Birkhäuser

Misha Gromov

Great Circle of Mysteries

Mathematics, the World, the Mind

Birkhäuser

Misha Gromov
Institut des Hautes Études Scientifiques
Université Paris-Saclay
Bures-sur-Yvette, France
and
Courant Institute of Mathematical Sciences
New York University
New York, USA

ISBN 978-3-030-09605-2 ISBN 978-3-319-53049-9 (eBook)
https://doi.org/10.1007/978-3-319-53049-9

Cover illustration: homographic sequences: projective dynamics © Benoît Kloeckner, http://perso-math.univ-mlv.fr/users/kloeckner.benoit/posts/2012-04-23-SuitesHomographiques.html

Printed on acid-free paper

This book is published under the imprint Birkhäuser, www.birkhauser-science.com by the registered company Springer Nature Switzerland AG part of Springer Nature.
The registered company address is: Gewerbestrasse 11, 6330 Cham, Switzerland

Contents

Introduction vii

Quotations and Ideas

1 Beautiful Elsewhere 1
2 Science 2
3 Numbers 4
4 Laws 9
5 Truth 26
6 Life 30
7 Evolution 37
8 Brain 58
9 Mind 62
10 Mysteries Remain 68

Memorandum Ergo

1 Brain, Ergo-Brain, and Mind 73
2 Ergo Project 78
3 Formality and Universality – Meaning, Folding, and Understanding 80
4 Universality, Simplicity and Ergo-Brain 89
5 Freedom, Curiosity, Interesting Signals, and Goal Free Learning 92
6 Information, Prediction, and a Bug on the Leaf 94
7 Stones and Goals 100
8 Ego, Ergo, Emotions, and Ergo-Moods 102
9 Common Sense, Ergo-Ideas and Ergo-Logic 104
10 Ergo in the Minds 107

11 Language and Languages 111
12 Meaning of Meaning 115
13 Play, Humour, and Art 120
14 Ergo in Science 122
15 Unreasonable Men and Alternative Histories 125
16 Mathematics and its Limits 128
17 Numbers, Symmetries, and Categories 132
18 Logic and the Illusion of Rigor 137
19 Infinite Inside, Finite Outside 141
20 Small, Large, Inaccessible 144
21 Probability: Particles, Symmetries, Languages 149
22 Signal Flows from the World to the Brain 156
23 Characteristic Features of Linguistic Signals 161
24 Understanding Structures and the Structure of Understanding 164
25 Sixteen Rules of an Ergo-Learner 168
26 Learning to Understand Languages: From Libraries to Dictionaries 171
27 Libraries, Strings, Annotations, and Colors 174
28 Teaching and Grading 176
29 Atoms of Structures: Units, Similarities, Co-functionalities, Reductions 178
30 Fragmentation, Segmentation, and Formation of Units 184
31 Presyntactic Morphisms, Syntactic Categories, and Branched Entropy 186
32 Similarities and Classifications, Trees and Coordinatizations 191
33 Clustering, Biclustering, and Coclustering 193
References 201

Introduction

© pixelparticle / Getty Images / iStock

Geometry is one and eternal shining in the mind of God.

JOHANNES KEPLER

God's plan for mathematics is beyond human reach but mathematics is the only light that can illuminate the mysteries of this world.

What we call the *science of physics* is filled with the radiance of mathematics; the fog of ignorance around what we now call LAWS OF NATURE receded in the brilliance of this physics.

But the beam of this light has barely touched the edges of the kingdom of LIFE and the face of PRINCESS THOUGHT remains hidden from us in the shroud of darkness.

And unless we know the ways of THOUGHT we cannot understand what mathematics is.

The first part of the book – *Quotations and Ideas* – is sprinkled with the ideas of those who saw sparks of light in the dark sea of the unknown. In the second part – *Memorandum Ergo* – we reflect on what in mathematics could shed light on MYSTERY THOUGHT.

Quotations and Ideas

1. Beautiful Elsewhere

Round us, near us, in depth and height,
Soft as darkness and keen as light.

ALGERNON SWINBURNE, LOCH TORRIDON

"*Mathématiques, un dépaysement soudain,*" was an exhibition organized in 2011 by *Fondation Cartier pour l'art contemporain* in Paris. It featured, among other things, *The Library of Mysteries*:

THE MYSTERY OF PHYSICAL LAWS
THE MYSTERY OF LIFE
THE MYSTERY OF THE MIND
THE MYSTERY OF MATHEMATICS

These were presented as quotes from writings of great scientists in a film made by David Lynch – an artist's visualizations of the ideas of Time, Space, Matter, Life, Mind, Knowledge, and Mathematics.

M. Gromov, *Great Circle of Mysteries*, https://doi.org/10.1007/978-3-319-53049-9_1

Michel Cassé and Hervé Chandés have persuaded me to try to do a *Naive Mathematician's* version of what David Lynch has done – to project a vision of these ideas to the invisible screen in our mind, an image illuminated by the eternal shining of mathematics rather than by the reflection of the beauty of the goddess of arts.

I knew it would be impossible but I tried anyway. Much of what I wrote was done in discussions with Giancarlo Lucchini and some of my English was corrected by Bronwyn Mahoney. Below is a modified version of what came out of it.

2. Science

Nothing exists except atoms in the void;
everything else is opinion.

DEMOCRITUS OF ABDERA (?), 460–370 BCE

All men by nature desire knowledge. Thinking is the talking of the soul with itself. ... knowledge is the one motive attracting and supporting investigators ... always flying before them ... their sole torment and their sole happiness. To perceive is to suffer.

Common sense is the collection of prejudices acquired by age eighteen.

... Science ... [is] *asking, not whether a thing is good or bad ... but of what kind it is?*

Science is the belief in the ignorance of experts. What we know already ... often prevents us from learning. Our freedom to doubt was born out of a struggle against authority in the early days of science.

Science is no more a collection of facts than a heap of stones is a house. A fact is valuable only for the idea attached to it, or for the proof that it furnishes. He who does not know what he is looking for will not understand what he finds. The investigator should have a robust faith – and yet not believe.

Science increases our power in proportion as it lowers our pride. Ignorance more frequently begets confidence than does knowledge. Whoever undertakes to set himself up as a judge of Truth and Knowledge is shipwrecked by the laughter of the gods.

The gods are fond of a joke – the universe is not only queerer than we suppose, but queerer than we can suppose. And the most incomprehensible thing about the world is that it is comprehensible.

There is geometry in the humming of the strings, there is music in the spacing of the spheres. A hidden beauty is stronger that an obvious one. It is godlike ever to think on something beautiful and on something new.

The most beautiful thing we can experience is the mysterious. It is the source of ... art and ... science – branches of the same tree. He to whom this emotion is a stranger ... is as good as dead: His eyes are closed.

When it comes to atoms, language can be used only as in poetry. Poetry is nearer to vital truth than history. Knowledge is limited. Imagination encircles the world.

But put off your imagination, as you put off your overcoat, when you enter the laboratory. Put it on again when you leave.

The objective reality of things will be hidden from us forever; we can only know relations. Everything we call real is made of things that cannot be regarded as real. The internal harmony of the world is the only true objective reality.

It is not nature that imposes time and space upon us, it is we who impose them upon nature because we find them convenient. ... the distinction between past, present, and future is a most stubbornly persistent illusion.

A human being is a part of a whole, called by us "universe" – a part ... restricting us to our personal desires and to affection for a few persons nearest to us. Our task must be to free ourselves from this prison by widening our circle of compassion to embrace all living creatures and the whole of nature in its beauty.

PYTHAGORAS 1	CHARLES DARWIN	NIELS BOHR
HERACLITUS 1	CLAUDE BERNARD	ALBERT EINSTEIN
PLATO 1	JAMES CLERK MAXWELL	JOHN HALDANE
ARISTOTLE 1	HENRI POINCARÉ	RICHARD FEYNMAN

What these people think and how they write enlightens your mind and elevates your spirit but these thoughts have little life of their own. They do not grow, they do not transform, they do not shoot new green sprouts – they luminesce as crystals of fiery flowers frozen in eternity. They are not quite what mathematicians call *ideas*, they are halfway between *ideas* and *opinions*.[1]

Great scientific *ideas* are different – they are alive, they ignite your soul with delight, they invite you to fight and to contradict them. Unlock your spirit from the cage of the mundane, let your imagination run free, start playing with such ideas as a little puppy with its toys – and you find yourself in the world of *Beautiful Elsewhere* – that is called *Mathematics*.

[1] An OPINION about X is a function, say $OP_X = OP_X(p)$, that assigns YES (agree) or NO (disagree) to a person p who is dying to say what he/she thinks about X; e.g., about the existence (non-existence) of *vacuum*. Democritus did not care about specific YES/NO-values of $OP_X(p)$, except for p being his best friend, maybe. But the philosopher would love to see a correlation between $OP_X(p)$ for $X =$"*vacuum exists*" and the distance from the house of p to Abdera.

3. Numbers

© Siim Sepp, http://www.sandatlas.org

All the mathematical sciences are founded on relations between physical laws and laws of numbers.

JAMES CLERK MAXWELL (1856)

The numbers named by me exceed the mass equal in magnitude to the universe.

ARCHIMEDES, THE SAND RECKONER, 250 BCE (?)

Archimède … dont la vie est une époque dans l'histoire de l'homme, et dont l'existence parait un des bienfaits de la nature.

NICOLAS CONDORCET

Archimedes estimated the diameter of the universe at about 2 light-years that is $\approx 2 \cdot 10^{13}$ km – twenty thousand billions kilometers, about half, as we know today, of the distance to the nearest stars – the binary system Alpha Centauri A & B.

Then Archimedes invented an exponential representation of large numbers and evaluated the number of sand grains or rather of ≈ 0.2 millimeter poppy seeds needed to fill it at less than 10^{63} in modern notation. (I took these numbers from the Wikipedia article. In fact the $2 \cdot 10^{13}$ km cube has volume $8 \cdot 10^{57}\,\mathrm{mm}^3$ that makes 10^{60} cubes of 0.2 mm.)

If a philosopher would not be impressed and say that

a good decision is based on knowledge and not on numbers,

Archimedes might respond that decisions may be left to our mighty rulers but that *numbers* are the guardians of our true knowledge.

Large numbers are everywhere. Even Socrates, Plato, and Aristotle would admit that *knowledge of thyself* is incomplete if you are unaware of the roughly 10^{14} (100 000 billion) bacteria in your guts – several bacteria per each cell of your own body.

(Bacteria are roughly 1 μm – one thousandth of a millimeter = one millionth of meter – in size, a few thousand times smaller than your own cells volume-wise. If you had a bacterium inside every one of your cells you would hardly notice this – you will be safely dead by that time.)

A single bacterium, if there are enough nutrients, can divide every 20–30 minutes and, in 24 hours you may have a blob $10^5\mu$m = 10 cm in size with about $2^{50} = (2^{10})^5 = 1024^5 \approx 1000^5 = 10^6 \times 10^9$ (million billions) bacteria in it.

A schoolboy now computes:

Day 2. The blob contains $10^{15} \times 10^{15} = 10^{30}$ bacteria and is $10^{10}\mu$m = 10 km in diameter, about 1 kg of bacteria per every square meter of the surface of the Earth.

Day 4. It increases to $10^{20}\mu$m = 10^{11} km in diameter and reaches the outermost regions of the Solar system. It will engulf the Sun ($\approx 1.5 \times 10^8$ km from Earth), all planets including Pluto ($\approx 6 \times 10^9$ km) but not fully the orbit of Sedna at its farthest point from us ($\approx 1.4 \times 10^{11}$ km).

(Aristotle, who maintained that

the shape of the heaven is of necessity spherical,

would feel relieved if he knew that the bacteria are still contained by the *spherical heavenly shell* of the hypothetical *Oort cloud* of comets around the Sun about a light-year away.)

Day 7. The blob contains $10^{15\times 7} = 10^{105}$ bacteria and has $10^{5\times 7}\mu\text{m} = 10^{26}$ km $\approx 10^{13}$ light-years in diameter, a hundred times the diameter of the observable universe ($\approx 10^{11}$ light-years).[2]

Bacteria have been around for billions of days, but do numbers like $10^{100\,000\ldots}$ make any sense at all? The answer is *yes and no.* They cannot be reached by counting 1, 2, 3, ..., at least not in our space-time *continuum*, nor be represented by *collections of physical objects* of any kind. However, unrealistically large (and unrealistically small) numbers are instrumental in our treatment of NATURE'S LAWS that are manifested in observable properties of *objects* in the Universe.

How does Nature, who, as Einstein says, *integrates empirically,* manage to satisfy these laws?

Is it because she has something *much bigger* than space/time (kind of quantum fields?) at her disposal where *empirical integration* is possible?

Or is there a *secret logical something* built into Nature and she proceeds by *mathematical induction* as mathematicians do?

Or had she found a *simple logical bypass* for arriving at these laws but we cannot reach it being bound to the mental routes available to our brains?

---

[2]Darwin computed the numbers of descendants of a couple of elephants and arrived at the number 15 000 000 after 500 years. This is much exaggerated, but in 5 000 years the Earth would be covered several times over by more than 10^{15} elephants and the whole Universe would not contain 10^{90} elephants after 30 000 years. (30 000 is yesterday on the Earth geological time scale, it is less than 0.02% of (about 200 000 000 years of) the evolution time of mammals on Earth.)

These questions, probably, make no sense; it is frustrating being unable to formulate a good one.

Yet, a mathematician may find a consolation in trying to estimate the number N_{can} of different logical arguments (brain routes), say in L words, that a sentient brain *can*, in principle, generate. Probably, if somebody told our mathematician what the words *can* and *in principle* signify, he/she would bound N_{can} by something like $\sim L \log L$, or even less than that – far from the number $N_{all} \sim 2^L$ of *all* such arguments, well behind what bacteria can do.

This might hurt his/her pride but then the mathematician will soon realize that numbers that linger behind his/her own logic/language beat the fastest replicating bacteria.

Indeed, *think* of Schrödinger's cat. The body of a cat is, roughly, composed of $N \approx 10^{26}$ molecules, those of water and small molecular residues in macromolecules. Suppose each molecule can be in two states. Then there are $S = 2^N \approx 10^{0.3N}$ *states of a cat.* Some of these states are judged being *alive* and some are classified as *dead.* The number, say $\mathcal{CAT}$, of possible judgements/opinions is

$$2^S = 2^{2^{10^{26}}} > 10^{10^{10^{25}}}$$

How does one decide, how can one select a sound judgement from this super-duper-universe of possibilities? Mathematicians do not understand how it works, but a cat, if he/she is alive and oblivious of math., somehow manages, makes a right choice and ... stays alive.

Some courageous people play with unimaginably greater numbers, the descendants of *Gödel's incompleteness theorem.* If you meet such a number on your path of reasoning about "real world" your logic is as good as dead.[3] Fortunately, you do not meet them in "real life" unless you call these misshapen monsters by their names.

THE MONSTER OF **STOP**. If your computer has M bits of memory, say with $M = 10^{10}$, then whatever you "ask" the computer to do, it either *stops* after $< 2^M$ steps or it goes into a cycle and runs forever. (You can use convenient *time units* instead of "steps"; the number $2^{10^{10}}$ is so big, it makes little difference if you take nanoseconds or billions of years for these units.)

Here, what we call a *question* or a *program* that you "ask" your computer to perform, is a sequence of *letters* that are the names of keys on the keyboard you have to press in order to activate this *program.*

Let us leave the "real world" and allow your computer to have an *infinite* (unbounded) memory. As in the finite case, the computer may stop after finitely many steps or run for infinite time *depending on your question*, (and on the design of the hardware and the operating system in your computer) except that *infinite* does not have to be *cyclic* for infinite memory. For example, if asked to find a file

[3]These numbers lie at the heart of *Turing's halting theorem* and of Kolmogorov–Chaitin complexity. A whiff of either of the two is poisonous for any *scientific* theory.

named *cell-phan-nimber-Bull-Gytes* the computer either finds it and stops, or, if there is no such file, runs forever.

(A fundamental ability of your *slow* brain, not shared by the *fast* Windows search system, is an *almost instantaneous* NO SUCH FILE response to this kind of inquiry. A simple, yet structurally nontrivial, instance of this is BITTER on your tongue for almost anything chemically away from potential nutrients, with a few mistakes; e.g., SWEET for saccharin.

One may speculate together with Robert Hooke, Charles Babbage, and Alan Turing on plausible brain memory architectures that makes this possible. There is no experimental means for checking non-trivial conjectures of this kind but there is a mathematically attractive model of such a memory suggested by Pentti Kanerva, called *sparse distributed memory*.)

Pick up a (moderately large) number L and take *all those programs* in L letters for which the computer *eventually stops*. Since the number of these programs is finite ($< 100^L$ if there are < 100 different letters at your disposal for writing programs) the longest (yet, finite!) time of executing such a program, say, the time measured in years, is also *finite*; call this time **SToP**(L).

Although finite, this number of years, even for moderate L, say **SToP** = **SToP**(50 000), (a program in 50 000 letters takes a dozen pages to write down) is virtually indistinguishable from infinity in a certain precise logical sense.

© 35007 / Getty Images / iStock

Our Universe is pathetically small not only for containing anything of this size but even for holding a *writing* of an *explicit formula* or of an *explicit verbal description* of a number of such size, even if we use atoms for letters in such a writing. (The way we have just described this number does not count – distinguishing stopping times from infinity without indicating a specific "experimental protocol" for this is not what we call *explicit*.)

By comparison, CAT appears tiny – the corresponding exponential formula can be expressed with a few dozen binary symbols (and *physically* written down with a few thousand atoms manipulated by means of an *atomic force microscope*).

To be fully honest we must admit there was something missing from our definition of **STOP**(L).

Imagine, for example, that the memory of our computer does contain the string *cell-phan-nimber-Bull-Gytes* but it is positioned so far away that it cannot be reached in less than T time units. Since you can choose T as large as you wish, the logic of our definition inevitably makes **STOP**(L) equal infinity, even if you limit the length L of admissible programs to something quite small, say, less than one thousand letters.

Somehow one has to prohibit such a possibility, by insisting that all "memory cells" in you computer that *cannot* be reached in less than, say, 10^{10} time units are empty, nothing is written in these "far away cells". Moreover, the computer is supposed to know when it crosses the boundary of "non-empty space" and will not spend any time in searching the empty one.

On the other hand, the computer is allowed, in the course of a computation, to write/erase in these "far away cells"; this is what may eventually generate an enormous volume of occupied memory cells and make its reading excruciatingly long.

With this provision, the definition of **STOP**(L) becomes correct, it does give you something *finite*, PROVIDED you have *precisely* defined what "far away cells" and "reaching something in the memory" mean.

But can one *explicitly* describe in *finitely many* words an *infinite* memory along with a description of a search program through this memory?

A commonly accepted solution articulated by Turing, is to assign memory cells/units to *all* numbers $1, 2, 3, 4, \ldots, 1000\ldots$ with labels "*empty*" / "*non-empty*" attached to them, and with the symbols 1 and 0 written along with all "*non-empty*" marks. Then an individual step of a memory search is defined as moving from an i-cell to $i+1$ or $i-1$, where each cell is labeled "*non-empty*" after having been visited.

If you are susceptible to the magic of the word "all" and believe this truly defines the infinity of numbers, then you will have a "mathematically precise" definition of **STOP**(L).[4]

All this being said, there is something not quite right with **STOP**: it is a byproduct of the machinery of formal logic – not a *true mathematical* number. But about thirty years ago, **pretty** *HUGE* numbers were discovered; e.g., the time Hercules needs to slay Hydra in a mathematical model of this regrettable parricide. But all such *GIANT BEAUTIES* are incomparably smaller than the ugly looking **STOP**.

[4]There are, probably, 5–10 mathematicians and logicians in the world who think that the monsters like **STOP** are indicators of fundamental flaws in our concepts: "number", "finite", "infinite".

4. Laws

Lex 1: *Corpus omne perseverare in statu suo quiescendi vel movendi uniformiter in directum, nisi quatenus a viribus impressis cogitur statum illum mutare.*

Isaac Newton, Philosophiæ Naturalis Principia Mathematica, 1687[5]

... les lois générales, connues ou ignorées, qui règlent les phénomènes de l'univers, sont nécessaires et constantes.

Nicolas Condorcet

So much as we know of them [the laws of nature] *has been developed by the successive energies of the highest intellects.*

Michael Faraday

The most astounding single event in the history of science, a physicist argued, was the discovery of *general relativity* in 1916 – this would have been postponed by 20–30 years if not for Einstein, twice as long as any other scientific discovery, according to our physicist's calculation on the back of an envelope.

Our physicist, a theoretician, could not confirm his calculation by an experiment, but history has done it for him and ... proved him wrong.

© baphotte / stock.adobe.com

The true logic of this world is the calculus of probabilities.

James Clerk Maxwell

[5]Historically, the *first recorded* LAW – the velocity/force formula for *motion in a viscous medium* – was stated by Aristotle. (Later versions of this were suggested by Newton, 1687, and by Stokes, 1851.) A century after Aristotle, Archimedes discovered basic laws of *statics of mechanical systems*: The laws of the lever and of equilibrium of solid bodies in liquids.

Mendel's theory of genes – units of inheritance – was published in 1866, where Mendel derived their very existence and essential properties from

the striking regularity with which the same hybrid forms always reappeared

in thousands of his experiments with pea plants.

The discovery of genes was the greatest event in biology since the discovery of *cells* by Robert Hooke in 1665 and of *infusoria* by Antonie van Leeuwenhoek in 1674.

Mendel's methodology of

combinatorial design of multistage interactive experiments

+

extracting specific structural information from statistics of observable data by mathematical means

was novel for all of science. This is why Mendel's paper was ignored by biologists for about thirty years.

When similar data were obtained and analyzed by de Vries, Correns, and Tschermak at the turn of the 20th century, biologists returned to Mendel's paper. Many, including Alfred Russel Wallace, were appalled by Mendel's ideas, while even the most sympathetic ones were confounded by Mendel's "counterintuitive" and "biologically unfeasible" algebra.

In 1908 a leading English mathematician G.H. Hardy and independently a German physician Wilhelm Weinberg spelled out this *counterintuitive* as

$$\frac{\left[(p+q)^2+(p+q)(q+r)\right]^2}{\left[(p+q)^2+(p+q)(q+r)\right]\cdot\left[(p+q)(q+r)+(q+r)^2\right]}=\frac{(p+q)^2}{(p+q)\cdot(q+r)}$$

and MENDEL'S LAWS OF INHERITANCE were accepted by (almost) everybody.

(Hardy called this *a mathematics of the multiplication table type*. He overlooked the mathematical beauty of *Mendelian dynamics* of the *next generation maps* M in the spaces of *truncated polynomials* – the maps that represent transformations of *distributions of alleles* in populations under *random matings*. In a simple instance, such an M applies to matrices $P=(p_{ij})$ by substituting each (i,j)-entry p_{ij} by the product of the sums of the entries in the i-row and the j-column in P,

$$(p_{ij})=P\overset{M}{\mapsto}P^{\text{next}}=(p_{ij}^{\text{next}})\text{ for }p_{ij}^{\text{next}}=\sum_i p_{ij}\cdot\sum_j p_{ij}.$$

It is amazing, albeit obvious, that $M(M(P))=\text{const}\cdot M(P)$ for $\text{const}=\sum_{ij}p_{ij}$, where this amounts to the above (p,q,r)-formula for symmetric 2×2 matrices in Hardy's notation.)

> *... are we justified in regarding them [genes] as material units; as chemical bodies of a higher order than molecules? ... It makes no difference in ... genetics. Between the characters that are used by the geneticist*

and the genes his theory postulates lies ... embryonic development.

THOMAS HUNT MORGAN, 1934

In 1913, almost half a century after the publication of Mendel's paper *Versuche über Pflanzen-Hybriden*, 21-year-old Alfred Sturtevant made the next step along Mendelian lines of logic and determined *relative positions* of certain genes on one of the chromosomes of Drosophila by analyzing frequencies of specific morphologies in generations of suitably interbred flies.

Just think about it. You breed fruit flies, you count how many have particular *combinations* of certain features; i.e., you record the distribution of occurrences of the following eight $(2 \times 2 \times 2)$-possibilities:

[striped bodies]/[*yellow* bodies]
[red eyes]/[*white* eyes]
[normal wings]/[*smallish* wings]

Then you distinctly see with your mathematically focused mind's eye (even if you happened to be as color blind as Sturtevant) that the corresponding genes – abstract entities of Mendel's theory – that are associated with these features, are all lying in definite relative positions on an imaginary line, where, as for Mendel's genes, the existence of this line is derived from how *hybrid forms reappear*. In particular, you assign the "eyes gene" a position *between* the "body gene" and the "wing gene", since

[smallish wings] + [yellow bodies] "**implies**" [white eyes]

in descendants of certain parents with an abnormally high probability.

Decades later, molecular biology and sequencing technology demystified "Sturtevant's line" by identifying it with a DNA string segmented by genes, but a mathematical unfolding of Sturtevant's idea still can be seen only in dreams.

ABOUT ROBERT HOOKE, ANTONIE VAN LEEUWENHOEK, DROSOPHILA MELANOGASTER AND THE IDEA OF STURTEVANT.

Hooke's name is associated with *Hooke's elasticity law* but that was only one of many of his experimental discoveries, original concepts, and practical inventions. For example, he recognized fossils as the remains of *extinct species*, he developed an almost modern *model of memory*, he proposed (1665?) the construction of *the spring balance watch* (Huygens' description of his own construction dates to 1675) and he suggested (1684) a detailed design of an *optical telegraph with semaphores.* (The first operational system with a network of ≈ 500 stations was built in 1792 in France.)

Leeuwenhoek found out how to obtain small glass balls for the lenses of his microscopes but he made others to believe he was grinding tiny lenses day and night by hand. Besides infusoria, he observed and described crescent-shaped bacteria (*large Selenomonads*), cell vacuoles, and spermatozoa. The secret of Leeuwenhoek's microscopes was rediscovered in 1957.

Drosophila melanogaster – the pomace fly ($\approx 2.5\,\text{mm}$ long, normally with red-eyes and stripped bellies) – were introduced as a major model organism in genetics by Thomas Hunt Morgan. He and his students were counting the mutant characteristics of thousands of flies and studied their inheritance. Analysing these data, Morgan demonstrated that genes are carried on chromosomes; also he introduced the concepts of *genetic linkage* and of *crossing over*.

Sturtevant mathematically "synthesised" his string of genes from the "substrate" of results and ideas taken from the work of Morgan almost as Kepler "crystallized" elliptical orbits from Tycho Brahe's astronomical tables. Sturtevant recalls the great moment as follows.

> *I suddenly realized that the variations in strength of linkage, already attributed by Morgan to differences in the spatial separation of the genes, offered the possibility of determining sequences in the linear dimension of a chromosome. I went home and spent most of the night (to the neglect of my undergraduate homework) in producing the first chromosome map, which included the sex-linked genes y, w, v, m, and r, in the order and approximately the relative spacing that they still appear on the standard maps.*

Sturtevant's idea of (re)construction of the (a posteriori linear) geometry of the genome is similar to Poincaré's suggestion for how the brain (re)constructs the (a posteriori Euclidean) geometry of the external world from a set of samples of retinal images.

Grossly oversimplifying, an unknown geometric (or non-geometric) structure S from a given class $\mathcal{S}$ on a set X under study – be it the set of (types of) genes in the genomes (of organisms) of a given species or the set of photoreceptor cells in the retina – is represented by some probability measure on the set of subsets Y of X. What is essential, this measure is supported on $\mathcal{S}$*-simple* (*special*) *subsets* Y that admit *short descriptions in the language of* $\mathcal{S}$; this allows a reconstruction of S from relatively few samples.

This sampling is far from random. In genetics, such a $Y = Y(O)$ is the subset of genes of particular allele versions in the genome of an individual organism O, where these O are obtained via a *controlled* breeding protocol that had been specifically designed by an experimenter.

In vision, such a $Y = Y(t)$ is the set of excited photoreceptor cells in the retina of your eye at a given moment t, where more often than not the variation $Y(t)$ with t is due to a motion of an object relative to your eye; the brain, which commands the muscles that move your eye, has an ability to design/control such variations.

The hardest step in finding S is *guessing* what $\mathcal{S}$ is. After all, what is a *structure*?

Mendel's laws are no more than a Platonic shadow – a statistically averaged image of the workshop of Life on the flat screen of numbers. The molecular edifice of the cell is crudely smashed on this screen and its exquisite structure cannot be reconstructed from this image by pure thought. Hundreds (thousands?) of ingenious experiments are needed to recover the enormousness of information that had been lost.

> *. . . I hold it true that pure thought can grasp reality,*
> *as the ancients dreamed.* ALBERT EINSTEIN

Contrary to what we see in biology, the mathematical image of the basic machinery running the world of physics retains the finest details of this machinery. It even may seem, probably, only to our *Naive Mathematician*, that the less you know the better you understand how the Universe is run.

For example, forget about velocities, forces, accelerations. Imagine a world exclusively populated by wandering watches that have no perception of speed and force. But when two watches meet, they can recognize each other and compare their records of the intervals of times between consecutive meetings.

A watch-mathematician would sum up what he/she believes he/she "sees" in the watch-world as *self-obvious axioms* and, after pondering for a few centuries on what they imply, he/she will figure out that there is a unique simplest most symmetric *watch space* of every given dimension. It is the *Lorentz–Minkowski time-space*, that is four-dimensional in the Universe in which we happened to exist.

The mathematician will be delighted by this marvelous space-idea; yet perplexed, since his/her mental picture of the world does not explain

> *why the watches that have no physical contact apart from their meeting points remain synchronous.*

(Here, on Earth, it is not this incredible synchronisation but its violation that is regarded as paradoxical.)

But then his friend physicist conceives the idea of speed and his colleague experimentalist designs fast traveling watches. The mathematician sighs with relief: the formulas of his/her theory (that is called *special relativity* on Earth), are perfectly right and desynchronisation is clearly seen for watches traveling with mutual relative speeds close to 1. (On Earth, this 1, that is the speed of light, is elegantly expressed as 299 792 458 ... × another unit of speed the meaning of which no watch-mathematician has ever been able to grasp.)

> *Lines of force convey a far better and purer idea ...*
>
> MICHAEL FARADAY, 1833[6]

> *Thy reign, O force! is over. Now no more*
> *Heed we thine action;*
> *Repulsion leaves us where we were before,*
> *So does attraction.*
>
> JAMES CLERK MAXWELL, 1876

> *The theory of relativity by Einstein ...*
> *cannot but be regarded as a magnificent work of art.*
>
> ERNEST RUTHERFORD[7]

---

[6] *Faraday must have grasped with unerring instinct ... spatial states, today called fields, ...*
ALBERT EINSTEIN, 1940

[7] Rutherford, named the Faraday of nuclear physics, experimentally identified alpha, beta and gamma emissions, discovered atomic nuclei and proposed protons and "nuclear electrons" for their constituents. (Neutrons were discovered and fit into the model of the nucleus about 20 years later.) *"All science is either physics or stamp collecting"* is a saying attributed to him. He, apparently, regarded chemistry as a part of physics and he did not live to see the birth of molecular biology from the seed of the Mendelian-chromosome theory of heredity.

Next our mathematician will look not for a most symmetric space but for the most symmetric *law* of "motion" applicable to all imaginable to him/her watch-spaces filled/strained by *lines/fields of force* à la Faraday.

First he/she thinks that no such distinguished law is possible, but then some terms in his/her calculation miraculously cancel each other and a beautiful equation pops out. He/she, undoubtedly, will call it the *Einstein vacuum equation*. (This was so derived by David Hilbert.)[8]

Then his friend physicist will bring in *energy/matter* and, to everybody's satisfaction, an experimentalist/cosmologist will show that the universe behaves as predicted by the resulting equation – the simplest mathematically imaginable description of [watch/space]-worlds – *general relativity theory*.

The greatest change in ... our conception of the structure of reality – since Newton ... was brought about by Faraday's and Maxwell's work on electromagnetic field phenomena.

ALBERT EINSTEIN, 1931

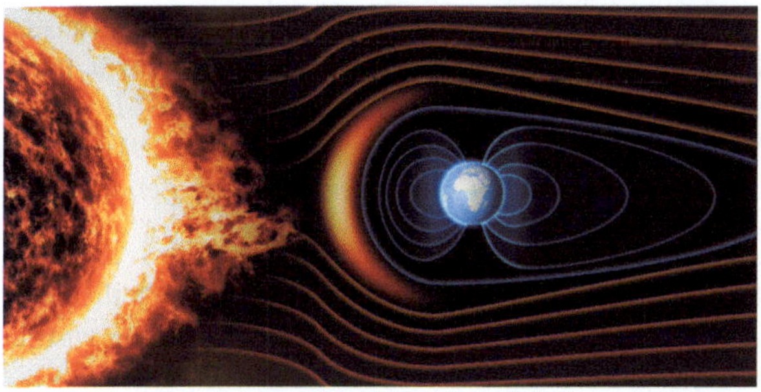

© Naeblys / Getty Images / iStock

... ten thousand years from now,
the most significant event of the 19th century
will be judged as Maxwell's discovery of the laws of electrodynamics.

RICHARD FEYNMAN, LECTURES ON PHYSICS, 1964.[9]

The *watch world* symmetry, that was acclaimed by the terrestrial physicists of the 20th century, originated in the work by Maxwell who, over 1855–1873, thought out (a system of twenty) differential (wave) equations that logically embrace and unify the *Ampere law* (1826) on magnetic effects of electric currents and *Faraday's law of induction* of electric currents by moving magnetic fields (1831).

---

[8]Hilbert was the one who, along with Poincaré, guided the passage from the 19th to 20th century mathematics.

[9]The 120th century seems far off – *another age must be the judge* – Charles Babbage's writes in his *Ninth Bridgewater Treatise* (1837). His design of *analytic engine* (that implemented what is now called the *universal Turing machine*) will be the most likely number one on the 19TH CENTURY SIGNIFICANT EVENTS list of a robot of the 22nd century.

Formally, these are *Lagrange–Hamilton equations* dictated by *Maupertuis' principle* for electromagnetic fields; they possess a remarkably high – what is now called *Lorentzian* – symmetry that, as Einstein wrote in 1953,

transcended its connection with Maxwell's equations.

(The historically first *wave equation* – the equation of a vibrating string, was written down and studied by d'Alembert in 1747; apparently, its *Lorentzian symmetry* went unnoticed until much later.)

The prima donna role for this symmetry in physics hinted at by FitzGerald in 1889 and suggested by Larmor in 1897 and by Lorentz in 1899, was consummated in two 1905 papers:

Sur la dynamique de l'électron[10] by Poincaré

and

Zur Elektrodynamik bewegter Körper by Einstein.

Two years later, Hermann Minkowski, who was usually preoccupied (besides *convex bodies*) with multidimensional geometry of *quadratic equations in many variables* (quadratic forms) and with their transformation groups, suggested a four-dimensional geometric realization of the (Poincaré)-Lorentz symmetry. The Minkowski 4-D geometry was extended by Einstein in 1916 to (*Einstein*)-*Lorentz spaces* that he used as a mathematical framework for *general relativity theory.*

Now, imagine a technologically backward watch-civilization where no high energy experiment is available. Here a mathematician will have to invent some mystical *absolute time* that will synchronize non-interacting watches as windowless *monads* of Leibniz.

Absolute, true and mathematical time, of itself, and from its own nature flows equably without regard to anything external, and by another name is called duration. ISAAC NEWTON

Then mathematician's friend physicist breaks a window somewhere from where he could see motions of other watches and the mathematician will realize that granting an *absolute time* to a Lorentz space necessarily implies an *absolute space* where motion is possible.

Eventually the two – the mathematician with his physicist friend – such a couple is called *Isaac Newton* on Earth – will arrive at the THREE LAWS and, when they come to Earth, they incorporate INVERSE SQUARE LAW for gravitation into their theory, where "square" is suggested (implied?) by the geometric

[10]where the *Lorentz group* materialized in the words

... the equations of the electromagnetic field are not altered by a certain transformation (which I will call by the name of Lorentz) of the form: ...

The *principle of relative motion* was discussed by Poincaré in earlier papers but that was not quite what we call *special relativity theory* as formulated by Einstein.

AREA $\sim R^2$-LAW: the area of the surface of the R-sphere is proportional to R^2 where the exponent 2 comes as $2 = 3 - 1$ for 3 being the dimension of our physical space.

(Newton mathematically *proved* that *only* the gravitation $\sim 1/R^2$-LAW agrees with astronomical observations,[11] but it is unclear who was the first to *conjecture* $\sim 1/R^2$. Robert Hooke claimed to be the one. Indeed, in his communication to the Royal Society of 1666 he states:

> *... heavenly bodies ... mutually attract each other within their spheres of action. ... so much the greater as the bodies are nearer.*

But others, including Galileo, Kepler, and Newton himself, expressed similar ideas on gravitation.

Newton regarded the proof of $\sim 1/R^2$ as by far more difficult than guessing it. Since he was the only person on Earth, if not in all of the astronomically observable[12] Universe, qualified to approach the problem and to evaluate its mathematical difficulty, we accept his judgement.)

But despite the remarkable agreement of the theory with recorded planetary orbits over short time intervals (thousands of years) a disturbing question will nag our friends.

Are these laws – the laws of classical mechanics + the inverse square law of gravity – *consistent* with the apparent stability of the solar system on the time scale of millions and hundreds of millions of years?

Newton himself believed the answer is NO and that planets would collide with the Sun if not for divine intervention from time to time. But about 250 years after Newton an optimistic *maybe* was suggested by a counterintuitive mathematical theorem, usually referred to as *KAM*, that (very roughly) says that *quite a few* physically significant dynamical systems may behave "asymptotically rather periodically", contrary to what physicists (as well as mathematicians) had always believed. (Everybody's intuition suggested "asymptotically chaotic" behavior of the *predominant majority* of mechanical systems with more than two degrees of freedom.)

This theorem depends on a hidden *symplectic symmetry* between the kinetic and potential energies as seen on yet another mathematical screen discovered by Hamilton ≈ 100 years after the death of Newton.

---

[11]Planetary motions to which Newton's equations apply are *not* directly observable. It took the *successive energies of the highest intellects* of (not only) Copernicus, Tycho Brahe, and Kepler to make the observable data amenable to mathematical analysis.

[12]"Observable" here, rather parochially, refers to an observer centred somewhere in the vicinity of *the Milky Way galaxy.*

Quantum electrodynamics describes Nature as absurd
from the point of view of common sense.
And it fully agrees with experiment.

RICHARD FEYNMAN, QED: THE STRANGE THEORY OF LIGHT AND MATTER, 1985

Nothing is too wonderful to be true,
if it be consistent with the laws of nature;
and in such things as these experiment
is the best test of such consistency.

MICHAEL FARADAY, LABORATORY JOURNAL ENTRY, 1849

Is the running of "watches" themselves governed by classical/relativistic mechanics?

Probably it is not hard to "prove" by an argument in the spirit of *Zeno's paradoxes* that no mechanical/electromagnetic model of Newton+Maxwell+Einstein can be compatible with the properties of *matter* we see everywhere around us. Apparently, our beautiful physical laws are "just" Mendelian kinds of images of something else, where this "something else" is expected by physicists to live in the *quantum world*.

Mathematical fragments of *quantum* may be accessible to us but when we try to imagine it as a whole our mind revolts, and if we insist it becomes giddy in a tangle of paradoxes and ambiguities. And little consolation can be found in what physicists keep repeating after Niels Bohr:[13]

If anybody says he can think about quantum physics without getting giddy, that only shows he has not understood the first thing about them.

You can hardly see a clear picture of anything while being giddy, but you would feel better if you could prove mathematically that, in principle, no common sense (including rigorous mathematics) model of *anything* resembling the *physical world* is possible.

"Thinking quantum" is incompatible with our deeply ingrained intuition of "reality". As Wolfgang Pauli[14] says:

The layman always means, when he says "reality" that he is speaking of something self-evidently known; whereas to me it seems the most important and exceedingly difficult task of our time is to work on the construction of a new idea of reality.

He also writes:

---

[13]In 1913, Bohr invented the concept of *quantization of a physical system* in the context of the *planetary model of the atom* that was suggested by Rutherford in 1911.

[14]Pauli (1900–1958) introduced *spin* into quantum mechanics, formulated the *Pauli exclusion principle*, and he predicted the existence of *neutrinos*.

Just as in the theory of relativity a group of mathematical transformations connects all possible coordinate systems, so in quantum mechanics a group of mathematical transformations connects the possible experimental arrangements.

This is only a hint about reality being a property of some transformation group, but a realistic idea of "physical reality" remains beyond our wildest dreams.

And quoting Pauli again,

The best that most of us can hope to achieve in physics is simply to misunderstand at a deeper level.

Yet, maybe, there remains a hope of seeing the image of the world in the mirror of mathematics.

ἀγεωμέτρος μηδελταεὶς εἴστω.
Let no one ignorant of geometry enter.

This is associated with the name of Plato who did not think much of those unfortunates who were

ignorant that the diagonal of a square is incommensurate with its side.

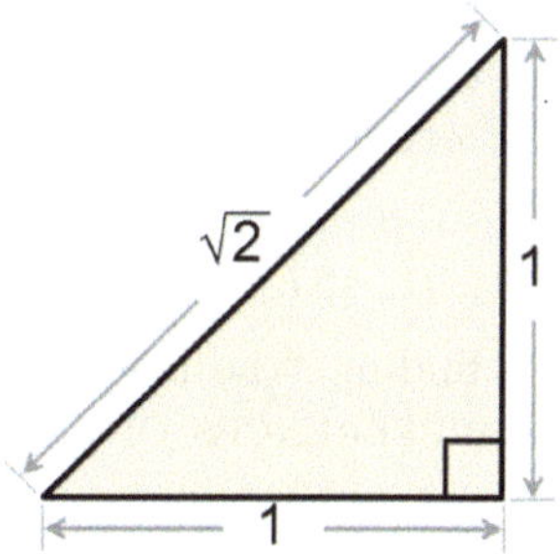

Fredrik / Wikimedia Commons / Public Domain

Twenty four centuries later, certain of Plato's disciples went further by suggesting that *true human consciousness* is distinguished by the ability of its possessor to grasp validity, meaning and implications of "incommensurability" between certain *theories* of numbers (rather than between certain individual numbers) that follows from *Gödel's incompleteness theorem*.

According to this supposition, any conceivable *imitation* of human mind, be it by an electronic implementation of a Turing universal computer or by a bio-robot, would be instantaneously recognizable by the flagrant absence of this ability.

However, no significant data on a presence/absence of this ability in human populations have been collected despite the renewed interests of neuropsychologists in the problem of consciousness.)

Whoever despises the high wisdom of mathematics
nourishes himself on delusion.

LEONARDO DA VINCI

Every new body of discovery is mathematical in form,
because there is no other guidance we can have.

CHARLES DARWIN

To those who do not know mathematics
it is difficult to get across a real feeling
as to the beauty, the deepest beauty, of nature.

RICHARD FEYNMAN

When does a fragment of science qualify for a LAW OF NATURE? Why does finding these laws require *successive energies of the highest intellects*?

Isn't a LAW *just* a *compression of information*, a *record of systematic correlations* between *facts* where these correlations can be found by *analyzing* your *observations*?

A "Yes" or "No" answer to this question depends on how you understand *just*, *information*, *systematic*, etc., where an essential (but not the only) difficulty facing a "law maker" is seen in the following example.

Imagine, that your *facts* or *events* are numbers, where the ones you happened to observe are

$$7,\ 19,\ 37,\ 56,\ 61,\ 91,\ 127,\ 189,\ 208,\ 296,\ 342,\ 386.$$

The only effective approach to finding "the law" behind these numbers is ... to *guess*: THESE NUMBERS ARE DIFFERENCES OF CUBES[15],

$$1 = 1^3,\quad 8 = 2^3,\quad 27 = 3^3,\quad 64 = 4^3,\quad 125 = 5^3,\quad 216 = 6^3,\quad 343 = 7^3, \ldots.$$

For instance,

$$91 = 216 - 125,\quad 127 = 343 - 216,\quad 189 = 216 - 27,\quad 386 = 9^3 - 7^3.$$

(This is kind of how it happened in Bohr's rendition of the *Balmer–Rydberg formula* for the wavelength λ of the Hydrogen spectrum, e.g., observed in the stars. This formula, in suitable units, says that $\frac{1}{\lambda} = \frac{1}{m^2} - \frac{1}{n^2}$, with the rationale behind this provided by Bohr's quantized model of the Hydrogen atom.)

In general, a mathematician may say that a law $\mathcal{L}$ is a "simple" map (function) from some "small and simple" *parameter set* P, (e.g., the set of pairs of cubes of integer numbers) *onto* a "complicated" subset *Obs* of observable events (differences of cubes of integers) within another "simple but possibly large" set *Ima* of all imaginable events (e.g., of all integers or, if you wish, of all real numbers that make a much larger, yet logically equally simple set).

[15] *A little mathematics of the multiplication table type* tells you that the predominant majority of numbers are *not* differences of cubes. This makes this $(m^3 - n^3)$-*law* quite restrictive; hence, informative.

The number of conceivable *simple laws* $\mathcal{L}$, e.g., of those describable in a few dozen words, is enormous – it grows exponentially with the number of words/symbols encoding an $\mathcal{L}$. There is no (?) general rule for guessing such an $\mathcal{L}$, from a realistic number of samples from *Obs* in each particular case.[16] And *he who does not know what he is looking for will not understand what he finds* – as Claude Bernard says.

(One may object by pointing out that an animal/human brain builds up a coherent mental picture of the external world by *systematically* finding correlations and cutting off redundancies in the flows of electric/chemical signals it receives from other parts of the body.

It is hard to argue since we do not know what is the *system* used by the brain, but, in any case, this system was designed by Nature *not at all* for discovering her inner workings.)

If you are a biochemist rather than a mathematician, you may visualize *a Law of Nature* as an intricately shaped "molecule", a kind of "logical enzyme" for "catalytic synthesis of theories" out of "solution of empirical facts", where the role of solvent – the supporting matrix of the process – is played by mathematics.

The *architecture of logic* may vary from law to law and the corresponding *chemistries of facts* may have nothing in common (as statistical mechanics and classical genetics, for example) but the principles of *mathematical catalysis* remain the same for most (all?) *enzymatic laws* worthy of the name LAW OF NATURE.

But beware: A shorthand expression of a law, e.g., of inertia – *Corpus omne perseverare* ... – a string of a hundred characters, tells you as little of the structure and the function of a law as a sequence of hundred amino acids says by itself about the physiological role of the corresponding protein. In order to read the message encoded by such a string you need to have a fair insight on the mathematical nature of the *solvent matrix* as much as on the *natural chemistry of facts*.

And if no mathematics in your head is available for this purpose, you shall not understand much of the corresponding law if at all.[17]

This was what happened to Mendel's ideas – apparently, mathematics and physics were not in the biology curriculum of that time. If Mendel, instead of Nägeli, had corresponded with people like Boltzmann, Guldberg and Waage, or van 't Hoff, the timetable of genetics might have been shifted by a quarter of a century.

---

[16] *Making* a correct guess is, in most cases, much harder than *verifying* one. There are several *conjectures* in mathematics, e.g., $P \neq NP$, that attempt to rigorously express this idea but, probably, we do not know yet how to *properly formulate* the problem.

[17] The laws of mechanics are special in this respect – they are built into the motor systems of the brain. *Instinctively*, we always try to find mechanical explanations (models) of everything around us. However, these "brain's laws" are, by necessity, non-Newtonian – true Newtonian mechanics is *instinctively* rejected by most (all?) people even if they formally understand it. (One wonders if this can be tested by experimental psychologists.)

Nägeli, Boltzmann, Guldberg, Waage, van 't Hoff

Karl Wilhelm von Nägeli (1817–1891), a leading botanist of the 19th century, introduced the concept of *meristem* – a group of plant cells capable of division; he also realized the role of sequences of cell divisions in plant morphology. But his fame rests on him being the first on the list of people who failed to understand Mendel's work.

Ludwig Eduard Boltzmann (1844–1906) received his PhD on kinetic theory of gases in 1866 – the year of publication of Mendel's *Versuche über Pflanzen-Hybriden.*

Cato Maximilian Guldberg (mathematician) and *Peter Waage* (chemist) proposed in 1864 the *law of mass action* in chemical kinetics that was logically similar to Mendel's laws; their work went unnoticed until it was rediscovered by van 't Hoff in 1877.

Jacobus Henricus van 't Hoff (1852–1911) was the first winner of the Nobel Prize in Chemistry for his discoveries in chemical kinetics, chemical equilibrium, osmotic pressure, and stereochemistry.

The significance of Mendelian hybridization data, that are similar to the *simple volumes ratios property*, also called *the law of combining volumes* of reacting gases, discovered by Gay-Lussac and reported in his 1808 article

Sur la combinaison des substances gazeuses, les unes avec les autre

would have been apparent to these people. (Gay-Lussac's law says, in Avogadro's words, that

> *les combinaisons des gaz entre eux se font toujours selon des rapports très simples en volume, et que lorsque le résultat de la combinaison est gazeux, son volume est aussi en rapport très simple avec celui de ses composants.*)

Boltzmann especially, a proponent of atomic theory of matter and energy, would have been delighted by Mendel's idea of *atoms of inheritance* conceived with logic similar to that involved in Avogadro's derivation of the Gay-Lussac law from his atomic conjecture:

> *... the number of integral molecules in any gases is always the same for equal volumes, or always proportional to the volumes.*

This is nowadays called the Avogadro Law. It goes along well with the Gay-Lussac law if you assume that chemical substances are composed of distinct mutually identical units (atoms or molecules) but one cannot a priori exclude these units from being *infinitesimally small* in the sense of Leibniz and, accordingly, the *number* of molecules in a finite volume of gas being *infinitely large.* (Such *infinitely large/small* numbers are describable nowadays in the language of *non-standard analysis.* For instance, the number $\mathbf{STOP}$ that we met may be regarded as infinitely large and particles of mass $\mathbf{STOP}^{-1}$ as infinitesimally small.)

But this *number* (defined nowadays as the number of atoms in 12 grams of pure carbon ^{12}C) happened to be finite and not terribly large: the (*Avogadro*) number N_A of molecules in $\approx$ 22.4 liters of a gas at 0°C and 1 atm pressure is $N_A \approx 6 \cdot 10^{23}$. This also $\approx$ equals the number of molecules in 18 g of water H_2O of molecular weight $\approx$ 18. If atoms were one-millionth as large, with $N_A = 6 \cdot 10^{41}$ instead of $6 \cdot 10^{23}$, the idea of atoms could forever remain only an idea.[18] (If atoms were so small, then the age of the Universe – a few billon years – would be too short for the 1 μm cell and, consequently, for organisms of our size, to evolve. Then even the *idea* of atoms could not be possible – no brains no ideas ... unless small atoms would allow brainy organisms of 1 μm in size.)

Avogadro's ideas on atoms had no better luck than those of Mendel on genes – they were generally accepted only years afterwards. But if the problem with Mendel was because his contemporary biologists were not up to his level of science and mathematics, the physicists of the 19th century were skeptical about atoms because of their (physicists') high standards for acceptance of new ideas. Faraday, for example, who understood the problem of atoms as much as Avogadro did, says:

> *it is very easy to talk of atoms, it is very difficult to form a clear idea of their nature, especially when compounded bodies are under consideration.*

(The logical/philosophical problem of atoms had also been very much on the mind of Lavoisier; today, we know that the concept of atoms is self contradictory in the context of classical (non-quantum) physics. But in the natural sciences, unlike how it is in mathematics, a contradiction does not necessary invalidate an idea.)

Atoms and molecules were always on physicists' minds. For example, Maxwell writes:

> *Assuming that the volume of the substance, when reduced to the liquid form, is not much greater than the combined volume of the molecules, we obtain from this proportion the diameter of a molecule. In this way Loschmidt, in 1865, made the first estimate of the diameter of a molecule. Independently of him and of each other, Mr. Stoney in 1868, and Sir W. Thomson, in 1870, published results of a similar kind, those of Thomson being deduced not only in this way, but from considerations derived from the thickness of soap-bubbles, and from the electric properties of metals.*
>
> *According to the table, which I have calculated from Loschmidt's data, the size of the molecules of hydrogen is such that about two million of them in a row would occupy a millimeter, and a million million million million of them would weigh between four and five grammes!*

Maxwell might not have been 100% certain *if there is any truth in the dynamical theory of gases ...* but eventually, atoms won under efforts by Boltzmann, and when Einstein and Smoluchowski wrote down an equation for describing the

[18] 10^{-6} in linear size makes 10^{-18} in volume and therefore in mass the exponent is $18 = 41 - 23$.

stochastic, called *Brownian*,[19] motion of microscopic particles suspended in a fluid that allowed an evaluation of "the size of atoms".

The idea behind this equation is old and simple:

... small compound bodies ...
are set in perpetual motion
by the impact of invisible blows ....
The movement mounts up from the atoms
and gradually emerges
to the level of our senses.

TITUS LUCRETIUS, 50 BCE (?)

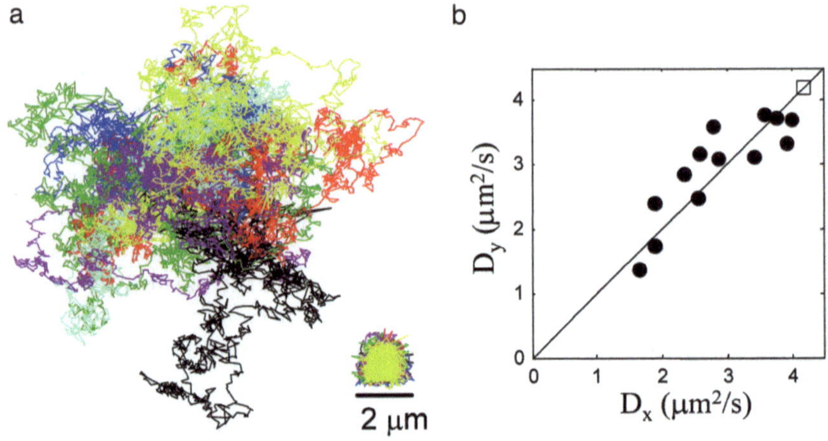

© Adam E. Cohen

But it took nearly two thousand years to turn poetry into science by translating these lines to mathematical language.

This "translation" was performed in different contexts independently (?) by: *Thiele* (1880), *Bachelier* (1900), *Einstein* (1905), *Smoluchowski* (1906), *Wiener* (1923), where the mathematics of this translation was not terribly far from *the multiplication table type*.

(The idea would have been apparent, to Pascal and to Buffon, while Euler and/or Laplace, would have had no problem performing the detailed computation, IF either of them had put his mind to it.)

Is it sheer luck or is it a consequence of the *weak anthropic principle* that Brownian displacement of particles is visible with an optical 1000× microscope?

[19]Robert Brown (1773 –1858) did not discover Brownian motion but he gave a detailed description of a cell nucleus and of other cellular structures. Brownian motion was systematically studied for the first (?) time by Jan Ingenhousz around 1785. (Ingenhousz also pinpointed the role of sunlight in photosynthesis.)

Molecules are too small to be detectable by the human eye in the visible light spectrum, but their impacts are felt by objects of intermediate size.

Say, water molecules, $\approx 0.3 \cdot 10^{-3}\mu$m, are *only* (!) 1000 times smaller (twice on the logarithmic scale normalized by our eyesight) than the smallest optically distinguishable objects/displacements, $\approx 0.2\,\mu$m. A 1 μm bacterium, for example, scaled by the factor 1 000 000 to human size (1 m $= 10^6\mu$m) would "see" molecules of water as grains of sand.

And water molecules move fast: Their (square) average speed is approximately $650\,\mathrm{m}/s \approx 1000^3 \cdot 0.5\,\mu\mathrm{m}/s$ at the room temperature. (Muzzle velocity of a bullet varies 200–1200 m/s for most guns.)

What one observes in reality are displacements of particles suspended in water under *multiple* collisions with water molecules where disproportionally many of them move in a certain direction. The frequency and average size of these displacements is related to the Avogadro number (or, essentially equivalently, to *Boltzmann constant*) via the Einstein–Smoluchowski formula (the reciprocal to the *Avogadro number* corresponds to "the volume size of an atom.")

Performing a measurement and an accurate evaluation of the Avogadro constant was by no means simple – this was achieved by Jean Baptiste Perrin and his team around 1913–1914 with the resulting value for the Avogadro number: $6.03 \cdot 10^{23}$. (The currently accepted value is $6.0221 \cdots \cdot 10^{23}$ that is obtained with X-ray crystallography of silicon single crystals.)

Do the laws of classical (*non-molecular*) *genetics*, and/or the *classical laws* of physics and physical chemistry of the 19th century still have something new and interesting to offer to us?

A scientist would find it unlikely but a mathematician may think otherwise – we do not feel that anything is understood, unless it is expressed in the language of 21st-century mathematics. But even 20th-century mathematical rendition of something as innocuous as the 18th-century *ideal gas laws*, (e.g., in terms of Riemannian and/or integral geometry) seems non-trivial. (The 20th-century mathematics "dissolves" the law of inertia in the "matrix" of geodesic flows over spaces with affine connections; we expect something more interesting from such a "matrix" in the 21st century.)

Mathematics also serves as a *cord* that ties laws to observations – the *empirical truth* of a law is not seen in passively observed *facts*. Mathematical ideas enter, often implicitly, into design and interpretation of experiments.

Nobody, for example, ever saw a *body* moving by itself with constant velocity along a straight line – the law of inertia stands in stark contradiction with what you observe. Yet, you cannot draw a mathematically consistent/elegant picture of mechanics without this law; e.g., you cannot write down an esthetically acceptable (say, given by analytic functions) formula compatible with Galileo's rolling balls on inclined planes without *taking* the balls velocity being *constant* for the inclination angle *zero*.

5. Truth

La pensée ne doit jamais se soumettre,
ni à un dogme, ni à un parti, ni à une passion,
ni à un intérêt, ni à une idée préconçue.

HENRI POINCARÉ

When you point out something naive or just plain silly in writings by a scientist of the past, one may say that you yourself are being foolish by judging an idea of yesterday from the standpoint of today.

Well ... this would be convincing if not for people like Lavoisier, Claude Bernard, Faraday, Poincaré. They had fine ears for the ringing of scientific truth – they never (almost never?) said anything senseless; mistaken – maybe, but never trivial, vacuous, or intrinsically inconsistent.

Qu'un objet change d'état ou seulement de position,
cela se traduit toujours pour nous de la même manière
par une modification dans un ensemble d'impressions.

HENRI POINCARÉ, LA SCIENCE ET L'HYPOTHÈSE, 1902

Everything shined at the touch of Poincaré's mind, be it mathematics, physics, or anything else. His analysis of the space perception in Chapter IV of *La science et l'hypothèse* remains unrivaled in its depth and clarity. His conjecture on how the brain reconstructs the *a posteriori Euclidean* geometry of external space with its rotational (orthogonal[20]) symmetry from the input of moving retinal images, that amazingly agrees with discoveries of 20th-century neurophysiology, stands as a challenge for experimental (mathematical?) psychologists of the 21st (22nd?) century.

... science d'observation est une science passive.
Dans les sciences d'expérimentation, l'homme ...
provoque à son profit l'apparition de phénomènes,
qui ... se passent toujours suivant les lois naturelles,
mais dans des conditions que la nature
n'avait souvent pas encore réalisées.

... l'astronome fait des observations actives,
c'est-à-dire des observations provoquées par
une idée préconçue sur la cause de la perturbation.

CLAUDE BERNARD, INTRODUCTION À L'ÉTUDE
DE LA MÉDECINE EXPÉRIMENTALE, 1865

---

[20]The *orthogonal group* $O(3)$ of rotations of 3-space is the basic *non-commutative atom* of our world geometry. Mathematicians' penetration into its amazing structure took two-and-a-half millenia – from *Pythagoras' theorem* to the theory of *Lie groups* developed at the turn of the 20th century and the present day Euclidean (elliptic) gauge theory over 4-D spaces. Amazingly, the brains of all actively moving animals on Earth have developed (by learning?) fair models of this structure.

If you quote a writing by Claude Bernard in an audience of mathematicians they will assume this must be Poincaré, physicists would think of Einstein or Feynman, and biologists of Darwin. But some may recall having seen these very words in a memorandum signed by several Nobel laureates in the last issue of Nature.

One hardly can add anything of substance to what Claude Bernard had already said about experimental science and what he says applies to theories as well, since a design of a logical argument is similar to that of an experiment.

Claude Bernard insisted that an *active science* depends on experiments and these must *control rather than confirm* our ideas. He could have continued by saying that an *active logic* must similarly control rather than confirm our ideas. We should not decide beforehand where the argument will bring us; yet, the rules of our logic must be set in advance. Then we go through an argument step by step and accept whatever comes out of it even if this contradicts our original ideas.

It is almost impossible to design such arguments without mathematics, where *going through an argument* being furnished by something like a proof of a theorem or "just" a computation; it is amazing how often the result of a simple computation may contradict your intuition.

An instance of that is the (Mendel)–Hardy–Weinberg formula we met before that expresses the (counterintuitive) idea that (naively understood) *evolution by natural selection* may stabilize on the second round of reproduction.[21]

Let us give another example that simultaneously illustrates Claude Bernard's remark that *averages confuse, while aiming to unify.*

Suppose that the average success of a certain brain surgery performed in two hospitals H_1 and H_2 depends on whether the father of a patient was right or left handed.

The *right* patients are sent to H_1, since

H_1 has **better average success** for ***right***.

The *left* patients are also sent to H_1, since

H_1 has **better average success** for ***left*** as well.

But if you do not know whether you are in the *left* or *right* category you are sent to H_2, since

H_2 has **better average success** for ***right*** + ***left*** together.

Perplexed?[22] Then look at the following formula for the average success rate S_{r+l} for *right* + *left* together in terms of those for *right* and *left* separately, denoted

[21] The mathematics of this formula necessitates the passage from Darwinian idea of *variation* to the more abstract (and more adequate) concept of *mutation* that was envisioned by Maupertuis in his picture of Life. (The term *mutation* for suddenly appearing variations was introduced by Hugo de Vries in his two-volume *The Mutation Theory,* 1900–1903.)

[22] All mathematicians I know refused to believe this could ever happen, but those of my friends who were not certain about their math pretended they could admit such possibility.

S_r, and S_l, where N_r and N_l are the numbers of the *right* and *left* patients who have undergone this surgery,

$$S_{r+l} = \frac{N_r S_r + N_l S_l}{N_r + N_l}.$$

Now you see how this *can* happen in certain cases: S_{r+l} depends not only on S_r and S_l but also on N_r and N_l, where these numbers may be quite different in the two hospitals.

Rephrasing Einstein – *the laws of mathematics are certain even when they refer to reality; it is which reality they refer to that is not certain.*

... à ne chercher la vérité que dans l'enchaînement naturel des expériences et des observations, de la même manière que les mathématiciens parviennent la solution d'un problème ...

ANTOINE LAVOISIER, DISCOURS PRÉLIMINAIRE AU TRAITÉ ÉLÉMENTAIRE DE CHIMIE, 1787

Je vous dénonce le coryphée des charlatans, le sieur Lavoisier.

JEAN-PAUL MARAT, L'AMI DU PEUPLE, JANVIER 1791

What unworthy motives ruled for the moment under high sounding phrases.

MICHAEL FARADAY

Lavoisier was arrested in November 1793, accused of selling watered-down tobacco and guillotined on 8 May 1794 at the age of 50.

Lagrange[23] commented on Lavoisier's execution:

Cela leur a pris seulement un instant pour lui couper la tête, mais la France pourrait ne pas en produire une autre pareille en un siècle.

Claude Bernard was born in 1813 and Henry Poincaré in 1854.

But who was Jean-Paul Marat?. Marat performed numerous experiments and published work on fire, heat, electricity, and light. This was not taken seriously by the contemporary French "academic establishment", in particular, by Lavoisier and Condorcet.

Some historians sympathetic to Marat argue that, according to certain evidence, he was a knowledgeable and dedicated scientist who did not get a fair hearing at the French Royal Academy and that a comparison of Marat to somebody like Mesmer is unfair. (In 1784, Lavoisier and Benjamin Franklin, who were heading *la Commission royale d'enquête sur le magnétisme animal*, conducted the historically first controlled clinical trial and disproved Mesmer's claims.)

This might be true, but, apparently, Marat, however *knowledgeable*, was not appreciative of the extent of his *non-understanding*. Thus, he published a volume of a few hundred pages on his study of electricity. Lavoisier, on the other hand, who believed that

l'électricité ne jette pas seulement de la lumière sur les effets du tonnerre, elle sert encore à expliquer un grand nombre des opérations de la nature,

and who had been pondering on electricity most of his life, had published virtually nothing on it. Lavoisier thought at some point that

le fluide éctrique et le fluide magnétique sont eux-mêmes composés, comme l'eau, de deux fluides plus subtils encore, qui sont susceptibles de se réunir et de se séparer, et, en quelque façon, de se neutraliser l'un par l'autre,

but he could not find sufficient experimental evidence for his ideas and remained uncertain on the nature of electricity; but Marat believed he understood what electricity was. One tends to agree with Lavoisier's assessment of Marat's work as being insignificant.[24]

Marat refused to accept Lavoisier's judgement and hated Lavoisier, but he was not, however, *directly* involved in Lavoisier's condemnation: Marat was assassinated in July 1793.

---

[23]Lagrange (1736–1813), born Giuseppe Lodovico (Luigi) Lagrangia, was a mathematician. The imprints of his analytic rendition of the *least action principle* (more precise than that by Maupertuis and more general than that by Euler) are pronouncedly visible as *Lagrangians* (instead of Newtonian forces) *of dynamical systems* in many segments of mathematical physics along with their younger sisters *Hamiltonians*.

[24]Lavoisier could have said what he thought of Marat in Pauli's words: *I don't mind your thinking slowly; I mind your publishing faster than you think.*

6. Life

... strutture di ossa per uomini, cavalli o altri animali, che potessero sussistere e far proporzionatamente gli uffizii loro, mentre tali animali si dovesser agumentare ad altezze immense .

GALILEO GALILEI, DISCORSI E DIMOSTRAZIONI MATEMATICHE INTORNO À DUE NUOVE SCIENZE, 1638

The science of *allometry*, the birth of which you witness reading the above lines, still mystifies us by some of its findings. Why, for example, does the metabolic rate R of an animal of mass M closely follow

Kleber's 3/4-law: $R \sim M^{3/4}$ *?*

La Nature contient le fonds de toutes ces variétés, mais le hasard ou l'art les mettent en oeuvre.

PIERRE-LOUIS MAUPERTUIS, VÉNUS PHYSIQUE, 1745

Maupertuis pinpointed the role of natural selection in evolution, outlined a (not quite Mendelian) picture of heredity, sketched the theory of mutations and suggested a germ plasm mechanism close to that of Weismann. Since he was a mathematician, biologists are justified in having ignored his ideas.

... *que de machines ... sont renfermés dans cette petite partie de matière qui compose le corps d'un animal! ... combinaisons ... de principes, que nous ne connaissons que par des résultats si difficiles à comprendre, qu'ils n'ont cessé d'être des merveilles que par l'habitude que nous avons prise de n'y point réfléchir!* [25]

GEORGES-LOUIS BUFFON, HISTOIRE GÉNÉRALE DES ANIMAUX, 1749

Buffon applied integral calculus to measure chances of geometric random events. For example, if one throws a needle of unit length to the plane divided into parallel strips of unit width, then the probability that the needle will cross a line between two strips equals $2/\pi$ for $\pi = 3.14\ldots$ by Buffon's formula.

(This brought to life the fields of integral geometry and of geometric probability; also this inspired the currently accepted mathematical foundation of probability theory laid out by Kolmogorov[26] in 1933.)

Buffon designed efficient lenses for lighthouses.

Buffon suggested a construction of concave mirrors that has been in use for two centuries afterwards.

[25] Was Buffon able to see this because of his mathematical background? Was he the first to realize it? Was it possible at all to appreciate Life's complexity and our powerlessness in structural representation of live entities prior to the development of molecular biology in the last few decades?

[26] Beside modern probability, several other branches of 20th-century mathematics and mathematical physics grew out of work by Kolmogorov including *KAM theory* in classical (Hamiltonian) mechanics, a stochastic theory of *turbulence*, and *entropy* theory of dynamical systems.

Buffon offered the first scientific scenario of formation of planets, namely from a collision of a comet with the sun.

Buffon evaluated the age of Earth at several (hundred?) million years on the basis of sedimentation rates and argued that at least $\approx 75\,000$ years were needed for the Earth to cool down from the molten state, as he calculated by "rescaling" the results of heating and cooling iron balls.

(The idea of hot molten Earth was present in the writings of Descartes and Leibnitz.)

Buffon gave a definition of species:

> *One should consider as being of the same species that which by means of copulation perpetuates itself and preserves the similarity of that species ... If the product of such mating is sterile, as is the mule, the parents are of different species. Any other criterion, particularly resemblance, is insufficient ... because the mule resembles the horse more than the water spaniel resembles the greyhound.*

This is rather different from how it was formulated by the natural philosopher John Ray who defined species by their

> ... *distinguishing features that perpetuate themselves in propagation from seed.* (1686)

Apparently, Buffon, a mathematician in the depth of his heart, must have been mystified and fascinated by the remarkable fact that the obstruction to interbreeding shows in the *second* generation, defined an *equivalence relation* between groups of organisms that allowed an introduction of the concept of species.[27]

Buffon would be happy to see that his kind of mathematically active scientific thinking returned to taxonomy in biology at the end of the 20th century with the comparative study of genome sequences.

Buffon applied his definition to prove the unity of humanity:

> *The Asian, European, and Negro all reproduce with equal ease with the American.*

Buffon writes:

> ... *one could equally say that man and ape have had a common origin like the horse and donkey that each family among the animals and vegetables have had but a single stem, and that all animals have emerged from but a single animal which, through the succession of time, has produced by improvement and degeneration all the races of animals.*

---

[27]In 1942, an influential evolutionary thinker Ernst Mayr reiterated Buffon's definition as

> ... *interbreeding natural populations, which are reproductively isolated...*

What would Buffon make of the 20th-century "evolutionary thinkers" who were unaware of mathematics of *clusterization* and who authoritatively stated that having a progeny needs mutual proximity of participating animals? (Never mind plants.)

Buffon, who had already had enough trouble with ecclesiastical authorities, rejected this idea in writing, since it contradicted the biblical version of creation.

Buffon would find it ironic that certain post-Darwinian evolutionary thinkers who have freely borrowed from his ideas, have, nevertheless, inherited a pre-Cartesian respect for *authority* and submission to *ideology*[28] from his ecclesiastical opponents. These thinkers have not admitted Buffon to membership in their club as a *non-believer* in Darwinism.

Buffon overhauled much of the knowledge of his time and developed a view on Nature and Life from a broad scientific perspective. But he wrote down only 36 volumes of his *Histoire naturelle, générale et particulière* out of planned 50 before he died in 1788.

(Buffon, like Einstein, complained that his natural laziness had precluded him from achieving more in science.)

Buffon's idea of evaluating the age of Earth was taken up by William (Kelvin) Thomson in 1862 who calculated the rate of thermal diffusion from the Earth's crust of its initially molten state and, thus, estimated Earth being at most of order 100 million years old. (This was in contradiction with the sedimentary geological estimates by Maillet, Lyell, and Darwin.)

Thomson is famous for the concept of *absolute zero*, $\mathbf{0}^\circ\,\mathrm{K} \approx -273.15^\circ\mathrm{C}$ and for the idea of the *heat death of the Universe*:

> *... a state of universal rest and death, if the universe were finite and left to obey existing laws. ... science points rather to an endless progress, ... involving the transformation of potential energy into palpable motion and hence into heat ....*

As an engineer Thomson had greatly contributed to the transatlantic telegraph project.

But Thomson is also renown for his:

> *heavier-than-air flying machines are impossible*

and the characteristically late Victorian:

> *There is nothing new to be discovered in physics now.*

This contrasts with

> *... we have no right to think thus of the unsearchable riches of creation ....*

by Maxwell (who is often misquoted at this point).

---

[28]We value in ideas their *beauty* and their *novelty*; we appreciate in people the depth and originality of their *understanding* of non-trivial ideas. Carriers of an ideology, on the other hand (often recognizable by *ist* attached to their tails), may be tolerant to those who are confused by what they (the carriers) proclaim – these are potential converts, but they are indignant against anyone who *clearly sees and rejects* their views. Faraday's

> *Nothing* [is] *quite as frightening as someone who knows they are right*

is about these people.

Buffon's direct impact on physics was, probably, rather limited but the flow of ideas emanating from his "Natural History" had been shaping the minds of generations of biologists and natural philosophers up to the middle of the 20th century.

> *It is in relation to the statistical point of view that the structure of the vital parts of living organisms differs so entirely from that of any piece of matter that we physicists and chemists ...*
>
> *... living matter ... is likely to involve "other laws of physics" hitherto unknown, which however, once they have been revealed, will form just as integral a part of science ....*
>
> ERWIN SCHRÖDINGER, WHAT IS LIFE? 1944

> *The principle of continuity renders it probable that the principle of Life will hereafter be shown to be a part, or consequence of some general law.*
>
> CHARLES DARWIN, LETTER TO GEORGE WALLICH, 1882

Whereas Darwin dreamed of some *mathematical/philosophical* PRINCIPLE of LIFE, Schrödinger wanted to make sense of LIFE in the light of PHYSICAL LAWS.

But LIFE breaks physical symmetries/regularities and creates new ones[29] of quite different nature, e.g.,

the striking regularity with which the same hybrid forms reappear.

The laws of physics, obviously, restrict possible structures/behaviours of living systems as much as the rules of chess restrict the moves of pieces on a chess board.

But moves of a master in a chess match are no more accountable for by the rules of chess than mathematical theorems are representable by reflections of the laws of logic.

These rules, one might argue, well account for the statistics of the moves of somebody who plays chess for the first time in his/her life but not of a player with a couple of years of experience. This is how it is with most *physical systems* – they start their games essentially from level zero, where the laws of physics apply.[30]

But the game of life on Earth has been played by Nature for a few billion years. The laws of physics, as Galileo pointed out, tell you that there are no flying elephants but they cannot help you to draw the path that brought *walking* elephants to Earth. Compatibility with the laws of physics is a minuscular part of the reason for existence in biology.

---

[29] It is not always clear what should be called "breaking" and what is "creating" a symmetry. Does, for example, selection of a particular *chirality* by biological systems break or create stochastic symmetry in ensembles of molecules?

[30] The second law of thermodynamics is an exception: it applies, as Boltzmann pointed out, only to *prepared states* of physical systems, prepared by their *history*. This law appears to *Naive Mathematician* as a physical corollary to a general mathematical theorem, a theorem nobody has managed to formulate yet.

© AaronAmat / Getty Images / iStock

It is not accidental that when Buffon and Schrödinger try to speak of Life in the language of physics they start speaking poetically.[31] The great painting of Life is not about the physics of the canvas and the chemistry of dyes.

But don't physical and biological theories share the hallmark feature of "true science": *falsifiability*?

> *You don't need 100 famous intellectuals to disprove a theory. All you need is one simple fact.*
>
> ALBERT EINSTEIN

For instance, think of *Galileo's law of falling bodies*:

if you let go down some objects ✱, ✺ *and* ❀ *you hold in your hand, say of* 300, 30 *or* 3 *grams of weights, they all reach the floor in* $\approx \frac{1}{4}$ sec *essentially independently of their weights.*

Would the edifice of classical mechanics collapse if, in some experiment on the sea level of planet Earth, one of these objects, for instance ✺ of 30 grams of weight, had remained in the air for 10 seconds before reaching the ground?

The answer could be "*Yes, it would*" IF Earth were bare of life. And to get the correct answer, no 100 intellectuals are needed, a single *healthy adult sparrow* for ✺ will do.

But there is no way to say what "(bare of) *life*" signifies in the language of physics.

The main impediment to "*reducing Life to Physics*" is not that we lack some "Law of Nature", but incompatibility of the two contexts underlying the languages appropriate for description of physical and of biological phenomena. What we call *physics*, or rather *theoretical physics*, is not only a collection of mathematical models, but also a set of *tacit* rules of when, where, and how these models should be related with results of experiments.

[31] This is reminiscent of Trevize in Asimov's *Foundation and Earth* who sees the Milky Way Galaxy as an unfolding of Cosmic Life.

One can imagine a universe with physical laws very different from ours; yet, inhabited by somebody very much like ourselves, e.g., a virtual world of a sophisticated computer program. Probably, an outline of the ways of Life can be seen in how *mathematical order* (*mysteriously*) "assembles" from simple "logical pieces", not in the pieces themselves.

But is there an abstract notion of *system* suitable for biology?
Can a cell be regarded as a *single* physical system?
What is a true *Grothendieck-style*[32] definition of "*single unit*" in biology?

A few facets of life can be described in the language of probabilities that is commonly used in physics, but these descriptions are dissimilar to the pictures that we see in physics and in chemistry as Schrödinger says. *Statistics* of LIFE, unlike *statistics* of *non-Life*, is characterized by

- *Improbably* **high multiplicity** *of occurrence of supposedly rare seemingly independent events,*
  such as the presence of thousands of *nearly identical* randomly looking *complex* protein molecules in a cell. (But hardly *two complex* snowflakes found in Antarctica are ever completely alike.)
  On a different scale, and for a somewhat different reason(s), there may be trillions of *copies* of *thousands residues long* polynucleotide molecules (DNA and RNA) and/or of viral particles in a pond of water, not to mention seven billion enormous multi-molecular aggregates of almost indistinguishable compositions and shapes, that bipedally locomote themselves on the surface of Earth.

This *improbable multiplicity* phenomenon is partly due to

- *Amplification of rare and/or low energy events*, e.g., of advantageous mutations in populations and of chemical signals in cells. (The latter would look similar to a chain reaction of combustion ignited by an accidental spark, if not for *key/lock mechanisms* ever-present in biological systems.)

This amplification is also responsible for

- *Gross discrepancy between "statistically averaged", "significant", and "typical" events/behaviours.*
  Processes in Life often run as lotteries – no one's gain is close to the average – in stark contrast with comparably large *stochastically homogeneous* physical systems, such as the bulks of gases and liquids where *typical* tends to be close to *average*.

In many (but not all) cases the above features are associated to *information* that is *encoded* in a biological system and that can be *transmitted* from one biological (sub)system to another.

---

[32] Alexander Grothendieck (1928–2014) is the originator and crystallizer of a body of fundamental concepts of 20th-century mathematics.

In fact, "information" is a commodity more valuable for Life than energy: The transfer of *sequential information* in the course of replication

$$\text{DNA} \to 2 \times \text{DNA}$$

is a defining factor of Life on Earth.

And the "*circulation of information flow*(*s*)" that makes the cell tick starts with the sequential transfers

$$\text{DNA} \to \text{RNA} \to \textit{polypeptides}.$$

These transfers are "formal" and (essentially) expressible in terms of Shannon's style of information theory. But there would be no Life if not for *macromolecular folding*. In a way, Life is created by

$$\textit{polypeptides} \overset{\text{folding}}{\rightsquigarrow} \textit{proteins}.$$

Albeit folding is an essentially physical process,[33] physics per se cannot tell how and why it works, because, besides lacking precise "physical/chemical formulae" for amino acid residues' interactions in the aqueous environment,[34] folding applies only to those *rare* polypeptide chains that were "selected by Life" for their specific roles in the cell.

Neither physics nor present day mathematics can identify and formally describe polypeptide sequences that would fold to "*potentially live-cell proteins*".

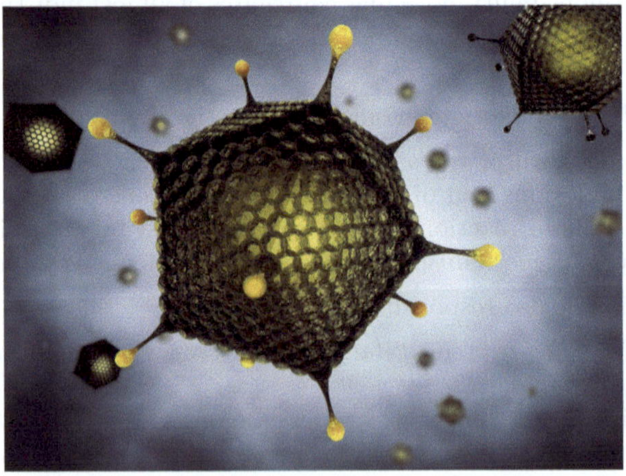

Even worse conceptually, we possess no mathematical framework where we could formulate the idea of "*creative reduction of information by folding*".[35]

[33] Folding of some proteins needs "biological assistance" in a cell.

[34] There is not even a satisfactory model of water itself on the nanoscale.

[35] Apparently, the bulk of the message carried by the amino acid sequence tells a protein P how it should fold rather than what it is supposed to do upon folding. This "folding information"

Similarly, there is no available mathematical formalism for the expression of the "*economy of information principle*" that dictates the icosahedral symmetry of viral particles (such as in the picture blow), where this information is what is needed to genetically encode an assembly of such particles in the rotationally symmetric physical space.

And evolving the ability of termites to build skycrapers was, probably, grossly facilitated by the possibility of a *symmetric* implementation of the building program in *identical* genomes of the builders.

Is this kind of *information* a precursor of a non-trivial mathematical concept or it will forever remain a poetic metaphor?

Physics has no means for addressing these kinds of questions. But why – one asks with an insinuating smile – were there fundamental contributions to biology by physicists and not so many, if at all, by mathematicians?

I could give you an answer, but this would be ... opinion.

7. Evolution

Nothing in Biology Makes Sense Except in the Light of Evolution.[36]

This is the title of a 1973 article by THEODOSIUS DOBZHANSKY.

We now may be able to understand it [evolution] *in biology.*

JACQUES MONOD, ON THE MOLECULAR THEORY OF EVOLUTION, 1975

We have an excellent idea of the core genetic makeup
of the last common ancestor of all bacteria
that probably lived more than 3.5 billion years ago.

EUGENE KOONIN, THE LOGIC OF CHANCE, 2011

The latter is not another excellently contrived JUST SO STORY but an outcome of statistical analysis (of *multiple alignments*) of genomes from data bases that have been accumulated at a *petabyte* (10^{15} bytes) rate for the last ten years. From the petabyte hight, Dobzhansky's *Light* and Monod's *understand* appear as dim as the early 20th-century picture of evolution, often called *modern evolutionary synthesis*, gauged by the molecular standards of the 1970s.

And the ideas of the 18th and 19th centuries serve for molecular biologists only as references to the names behind them. Yet, such concepts as *reproduction – inheritance – selection – competition* seem to carry still undeciphered messages that we would be able to read if we could translate *natural poetry* of olden days to our mathematical language, preferably, not exactly of *multiplication table type*.

is, for the most part, invisible on the exposed surface of a folded P but the sequence can be recovered when P is unfolded. Thus, the term "reduction" is not quite appropriate.

[36]In mathematical terms, LIFE ON EARTH becomes *connected*; thus, a topologically *non-trivially* structured entity *only* in the presence of the *time* coordinate. Is this connectivity or the *full* topology (geometry?) of LIFE ON EARTH what Dobzhansky calls *sense*?

First forms minute, unseen by spheric glass,
Move on the mud, or pierce the watery mass;
These, as successive generations bloom,
New powers acquire and larger limbs assume;
Whence countless groups of vegetation spring,
And breathing realms of fin and feet and wing.

ERASMUS DARWIN, THE TEMPLE OF NATURE, 1802

... difficult to believe ... that the more complex organs and instincts should have been perfected, by the accumulation of innumerable slight variations. Nevertheless, ... all organs and instincts are, in ever so slight a degree, variable ... a struggle for existence ...

CHARLES DARWIN, THE ORIGIN OF SPECIES, 1859

Some pages later Darwin writes:

... it is as easy to believe in the creation of a hundred million beings as of one; but Maupertuis' philosophical axiom "of least action" leads the mind more willingly to admit the smaller number.... Analogy would lead me one step further, namely to the belief that all animals and plants have descended from one prototype.

This sounds almost like a mathematical proposition! Pythagoras would have loved such a quintessential formulation of the idea of evolution. He might have learned this idea from Anaximander and he would have stated it almost as Darwin put it:

Theorem 1. *Every two organisms on Earth, be they plants, animals or humans, have a common ancestor.*

But only Pythagoras' corollary reached us across the chasm of time:

As long as man continues to be the ruthless destroyer of lower living beings he will never know health or peace. For as long as men massacre animals, they will kill each other.

Darwin, who hated slavery of any kind, adds:

Animals, whom we have made our slaves, we do not like to consider our equal.

Lamarck, Darwin, and Wallace had scrutinised and systematised vast amount of material, in part collected by themselves, that led them to believe in transmutation of species and evolution in general. Darwin and his followers chose to present their ideas to the general public in the language of political economy: *competition for resources, struggle for existence*, etc. That was well taken by their Victorian audience acquainted with *The Wealth of Nations* by Adam Smith and prepared for evolutionary ideas by *Vestiges of the Natural History of Creation* published anonymously in 1844 by Robert Chambers.

But **Theorem 1** was truly *proved* only a century later with the advent of molecular biology and sequencing techniques that uncovered *non-ambiguous* structural similarity between molecular architectures of all living cells.

Everyone who is able to *understand* Pythagoras' theorem will also *understand* that such a degree of similarity could be neither accidental nor come from any kind of *convergent evolution.* You do not have to *convince* anybody anymore and can discard *struggle for existence, fittest survives,* and *creative power of selection.* Granted the accumulated sequential and molecular structure data, all you need is *a little mathematics of the multiplication table type* in Hardy's terms.

(The Darwinian *metaphor* of *struggle for existence* applies, with no possibility of being taken literally, to *numbers* $R_1 > 1$ and $R_2 > 1$: The greatest of the two numbers survives in the formula $\sqrt[T]{R_1^T + R_2^T}$ for time T being large. But when struggle for existence is applied to animals the reproduction rates of which are represented by these numbers, it may give you wrong ideas.)

Yet, there are two problems with **Theorem 1**.

1. It logically (and obviously) implies that there was a first *proto-cell* from which all other cellular organisms (viruses?) descended, where *proto* is a shorthand for *I have no idea what the Hell I am talking about.*[37]
2. The second question reads: Where is the hidden beauty? What are the physical/chemical/biological mechanisms besides heredity that drive evolution?

The idea (opinion?[38]) of NATURAL SELECTION can be condensed to a single word:

NONE[39]

The *potentially exponential* rate of growth of populations allows *random* variations to cover *all* possibilities; all Nature/enviroment has to do is to *select* those she favors.

Darwin also insisted that consecutive steps of evolution are implemented by *small* variations.[40] These, he argued, are more likely to be non-deleterious; besides, this increases the chance of a (more or less) simultaneous occurrence of several, say two, variations that are advantageous only if they come together.

---

[37]The Earth is about 4.5 billion years old and *great oxygenation* of Earth's atmosphere happened ≈ 2.5 billon years ago. There are ≈ 2 billion year-old fossils of multicellular organisms and traces of pre-oxygenic bacteria ≈ 3.5 billion years old. But when and what was the age of proto-life?

[38]An assertive expression of an idea that has been around for a few decades classifies as an *opinion.*

[39]This answer brings to mind the famous apocryphal response of Laplace to Napoleon:

Sire, je n'ai pas eu besoin de cette hypothèse.

[40]The Western idea of gradualism can be traced back to Milo of Croton, an associate of Pythagoras, who, around 540 BCE, gradually developed his strength by having carried a calf growing to bull on his shoulders.

(Major transitions in evolution are, contrary to what Darwin would think, associated not with the accumulation of *point mutations* but with *abrupt* and significant genome rearrangements, including *duplications of genes and of whole genomes*. For instance, the divergence of the lineages of humans and chimpanzees, that occurred about 5 million years ago, was, most likely, triggered by the *fusions of two chromosomes*: great apes have 24 pairs of chromosomes while humans have only 23 pairs.)

Besides, Darwin suggested several scenarios of how particular patterns of evolution, e.g., that of the vertebrates' eyes, could have been gradually achieved by selection.

But Pythagoras would point out that *X is not impossible* does not imply *X is true*, e.g., for *X* being *selection and nothing else*,[41] and that the true biological problem is concealed by dismissive *besides* attached to *heredity*.

In fact, until the development of cellular/molecular biology that started at the end of 19th century, nobody was close to imagining the immensity and the beauty of the hidden structural complexity of Life; hence, of heredity. Even today, some may argue, we have not reached the point where *I know that I know nothing* coming from a biologist will be not just words. But as recently as at the beginning of the 19th century, some, e.g., Lamarck (but not Maupertuis) earlier, believed that spontaneous generation of worms from dirt was plausible.

There was a logical reason for this belief: A presence of rich flora of parasitic worms in the intestine of Adam would contradict the idea of Eden. But it became clear to (almost) everybody by the end of the 19th century that "spontaneous generation" of a single cell of a worm is no more probable, than spontaneous emergence of *great pyramid of Giza* – the first of the seven wonders of the world, out of stones, sand, and dirt by some natural physical process some forty-five centuries ago.

Eventually, accumulation of data on *cell division*, in particular, the discovery of *meiosis* (cell division for production of *gametes*, e.g., sperm and egg cells in animals) by Oscar Hertwig in 1876, led to the *germ plasm* idea proposed by August Weismann around 1890 that turned evolutionary theory toward *active science* in the sense of Claude Bernard.

In modern terms, the *Weismann principle* says:

genomes vary but organisms are selected.

(To confirm his idea and to disprove Lamarck, Weismann was chopping off the tails of several hundred mice for ≈ 20 generations and recorded no mouse born without a tail.[42])

---

[41]Darwin himself wrote in the 1872 edition of *The Origin of Species* that he was

> ... *convinced that natural selection has been the main but not the exclusive means of modification.*

This, however, like prophecies by the oracle of Delphi, is open to many interpretations.

[42]Lamarck would point out that if Weismann had done it to a hundred million mice for a hundred million generations, then the tails would degenerate as do the eyes of fish living in dark caves. He

Soon thereafter, Mendel's ideas was rediscovered and an explosion of genetics followed by that of molecular biology began. The focus of the evolution problem shifted from organisms to genomes.

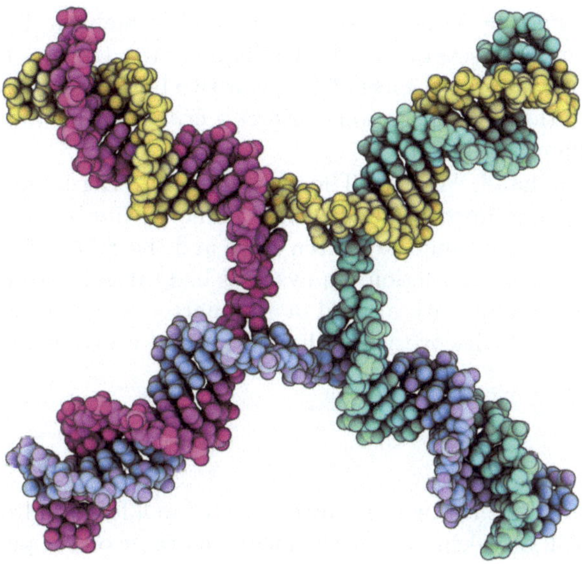

© Laguna Design / Science Photo Library

We want to learn how genomes "fly", how their "engines" work. This, we believe, will bring us closer to understanding how they fare in the adverse winds of competition and selection. (But doesn't this metaphor go too far when applied to natural selection? Granted, for instance, that the main factor for having a safe flight is a *selection* of an air company, can you seriously say that this selection is the main factor that keeps an airplane safely in the air? If not, how can you take natural selection as the *main* factor in evolution as Darwin says?)

A rough quantitative picture of genome evolution can be seen with simple mathematics of *biased* (non-symmetric) *random walks* on *finite* (but very large) *graphs* with *absorption*, where the vertices of the graphs represent genomes and absorption corresponds to extinction/selection. But we are still far from *formulating* **Theorem 2**.

An amazing thing about old fashioned naturalists, especially Darwin, is that with *no support by an experiment* and/or with *no quantitative reasoning*,[43] they

would also conjecture that if Weismann had done it to a hundred million generations of lizards then, instead of regenerating their tails, they would start growing second heads with one not being enough to comprehend the scientific significance of such experiment.

[43]Because of these shortcomings, the Darwinian scheme of evolution may appear exceedingly naive. For instance, biochemist Ernst Chain (1906–979), one of the main contributors to isolating and applying penicillin, writes: *... variants do arise in nature and ... their emergence can and*

could built consistent pictures of large fragments of Nature and sometimes could do it (literarily) a hundred times better than the most luminous physicists, astronomers, and mathematicians of their time.

For example, Darwin and his friend, geologist Charles Lyell, following ideas originated in the work by James Hutton, estimated (somewhat differently) the age of Earth to be at least several hundred million years. (About 150 years earlier Benoît de Maillet evaluated the age of Earth as two billion years by estimating the rates of sedimentation and formation of Earth's crust, but his argument is judged as having been poorly founded.)

On the other hand, William Thomson (Kelvin), Hermann Helmholtz, and Simon Newcomb came up with about thirty million years by evaluating the time needed for Earth to cool from the molten state and the Sun to be heated and sustained by gravitational contraction. If physicists had taken Hutton, Lyell, and Darwin seriously, they might have arrived at the matter-energy idea several decades before Becquerel's 1896 discovery of radioactivity and Einstein's $E = mc^2$ of 1905.

And pondering over

It is not the strongest of the species that survives, nor the most intelligent that survives. It is the one that is the most adaptable to change.

one starts wondering if these words may come true in a few hundred years from today: The Earth will be taken over by the most adaptable ones – protozoa, bacteria, and viruses after Man will have done with multicellular life.

(The reader may be relieved: The above is a common *misquotation* of Darwin; he was neither that stupid nor *that* gloomy on the future of mankind.)

About Hutton, Lyell, Helmholtz and Newcomb

James Hutton (1726–1797) recognized the role of subterranean heat in the creation of new rock followed by the gradual process of weathering and erosion on a very long geological time scale – *la route éternelle du temps* as Buffon says in *les Époques de la Nature*, published in 1778. (Many of these ideas, including evolution, appear in *Protogaea*, written by Gottfried Leibniz between 1691 and 1693, and published in 1749. For example, Leibniz writes: *The globe of the earth ... has hardened from liquid, light or fire being the motive cause.*)

Hutton thought of Earth geo-dynamics as being cyclical, with time stretched without limit into past and future. Concerning evolution of life, he accepted the dominant role of selection in what we now call *microevolution* but not, as Darwin and Wallace did, in *macroevolution.*

Charles Lyell (1797–1875) was a proponent of gradualism in geology whose uniformitarian ideas influenced Darwin.

---

does make some limited contribution towards the evolution of species. The open question is the quantitative extent and significance of this contribution.

Hermann von Helmholtz (1821–1894) invented the *ophthalmoscope* for examining the inside of the eye and the *Helmholtz resonator* to identify frequencies of sound waves.

Helmholtz measured the speed of propagation of nerve impulses and he developed mathematical and empirical theories on depth, color, sound, and motion perceptions.

Helmholtz formulated the *law of conservation of energy* in his mechanical foundation of thermodynamics, where he also introduced *Helmholtz free energy.*

Being a rare scientist whose discoveries had found immediate uses, he nevertheless states:

> *Whoever in the pursuit of science, seeks after immediate practical utility may rest assured that he seeks in vain.*

In Maxwell's words:

> [Helmholtz] ... *Who prosecutes physics and physiology, and acquires therein not only skill in developing any desideratum, but wisdom to know what are the desiderata.*

Simon Newcomb (1835–1909) conducted a precise measurement of the speed of light. He discovered what is now known as *Benford's law*: More numbers, taken from "real life" data, will begin with 1 than with any other digit. Newcomb believed that the astronomy of his time was nearing the limit and, like Thomson, he was skeptical about *flying machines*.

> *... those which depart most from the best adapted constitution, will be the most liable to perish, while, ..., those organized bodies, which most approach to the best constitution for the present circumstances, will be ... multiplying the individuals of their race.*
>
> James Hutton, An Investigation of the Principles of Knowledge and of the Progress of Reason, from Sense to Science and Philosophy, 1794

> *And love Creation's final law*
> *Tho' Nature, red in tooth and claw*
>
> Alfred Tennyson, In Memoriam A.H.H., 1849

> *The powerful retractile talons of the falcon- and the cat-tribes ... survived longest which had the greatest facilities for seizing their prey.*
>
> Alfred Russel Wallace. On the Tendency of Varieties to Depart Indefinitely From the Original Type, 1858

Teeth, claws, talons, nails – there are gentler designs of Mother Nature that are also more important for survival of her children.

The mortality rate for all animals is the highest before they reach maturity. If you are a bird or mammal, your survival depends 100% on the care of your parents. Not enough mummy's milk – you are dead long before you learn what your talons are for.

© zmeel / Getty Images / iStock

But it is not easy to figure out how Nature managed the *simultaneous* evolution of several *intrinsically uncorrelated* (?) functions, e.g., physiology + psychology of a kitten and then of the *same* animal in the role a mother cat where everything must function *in concert.* (Darwin's appeal to gradualism of evolution with no support by detailed/quantitative analysis of data is hardly acceptable for more than two "elementary functions".[44])

But talons or no talons, Wallace was a great (the greatest?) 19th-century naturalist. He collected more than 100 000 specimens in Malaysia and Indonesia and discovered more than a thousand new species, e.g., *Wallace's flying frog.*

As much as Darwin, he had been thinking on how and why species transform into new species and why the separation between different species is rather sharp.[45] He arrived at the same (but not quite) natural selection theory as Darwin, but, apparently, he was more ecologically minded than Darwin. A witness to that is Wallace's self-regulation principle in animal populations:

> *The action of this principle is exactly like that of the centrifugal governor of the steam engine, which checks and corrects any irregularities almost before they become evident; and in like manner no unbalanced deficiency in the animal kingdom can ever reach any conspicuous magnitude, because it would make itself felt at the very first step, by rendering existence difficult and extinction almost sure soon to follow.*

---

[44]An outcome of any kind of such analysis would depend, in particular, on the comparative size of atoms and molecules versus cells, for example.

[45]This sharpness, (that is, *disjointness*) of the corresponding *attractors in the dynamics of evolution*, probably, is determined by a presence of feedback loops in this dynamics. Negative feedbacks constrain the spread of attractors (that is, the intraspecies variation), whereas the positive feedback loops make different attractors (representing different species) drift apart.

I do not know if either he or Darwin realized that the *self-regulated* (negative feed-back) equilibrium can be oscillatory as in the *Lotka–Volterra equation.*[46] But Wallace gave a devastating analysis of how the ecology of St. Helena island (famous for Napoleon who had *not* studied it) was interfered with by European colonists and what happened to the equilibrium afterwards.

Disagreeing with Darwin, Wallace rejected the superficial similarity between artificial selection and natural selection in the wild:

> *... varieties produced in a state of domesticity are more or less unstable, and often have a tendency, if left to themselves, to return to the normal form of the parent species; and this instability is considered to be a distinctive peculiarity of all varieties, even of those occurring among wild animals in a state of nature, and to constitute a provision for preserving unchanged the originally created distinct species.*

The current view (if I get it right) is that *instability of domesticity* and intraspecies variation in general are mainly due (besides ever-present Gaussian bell-shaped variations[47]) to genome *crossover recombination* during meiosis that, unlike mutation, is (quasi)reversible.

Wallace, who could not know anything of this, gave a 19th-century explanation:

> *If turned wild on the pampas, such* [domesticated] *animals would probably soon become extinct, or under favorable circumstances might each lose those extreme* [artificially selected] *qualities which would never be called into action, and in a few generations would revert to a common type.*

Isn't it amazing that the two explanations have hardly anything in common?

The first one was unthinkable a hundred and fifty years ago. It depends on cellular/molecular data that have been obtained by many technically involved experiments the outcomes of which were by no means predictable a priori.

On the other hand, Wallace's common sense argument could have come from Lamarck with *adaptation* instead of *selection* (implicit in the above quote) and even from Aristotle whose choice of words would be *openly* teleological.[48]

---

[46]This differential equation was used as a model of the predator-prey systems in the mid 1920s by the mathematical chemist Alfred Lotka and independently by the mathematician Vito Volterra. Prior to that, a similar equation was introduced in 1838 by the mathematician Pierre Verhulst for describing the number of individuals that an environment can support. Even earlier, in 1766, Daniel Bernoulli, stimulated by the inoculation controversy, solved an equation of this type in his study of smallpox epidemics.

[47]These variations, as was pointed out by Schrödinger and ≈ 150 years earlier by James Hutton, may be purely phenotypical; thus, *non-inheritable*, contrary to one of the main premises of pre-Weismann Darwinism.

[48]Aristotle categorized "explanatory causes" into four classes:

> *the material cause, the formal cause, the efficient/moving cause*, and *the final cause*,

(Teleology is built into human language and it inadvertently pops up in our reasoning when we least expect it, especially in evolutionary biology and in psychology.[49] Even Darwin, who fought teleological thinking in biology for years, writes:

> *Natural selection cannot possibly produce any modification in a species exclusively for the good of another species.*

By logic, every explanatory non-teleological sentence about evolution with *for* in it must be either vacuous or self contradictory.)

Wallace was also skeptical about Darwin's powerful idea of *sexual selection*. He, apparently, found it too powerful. You can explain almost everything with it:

> *a certain feature evolves because the opposite sex happened to like it.* (*This dispenses with the necessity of reflection* – Poincaré would say.)

Thus, for example, Wallace identified the true role of what is now called *warning coloration in animals*, that Darwin originally attributed to sexual selection.

An essential point of divergence between Wallace and Darwin was in the dominant role of selection, in particular of sexual selection, in the *human* evolution that took merely 200 000–2 000 000 years depending on from where you count.

> *Man was generated from all sorts of animals,*
> *since all the rest can quickly get food for themselves,*
> *but man alone requires careful feeding for a long time;*
> *such a being at the beginning could not have preserved his existence.*[50]

It is hard to disagree that sexual selection *is* the most probable cause of, say, sophisticated courtship of birds – just think of a peacock burdened by his

---

where the latter corresponds to something like our *teleological*. It would be presumptuous however, even to raise it as a *question* what *exactly* Aristotle meant by these "causes" and what would be his vision on their respective roles in the performance of evolution; yet, this classification is inspiring anyway. For example, it brings the modern stochastic genome dynamics perspective on evolution to the same "material/efficient category" as Lamarckian ideas on means and causes of evolutionary modifications of organisms whereas "explanations by selection" go to the "formal category".

[49] Rephrasing von Brücke and/or Haldane,

> *Teleology is like a mistress to an evolutionary biologist: he cannot live without her but he's apprehensive of being caught up in her company by a mathematical physicist.*

Ernst Wilhelm Ritter von Brücke (1819–1892), as much as Claude Bernard and Hermann Helmholtz, can be regarded as a father of modern physiology. Among many other things, he studied the change of color in chameleons and the ways sounds of European and Oriental languages are produced.

John Burdon Sanderson Haldane (1892–1964) was one of the founders, along with Ronald Fisher and Sewall Wright, of *population genetics* – that is Mendelian dynamics in a *microevolutionary setting*. His idea (1941) of checking *semiconservative mode of DNA replication* with the use of ^{15}N (the heavy isotope of nitrogen), was implemented in 1957 by Matthew Meselson and Franklin Stahl in one of the most logically beautiful experiments in biology.

[50] Attributed by (Pseudo?)-Plutarch (2nd century?) to Anaximander, 610–546 BCE.

"useless" tail. Yet, only (adult?) males may be expendable; precarious childbirth and delayed maturity seems too high a price for the big brain in humans.

Apparently, the brain + language evolution went through a positive feedback loop. Such loops, be they positive or negative, must be abundant in the organisms/environment systems since organisms modify/shape environment. (The interactions between opposite sexes of the same species is a basic instance of that, where positive loops are more easily recognizable as they enhance sexual selection.[51])

An obvious environment of an individual in the early human brain evolution, besides one's mates, was one's *tribe/clan identified by common language*:

in a community of speakers and listeners,
selection favors the most articulate ones.

Besides, these tribes themselves became *units of selection*, where evolution speeds up by a significant factor for a population that is divided into N competing tightly knit groups. (*A little mathematics of the multiplication table type* suggests this factor may be almost as large as N but I am not certain about it.)

Wallace himself maintained that there must be *something transcendental* responsible for the emergence of higher cognition in humans. This does not look *scientific*, unless *transcendental* is read as

a simple yet subtle abstract structure that underlies human cognition and that is evolutionarily accessible.

A remarkable instance of this kind of a structure – *imprinting in young animals* – was described in an 1872 paper by Douglas Spalding. This marked the birth of psychology as a *science* separated from neurophysiology and ... Spalding's results shared the fate of those by Mendel of being ignored for several decades.

© Rike_ / Getty Images / iStock

An attractive feature of this structure for a theoretically inclined *Naive Mathematician* as well as for practically minded Mother Nature with her unsophisticated

[51] *Sexual preference for such developments in the female, must thus advance together, and so long as the process is unchecked by severe counterselection, will advance ... in geometric progression.*
RONALD FISHER

strategy of evolution by selection, is the *universality* of imprinting: A baby animal (of a certain class of species) takes the first moving object, *whatever it is*, for its mother. (Does this very simplicity of imprinting disclosed by Spalding's experiments that excites mathematicians hold it in low esteem among psychologists?)

Possibly, many (all?) fundamental patterns/units of human/animal psychology/behaviour are comparably mathematically simple/universal; thus, evolutionarily accessible. But they may have no apparently straightforward manifestations and be hard to detect by direct experiments.[52]

> *Les développements de la vie, la succession de ses formes, la détermination précise de celles qui ont paru les premières, la naissance simultanée de certaines espèces, leur destruction graduelle, nous instruiraient peut-être autant sur l'essence de l'organisme.*
>
> GEORGES CUVIER, RECHERCHES SUR LES OSSEMENS FOSSILES DES QUADRUPÈDES, 1812

Cuvier was a magician of palaeontology and comparative anatomy:

> *... after inspecting a single bone, one* [he modestly speaks of himself] *can often determine the class, and sometimes even the genus of the animal to which it belonged, above all if that bone belonged to the head or the limbs. ... This is because the number, direction, and shape of the bones that compose each part of an animal's body are always in a necessary relation to all the other parts, in such a way that – up to a point – one can infer the whole from any one of them and vice versa.*

Cuvier's analysis of the fossil data was the primary source of 19th-century evolutionary theories, but Cuvier rejected the idea of gradual transmutation of species proposed by Lamarck, since fossil records indicated abrupt rather than gradual changes, and he scorned the mechanisms of evolution suggested by Lamarck.

Would Cuvier accept natural selection as a scientifically valid solution to the problem of evolution? Would Huxley – *Darwin's bulldog* as he was called – stand a chance against arguments that Cuvier could master?[53]

(Thomas Henry Huxley, probably, the second great comparative anatomist after Cuvier, beat bishop Wilberforce in a public evolution debate. But that was rhetoric, not science.)

---

[52]Another such pattern, *Hawk/Goose effect*, is a habituatory response of a baby animal who learns not to fear *frequently* observed shapes sliding overhead.

The structure of *imprinting* and of *Hawk/Goose* does not look simple if you go to the bottom of it. To see the problem, try to honestly describe these in *brain language* – in terms of retinal images and/or of properties of flows of signals received by the vision processing centres in the brain.

[53]Cuvier would have no problem with the (pre-Darwinian) concept of *purifying selection*, that is similar to weathering that smoothes out the shapes of old mountains and hills, but *creative power of selection in formation of species* would sound to him as bizzare as the *creative power of erosion in the formation of mountains*.

The discontinuity of fossil records that puzzled Cuvier remains a puzzle and his idea of the essential role of *catastrophes* in shaping evolution may be correct. Who knows, if not for the catastrophes, if the Earth were moving through time without hitting potholes and bumps on the way – nice, smooth, and continuous ride – then Nature would be more than satisfied with Precambrian jelly fish – there would be no need for the luxury of "higher" plants, animals, and people.

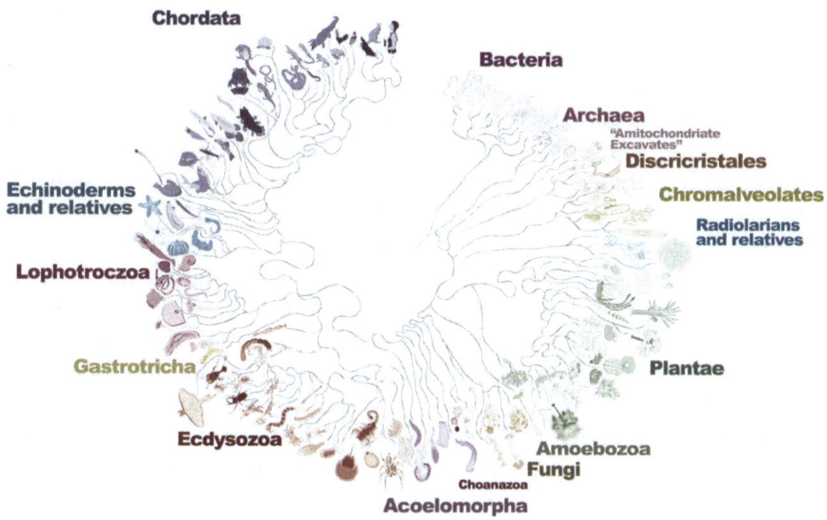

© Nemo Ramjet / Science Photo Library

Les végétaux sont des corps organisés vivants,
jamais irritables dans leurs parties, ne digérant point,
et ne se mouvant ni par volonté, ni par irritabilité excessive.

JEAN-BAPTISTE LAMARCK, PHILOSOPHIE ZOOLOGIQUE, 1809

A definition worth its name – mathematicians learned this from Alexander Grothendieck – is not a concise wording of what everyone knows, but a pointer toward unknown. Many unexpected fruits have been harvested from what grew from the seeds of ideas in his definitions.

Lamarck had never read Grothendieck. Not only did his definition miss moving carnivorous plants, such as *the Venus Flytrap* (undoubtedly known to Lamarck), but his criteria are inapplicable to most of the tree of life. The two minor branches of this tree – plants and animals – are poor representatives of (mostly unicellular) life on Earth.

The tragedy of Lamarck was that he put wrong signs at the portals of the edifice of the evolution theory he had erected. Like Columbus, who misread *America* for *Indies* on the coastline of the new land, Lamarck misspelled *selection* for *adaptation*.

Worse than that, unlike his evolutionary successors, Lamarck had proposed several biological mechanisms underlying evolution. His ideas had the "drawback" of having scientific (*material* and/or *moving* in Aristotelean terms) ingredients in them that made these ideas experimentally verifiable; they turned out to be (essentially but not fully) incorrect.[54]

If you take for granted the miracle of adaptation of organisms to environment during their lifetimes, you may equally accept the Lamarckian idea of adaptation on the evolutionary time scale via some non-teleological(!) mechanism that allows an environmentally channeled influence of the potential future on certain internal processes in an organism that could benefit descendants of this organism.

This idea creates no logical problem in many instances where *selection* and *adaptation* are interchangeable in "explaining" evolution; yet, as Darwin points out, no specific environmentally induced mechanism (e.g., what was suggested by Lamarck) seems plausible in *evolution of the mode of reproduction* – this most essential feature of any species. (An instance of such "non-Lamarckian" feature is the *stability of* 1 : 1 *sex ratio* for such species as elephant seals with harems of several dozen females.) Another "non-Lamarckian picture" is seen, as was also indicated by Darwin, in *evolution of social insects*. (There is no essential disagreement, however, between Lamarck and Darwin about the folk idea on inheritance of acquired traits.)

Geometrically speaking, the connectivity of LIFE ON EARTH along the time coordinate via the heredity threads between organisms is rather tenuous unlike the full-fledged spacial connectivity/unity of individual organisms.

But is it possible *to directly* disprove, Lamarck's idea of adaptation and show that beneficial mutations take place *prior* to the change in the environment where they turn out to be beneficial? You cannot do this by staring at fossilized remnants of extinct animals.

Yet, in 1943, Salvador Luria (biologist) and Max Delbruck (physicist) thought of an experiment the logic and the beauty of which would delight Gregor Mendel. This goes, roughly, as follow.

Grow a colony with, say, about a billion bacteria, starting from a single cell, e.g., of E. coli, a bacterium adored by bacteriologists. (E. coli colonizes your guts within several hours of birth, where it switches from aerobic to anaerobic life and it adheres to the mucus of your large intestine until you die.)

Suppose, that when we apply some factor X to such a colony, say a virus (e.g., Bacteriophage T1 – Escherichia coli's best friend) which is deadly for the cells, there is, nevertheless, a small probability, say $p_1 = 0.02$, that some cell in the colony survives. In fact, such a survivor is a result of a mutation that brings a certain new property Π to the mutated cell that is manifested in tolerance of the mutant to the X factor. Moreover, and this is essential, Π passes on to all descendants of our Π-cell.

---

[54]Some Darwinists rejoice in finding inconsistencies in Lamarck as if this makes their own theories more substantial.

If you apply this procedure, to, say, one thousand different colonies, then about twenty will have survivors in them. Discard dead colonies and look how many among the remaining twenty have *two* (or more) surviving bacteria. There are two conjectural alternatives.

1. *Lamarckian Adaptation.* If property Π develops in response to the factor X *after* it has been introduced, then the probability of having two survivors will be $p_2(\text{adap}) = 0.0004 = p_1^2$ being the results of simultaneous occurrence of two *independent* p_1-events. In this case

a presence of three colonies with two survivors is highly unlikely.

2. *Sheer Luck.* If some bacteria mutate *prior* to introduction of X, then this could happen before the last round of division with *half* billion bacteria on the plate with probability $0.01 = p_1/2$. This makes $p_2(luck) \geq 0.01$ and there must be something like

from seven to thirteen colonies with two survivors in them.

This is what Claude Bernard calls *active science* where *you make experiments not to confirm your ideas, but to control them.*

When you check what actually happened, you find out that, unequivocally, **2** is true and **1** is false: About half of not fully dead colonies have two or more active bacteria in them.

(To Lamarck this would be no more in a contradiction with his idea of evolution, than, say, a guillotined head could be a counterexample to how the use/disuse of an organ implies an adaptive evolution of this organ in a *slowly* changing environment.)

Stop! How can you tell how many bacteria out of *billion* are alive?

Elementary, my dear Watson. Shake your cells in a liquid to randomly mix them and spread the culture on a nutrient plate. (Technically it is better first to spread the fully alive culture and then apply X.) Then each survivor – each Π-cell – starts dividing and in a short while you see as many colonies on the plate as you had Π-cells to start with.

(If we do not introduce X and keep the colony alive at a limited size with a constant flow of nutrients, then, some time afterwards, only a few cells will have surviving descendents and, eventually, only one cell. This follows from Anaximander's **Theorem 1** when it is empowered by an estimate of the *probability of extinction* that was calculated by Francis Galton and Henry William Watson in their 1874 paper. There is hardly one chance in a million that the Π-cells from the original colony would have descendants in a few years.

The *original* resistance to X is not preserved[55] and none is visible in the colony; yet, the "dormant guardians" are there and they "fight back" when the

[55] It can be preserved, if the colony was exposed to the virus: CRISPR-based adaptive immunity in bacteria is heritable. This justifies Lamarckian ideas if these are applied to evolution of immune systems in prokaryotes.

colony is attacked by X. Lamarck would say that colonies do adapt by "voting for survival" of their luckiest members and that a female egg-cell might be similarly selecting the fittest contender out of hundreds of millions of available spermatozoids; thus, ensuring adaptation of the progeny. Let Darwin and Weismann disprove this.)

Another idea of Lamarck is the presence of a *complexifying force* that *drives* organisms from simple to complex forms with environmental "pressure" being superimposed on this force. You cannot discard this by brandishing the banner of natural selection in the air, so try something cleverer.

Probably, such a "force", if it exists, resides in the *logic of chance* in the evolutionary game (frame?) of LIFE, possibly, expressible in a language of a Grothendieck-like mathematics. But deciding one way or the other seems more difficult, than, for example, (quasi)*rigorously* "deriving" irreversibility in thermodynamics from the reversible laws of physics where the present day mathematical rendition of Boltzmann's arguments does not seem satisfactory.

In either case *mathematics of the multiplication table type* that suffices for the Luria–Delbruck experiment seems of little help to you. (But maybe the main problem is that we do not understand *multiplication table*.)

... there perished many a stock, unable
By propagation to forge a progeny.
For whatsoever creatures thou beholdest
Breathing the breath of life, the same have been
Even from their earliest age preserved alive
By cunning, or by valour, or at least
By speed of foot or wing.

TITUS LUCRETIUS, ON THE NATURE OF THINGS, 50 BCE(?)

... et ces espèces que nous voyons aujourd'hui ne sont que
la plus petite partie de ce qu'un destin aveugle avait produit.

PIERRE-LOUIS MAUPERTUIS, ESSAI DE COSMOLOGIE, 1750

... just in proportion as this process of extermination
has acted on an enormous scale, so must the number of
intermediate varieties, which have formerly existed,
be truly enormous.

CHARLES DARWIN, ORIGIN OF SPECIES, 1859

Dramatic cut-off of exponentially growing functions in a bounded space, called by Darwin *natural selection*, is an apparent logical necessity, not an intrinsically biological property of living systems.

The principle of natural selection no more "explains" evolution than differential equations "explain" mechanical motions, but this principle provides a conceptual framework and suggests a language for possible mathematical models of evolution.

© vege / stock.adobe.com

This must have been obvious to Maupertuis and to anybody who had ever pondered on population growth and who could fathom the immensity of $\exp T$, e.g., to Buffon, who along with Maupertuis, was wondering

> *What purposes then are served by this immense train of generations, this profusion of germs, many thousands of which are abortive for the one that is brought to life?*

to Benjamin Franklin, who wrote on some occasion in 1751:

> *no bound to the prolific nature of plants or animals, but what is made by their crowding and interfering with each others means of subsistence,*

to Euler, who, in *Introductio in analysin infinitorum* (1748), illustrates the exponential function by examples from population dynamics, and to Fibonacci who savors numerical subtleties of an idealized rabbit replication model in his *Liber Abaci* (1202).

An imbalance between the growth of the population and of resources also must have been known for centuries. The problem and an approach to it are discussed, for example, in Giovanni Botero's treatise *Delle cause della grandezza delle città* (1588).

A humanistic perspective on the solution of the overpopulation problem was proposed, in almost modern terms, by Nicolas de Condorcet in *Esquisse d'un tableau historique des progrès de l'esprit humain* that was published posthumously in 1795. But there is nothing humanistic in how Nature has been handling the (over)population problems for millions of years.

Condorcet and Malthus

Condorcet is remembered by mathematicians for his 1785 essay

> *sur l'application de l'analyse à la probabilité des décisions rendues à la pluralité des voix*

where he proves his *jury theorem* on probability of a group of individuals to arrive at a correct decision and where he analyses the (*voting*) *paradox* of non-transitivity of the *order relation on decisions* defined by collective preferences.

His passionate manifesto

Esquisse d'un tableau historique des progrès de l'esprit humain

was written between October 1793 and March 1794, when Condorcet was in hiding having been condemned to death by the Robespierre government for his humanistic political views.

(About a century afterwards, a monument was erected in Paris with the torso of Lavoisier, who was guillotined forty days after Condorcet was found dead in his jail cell, and the head of Condorcet. This was not done on purpose.)

Thomas Malthus, in *An Essay on the Principle of Population*, published in 1798, sides with Nature in her solution to the overpopulation problem and expresses his skepticism on the social practicability of the solution proposed by Condorcet.

Malthus' influence is due to two facts.

1. In the 19th century, the number of readers who were willing to pay for a book with $\exp T$ in it reached the critical mass needed to make the publication of such a book profitable.
2. Darwin and Wallace were among these readers.

But the idea that technological progress can compensate for Malthusian $\exp T$ remains as much in contradiction with the multiplication table today as it was at the time of Malthus. The rules of arithmetic do not change, at least not on so short a time scale.

Entropy versus Energy Barriers in the Development of Science

The world is full of magical things patiently waiting for our wits to grow sharper.

Bertrand Russell

There is a fundamental difference in the logical structure of principles of physics, such as *Newton's second law*, and those of evolutionary biology, such as *adaptation by selection*, that is reflected in our perception of these principles.

Learning and understanding physics is hard; one needs a firm mathematical background and a non-trivial intellectual effort to comprehend the meaning and the consequences of something like the second law, for example. Only a small percent of "educated people" even understand what it means to UNDERSTAND the second law.[56]

On the other hand the idea of *evolutionary adaptation by selection* strikes anybody, who is aware of the enormousness of exponential growth, as obvious.

[56]This is why the members of *flat Earth societies* do not argue against the second law.

(Realizing insufficiency of this idea in its original Darwinian form needs a higher level of mathematical sophistication.[57])

But why hadn't the full-fledged idea of evolution by selection sprouted from the brilliant mind of some mathematician such as Daniel Bernoulli or Leonhard Euler?

An idea in science, as a molecule in a chemical reaction, in order to reach the critical maturation point may need not to climb over a mountain of high energy, but rather to follow a narrow pathway in a labyrinth of tunnels through this mountain. Once such a pathway is found, everybody can follow effortlessly.

There may yet be other simple ideas in science waiting to be discovered that are as illuminating as the idea of the pivotal role of the exponential function in evolutionary dynamics. If you think such ideas are "obvious", pinpoint a single one of them.

> *... preservation of favorable variations, and the destruction of injurious variations, I call Natural Selection, or the Survival of the Fittest.*
>
> CHARLES DARWIN, ORIGIN OF SPECIES 5TH EDN. 1869

It is not this survival rhetoric however that had been mainly occupying Darwin's mind but a persistent idea of turning the *cut-off effect on the exponential growth in biology* into a *principle of selection* (of *continuity*?), something like Maupertuis' *principle of least action.*[58]

Maupertuis, who had been working over his principle for two decades, believed that Nature always minimizes/optimizes whatever she does and he must have tried to figure out the formula for "evolutionary action" that is minimized by Nature in the course of selection. Part of this must be *time*, he would assume, since not so much *perfection* of your fitness makes you the winner, but how fast, lets imperfect, your fitness can be evolutionarily achieved. But he, probably, could not guess and write down other terms in this "action".

Besides, Maupertuis would observe, that mutability and the reproduction rate must be sufficiently high in order for *evolution by selection and nothing else* to be logically/mathematically feasible.

Very roughly, R^T for R being the reproduction rate, must "beat" something like 2^P where P is the number of "variable parts" of an organism and where a presence of selection, that channels evolution by pruning off the majority of mathematically conceivable branches of development, makes the estimated time T needed for evolution longer rather than shorter.

On the other hand, the mutation rate cannot be too high; otherwise, deleterious mutations, that are in a dominant majority, would result in extinction. The

[57]On the other hand, mathematically insensitive people may be driven against the very ideas of evolution and natural selection by some (Freudian?) psycho-sociological mechanism.

[58]Maupertuis identified a certain quantity, called *action* $A = A(motion)$, such that a *moving* physical system S minimizes this A, or rather, the *motion* of S, thought of as a spatial curve, satisfies the corresponding *Lagrange–Euler* (*differential*) *equations.* Lagrangian and Hamiltonian incarnations of this principle are present in all branches of mathematical physics.

most dangerous are *dormantly deleterious mutations*, e.g., those that increase the mutation rate).

Being a student of Newtonian (non-Aristotelian) mechanics (with impulses as coordinates), Maupertuis would assume that the main *biological observable* (feature) that evolves by selection is "mutability", or rather its opposite – *fidelity of reproduction*. (All Nature knows is to keep the mutation rate as low as she possibly can – evolution is a random dance on the razor's edge between stagnation and extinction.[59])

Finally, Maupertuis, a mathematical physicist rather than a pure mathematician, would try to match rough numerical estimates of this kind with the schedule of the fossil data.

Well ... Maupertuis, had done none of this and neither had Darwin who remained dissatisfied with the selection theory of evolution, judging by how insistently he was *convincing* himself that natural selection explains evolution and by how eloquently he was declaring that he came to *believe* in the validity of the idea. It is for myths, not for science, to *explain* the world and life and it is up to preachers and politicians, not for scientists, to *convince* people of anything.

Darwin wanted more of his selection theory than being merely a *convincing explanation* of the evolution of Life on Earth but mathematics of the 19th century had nothing to offer for substantiating his vision.

A cautious 21st-century mathematician would not expect a single clear cut mathematical theorem/theory fulfilling Darwin's dream but he/she may still envisage a possibly vague, yet, *truly mathematical* setting embracing the idea of LIFE.

An instance of such a not quite precise but purely mathematical idea often used in physics is that of representing a state of a dissipative dynamical system, e.g., of a quasi-stationary flow of a viscous fluid, as an *ensemble of attractors* in the corresponding phase space.

But no such concept, nothing of what one expects to find in 20th-century dynamics, seems suitable for expressing much of the whole idea of LIFE.

(There may be dynamic-theoretic models of some of LIFE'S fragments. For instance, one can possibly describe the multi-scale time feature of *evolution-by-selection*, with different classes of *units of selection*[60] on different time scales, in the language of "multistage dynamical systems" where attractors themselves are subject to dissipative-like *predominantly contractive* dynamics.)

---

[59]Nature faces the same problem as a public education system – it should be *almost but not fully* perfect: If students 100% comply with the demands of their teachers, society stagnates; but too little rigour results in *error catastrophe*, cancer of corruption and extinction.

What is amazing (for non-Lamarckians) is that besides sheer sloppiness, there are provisions by Nature for (quasi-random) "desirable mutations", e.g., in specified loci of genomes of parasites evading the immune systems of their hosts and in genomes of bacteria under stress.

[60]Mathematically speaking, units of selections are *particular* biological observables but it is unclear *particular* in what sense. For example, can *fidelity of reproduction* – one of the basic and evolutionarily slowest observables – be taken for such a "unit"?

It seems that a satisfactory understanding of genomes' stochastic dynamics and transforming the mathematical poetry of the Darwinian idea of evolution by selection to a hard-boiled scientific theory lies a long way off.

> ... *our* [19th] *century will be called ... the century of the mechanical view of nature, the century of Darwin.*
>
> LUDWIG BOLTZMANN

The abstractly logical – *formal* in the Aristotelian sense – Darwinian SELECTION PRINCIPLE may be alluring to mathematicians and to *mathematical* physicists like Boltzmann; yet, this *formality* is disconcerting. How can some property of the logic/syntax/mathematics of the language that one uses stand for the *main reason* of something *physically* or biologically occurring?

Refined mathematical concepts may seem no more helpful for understanding the harsh truths of the real world than poetic metaphors for this purpose. Only *Naive Mathematician* may take, for instance, the summation rules of infinite series for "explanation" of *Zeno's paradoxes* or the general theory of (classical or/and quantum) Einstein–Lorentz spaces for a full physical model of SPACE-TIME. But *Naive Mathematician* may protest by pointing out, for example, that when Laplace writes about

> *Une intelligence qui, pour un instant donné, connaîtrait toutes les forces ... embrasserait dans la même formule les mouvements des plus grands corps de l'univers et ceux du plus léger atome ....*

he hardly thinks that the solution of the problem of determinism may be advanced by a study of the mind of this metaphoric *intelligence* by an experimental neuropsychologist (as some 21st-century thinkers suggest), but he would accept that the (quasi)deterministic behaviour of planetary motions might be explicable in the light of *unique solvability of differential equations* and/or KAM-like theorems.

The skepticism of the hard core scientists should not divert us from a quest for mathematics that would bring "light of sense" to evolutionary biology.

We want to have maximally general, yet biologically sound mathematical concepts of

gene, organism, population, unit of selection, environment, adaptation

so that we can make sense of the following questions.

- What are probability (?) spaces where selection works?
- What defines (dis)connectivity of genomes, organisms, populations?
- Why are adaptations represented by (quasi)fixed points of (gradient?) evolutionary dynamics in *different* spaces, e.g., of genes in the environments of other genes and of organisms themselves in the "natural" environments, leads to *similarly looking* "coarsely grained adaptation" of populations, as we saw in the example of domesticated species reverting to wild type?

8. Brain

Tell me where is fancies bred, Or in the heart or in the head.

SHAKESPEARE, THE MERCHANT OF VENICE
(Probably, written between 1596 and 1598)

... mental activities are entirely due to the behavior of nerve cells, glial cells, and the atoms, ions, and molecules that make them up and influence them.

FRANCES CRICK, THE ASTONISHING HYPOTHESIS(1994)[61]

Shakespeare would not have been intimidated by the scientifically sounding "*mental activities*" instead of his "*fancies*"[62] and *the Hypothesis* would not have stricken him as especially astonishing.

After all, it has been known since about 2500 BCE that different types of head injury produce different symptoms, as it was recorded in *Edwin Smith Surgical Papyrus*, of about 1500 BCE where the idea of BRAIN appeared for the first time.[63]

Shakespeare could not have been acquainted with the *Surgical Papyrus* – this was discovered in the middle 1800s, but he might have been aware of the following.

The seat of sensations is in the brain. ...
All the senses are connected in some way with the brain ...
This power of the brain to synthesize sensations
makes it also the seat of thought.

[Attributed to] ALCMAEON OF CROTON ($\approx$ 450 BCE)

the source of our pleasure, merriment, laughter, and amusement
as of our grief, pain, anxiety, and tears, is none other than the brain.

HIPPOCRATES (?), ON THE SACRED DISEASE, ($\approx$ 425 BCE)

---

[61] In the early 1950s, Frances Crick and James Watson worked out, with the X-ray crystallography data contributed by Rosalind Franklin, a correct helical model of DNA. (An incorrect triple-stranded helix structure was earlier suggested by Linus Pauling). They concluded:

it therefore seems likely that the precise sequence of the bases is the code that carries the genetical information.

Crick also proposed THE CENTRAL DOGMA OF MOLECULAR BIOLOGY: DNA→RNA→Proteins.

This ... *deals with the detailed residue-by-residue transfer of sequential information. It states that such information cannot be transferred back from protein to either protein or nucleic acid.*

From age 60 until his death (2004) Crick worked on the Brain; in particular, he tried to identify specific neuronal processes responsible for consciousness.

[62] No matter how hard you try, you cannot say much about the arrow [brain]$\rightsquigarrow$[mind] without resorting to metaphors. (Poetic "*bred*" prompts in you a surge of ideas, while dry "*entirely due*" transmits zero positive information.)

[63] Surgical Papyrus is an incomplete copy of a text attributed to *Imhotep* ($\approx$ 2650–2600 BCE), called "*inventor of healing*", who was a high priest of the Old Kingdom of Egypt as well as chief builder and chief physician in his time. He is the first physician known by name.

... sense-perception in sanguineous animals is the region of the heart ...

ARISTOTLE, SLEEP AND SLEEPLESSNESS ($\approx$ 350 BCE)

It is hard to believe that clinical observations by Alcmaeon and Hippocrates did not convince Aristotle, who had not accepted "the Astonishing Hypothesis" because:

(1) the heart, unlike the brain, connects with all the sense organs;[64]
(2) the heart is more centrally placed;
(3) the heart in embryos develops before the brain;
(4) invertebrates, who have hearts but no brains, have sensations;
(5) the heart but not the brain is affected by emotions;
(6) the heart is warm but the brain is cold;
(7) the heart but not the brain is essential for life.

Half a century later, Alexandria's anatomist Herophilus (335–280 BCE) discovered that nerves spread from the brain throughout the body in agreement with the idea that the brain was the controlling organ in Man.

His younger colleague Erasistratus (304–250 BCE), who believed that the psychic pneuma was transmitted through motor nerves to muscles, appreciated the separate neural pathways for motor and sensory functions. He also suggested that the degree of intelligence in animals is correlated with how much cerebral hemispheres are convoluted.

But the experimental evidence for the brain control over the body came four centuries later, when Galen of Pergamon (129–200?) had shown that cutting the recurrent laryngeal nerves which innervate the larynx makes a pig stop squealing but not struggling.[65]

Here is his reaction to those who were not convinced of the dominance of the brain after having attended this experiment.

> *When I heard this, I left them and went off, saying only that I was mistaken in not realizing that I was coming to meet boorish skeptics; otherwise I should not have come.*

Ancients had no idea of a cell and could not fathom how the brain works. And when the cellular structure of the brain started unravelling itself in all its beauty, founders of *neuronal theory* were inspired to speak of the brain in poetic language:

> *The brain is waking and with it the mind is returning.*
> *It is as if the Milky Way entered upon some cosmic dance.*
> *Swiftly the head mass becomes an enchanted loom*

[64] Nerves are less prominent than blood vessels.

[65] Besides founding experimental medicine including experimental neurophysiology, Galen greatly advanced anatomy, physiology, pathology, and pharmacology of his time. For example, he showed that urine comes from the kidneys, he demonstrated that the larynx generates the voice and he recognized the essential difference between venous and arterial blood. He also developed several surgical techniques, including one for correcting cataracts.

where millions of flashing shuttles weave a dissolving pattern,
always a meaningful pattern though never an abiding one;
a shifting harmony of subpatterns.

CHARLES S. SHERRINGTON, MAN ON HIS NATURE, 1942

To know the brain ... is equivalent to ascertaining the material course of thought and will, to discovering the intimate history of life in its perpetual duel with external forces.

SANTIAGO RAMON Y CAJAL, RECUERDOS DE MI VIDA 1917

"Structure of the Mammalian Retina" © 1900 By Santiago Ramon y Cajal / Wikimedia Commons / Public Domain

Cajal and Sherrington, apparently, speak of *the human* brain but experimental neuroscience and, in part, brain anatomy relied on the study of animal brains, starting from ox and pig brains dissected by Galen and continued with a microscopic study of insect brains since the 18th century.

... the brain of an ant is one of the most marvelous atoms of matter in the world, perhaps more so than the brain of a man.

CHARLES DARWIN (1859)

For instance, the collective mind of ants is able to achieve shortest paths between locations in a rugged terrain:

a busy ant highway between an anthill and a source of food usually implements a nearly shortest possibility.

ANTS, CHEMISTRY AND LOGIC

There are more than 10 000 different species of ants and variations within a species may also be high. On the average (?), there are about a quarter of million brain cells in an ant. But there are ants and ants with their weights ranging from 0.01 mg to half a gram, where, in small ants, the brain makes more than 10% of the animal body weight – comparable to that in a newborn infant and by far more than 2% in adult humans.

If my opinions are the result of the chemical processes going on in my brain, they are determined by the laws of chemistry, not those of logic.

JOHN HALDANE, THE INEQUALITY OF MAN (1932)

Ants mark their trails with pheromones and themselves tend to choose the routes that have stronger pheromone odors. All things being equal,

the number of ants that pass back and forth on some track, say during 1 h, is inversely proportional to the length of this track;

© Benz3536 / Getty Images / iStock

hence, the shortest track becomes the smelliest one, thus, eventually preferred by the ants.

What has made this algorithm evolutionarily attainable is its simplicity and universality. Probably, basic programs running within our minds, just in order to exist at all, must be comparably simple and universal. But there is not any "law" in sight[66] that determines ants' opinions of where to go.

Cajal and Sherrington

Cajal studied nervous and brain tissues with a *silver staining technique* discovered by Camillo Golgi in 1873 and established that the *neurones* are basic units of nervous structure. This theory was completed with the introduction of the concept of *synapse* by Sherrington who writes in his 1906 book The Integrative Action of the Nervous System:

> *At the nexus between cells if there be not actual confluence, there must be a surface of separation ... In view, therefore, of the probable importance physiologically of this mode of nexus between neurone and neurone it is convenient to have a term for it. The term introduced has been synapse.*

[66] It is doubtful whether the expressions "*laws of chemistry*" and "*laws of logic*", unlike those of "*laws of classical mechanics*", make any sense at all.

9. Mind

The energy of the mind is the essence of life.

ARISTOTLE

As followers of natural science we know nothing of any relation between thoughts and the brain, except as a gross correlation in time and space.

CHARLES SHERRINGTON

WHAT IS THE MIND? WHAT ARE THOUGHTS?

How much have we to know about the brain to understand the mind?

Imagine that by the middle of the 21st century we will have learned as much about the brain as we know to-day about the fundamental quantum mechanical laws of matter and energy or about the nanoscale dynamics of the cell. Would it help?[67]

Or, will we only more clearly realize our helplessness and, following Pauli, will say that "mind", like "reality" is

> *something self-evidently known* ... [but it is] *an exceedingly difficult task ... to work on the construction of a new idea of* mind.[68]

But is it possible at all to translate the following poetic mage of brain/mind to the language of science?

> *The eye sends ... into the cell-and-fibre forest of the brain throughout the waking day continual rhythmic streams of tiny, individually evanescent, electrical potentials.*
>
> *This throbbing streaming crowd of electrified shifting points in the spongework of the brain bears no obvious semblance in space pattern, and even in temporal relation resembles but a little remotely the tiny two-dimensional upside-down picture of the outside world which the eyeball paints on the beginnings of its nerve fibers to the brain.*
>
> *But that little picture sets up an electrical storm. And that electrical storm so set up is one which affects a whole population of brain cells. Electrical charges having in themselves not the faintest elements*

[67]The molecular machinery of a living cell *transcribes + translates* each *codon* – triple of four basic nuclear acids in DNA (except for three stop codons) – to one out of twenty standard amino acids (with one stop codon sometimes coding for *Selenocysteine)*. The particular coding/translation rule used by Nature, that is *the same for* (almost) *all* organisms on Earth and that is called the *genetic code*, is, for an experimentalist, the most fundamental law of biology. On the other hand, nothing (?) in life would visibly change if the code was somewhat different. What is essential in biology from a mathematician's perspective is the *principle* of coding, rather than the *specificity* of the code being used.

The brain may harbor an immensity of such "coding arbitrariness"; an accumulating avalanche of the detailed knowledge of this "code" may block out rather than promote our understanding of the arrow [brain] $\rightsquigarrow$ [mind].

[68]The idea of "reality" is inseparable from "mind": *Cogito ergo sum*.

of the visual – having, for instance, nothing of "distance," "right-side-upness," nor "vertical," nor "horizontal," nor "color," nor "brightness," nor "shadow," nor "roundness," nor "squareness," nor "contour," nor "transparency," nor "opacity," nor "near," nor "far," nor visual anything – yet conjure up all these.

A shower of little electrical leaks conjures up for me, when I look, the landscape; the castle on the height, or, when I look at him approaching, my friend's face, and how distant he is from me they tell me.

Taking their word for it, I go forward and my other senses confirm that he is there.

SHERRINGTON, MAN ON HIS NATURE

What does it mean to understand mind? What are the right questions to ask?

It seems there is an invisible interface between the brain and what we perceive as our mind – an interface that is comparable in its structural complexity to the *machinery of embryological development* that, as Thomas Hunt Morgan says, effectuates the transformation of what is "written" in the genes into [phenotypic] *characters that are used by the* [classical] *geneticist.*

What I cannot create, I do not understand.

RICHARD FEYNMAN

Here we speak of creating *abstract models* not necessarily implemented by physical devices,[69] and some of them may not need a knowledge of the brain, at least for modelling "parts" of the mind, such as *memory*, for instance.

© Di_Studio / Getty Images / iStock

[69]Ability to create does not imply understanding. Animals (as well as plants and bacteria) can create their approximate copies but even the smartest of the human stock have only the vaguest images in their minds of the ways of embryological development of their progeny.

Memory ... an Organ, as the Eye, Ear, or Nose,
where the Nerves from the other Senses concur and meet. ...
... I conceive to be nothing else but a Repository
of Ideas formed partly by the Senses, but chiefly ...
... receiving, and being excited by such Impressions,
they do again renew their former Impression ....

ROBERT HOOKE, ON HYPOTHETICAL EXPLICATION OF MEMORY, LECTURE TO THE ROYAL SOCIETY, 1682

This looks almost like a verbal description of a *mathematical model* of memory, but some do not like such an idea.

Intellectual snobbery makes them [mathematicians?] *feel they can produce results that are mathematically both deep and powerful and also apply to the brain.*

FRANCES CRICK, WHAT MAD PURSUIT (1988)

However, new ideas on mind/brain came from the mathematical side along with the concepts of *information flows* and of *Turing machines* that allowed formulations of new kinds of questions, such as:

Can machines do what we (as thinking entities) can do?

ALAN TURING, COMPUTING MACHINERY AND INTELLIGENCE (1950)

Attempts to disprove Turing's reasoning that a digital computer can imitate human intelligence are in line with the argument that people could not have come from the same stock as monkeys, since monkeys lack moral virtues and do not abide by traffic laws.

And it is not only apes and monkeys – also bacteria in our guts are our remote cousins as far as biology is concerned. The unity of life on Earth lies deeper then similarity in anatomy and physiology. Something of this kind must be true about "intelligence" except we do not know what it is.

The difficulty in understanding the human mind is that it is *unique of its kind*; hence, there is no language to speak about it. The brain, on the contrary, is shared by almost all animals on this planet. This is the way neuroscience advances but the scientific theory of mind has stayed still for several millennia.

And how on Earth can one disprove Turing? We *are* biological machines as much as, say, ants are, albeit, individually, we have more neurons in our brains:

Man is so complicated a machine that it is impossible
to get a clear idea of the machine beforehand,
and hence impossible to define it.

JULIEN OFFRAY DE LA METTRIE, MAN A MACHINE. 1748

But maybe, what happens in our minds with wheels within wheels within wheels in them is too complicated to be representable by mere math? This is what Edgar Allan Poe thought in connection with chess.

Arithmetical or algebraical calculations are, from their very nature, fixed and determinate … [But] *no one move in chess necessarily follows upon any one other.*

EDGAR ALLAN POE, MAELZEL'S CHESS-PLAYER, APRIL 1836

Of course, Poe could not have known of Turing, but he was aware of the calculating machine of Babbage and argued that playing chess *cannot* be modelled on such a machine. And although what he was saying was formally incorrect, a fundamental difficulty[70] he pinpointed in such a modelling remains unresolved up to the present day.

Another problem with understanding human mind is that our intuitive concept of *intelligence* shaped by the *existential ego* of Cartesian *cogito ergo sum*, is buried in the multilayered cocoon of *teleology – purpose, function, usefulness, survival.* Turing purposefully avoids the question – *What is intelligence?* – since you cannot answer this question unless you develop an appropriate language. (It is the same as with the question – *What is* LIFE? – that cannot be answered in the language of LIFE ON EARTH.)

However, Turing suggests that a "human-like intelligence" can be built in a machine by subjecting it to education. Indeed, *learning* is a more intelligent

[70]As one tries to analyze a position many moves ahead, one becomes overwhelmed by the exponentially growing numbers of branches of possible game strategies.

concept than *intelligence*,[71] but one will not go far with it if one assumes that the child brain is *something like ... lots of blank sheets.*

> *Instead of trying to produce a programme to simulate the adult mind, why not rather try to produce one which simulates the child's?*
>
> *If this then were subjected to an appropriate course of education one would obtain the adult brain.*
>
> *Presumably the child brain is something like a notebook as one buys it from the stationer's. Rather little mechanism, and lots of blank sheets.*
>
> ALAN TURING, COMPUTING MACHINERY AND INTELLIGENCE

An unbridled delight in the brilliance of our "intelligence" blinds us to seeing the essence of the mind but modelling the learning process by a child may bring the light into the picture.

> *Man is most nearly himself when he achieves*
> *the seriousness of a child at play.*
>
> HERACLITUS

All a child knows is play and this is how he/she learns. The mathematics of this *playful learning process*, that is a transformation of the flow of electric/chemical signals the brain receives into a coherent picture of *external world* in the first two years of human life, is as intricate and mysterious as that of emergence and evolution of *live structures*.

And it is, so to speak, *free learning*: if you subject a human (or animal) infant to "education" you only impair the learning, since you have no idea of what happens in the brain/mind of a child. (Not that you know better how your own mind works.) Applying your "adult intelligence" to aid an infant's development is like a use of forceps in helping an amoeba to divide. No surprise, learning programs that have been developed until today are a far cry from Turing's dream.

In order to progress, instead of inventing more and more "intelligent" definitions of *intelligence*, *we must recognize our ignorance* as Feynman says, forget/rectify[72] how we "naturally" see ourselves and start with potentially answerable questions.

---

[71] *Learning* brings along a relatively slow *time coordinate* in the description of *mind* similarly to how *evolution* brings such a coordinate in the picture of *Life on Earth* where *evolutionary learning* is described in the language of *natural selection*.

[72] *L'esprit humain se plie à une manière de voir* – Lavoisier says in his paper on *phlogiston* – the common sense idea of *warmth* disguised as a scientific concept in pre-Lavoisier chemistry.

If you follow Lavoisier's path of reasoning, you will find his words applicable in connection with "intelligence" and "consciousness" as well. These notions are helpful in practical psychology but no common-sense definition of *intelligence* and/or *consciousness* can be taken for anything *scientific* in Lavoisier's understanding of the idea of science.

What are the *logical* (*molecular-like*) *units*[73] that *abstract intelligence* is composed from?

An absence (presence?) of what structure(s) in the mind of a baby ape makes its ability and willingness to learn after the age ≈ 1.5 years to fall below that of a human child but still keeps its learning gradient above that of an adult human?

What is the mechanism(s) responsible for rarely occurring structurally elaborate patterns within "human intelligence" such as exceptional musical and mathematical abilities? Are there conterparts for these in animals?

Do these patterns, that are harmful rather than helpful for *survival in the wild*, come by way of natural selection and if so what is the pool of possibilities/variations of these patterns?

What is a non-teleological description of the role played by *predictable future* in the functioning of an "intelligent system"?

What is a simple/short list of questions (or, rather an interactive algorithm generating questions depending on the history of a conversation) that would be easily answerable by a human of any culture, e.g., by a Cro-Magnon child (IF the language problem is somehow taken care of), but would cause an insurmountable difficulty for currently available "imitation programs".[74]

We need to design arguments and/or experiments for finding the answers that would serve to control rather than to confirm our ideas, being most satisfied with an outcome if it does not resemble anything we had imagined. Thus we shall be able to trace and to seal the route by which the self-gratifying illusion of our "intelligence" came from and to start a walk on a road toward understanding what the mathematical essence of mind/learning/intelligence is.

Quite likely, a full instruction for making a human-like learning system can be written down on a couple of hundred pages.

But even if you are on the right track for doing so, it may be a long way to go. It took Nature half a billion years and quadrillions of tries to arrive at the present design of our nervous system and we may need quadrillions of man/computer hours for designing a comparable system ourselves.[75] And the detailed knowledge of the wiring of the human brain would not necessarily help because of the indecipherable stochastic redundancy in its architecture.

---

[73]These units must be functions of *abstract context/purpose free* "variables", like *imprinting*, that applies to [*first/second* ...] (moving object) and of *Hawk/Goose* that depends on [*frequently/rarely*] (occurring event).

[74]Such questions must refer to (and/or recycle) phrases that had been already used in the course of a dialog. Probably, the length L of the (suitably defined) *shortest naive* imitation program that is undetectable by a clever algorithm after an exchange by n sentences must grow at least exponentially, $L \sim 2^n$.

[75]One may accept this as a possibility if one believes that $NP \neq P$.

10. Mysteries Remain

Ignoramus et ignorabimus.

EMIL DU BOIS-REYMOND (1872)[76]

In mathematics there is no ignorabimus.

DAVID HILBERT (1900)

Wir müssen wissen – wir werden wissen!

DAVID HILBERT (1930)

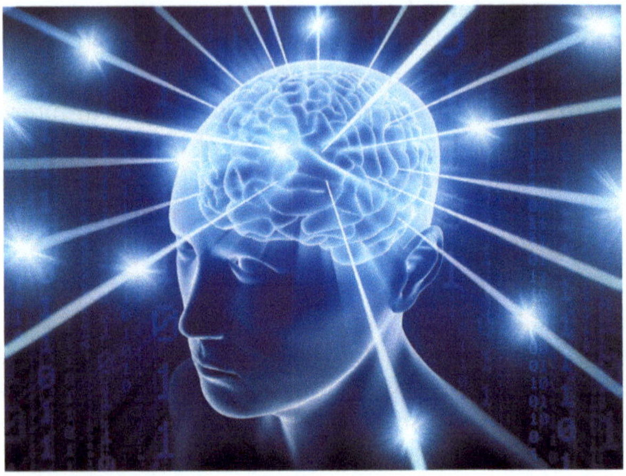

© Henrik5000 / Getty Images / iStock

Du Bois-Reymond suggested[77] that, possibly, humans shall *never* understand the following:

1. *Nature of matter and force.*
2. *Origin of motion.*
3. *Origin of Life.*
4. *Emergence of seemingly purposeful organisation in Nature.*
5. *Generation of conscious sensations by unconscious nerves.*
6. *Source of intelligent thought and Language.*
7. *Non-determinism of free will in deterministic Universe.*

Do you agree with what Du Bois-Reymond thinks of these issues?

To answer this, you need to find mathematical counterparts to the words there that are used (implicitly) metaphorically.

[76] Du Bois-Reymond discovered *nerve action potential* and proposed the chemical nature of synaptic transmission.

[77] *Über die Grenzen des Naturerkennens,* 1872 and *Über die Grenzen des Naturerkennens: Die sieben Welträtsel,* 1891.

For instance what is "DETERMINISTIC Universe"?

Most likely, Du Bois-Reymond carries in the back of his mind Laplace's idea of *une intelligence qui, pour un instant donné, connaî trait toutes les forces …* that models the universe by a system of ordinary differential equations.

This *model* is deterministic due to the unique solvability of such systems under due regularity conditions.

Of course, this is absurd, nobody would literally think that the state (?) of the Universe is given by a set $u(t) = (u_1(t), \dots, u_k(t)) \in U = \{u_1, \dots, u_k\}$ at a moment (?) t by EXACT values of some real parameters u_i and that determinism amounts to some technical uniqueness theorem.

However, one can make sense of it by formulating the question in the following terms.

Selective Instability of Dynamical Systems. Let $F = F_1$ be the "future" map from U to itself, for $F : u(t = 0) \mapsto u(t = 1)$. Given a point u in U define its *instability profile* $IP(u)$ as the set of those u' in the "vicinity" of u for which "essential parameters" of $F(u)$ in U undergo "improbably large" (nearly discontinuous) variations under "small variations" of u in $IP(u)$.

Selective Instability of a state u signifies in this terms some particular mathematical property of the set $IP(u)$ that corresponds to the vague[78] idea of being "structurally interesting".

States u of the Universe that depict *physical* systems $P = P(u)$ are (tacitly) assumed NOT to be *selectively unstable.*[79]

But if there is something *alive* about u, e.g., a protein molecule $M(u)$, a cell $C = C(u)$ or, even better, an "intelligent" humanoid ape $A = A(u)$, then $IP(u)$ does look very interesting. And it may (or may not) be possible to correlate particular features of $IP(u)$ with what $A(u)$ calls "my free will" thus refuting what Du Bois-Reymond says about it.

In order to make the above "rigorous" one needs to specify various terms used: "small variations", "improbably large", etc., which is done by "adjusting logical parameters" with a use of a representative set of examples. Below is one.

Unlocking The Door. Let your universe be reduced to a room with a door and a lock in it, with a key K that may be in any position in this room and a mechanical contraption/apparatus A with say 100 degrees of freedom and such that, according to your record, it is equally likely to occupy all its own states.

If A is "non-intelligent" and, at $t = 0$, the key in a state u is positioned, say in the middle of the room, then the probability of the door being unlocked at the moment $t = 1$, if it was locked at $t = 0$, is virtually zero.

[78] If you are a student of ergo-logic, there is nothing vague for you in the word "interesting".
[79] Yet, some physical systems exhibit some "selectively unstable features", such as localisation/concentration of energy to a few (selective) degrees of freedom.

But if you observe that this probability greatly increases whenever some appendage of A holds K, then, you bet, A is intelligent.[80]

"Nature of mathematics" is absent from Du Bois-Reymond's list. Probably, he did not realize there was "ignorabimus problem" in mathematics, while Hilbert who had thought about it emphatically stated several times that all mathematical problems will be *eventually* solved, and he even suggested a program for *mathematically* proving this assertion.

WHERE DO WE STAND TO-DAY?

The twentieth century taught us humility. We are not even certain any more what exactly "eventually" means for human civilization. Our time may be short.

And the fascinating depth of the present day *non-understanding* of the physical world – matter, force, energy, motion – was inconceivable to people of the 19th century, who were oblivious of the ideas of relativity and quantum fields and who possessed no knowledge of the large scale dynamics of the observable Universe.

Also, contrary to what Hilbert believed, his naively "optimistic" conjectures on the logical organization of mathematics were mathematically *disproved.*[81]

Mathematics is not a "logical bore" as Hilbert thought, but a miraculous structure that has no right to exist neither intrinsically nor in how it is intertwined with fundamental physics and with human minds.

(Inexplicably, despite the failure of Hilbert's program, the questions we ask in mathematics are getting answered almost instantaneously – often in the course of significantly less than a million of man-hours of people seriously thinking on these problems. But the answers are, sometimes, far from the expected ones.

Probably, it is not hard to design a random device generating mathematical problems that admit simple solutions but these will be not only unfindable but *unfathomable* in realistic time, say in a billion years of intense research by a prosperous Galactic civilisation with a quadrillion of active mathematicians in it. But the human brain is, apparently, unable to formulate such problems at will.[82])

And we believe that the fundamental questions about Nature cannot be formulated but in mathematical terms. The deepest nature of the two arrows

$$\text{SPACE/TIME/MATTER/ENERGY} \underset{?_1}{\rightsquigarrow} \text{LIFE/BRAIN}$$

[80] All this suggests the existence of a rigorous, yet remaining interesting, theory of *selective instability*; possibly, the other of the above seven questions (with possible exception of 1 and 2), if you concentrate hard, may also direct you toward non-trivial mathematics.

[81] Gödel has shown (1931) that there are unquestionably *true but unprovable statements* in mathematics. More dramatically, Paul Cohen proved (1963) that there are several parallel worlds of mathematics, such that some statements that are *true in one of the worlds may be false in another one.*

[82] Maybe, the P≠NP problem is of this kind.

and

$$\textrm{Life/Brain} \underset{?_2}{\rightsquigarrow} \textrm{Mind/Thought}$$

cannot be comprehended but in the ambience of MATHEMATICS.

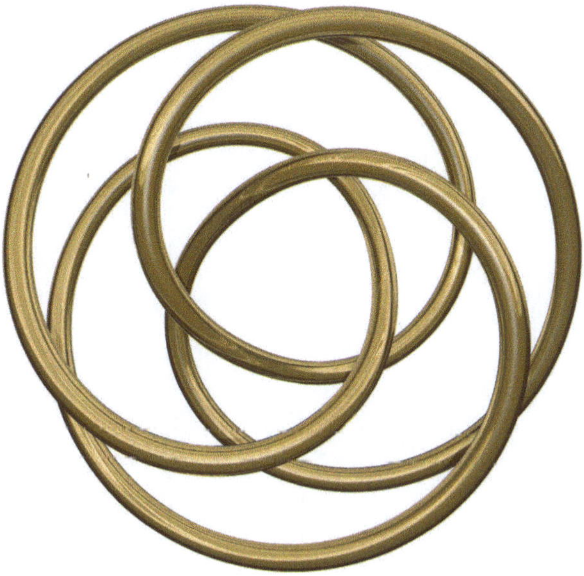

© Winthrop Brookhouse — Dreamstime.com

The Great Circle of Mysteries is closed with

$$\textrm{Brain/Mind/Thought} \underset{?_3}{\leftrightsquigarrow} \textrm{Mathematics}$$

and

$$\textrm{Mathematics} \overset{?_4}{\leftrightsquigarrow} \textrm{Space/Time/Matter/Energy}.$$

We know something about $?_4$ – mathematical physicists tell us story after story about it. But no mathematics available to us elucidates the mysteries of $?_1, ?_2$ and $?_3$. The time of this mathematics is yet to come.

Memorandum Ergo

1. Brain, Ergo-Brain, and Mind

The universe is built on a plan the profound symmetry of which is somehow present in the inner structure of our intellect.

PAUL VALERY

Decoding the Mind is impossible without creating a broad (semi)mathematical context allowing one to consistently speak of *mind-like structures*.

But

what kind of mathematics do we need to speak about the Mind?

Should we stick to *mathematics of numbers* – the language physicists speak about their World?

Some think that no radical departure from physics is needed. Frances Crick,[1] for instance, believed – we mentioned this earlier – that much of the mind may be understood in terms of the physiology of the brain:

a person's "mental activities are entirely due to the behavior of nerve cells, glial cells, and the atoms, ions, and molecules that make them up and influence them."[2]

Nobody argues that the sole source of your thoughts *is* your brain – this idea has been around for more than 4000 years.[3] But no matter how much you adorn this idea with persuasive words, everything you say about the arrow

$$[\text{BRAIN}] \rightsquigarrow [\text{mind}]$$

remains metaphoric. No sentence of a kind: "*The mind is*

caused/produced/generated/constructed or *determined/controlled/run*

About two thirds of this "memorandum" is compiled from slightly edited extracts from the two our earlier ergo-articles.

[1]Francis Harry Compton Crick (1916–2004), who greatly contributed to molecular biology, was educated as a physicist.

[2]This is written in Crick's 1994 book *The Astonishing Hypothesis*, where he promotes what he believes is a scientific approach to the problem of consciousness.

[3]A witness to this is *Edwin Smith Surgical Papyrus* (≈ 1500 BCE) – an incomplete copy of a text from the Old Kingdom of Egypt (circa 2686–2181 BCE).

M. Gromov, *Great Circle of Mysteries*, https://doi.org/10.1007/978-3-319-53049-9_2

by the brain" sheds any light on the nature of this arrow. We cannot but accept[4] what Charles Sherrington grudgingly acknowledged nearly a century ago:[5]

> *"As followers of natural science we know nothing of any relation between thoughts and the brain, except as a gross correlation in time and space."*

But what is wrong with this? What other relations do you want?

Your own brain, or rather, what we call *ergo-brain*, reconstructs the whole world in all its grandeur from space/time correlations between different events.[6] With a little mathematics, we may try something similar as follows.

Different types of brain injury produce different psychological impairments,[7] and experimental neurophysiology (ideally) delivers a correspondence between states of mind and collections of neurons in the brain that are active in the presence of such states. Since the anatomy of the brain is, roughly, the same for all people, this allows an objective comparison of similar collections in different individuals.

For instance, if experiencing a particular color, such as ●, were universally identifiable by records of excited neurons in the brains of a representative[8] group of people (animals), one would be justified in attributing the "predicate of *existence*" to *the quale* of this color.

More interestingly, the natural combinatorial distance, called *Hamming metric* between different collections of neurons in the brain[9] gives us a way to measure distances between states of mind.

If such a distance/metric were a reality, psychology would be equated with "*geometry of the mind*(*s*)" and albeit no such metric is available with the current state of knowledge, its very idea suggests the possibility of a mathematical approach to the study of the mind.

But, on the other hand, it seems there *cannot be* any "mathematics of the Mind": no matter how much you try you cannot discern anything "mathematical" in what you consciously perceive as "my Mind"; it is too loosely organized with no structurally significant patterns visible in it.

Well ..., if you watch soap operas on the screen of your computer you do not see much structure either. You have to look somewhere else.

> *It is in the admission of ignorance and the admission of uncertainty that there is a hope for the continuous motion ... in some direction... .*
>
> RICHARD FEINMANN

---

[4]Crick would disagree.

[5]This quote, which deserves being presented for the second time, is taken from Sherrington's book *Man on his Nature* based on his Gifford Lectures (1937–1938) where the scientist expounds his philosophical ideas on man's place in the universe from the point of view of the natural sciences.

[6]Space-time itself, as it is represented by the ergo-brain, results from such correlations.

[7]This had been already recorded in the Surgical Papyrus.

[8]Very likely, the so-defined quale of WHITE for the Inuit in the arctic regions of Greenland would correspond to GREEN for the Pirahã people of Amazonia.

[9]This distance is defined as the number of neurons that belong *only to one* of the two collections.

Ergo-Brain Conjecture. There exists a certain elaborate mental entity, we call it *ergo-brain*, that mediates between the electrophysiological dynamics of the brain and the thought processes in the conscious mind.

Ergo-brain is responsible for *deep learning* by humans, in particular for learning mother tongues by children and mathematics by future mathematicians.

Little of the ergo-brain is accessible to introspection. Yet, "ergo-patterns" are recognizable in natural languages and within mathematics.

Ergo-Structures/Ergo-Systems Conjecture. There are particular mathematical, essentially combinatorial, structures, call them *ergo-structures*, and a class (mathematical category?) of mathematics objects, called *ergo-systems*, that carry within themselves such structures. Ergo-brain is a particular instance of an ergo-system.

Our ultimate goal is developing the theory of ergo-structures that would bring *mathematical* means for *analysing and synthesising universal learning systems.*[10]

We imagine such a system $\mathcal{LERNER}$ that interacts with *an incoming flow of signals* similarly to a photosynthesizing plant growing in a stream of photons of light or to an amoeba navigating in a sea of chemical nutrients and/or of smaller microbes: $\mathcal{LERNER}$ recognizes and selects what is *interesting* for itself in such a flow and uses it for *building* its own structure.

This analogy is not fully far fetched. There is no *significant* difference between human activities and those by amoebas and even by bacteria, well, ... on the GRAND SCALE. Say, the probability of finding the first million digits of the number $\pi = 3.14159265359\ldots$ "written" at some location of an imaginable UNIVERSE increases by more than a billion-by-billion-by-billion factor if you find a bacterium machine feeding on a source of almost amorphous *free energy* at a point within many thousand light years of this location.

Ergo-logic: this is a particular way of thinking that is needed to approach our conjectures.

Ergo-logic sharply contrasts with everything we take for granted about what we are and what happens in our heads. We reject such ideas as

intelligent – rational – intuitive – important,

as far as ergo-brain and ergo-learning are concerned, and replace them by

interesting – curious – funny – informative.

Albeit counterintuitive, the manifestations of this logic are seen in the depth of mathematics and also in *molecular structures* of living systems uncovered by biologists in the last 50 years.

Also the idea of ergo-brain comes by assessing *limitations of natural selection* in the evolution of Man.

The structural patterns we find in the ergo-brain, although being of evolutionary origin, cannot be accounted for by the naked survival/selection mechanism, but rather by inevitable constrains on possible ergo-system architectures. These

[10]Such a theory may also elucidate the nature of mathematics.

are, essentially, mathematical constraints, and, seemingly paradoxically, they make ergo-brain more likely to be evolutionarily accessible than the "amorphous" human Mind.

And inspired by *the history* of evolution theory, where the insight by Darwin and Wallace was *not so much in biology per se but in realization of potentiality of the exponential function*,[11] we search for the key to the mystery of the Mind *in mathematics rather than in neurophysiology*:

What stands on the way of relating the world of thoughts in your mind to that of neurons in your brain is the disparity of the two structures and incompatibility of the languages describing these structures;[12] amassing data on the Brain will be of little help.[13] The arrow [BRAIN] $\rightsquigarrow$ [MIND] is as defiant of all attempts to contain it in chains of clever words as its younger sister, the arrow

[MATTER/ENERGY] $\rightsquigarrow$ [life].

One *cannot assert* (as Crick would undoubtedly do) that all life processes are *entirely due* to the interactions between atoms, ions, and molecules that make biological entities.

Of course, physicists disagree: Richard Feynman says in *Six Easy Pieces* of his famous *Lectures on Physics* that

> ... *there is nothing that living things do that cannot be understood from the point of view that they are made of atoms acting according to the laws of physics.*

However, the laws of physics are not suspended in a logical vacuum, they are immersed in a mathematical framework. Physics practiced by humans is a "*network of ideas*"[14] within this framework where some "nods" are taken for "laws of physics".

The "spirit of physics" resides in *the combinatorial architecture* of this network that is constrained and directed by many conventions, instructions, assumptions, such as

symmetry, infinitesimal linearity, stability, genericity.

But Life, albeit *constrained* by "physical laws", *excels in violating* "physical conventions and assumptions" – this is what makes Life Life.

---

[11]The enormousness of the exponential growth of unrestricted populations was obvious to mathematically minded people from the time of antiquity. But this might have been a revelation to the biologists of the 19th century who were not well versed in math – Darwin himself, who has a fine intuitive feeling for large numbers, was unable to correctly evaluate the number of descendants of a couple of elephants after 500 years.

[12]This is reminiscent of the *collapse of quantum states arrow* that stands for a (still unavailable) translation of the "quantum language" to the classical one.

[13]This would be like trying to achieve understanding of *proteins* – of their 3D structures and functions in the cell – by accumulating data on chemistry of *polynuclear acids* – DNA – that direct the synthesis of these proteins.

[14]"Idea" may stand for a record of an observation or an experiment as well as a recipe/rule for designing, conducting, and interpreting experiments.

Think, for instance, what happens to a 100 kg, BODY colliding with something tiny, something that weighs less than one billionth of a gram.

Nothing, obviously, but ... let BODY be the body of a *predator*. Let your "something" be a few billion molecules that depart from *the scent glands* in the *body* of a potential *pray* and "collide" with the *olfactory epithelium in the nasal cavity* of BODY.

Would you solely rely on the *Law of conservation of momentum* for predicting the time evolution of the distance between BODY and *body* especially if this second *body* happens to be yours?[15]

The idea of "mathematics of the Mind" is not new. "Algebra of thought" was conceived by Leibniz around 1676.

In 1869, William Jevons[16] built a mechanical *Logic Piano*, that, in his words, represented

> *a mind endowed with powers of thought, but wholly devoid of knowledge.*

In 1887, Charles Peirce[17] was asking how much

> *the business of thinking a machine could possibly be made to perform.*

In 1950, this idea was expounded by Alan Turing in the article *Computing Machinery and Intelligence* where he argues that nothing stands on the way of

BUILDING MACHINES THAT CAN THINK.

But what is *the logical structure* in your mental processes that can be mathematically modelled and implemented on a machine?

The structurally rich *neurophysiology of the brain* is too far removed from what we want to simulate, e.g., the learning process of the mother tongue by a child, but *the flows of your conscious thoughts* are void of interesting structures.

Our suggestion is to switch the focus from *dynamics of the brain* and *logic of thoughts* to

invisible and apparently illogical undercurrents of thoughts

that we collectively call *ergo.*

The core structure of this is determined by the *mathematical* necessity of simplicity and universality, and the shape "ergo" takes in the human mind is influenced by the constraints of neuronal organisation of the brain and by (conjectural) limitations of evolutionary selection.[18]

---

[15]Indian leopards (40–80 kg) and more rarely tigers (150–300 kg) may attack men.

[16]William Stanley Jevons (1835–1882) was an economist and logician. His book *A General Mathematical Theory of Political Economy* (1862) was a start of the mathematical method in economics.

[17]Charles Sanders Peirce (1839–1914), "the father of pragmatism" and the founder of *semiotic*, was an innovator in mathematical logic, philosophy, and statistics.

[18]Think of the shape of a cucumber grown in a bottle.

In the first twenty-one sections of this "memorandum", except maybe for "formalistic" Section 3, we informally discuss what "ergo" might be and in the remaining more technical sections we turn to a concrete analysis of the two basic questions about the Mind, that are:

Is there enough structural universality in the process of "thinking" to allow a mathematical modelling of this process?

What, conceivably, could serve as
THE MATHEMATICS OF THE ERGO-BRAIN?

2. Ergo Project

It is not knowledge, but the act of learning, . . .
which grants the greatest enjoyment.

CARL FRIEDRICH GAUSS

The ultimate aim of the *ergo project* is designing a *universal learning program* that upon encountering an *interesting flow of signals*, e.g., representing a natural language, starts *spontaneously interacting* with this flow and will eventually arrive at *understanding* of *the meaning* of messages carried by this flow.

We do know that such programs exist, we carry them in the depths of our MIND, in what we call *ergo-brain*, but we have no inkling of what they are.

Prior to embarking toward the design of such programs we must proceed with

- assessing *flows of signals* commonly encountered in life from a mathematical perspective and formalising what we find *interesting* about them;
- describing, let it be in general, yet, mathematical, terms, what the words *learning, understanding, meaning* signify;
- working out general conceptual guidelines for ergo-learning.

When approaching these issues, one should follow the principles of what we call *ergo logic*, thus *distancing ourselves* from *common sense* ideas about the Human Mind that are dominant in our *ego-mind* and that are pervasive in our (*popular*) *culture.*

Ego-mind is a part of a greater MIND; in fact, it is the part you normally perceive as your mind, but MIND, as we understand it, also contains ergo-brain that, unlike ego-mind, is inaccessible to your mind's eye.

Schematically, MIND is a *finite connected graph* – network of ideas – that is composed of two subgraphs (very roughly) corresponding to the ergo-brain and the ego-mind,

$$\text{MIND} = M_{ergo} + M_{ego}$$

where M_{ergo} is a kind of a *core* of the MIND, that is a *union of cycles* and M_{ego}, a *periphery*, is a disjoint union of *trees*[19] such that each of these trees meets M_{ergo} at a single vertex – the root of this tree from where it grows.[20]

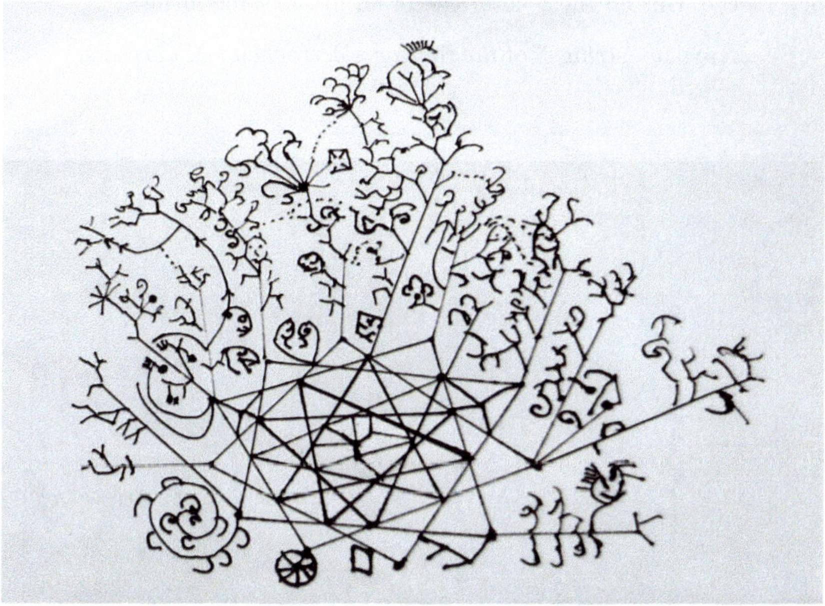

Ego is *rational*: common sense – the logic of "ego" carries accumulated evolutionary wisdom needed for our personal survival and that of our genes. Common (popular) culture is a kind of collective ego-mind.

Human ergo is *irrational*.[21] It is after beautifully interesting structures in the world, not practically useful ones, it is enthralled by play, art, science. Science and mathematics are at the core of our *collective ergo.*

Common-sense ideas and opinions, unlike the ideas of science, are unquestionably self-evident; you are not suppose to overrun them. For instance,

> *if something heavy falls on you – dodge out of the way as fast as you can: Heavy objects fall faster than the light ones and they hit you harder.*

This is great for your survival. But nothing of this kind – nothing suggested by your common sense – is good as an idea in science.

We do not claim we know which model of the human mind is nearest to the truth, but it must be as dissimilar from what intuition and common sense whisper in your ear as one is capable to imagine.

---

[19] *Trees* are connected graphs without cycles.

[20] Of course, all finite graphs decompose this way.

[21] Man is least rational of all animals. No matter what a cockroach does, even if it gets killed in the process, its behaviour is 100% rational. You cannot say this about people.

3. Formality and Universality – Meaning, Folding, and Understanding

Sources and channels of flows of signals we perceive may be different – visual, auditory, tactile. But all these enter the brain in the same form:

arrays of "strings" of fluctuating electrochemical currents.

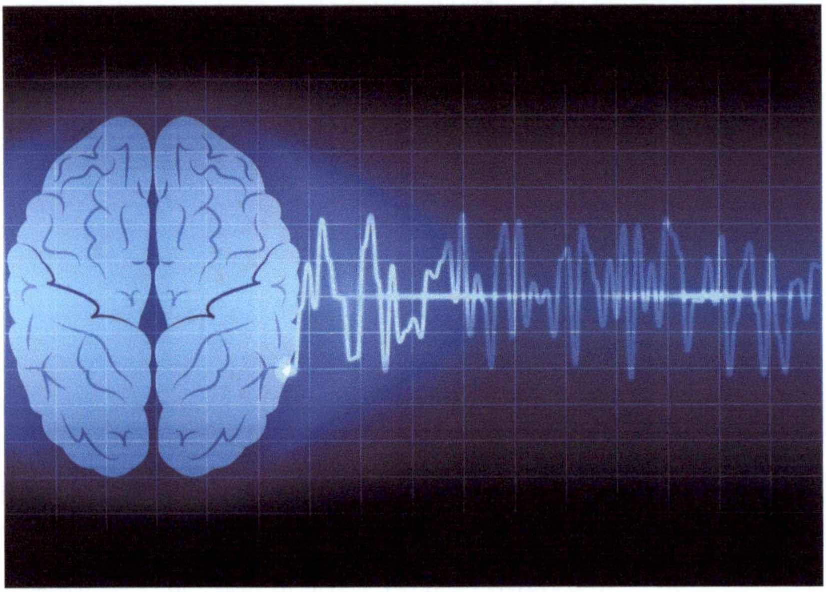

SIGNALS → BRAIN → MIND → MEANING QUESTION. How much of the *meaning* encoded in such "strings" can be recaptured and *understood* by *formal symbolic* manipulations following some *universal mathematical* rules?

Meaning and *understanding* are elusive concepts – there is no definition of them suitable for approaching this question. But drawing a formal parallel with the central problem in *molecular genetics* is helpful.

GENOTYPE ⇝ PHENOTYPE QUESTION. How much of the *anatomy, physiology, behaviour* of an organism – bacterium, plant, or animal – can be reconstructed by a *formal* analysis of the *DNA sequence* of the genome of this organism?

Life effectuates the green arrow "⇝" by several developmental processes of organisms of which we single out the following two.

⋆ *Protein folding* and ★ *Embryonic development.*

Concerning ⋆, recall, that proteins are synthesised in cells as *polymer chains* from 20 (sometimes 21) molecular units – *common amino acids*.[22] (The smallest amino acid, *glycine*, is composed of 10 atoms and the largest one, *tryptophan*, has 27 atoms.)

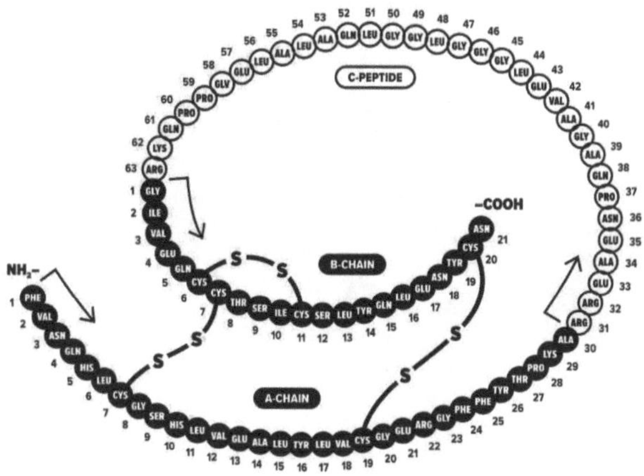

Then such a chain "folds" and thus arrives at a specific (often "potato shaped") *spatial structure* (*conformation*)[23] that determines the *physiological function of the protein.* This function is *reason d'etre* of protein, its *meaning* in the eyes of the cell.

The (stochastic) dynamics of the folding of a loosely positioned flabby string $\mathcal{S}$ of amino acids takes place in a watery environment E, where it is propelled by interactions of $\mathcal{S}$ with randomly moving water molecules[24] and also with other molecules and ions whenever these are present in a significant concentration in E.

But what mainly defines the final compact protein form[25] P is the interactions, kind of (but not quite) *attraction* forces, between constituents of $\mathcal{S}$ when they come in close contact one with another.

Metaphorically, the physical chemistry of the environment *understands* the sequence of $\mathcal{S}$ and extracts its *meaning* by forming P.

---

[22]The synthesis of proteins effectuates *translation* of what is written in four letters on mRNA's, that are, essentially, replicas of fragments of DNA, to the 20-letter proteins language. This, probably the most complicated chemical process in the Universe, is accomplished by large proteins + RNA aggregates, called *ribosomes*, that are assisted by dozens of other proteins and RNA molecules.

[23]This applies to (many but not all) moderately short chains, say of 50–300 units, where some protein molecules are formed by several chains. For instance the molecule of *haemoglobin* (in the blood of adult humans) incorporates two identical 141-chains and two 146-chains – 574 amino acids altogether.

[24]The (quadratic) average speed of water molecules at the room temperature is about 650 m/s.

[25]This form, unbelievably for a mathematician, is (essentially) unique and it is *determined*, up to a controllable stochastic error, *by the amino acid sequence*, for most proteins in living organisms.

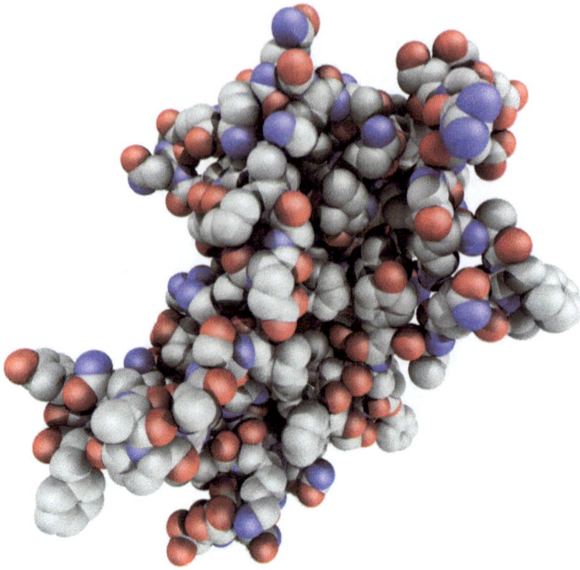

Then the physiological environment of the cell implements the message encoded in $\mathcal{S}$ by letting P perform the protein P function that the amino acid sequence of P was fashioned for.

Physically, folding can be divided into elementary steps/motions – bending, twisting, and stretching of chemical bonds of an $\mathcal{S}$-molecule. These go in parallel at different locations of $\mathcal{S}$, up to 10^{14} motions per second at each location, and happen simultaneously in many copies of $\mathcal{S}$ in the cell.

But the mechanical speed of the molecular dynamics would not suffice for fast folding of proteins that is observed in the cell – seconds or less, if not for a helping hand of evolutionary biology: *Native* protein sequences are "designed", among other things, for fast folding under physiological conditions.

Computationally, *ab initio* protein structure prediction (granted full knowledge of intermolecular interactions[26]) remains well out of range of the fastest computers.

Experimentally, if – this is a big IF – one succeeds in crystallization of a protein, then its structure can be determined (this is non-trivial) on the basis of *X-ray images of the crystals.*

Bioinformatically, a protein $\mathbf{P}_{\text{new}}$ is studied by comparing its sequence with those of related proteins the structures of which have already been resolved, where the final determination of the structure of $\mathbf{P}_{\text{new}}$ depends on *stereochemistry* of proteins.

---

[26]The available data on these is far from being complete and accurate.

Logically, the arrow

$$\text{sequence} \mapsto \text{structure}$$

transfers the sequential information in $\mathcal{S}$ to that encoded by the geometry of the *outer surface* ∂P of P.[27] This goes with a significant loss (some say "compression") of information: Different sequences may result in essentially the same protein structures.[28]

Although even the *concepts* of a protein P as "a physical body" is not expressible in the language of $\mathcal{S}$-sequences if P is taken *in isolation*, the space $\mathscr{PB}$ of all such "protein bodies", is describable in this language as

> a *subquotient* (quotient set of a subset) of the set of sequences in 20 letters defined via an *equivalence relation* on the subset of the sequences that properly fold to well shaped proteins, where *two sequences are declared equivalent if the corresponding surfaces ∂P have essentially the same geometries.*

Commentaries

• The conditions "properly-folded" and "well-shaped" may vary and define different classes of sequences, but all of these classes must contain the majority of the native (globular?) proteins.

• A "well-shaped" protein with a "properly-folded" sequence does not *generically* perform what can be called "a function" in any living cell; it is *biologically meaningless*, similar in this respect to a grammatically correct but meaningless (this is a generic property) sentence in a natural language.[29]

On the other hand, among "generic properly-folded" sequences, there are many (say, $> 2^{200} > 10^{60} < 20^{300}$) *non-native* ones; i.e., those that have never been and never will be found in nature, but that, upon folding, would perform certain functions as well as – if not better than – native proteins.

• Let the above "essentially" be understood as "approximately". Then our space of "protein bodies" will depend on a kind of approximation being used and, instead of a single $\mathscr{PB}$, we arrive at an ensemble (*small category*?) $\{\mathscr{PB}_a\}$, parametrised/indexed by possible approximations a, where the "ideal space" $\mathscr{PB}$ emerges as a (*projective*?) limit of these $\mathscr{PB}_a$.

Granted a formal *definition* of "protein folding", one may speak of a formal *resolution* of the *protein folding problem* as a set $\mathscr{R}$ of mathematical rules, that,

---

[27] "Geometry" here means that of a "colored" surface ∂P, where "colors" represent the physical/chemical properties, such as *polarity and hydrophobicity*, of the amino acids forming this surface.

[28] It debatable if the mapping *sequence* $\mapsto$ *structure* and/or the corresponding *information 3D-transfer maps* possess some kind of (stochastic?) continuity.

[29] Grammar of folding and semantics of functions of proteins in different cells are closer to each other than grammar and semantics of sentences in different human languages.

using the (sample) set $\mathcal{N}$ of sequences of native proteins that are known (or believed) to properly fold as an input, would come up with (something close to) the ensemble $\{\mathcal{PB}_a\}$.

Such a set $\mathcal{R}$, even one of a moderate size, may exist. But since $\mathcal{R}$ must depend on physical/chemical specifics of intermolecular interactions and due to the computational complexity inherent in these interactions, such an $\mathcal{R}$ hardly can be found on the basis of *universal mathematical* principles *with no a priori idea* of geometry and physics of folding and/or with no knowledge of the origin of $\mathcal{N}$.

The solution of the protein folding problem by purely formal mathematical reasoning is impossible.

However, certain fragments (mathematically speaking, *subquotients*) of the ensemble of spaces $\{\mathcal{PB}_a\}$ that represent some "meaningful features" of proteins, e.g., persistence of certain sequential patterns, say, those corresponding to *α-helices*, are easily formally detectable. And a comprehensive mathematical model of proteins open to an input of experimental data may furnish a solution of the folding problem.

The colors of logic change in a variety of ways when we look at the arrow

protein structure → protein function.

(⋆) *Collectivity of Protein Functions.* The function of a given protein P in a cell makes sense only in the ambience of other proteins that interact with P. (This is formally similar to the *function* or *semantic meaning* of a sentence S in a text: Unlike the syntax of S, the meaning cannot be defined out of an extended lexical context.)

Accordingly, it is the space of *(macro)molecular dynamics of the cell* – $\mathcal{MDC}$ in the role of $\mathcal{PB}$ – not the spaces of functions of individual proteins, that is amenable to a formal representation, where $\mathcal{MDC}$ must be represented by a subquotient of the set of DNA sequences of (real and imaginary) cell genomes.

(⋆⋆) *Predominance of Biology over Physics.* Certain (not all) specific functions of a protein P depend on the "colored shape" of the surface ∂P that is responsible for direct *physical interactions* of P with other molecules, e.g., proteins, that are similar to the attractive interactions between different loci in the molecular chain of P.

But evolutionarily "designed" organisation(s) of cellular functions of proteins in the cell is very much *unlike* what is seen in natural *physical* systems. There is no physical embodiment of this organisation, nothing similar to "protein body" P.

(Some protein functions are reflected, albeit in a limited way, in *protein interaction networks* that record systematic close encounters between proteins molecules in a cell.)

Let us move up the complexity ladder and try to understand in formal terms how the physiological *meaning* encoded in the genome of an organism, that is the *phenotype* of this organism, "unfolds" in the course of embryogenesis.

A possible mathematical format for such understanding can be, as earlier, a representation of the "space" $\mathcal{APO}$ of *anatomies + physiologies* of *multicellular organisms* as a subquotient of the space $\mathcal{DNA}$ of (real and imaginary) genome sequences.[30] Achieving this, however, doesn't seem realistic because of the following.

☹ The "space" $\mathcal{APO}$, unlike the (true) protein space $\mathcal{PB}$, is unknown in its entirety and the spaces $\mathcal{APO}_\alpha$ of *accessible attributes* α of organisms, by no means approximate (the largely unknown) $\mathcal{APO}$.[31]

☹☹ One does not know, not even in general terms, what are the rules of *intercellular interactions*, including *cell signalling*, that describe the *biology of morphogenesis.*

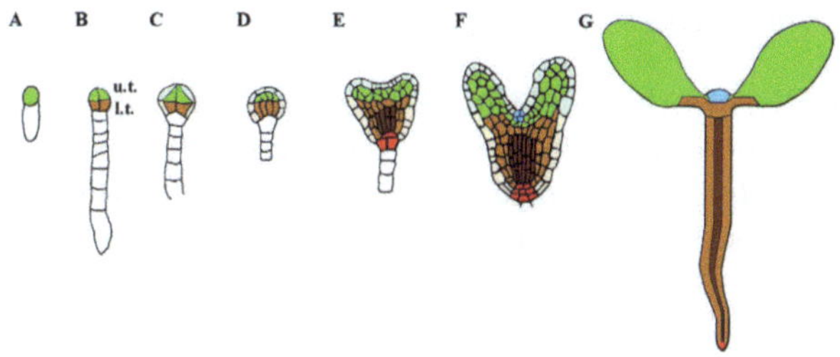

Republished with permission of American Society of Plant Biologists, from The Arabidopsis Book 7, Embryogenesis: Pattern Formation from a Single Cell, A. Capron, S. Chatfield, N.Provart, and Th. Berleth, 2009

☹☹☹ No known mathematical model is able to reproduce essential features of morphogenesis, e.g., the resilience of the forms of animals (say, mammals and birds) as they grow from childhood to adulthood.

While the body of a humble bug reads and understands the messages encoded in the bug's genome and implements their *meaning* by building and sustaining itself, our enlightened mind miserably fails to discern the *meaning* in these messages.

And if we cannot match the intelligence of a bug's body do we stand a chance to understand a human brain?

Can we succeed in emulating *the brain's way* of extracting *meaning* from the signals it receives and assigning *meaning* to the signals it generates?

This looks hopeless. The brain's *meaning*, unlike the cell's *meaning* of proteins and of intercellular signals, does not even possess a physical or physiological embodiment – it is a whim of the brain's imagination.

---

[30]This space $\mathcal{DNA}$ may need to be augmented by additional *epigenetic* information (e.g., by what is molecularly "encoded" in the maternal egg cell), and, possibly, by evolutionary data on organisms and their genomes.

[31]The structure of the ensemble $\{\mathcal{APO}_\alpha\}$, even of the parts of it that have been understood, cannot be represented by anything as logically simple as a mathematical category.

But, paradoxically, this "non-physicality" of *meaning* makes a formal mathematical model of the brain-style understanding of natural signals as well as of unnatural ones – generated by other brains – feasible.

Optimism stems from the apparent *limitations* of the human brain. Signals perceived and emitted by the brain lack fine-tuned specificity of intercellular chemical signals. The brain, unlike the live cell, does not come into direct contact with the environment, it has no internal knowledge of the outside world, nor does it have built in faculties for physiological modelling of the external physical world and/or the world of human relations.[32]

> *The "logical folding" of "sequences" of electrochemical "symbols" received as well as generated by the brain into meaningful shapes – shadows of events and objects in the "real world" – is, by necessity, (almost) entirely a formal process.*

Mathematically, we think of the *meaning* resulting in this "folding" as an ensemble of subquotients $\mathscr{M}_\alpha$ of the space of (flows of a particular class of) signals $\mathscr{S}$, where *understanding* is a kind of *operational representation* of the arrow that symbolises taking these subquotients,

$$\mathscr{S} \rightsquigarrow \mathscr{M}_\alpha.$$

This is, of course, by no means a proposition of a *solution* to the problem of *meaning* but a suggestion of a possible *language for formulating* the problem. If we accept such a language, then we may search for a solution as a description of a *particular mathematical structure* in the ensemble $\mathscr{M}_\alpha$, and in the ensemble $\mathcal{A}$ of *attributes* α themselves.[33]

The structures of these $\mathcal{A}$ and $\mathscr{M}_\alpha$ as well as of the *arrow of understanding* "$\rightsquigarrow$" depend on a sample $\mathcal{S}$ of signals used for their construction. These structures are elaborately messy – anything but simple and universal.

What is simple and universal, we believe, is the *set of rules of learning* that lead from $\mathcal{S}$ to "$\rightsquigarrow$" and thus, to $\mathscr{M}_\alpha$.

Universality is the most essential property we require from the learning systems/programs which we want to design – these programs must *indiscriminately* apply to *diverse classes* of incoming signals regardless of their "meanings" using the same toolbox of rules for learning languages, chess, mathematics, and tightrope walking.

---

[32]The brain is able to map the geometry of the external world to its own internal geometry. Also, the control of motion of (the parts of) the body, e.g., turning the eyeballs and skeletal joints, may be a key factor for the brain's ability to apprehend the *rotational symmetry* of 3-space. But these *non-symbolic representations* hardly lie at the core of human higher cognitive abilities such as language and sequential reasoning.

[33]We say "ensemble" and "attribute" in order not to commit ourselves prematurely to precise mathematical notions, such as *set* for instance.

Without universality there is no chance of a non-cosmetic use of mathematics;[34] and only "clever mathematics" may furnish universality in learning.[35]

But in most of what follows we explain the ideas of ergo-learning with no appeal to "abstract" mathematics.

TERMINOLOGY: *Quotient, Reduction, Categorization, Compression*

Given what mathematicians call a *set* S, that is, a well-defined[36] "collection" of "objects" a *quotient set* $\underline{S}$ of S is obtained by identifying a kind of "gluing" or "binding together" of certain elements in S, where the "rule of gluing", of pairs of members of S, customary written as $s_1 \sim s_2$, is called the *equivalence relation* that defines $\underline{S}$.

(A simple example is where S consists of strings, say of length 50 in 40 symbols: 26 letters, 13 punctuation marks and the "space" symbol, where two strings are declared equivalent if they differ only at the last 5 places.)

Symbolically, the passage from S to $\underline{S}$, called *factorization*, or the *quotient map*, is represented by an arrow $Q : S \to \underline{S}$, also written as $s \mapsto \underline{s} = Q(s)$, where such a Q sends S *onto all* of $\underline{S}$. Pictorially, $\underline{S}$ is a kind of shadow of S defined by the arrow of the rays of light.

(In the above example of strings, the quotient set $\underline{S}$ consists of all strings of length 45 and applying Q amounts to dropping the last 5 letters in the strings.)

Conversely, given an onto map Q from S to another set T, gluing or identification of s_1 and s_2 in S can be expressed by the equation $Q(s_1) = Q(s_2)$ and then T can be regarded as a quotient set of S for the equivalence relation

$$s_1 \underset{Q}{\sim} s_2 \Leftrightarrow Q(s_1) = Q(s_2).$$

A subquotient of a set S is a quotient of a part (subset) P of S. That is, the equivalence relation and the corresponding quotient map are defined only on P.

(For instance, P may be the "set" of *meaningful* English sentences expressed by strings of fifty symbols in the $26 + 13 + 1$ "alphabet", where two such strings are declared *equivalent* if they carry *the same meaning*. Here, the ambiguity of "meaningful", and "same meaning" necessitates considering not individual P and Q but rather a family of maps $Q_{\alpha,\beta}$ defined on subsets P_β in S.)

[34]This is meaningless unless you say what kind of mathematics you have in mind. Mathematical creatures, such as, for example, *Turing machine* and *Pythagorean theorem*, differ one from another as much as a single-stranded RNA virus form a human embryo.

[35]Our objectives are different from those taken by *mathematical psychologists* (e.g., Robert Duncan Luce and James Tarlton Townsend) – *Logical and Mathematical Psychology: Dialectical Interpretation of Their Relations* by Nicolae Mărgineanu, Presa Universitară Clujeană, 1997; *Mathematical Psychology: An Elementary Introduction*, by Clyde H. Coombs, Robyn M. Dawes, and Amos Tversky; Englewood Cliffs, N.J.: Prentice Hall, © 1970. and *Mathematical Psychology: Prospects for the* 21*st century* by James T. Townsend (2008). Also see http://www.indiana.edu/~psymodel/publications_all.shtml as we are not so much concerned with modelling Human Mind but rather the "invisible" processes that shape the Mind.

[36]This "well-defined" makes it problematic, for instance, to say that *meaningful sentences* in a language constitute a set.

© ilbusca / Getty Images / iStock

In physics, factorization maps between *spaces of states*, e.g., the maps obtained by dropping some coordinates describing "states", are called *reductions.*

In linguistics, one says *classification* or *categorisation* and subsets of mutually equivalent elements s in S are called *classes* or *categories.*[37]

In our context, the arrow $S \to \underline{S}$ can be called, albeit only metaphorically, *compression of information* or *suppression of redundancies.*

This "suppression", however, does not mean that redundancies are any kind of nuisance; on the contrary, they are crucial for "logical folding" of natural flows of signals similarly to how protein folding relies on redundancy of information encoded by residue sequences. No meaning can be extracted from a non-redundant flow.[38]

But, unlike how it is with proteins, we have no clear idea of what this "logical folding" is. And it is not just technicalities – what we still miss is the *essence* of mathematics responsible for *understanding* and for *learning to understand meanings* carried by flows of signals.

[37]These have nothing to do with *mathematical categories.*

[38]Thus a logically perfect formalization of a mathematical idea, that redirects mathematical flows from the comfortable channels shared with natural languages to the narrow ditches of thought dug by logicians, impairs understanding of the idea.

4. Universality, Simplicity and Ergo-Brain

Out of chaos God made a world,
and out of high passions comes a people.

BYRON

Our fascination by learning systems comes from what may seem as an almost godlike ability of human (and some animal) infant brains of building a *consistent* model of *external world* from an *apparent* chaos of *flows of electric/chemical signals* that come into it.

Imagine that you see on a computer screen what a baby brain "sees": A *throbbing streaming crowd of electrified shifting points* encoding, in an incomprehensible manner, a certain never seen before, not even imaginable, "reality". Would you reconstruct anything of this "reality"? Would you be able to form such concepts as *shadow*, *roundness*, *squareness*?

Could you extract any *meaning* from a Fourier-like transform of the sound wave the brain auditory system receives?

No. This ability is lost by "mature minds". One cannot even recognize two-dimensional images by looking at graphical representations of the illumination levels, which is a much easier problem. What a baby chimpanzee's brain does is more "abstract" and difficult than the recently found solution of *Fermat's last theorem*.

Yet, we conjecture that an infant's ergo-brain operates according to

a universal set of simple learning rules.

The ergo-brain *extracts structural information* "diluted" in flows of signals following these rules and continuously *rebuilds itself* by incorporating this structure.

(It would be unrealistic to make any conjecture on how such rules could be implemented by the neurophysiology of the human brain, although it seems plausible that they are incorporated into the "architecture of pathways" of signal processing in the brain. But we shall try to guess as much as possible about these rules by looking at the universal learning problem from a mathematical perspective.)

At the moment, one may only speculate in favour of universality by appealing to "evolutionary thrift of Nature" and to "brain plasticity" where, besides propensity for learning languages, convincing evidence for simplicity and universality of the performance of ergo(brain) is the human ability to learn mathematics.

It may strike you as paradoxical that something as *complex* as learning 1000 pages of math and coming to understand the proof of Fermat's last theorem can be effectuated by a *simple* program in your ergo-brain. But a specialized and/or complicated *evolutionarily designed* learning program could hardly do mathematics that is far removed from the mundane activities the brain was meant for.

Ultimately, we want to write down a *short* list of *general* guidelines for "extracting" *mathematical structures* from *general* "flows of signals". And these flows may come in many different flavours – well organized and structured as mathematical deduction processes, or as unorderly as "*a shower of little electrical leaks*" depicted by Charles Sherrington in his description of the brain.

Of course, *nontrivial* structures can be found by a learning system (be it universal or specialized), only in *interesting* flows of signals. For instance, nothing can be extracted from fully random or from constant flows.

But signals that are modulated by *something meaningful* from "the real world" carry within them certain *mathematical structures* that the brain of a human infant can detect and reconstruct.

Universal Patterns in Animal Behaviour

When, as by a miracle, the lovely butterfly
bursts from the chrysalis full-winged and perfect, ...
it has, for the most part, nothing to learn,
because its little life flows from its organisation
like melody from a music box.

Douglas Spalding[39]

Why Universal? It would be unrealistic to expect that evolution had time enough to select for many long sequential learning programs specific to different goals, but "clever" universal programs may accommodate several different situations.
Example: "Hawk/Goose" Effect. A baby animal, as we described this earlier, distinguishes *frequently* observed shapes sliding overhead from those that appear *rarely.*

The former eventually stops soliciting HIDE! responses, but every unusual kind of a shape, e.g., that of a hawk, makes an animal run for cover.

Such universal programs develop in the environment of evolutionary older "*general ideas*" that are kinds of *tags*, such as dangerous/harmless, edible/useless, etc., selectively associated in an animal/human brain with a variety of *particular* "real somethings" in the world.

NO universality, however, in biology is mathematically perfect – "laws of biology" are no brothers of laws of physics.

The most general law of Life from a biologist's point of view is *the genetic code* – a *specific* correspondence between two sets accidentally frozen in time, one is comprised of 61 out of 64 triplets of four bases – adenine, thymine, cytosine,

[39] Douglas Spalding (1840(41?)–1877), the founder of *ethology*, was arguably, along with Gregor Mendel, the most original biology (and psychology) thinker of the 19th century. He discovered *imprinting* in baby animals (popularised by Konrad Lorentz in the 1930s) and he began the study of anti-predator reactions. Unlike Darwin and Freud, he had not exposed his ideas to laymen and his name remained unknown to the general public.

and guanine – in DNA –, and the second one is the set of 20 basic amino acids in proteins.[40]

$$\{AAA, AAT, \ldots, TTT, AAC, \ldots, GGG\}$$

⤳ {❀, ✲, ❈, ✾, ❁, ✽, ❊, ✸,✹, ✤,✺,✷, ✥,❉,✶, ❃,❆,❂, ✱,✿,✖}.

But the primal universality in biology for an ergo-minded mathematician is seen in

⋆ *one-dimensionality* of polynucleotides and polypeptides,

⋆⋆ *digital* nature of the genetic code,

⋆⋆⋆ *information 3D-transfer principle* implemented by *folding of biological heteropolymers.*

Evolution, Universal Grammar and Chomskyan Theory

> ... *ideas which consist of "symbolic images". The first step to thinking is a painted vision of these inner pictures ... which are produced by an "instinct to imagining" and ... re-produced by different individuals independently ...* Wolfgang Pauli

According to Chomsky, Lenneberg, and their followers, *poverty of stimulus* (i.e., limited data) would prevent children, who have amazing innate ability for language acquisition, from learning mother tongues as quickly as they do unless they have *universal grammar* "unscripted" in their LAD – *Language Acquisition Device* – a "language organ" in the brain or rather a module of the human mind that emerged in the course of human *evolution* as a result of some peculiar *mutation.*

Linguists usually do not bother furnishing any specific genetic or neurophysiological data on this "mutation" but rather operate with concepts of "evolution" and "mutation" *metaphorically.*

It is more likely, contrary to what Chomsky insists upon, that

> *language...* [is] *not a wholesale innovation, but ... a ... reconfiguration of ancestral systems.*[41]

And from our ergo perspective, "universal deep learning mechanisms" are not limited to language, but also carry a variety of other functions, such as

- learning to read and to write by ~95% of (non-dyslexic) people on Earth;
- learning mathematics by mathematically inclined students that make 10%–30% of all people;[42]
- learning to play chess by particularly gifted children (this is said about Morphy, Capablanca, Tal, Waitzkin) by observing adults play.

---

[40]Three nucleotide triplets: TAA, TAG, and TGA are *stop codons* that do not correspond to amino acids. Yet TGA may represent the 21st amino acid: *selenocysteine.*

[41]*The eloquent ape: Genes, brains and the evolution of language*, Fisher & Marcus [15].

[42]It may be close to 30%(100%?) at ages 3–5; eventually it declines, probably, below 1%, partly under the pressure (unconsciously) exerted by "mathematically dyslectic" parents and teachers.

The presence of these abilities in human populations forcefully denies naive arguments of Darwinian[43] adaptive evolution.

Technical (Im)Practicality of Universality. Multi-purpose gadgets are not among the Greatest Engineering Achievements of the Twentieth Century: *flying submarines*, if they were a success, then only in James Bond movies.[44] On the other hand, 20th century machine computation has converged to universality; basic machine learning will, most probably, follow this path in the 21st century.

5. Freedom, Curiosity, Interesting Signals, and Goal Free Learning

The essence of mathematics lies in its freedom.

GEORG CANTOR

These words by Georg Cantor equally apply to learning in place of mathematics. Universality necessitates the *non-pragmatic* character of learning. Indeed, *formulating* each utilitarian goal is *specific* for this goal – there is no universal structure on the "set of goals". Thus,

the essential mechanism of learning is goal-free and independent of external reinforcement,[45]

where the primary example of free learning is the first language acquisition.

The ability of native learning systems to function with

no purpose, no instruction, no reinforcement

is no more paradoxical than, say, a mechanical system moving in the absence of a force.

External constraints and forces change the behaviour of such systems, but *inertia* remains *the source* of motion. (The use of metaphors in science leads to confusion. The force of gravity is what *makes* things fall but it can hardly be called the *source* of motion of the Earth around the Sun.)

Closer to home, think of your digestive system. The biochemistry of metabolic networks in the cells in your body needs no teacher instruction, albeit hunger initiates the digestive process.

---

[43]More often than not, "Darwinian" is used synonymously to "truly scientific". But our "Darwinian" refers to how evolution was understood prior to the genetic revolution in the last decades of the 20th century.

[44]There are sea birds, e.g., *pelagic cormorants* and *common murres*, who are (reasonably) good flyers and who also can dive, some up to more than 50 (150?) m. The technology for building comparably universal/adaptable machines may be waiting ahead of us.

[45]The feeling of pain when you fall down or bump into something may be helpful in learning to run – this is debatable; but contrary to what a behavioristically minded educator would think, *reward/punishment reinforcement* does not channel the learning process by *reinforcing* it, but rather by *curtailing and constraining* it. Compare [33], [38].

Similarly, you may start learning to play chess or to walk a tightrope in order to impress your peers, but the learning program(s) in your (ergo)brain carries no trace of this purpose.

Ergo-Systems. These are universal learning systems that we want to design. They also must be self-propelled learners that learn spontaneously with no need for instructions and reinforcement. (Strictly speaking, our concept of ergo-system is broader, in particular it does not exclude native ergo-brain learners.)

Curiosity as Intrinsic Motivation. The idea of what we call *ergo-systems* is close to what was earlier proposed by Schmidthuber [29] and by Oudeyer, Kaplan, and Hafner [25] under the name of *Intrinsically Motivated Curiosity Driven Robots.*

This "motivation" is implemented by a class of *predictor programs*, that depend on a parameter B that is coupled with (e.g., by being a function of) the behaviour of robots.

These programs $\mathrm{Pred} = \mathrm{Pred}(H, B)$ "predict" in a certain specified way incoming signals on the basis of the history H, but robots (are also programmed to) optimize (in a specific formally defined way) the quality of this prediction by varying B.

Thus, "freedom" for an ergo-brain is not just a possibility to generate any kind of signals it "wants", but rather to have "interesting" environmental responses to these signals.

For instance, a bug crawling on an *infinite* leaf has *zero* freedom: No matter where it goes it learns nothing new. But an accessible edge of the leaf, adds to the bug's "freedom".

Similarly,[46] an ergo-brain comes to "understand" the world by "trying to maximize" its "predictive power" but what exactly it predicts at every stage depends on what structure has been already built. The "architecture of understanding" in the human mind is built from "bricks of predictability" that come in all shapes and colors. (This may be hard to reconcile with Rene Thom's "*Prédire n'est pas Expliquer*".)

In order to maximise anything, one needs some freedom of choice, e.g., your eye needs a possibility to run along lines/pages or, in a chess game, you can choose from a certain repertoire of moves.

When this repertoire becomes constrained, the ergo-brain feels *bored* and *frustrated.* This happens to you when a pedantic lecturer curbs your curiosity by displaying slides on the screen line by line, preventing you from seeing the whole page.

And a most dramatic instance of being prevented from learning is described by Helen Keller:[47]

[46]The distribution density of a bug's positions crawling on a leaf is similar to that of your eye scanning this very leaf.

[47]Helene Keller, who lost her sight and hearing at the age of 18 months, was not exposed to *tactile sign language* until nearly age seven.

> *Once I knew only darkness and stillness ... my life was without past or future ... but a little word from the fingers of another fell into my hand that clutched at emptiness, and my heart leaped to the rapture of living.*

The idea of "*interesting*" – that is, the feature of a structure that excites "curiosity" of a learner – can be best grasped by looking at the extreme instances of *uninteresting* flows of signals – the constant ones:

●●

○○○

There is (almost) nothing to predict here, nothing to learn, there is no substance in this flow for building your internal ergo-structure. (If you were deprived of freedom to learn by being confined to an infinite flat plane with no single distinguished feature on it, you would soon be mentally dead; boredom cripples and kills – literally, not metaphorically.)

And random *stochastically constant* sequences do not look significantly more interesting.

○●○●●○○●●○●○○○●●○○●●●○●●●○○●○○●●○○○○●●●○●○○●●○●○●○●●●●○●○

This appears "non-interesting" because one loses control over incoming signals, but there is much to learn from the following string that makes your curious ergo much happier.

Our ergo idea of "interesting" is suspended in balance between *maximal novelty* of what comes and *being in control* of what happens.

(Pure randomness looks boringly uneventful to your eye but your *vestibular* and the *proprioceptive/somatosensory systems*[48] would enjoy propelling your body through a rugged terrain with occasional random jumps from one rocky stone to another making the trip enjoyably dangerous.[49])

6. Information, Prediction, and a Bug on the Leaf

The Optimal Prediction idea of Schmidthuber–Oudeyer–Kaplan–Hafne is central in our thinking on ergo-systems but we emphasize "structure" instead of "behaviour", with *degree of predictability* being seen as a part of the structure of flows of signals within and without an ergo-system.

This "degree" is defined as a function in three (groups of) variables:

---

[48] These sensory systems tell you what are the current (absolute and relative) positions, velocities, and accelerations of your body and of its parts, with most accelerations being perceived via stresses in your skeletal muscles.

[49] *Irrationality* is a hallmark of humanity. Only exceptionally, grown-up *non-human* animals are able to derive pleasure from doing something that carries no survival/reproduction value tag attached to it.

the learner system LEARNER *and two fragments, say* $\overleftarrow{past}$ *and* $\overrightarrow{future}$ *in the flow of signals, where* LEARNER *predicts "something" from* $\overrightarrow{future}$ *on the basis of its knowledge of* $\overleftarrow{past}$.

This "something" refers to the result of some reduction – a kind of simplification procedure applied to $\overrightarrow{future}$ *where such a reduction may be suggested by* LEARNER itself or by another ergo-system, e.g., by a human ergo-brain.

An instance of this is predicting a *class of a* [*word*] in a [*text*] on the basis of several preceding words or classes of such words. Such a class may be either syntactic, such as *part of speech*: verb noun . . . , or semantic, e.g., referring to

vision, hearing, motion, animal, inanimate object,

or something else.

And "degree of predictability" of a class of [*word*] derived from correlations of this class with words that *follow as well as precede* [*word*] is also structurally informative.

(The *proper direction*, that is "follow" versus "precede" relation, is *not intrinsic* for (a record of) a flow of speech: it is non-obvious if strings have to be read left to right or right to left in an unknown language.

But, possibly, the direction can be reconstructed via some *universal* feature of the "predictability (information) profile" of such a flow *common to all* languages[50], similarly (but not quite) to how the arrow of time is derived from the evolution of macroscopic observables of large physical ensembles.)

© Kathy Pilato,
http://katpilato.blogspot.ch/2011/02/lady-of-different-sort.html

Let us apply the prediction idea to a bug crawling on a leaf or of your eye inspecting a *green* spot on a *brown* background.

[50]Phonetics of recorded speech suggests an easy solution but it would be more interesting to do it with deeper levels of the linguistic structures. In English, for instance, the correlations of *short* words with their neighbours are stronger for the neighbouring words that *follow short-words* rather than precede them; but this may not be so in other languages.

We assume (being unjust to bugs) that all the bug can perceive in its environment are two "letters" G and B – the colors (textures if you wish) of its positions on the leaf, where the bug has no idea of color (or texture) but it can distinguish green G-locations from the brown ones, B.

The four "words" our bug (eye) creates/observes on contemplating the meaning of its two consecutive positions are GG, GB, BG, BB

But can the bug tell GB from BG?
Is GG "more similar" to GB than to BB?

An algebraically minded bug will translate these questions to the language of *transformations* of the "words in colors":

1. *Switching the colors*: GG ↔ BB, GB ↔ BG.
2. *Interchanging the orders of the letters*: GB ↔ BG with no action on GG and BB.[51]

These transformations do not essentially change the "meaning" of the words: a green square on a brown background is identical, for most purposes, to a brown square on a green background. (Is this "equality" recognized by animals?)

Alphabet of Bug's Moves. Besides perceiving/distinguishing colors, our bug (as well as our eye) has a certain repertoire of moves but it knows as little about them as it knows about colors. Metaphorically, a bug moves by pressing certain "buttons" and then records the colors of the locations these moves bring it to.

The bug does not know its own position on the leaf and the *equality of two moves* – pressing the *same button* – translates in the bug's mind to the constancy of its direction and speed of motion.

(The eye, unlike the bug, "knows" its position s and, in order to repeat a move, it needs to "forget" s. Besides, the eye has several independent arrays of buttons corresponding to different modes of eye movements, some of which are stochastic.)

Amazingly, albeit obviously, this is exactly what is needed for reconstruction of the *affine geometry* of the Euclidean plane that tells you which triples of points • • • in the plane are positioned on a straight line and, moreover, when one of the points, say •, is positioned exactly *halfway* from • to •.

And if the bug can "count" the numbers of repetitions of identical moves, it can evaluate distances and, thus, reconstruct the full *Euclidean* (*metric*) *structure* of the ambient space, that is the 2-plane in the present case.

Which buttons has the bug to press in order to efficiently explore the leaf and learn something about its meaning – the shape of the leaf?

---

[51] There is yet another transformation: changing the color of the first letter: GG ↔ BG, BG ↔ GG. This, together with the positional transformations generates a *non-commutative group* with 8 elements in it, called the *wreath product* $\mathbb{Z}_2 \wr \mathbb{Z}_2$; the role this group plays in the life of insects remains obscure.

The bug feels good at the beginning being able to predict that the color usually does not change as the bug makes small moves. (The eye, unlike the bug, can make fast large moves.) But it becomes bored at this repetitiveness of signals, until it hits upon the edge of the leaf. The bug becomes amazed at the unexpected change of colors and it will try to press the buttons that keep it at the edge.

(Real bugs, as everybody has had a chance to observe, spend a disproportionally long time at the edges of leaves. The same applies to human eyes.)

In order to keep at the edge, the bug (this is more realistic in the case of the eye) needs to remember its several earlier moves/buttons. If those kept it at the edge in the past, then repeating them is the best bet to work so in future. (This does work if the edge is sufficiently smooth to be close to straight on the bug's scale.)

Thus, the bug learns the art of the navigation along the edge, where it enjoys twice the predictive power of what it had inside or outside the leaf: The bug knows which color it will see if it pushes the "left" or the "right" buttons assuming such buttons are available to the bug. (The correspondence "left"↔"right" adds yet another involution to the bug's world symmetry group.)

Amazingly (accidentally?), this tiny gain in predictive power, which makes the edge interesting for the bug, goes along with a tremendous information compression: The information a priori needed to encode a leaf is proportional to its area, say $A \cdot N^2$ bits on the N^2-pixel screen (where A is the relative number of the pixels inside the leaf) while the edge of the leaf, a curve of length l, can be encoded by $\mathrm{const} \cdot l \cdot N \cdot \log N$ bits (and less if the edge is sufficiently smooth). Unsurprisingly, edge detection is built into our visual system.

(The distribution of colors near the edge has a greater entropy than inside or outside the leaf but this is not the only thing that guides bugs. For example, one can have a distribution of color spots with essentially constant entropy across the edge of the leaf but where some pattern of this distribution changes at the edge, which may be hard to describe in terms of the entropy.)

Eventually the bug becomes bored traveling along the edge, but then it comes across something new and interesting again, the tip of the leaf or the T-shaped junction at the stem of the leaf. It stays there longer playing with suitable buttons and remembering which sequences of pressing them were most interesting.

When the bug starts traveling again, possibly on another leaf, it would try doing what was bringing ut to interesting places before and, upon hitting such a place, it will experience the "déjà vu" signal – yet another letter/word in the bug's language.

We have emphasized the similarities between eye and bug movements but there are (at least) two essential differences.

1. The eye moves much faster than the bug does on the neurological time scale.
2. The eye can repeat each (relatively large) "press the button" move only a couple of times within its visual field.

3. Besides "repeat", there is another *distinguished* move available to the eye, namely *reverse.*[52] Apparently, approximate back and forth movements of the eye appear disproportionally often, especially when comparing similar images.

But the problems faced by our bug are harder than evaluating a metric in a *given* space.

Imagine that you are such a bug at a keyboard of buttons about which you know nothing at all. When you press a button, either nothing happens – the color does not change – or there is a blip indicating the change of color.

> *Can you match these buttons with moves on the plane and the blips with crossing the boundaries of monochromatic domains?*
> *What is the fastest strategy of pressing buttons for reconstruction of the shape of a domain?*

The answer depends, of course, on the available moves and the shapes of the domains: You need a rich (but not confusingly rich) repertoire of moves and the domains must not be too wild.

What you have to do is to create a language, with the letters being your buttons – blips, such that the geometric properties of (domains in) the plane would be expressible in this language of sequences of pressing the buttons marked by blips. If in the course of your experiments with pressing the buttons, you observe that these properties (encoded by your language) are satisfied with significant (overwhelming?) probability, you know you got it right.

But what the bug has to do is even more difficult, since, apparently, there is *no a priori idea of spatial geometry* in the bug's brain.[53] Bug's geometry is the grammar of the "button language".[54]

Because of this, the bug's brain (and an ergo-brain in general) cannot use a strategy tailored for a particular case, but must follow *universal* rules, as the real bugs, we believe, do. Success depends on the relative simplicity/universality of plane geometry, more specifically on the group(s) of symmetries of the plane. (This symmetry is broken by "colored" domains in it, and, amusingly, *breaking* the symmetry makes it perceptible to an "observer" – the bug or the eye at the keyboard.)

And the bug is able to make an adequate picture of the world, because, incredulously, *universality is universal*:

---

[52]Geometric incarnations of "reverse" are more amusing than the affine spaces associated with "repeat". These are *Riemannian symmetric spaces.*

[53]Some animals, e.g., mice, have *map making programs* in their brains.

[54]This is similar in spirit but dissimilar in every single detail from axiomatic representations of geometries by mathematicians.

The mathematical universality of a bug's strategies matches the universal mathematical properties of the world.

This universality shapes the mechanisms of your mental processes as much as those of a bug. Your eye spends more time focused at the *edges* of images as much as a bug that crawls along the edges of leaves, and you pay most attention to the ends of words and it usually doesn't mcuh mttaer in waht oredr the ltteers in a wrod are.

> *Eye movements reflect the human thought processes; so the observer's thought may be followed to some extent from records of eye movements... The observer's attention is frequently drawn to elements which do not give important information but which, in his opinion, may do so. Often an observer will focus his attention on elements that are unusual in the particular circumstances, unfamiliar, incomprehensible, and so on.*
>
> Alfred Yarbus.

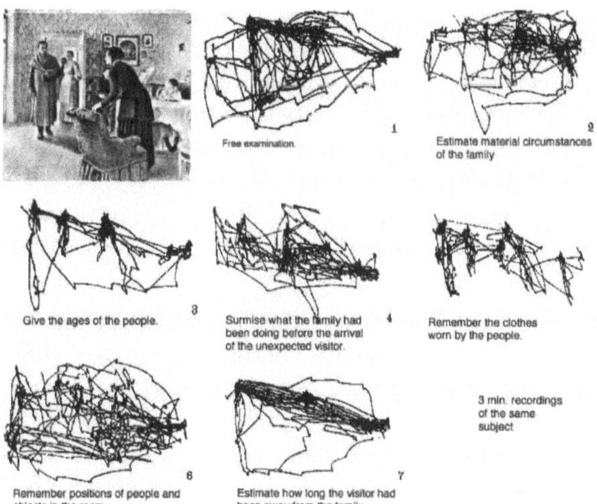

Eye Movements and Vision, Eye Movements During Perception of Complex Objects, 1967, page 174, Alfred L. Yarbus (Springer-Verlag US 1967)

7. Stones and Goals

Reach high, for stars lie hidden in you.

RABINDRANATH TAGORE

Some maintain that mastering accurate throwing, a uniquely human[55] capacity, could have been, conceivably, a key factor in early hominid brain evolution.[56] According to *the unitary hypothesis*, the same neural circuitry may be responsible for other sequential motor activities, including those involved in speech production and language [5], [39].

We cannot judge the neurophysiological plausibility of this conjecture[57] but there is a chasm of differences between learning to throw and learning to speak as far as mathematics of the two learnings is concerned.

Achieving precision throwing is a *single layer problem*. Any conceivable algorithm for it, no matter how naive – that may not even need the knowledge of the laws of mechanics – is going to work. And nowadays, it is no big deal to build a mechanical contraption that will beat any human in a throwing contest many times over, the best of Palaeolithic hunters included.

It is more difficult, but probably feasible, to design a similar program that would imitate whatever controls your tongue and larynx for proper production of sounds. But language is a multilayer structure, wheels within wheels. No one is close to designing "a speaking algorithm" that would come close to a silliest of human conversation.[58]

The unitary hypothesis, regardless whether it is right or wrong, is hardly (?) relevant to our "ergo" but looking at throwing from a position of a goal free learner is instructive.

For a thrower, the most important is his/her *aim*, that must be achieved with a correct *initial condition* – the velocity vector of a stone – that then will follow the trajectory toward a desired target. You may (and you better do) fully forget the laws of Newtonian mechanics for this purpose.

But from a physicist's point of view, it is the *second law* + *the force field* (gravitation and air resistance) that determine the motion – the initial condition is a secondary matter and the destination point is even less so.

A mathematician goes a step further away from the ancient hunter and emphasizes the general idea of time-dependent processes being described/modelled by *differential equations*.

[55]Elephants may be better than humans at precision throwing.

[56]500 000-year-old hafted stone projectile points, 4–9 cm long, were found in the deposits at Kathu Pan in South Africa, http://www.newscientist.com/article/dn22508-first-stonetipped-spear-thrown-earlier-than-thought.html

[57]There is a parallelism between spontaneous drives to learn to speak and to learn to walk/run/jump by children, but precision throwing is not in the same basket.

[58]An exception is what a patient expects to hear during a seance of psychoanalysis: ELIZA – a program written by Joseph Weizenbaum in the mid 1960s – successfully imitates a psychotherapist.

We – physicists and mathematicians with all our science would not stand a chance against *Homo heidelbergensis*[59] in a spear throwing contest; however, for instance, we, at least some of us, shall do better in mathematically designing *gravity-assist trajectories* from Earth to other Solar System bodies.

But from the position of Homo heidelbergensis[60] it would be unreasonable, not to say plain stupid, to aim at an inedible target.

Well, let us make it clear, *goal-free learning* is far from being "plain and reasonable" but it rather follows a mathematical physicist in his view on mechanical motion: There is nothing special, nothing *intrinsically* interesting, *neither in the hunter's aim* no matter how hungry he/she is, *nor in the initial condition*, although much skill is needed to achieve it. But the *transformation*

$$\textit{initial position} \mapsto \textit{aim}$$

that incorporates the laws of motion expressed by differential equations, is regarded as something *universal* and the most essential from our point of view.

There are many possible aims and initial conditions but not so many fundamental laws L and of transformations *initial position* $\mapsto$ *aim* associated to them.[61] This what makes these laws so precious in our eyes.

Similarly, one may think of learning as a transformation of an *initial input* and/or of a *learning instruction* to the *final aim* of learning.

Here we are even in a poorer position than the ancient hunter: We have hardly an inkling of what the corresponding "transformation by learning" does as it brings you from the initial input/instruction to your aim:

What is the "space" where all this happens?
What is the structure of "the trajectory" for *initial input* $\underset{?}{\rightsquigarrow}$ *aim of learning*?

And, unlike a teaching instructor, we are not concerned with *observable* inputs and aims but with mathematical models of *invisible* intrinsic structures of transformations *inputs* $\to$ *aims* that are built according to "universal laws of learning".

It is not that we deny the importance of goals, instructions, and external stimuli for learning, but we relegate them to secondary roles in the "transformation formula" that is responsible for the arrow *inputs* $\to$ *aims*. We try to understand learning processes regardless of their specific aims, or, rather, we want to see general *aim generating mechanisms* within the "universal laws" of learning.

---

[59] *Homo heidelbergensis*, a probable ancestor of Homo sapiens as well as of Neanderthals and of Denisovans, lived in Africa, Europe, and western Asia between 1 000 000 and 200 000 years ago.

[60] This position is articulated by Lev Tolstoy in an essay on science where he speaks for *a plain and reasonable man*. But of course not all (if any) of Heidelberg men were plain and reasonable. Those who were have returned to the trees.

[61] This stands in sharp contradiction with *Cantor's theorem*: There are more *logically conceivable* functions $f : x \mapsto y = f(x)$ than arguments x. But logic should not be taken literally when it comes to "real life mathematics".

8. Ego, Ergo, Emotions, and Ergo-Moods

One may understand the cosmos, but never the ego;
the self is more distant than any star.

GILBERT K. CHESTERTON

Our main premise is that learning mechanisms in humans (and some animals) are *universal, logically simple, and goal-free.* An organized totality of these mechanisms is what we call *ergo-brain* – the essential, albeit nearly invisible, "part" of human mind – an elaborate mental machine that serves as an *interface* between the neurophysiological brain and the (*ego*) mind.

Metaphorically, this "invisible" is brought into focus by rewriting the Cartesian

I THINK therefore **I** AM

as

cogito ERGO sum.

"I think" and *"I am"* are what we call *ego-concepts* – structurally shallow products of *common sense.* But ERGO – a mental transformation of the seemingly *chaotic flow* of electric/chemical signals the brain receives into a coherent picture of a *world* that defines your personal idea of existence – has a beautifully organized *mathematical structure.*

Our MIND, as depicted on p. 79, contains two quite different separate entities, that we called *ego mind* and *ergo-brain.*

Ego-mind is what you see as your personality. It includes all that you perceive as your conscious self – all your thoughts, feelings and passions, with subconscious as a byproduct of this *ego.*

The rational and intelligent "ego", shaped by the *evolutionary selection* that acted on tens of millions of generations of our animal forebears, and that serves your survival and reproduction needs, also carries imprints of *the popular culture* of the social group an individual belongs to.

Ego-processes are observed in the behaviour of humans and animals and some are perceived by retrospection.

Ego-mind is "real", large, and *structurally shallow.* Most (all) of what we know of ego-mind is expressible in *the common sense* language that reflects the logic of ego-mind. This language is adapted to our social interactions; also it suffices for expressing ideas in *the theory of mind* of *folk psychology.*

Ego-mind is responsible for WHYs about your thoughts; if you want to understand HOWs you must turn to the *ergo-brain.*

Ergo-brain, logically, mediates between electrochemical dynamics of neuronal networks in the brain and what we perceive as our "thinking".

Ergo-brain is something abstract and barely existing from ego's point of view. Ultimately, ergo-brain is describable in the language of what we call (mathematical universal learning) *ergo-systems*, but it is hard to say at the present point what ergo-brain truly is, since almost all of it is invisible to the conscious (ego) mind. (An instance of such an "invisible" is the mechanism of *conditional reflexes* that is conventionally regarded as belonging with the brain rather than with the mind.)

Ergo-brain, unlike ego-mind, is a structural entity, which underlies deeper mental processes in humans and higher animals; these are not accessible either to retrospection or to observations of behaviour of people and/or animals. This makes the ergo-brain difficult (but not impossible) for an experimental psychologist to study. (Folk psychology, psychoanalysis and the like are as unsuitable for looking into the depths of the mind as astrology is for the study of the synthesis of heavy atomic nuclei in supernovae.)

The ergo-brain and the ego-mind are autonomous entities. In young children, human and animals, the two, probably, are not much separated; a presence of *ergo in the mind* is visible in how children think about play.

As the ego-mind ("personality", in the ego-language) develops it becomes protected from the ergo-brain by a kind of a wall.[62] This makes most of ergo-brain's activity invisible.

In grown ups, ergo, albeit reluctantly, may comply with demands by ego:
"*Concentrate and solve this damn problem! – I need a promotion.*"
But the two can hardly tolerate each other.

Human ergo has a seriousness of a child at play. As a child, it does not dutifully follow your instructions and does not get willingly engaged into solving your problems. This irritates ego. From the ego perspective, what ergo does, e.g., composing utterly useless chess problems, appears plain stupid and meaningless.

Reciprocatory, utilitarian ego's activity, e.g., laboriously filing in tax return forms, is dead boring for ergo.

Certain aspects of ergo may be seen experimentally, e.g., by following *saccadic eye movements*, but a direct access to ergo-processes is limited.[63]

But there are properties of the working ergo in our brain/mind that are, however, apparent.

For example, the *maximal number* $N_\circ$ of concepts our ergo-brain can manipulate *without structurally organizing them* ("chunking" in the parlance of psychologists) equals three or four.[64] This is seen on the conscious level but such a bound is likely to apply to all signal processing by the ergo-brain.

For instance, this $N_\circ$ for (the rules of) chess is between three and four: The three unorganized concepts are those of "rook", "bishop" and "knight", with a weak structure distinguishing king/queen.

---

[62] Dramatic effects of accidentally breaking this wall are described in [10], [2], [26], [34], [35].

[63] This is similar to how it is with cellular/molecular structures and functions, where the "ergo of the cell", one might say, is the machinery controlled by the *housekeeping genes that is not directly involved in any kind of production by the cell.*

[64] Some people claim their $N_\circ$ is as large as (Miller's) "magical seven" but this seems unlikely from our mathematical perspective; some psychologists also find the number four more realistic.

Similar constraints are present in the structures of natural language where they bound the number of times operations allowed by a generative grammar may be implemented in a single sentence.[65]

Animal (including human) emotional responses to external stimuli are rather straightforward with no structurally elaborate ergo mediating between neuronal and endocrine systems.

Think of emotions as colors or typefaces – a few of dozen different kinds of them, which the brain may choose for writing a particular message, such as

run! run! RUN! RUN!

Ergo-moods also come in different colors:

curious, interested, amused, amazed, *perplexed, bored,*

serve as indicators as well as dynamic components, of the activity of the ergo-brain. These indicators tell us how far our ergo-brain is from animal rationality.

Our visual system is *amused* by optical illusions, *amazed* by tricks of magicians, *fascinated* by performances of gymnasts.

Our auditory system is *enchanted* by music.

Our olfactory system is attracted by exotic perfumes.

Our gustatory system is hungry for strange and often dangerously bitter foods.

Our motor/somatosensory system plays with our bodies making us dance, walk on our hands, perform giant swings on the high bar, juggle several unhandy objects in the air, climb deadly rocks risking our lives, play tennis, etc.

Ergo-moods, being independent of the pragmatic content of the signals received by the ergo-brain, serve as universal signatures/observables of ergo-states.

These moods are apparent as reactions to *external* signals by the ergo-brain; we conjecture that similar signatures mark and guide *the internal ergo-processes* as well.

9. Common Sense, Ergo-Ideas and Ergo-Logic

Einstein, when he says that

common sense is the collection of prejudices acquired by age eighteen

does not try to be intentionally paradoxical. There is a long list of human conceptual advances based on *non-trivial* refutations of the *old way which is also the common-sense way.*[66] The first entry on this list – *heliocentrism* – was envisioned

---

[65] An often repeated statement that "one *can* potentially produce an *infinite* number of sentences in any language" is, to put it politely, a logical misdemeanour.

The only meaningful concept of "infinite" belongs with mathematics and there is no room for the concept of "can" within mathematics proper. (Hiding behind "potentially" or appealing to such definition as "a language is a set of strings ..." does not help.)

[66] This is Lev Tolstoy's idea of how *a plain, reasonable working man* should think.

by Philolaus, albeit not quite as we see it today, twenty four centuries ago. The age of enlightenment was marked by the counterintuitive idea of Galileo's *inertia*, and the 20th century contributed *quantum physics* – *absurd from the point of view of common sense* – in Richard Feynman's words. (Amusingly, Einstein sided with common sense on the issue of *quantum*.)

It is hard to withstand the command of common sense with its incessant buzz in your head and murmur in your guts. The patulous tree of ego-wisdom in your mind, as much as your emotional self seeping with "noble feelings", is the net result of millions years of evolutionary pruning – "*the fittest survives*". And the final form of your ego-self was moulded by equally brutal cultural pressure of the last few thousand years. Unsurprisingly, a pragmatically teleological ego-centred mode of thinking that was installed by evolution into our conscious mind along with the cauldron of high passions seems to us intuitively natural and logically inescapable. But this was selected by Nature not at all for a structural modelling of the world including the mind itself.

Ergo-ideas flow from a different source and their orientation is anything but pragmatic. Ergo, unlike ego, was not specifically targeted either by evolutionary selection nor by the pressure of a popular culture – it was adopted by evolution out of sheer logical necessity. (This is similar to the *one-dimensionality* of DNA molecules that is not a result of any kind of selection as it has never been in competition with, say, two-dimensionality.) Non-accidentally, ergo is often in discord with our intuition and with the dominant cultural traditions of our social environment.

> *The first principle is that you must not fool yourself –*
> *and you are the easiest person to fool.*
>
> RICHARD FEYNMAN

The self-gratifying ego-vocabulary of

> *intuitive, intelligent, rational, serious, objective,*
> *important, productive, efficient, successful, useful*

will lead you astray in any attempt at a rational description of processes of learning; these words may be used only metaphorically. We *cannot*, as Lavoisier says,

> *improve a science without improving the language or nomenclature that belongs to it.*

The intuitive common sense concept of *human intelligence* – an idea insulated in the multilayered cocoon of *teleology* – purpose, function, usefulness, survival – is a persistent human illusion. If we want to understand the *structural essence* of the mind, we need to break out of this cocoon, wake up from this illusion, and pursue a different path of thought.

It is hard, even for a mathematician, to accept that your conscious mind, including basic (but not all) mathematical/logical intuition, is run by a blind evolutionary program resulting from "ego-conditioning" of your animal/human ancestor's minds by a million years of "selection by survival" and admit that mathematics is the only valid alternative to common sense.

Yet, we do not fully banish common sense but rather limit its use to concepts and ideas *within* mathematics. To keep on the right track we use a semi-mathematical reasoning – we call it *ergo-logic* – something we need to build along the way. We use, as a guide, the following

ERGO LIST OF IDEAS:

interesting, meaningful, informative, funny, beautiful,
curious, amusing, amazing, surprising,
confusing, perplexing, predictable, nonsensical, boring.

These concepts, are neither "objective" nor "serious" in the eyes of the ego mind, but they are *universal.* By contrast, such concept as "useful", for instance, depends on what, *specifically,* "useful" refers to.

Hopefully ergo-logic and ergo-ideas will direct us toward developing ergo-programs that would model learning processes in children's minds. After all, these minds can hardly be called serious, rational, or objective.

It is difficult to bend your ego-mind to the ergo-way of thinking. This, probably, is why we have been so unsuccessful in resolving the mystery of the Mind.

CHIMPANZEE MODEL

We cannot learn much about ergo by a study of animal behaviour,[67] but our egos are similar to those of animals. This is seen in the following experiment performed by Sarah Boysen more than 20 years ago.

X (Sarah) and Y (Sheba) were chimpanzees who learned the concepts "more than" and "less than" and who adored gumdrops, the more the better.

While Y watched, X was asked to point to one of the two plates on the table: A "large" one, with many gumdrops and a "small" one, with few of them. Whichever plate X pointed to was given to Y.

Try after try, X pointed to the "large" plate and received only a few gumdrops. Apparently X realized it was behaving stupidly but could not override the "grab what you can" drive.

Then gumdrops were replaced by plastic chips. Now X invariably pointed to the "small" plate and thus received more gumdrops than Y.[68]

---

[67]There are exceptions. Orangutans, for instance, have a propensity for 3*D topology.* They may enjoy playing with knots as much as human mathematicians do.

[68]Abstract "more" and "less" are not ingenious of ergo-brain as we shall argue later on. Chimpanzees' "more"/"less" for food (depending on the intensity of the smell?) and for non-edible items might be located in mutually disconnected parts of their brains/minds.

We invite the reader to find yet another interpretation of this experiment besides this and the obvious one.

10. Ergo in the Minds

Those who dance are often thought mad
by those who hear no music.

TAO TE CHING

The most dramatic evidence for the existence of an unbelievably powerful survival indifferent mental machinery in our heads comes from the rare cases where *the ergo insulating wall* has "*leakages*".

The legacy of evolution keeps "ergo-power" in our minds contained by an *ego-insulating wall*: A hunter-gatherer whose ergo-brain had overrun his/her pragmatic ego-mind did not survive long enough to pass on his/her genes.

But if in olden times, people with such "leakages" in their ergo-brains had no chance for "survival", in today's civilized societies they may live; they shine like "mental supernovas" unless their fire has been stifled by educational institutions.

Srinivasa Ramanujan (1887–1920) was the brightest such supernova in the Universe of Mathematics; only accidentally, due to the intervention of Godfrey Harold Hardy, did he get a chance to become visible.

When he was 16, Ramanujan read a book by G.S. Carr, "A Synopsis of Elementary Results in Pure and Applied Mathematics", that collected 5000 theorems and formulas. Then in the course of his short life, Ramanujan wrote down about 4000 new formulas, where one of the first was

$$\sqrt{1+2\sqrt{1+3\sqrt{1+4\sqrt{1+5\sqrt{1+\cdots}}}}}=3.$$

During his life, Ramanujan recorded his discoveries in four notebooks. The fourth notebook – a bunch of loose pages – the so-called "lost notebook", with about 650 of Ramanujan's formulas, most of them new, was rediscovered in 1976 by George Andrews in the library at Trinity College.[69]

Judging phenomena of "Ramanujans" and "Mozarts" *statistically insignificant*[70] is like signing off *explosions of supernovae* to *mere accidents*, just because only a dozen supernovae were recorded in our galaxy with billions of stars (none since 1680).

The hidden mental power of everybody's (ergo)brain, not only of Ramanujan's brain, must be orders of magnitude greater than what is available to the ego-mind, since *rare* mental abilities could not have been evolutionarily selected

[69]George Andrews and Bruce Berndt collaborated on publishing four volumes, appearing in 2005, 2009, 2012, 2013, of the proofs of Ramanujan's formulas included in the lost notebook. http://www.math.uiuc.edu/~berndt/lostnotebookhistory.pdf.

[70]Ergo-logic, unlike insurance companies, assigns *significant* weights to *miraculously improbable* events.

for and structurally complex functional features (be they anatomical or mental) cannot come by accident.[71]

What kind of mathematical structure could adequately describe the "mysterious something" in the human brain/mind that caused the transformation from the flow of written symbols from Carr's book to the formulas written by Ramanujan?

Unless we develop a fair idea of what such a structure can be, we would not accept any speculation either on the nature of mathematics or of the human mind, be it suggested by psychologists or by mathematicians.

Further evidence in favour of ergo-brain – a universal mathematically elaborate machine hidden in *everybody's* head that is responsible for non-pragmatic mechanism(s) of learning can be seen in the following.

1. *Spontaneous learning mother tongues by children.*

Albeit human speech depends on our inborn ability to distinguish and to articulate a vast variety of phonemes, the structural core of learning mother Language goes according to some *universal rules* that are not bound to a particular physical medium supporting a "linguistic flow". Learning languages and writing poetry by deaf-blind people is evidence for this.

2. *Learning to read and to write.*

This, unlike learning to speak, has no evolutionary history behind it.

3. *Mastering bipedal locomotion.*

One is still short of designing bipedal robots that would walk, run, and jump in a heterogeneous environment.

4. *Human fascination by sophisticated body movements: dance, acrobatics, juggling.*

5. *Playful behaviour of some animal, e.g., human, infants during the periods of their lives when the responsibility for their survival resides in the paws of their parents.*

6. *Attraction to useless (from a survival perspective) activities by humans, such as climbing high mountains and playing chess.*

Albeit rarely, adult animals, e.g., dolphins, engage in similarly useless playful actions.

7. *Creating and communicating mathematics.*

Probably, several hundred, if not thousands or even millions, people on Earth have a mental potential for understanding *Fermat's last theorem,*[72] by reading a thousand-page *written proof* of it.

The following example demonstrates human ergo in all its illogical beauty.

---

[71]The development of the brain is a random process, where only its general outline is genetically programmed. Rare fluctuations of some average "connectedness numbers" can be further amplified by "Hebbian synaptic learning". To properly account for this one has to argue in terms not of individual ergo-brains but of *(stochastic) moduli spaces of ergo-brains.*

[72]No integers $x > 0, y > 0, z > 0$ and $n > 2$ satisfy $x^n + y^n = z^n$.

A 4–5 year-old child who sees somebody balancing a stick on the tip of a finger, would try to imitate this; eventually, without any help or approval by adults, he/she is likely to master the trick.[73]

What is the mathematics behind this?

A naive/trivial solution would be reformulating the problem in terms of classical mechanics and control theory. The balancing problem is easily solvable in these terms but this solution has several shortcomings:

- It does not apply where the external forces are unknown.
- It does not scale up: No such robot came anywhere close to a healthy human in its agility.
- It suggests no universality link between balancing sticks and Ramanujan's $\sqrt{1+2\sqrt{1+3\sqrt{1+\cdots}}}$.

A better (ergo-style) solution of the balance problem with a single degree of freedom – the inclination angle α, may be obtained by following $grad_v(T)$ for $T = T(\alpha, \alpha', v)$ being the empirical "falling time" where α' denotes the angular velocity and v is the control parameter – the (horizontal) velocity of the support.

But even this distracts you from the key issue:

What on Earth drives children to try to perform such tricks?
What drives Ramanujans to invent impossible formulas?

(Younger children at play enjoy putting pencils vertically on their non-sharp ends on the table. And if captured and caged by an extraterrestrial being unable to write Ramanujan-style formulas, you would have to prove your "non-animal level of mentality" by putting a stick vertically in the *centre* of your cage.)

---

[73] Possibly, let it be rarely, a baby ape may also try to do it, but a *reasonable* human or non-human grown-up animal would have none of this nonsense.

About Stars

There are between 100 billion and 1000 billion stars in our galaxy with less than 10 thousand visible by the naked eye from Earth. One estimates that there are 2–4 supernovae explosions per century in our galaxy.[74]

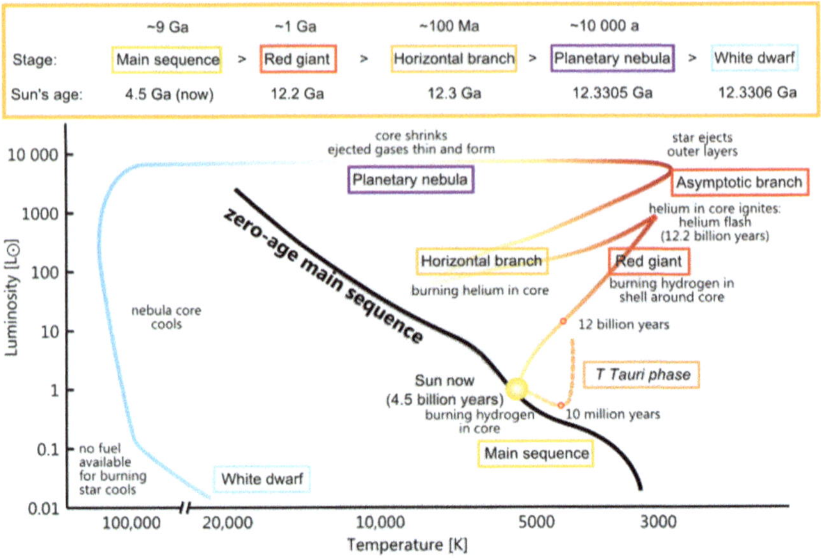

The majority of stars do not become supernovae. For instance, the life expectancy of Sun-like stars is about 10 *billion* years and their ends are *relatively* peaceful.

Stars that are ten times bigger than the Sun shine 10^4 times brighter and live only $\approx$ 10 *million* $(=(10 \cdot 10^{10})/10^4)$ years. They end up exploding as supernovae. Also some stars in binary systems turn to supernovae by accreting matter from their companions.

During several weeks a supernova radiates with an intensity of 1 billion–100 billion Suns.

[74]The brightest supernova in the 19th century sky of science, as it is seen from the position of the 21st century, was the 1866 article *Versuche über Pflanzen-Hybriden* by Gregor Mendel, who derived the *existence of genes* – atoms of heredity by a statistical analysis of the results of his experiments with pea plants. The world remained blind to the light of this star for more than 30 years.

11. Language and Languages

... may well have arisen as a concomitant of structural properties of the brain.

NOAM CHOMSKY

Most of what we know about the structure of the human ergo-brain is what we see through the window of human language. This "window" is brain's own invention, kind of a *half-step* toward *meaning* of signals it receives and it produces. Formally, in the sense of Section 3, this is a decomposition of the (subquotient) signals/meaning arrow $\mathscr{S} \rightsquigarrow \mathscr{M}_\alpha$ as

$$\mathscr{S} \rightsquigarrow \mathscr{L}\mathscr{A}\mathscr{N}\mathscr{G} \rightsquigarrow \mathscr{M}_\alpha.$$

But what is LANGUAGE in simple terms? Is it *conversing, writing, reading?*

What are essential *mathematical* structures characteristic for languages?

What would make signals coming from space classified as "*language*" rather than "*music*" or *a record of an elaborate chess-like game*?

We forfeit any idea of definition[75] but rather sketch in a few words a picture of languages of the world.

One counts ≈ 7000 currently spoken[76] (or having recently become extinct[77]) languages where *distinct languages* (rather than distinct dialects) are defined by linguists as *clusters* of vertices in the graph $\mathcal{L}_{Earth}$ where the nodes of $\mathcal{L}_{Earth}$ represent "something" (parlances) spoken by small communities of people and where the (weighted) edges of $\mathcal{L}_{Earth}$ express (degrees of) *mutual intelligibility* of these "somethings".[78]

Languages with traceable common origins are organized into ≈ 150 *language families.*

Whistled languages. There are several dozen regions in the world where the speakers of native languages can also communicate by whistling, thus transmitting messages over distances of several distances.

Silbo, based on Spanish, is practiced by 20 000 inhabitants on the island of La Gomera in the Canary Islands.

Silbo replaces Spanish phonemes with whistling sounds distinguished by pitch and continuity. (Two whistles replace the five Spanish vowels, and four whistles used for consonants.)

---

[75] A definition of *language* with a reference to human activity is as helpful for us as defining *plants* by their cooking recipes for a molecular biologist.

And defining *language* as a set of strings of symbols is no better than defining *plant* as a set of atoms.

[76] There are data on writing systems for ≈ 3500 languages (not all of them are widely used) and ≈ 700 languages are known to be unwritten.

[77] There are ≈ 1000 languages each spoken by $\lesseqqgtr$1000 people with about 20–30 languages disappearing every year.

[78] Such an $\mathcal{L}_{Earth}$, of course, is only a dream of a linguist; besides, identifying languages as clusters depends on a *clusterization algorithm* in use.

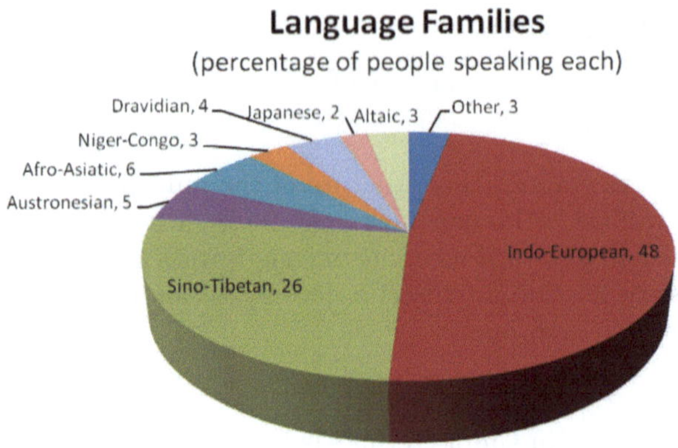

MRI monitoring shows that whistling sounds are processed in the same localities of the brain as Spanish sentences.

Mazatecan languages, spoken by 200 000 people in southern Mexico, are *tonal*; thus, well adapted to whistling. Whistled communication is used predominately by men but understood also by women in Mazatecan speaking communities.

Pirahã – the language of 200–300 Pirahã people, living in the Amazon rainforest in Brazil, can be whistled, hummed, or encoded in music.

The Pirahã language, unrelated to any other living language, has been studied by Keren and Dan Everett who lived with Pirahã people for nearly 10 years.

According to the Everetts there are no specific names for colors, no plural, and no concept of *number* in Pirahã, possibly not even for "*two*". Many other features of Pirahã, such as sentences not being produced according to Chomskyan-style *transformational grammar,* also remain controversial.

Pidgin and Creole Languages. Pidgins that serve for communication between people, e.g., traders, having no common language, are built from words and other units, of several other languages.

Children who learn a pidgin as their first language create in the process learning a creole language with as elaborate a grammatical structure as those of natural languages and that are missing from pidgins.

Sign languages. There are more than 100 different sign languages in deaf communities in the world. They are mainly independent of spoken languages; their grammars have little (if any) resemblance to that of spoken languages in the same areas. For instance, *British Sign Language* (more than 100 000 users) and *American Sign Language* (more than 300 000 users) are mutually unintelligible.

Sign languages are structurally in the same league as spoken ones, but they have a high *non-sequential* component: Many "phonemes" are produced simultaneously by combining shapes, orientations and movements of the hands, arms, and body, as well as facial expressions. This makes developing writing forms of sign languages quite difficult; and most sign languages have no written counterparts.[79]

Learning and developing sign languages follow ergo-routes similar to those of spoken languages and a well-minded ergo-blind teacher intervention only serves to block the learning – even more so the language creation – process.

ISN. A unique (?) instance of the emergence of a new language is provided by *Nicaraguan Sign Language* (ISN) that was developed by about 400 deaf children in Nicaragua in the1980s, after attempts to teach children "sign Spanish" had failed, and children became linguistically disconnected from their teachers.

This is considered by some as the strongest evidence for *innate human language capacity.*

Language Acquisition by Deaf-blind Children

Deafness is a much worse misfortune [than blindness]. *For it means the loss of the most vital stimulus – the sound of the voice that brings language, sets thoughts astir and keeps us in the intellectual company of man.* Helen Keller

The bulk of information we receive enters the brain through the eyes and the ears[80] (with more than 50% of the human cortex dedicated to vision), while the external world reconstructed by the brain of a deaf-blind, the world that is defined by what and who he/she touches, must be dissimilar to the world of those who see and hear.

However, given an opportunity, deaf-blind children master languages up to the point of writing poetry. This is amazing.

Life is either a great adventure or nothing. Helen Keller

The great adventure of learning by Helen Keller (1880–1968), who was left blind and deaf at the age of 18 months, commenced when she was six under the guidance of 21 year old Anne Sullivan.[81]

After an initial painful failure to understand the meaning of tactile signs, a dramatic breakthrough came when Keller connected the feeling of cool water running on her hand with the Braille sign for water.

---

[79] Probably, there is an (ergo)algorithm encoding flows of signs by sequences of "phonemes"; thus, delivering *phonetic* representations of sign languages, such that children would be able to learn languages by listening and reproducing such "flows of phonemes"; but hardly an experiment is possible. Compare [18].

[80] Babies and young children are hungry for *tactile information* received from the lips, the tongue, and the hands.

[81] Anne Sullivan (1866–1936) – a brilliant educator, dubbed "miracle worker" by Mark Twain – was herself visually impaired. Born in a poor illiterate family of Irish immigrants she contracted trachoma at the age 7 and nearly fully lost her sight. Anne's education began in 1880, first by learning to read and write and to use the manual Braille alphabet. At that time, she had undergone several eye operations, which improved her sight. Soon after her graduation in 1886 she became the tutor of blind-deaf Helen.

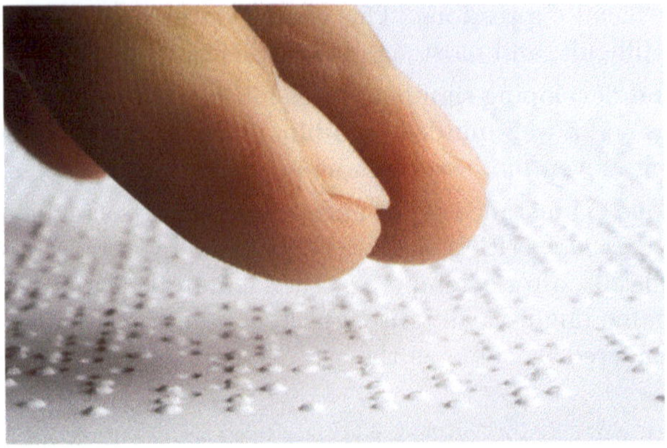

© LuisPortugal / Getty Images / iStock

Supported by Anne's creative teaching adapted to her needs, Helen had learned in a few months 600 words in Braille and the multiplication table. Eventually, Keller had mastered touch-lip reading, typing, and finger-spelling. Later on she learned to speak.

In the course of her life Helen wrote a dozen books as well as a multitude of essays: on faith, on blindness prevention, birth control, the rise of fascism in Europe, and atomic energy. Also she gave many public speeches campaigning for women's suffrage, labor rights, and social equality.

Poetry by Deaf-Blind People

The following gives an idea of how deaf-blind people perceive the world.

....

In the realms of wonderment where I dwell
I explore life with my hands;
I recognize, and am happy;
My fingers are ever athirst for the earth,
And drink up its wonders with delight, ...

From *A Chant of Darkness* by Helen Keller[82]

....

My hands are ...
My Ears, My Eyes, My Voice ...
My Heart.
They express my desires, my needs
They are the light
that guides me through the darkness

....

[82] http://www.deafblind.com/hkchant.html

With my hands I sing
Sing loud enough for the deaf to hear
Sing bright enough for the blind to see . . .

From *My Hands* by Amanda Stine[83]

The free use of ordinary language by deaf-blind people, whose internal model of the external world is quite different from the rest of us, points toward a significant independence of language from non-linguistic stimuli in accordance with ideas of Chomsky.

Language may occupy a tiny part of the brain compared to that committed to vision but, for a human being, as Ludwig Wittgenstein says,

The limits of my language mean the limits of my world.

And the inexplicable ability of children to learn and to use language is most astounding in deaf blind people. But only two lines out of 800 pages in the Handbook of Linguistics[84] are dedicated to this miraculous phenomenon:

Even a child like Helen Keller, who has lost both hearing and sight, can still acquire language through symbols expressed in touch and motion.

In the words of Helen Keller,

The only thing worse than being blind is having sight but no vision.

12. Meaning of Meaning

Everything we call meaningful is made of things that
cannot be regarded as meaningful.
. . . "meaning" is . . . a word which we must learn to use correctly.

NIELS BOHR, MISQUOTED[85]

Meanings of words are determined to a large extent
by their distributional patterns.

ZELIG HARRIS

The idea of "*meaning*" advocated by Harris is quite different from the common usage of the word "*meaning*" that invariably refers to "the real world" with "meaningful" being almost synonymous to what is advantageous for preservation and propagation of (observable features encoded by) your genes. (The speakers of the word are usually blissfully unaware of this and they become unhappy if you suggest such an interpretation of the meanings of their actions.)

The former is a *structural meaning* the full extent of which may be discerned only in the dynamics of the learning processes in humans, but the latter, the concept *pragmatic meaning*, is shared by all living organisms, at least by all animals

[83] http://www.deafblind.com/myhands.html
[84] Blackwell Publishing (2007), p. 466.
[85] In the original, one has "real" instead of "meaningful" and "reality" instead of "meaning".

from insects on.[86] This idea of meaning – the commandment to survive – was firmly installed in our brain hardware by evolutionary selection several hundred million years before anything resembling humans came into existence.

A possible way to look beyond a survival-oriented mode of thinking is to turn your mind toward something like chess, something that does not (contrary to what Freudians say) carry a significantly pronounced imprint of the evolutionary success of your forefathers.

But even if you manage to switch your mind from ego- to ergo-mode, you may remain skeptical about (ergo)chess telling you something nontrivial about learning languages and understanding their meanings.

Superficially (this is similar but different to what was suggested by Wittgenstein), one may approach a dialog in a natural language as a chess-like game that suggests an idea of (ergo) meaning: The *meaning* of an utterance UTT is derived similarly to that of the meaning of a position POS in chess. The latter is determined by the combinatorial arrangement of POS within the ergo-structure $\mathcal{CHESS}_{\text{ergo}}$ of "all" ergo-interesting chess positions/games and the former is similarly determined by its location in the architecture of $\mathcal{TONGUE}_{\text{ergo}}$ of a language.

More generally, we want to entertain the following idea that is an elaboration on the "formal discussion" in Section 3.

> *The meanings assigned by ergo-structures (e.g., by our ergo-brains) to signals are* ***entirely*** *established by patterns of combinatorial arrangements and of statistical distributions of "units of signals", be they words, tunes, shapes, or other kinds of "units".*
>
> *Understanding is a* ***structurally organized*** *ensemble of these patterns in a human/animal ergo-brain or in a more general ergo-system.*

But even leaving aside the lack of precision in all these "pattern", "arrangement", etc., one may put forward several objections to this idea.

The most obvious one is that words, and signals in general, are "just names" for objects in the "real world"; the "true meaning" resides in this world. But from the brain perspective, the only "reality" is the interaction and/or communication of the brain with incoming flows of signals. The "real word" is an abstraction, a model invented by the brain, a conjectural "external invisible something" that is responsible for these flows. Only this "brain's reality" and its meaning may admit a mathematical description and be eventually tested on a computer.[87]

(There are many different answers to the questions "What is meaning?" and "What is understanding?" offered by linguists, psychologists, and philosophers.[88]

---

[86]Semiotically minded *vervet monkeys* would not hesitate to say that the meaning of the *word-signs* of their language resides in (dangerous) object-events as these come to their fields of vision: a leopard, an eagle, a python, a baboon.

[87]We do not want to break free from the *real world*, but from the hypnosis of the words EXISTENCE/NON-EXISTENCE coming along with it.

[88]References can be found on the corresponding pages of Wikipedia.

We, on the other hand, do not suggest such an answer, since we judge our understanding of the relevant ergo-structures as immature. The expression "*structurally organized ensemble*" is not intended as a definition, but rather as an indication of a possible language where the concept of *understanding* can be productively discussed.)

Another objection may be that learning chess and understanding its meaning, unlike learning native languages by children, depends on specific verbal instructions by a teacher.

However, certain children, albeit rarely – Paul Morphy, Jose Raul Capablanca, Mikhail Tal, and Joshua Waitzkin – as we said earlier, learn chess by observing how adults play. And as for supernovas, it would be foolish to reject this evidence as "statistically insignificant".

More serious problems that are harder to dismiss are the following.

(∘) The structures $\mathcal{TONGUE}_{\mathrm{ergo}}$ of natural languages are *qualitatively* different from $\mathcal{CHESS}_{\mathrm{ergo}}$ in several respects.

Unlike how it is with chess, the rules of languages are non-deterministic, they are not explicitly given to us, and many of them remain unknown. Languages are bent under the load of (ego)pragmatics and distorted by how their syntactic tree-like structures are packed into one-dimensional strings.

SELF AND TIME. The most interesting feature of natural languages is the *self-referentiality* of their (ergo)syntax (e.g., expressed by *pronouns* and/or by certain *subordinate clauses*) that allows languages to *meaningfully* "speak" about themselves.

This is present in most condensed form in *anaphoras* such as in

X thinks he is a good chess player,

and related features common to all human languages are seen in *deixis*, such as in

but I am afraid you may be disappointed by the naivety of his moves,

along with various forms of *grammatical aspects* linked to the *idea of time.*

(It is hard to say how much of *time in the mind* is necessitated by the time dynamics of the neurobrain, what had been installed by the evolution and what comes with flows of incoming signals. And it is unclear if *time* is an essential structural component of the ergo-brain and if it should be a necessary ingredient of universal learning programs.)

None of these have counterparts in chess[89] or in any other non-linguistic structure, e.g., in music. Yet, self-referentiality is seen in mathematics on its borders with a natural language, e.g., when it is communicated from mind to mind and in its logical foundations such as *Gödel's incompleteness theorem.*

---

[89] Does the "meaning" of the following sentence reside in the game being played or in the conjunction of syntactic self-referentiality loops in there?

I thought I understood why X's white knight was placed on square a1 *but his next move caught me by surprise.*

($\circ\circ$) The internal combinatorics of $\mathcal{TONGUE}_{\rm ergo}$ may be insufficient for the *full* reconstruction of the structure of the corresponding language.

For example, linguistic signals a child receives are normally accompanied, not necessarily synchronously, by what come via all his/her sensory systems, mainly *visual* and/or *somatosensory* signals – feelings of touch, heat, pain, sense of the position of the body parts, as well as *olfactory* and *gustatory* perceived signals.

The full structure of $\mathcal{TONGUE}_{\rm ergo}$ and/or the meaning of an individual word may depend on (ergo)combinatorics of $\mathcal{VISION}_{\rm ergo}$ coupled with $\mathcal{TONGUE}_{\rm ergo}$, not on $\mathcal{TONGUE}_{\rm ergo}$ alone.

$\mathcal{VISION}_{\rm ergo}$ is vast and voluminous – more than half of the primate (including human) cortex is dedicated to vision, but the depth of the structure of "visual" within $\mathcal{TONGUE}_{\rm ergo}$ seems limited and cannot by itself support linguistic structures. This is in concordance with the ability of deaf blind people to learn to speak by essentially relying on their *tactile* sensory system; that is, their feeling of touch.[90]

The role of *proprioception* (your body/muscle sense) and the *motor control system* in learning (and understanding?) language is more substantial than that of vision, since production of speech is set in motion by firing motor neurons that activate muscles involved in speech production – *laryngeal muscles, tongue muscles* and hordes of other muscles (hand/arms muscles of mute people); thus, an essential part of human linguistic memory is the memory of sequential organization of these firings.

(Proprioception, unlike vision, hearing, and olfaction, has no independent structural existence outside your body; also it is almost 100% interactive – you do not much feel your muscles unless you start using them. The internal structure of proprioception is quite sophisticated, but, probably, it is by no means "discretized/digitalized" being far remote from what we see in language. It is hard to evaluate how much of language may exist independently of $\mathcal{PROPRIOCEPTION}_{\rm ergo}$, that may include $\mathcal{TACTILE}_{\rm ergo}$, coupled with the motor control system, since a significant dysfunction of these systems at an early age makes one unable to communicate.)

The above notwithstanding, (ergo)programs (as we see them) for learning chess and a language, and accordingly, the corresponding ideas of *meaning* and *understanding*, have much in common.

To imagine what kinds of programs these may be, think of an *ergo-entity*, call it $\mathcal{EE}$, from another Universe, to whom you want to communicate the idea/meaning of chess and with whom you want to play a game.

A preliminary step may be deciding whether $\mathcal{EE}$ is a *thinking* entity; this may be easy if $\mathcal{EE}$ possesses an ergo-brain similar to ours, which is likely if ergo is universal.

[90]There are most intriguing differences in the first language acquisition by sighted and blind children, see [3]. Probably, comparable peculiarities would be present in the language acquisition by a sighted child born in a society of blind people who were unaware of their blindness.

For example, let $\mathcal{E}\mathcal{E}$ have the mentality of a six-year-old Cro-Magnon child, where this "child" is separated from you by a wall and where the only means of communication between the two of you is by tapping on this wall.

Could you decide if the taps that come to your ears are produced by a possessor of an ergo-brain – more versatile than yours if you are significantly older than six – or from a woodpecker?

If you also happen to be six years old, the two of you will develop a common tap language-game and enjoy *meaningfully* communicating by it, but possessors of two mature human minds separated by a wall will do no better than two adult woodpeckers.

To be a good teacher of chess (or of anything else for this matter), you put yourself into $\mathcal{E}\mathcal{E}$'s shoes and think of what and how you could learn from (static) records of games and how much a benevolent and dynamic chess teacher would help. You soon realize that this learning/teaching is hard to limit to chess as it is already seen at the initial stage of learning.

Even the first (*ergo-trivial*) step – learning the *rules* how pieces move on the board will be virtually insurmountable in isolation, since these rules cannot be guessed on the basis of a non-exhaustive list of examples, say, a thousand samples, unless, besides ergo, you have a simple and adequate representation of the geometry of the chess board in your head.

If your are blind to the symmetries of the chessboard, the number of possible moves by a white rook ♖ in the presence of the white king ♔, that you must learn (in $64 \cdot 63$ positions), is $\approx 64 \cdot 63 \cdot 13 \approx 50\,000$. "Understanding" space with its symmetries, be this "understanding" preprogrammed or acquired by *a learning process of spatial structure*(s), is a necessary prerequisite not only for learning chess but also for communication/absorption of the rough idea of chess.[91]

But if you have no ergo counterparts to such concepts as "*some piece on a certain line*"[92] in your head, you'll need to be shown the admissible moves of the rook in *all* ($> 10^{45}$) possible chess positions.

And the more you think about it the clearer it becomes that the only realistic way to design a chess learning/understanding program goes via some general/universal mathematical theory equally applicable to learning chess and learning languages.

[91]The geometry of the board can be reconstructed from a moderate list of sample chess games with *Poincaré–Sturtevant space learning algorithms* but these algorithms are slow.

[92]Such "abstractions" are probably acquired by the visual ergo-system of a child well before anything as "concrete" as *white knight in a particular position on a chessboard*, for example.

13. Play, Humour, and Art

© Lynn_Bystrom / Getty Images / iStock

Without play and "playful thinking" we would not be human.

Children carry magic lanterns within themselves – the world projects onto the playground screens in their minds. And a similar *play-mode behaviour* of kittens and puppies is familiar to all of us.

Most young mammals play and also some birds, e.g., crows and ravens. (It is not always clear what behavior can be classified as "play")

Playfulness is retained into adulthood in humans and dogs, and goes along with other neonatal characteristics. Some adults animals in the wild also play, e.g., *dholes.*[93]

A Bear and a Dog[94]

> [the dog] wagged his tail, grinned, and actually bowed to the bear, as if in invitation. The bear responded with enthusiastic body language and nonaggressive facial signals. These two normally antagonistic species were speaking the same language: "Let's play!" The romp was on. For several minutes dog and bear wrestled and cavorted.

There is no accepted adaptive evolutionary explanation for the play. Apparently, patterns of play programs reflect some facets in the mental architecture of the ergo-brain that came about *despite*, not because of, selection.[95]

Ego and Ergo in Play. The drive *to win* originates in ego, but "winning/losing" is, structurally speaking, a trivial component of play.

A pure ergo-system would not try to win but rather adjust to a weaker player to make the play/game *maximally interesting.*[96]

One's ego-mind approaches the problem of play (as much as everything else) with "why"-questions; purpose-oriented solutions are welcomed by ego and such "explanations" as *Oedipus complex* for chess are acceptable.

---

[93]Dholes, also called *red Asian dogs* and *whistling dogs*, are agile and intelligent animals, somewhat one-sidedly depicted by Kipling in *The Jungle Book*.

The systematic killing of dholes was conducted by locals and promoted by British sport hunters during the British Raj. Later, some European "naturalists" called for extermination of dholes, because dholes had no "redeeming feature" but rather hair between their toes. Despite recent measures protecting dholes their population (¡2000) keeps declining.

[94]Taken from: http://www.onbeing.org/program/play-spirit-and-character/feature/excerpt-animals-play/1070.

[95]Eventually, selection may win out and populate Earth exclusively with bacteria which would have no risk-prone inclination for play. This would be the most stable/probable state of the biosphere of an Earth-like planet, granted an "ensemble" of $10^{10^{10}}$ such planets.

[96]This may not be very *interesting* to the second player, e.g., in the cat and mouse game.

We – students of ergo – on the other hand, admit that we do not understand the deep nature of play, but we reject the very idea of any common sense (teleological) explanation.

For instance, contemplating on the "meaning" of a chess-like game, we do not care what drives one to win, but rather think of the architecture of an elaborate network of *interesting game positions.* Algorithms for representation of such networks lie at the core of *universal learning.*

> *The ... phrase that heralds new discoveries is not "Eureka!" but "That's funny".* ISAAC ASIMOV[97]

Sense of humour, laughing at "funny", is closely associated with play – this is apparent in children. This "sense" is an instance of what we call an *ergo-mood* – a reaction of the ergo-brain to "*funny arrangements of ideas*".

Making a universal style program recognizing such "funny arrangements", say on internet pages, seem easier than recognizing *interesting arrangements of pieces* on the chessboard.

Performing as well as fine Arts – theatre, dance, music, along with painting, sculpture, poetry – grew out of child's play on the ego-soil of the Human Mind, where *aesthetic perception* – feeling of beauty of nature and of artistic beauty – is shared within our minds with the *sense of opposite-sex beauty.*

Music, poetry, the architecture of plants, animals and cathedrals, kaleidoscopic symmetry of peacock's tails – all with no "reproduction" tag on them,[98] – trigger in us a feeling similar to that caused by attraction to the opposite sex.

But this may only distract you from what we want to understand. For instance,

> what is in arts which is of "ergo" rather than of "ego", what are universal structures in there?

Well, ... the formal study of arts, especially of music, goes back to Pythagoras.

Also, there are active fields of *neuroscience of art, neuroaesthetics, cognitive neuroscience of music*, with many publications openly accessible on the web.

For instance,

> *"positron emission tomography scanning, combined with psychophysiological measures of autonomic nervous system activity*

shows that

> *endogenous dopamine release in the striatum at peak emotional arousal during music listening".*[99]

And an avalanche of superlatives that music lovers pour on you when they speak of music[100] tells you something about the levels of endorphins released into their blood, but does not help answering the following kind of questions.

---

[97]Fifty-fifty maybe?

[98]Peacock's tails are sexually significant for peahens.

[99]See http://www.ncbi.nlm.nih.gov/pubmed/21217764.

[100]This is also how mathematicians speak of their beloved science.

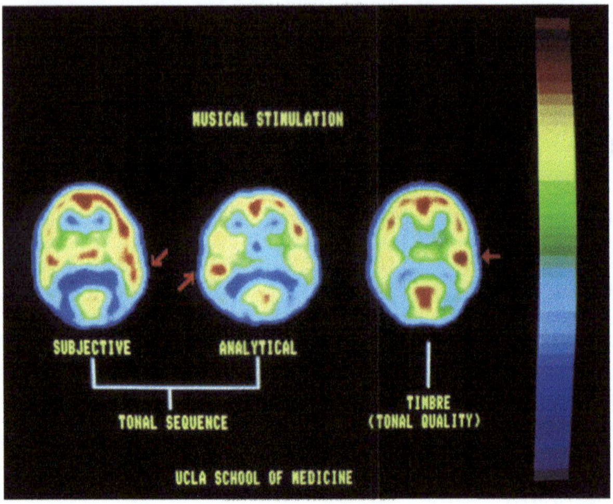

© Dr. John Mazziotta et el / Science Photo Library

What is the starting level of complexity an ergo-system must have, such that, upon unsupervised learning, it will achieve the ability to "correctly" assign aesthetic values to pieces of art?

Probably, this level need not be prohibitively high, if such a "value" is represented *not by a number* $V = V(A)$ assigned to a piece of art A but by a (partial?) order relation

$$V(A_1) >_c V(A_2), c \in C,$$

that depends on c taken from a set C of groups of art critics c.[101]

14. Ergo in Science

... a scientist ... is a curious man looking through the keyhole of nature.

JACQUES YVES COUSTEAU

It is the harmony of the diverse parts, their symmetry,
their happy balance; in a word it is all that introduces order,
all that gives unity, that permits us to see clearly and
to comprehend at once both the ensemble and the details.

HENRI POINCARÉ

Seduction by an unadulterated beauty of the world overrides the pragmatic dictum of evolution and lures us to chess, to arts, and to the ultimate human game: The mad pursuit of harmonious structures in science and mathematics.

[101]Owners of art galleries routinely solve the problem of assigning consumer-dependent prices $P_c(A)$ to pieces A of modern art.

This is "ergo" in the human character that shapes the mental set-up of a scientist. "Ergo" makes the very existence of science possible.

Henri Poincaré articulates this as follows.

The scientist does not study nature because it is useful to do so. He studies it ... because it is beautiful. [It is] *intimate beauty which comes from the harmonious order of its parts, and which a pure intelligence can grasp.*

But – one objects – Poincaré was a high priest of pure thought. Would experimentalists agree?

The experimental scientist who single-handedly contributed most to our electricity-hungry industrial civilisation was Michael Faraday.[102] He writes:

It is the great beauty of our science, chemistry ..., that ... opens the doors to further and more abundant knowledge, overflowing with beauty and utility.

Yet, medical researchers – doctors and inventors of drugs – were not playing "scientific curiosity games" but were driven by the concern for the well-being of their fellow humans. Weren't they?

Let us listen to what Alexander Fleming, who discovered penicillin, and Howard Florey, who brought penicillin to therapeutic use, say.

Fleming: *I play with microbes. There are, of course, many rules in this play ... but ... it is very pleasant to break the rules ....*

Florey: *This was an interesting scientific exercise, and because it was of some use in medicine is very gratifying, but this was not the reason that we started working on it.*

We are in no position to brush aside what these people say about science. Penicillin has saved about 100 million human lives.[103] Without either Fleming or Florey half of us would not be alive today and younger ones would not be even born.

Science produces ignorance, and ignorance fuels science.

Stuart Firestein[104]

Science never solves a problem without creating ten more.

Bernard Shaw

Shaw meant this as a mockery, but science *is* an art of *not* understanding. We strive to understand but we are not satisfied with kindergarten explanations like: *Teeth are for chewing* and *wings are for flying*. We search for *new, unknown,*

[102] Without him on the scene, world history would be shifted by a few years backward and would be, of course, quite different from what we know as our world today.

[103] By comparison, the number of victims of 20th-century "fighters for people's happiness" is estimated at 180 million–220 million.

[104] Firestein is a biologist. In his lab, they study *the vertebrate olfactory receptor neuron as a model for investigating general principles and mechanisms of signal transduction.*

invisible to understand[105] what we see in front of our eyes.[106] We are happy to discover ten new problem where originally we could discern only a single one.

A four-year-old asks:

> *Why is the grass green? Why do we breathe? Why is the water wet and stones are hard? Why do we not see in the dark? Why do we not fall upward?*

A *plain and reasonable man* of Lev Tolstoy would smile at the naivety of these questions but a 21st-century scientist would readily admit that the he/she understands none of these either, but he/she may continue with further and better-focused questions:

> *How does chlorophyll-assisted photosynthesis work?*
> *When and exactly how did the great oxygenation event occur?*
> *How does an animal cell transfer chemical energy of oxidation into mechanical energy?*
> *What is a workable microscopic model of liquid water?*
> *What is the nature of divergences in the quantum electrodynamics model of light and matter?*
> *Is there a self consistent theory of quantum gravity? ...*

Of course, *a plain and reasonable man* would have none of this. If anything, he would like to understand Nature in "a few simple words".

Well, the way the world is run may be beautifully simple, but our mind was not designed by Nature for contemplating her beauty. Perceiving this beauty requires an utmost intellectual effort on our part.

Even most familiar and apparently simple things in science are intuitively hard to accept, such as Newton's second law of motion that presents a manifestly mathematical (ergo)way of thinking about motion:

Lex II: Mutationem motus proportionalem esse vi motrici impressae, et fieri secundum lineam rectam qua vis illa imprimitur.

This law, even more so than the first law, runs against how our visual and somatosensory (mainly *proprioception* – the body sense) systems represent properties of motion in our mind.[107] The majority of us, even if we can correctly recite the three laws of motion, do not believe in these laws. We intuitively reject them

---

[105] Here, "to understand" refers to what is called "understanding" by students of natural science. This is different from "to understand" of mathematicians and has little in common with "to understand" of humanities scholars.

[106] A rare for humanities instance of an explanation of known by unknown, be it only a *conjectural* one, is that by Julian Jaynes who suggested that the major Mediterranean religions had resulted from *the Breakdown of the Bicameral Mind* about three thousand years ago. Jaynes' bicameral mind conjecture is, in principal, falsifiable – it may be either true or false. (One cannot assign truth value to "teeth are for chewing" theories of religion or of anything else for this matter.) But conducting an actual experiment verifying Jaynes' conjecture is ethically prohibitive.

[107] The essential logic of this reconstruction is of ergo but it serves the survival of our ego and serves it well, better than a mathematical Newtonian model would do.

in view of the apparent inconsistency of these laws with much of what we see with our own eyes, such as the motion of a pendulum that visibly contradicts the law of *conservation of momentum.*

But the main reason our brains resist absorbing scientific knowledge is the complexity of the combinatorial organisation of this "knowledge" that a brain needs to represent within itself.[108] Probably, the number of synaptic connections needed for *understanding* "N units of knowledge", of, say, *string theory* in mathematical physics grows (at least) as N^2 rather than const $\cdot N$ which is required for *absorbing*[109] the same number of "units of knowledge", say, in *cultural anthropology* that conveniently fit into pre-prepared niches in your ego-mind.

15. Unreasonable Men and Alternative Histories

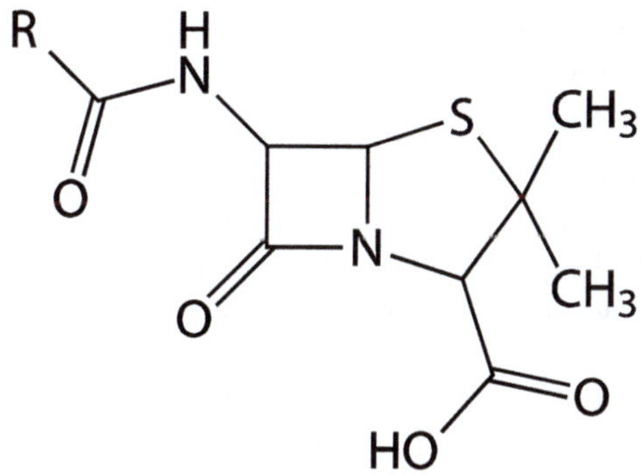

Graphic formula of penicillin, where R is the variable group.

> *It* [science] *triumphantly tells him: how many million miles it is from the earth to the sun; at what rate light travels through space; how many million vibrations of ether per second are caused by light, and how many vibrations of air by sound; it tells of the chemical components of the Milky Way, of a new element – Helium – of micro-organisms and their excrements, of the points on the hand at which electricity collects, of X-rays, and similar things.*
>
> *But I don't want any of those things, says a plain and reasonable man – I want to know how to live.*
>
> LEV TOLSTOY

---

[108]Partly, the difficulty in understanding "abstract" mathematical ideas is due to the protective wall separating ego from ergo.

[109]Memory in the brain is not straightforward unlike that on the magnetic tape. For instance, remembering long loosely structured (quasirandom) sequences, such as pages of telephone directories and arrays of the dates of "great historical events", is difficult – *excruciatingly difficult* for the mathematically inclined among us.

Nothing about penicillin – the miracle drug that had cured millions over millions of people was ever reasonable. And those who had discovered and developed this drug were anything but *plain and reasonable people.*

The modern chapter of penicillin's history starts with a *Staphylococci plate* in Fleming's lab that, between 27 July and 6 August of 1928, was unreasonably and unaccountably contaminated by an unusual (in its antibacterial activity) strain of *Penicillium notatum*, probably by the spores that escaped from the lab of La Touche – the mycologist at St Mary's Hospital working in the room below Fleming.

In 1928–1929, Fleming determined that the secretion from the mold *Penicillium notatum*, that he called *penicillin*, was effective against many bacteria.

(Penicillin suppresses growth of the so-called *gram-positive* bacteria, e.g., *streptococci* and *staphylococci*, that have no outer cell protective membrane, by blocking cell wall growth when bacteria replicate.)

Fleming suggested that penicillin could serve as a low toxicity[110] disinfectant; also he indicated the laboratory use of penicillin for the isolation of *Bacillus influenzae*. But despite a few therapeutic successes,[111] he became disappointed with instability of penicillin and turned to his other projects.[112]

In 1938, Ernst Chain (1906–1979), upon reading the 1929 paper by Alexander Fleming, proposed the study of penicillin to Florey.

In 1939, Howard Florey (1898–1968) created and directed a team of scientists for the study of anti-bacterial substances that are produced by mould. Using the sample of Penicillium notatum preserved by Fleming, they extracted and purified *penicillin* – the active antibacterial agent in the mould – and produced therapeutically significant amounts of it (1940), with the key roles played by three biochemists: Chain, Heatley,[113]and Abraham.[114]

---

[110]Penicillin kills bacteria but it is harmless for humans. However it is toxic to a few mammalians, e.g., to Guinea pigs, whose intestines are inhabited by gram-positive bacteria. Luckily, Fleming and people from the Florey's team, Chain and Heatley, tried penicillin on mice, apparently because they had limited amounts of the drug.

[111]In 1930, Cecil George Paine, treated a gonococcal infection in infants and achieved the first recorded cure with penicillin. He then cured four additional patients of eye infections, and failed to cure a fifth.

[112]Alexander Fleming (1881–1955) was a member of the research department at St Mary's Hospital in London that was organized and directed by Almroth Edward Wright (1861–1947).

In the first year of Wold War 1, Wright and Fleming worked out a treatment of infected wounds; ever since Fleming had been searching for antibacterial agents. In 1921, prior to penicillin, he discovered *Lysozyme* – an enzyme, present in tears, saliva, human milk, and mucus – that protects from gram-positive pathogens.

[113]Norman Heatly (1911–2004) devised the main steps for producing therapeutic quantities of penicillin. This, combined with know-how in fermentation technology of organic acids, has led to fast development of production of penicillin on an industrial scale.

[114]In 1943, Edward Abraham (1913–1999) determined the structure of penicillin which involved a *beta-lactam ring*; this was confirmed in 1945 by Dorothy Hodgkin by X-crystallography. In the 1950s, Abraham essentially contributed to isolation and development of *cephalosporin*, that kills penicillin-resistant bacteria.

Does the Flap of a Butterfly's Wings in Brazil set off a Tornado in Texas?

EDWARD LORENZ

There had been several decisive moments in history when ideas, insights, and decisions by such people as Fleming, Florey, Chain, Heatley, and Abraham were turning the path taken by humanity to its present course.[115]

Below are two other, even more vivid instances of (potential) instability of the course of the human history.

1. In 1896–1897, Ernest Duchesne (1874–1912) conducted the first(?) scientific study of antibacterial properties of mould. His results on *Penicillium glaucum*, similar to what was observed by Fleming, were recorded in his 1897 thesis: *Contribution à l'étude de la concurrence vitale chez les micro-organismes: antagonisme entre les moisissures et les microbes*, that he sent to *the Pasteur Institute.*

If Duchesne's work had been taken seriously by people (a single person?) at Pasteur Institute the development of antibiotics (and of the world pharmacological industry, in general) could have started a few decades earlier.[116]

It is hard to imagine what kind of world we would be then living in today.

[115]The perturbative effects of these people on human history were more subtle than of those depicted by Stefan Zweig in *Sternstunden der Menschheit*.

[116]This may be compared to what could have happened if Nägeli had understood Mendel. Possibly, nothing would have changed in both cases: The whole of the scientific community, stabilized by inertia, was no ready for these ideas.

2. A boy fails to recall the name of the river the city Berlin is located on and as a result he is denied entrance to a gymnasium in Odessa. Seeking education, he is compelled to move to the United States, where a few decades later he starts research on soil bacteria. Around 1940 this "boy" develops a comprehensive program for screening and testing *actinomycetes* for antibacterial activity. This leads to the discovery of a dozen antibiotics including, in 1943, *streptomycin* – a drug harmful to *gram-negative* bacteria, the first one effective against *mycobacteria* that cause tuberculosis.[117]

The name of the "boy" was Selman Waksman (1888–1973).[118] Streptomycin was discovered by Albert Schatz (1920–2005) who worked in Waksman's group.

Were Waksman's examiner in Odessa less pedantic, the discovery of streptomycin and other antibiotics could have been delayed by several years and hundreds of thousand people would have died of tuberculosis during this time.

A Few References On Antibiotics and their Discoverers

Gwyn Macfarlane, a haematologist who worked with Florey, wrote down two masterful accounts on the lives and works of the discovers of penicillin, [21, 22].

More can be found in [4, 6–8, 11, 12, 17, 24, 28, 30, 31, 36, 37].

16. Mathematics and its Limits

... *the object of pure mathematics that of unfolding the laws of human intelligence.*

James Joseph Sylvester

© William van der Fliet

[117]A hundred years ago, 10–15% deaths in Europe were inflicted by tuberculosis. In 2013, about nine million people in the world fell ill and 1.5 million died from TB.

[118]A significant factor turning Waksman's interest toward antibiotics, besides the achievements by Florey's group in Oxford, was the work of René Dubos (1901–1982) who, in 1939, under the influence of Oswald Avery (1877–1955) and with the help of Rollin Hotchkiss (1911–2004) at Rockefeller University in New York, isolated a bactericidal substance from the spore-forming bacillus of the soil, that he called *gramicidin*.

We share our inborn[119] ability to count with *pigeons* and *vervet monkeys*, but it is a long way from *quantity and shape perception* by animals (humans included) to something like *Ramanujan's mysterious formula*

$$\frac{1}{\pi} = \frac{2\sqrt{2}}{9801} \sum_{k=0}^{\infty} \frac{(4k)!(1103 + 26390k)}{(k!)^4 396^{4k}} = \frac{2\sqrt{2}}{9801} \left(1103 + \frac{24 \cdot 27493}{396^4} + \cdots\right)$$

This formula, similarly to the equally incredulous but more familiar *Leibniz formula* $1 - \frac{1}{3} + \frac{1}{5} - \frac{1}{7} + \frac{1}{9} - \cdots = \frac{\pi}{4}$, relates the *geometrically* defined number $\pi = 3.14159265\ldots$ to an *arithmetically generated* infinite sum, that in the Ramanujan case is comprised of impossibly complicated terms.[120]

What allows such miracles in mathematics?
What is mathematics from the ergo perspective?
What is the mathematics that underlies ergo?

The relations between mathematics and ergo, that are by necessity circular, may be summarized as follows.

Mathematics at its core, is "just" an instance of an ergo-structure.
Mathematics is "just" a fragment of the collective[121] human ergo.
Mathematics is the tool and the language for a study of ergo-structures where the latter are "just" particular mathematical structures.

Let us say a few more words about these.

The shape of heaven is of necessity spherical. ARISTOTLE

1. *Core mathematics* is all about *amazing structures* clustered around *symmetries*: perfect symmetries, hidden symmetries, supersymmetries, partial symmetries, broken symmetries, generalized symmetries, linearized symmetries, stochastic symmetries. Two thirds of this core along with most theoretical physics would collapse if the symmetry axes were removed.

The main transitions in the evolution of mathematics were not achieved by reduction of unknown to known but, contrary to common sense, by inventions of "irreal objects" such as *negative* and later *complex imaginary* numbers, *infinitesimals, ideal numbers, n-dimensional spaces*, etc., similarly to how progress in physics was driven by "irreal ideas" of *atoms, wave functions, quantum fields.*

---

[119] Inborn? – not quite, at least not quite inborn in *humans*. Our number sense is intertwined with the language learned in the cradle.

[120] The series $1103 + \frac{4!(1103+26390)}{396^4} + \frac{8!(1103+26390\cdot 2)}{(2!)^4 396^8} + \frac{12!(1103+26390\cdot 3)}{(3!)^4 396^{12}} + \cdots$, unlike Leibniz' $1 - \frac{1}{3} + \frac{1}{5} - \cdots$, converges exponentially fast; this allows a practical computation of the decimals of π with Ramanujan's formula.

[121] We mathematicians are a tiny community, something of order 0.001% of the total population, and, probably only a couple of hundreds among us (myself is not in the club) understand the Ramanujan formula.

It is amazing how mathematics manages to contain these symmetries and to represent, for example, "*roundness*" immanent in something like $\pi = 3.14159265\ldots$ by a Ramanujan-like, let it be infinite; yet, *arithmetic* formula or in the combinatorial, nearly digitalized, form of axioms, lemmas, theorems, and proofs.

This may seem not surprising, since the (collective) ergo-brain that created mathematics – represents the external world in this manner. But it may be *an endogenous* property of mathematics as well.[122]

This "combinatorial" nature of mathematics may also be compared to that used by Life for encoding shapes of organisms by DNA sequences, except that there is no (?) mathematical counterpart to *transfer of information by 3D-folding*. (A primitive form of "embryonal development" may be discerned in organisation of some mathematical proofs.)

A mathematician is an ergo-brain's way of talking to itself.

NIELS BOHR [misquoted]

2. Mathematics is the last born child of the ergo-brain, its development is guided by our ergos.

Mathematics *shines in the mind of God*, as Kepler says, but we are no gods and our minds are not pure ergo, our thinking is permeated by ego that makes it hard for us to tell "true and interesting" from "important" and that makes the (ergo) right choices difficult.

In the eyes of the ego-mind, much of mathematics appears *abstract and difficult*, but what you see in front of your eyes is *simple and concrete*.

But this simplicity is deceptive: What your eyes "see" is *not* simple – it is an outcome of an elaborate image building by your visual ergo-system that is, probably, more abstract and difficult than most of our mathematics.

Compelled by our "ergos", we search for another kind of "simplicity" that is *beautiful and interesting* – not at all trivial; *trivial* bores us to death. We are thrilled when ego-mind's "simple and apparent" is explained in terms of "abstract and difficult", that may not, a priori, even exist.

Our mathematical diamonds have been polished and their edges sharpened – century after century, by scratching away layers of ego from their facets, especially for the last fifty years. Some of what came out of it may appear as "abstract nonsense" but, as Alexander Grothendieck says,

The introduction of the cipher 0 or the group concept was general nonsense too, and mathematics was more or less stagnating for thousands of years because nobody was around to take such childish steps.

[122]The proposition – *mathematics exists as an independently entity* – may be understood only metaphorically. No conceivable experiment or argument would make it more or less feasible. But ... you cannot do mathematics if you do not believe in it. And this is also the way that physicists take *reality of the physical word.*

3. In building a mathematical frame for "ergo" we need to recognize what of our mathematics is ready to serve as "parts" of ergo-systems, what should be rejected and what needs to be made anew. And remember that

you cannot apply mathematics as long as words still becloud reality.

as Hermann Weyl said.

Our choice of the components of a logical reconstruction of human ergo-brain, call it $\mathcal{HEB}$, follows the criteria used everywhere in mathematics:

NATURALITY, UNIVERSALITY, LOGICAL PURITY, CHILDISH SIMPLICITY.

Mathematical universality of $\mathcal{HEB}$, in particular of its learning strategies (programs) may be seen in how we *enjoy and learn many different* logically complicated games. Thus, for instance a chess learning program in somebody's $\mathcal{HEB}$ must be a *specialization of a universal* learning program.

But why should such programs be simple? After all

The human brain is the most complicated object in the Universe. Isn't it?

The answer is that most general/universal theories are logically the simplest ones.[123] What is not simple is discovering and formulating such theories.

As mathematicians we are ready to accept that we are a hundred times stupider than evolution is but we do not take it for the reason that evolution is able to make miracles, such as a logically complicated brain at birth.

Believers in simplicity, we seek our own solution to the *universal learning problem* by adapting the *purest* kind of mathematics to the "dirty world" of *flows of signals* and their "*transformations*" by our (even if conjectural) ergo-brains.

But beware, mathematics that directs your thoughts toward "logically perfect structures", may mislead you when it comes to "real life".

(The arithmetic of numbers is seductively beautiful, but the surface temperatures of the sixteen named stars in the constellation of the Great Bear, even if taken in Kelvin, are not meant for addition and multiplication.)

The structures of "ergo" grow on the soil of mathematics but they diverge unrecognisably far from their pure mathematical prototypes.[124]

And it may appear – this has been the dominant philosophy in mainstream AI – that it is not so much mathematics proper, but rather such concepts as *axiomatic systems, automata, Turing machines, Gödel's theorems, etc.*, that are essential for understanding mental processes.

However, these ideas – this is witnessed by their poor record in implementing Turing's original program – are as insufficient for understanding the nature of human thinking processes as for illuminating the nature of human mathematics.

---

[123]The simplicity of a universal idea, e.g., of *Gödel's incompleteness theorem*, may be obscured by a plethora of technical details.

[124]This is not so in theoretical physics but similar to the position of mathematics in shaping (mostly unknown) fundamental principles of biology.

17. Numbers, Symmetries, and Categories

© ngaga35 / stock.adobe.com

The existence of Mathematics as we know it strikes one as improbable as the emergence of Life on Earth. Nothing in the foundation of mathematics suggests such a thing is possible, like nothing in Earth's chemistry suggests it can beget Life.

One may say that mathematics starts with *numbers*. We are so used to the idea that we forget how *incredible* properties of *real numbers* are. The seamless agreement of several different structures – *continuity, order, addition, multiplication, division* – embodied in this *single* concept is amazing.

Unbelievably perfect *symmetries* in geometry and physics – *Lie groups, Hilbert spaces, gauge theories*... – emerge in the world of numbers from the seed of the Pythagorean theorem. Mathematics and theoretical physics are the two facets of these symmetries that are both expressed in essentially the same mathematical language.

As Poincaré says,

> *... without this language most of the intimate analogies of things would forever have remained unknown to us; and we would never have had knowledge of the internal harmony of the world, which is, as we shall see, the only true objective reality.*

In the "harsh real world", away from pure mathematics and theoretical physics, the harmony of the full "symmetry spectrum" of numbers comes into play only rarely. It may even seem that there are several different kinds of numbers: some may be good for *ordering* objects according to their size and some may be used for *addition* of measured quantities. Using the all-powerful real numbers for a limited purposed may strike you as wasteful and unnatural.

For example, *positive* numbers appear in classical physics as *masses* of bulks of matter but electric charges represent positive and negative numbers. The relevant *operation* with these numbers is *addition*, since mass and electric charge are naturally (nearly perfectly) additive: $(a, b) \mapsto a + b$ corresponds to bringing two physical objects together and making a single $(a + b)$-object out of the two corresponding to a and to b.

But there is no comparably simple implementation of, say, $a \mapsto 2a$ – one cannot just copy or double a physical object.[125] And writing $2a = a + b$ for $a = b$ does not help, since mutually equal macroscopic physical objects do not come by themselves in physics.

In contrast, doubling is seen everywhere in Life. All of us, most likely, descend from a polynucleotide molecule that had successfully doubled about four billion years ago. Organisms grow and propagate by doubling of cells. Evolution is driven by doublings of genomes and of significant segments of whole genomes (not by so-called "small random variations").

A true numerical addition may be rarely (ever?) seen in *biology proper* but, for example, additivity of electric charges in neurons is essential in the function of the brain. This underlies most mathematical models of the neurobrain, even the crudest ones such as neural networks. But the ergo-brain has little to do with additivity and linearity.[126]

The apparent simplicity of *real numbers* represented by points on an infinite straight line is as illusory as that of visual images of the "real world" in front of us. An accepted detailed exposition (due to Edmund Landau) of real numbers by *Dedekind cuts* (that relies on the order structure) takes about a hundred pages. In his book *On Numbers and Games*, John Conway observes (and we trust him) that such an exposition needs another couple hundred pages to become complete.

To appreciate this "problem with numbers", try to "explain" real numbers to a computer, without ever saying "obviously" and not resorting to anything as artificial as decimal/binary expansions. Such an "explanation computer program" will go for pages and pages with a little bug on every second page.

We shall not attempt to incorporate the full theory of real numbers in all its glory into our ergo-systems, but some "facets of numbers" will be of use. For example we shall endow an ergo-learner with the ability of distinguishing frequent and rare events, such as seen in the behaviour of a baby animal that learns not to fear *frequently* observed shapes.

---

[125]Certain power/energy quantities, e.g., those measured in *decibels*, are described on a multiplicative (logarithmic) scale, and their doubling is perfectly meaningful.

[126]"*Non-linear*" customarily applies to systems that are set into the framework of numbers with their *addition structure* being arbitrarily and unnaturally contorted.

On the other hand, while describing and analyzing such systems we shall use real numbers as much as we want.

Numbers are not in your ergo-brain but the idea of symmetry is in there. Much of it concerns the symmetries of our (Euclidean) 3-space, the essential ingredient of which – the group of the (*three-dimensional Lie*) *group* of all possible motions, call them *rotations* of the Euclidean *round* 2-*sphere* within itself – has been fascinating mathematicians and philosophers for millennia. And not only the haven of Aristotle but also your eyes and some of your skeletal joints that "talk" to the brain are spherical; hence, *rotationally symmetric*.

Building and identifying symmetries *within itself* serves as an essential guideline for an ergo-learner. These are created, seen from outside, by a *statistical analysis*[127] *of signals* that *break* spatial and temporal symmetries.

For instance, the input of the visual system may be represented by the set of samples of a distribution of (not quite) a probability measure on the set of subsets of the light receptors in your retina. *In principle*, this is sufficient for reconstruction of Euclidean geometry similarly to how Alfred Sturtevant obtained the partial genetic map of a drosophila X chromosome on the basis of distributions of phenotype linkages.

However, your brain would not (?) be able to "map" the 3D-space without receiving along with visual signals also signals coming from the firing of the motor neurons controlling your motions, especially those of your eyes, which, being spherical, enjoy full rotational symmetry.[128] The movements of the eye, have, apparently, prepared the brain for the idea of spatial symmetries and have helped the brain to learn how to identify images that move in the visual field.[129]

Our ergo-brain is also sensitive to *arithmetic symmetries* that issue from prime numbers as is seen in the recurrence of the *magical pentagram figure* depicting what mathematicians called *five-element Galois field* that can be visualized as the set of the five vertices of ⬠ with 20 transformations acting on it, where only 10 of them are geometrically apparent – [5 rotations] × [2 reflections]. But there is an extra one, that is there because 5 is *a prime* number and that can be depicted as ⬠ ↦ ⛤.

A fantastic vision, unimaginable to ancient mystics and to mediaeval occultists, emerges in the *Langlands correspondence* between arithmetic symmetries and the *Galois symmetries* of algebraic equations, where much of it is still in the

[127] The initial learning programs we envisage contain no counting mechanism, but an ergo-learner must be able to distinguish "significant/persistent" signals and ignore "accidental" ones.

[128] The eye shares spherical symmetry with the shoulder and hip joints that also have three degrees of freedom, while the cylindrical knee joint allows only circular motions. And the elbow hinge-joints is "designed" with *exactly two* degrees of freedom.

[129] This is explained in ○IV of Poincaré, *La science et l'hypothèse*.

© SimonHS / Getty Images / iStock

clouds of conjectures. It is tantalizing to trace the route by which the ergo-brain has arrived at comprehension of these kinds of symmetries.[130]

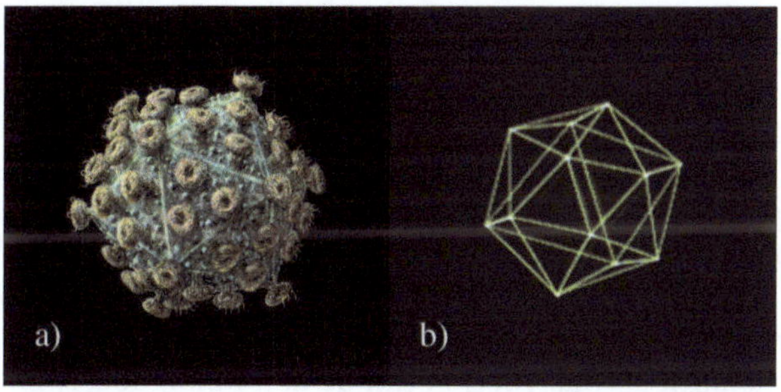

a) © Russell Kightley / Science Photo Library and
b) © maximmmmum / stock.adobe.com

Categories, Functors, and Meaning. Mental ergo-objects, e.g., sentences of a language, rarely (ever?) possess perfect internal symmetries, that move objets within

[130]Seen by an outsider, the symmetry in mathematics is diluted to the point of invisibility by *useful formulas, difficult computations, efficient algorithms, logical axioms, reliable* (*or unreliable*) *statistics* ....

themselves, such as the rotations of spheres, pentagons, or of icosahedra[131] preserving their geometric structures. (An icosahedron admits 120 transformations where – this is not fully accidental – $120 = 5! = 1 \times 2 \times 3 \times 4 \times 5$.)

However, certain transformations of such objects, e.g., of sentences, can be depicted, albeit only approximately, in *mathematical category theory.*

The simplest, geometric rather than syntactic, transformations are *insertions* of strings of letters into longer strings positioned somewhere else in a text,

$$\ldots ABC \ldots \longrightarrow \ldots DEABCMN \ldots.$$

Mathematical categories extends the concept of symmetries by allowing the kind of transformations that "move" one object into another one.

Such transformations are depicted by arrows between objects, e.g.,

$$\text{string}_1 \to \text{string}_2, \text{ or symbolically } \circ \xrightarrow{r} *,$$

where the structure of a particular category is given by a *composition rule* between incoming and outgoing arrows for all objects in this category.

Namely, one distinguished certain triples of arrows, say (r, s, t) between triples of objects, call them $(\circ, *, \star)$, such that $\circ \xrightarrow{r} *$, $* \xrightarrow{s} \star$ $\circ \xrightarrow{t} \star$, for which one declares that t equals the composition of r and s and one depicts this by the following diagram.

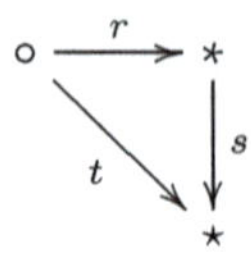

(In geographic terms, objects are "locations" and arrows between pairs of "locations" – there may be many of these – are possible routes from one location to another one, where composition of routes is the route that is obtained by consecutively following these routes, say, first r from $\circ$ to $*$ and then s from $*$ to $\star$.)

The combinatorics of (large) arrangements made from triangles of arrows carries an unexpectedly rich amount of information about the *internal* structures of the objects in our category. For instance, a seemingly vague idea of "*naturality*" of a certain mathematical construction can be non-ambiguously expressed in terms of *functors* between categories.

When applied to a language, this leads to a workable (sometimes called "holistic", see p. 242 in [20]) definition of "meaning" of a text without any reference to "real meaning" along the ideas maintained by Zelig Harris: *The meaning of a word is determined by statistics of distributions of accompanying words.*

---

[131] Viruses – they are not bound by ergo – love icosahedral symmetry, since this minimises the area of their protein shells that must contain all of the DNA encoding these very proteins.

This is how Life works: Information and geometry (of physical matter) go hand in hand. But mathematicians lag far behind viruses in the solution of *inforimetric problems.*

18. Logic and the Illusion of Rigor

As we aim at the very source of mathematics – ergo-brain itself – and try to develop a theory of ergo-systems, the purity and simplicity of the building blocks of such theory becomes essential. It is not logical rigor and technical details that are at stake – without clarity you miss diamonds – they do not shine in the fog of an ego-pervaded environment.

The evolution of mathematical concepts in their convergence to the clear shapes they acquired in the 21st century suggests how one may design ergo-systems. Yet, not all roads we explored had lead us to the promised land; understanding what did not work and why may be more instructive than celebrating our successes.

The problem with our mathematical ideas is not that they are too abstract, too difficult. or too farfetched, but that we lack imagination for pulling abstract difficult and farfetched ideas out of thin air. Nor do we have the foresight for predicting how an idea will develop.

> *Contrariwise, if it was so, it might be;*
> *and if it were so, it would be;*
> *but as it isn't, it ain't.*
> *That's logic.* LEWIS CARROLL

According to the *logicism* of Frege, Dedekind, Russell, and Whitehead, mathematics is composed of atomic *laws of thought* dictated by formal logic and the rigor of formal logic is indispensable for making valid mathematical constructions and correct definitions.

Admittedly, logicians participated in dusting dark corners in the foundations of mathematics but ... most mathematicians have no ear for tunes of formal logic.[132] We are suspicious of "intuitive mathematical truth" and we do not trust *meta*mathematical rigor[133] of formal logic.

(Logicians themselves are distrustful one of another. For example, Bertrand Russell pointed out that Frege's *Basic Law V* was self-contradictory, and in Gödel's words,

> [Russell's] *presentation ... so greatly lacking in formal precision in the foundations ... presents in this respect a considerable step backwards as compared with Frege.*

---

[132]Not everything in logic is collecting, cleaning, and classifying morsels of common sense – it is hard to believe but logical thinking can be creative. But this "creative logic" is what we call *mathematics*. We happily embrace *model theory, set theory, theory of algorithms*, and other logical theories that became parts of mathematics.

[133]The concept of *logical rigor* unlike that of *mathematical rigor* cannot be defined even with minimal requirements for being precise and rigorous.

Russell's words

> *Mathematics may be defined as the subject in which we never know what we are talking about, nor whether what we are saying is true.*

apply to formal logic rather than to mathematics.)

The cleanness of things does not make them beautiful in the eyes of a mathematician. We care for logical pedantry as much as a poet does for preaching of grammarians.

The soundness of mathematics is certified by an ***unbelievably equilibrated*** *harmony* of its edifices rather than by the strictness of the construction safety rules. The miracle of the Leibniz formula $1 - \frac{1}{3} + \frac{1}{5} - \cdots = \frac{\pi}{4}$ (of 1682) achieved by an appeal to infinitesimals, makes the criticism of insufficient rigor in mathematics by George Berkeley (1734) as well as the idea of "redemption" of Leibniz' calculus by Abraham Robinson (1966) look puny.[134]

Historically, the system of calculus rolling on fuzzy wheels of *infinity* and *infinitesimals*, has been the main intellectual force driving the development of mathematics and science for more than three hundred years. But just a step away from mathematics, volumes of philosophical speculations on the "true nature" of infinity remain on library shelves covered in dust year after year.

(Unbelievably, as recently as at the beginning of the 20th century, Florian Cajori, then a leading historian of mathematics, hailed *The Analyst* – the treatise by George Berkeley who had lambasted "non-rigorous infinitesimals" – as *the most spectacular event of the* [18th] *century in the history of British mathematics.*

The landscape of 18th-century British mathematics was, indeed, so bleak that even *The Analyst* was noticeable. But there were, however, two English mathematicians who, unlike Berkeley, had left non-trivial imprints on 18th-century science – Thomas Bayes, who suggested what is now called a *Bayesian* approach to empirical probability,[135] and Edmond Halley, famous for computing the orbit of *Halley's Comet.*[136])

We cannot take seriously anything like $(a, b) := \{\{a\}, \{a, b\}\}$.[137] To get "convinced" that this definition is worth making, you must accept logicians' appeal to metamathematical intuition; however, *nothing* of meaning can be communicated *without a use of natural language*, the metaphoric essence of which dissolves logicians' idea of "perfect rigor".

[134]The achievement of Robinson from a working mathematician's perspective was not so much in justification of Leibniz' idea of infinitesimals but rather in a vast and powerful extension of this idea.

[135]The Bayesian approach relies on continuous updating of conditional probabilities of events rather then on integrated frequencies; it is systematically used nowadays in *machine learning*.

[136]Halley's is the only short-period comet that is clearly visible from Earth when it returns to the inner solar system, at approximately 75 year intervals.

[137]This is the 1921 definition of an *ordered pair* by Kuratowski.

We communicate mathematics not by copying scores of logical symbols from one mind to another but by making another's ergo-brain resonate to the tunes we hear within ourselves.

And we do not bow down before "power of intuition" so extolled by logicians and mathematicians but rather try to understand the source of this intuition in the human ergo-brain/mind.

Who argues establishing "truths" of certain kind needs perfect precision, all 200% of it; below is an example of such a "truth".

$$4579 + 8763 - 3459 + 4686 - 6537 + 7763 - 4579 + 1099 - 8765 + 1238 - 3677 = 1111.$$

But – this is a miracle from a logician's point of view – a formally imprecise outline of the "core idea" of a *meaningful* statement, even of a relatively structurally plain one such as *Gödel's first incompleteness theorem*, ascertains its validity in the eyes of a mathematician.

(It is how "*something alive*!" in an incomplete and distorted image of an animal, that may be even unknown to you, catches your eye.)

(The century-old foundational dust finds its way to our textbooks under the pretext of rigor,[138] e.g., in the following definition of a *graph* G as

an ordered [by whom?] *pair* $G = (V, E)$ *comprising a set* V *of vertices* ...

Brr ..., there is more sense in simply drawing a graph than giving such a definition.)

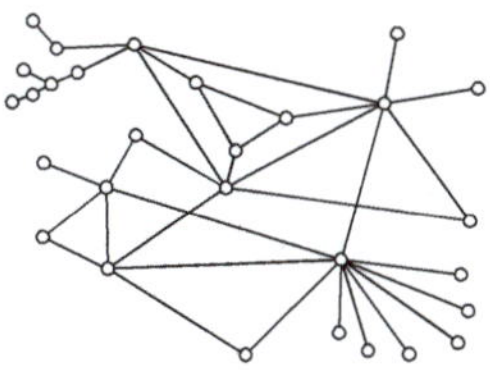

[138]The persistent urge for "rigor" in certain minds begs for an explanation by a Freudian style psychologist.

Problem with Order

The concept of *logical rigor* doesn't fare well in the ergo environment. For instance, the abstract concept of *order* cannot be defined logically rigorously but only with reference to a preexistent order in the physical or psychological medium where the idea of order is expressed.

Ordering a pair involves breaking $-1 \leftrightarrow +1$ symmetry similar to that encountered by *Buridan's Ass.* (To see it picturesquely, switch from the habitual $A < B$ to ▲ ❖ ▼ that is graphically but not contextually invariant under rotation by $180°$.[139]

Besides the mathematical impossibility of algorithmically resolving the "order problem" (the reader is invited to count the number of logical flaws in the above "solution" $(a,b) := \{\{a\},\{a,b\}\}$) there is good evidence that there is no innate "idea of order" in our ergo-brains. For instance, children often use mirror images of letter-signs when they begin writing and mathematicians tend to reverse (mentally as well as graphically) the directionality of their inequalities.

Accordingly, we should not (?) postulate *a primary idea* of order in our design of ergo-systems.

In general, we need to be choosy in our terminology and in assigning basic concepts/operations to our learning systems:

> *no elementary structure, no matter how simple looking and "obvious", can be taken for granted.*

For instance, a child who has learned to count on fingers and can figure out that $2+3=5$ and $2\times 3=6$ resists $3=3$ and $1\times 5=5$. If you think the child should be instructed to accept $[3=3]=[3+2=5]$, this is yourself, not the child, who lacks proper education: $[3=3] \neq [3+2=5]$ – mathematics has arrived at the point of accepting child's attitude and developing means (around *category theories*) for differentiating between various kinds of "equalities". And a *hierarchy of equalities* is essential for our models of ergo-brain.

Logic in Science

Mathematical rigor and logical certainty are absent not only from the logical foundations of mathematics, but also from all natural sciences even from theoretical physics. Einstein puts it in words:

> *As far as the laws of mathematics refer to reality, they are not certain; and as far as they are certain, they do not refer to reality.*

But "the physical level of rigor" is higher on certainty than the logical one, since reproducible experiments are more reliable than anybody's, be it Einstein's or Gödel's, intuition.

---

[139] One may only wonder what face of mathematics one would see in the world where "formulas" were invariably represented by *symmetric* combinatorial arrangements of symbols on the plane.

19. Infinite Inside, Finite Outside

If any philosopher had been asked for a definition of infinity, he might have produced some unintelligible rigmarole, but he would certainly not have been able to give a definition that had any meaning at all.

BERTRAND RUSSELL

Mathematics is the abode of infinity. Almost all our theorems are about infinite objects, e.g., associated with *real numbers*, such as *Lie groups*[140], or about infinities of finite objects, e.g., *prime numbers.*[141]

And basic properties of finite objects such as prime numbers can be understood only in the ambience of transcendental infinities of real as well as *p-adic numbers*[142] and of uncountably infinite *adelic groups*[143] represented by rotations of *infinite-dimensional geometric* (*Hilbert*) *spaces.*

Also this kind of infinity is indispensable for how we describe the physical universe and its fragments. Most (all?) physical systems are modelled by (uncountably) infinite sets of *real numbers.* (Physical constants, even dimensionless ones such as 0.00729735257[144] ..., are, of course, not quite numbers, but it is hard to pinpoint what exactly this quite is.)

But looking from outside, the whole body of mathematics, call it $\mathcal{M}$, is a humble mathematical object describable in *finitely many* words.[145] These words

[140]The basic example of such a group is that of *rotations of an imaginable rigid body* in 3D-space.

[141]Even individual (especially transcendental) real numbers, such as
$\pi = 3.1415926535897932384626433832795028841971693993751058209749445923078164062862\ldots$
would barely exist without an ocean of infinity that surrounds and supports them.

[142]These are limits of rational numbers where the usual rule $\varepsilon^i \to 0$ for $|\varepsilon| < 1$ and $i \to \infty$ is formally replaced by $p^i \to 0$ for a given prime number p and $i \to \infty$. Amazingly, this leads to a meaningful concept of a "number".

[143]These groups are infinite products of certain Lie groups with all their p-adic counterparts, $p = 2, 3, 5, 7, 11, \ldots$.

[144]This is the *fine-structure constant* $\alpha = \pi \cdot [\textit{elementary charge}]^2/hc$, introduced by Sommerfeld in 1916 to account for the spectrum of the hydrogen atom.

[145]From a non-mathematical scientist's point of view, such an $\mathcal{M}$ is *a mathematical model* (not of a kind defined by mathematical logicians) of "real mathematics" that is practiced by human mathematicians on the planet Earth.

however, generate a language that is represented by something *infinite*, call it $\mathcal{M}'$ – a "fragment" of $\mathcal{M}$ taken from it and positioned outside of it.

Thus, $\mathcal{M}$ starts looking like a giant kaleidoscope full of tiny mirrors $\mathcal{M}'$ each with the ability to fully reflect all of $\mathcal{M}$ with the mirrors, including $\mathcal{M}'$ itself, in it.

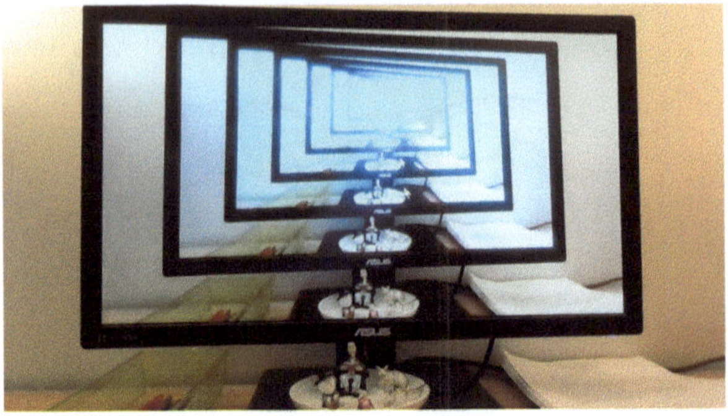

© Vadim Kaimanovich

The coexistence of "equivalence" $\mathcal{M}' \sim \mathcal{M}$ with *strict* "inclusion" $\mathcal{M}' \subsetneqq \mathcal{M}$ allowed Gödel to untangle self-referentiality within the *Liar Paradox: I am unprovable is unprovable* and, thus, proves his theorem on the existence of *formally unprovable mathematical propositions.*

(Gödel's $\mathcal{M}'$ is an infinite set of strings in finitely many, say 10, symbols/letters: These strings represent proofs in $\mathcal{M}$ described by sentences written in a language with a ten-letter alphabet. This $\mathcal{M}'$ "embeds" into $\mathcal{M}$ by means of *Gödel's enumeration*, where the "letters" are depicted as 0, 1, 2, 3, 4, 5, 6, 7, 8, 9; thus, "translating" formulas in $\mathcal{M}'$ to *decimally represented* numbers from $\mathcal{M}$ and properties of these formulas, notably their provability/non-provability, to arithmetic properties of these numbers.[146]

Gödel's proof depends on the "rigid" physical space-time of our Universe that serves as supporting background for *positional* representation of numbers. No such proof would be possible in a "Liquid World" with *nothing* "rigid" like our space-time in it.)

All of mathematics with all these *uncountable sets* that are incomparably greater than $\mathbb{N} = \{1, 2, 3, 4, \dots\}$ can be reflected in the countably large (small?) mirror of arithmetic according to a childishly simple theorem called *Skolem's paradox*:

> *"All" mathematical objects comprising uncountable sets can be "adequately represented" by countably many their "verbal descriptions".*

But this is still about "infinities"; in "reality", all these infinities are fictional: The number of mathematical statements that can ever be produced by intelligent

[146]Logicians, apparently to show their respect for Gödel, compose this with the map $\mathbb{N} \to \mathbb{N}$ defined by $\{a_i\} \to \prod_i p_i^{a_i}$ for the prime numbers $p_1 = 2, p_2 = 3, \dots, p_i, \dots$.

beings in our or any other conceivable universe is pathetically small[147] compared to the immeasurable infinity of "objects" allowed by the grammar of the mathematical language.

Also, since all physical, biological, and/or mental structures may function only on specific space/time/complexity scales, meaningful philosophical interpretations of Gödel's theorem seem difficult, if not impossible, to attain.

Most (all?) speculative (rather than purely mathematical) arguments about these structures (e.g., in the philosophy of artificial intelligence) with a whiff of a hint at Gödel's or similar "infinity theorems" such as *Turing halting theorem* or *Kolmogorov–Chaitin complexity* inevitably harbor gross misinterpretations of the concepts underlying these theorems.

Ergo-Logical Model of Mathematics

Out of an infinity of designs a mathematician chooses one pattern for beauty's sake and pulls it down to earth.

Marston Morse

A logician's model $\mathcal{M}$ of mathematic, e.g., *Peano arithmetic*, has an unattractively amorphous structure that may be compared to that of the morass of *all positions* of chess pieces on the board that may be obtained from the initial position by the rules of chess.

(Peano arithmetic is an axiomatic description of the structure of integers $n = 1, 2, 3, \ldots$ that is defined via the *successor operation* $n \mapsto n'$ (where, secretly, $n' = n + 1$),

$$\overset{1}{\bullet}\mapsto\bullet\mapsto\cdots,$$

where the basic axiom is that of *induction*:

If a "proposition (or a definition) expressed by a formula" F is valid for $n = 1$ and F-for-n implies F-for-n', then F is true for all n.

For instance the addition $m + n$ is *defined* by the formula

$$m + 1 = m' \;\&\; m + n' = (m + n)'.$$

There is no means in the traditional logic to see what is "interesting" and to describe the structure of "the set" $\mathcal{M}^{\text{ergo}}$ of all "interesting" theorems.[148]

One of the goals of ergo-logic is to achieve this end but we cannot say yet with a convincing degree of precision what *interesting* theorems (and/interesting chess positions) are, but there are illuminating examples.[149]

---

[147] The number of strings of symbols that can be generated during the life cycle of any conceivable universe can be safely bounded by $10^{10^{10}}$.

[148] "Interesting" theorems do not naturally comprise what we call "a set".

[149] It may also be worthwhile to borrow ideas from evolutionary biology (and immunology?) and pursue the ideas of *competition, selection, adaptation* (partly) responsible for the emergence of "interesting structures" in living systems and, on a smaller scale, in combinatorial games such as chess.

For instance, Harrington and Paris found out several bona fide mathematical propositions, such as *Goodstein's base change theorem* that are unprovable in Peano arithmetic.[150]

20. Small, Large, Inaccessible

Mathematicians treat all numbers on equal footing, be these

$$2,\ 3,\ 4,$$

or

$$10,\ 20,\ 30,\ 40,$$

or

$$1\,000,\ 10\,000,\ 100\,000,\ 1\,000\,000,$$

or

$$10^{10},\ 10^{20},\ 10^{30},\ 10^{40},$$

or

$$10^{10^{2}},\ 10^{10^{30}},\ 10^{10^{400}}$$

or

$$2^{3^{4^{\cdot^{\cdot^{\cdot^{399^{400}}}}}}}$$

But "democracy of numbers" breaks down in the "real world", be it the physical Universe or the human ergo-world.

Partly this is because properties of numbers (and formulas in general) depend on how their representations by symbols spreads over the background either in physical space/time or in the human ergo-mind.

Even quite small numbers, in fact, everything above four, unless they are represented by "structured entities" in some way, are not accessible to the human (ergo)mind.

Non-accidentally, the grammars of some languages, e.g., of Russian, distinguish the numbers 2, 3, and 4, while 5, 6, ... , 20, 30, 40, ... , 100, 200, ... (but not, say 23, 101, and 202) are, syntactically speaking, go in the same basket as infinity.[151]

Yet, some psychologists maintain the idea that a few larger numbers

••••• , •••••• and •••••••

[150]See *Brief introduction to unprovability*, by A Bovykin at https://www.cs.umd.edu/~gasarch/TOPICS/largeramsey/bovINTRO.pdf and the next section.

[151]Amusingly, there is also a chasm in essential properties between geometric spaces of dimensions 1, 2, 3, 4, and those of dimension 5 and more.

are also humanly perceptible but nobody(?) would claim he/she can immediately grasp

•• or even •••••••••••.

Habituation to the decimal notation deceives us. We have no idea what such numbers as 65 536, even less so 7 625 597 484 987, are[152] unless we write them as $2^{2^{2^2}}$ and 3^{3^3}.

Even recognition of "*seven trillion*" in 7 625 597 484 987, unlike that of "*million*" in 1 048 756$(= 2^{2^{2^2}+2^2})$, needs a few rounds of the mental counting algorithm that is not innate to the human mind.

On the other hand, the human (ergo)mind that feels uncomfortable with, say 177 148, readily accepts the same (?) number structurally organized as $3^{3^3+2}+1$.

Probably, this is because the "*combinatorial geometry of the* (*ergo*)*mind*" is nearer to that of the arrangements of exp-towers in the 2-plane than to the uneventful linearity of •••••••••••••••••••••••••... as well as of the ordinary positional depiction of numbers and formulas.

Look more closely at how it works. One instantaneously evaluates the cardinality of ••••, one needs a fraction[153] of a second to identify "(almost) unstructured five" •••••, it takes a couple of seconds for ••••••• (it is much faster if the symmetry is broken, for instance, as in ••• ••••) and it is impossible with

•••••••••••••••.

But a little structure helps:

And slightly larger numbers, such as

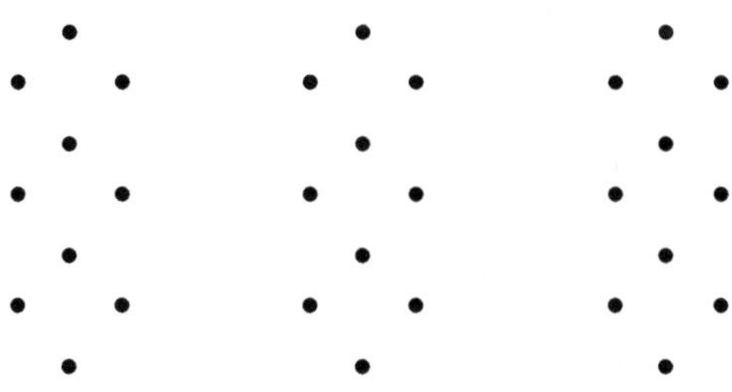

if perceived, then only through a lens of mathematics.

---

[152] Browsing through

$$\underbrace{\bullet\cdots\bullet}_{7625597484987}$$

point by point would occupy you for a couple of *hundred thousand* years.

[153] This may be a significant fraction, well above 200 milliseconds.

Our intuition does not work anymore when it comes to thousands, millions, billions. Answer fast:

Do you have more hairs on your head (assuming you are not bald) than the number of people an Olympic stadium may contain?[154]

What is greater, the number of bacteria living in your guts or the number of atoms in a bacterium?[155]

Below are ergo-relevant numbers.

• TIME. A hundred years contain < 3.2 billion seconds. With a rate three words per second you vocalize *less than ten billion* (10^{10}) words in the course of your life.

Ten billion garrulous individuals all together[156] will utter *at most*

$$10^{10} \times (3 \times 3.2 \cdot 10^7) \times 5 \cdot 10^9 < 5 \cdot 10^{27}$$

words until Sun turns into a *red giant* in about five billion years.

Speaking more realistically, humanity *cannot* come up with more than 10^{12}–10^{18} *different ideas* – poems, theorems, computer programs, descriptions of particular numbers, etc.[157]

10^{15} years of possible duration of the Universe is made of less than $10^{46} = 10^{15} \times 3 \cdot 10^7 \times \frac{1}{3} 10^{24}$ *jiffy-moments.*[158]

• BRAIN. The number of neurons in the human brain is estimated between ten and a hundred billion neurons with hundreds of synaptic connections per neuron, somewhere between 10^{12}–10^{14} synapses all together.

This gives an idea of the volume of the memory stored in the brain, that is comparable to that on a computer hard disk of about 10^{12}–10^{13} bits.

The (short time) brain performance is limited by the *firing rates* of neurons – something about 100 times per second.[159] Thus, say hundred million active neurons can perform 10^{10} "elementary operations" per second[160], which is what an average computer does.[161]

• LANGUAGE. There are 10^{22}–10^{25} grammatical sentences in five words in English.

---

[154] Both numbers are about 100 000.

[155] There about 10^{11}–10^{14} atoms in bacteria and more than 10^{12}–10^{13} bacteria living in your body, mainly in your guts.

[156] The human population on Earth today is slightly above seven billion.

[157] LIFE on Earth, in the course of its $\approx 3.9 \cdot 10^9$ year history, has generated a comparable number of "ideas" and recorded them in DNA sequences of organisms inhabiting the planet.

[158] *Jiffy* $\approx 3 \cdot 10^{-24}$ *s* is the time needed for light to travel a proton-sized distance.

[159] It takes 8 milliseconds for the brain of a *star-nosed mole* to decide what is edible and what is not. This is fifty times faster than it takes for a driver to decide whether to brake or accelerate on yellow.

[160] But the rate of learning is measured not in seconds but in hours, days, months, years. This is so, partly, because modification of the strength of synaptic connections is slow.

[161] The speed of modern *supercomputers* is measured in *petaflops* corresponding to 10^{15} (*floating point*) operations per second. This is achieved with particularly designed network architectures of processors that allow thousands (not millions as in the brain) of operations performed *in parallel.*

• SPACE. A *glass of water* contains about 10^{25} molecules, the *planet Earth* is composed of about 10^{50} atoms, and the *astronomically observable universe* contains, one estimates today, 10^{80} particles.[162]

Thus, there are (significantly) less than 10^{130} *classical* (as opposed to quantum) "events" within our space-time and this grossly overestimated number makes

> *an unquestionable bound of what will ever be achieved by any conceivable (non quantum) computational/thinking device of the size of the Universe.*

But ... there are at least $2^{10^{10}} > 10^{3\,000\,00000} \gg 10^{130}$ possible "texts" that you, a humble 21st-century human being, *can* (?) write in sequences s of 10^{10} bits on the hard disc of your tiny computer. Can't you?

How comes it that only a negligible percentage, less than $\frac{1}{10^{10^9}}$ of possibilities, can be actualized?

Worse than that, it is *impossible* to pinpoint a *single instance* of a non-realizable sequence s: Indicating an s will make this very s actual.

It is far from clear whether such inconsistency between "can" and "will" admits a clean mathematical reformulation or whether this belongs with the *paradox of the heap*. Yet, there are a few purely mathematical theorems and open problems that address this issue, albeit not satisfactorily.[163]

IS IT MATHEMATICS?

(1) It seems not hard to show that there exist lots of *provable* theorems T that can be formulated on a page or two in the standard mathematical language but such that their shortest proofs must be enormously long; hence, *humanly unprovable.* (To be unprovable one does not need "enormous" – "modest" 100^{100} will do.)
(2) On the other hand, pinpointing a *specific* a priori provable T that is *realistically unprovable* may be humanly impossible.
(3) A somewhat more problematic (and more interesting) possibility is that there exists a theorem T' that admits a hundred page proof but a search for the proof of T' would need at least 10^{50} human + computer hours; hence, being unfindable.

Explanation. Unlike (1) and possibly (2), that (almost) entirely belong with mathematics proper, this (3) hardly can be approached without making some conjectures on the innate resources of the human (ergo)mind/brain.

In fact, all kind of arguments and ideas – computations, proofs, etc. – that we can imagine and/or design, let this be done with a use of computers, are limited

[162] Archimedes, recall, evaluated the number of sand grains that would fill the Universe as $\approx 10^{60}$ where *exponential representation of numbers* was invented by him for this purpose.

[163] These turn around the following question. What makes it so difficult to locate/construct *individual* objects O with a property P from a given class C of similar objects, when we know that this P is satisfied by the *majority* of members of C?

to *compositions* of what is available in the "pool" of a *relatively small number* of innate atomic ideas evolutionarily installed by Nature into our (ergo)brain along with a collection of composition/selection rules of these ideas.[164]

Therefore, there exist "ideas" expressible in strings of 10^5 decimal symbols, that will NEVER be articulated in any form by humans on Earth.[165]

Shadows of Examples. By their very nature, *humanly inaccessible ideas* cannot be exemplified. But the following is suggestive.

($\star$) *"Colorless green ideas sleep furiously"* was probably meant by Chomsky as a *random grammatical*, hence, absurd sentence. But this is, in truth, a bona fide (sarcastic) English sentence due to pronounced, let them be negative, correlations between the words in it.[166]

(There is no 100% *random* option within your ergo-brain, but, given a list of words in front of you, you can (?) put your finger blindly on one of them.

NONSENSICAL SEVEN THUS CHOSEN ON RANDOM[167]

Illegal silly inflations mislead recklessly.
Extended meagre materials emphasize fortunately.
Fine joyous departments choose physically.
Wet coordinated articles complain coherently.
Wooden illustrious faults cost halfheartedly.
Wooden coordinated departments emphasize recklessly.
Illegal meagre departments complain halfheartedly.[168])

($\star\star$) To scientifically evaluate the intelligence of a monkey, a psychology professor suspended a banana high up under the ceiling, such that it could be reached only from a chair that had to be put on a table.

But the monkey, upon entering the room, discovered a better use of the professor's head than the professor himself: The animal jumped on the shoulder of the man, leaped up for the fruit using the man's head as a trampoline and safely landed on the table.

Apparently, this solution to the problem was absent from the toolbox of ideas under professor's skull.

---

[164]Some of these "compositions" have already been implemented in the course of human cultural history with incorporation of bits and pieces from the "information flows" from the "external world", but this does not change the essence of what we say.

[165]If – which is unlikely – the number of equivalence classes formed by "meanings" carried by these strings is (very much!) significantly less than 10^{10^5}, then this is objectionable.

[166]Possibly, this sentence came up looking meaningful because it is too short.

[167]These are taken from the pool of, what I guess, 5 000–10 000 most common words out of "full" English vocabulary of more than 150 000 words.

[168]Notice, that the last two sentences are the diagonals of the first five and observe that the longer you look at these sentences the better sense they make – the human brain is apt to detect non-existent meaning in all kind of nonsense.

21. Probability: Particles, Symmetries, Languages

Can one reconcile Maxwell's thesis that

> *The true logic of this world is the calculus of probabilities.*

with Naum Chomsky's assertion that

> *The notion of a probability of a sentence is an entirely useless one, under any interpretation of this term*?

Human languages carry imprints of the mathematical structure(s) of the ergo-brain and, at the same time, learning a natural (and also a mathematical[169]) language is a basic instance of the universal learning process of the human ergo-brain. We can hardly understand how this process works unless we have a fair idea of what LANGUAGE is. But it is hard to make a definition that would catch the *mathematical essence* of the idea of LANGUAGE.

But isn't a language, from a mathematical point of view, *just*

a set of strings of symbols from a given alphabet,

or, more generally,

a probability distribution on the set of such strings?

A linguist would dismiss such definitions with disgust, but if you are a mathematician these *effortlessly* come to your mind. Paradoxically, this is why we would rather *reject* than accept them:

> *Mathematics is shaped by definitions of its fundamental concepts, but there is no recipe for making "true definitions." These do not come to one's mind easily, nor are they accepted by everybody readily.*

A good definition must tell you *the truth, all the truth, and nothing but the truth*, but there is no common agreement, not even among mathematicians, about what constitutes a TRUE DEFINITION.

For example, the idea of an *algebraic curve* that is a *geometric* representation of

solutions of a polynomial equation $P(x_1, x_2) = 0$ *in the* (x_1, x_2)*-plane,*
e.g., of the equation $2x_1^2 + 4x_2^4 + x_1x_2^3 = 0$,

by something like , originated in the work of Fermat and Descartes in the 1630s, and these curves have been studied in depth by generation after generation of mathematicians ever since.

But what is now seen as the simplest and the most natural definition of such a curve – the one suggested by Alexander Grothendieck in 1950s in the language of *schemes* – would appear absurd to anybody a few decades earlier.

Defining "language" and/or "learning" is, non-surprisingly, more difficult than "algebraic curve", since the former have non-mathematical as well as purely

[169] *Mathematical language* for us is the language used for communication between mathematicians but not a mathematical language of formal logic.

mathematical sides to them. They are similar in this respect to the concept of *probability*, that by now is a well-established mathematical notion.

It is instructive to see how "*random*" crystallized to "*probability*", and what was gained and what was lost in the course of this "crystallization".

Also, we want to understand how much of "random" in languages in the (ergo)learning process (including learning languages) is amenable to what Maxwell calls "the calculus of probabilities".

The concept of *chance* is centuries old as is witnessed by some passages in Aristotle (384–322 BCE) and also in the Talmud.[170] And Titus Lucretius (99–55 BCE), a follower of Democritus, describes in his poem *De Rerum Natura* what is now called the *Einstein–Smoluchowski stochastic model* of Brownian motion[171].

But mathematics of "random" was originally linked to gambling rather than to science.

I of dice possess the science and in numbers thus am skilled

said Rituparna, a king of Ayodhya, after estimating the number of leaves on a tree upon examining a single twig. (This is from *Mahabharata*, about 5 000 years ago; also 5 000 year-old dice were excavated at an archeological site in Iran.)

What attracts a mathematician to random dice tossing and what attracts a gambler are the two complementary facets of *stochastic symmetry*.

Randomness *unravels and enhances* the *cubical symmetry* of dice (there are $3! \times 2^3 = 48$ symmetries/rotations of a cube) – this is what fascinates a mathematician.

But randomness also *breaks* symmetries: The only way for a donkey' ergo-brain (and ours as well) to solve Bouridan's ass problem is to go random.[172] Emanation of the "miraculous decision power of random" intoxicates a gambler's ergo.[173]

The first (?) documented instance of the *calculus* of probabilities – "*measuring chance*" by a European[174] appears in a poem by Richard de Fournival (1200–1250)

---

[170]Our sketchy outline of the history of probability relies on [23] [9], [14], [16], [32], [27] with additional *References for Chronology of Probabilists and Statisticians* on Ming-Ying Leung's page, http://www.math.utep.edu/Faculty/mleung/mylprisem.htm

[171]This is the collective random movements of particles suspended in a liquid or a gas that should be rightly called *Ingenhousz' motion*.

[172]No deterministic algorithm can select one of two points in an (empty) 3-space as it follows from the existence of the *Möbius strip*. And a general purpose robot that you can ask, for instance, *bring me a chair* (regardless of several available chairs being identical or not) needs a "seed of randomness" in its software.

[173]In the same spirit, the ABSOLUTE ASYMMETRY of an individual random ± sequence of outcomes of coin tosses complements the ENORMOUS SYMMETRY of the whole space S of dyadic sequences that is acted upon by the compact Abelian group $\{-1, 1\}^{\mathbb{N}}$ for $\mathbb{N} = \{1, 2, 3, 4, 5, \dots\}$ and by automorphisms of this group.

[174]Some "calculus of probabilities", can be, apparently, found in the *I Ching* written about 31 centuries ago.

who lists the *numbers* of ways three dice can fall. (The symmetry group in the case of n dice has cardinality $n! \times (48)^n$; that is, 664 552 for $n = 3$.)

Next, in a manuscript dated around 1400, an unknown author correctly solves an instance of *the problem of points*, i.e., of division of the stakes.

In 1494, the first (?) treatment of the problem of points appears in *print*[175] in Luca Paccioli's *Summa de Arithmetica, Geometria, Proportioni et Proportionalità.*[176]

Paccioli's solution was criticized/analized by Cardano in *Practica arithmetice et mensurandi singularis* of 1539 and later on by Tartaglia in *Trattato generale di numerie misure*, 1556.

Cardano

Gerolamo Cardano was after Vesalius the second most famous doctor in Europe. He suggested methods for teaching deaf-mutes and blind people, a treatment of syphilis and typhus fever. Besides, he contributed to mathematics, mechanics, hydrodynamics, and geology. He wrote two encyclopaedias of natural science, invented the *Cardan shaft* used in today's cars, and published a foundational book on algebra. He also wrote on gambling, philosophy, religion, and music.

The first (?) systematic mathematical treatment of statistics in gambling appears in Cardano's *Liber de Ludo Aleae*, where he also discusses the psychology of gambling, written in the mid 1500s, and published in 1663.

In a short treatise written between 1613 and 1623, Galileo, on somebody's request, effortlessly explains why upon tossing three dice, the numbers (slightly) more often add up to 10 than to 9. Indeed, both

$$9 \underset{1}{=} 1+2+6 \underset{2}{=} 1+3+5 \underset{3}{=} 1+4+4 \underset{4}{=} 2+2+5 \underset{5}{=} 2+3+4 \underset{6}{=} 3+3+3$$

and

$$10 \overset{1}{=} 1+3+6 \overset{2}{=} 1+4+5 \overset{3}{=} 2+2+6 \overset{4}{=} 2+3+5 \overset{5}{=} 2+4+4 \overset{6}{=} 3+3+4$$

have six decompositions, but $10 = 3+3+4 = 3+4+3 = 4+3+3$ is thrice as likely as $9 = 3+3+3$.

(If you smile at the naivety of people who had difficulties in solving such an elementary problem, answer, instantaneously,

> *What is the probability of having two girls in a family with two children where one of them is known to be a girl?*[177])

The formulation of basic probabilistic concepts is usually attributed to Pascal and Fermat who discussed gambling problems in a few letters (1653–1654) and to Huygens who in his 1657 book *De Ratiociniis in Ludo Aleae* introduced the idea of *mathematical expectation.*

---

[175]The first book printed with movable metal type was the Gutenberg Bible of 1455.

[176]Paccioli became famous for the system of *double-entry bookkeeping* described in this book.

[177]This would take half a second for Galileo – the answer is $1/3$ $(\pm\varepsilon)$.

© CharlieBfl / Getty Images / iStock

But the key result – *the Law of Large Numbers* (hinted at by Cardano) was proved by Jacob Bernoulli only in 1713.

This, along with the *Pythagorean theorem* and the *quadratic reciprocity law*[178] stands among the ten (±2) greatest mathematical theorems of all time.

"*Continuous probability*" was invented in 1733 by Buffon who thought of

> *a needle of unit length* (*instead of dice*) *randomly thrown on the plane, where this plane was divided into parallel strips of unit width.*

He proved that

> *the probability of crossing a line between two strips by the needle equals* $2/\pi$ *for* $\pi = 3.14\ldots$ *being one half the length of the unit circle.*

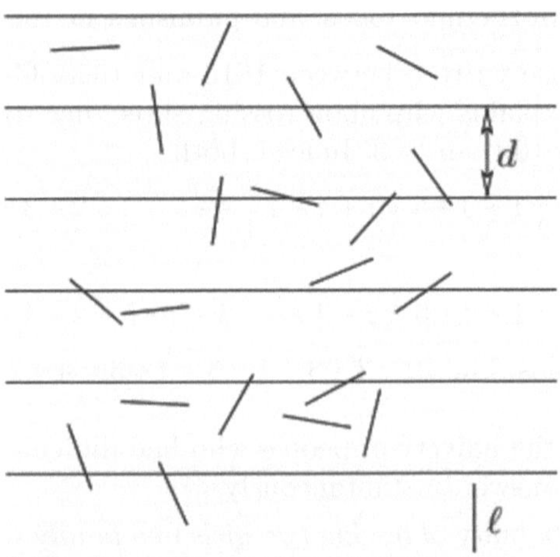

Das BUCH der Beweise, Das Nadel-Problem von Buffon, 2015, Seite 199, Martin Aigner & Günter M. Ziegler (Springer-Verlag Berlin Heidelberg 2015)

[178]Let p, q be odd primes and $q^* = (-1)^{(q-1)/2}q$. Then $n^2 - p$ is divisible by q for *some* integer n if and only if $m^2 - q^*$ is divisible by p for *some* m.

Georges-Louis Leclerc Buffon

Besides opening the fields of *geometric probability* and *integral geometry*, Buffon, who understood the physics of his time, contributed to theoretical and practical optics. As an experiment, he constructed a large (about 2 m) concave mirror built of 360 small ones that, by focusing sunlight, could melt iron at 10 m distance and ignite wood at 50 m.

But his major contribution, about which we wrote in Chapter 6 of Part I, was to what he called "natural history" – a development of a synthetic picture of Life on Earth, where he outlined many essential interactions between organisms and their environment, much of which now goes under the heading of "biogeography".

Buffon emphasized the preeminence of biological reproduction barriers between different groups of organisms over the obvious geographical ones that suggested a definition of *species* that has withstood the attempts to "improve" it by later natural philosophers including some 20th-century post Darwinian evolutionary thinkers.

Buffon was the first (?) who articulated the main premise of evolutionary biology – the concept of the *common ancestor of all animals*, including humans.

Buffon's view on Nature and Life, expounded in his *Histoire naturelle, générale et particulière* published between 1749 and 1789 in 36 volumes, became a common way of thinking among educated people in Europe for two centuries afterwards.

With the Buffon's needle, "random" merged with "analysis of continuum" and were empowered by "calculus of infinitesimals". This is what was hailed by Maxwell and exploited by generations of mathematicians and physicists after Buffon.

This calculus comes at a price: Probability is a "full fledged number" with the addition/multiplication table behind it. But assigning a *precise specific* numerical value of probability to a "random event" in "real life", e.g., to a sentence in a language, is not always possible.

On Symmetry in Randomness

The elegance and success of probabilistic models in mathematics and science (always?) depends on (often tacitly assumed and/or hidden) symmetry.

The essentiality of "equiprobable" was emphasized by Cardano and *parametrization* of random systems by "independent variables" has always been the main tenet of probability theory. Most (all?) of classical mathematical probability theory was grounded on (*quasi*)*invariant Haar*(*-like*) *measures* and the year 2000 was landmarked by the most recent triumph of "symmetric probability" – the discovery of (essentially) *conformally invariant* probability measures in spaces of planar curves (and curves in Riemann surfaces) parametrized by increments of *Brownian's processes* via the *Schram–Loewner evolution equation.*

Atoms and Bacteria. A bacterium-sized speck of matter may contain, say, $N_{AT} = 10^{12}$ atoms and/or small molecules in it, and the number N_{BA} of bacteria residing in your colon is also of order 10^{12}. If there are two possible states for everyone – be they atoms or bacteria – then the number of *conceivable* states of the entire system, call it S, is the monstrous

$$M = M(S) \geq 2^{10^{12}} > 10^{3\,000\,000\,000}$$

where its reciprocal,

$$\frac{1}{M} < 0.\underbrace{000\ldots0001}_{3\,000\,000\,000}$$

taken for the probability of S being in a particular state, is too small for making any experimental/physical/biological sense.

However, the assignment of the $\frac{1}{M}$-probabilities to the states is justified and will lead to meaningful results IF there is a symmetry that makes these tiny meaningless states "probabilistically equivalent", where the nature of such a symmetry, if it is present at all, will be vastly different in physics and in biology.[179]

But if there is not enough symmetry and one cannot *postulate* equiprobability (and/or something of this kind such as *independence*) of certain "events", then the advance of classical calculus stalls, be it mathematics, physics, biology, linguistic, or gambling.

On Randomness in Languages

Neither unrealistic smallness of probabilities, nor failure of "calculus with numbers", preclude a use of probability in the study of languages and of learning processes. And if you are too timid to contradict Chomsky, just read his

under any interpretation of this term as
under any interpretation of the term probability
you can find in a 20th-century textbook.

Probability applicable to languages must not be an assignment of *numbers* to events, but rather a "*functor*" from a "linguistic category" to a "small and simple" category, yet, in general, different from the "category of numbers".

And forfeiting numbers for probabilities in languages is unsurprising – numbers are not the primary objects in the ergo-world. Numbers are not there, but there is a visibly present *partial order* on "plausibilities" of different sentences in a language. This may not look like much, but a *hierarchical use* of this order allows recovery of many linguistic structures.

[179] It is not fully accidental that the numbers N_{AT} and N_{BA} are of the same order of magnitude. If atoms were much smaller or cells much bigger, e.g., if no functional cell with less than 10^{20} atoms (something slightly smaller than a Drosophila fly) were possible, then, most probably, LIFE, as we know it, could not have evolved in our short-lived Universe with hardly 10^{80} atoms in it.

Adaptation of probability to the needs of (ergo)linguistics will also require a revision of the concept of "event" the probability of which one measures.

The nowadays canonized definition of "event", suggested in 1933 by Kolmogorov in his *Grundbegriffe der Wahrscheinlichkeitsrechnung*, is essentially as follows.

> *Any kind of randomness in the world can be represented (modeled) geometrically by a subdomain Y in the unit square* ■ *in the plane. You drop points to* ■, *you count hitting Y for an* **event**, *and define the probability of this event as* $area(Y)$.

However elegant this *set-theoretic* frame is (with ■ standing for a *universal probability measure space*), it must share the faith of André Weil's *universal domains* from his 1946 book *Foundations of Algebraic Geometry*. The set-theoretic language introduced in mathematics by Georg Cantor that has wonderfully served us for almost 150 years is now being supplanted by a more versatile language of *categories* and *functors*. André Weil's varieties were superseded by Grothendieck's schemes, and Kolmogorov's definition will eventually go through a similar metamorphosis.

A particular path to follow is suggested by Boltzmann's way of thinking about statistical mechanics – his ideas invite a use of *non-standard analysis* as well as of a Grothendieck's style *category theoretic* language. But a mathematical interpretation of the idea of probability in *languages* and in *learning* needs a more radical deviation from (modification? generalization of?) this ■.

CARDANO, GALILEO, BUFFON. The very existence of these people challenges our vision on the range and extent of the human spirit. There is no apparent wall between the ergos and egos in the minds of these men.

Where are such people today? Why don't we see them anymore? Nobody in the last 200 years had a fraction of Cardano's intellectual intensity combined with his superlative survival instinct. Nobody since Buffon has made long-lasting contributions to domains as far-distant one from another as pure mathematics and life sciences. What needs to be done to bring Galileos back to us?

22. Signal Flows from the World to the Brain

© BargotiPhotography / Getty Images / iStock

The signals entering the ergo-brain via vision, hearing and olfaction are "written" on certain physical/chemical backgrounds, the structures and symmetries of which are mathematically quite transparent.

1. The visual signals are carried by the four-dimensional *space* + *time* continuum. Signals break the symmetry (and continuity) of the *space* + *time* but eventually, the ergo-brain reconstructs this symmetry.
2. The auditory signals are carried by the three-dimensional *time* + *frequency* + *amplitude* space. Ergo-brain, unlike mathematicians, does not seem to care about the underlying (symplectic) symmetry of this space; it is concerned with the "information content" of these signals and with *correlations* and/or *redundancies* in flows of the signals.
3. Structurally, olfaction has much in common with gustation as well as with perception of warm, cold, and pain, that are, however, more primitive than olfaction in most animals. But olfaction is fundamentally different from vision and hearing.

   Scents, unlike images and sounds, cannot be written and rewritten on a variety of backgrounds they are supported by physical/chemical substrates *specific to each of them.*

   The "letters of smells" are aromatic molecules entering our noses. The flows of these, *unless they are bound to a rigid substrate,*[180] have more internal symmetries than the flows of visual and auditory signals: N different kinds of aromatic molecules that freely float in a gas or in a liquid can be permuted in $N! = 1 \times 2 \times 3 \times \cdots \times N$ different ways.[181]

   No rewriting, no natural digitalization, no universal structuralization of smells comparable to that of images or sounds is possible.

[180]Scents attached to a rigid surface, allow, for instance, tracking by dogs.

[181]This is, of course, an exaggeration. Nothing on Earth has ever been permuted in $50! > 10^{64}$ ways.

Because of this smells cannot be positionally codified and olfaction inputs have significantly lower information contents than better organized flows of visual and auditory signals. (This is an apparent reason for why the olfactory perception depends on so many *different*[182] receptors.)

For instance, combinations of two different kinds of molecules, ⬡ and □, can encode only a few dozen distinct signals, because there is no perceptible order in how groups of these molecules, say from the following string, may enter your nose.

⬡□⬡⬡□⬡⬡□□□⬡□⬡□□□⬡□□⬡⬡□⬡□□□□⬡⬡□

Even though there are 2^{30} – more than a billion – of such binary strings, your nose can only perceive the relative amount of ⬡ and □. (This is a humble 13/15 for the ⬡□-string.)[183]

Possibly, bloodhounds, who have 200–300 million olfactory receptor neurons in their nasal cavities, may differentiate smells from a pool of size 10^6(?), 10^9(?), 10^{12}(?)[184], but they cannot beat a human visual system that can easily distinguish randomly-taken images from a pool of 10^{20} and with a little effort (training?)[185] up to 10^{1000}.

The internal library of smells is simpler than how we remember visual images, sounds, words, and ideas – no ergo is needed for organisation of this "library".

Besides, olfaction, unlike vision, does not depend on your muscles, it is disconnected from the proprioception system and we have no means of (re)producing scents at will, albeit we think we can recall them.

There aren't many clearly identifiable *universal smells* common to large groups of object. Non-surprisingly, languages (of urban populations?) have few specific names for smells – about ten in English:

musky, putrid, rotten, floral, fruity,
citrus, vegetative, woody, herbaceous, spicy.

(There are slightly more smell-words in certain languages, for example there are about fifteen of them in the *Kapsiki of Cameroon*.)

Natural languages do not waste words for naming *individual* objects/properties but rather exercise the art of giving the same name to *many* different things, very much as mathematical theories do. There is no grammar of scents, no books in the language of odors, no possibility to encode and "freeze" the flows of olfactory

[182] There are over 1000 different receptor proteins in these neurons, as was established by Richard Axel and Linda Buck.

[183] Variation of smell intensity in time does not contribute much, although this may work as a clock for a dog.

[184] Apparently, there is neither an unequivocal definition of "olfaction sensitivity" nor reliable data for a convincing estimation of the olfaction capacity of people and other animals.

[185] Strings of common words are easy to discriminate but see how long it takes to distinguish the following two.

अ ल ो ए व अ न अ गअ रई हअ तए ह अइ and अ ल ो ए व अ न अ अग रई हअ तए ह अइ.

perception.[186] (This is what stopped bloodhounds from developing a language based on olfaction.)

4. There is no apparent uniform (symmetric) spacial background for somatosensory and touch (haptic) perceptions but their information carrier potential is comparable to that of vision and hearing.

Seven Flows

... think of some step that flows into the next one, and the whole dance must have an integrated pattern.

Fred Astaire

© airindizain / stock.adobe.com

Incoming flows of signals are naturally classified according to the sensory receptors and pathways by which they enter the brain: *Visual, auditory*, and *somatosensory*, where the two relevant aspect of the latter are *proprioception*, the body sense, and *tactile*, i.e., *touch perception.*

But from an ergo-learner perspective, signals differ by how one learns their "meanings", how one interacts with them, how one arrives at an understanding of their structures.

1. *Spoken language* depends on the auditory and sensory-motor systems; ears to listen and sensory-motor systems to generate speech. However, deaf-mute people speak in sign language and deaf blind people communicate in tactile sign language.[187]

2. *Written language* (whenever it naturally exists) is likely to have a huge overlap with the spoken one in the human brain (of a habitual reader) but it also makes a world of its own. It is not inherently interactive, at least not so superficially,[188] and it is not *physically* bound to the flow of time.[189] Persistence of written literature is hard to reconcile with a naive selectionist view on co-evolution of language and the brain.

3. *Mathematics.* Learning mathematics is an interactive process but it is hard to say exactly in what sense.

[186] Perfumes do not count.

[187] Most amazingly, some deaf blind people can understand spoken language by picking up the vibrations of the speaker's throat.

[188] Writing and reading is kind of talking to oneself.

[189] The time arrow is implemented by *directionality* of what is written.

The images a mathematician generates in his/her mind are neither of language nor do they belong with any particular "sensory department". Thinking mathematics is like driving an imaginary bicycle or performing/designing a dance with elaborate movements entirely in your head. (This may differ from person to person.)

4. *Languages of games.* We are able to enjoy and to learn a variety of mental and physical games. Probably, these are divided into several (about dozen?) classes depending on how they are incorporated into our ergo-brains. Written language and mathematics may be particular classes of games.

5. *Music.* People gifted in music replay melodies in their minds and they can reproduce melodies vocally and/or with musical instruments; the rare few may generate new melodies. But melodies, unlike sentences in a language, cannot talk about themselves and there is no general context in which one can formulate what human (unlike that of birds) music is and/or what should be regarded as "understanding music".[190]

6. *Proprioceptive/somatosensory system.* Running over a rough unpredictable terrain is kind of talking to the road with the muscles in your body. This is much simpler than ordinary language but is still beyond the ability of computers that control robots. Neither is a present day robot able to sew a button on your shirt.

7. *Vision.* At least half of the neocortex in humans is dedicated to vision, but this may be mainly due to the sheer volume of the information that is being processed and stored, rather than to the structural depth of visual images.

[190]Recently, there was an attempt to understand what music does to one's brain: http://phenomena.nationalgeographic.com/2013/04/11/why-does-music-feel-so-good/ and http://www.zlab.mcgill.ca/home.php?1592876871.

This is probbaly, why vision impairment, even vision + hearing impairment, do not significantly affect human ergo. The ergo is robust and independent of particular sensory inputs as exemplified by people such as Hellen Keller.

Three flows among these, *Language, Mathematics, and Music*, have an essential feature in common: The receiver of such a flow $\overrightarrow{F}$ develops an ability, with no external reinforcement, to creatively generate a new flow $\overleftarrow{F}$ in the class of $\overrightarrow{F}$. (In the case of Mathematics and Music this happens rarely, but the miracles of this having happened in the brains of Mozart and Ramanujan outweighs any statistics.)

Modelling the transformation $\overrightarrow{F} \mapsto \overleftarrow{F}$ is one of the key aspects in our picture of the universal learning problem. (Possibly, there are counterparts of $\overleftarrow{F}$ for other incoming flows $\overrightarrow{F}$, but they may be kind of *internal*.)

The most studied among these is the learning of native languages by children, the (still unknown) mechanism of which must not be not far from how mathematics is learned by mathematicians.

The structure of a most sophisticated mathematics we build in our minds is likely to be simpler than that of natural languages (and smaller than that of vision), but it is still quite interesting, and the corresponding learning process may be more accessible, due, besides its relative simplicity, to a great variance in people's abilities in learning mathematics[191] and a presence of criteria for assessing its understanding.

Transformations and Reconstruction of Flows: Learning to Read by Learning to Speak

The original form of signals carried by the above seven flows is different from what arrives at your sensory systems. For instance, visual images result from 2D *projections* of *three-dimensional* patterns to the retina in your eyes; moreover, the brain's analysis of these projections is coupled with the activity of the brain's motor system that controls movements of the eyes that continuously modify these projections.

Similarly, the flow of speech as it is being generated in one's mind has, according to the tenets of *generative grammar*, a tree-like structure that is then "packed" into single time line.

Reconstruction of F_{orig} from the flow F_{rec} that you receive is an essential aspect of understanding the message carried by F_{orig}. For example, understanding a flow of speech is coupled with one's ability to speak, i.e., to reconstruct/generate F_{orig}, or something close to it, in one's ergo-brain.

This reconstruction can be expressed, albeit incompletely, as *annotation* to F_{rec}.

[191]Every sane person understands his/her mother tongue and has an adequate visual picture of the world. This uniformity makes understanding of these "understandings" as difficult as would be understanding motion in a world where all objects moved in the same way.

For instance, upon receiving a flat image on its screen (retina), an ergo-learner $\mathcal{L}$ must correctly resolve depth in *interpositions/occlusions* and/or "guess" relative values of the third coordinates at essential points of this image.

And the background tree structure in a (record of a) flow of speech can be indicated with *parentheses* properly inserted into sentences. (An annotation may also include additional syntactic and/or semantic comments concerning particular words and sentences.)

Such annotations performed by a human ergo-brain depend on an elaborate guesswork that is by no means simple or automatic. And besides annotating flows of signals, the ergo-brain augments them by something else.

For instance, the formation of a visual image in one's mind depends on the activity of motor neurons involved in eye movements, and "understanding" of these images depends on structural matching of this activity with similar actions of these neurons in the past.

This active process of perception can be seen as a conversation or a kind of a game of the ergo-brain with the environment. But such games, unlike anything like chess, are not easy to mathematically formalize.

23. Characteristic Features of Linguistic Signals

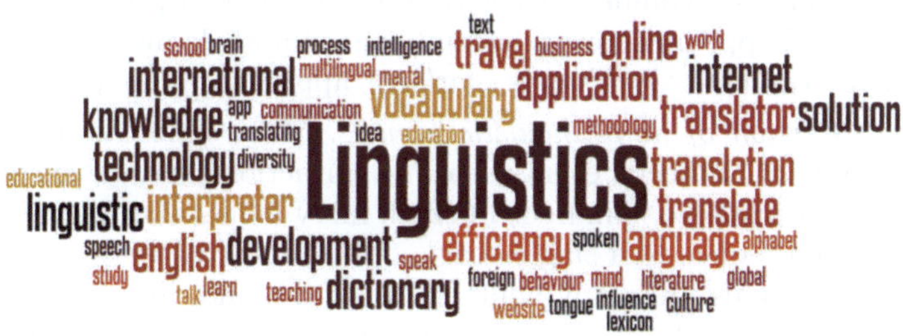

© kalpis / stock.adobe.com

Linguistic information entering the ergo-brain, unlike chemical messaging between cells, does not much depend on the physical carrier of this information and on the corresponding perceptual channels, be these auditory, visual (in reading and in sign languages), or tactile (in deaf-blind communication).[192] The properties of "verbal signals" reflect the properties of the underlying *syntactic ergo-structure*, the apparent attributes of which are as follows.[193]

(1) *Fast Language Specific Clustering.* Formally/physically *different* signals, e.g., sounds, are perceived/recognised as *identical* verbal units, e.g., phonemes, words, phrases, where this is achieved within *half-second* time intervals.

[192]Languages for a physicist are instances of flows of (very!) *low energy* signals that are able, however, to *strongly perturb* the dynamics of otherwise stable systems, such as the physiology of the human body.

[193]We do not speak here of *linguistic universals* such as *negative*, *interrogative*, and *antonymous* that are absent from non-linguistic signals.

The clusterization of *phonemes* (and, probably, of other, including non-auditory, basic verbal units) depends on a particular language, and the mechanism of learning these clusters by children (that deteriorates with age) is poorly (if at all) understood.

Yet, abstractly speaking, this is the easiest of our problems as is witnessed by the efficiency of (non-contextual?) speech recognition algorithms.

(2) *Formalized Division into Units.* Flows of speech are *systematically* divided (albeit non-perfectly) into (*semi*) *autonomous units*, where the basic ones are what we call "*words*".

This division, that is sharper than that of signals coming from "natural sources", is based to a significant extent on *universal* principles of *segmentation* that are applicable to all kind of signals, where the markers separating "segments" are associated with *pronounced minima* of the *stochastic prediction profiles* of signal flows, where determination of such a profile depends on structural patterns characteristic for a particular flow.

(3) *Medium and Long Range Structure Correlations.* There are more "levels of structure" in languages than in other flows of signals. This is seen, in part, in a presence of non-local "correlations" between different fragments in texts.

For instance, if a sentence starts with "*There are more ... in ...*" one may rightly expect "*than in*" coming next with abnormally high probability.[194]

And if "*Jack*" appears on every second page in a book and "*his eyes sparkled again*" than, you can bet, "*Jack's eyes sparkled*" appeared on the previous page.[195]

(4) *Verbal Reduction of non-Linguistic Signals.* Many different non-verbal signals, corresponding to objects, events, features, or actions may be encoded by the same word.[196] For instance, hundreds of small furry felines that have ever crossed your field of vision reduce to a single "cat".

Non-verbal signals are many while their word-names are few. The use of a language replaces the bulk of the "raw memory" in the brain by a network of "*understand*" links between individual items in this memory. This is why small children visibly enjoy the process of the verbal classification/unification of "natural signals" from the "external world" as they learn to identically name different objects.[197]

(5) *Imitation, Repetition, and Generation of Linguistic Signals.* Humans, especially children, have the ability to reproduce linguistic signals [*sign*] they receive, including those emitted by themselves, where, to be exact, *not signals* [*sign*] *themselves* are generated but members of the *same class/cluster* as [*sign*] and where the choice of a particular *classification rule* is not a trivial matter.

---

[194]Try: *there are more * in* on Google.

[195]A universal learning program faces here the difficulty of "understanding" *his* and *again*.

[196]This may be contrasted with the existence of *synonymous* words, but the multiplicities and significance of the latter are incomparable to the power of verbal reduction.

[197]Children of this age are close to being ideal ergo-learners – striving to learn and to understand is the main drive of ergo-systems.

One can hardly analyse languages without being able to generate them,[198] where the language generative mechanisms – called *generative grammars* – result from the repetitive nature of imprecise imitation.

(6) *Many Levels of Self-Referentiality.* No other flow of signals, and/or human medium of communications have the propensity of self-reference typical of language. The ergo-structures of languages contain multiple reflections of their own "selves", their internals "egos", such as

noun-pronoun pairs, allusions to previously said/written items, summaries of texts, titles of books, tables of contents, etc.

Understanding a language is unthinkable without the ability of generation and interpretation of self-referential patterns in this language.

(7) *Pervasive Usage of Metaphors.* Metaphors you find in dictionaries are kind of frozen reflections of their precursors in multiple colored mirrors of language (where such a precursor may have been long forgotten) that correspond to similar mirrors in human ergo-brains.

(Such "mirrors" in vision are, probably, implemented by projections from "deeper" regions of the brain to the primary visual cortex.)

Language in the ergo-brain

Mental representations of languages are described by *syntactic/linguistic ergo-structures* that are similar, but are in a way more *elementary and more abstract* than what is customarily studied by mathematicians and by linguists. (*Elementary, abstract, fundamental, rudimentary* are synonymous from the ergo-point of view.)

© Baris Simsek / Getty Images / iStock

We do not have a full idea of what these are but the *combinatorial core* of such a structure is visible as a *pair of multilayered colored networks*:

The first network is comprised of connectives between (syntactically or semantically) interacting (often spatially/temporally close) units in texts (conversations) and the second one is composed of various kinds of similarities between such units.

And also (non-group theoretic) symmetries implied by transformation rules within these structures bring to mind what one encounters in the *theory of n-categories.*

[198]Neuronal signal generation mechanisms play an essential role also in vision: much of what you "really see" is conjured by your own brain, but the details of this process are inaccessible to us.

24. Understanding Structures and the Structure of Understanding

If there was a parrot which could answer every question,
I should say at once that it was a thinking being.

DIDEROT, PENSEES PHILOSOPHIQUES, 1746

But ...

It never happens that it [an automaton] *arranges its*
speech in various ways, in order to reply appropriately
to everything that may be said in its presence,
as even the lowest type of man can do.

DESCARTES, DISCOURSE ON METHOD, 1637

Is Descartes justified in his belief that no machine can pass what is nowadays called *Turing Test*; i.e., *to reply appropriately to everything that may be said in its presence*?

Does passing such a test certify one as a THINKING BEING who UNDERSTANDS what is being said, as Diderot maintains?

What does it mean to UNDERSTAND, say, a language or any other flow of signals?

Diderot indicates a possible answer:

the continuity of ideas, the connection between propositions, and the links of the argument that one must judge if a creature thinks.

To go further, we conjecture that

- most (all?) structures we encounter in life, such as natural languages, mathematical theories, etc. are UNDERSTANDABLE;[199]
- this UNDERSTANDING universally applies to a large(?) class of *structural*, entities;
- UNDERSTANDING is an elaborate structural entity in its own right, thus, being subject to mathematical scrutiny.

Granted all this, we start searching for mathematical models[200] of UNDERSTANDING.[201]

---

[199]Over-optimistic? Yet, in line with the remark "... *mystery of the world is its comprehensibility*" by Einstein.

[200]When we say *mathematical structure* or *mathematical model* we do not have in mind any particular branch of the continuously growing, and mostly hidden from us, enormous tree that is called MATHEMATICS.

[201]Impossibility of resolving – not even formulating – the problems of UNDERSTANDING and of *thinking machine* in simple words does not abate one's urge to make the world know what one's *gut feeling* about these issues is – an incessant flow of publicised opinions on this subject matter is a witness to this.

Amusingly, the *gut feeling* itself, at least the one residing in a dog's guts, unlike the ideas propagated from human guts to human minds, was experimentally substantiated by H. Florey with his coworkers, with the results published in 1929 under the title *The Vascular Reactions of*

Answers to the following questions, let these be only approximate ones, may serve to narrow the range of this search.

QUESTION 1. What are essential (expected? desired?) features/architectures of mathematical models of structural understanding?

QUESTION 2. If such a model exists should it be essentially unique? In particular, must the hypothetical structures of understanding, say of a language and of chess, necessarily closely resemble one another?

QUESTION 3. How elaborate does such a model need to be, and, accordingly, how long should be a computer program implementing such a model?

QUESTION 4. What is an expected time required for finding such a model and writing down the corresponding program?

QUESTION 5. What percentage of this time may be delegated to machine (ergo)-learning with a given level of supervision?

QUESTION 6. How much of the supervision of such learning can be automated?

© iculizard / Getty Images / iStock

QUESTION 7. What are criteria/tests for performance of "I UNDERSTAND" programs?[202]

QUESTION 8. Can Turing-like tests be performed with *algorithmically* designed formal questions that would trick a computer program to give senseless answers?[203]

QUESTION 9. Are there simple rules for detecting senseless answers?

---

the Colonic Mucosa of the Dog to Fright. (This the same Florey who was responsible for bringing penicillin to the therapeutic use in 1940s.)

[202]A designer's own ability to pass a test is a poor criterion for designing such a test.

[203]Such questions must refer to an earlier part of a conversation. For instance, the examiner says "... *example* ..." at some point and then, responding to a question by the program, the examiner says; "*I have already given an example of this in my previous five word sentence.*" But to trick a sufficiently sophisticated program one must design a "logical *Russian doll*" of such questions.

QUESTION 10. Can the human learning (teaching?) experience be of use for designing clever learning algorithms?

QUESTION 11. Does ergo-logic help answer the above questions?

Apparently, UNDERSTANDING is composed of three ingredients:

$[\bullet]_U$ *a certain combinatorial (logical?)* **structure** $\mathcal{U}$ *in the understander's mind/ brain/program;*

$[\bullet]_{[IU]}$ *a* **process** $\mathcal{IMPU}$ *of implementation of* $\mathcal{U}$ *by an ergo-system representing "an understander";*

$[\bullet]_{[RU]}$ *the* **result** $\mathcal{RESU}$ *of such implementation,* $\mathcal{RESU} = \mathcal{IMPU}(\mathcal{FLOW})$, *where* $\mathcal{IMPU}$ *is seen as a transformation applied to flows of signals.*

We do not know for sure if *understanding* is a formalizable concept and, if "yes", there is no clear idea of what kind of structure this $\mathcal{U}$ could be.

The only convincing argument in favour of the existence of $\mathcal{U}$ would be designing a functional *thinking machine/program*, and the only conceivable NO might come from an incredible discovery of a hitherto unknown fundamental property of the live matter of the brain.

This precludes any speculation on how and where such a structure can be implemented. Besides such a structure is by no means unique but rather different $\mathcal{U}$ are organized as a structural community that can be partly described in category-theoretic terms.

However, we have a realistic(?) expectation of what space/time characteristics of this structure(s) could be: $\mathcal{U}$ is much smaller in volume content than the totality of the flow $\mathcal{FLOW}$ this $\mathcal{U}$ "understands", and implementation of $\mathcal{U}$, that is application of $\mathcal{IMPU}$ to a flow $\mathcal{FOW}$, is much faster than achieving understanding $\mathcal{U}$.

It takes, probably, $\approx l \log l$ elementary steps for learning $\mathcal{FLOW}$ of length l that translates to months or years when it comes to learning a language or a mathematical theory.[204] But when learning is completed, it takes a few seconds to realize, for instance, that a certain string of symbols in the language of your $\mathcal{FLOW}$ is completely novel and even less time to see that a string is meaningless.

On the other hand, the space/volume occupied by an understanding program $\mathcal{U}$ is a few orders of magnitude greater than a learner's starting program $\mathcal{PROG}$, where such a program is *universally* (independently of the total number of signals from $\mathcal{FLOW}$ received/inspected by a learner) bounded by something like 10^6 bits. Picturesquely,

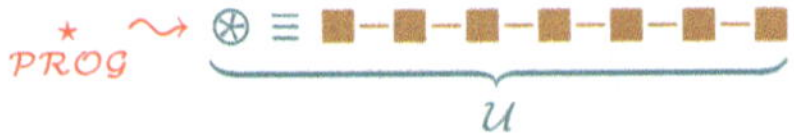

where ⊛ represents the "core understanding" – a few thousand page "dictionary + grammar" of $\mathcal{FLOW}$ – the flow of signals, where this ⊛ is augmented by several

[204]The true measure of time, call it *ergo-time*, should be multi-(two?)-dimensional, since it must reflect parallelism in programs modelling learning and other mental processes.

(tens, hundreds, or thousands depending on $\mathcal{FLOW}$) "volumes" ■ of loosely (imagine RAM on your commuter) organized "knowledge", while the available $\mathcal{FLOW}$ itself may number in tens or even hundreds of millions of nearly unrelated units – volumes, internet pages, images – a mess like this:

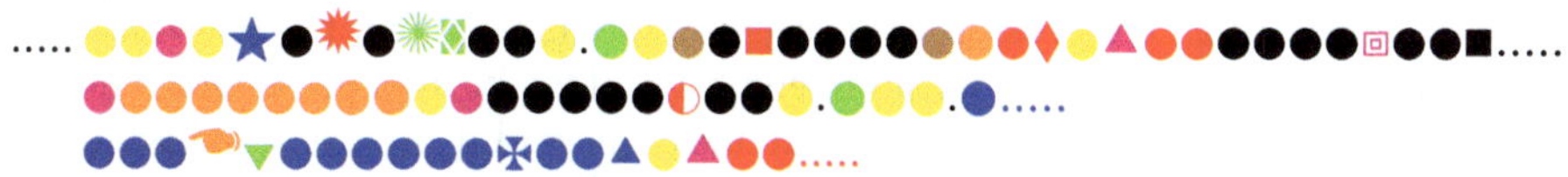

The branch of mathematics that would support, say human UNDERSTANDING, of language, must be quite elaborated – it is not grown yet on the "grand tree of math."

But sometimes things are simple, e.g., for the vervet monkey "*Alarm Call Language*" that matches a few (four?) word-signs – their alarm calls, with object-events, that are particular predators – leopards, eagles, pythons, baboons.

However, no monkey would think that a mathematical *one-to-one correspondence*, call it AlCaLa, between two 4-element sets *understands* the meaning of the alarm calls even if this AlCaLa is implemented by a monkey-shaped robot that properly reacts to predators by correctly emitting the corresponding calls and, thus, passes the vervet monkey Turing test.

Why then do Descartes and Diderot, not to speak of Turing himself, attach such significance to the Turing test?

Is there an *essential* difference between the correspondence "*questions*" ⟶ "*correct answers*" in a human language and the AlCaLa correspondence?

The answer is:

Yes, there is an essential difference, an ENORMOUS DIFFERENCE.

Operating with tiny sets, e.g., composed of four utterings – alarm calls of vervet monkeys and with correspondences between such sets needs no structure in these sets. But one CANNOT manipulate human utterings and even less so longish strings of utterings and/or written texts in a structureless way.

It is tacitly assumed by scientifically minded people – Descartes, Diderot, Turing ... – that the above correspondence "⟶" must be compatible with the essential structure(s) of the human language, call it $\mathcal{HL}$, used in a particular Turing test, where the basic (but not the only) structure in $\mathcal{HL}$ is that of an *exponential/power set*:

An uttering/sentence, say in thirty words, in a language with dictionary D is seen as a member of the HUGE *power set*

$$D^{30} = \underbrace{D \times D \times D \times \cdots \times D}_{30}.$$

Such structurality is indispensable for an implementation of a "thinking automaton" and/or the program running it in a *realistic* space time model [205] that

[205] The property of being physically realistic is often missing in philosophical discourses on artificial intelligence.

necessarily excludes, for instance, "imaginary programs" containing in their memories lists of more than, say N^{15}, sentences with the number N being comparable with the cardinality $card(D)$ of the dictionary.[206]

("Large sets", be they finite or infinite, have no independent existence of their own, but only as carriers of structures in them, similarly to how the space-time in physics makes no sense without energy-matter in it. This is not reflected, however, in the set-theoretic notation that may mislead a novice. For instance, it is rarely stated in elementary textbooks that the "correspondence" $x \mapsto y$ in the "definition" of a *real variable function* $y = f(x)$ is only a metaphor and that a function $f(x)$, if it claims the right to exist in mathematics, must *"respect" some structure* in the set of real numbers.[207])

25. Sixteen Rules of an Ergo-Learner

The general guidelines/principles suggested by *ergo-logic* for designing universal learning algorithms can be summarized as follows.

1. Flows of signals coming from the external world carry certain structures "diluted" in them.

Learning is a process of extracting these structures and incorporating them into the learner's own *internal structure.*

2. The essential learning algorithms are *universal* and they indiscriminately apply to all kind of signals.[208]

3. Universality is incompatible with any a priori idea of "reality" – there is no mental picture of what we call "real world" in the "mind" of the learner.

The only *meaning* the learner assigns to "messages" coming from outside is what can be expressed in terms of (essentially combinatorial) *structures* that are recognized and/or constructed by the learner in the process of incorporating these "messages" in learner's internal structure.

---

[206]This very sentence: "Such structurality is ... of the dictionary" contains forty words with roughly half of them being nouns, verbs, and adjectives. By varying these, one "can" generate more than $1000^{20} = 10^{60}$ grammatical sentences. Can one evaluate the number of meaningful ones among them? Would you expect *thousand* of them or, rather, something closer to *ten thousand*? Is it conceivable that "weakly meaningful" sentences number in 10^6, or are there more than 10^{10}, or even greater than 10^{18} of them?

[207]There is no accepted definition of "*function*" that would separate the *wheat*, $\sin x$, $\arctan x$, $\sqrt{x}$, $\log x$, Riemann's $\zeta(x)$, Dirac's $\delta(x)$, ..., from the *chaff*, such as the characteristic function of the subset of rational numbers.

[208]The learner's behaviour, that is learner's interaction/conversation with incoming signals, also depends on the learner's internal structure that has already been built at a given point in time. In particular, a prolonged exposure to a particular class of signals makes a learner's behaviour more specialized (more efficient?) but the learner's ability to absorb and digest different kinds of signals declines.

4. Universality also implies that the actions of the learner – building internal structures and generating signals, both within itself and/or released outside,[209] *are not governed by goals* expressible in terms of the external world.

The learning is driven by the learner's *"curiosity" and "interest"* in structural patterns the learner recognizes in the incoming flows of signals and in the learner's delight in the logical/combinatorial beauty of the structures the learner extracts from these flows and the structures the learner builds.

Essential ingredients of the learning process are as follows.

5. The learner discriminates between familiar signals and novelties and tries to match new signals with those recorded in its memory.

6. The learner tries to *structurally extrapolate* the signals already recorded in its memory in order *to predict* the signals that are expected to come.

7. Besides the signals coming from the external world, the learner perceives, records, and treats some *signals internally generated* by the learner itself.

8. The learner tends to *repeatedly imitate signals* being received, including some signals that come from within itself.

(The repetitiveness of their basic operations allows a description of learning processes as *orbits under some transformation in the space of internal structures of a learner.* The learning program that implements this transformation must be quite simple and the learning process must be robust. Eventually, "orbits of learning" stabilise as they approach approximately fixed points.)

9. The learner tends *to simplify signals* it tries to imitate.

10. The learner systematically makes guesses and "jumps to conclusions" by *making general rules* on the basis of regularities it sees in signals.

11. When the learner finds out that a rule is sometimes violated, the learner does not reject the rule but rather adds *an exception.*

12. The learner tends to use *statistically significant* signals for building its internal structure as well as for making predictions. But sometimes, the learner assigns significance to certain exceptionally rare signals and uses them as essential structural units within itself.[210]

13. The learner's probabilistic reasoning in uncertain environment is *yes-maybe-no* logic.

We impose the following restrictions on the abilities of our intended learner programs that are similar to those the human brain has.

14. The learner *does not accept unstructured sets* with more than four or five items in them; upon encountering such a set, the learner invariably assigns a certain structure to it.[211]

[209]These are the "actions" the human brain is engaged in.

[210]It is the rare words in texts that are significant, not the most frequent ones. For instance, *gifku mfink otnid* on three different pages of a book will impress you more than ooooo ooooo ooooo on twenty different pages.

[211]The partition of stars in the sky into constellations is an instance of this.

15. The learner has *no built-in ability of sequential counting* beyond 4 (maybe 3); we postulate that $5 = \infty$ for the learner.

In particular, the learner is not able to produce or perceive five consecutive iterations of the same process, unless this becomes a *routine* delegated from *cerebral cortex* to *spinal cord.*[212]

Our main conjecture is that

universal learning algorithms that converge to UNDERSTANDING *exist* and that, moreover, their formalized descriptions are *quite simple*.

The time complexity of such algorithms must be at most log-linear (with no large constant attached) and the performance of an "educated/competent program" must be no worse than logarithmic.

In fact, the essential features of (ergo)learning as we know it, make sense only on a roughly "human" time/space scale: Such learning may apply to flows of signals that carry $10^9 - 10^{15}$ bits of information altogether and one hardly can go much beyond this.[213]

Universality and Doublethink. If one expects an analysis of a *flow of signals*, e.g., of a collection of texts in some language $\mathcal{L}$, to be anywhere close to the TRUTH, and if one wants to design an algorithm for learning $\mathcal{L}$, one must, following ergo-logic, *disregard* all one a priori knows about this $\mathcal{L}$, *forget* this is a language, and *reject* the idea of meaning associated to it.

But the only way to evaluate the soundness of your design, prior to a computer simulation of it, is to compare its performance to that of the corresponding algorithms in a human head.

---

[212]Never mind *the kid that fought his dad that bought the car that struck the bike that hit the truck that brought the horse that kicked the dog that chased the cat that caught the rat that ate the bread.*

[213]The universal learning systems themselves, e.g., those residing behind our skulls, have no built-in ideas of *meaning*, of *time*, of *space*, of *numbers*. But any speculation on natural or artificially designed "intelligent" systems strikes one as *meaningless*, if *spatial* and *temporal* parameters of possible implementations of such systems are *not specified* and set within realistic *numerical* bounds.

26. Learning to Understand Languages: From Libraries to Dictionaries

© RobinsonBecquart / Getty Images / iStock

Mathematically, the process of *learning a language, call it $\mathcal{TONG}$* is represented by an *orbit* of the *universal learning program PROG* that acts on the *linguistic space of $\mathcal{TONG}$*, and where this orbit must eventually (approximately) converge to "*I understand $\mathcal{TONG}$*" state/program.

The principle of the existence of such a *PROG* is demonstrated by the linguistic performance of the brain of (almost) every child born on Earth that receives flows of *electro-chemical signals*, some of which come from linguistic sources, and the "meaning" of which the child's brain learns to "understand".[214]

A scenario closer to our experience is that of a visitor from another Universe[215] who attempts to "understand" what is written in some human "library" LIBR, e.g., on the English pages of the internet.

In either case, the process of what we call "understanding" is interpreted as assembling *an ergo-dictionary EDI* – a kind of "concentrated extract" of the combinatorial structure(s) that are present (but not immediately visible) in flows/arrays of linguistic signals.

Making a dictionary involves several *interlinked* tasks where a starting point is

Annotation & Parsing, that is identification and classification of ***textual units*** that are persistent and/or significant fragments in short strings s (say, up to 50–100 letter-signs) as well as attaching ***tags*** or *names* to some of these fragments.

Tagging may be visualized as coloring certain fragments in texts, where these fragments and the corresponding colors may overlap. Or, one may represent an

[214]Bridging linguistic signals to non non-linguistic ones is an essential but, probably, not indispensable ingredient of "understanding language" as it is suggested by the linguistic proficiency of deaf blind people.

[215]No imaginable Universe appears as dissimilar to ours as what the brain "sees" in the electro-chemical world where the brain lives.

annotation by several texts written in parallel with the original one, where the number of different color-words is supposed to be small, a few hundred (thousand?) at most, with a primitive "grammar" that is a combinatorial structure organizing them.[216]

One may think of such annotations as being written in strings positioned on several *levels*[217] over the original strings s, where the new tag-strings on the level l are written in the tag-words specific to this level and where the number of such l-tags (at least) decays exponentially fast with l.

An ergo-dictionary is obtained by several consecutive *reductions* – kind of *compressions* – applied to a library of annotated texts where the resulting combinatorial structure of the dictionary is more elaborate than that of the library of annotated texts. The overall reduction in volume is a thousandfold, roughly, from 10^9–10^{10} to 10^6–10^7 linguistic units.

The grammar of a language makes a part of EDI where the structural position of this grammar in EDI is supposed to imitate how it is (conjecturally) organized in the human mind.

A particular ergo-dictionary

$$EDI = EDI(\text{LIBR}) = EDI_{PROG}(\text{LIBR})$$

is obtained from a collection of texts, what we call a *library* LIBR, according to some universal (functorial?) process/program $PROG$ that drastically reduces the size of LIBR and, at the same time, endows what remains with a *combinatorial structure* – a kind of "network of ideas" – that is similar to but more elaborate than the structure of a *partially directed edge and vertex colored graph.*

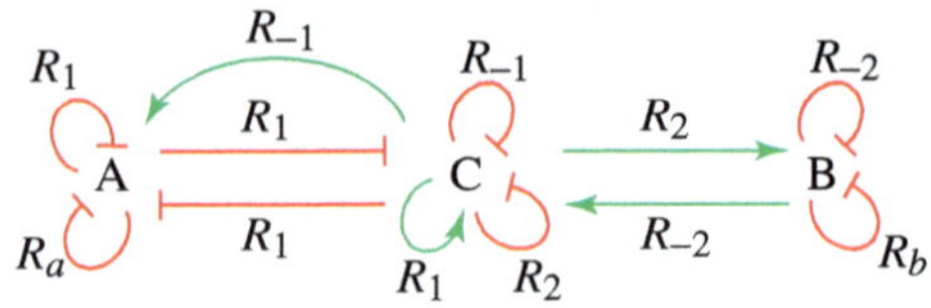

This EDI can be thought of as (a record of) *understanding* of the underlying language by the learner behind $PROG$. This understanding, call it U_t, is time-dependent, with EDI being an *approximate fixed point* of the learning process

$$U_{t_1} \underset{PROG}{\rightsquigarrow} U_{t_2},\ t_2 > t_1,$$

where, a priori, $PROG$ can be applied to "understandings" U that were not necessarily built by this $PROG$.

The essential problem here is finding a *uniform/universal* representation that can be implemented as a *coordinatisation* of "the space of understandings" U,

[216]An annotation may include references to *non-linguistic* signals but this would contribute to one's *knowledge* rather than to one's understanding.

[217]These levels l can be regarded as numbers $0, 1, 2, 3, \ldots$ with $l = 0$ corresponding to strings in the original text, where the number of level is small, something between 3 and 5. But, as we shall see later on, these levels are organized according to a structure that is not quite a linear order.

where a simple minded program *PROG* could act by consecutively adjusting "coordinates" $u_1, u_2, \dots$ of U, and where this space would accommodate incoming loosely-structured flows of signals encoded by libraries as well as rigidly organized dictionary structures.[218]

The apparent ingredients of ergo-dictionaries *EDI* that encode "understanding" and processes for assembling these dictionaries are the following.

- *Short range correlations,*[219] *segmentation and identification/formation of units in flows of linguistic signals.*
- *Memory, information, and prediction on different levels of structure.*
- *Similarities, equalities, contextual classification, cofunctionality, and coclustering.*
- *Local and non-local, links, and hyperlinks.*
- *Tags, annotations, reduction, classification, coordinatization.*
- *Structuralisation and compression of redundancies.*[220]
- *Ability and tendency for repetition and imitation.*
- *Fast recognition of known, unknown, frequent, significant, improbable, nonsensical.*
- *Evaluation of degree of "playfulness" or "metaphoricity" of words, phrases, and sentences.*[221]
- *Recognition of self-referentiality.*
- *Evaluation of parameters of ability/quality of predictions:*
- *Speed, precision, specificity, rate of success, the volume of memory, and the numbers of parallel and sequential "elementary operations" employed,*

  etc.

A program that would imitate a human conversing in a natural language and that is seen as "realistic" from the ergo-perspective must be within 10^9–10^{12} bits in length. If such a program would fool somebody like Diderot, then its level of structurality must necessarily be comparable to that of the human ergo-brain and one would be justified in saying that this program *understands* what is being said.[222]

More seriously, a validity of a particular program and the resulting linguistic competence of a learner in $\mathcal{TONG}$ can be certified only by comparing the outcomes of several programs, say comprised and executed by visitors from 100 different

[218]We know that such programs are fully operational in the brains of 2–4 year old children.

[219]Relative frequencies of "events" are essential for learning a language but such concepts as "probability", "correlation", "entropy", cannot be applied to languages, without reservation.

[220]The essence of understanding is not so much extracting "useful information" but rather understanding the structure of redundancies in texts. Non redundant texts, such as tables of random numbers and telephone directories do not offer much of what is worth UNDERSTANDING.

[221]Playfulness is the first visible manifestation of "ergo" in humans (and some animal) infants.

[222]Beware of ELIZA type programs that respond to everything you say by: "You are right, it is very profound what you say. You must be very intelligent".

Universes. Conceivably, 70 among them will not be able to communicate in $\mathcal{TONG}$ with anybody, but 30 will be able to talk to each other in $\mathcal{TONG}$ and find it interesting. These 30, by definition, UNDERSTAND $\mathcal{TONG}$.

(Competent in $\mathcal{TONG}$ and linguistically naive native speakers will judge differently.)

27. Libraries, Strings, Annotations, and Colors

Libraries LIBR we have in mind may contain 10^7–10^{11} *input units* that are possibly overlapping stings of letters starting from a fragment of a word to a paragraph with a few dozen words in it.[223]

Annotation may be seen as an assignment of "*colors*" to these strings as well as connecting some units with colored edges, where these colors depict "essential properties" of the nodes or edges they are assigned to and serve as *descriptors* of *input units* in (annotated) LIBR and later on in EDI.

A color on a node u may describe a *pre-syntactic* property of a string, such as

short string, median string, long string, and *frequent string*,

but syntactic/semantic features of strings will be assigned different classes of colors from different classes.

Similarly, a color on a connective may signify a type of a geometric/temporal relation between strings, such as

overlap, contained-one-in-another, close-one-to-another, far-one-from-another, next-to-each-other, in-between, begins-with,

etc., where these "colorful concepts" come in several subcolor-flavors similarly to (yet, differently from) how it is with lengths of strings.[224]

These connectives are tied together themselves by relations between them. For instance, *closeness* between two strings often comes as *simultaneous containment* of these string into a longer string.

This is essential for enlisting and keeping in memory (pre)syntactic insertions between strings, in particular all pairs of identical words w in L.[225]

And another class of colors such as for similarity $\leftrightsquigarrow$, and for reduction arrows, may refer to a particular classifier algorithm defining/producing these arrows.

[223]There are also larger units, call them "pages", "volumes", "shelves", but these play different roles.

[224]The ergo-brain, that processes incoming information in parallel, divides long strings into pieces and keeps track of their mutual positions by means of binary (and ternary such as ***in-between***) relations.

Although the linear order •••••••••••••••• can be *formally* reconstructed from ***begins-with***, a baby-like learner is unaware of formal logic and cannot indefinitely iterate a single relation.

[225]The number of pairs of (short) identical strings is uncomfortably large, being *quadratic*, but identifying identical words, say on a single page, goes *linearly* in time.

This is gratifying. Squares are unacceptable except of small quantities. We happily live through a *million* seconds that make up less than 12 days of our lives. But a *trillion seconds*, that is a million squared, stretch over more than 31 000 years.

Colors themselves are structural entities but much simpler ones than incoming units such as sentences. One may think of them as simple graphs, e.g., little trees with few branches.

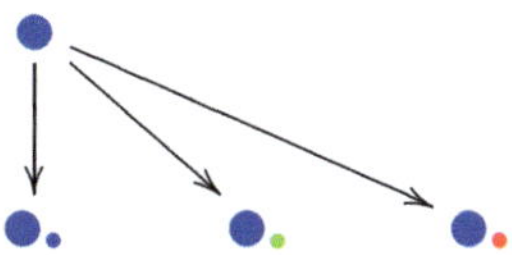

The library colors – about 100 in number – pass to *EDI*, our ergo-dictionary where they are organized into a more elaborate colored network with additional descriptor-colors, maybe up to 1000 of them.

A more difficult/interesting aspect of annotation is identifying "significant strings" (this is essentially what is called *parsing*) such as ***words***, ***phrases***, etc., and disregarding insignificant ones.[226] These also come in colors and subcolors such as $\textbf{\textit{word}}_{\text{rare}}$, for instance.

Collections, Ensembles, Sets

Collections/ensembles of linguistic units and of connections between them cannot be indiscriminately operated with, as we do with sets in mathematics for the following reasons.

1. The presence of a particular node in a network, e.g., of a particular phrase, in the long-term memory of a learner is often ambiguous.
2. Basic set-theoretic constructions, such as the union $X_1 \cup X_2$ and the Cartesian product $X_1 \times X_2$, cannot (and should not) be unrestrictedly performed in our networks.

The set theoretic language may lead you astray;[227] yet, we use fragments of this language whenever necessary.

Besides "localized" relations between strings and connectors between them, there is a large scale geometry in LIBR that is seen in the presence of (relatively large) *contextual units* such as ***pages*** and ***books***.

At the first stage of the annotation process these units are classified/colored by their size, where different classes must roughly fit into the corresponding frames of the *short-term, medium-term,* and *long-term memory.* Then the concept of "context" is modified and refined in the course of learning (not quite) similarly to how it happens to strings, where the true ***pages*** and ***books*** must be either sufficiently *statistically homogeneous*, or *structurally unified*, or have *pronounced boundaries.*

Coloring Colors. It seems not difficult to make a complete combinatorial description[228] of LIBR with a few dozen (about 100?) "colors" that are *descriptors* of *basic types* of units and of connectors between them in LIBR.

[226] An essential (but not the only) intrinsic motivation for doing this by an ergo-learner is economizing memory space.
[227] Bringing forth *random sets* and/or *fuzzy sets* may only aggravate the problem.
[228] Implementing this, i.e., achieving a proper annotation and/or parsing, is by no means easy.

But the principal issue is not so much LIBR per se but a construction of an adequate *colored network of descriptors* on the basis of a few, probably 4–8, "general rules". Eventually this network will be laid in the foundation of *EDI* – that is a *stationary model* of understanding $\mathcal{TONG}$ as represented by the library LIBR.

28. Teaching and Grading

© hurca.com / stock.adobe.com

A universal language-learning program *PROG* is supposed to model a mind of a child and it needs only minimal help from a "teacher", such as ordering texts according to their complexity[229] and allowing *PROG* a flexible access to texts.

On the other hand, evaluation of the *quality of understanding* by *PROG* is harder (albeit much easier than designing a learning program itself), since no one has a clear idea of what *understanding* is.

Our formal approach is guided, in part, by how it goes in physics, where an unimaginably *high level of understanding* is reflected in the *predictive power* of mathematically-formulated *natural laws* that encapsulate enormously *compressed data.*

This lies in a category quite different from what we call *"knowledge"*.

[229]One may also equip *PROG* with an ability, similar to that possessed by children up to age 2–3, to resist a "bad teacher" by rejecting environmental signals that are detrimental for learning. (This ability deteriorates with age as one has to adapt to the environment in order to survive.)

For instance, ancient hunters *knew* more of how planets wander in skies than most modern people do. But understanding of this wandering depends on "compression" of this knowledge by setting it into the slender frame of mathematically formulated laws of motion.[230]

Similarly, understanding languages depends on compression of structural redundancies[231] in flows of linguistic signals, albeit this compression is not as substantial as in physics.

Besides "sheer knowledge", *understanding* should be separated from *adaptation*. For instance, an experienced rodent (or a human for this matter) competently navigates in its social environment. But only metaphorically, one may say that the rodent (or human) "*understands*" this environment.

With the above in mind, we indicate the following two mutually linked attributes of what we accept as "*understanding*".

1. *Structural compression of "information"*.

2. *Power of prediction.*

These **1** and **2** can be quantified in a variety of ways. For example, one may speak of the degree of compression versus the "percentage" of structure lost in the course of compression, while the essential characteristics of a prediction is *specificity* versus *frequency of success.*

This kind of quantification may be used for *partially ordering* "*levels of understandings*" that may suggest tests for evaluating progress achieved by a learning program *PROG* in these terms.

Another attribute of "*understanding*" that is easy to test but hard to quantify is as follows.

3. *Ability to acquire knowledge.*

For instance, a program *PROG* minimally proficient in English, would "know", upon browsing through Encyclopaedia Britannica, that cows eat grass and cats eat mice.[232]

Also, the following can be seen as a hallmark of understanding.

4. *Ability to ask questions.*

(Those whose business is UNDERSTANDING – scientists and young children – excel in asking questions.)

Besides the ability to understand, learning programs *PROG* may be graded according to their "internal characteristics", such as the *volume of memory* a

---

[230] Ancients astronomers came to *understand* periodicity of planetary motions and were able to make rather accurate predictions.

[231] One cannot much compress "useful information" without losing this "information", but if we can "decode" the structure of redundancy it can be encoded more efficiently.

[232] Properly responding to "*Do black cats eat fresh mice?*" instead of the plain "*Do cats eat mice?*" would need a study of a more representative corpus of English than Encyclopaedia Britannica by *PROG*.

PROG has to use, the *number of elementary operations* and *the time* needed for it to make, for instance, a particular prediction.

Eventually, full-fledged ergo-systems must contain *self-control programs*, such as an evaluation of *increase* of quality of predictions as extra information becomes available to the learner. (Such control programs are much easier to design than the core ergo-programs.)

29. Atoms of Structures: Units, Similarities, Co-functionalities, Reductions

Much of learning and understanding consists of *structuralizing* incoming flows of signals you perceive that is achieved by identifying redundancies in these flows and representing "compressed flows" of these signals in a structurally efficient way.

It is a fundamentally unresolved problem in psychology to identify *mathematical classes of structures* that would *model mental structures* built by human brains that assimilate incoming "flows of signals".

We do not know what, specifically, these structures are but a few of their ingredients are visible.

Let us make a short (and incomplete) list of four "logically (quasi)atomic constituents" of (ergo)operations applied to flows of signals with no attempt at this point to give precise definitions of these "atoms", to justify their reality, and/or to explain how one finds them in flows of signals.

1. Discretization and Formation of Units

The first step in structuralizing flows of signals is identifying/isolating *units* in these flows, where the simplest (but not at all simple) process serving this purpose is *segmentation*: Dividing a flow into non-overlapping "geometrically simple parts". (The *internal* correlations/connections in these "parts" must be significantly stronger than *mutual* correlations/connections between different "parts".)

These may be small and frequently appearing signals, such as *phonemes, words* and *short phrases* in the flow of speech or basic visual patterns such as *edges* and *T-junctions*. Also these may be as long as *sentences, internet pages, chapters in books*, or intrinsically coordinated visual images of objects such as *animals, trees, forests, buildings, and mountains*.

And even if, say, paintings and ■'s, are both regarded as "units", they are unlikely to be filed by your (visual) ergo-system in the same "units-directory".

(Processing of linguistic and visual inputs by your (ergo)brain, probably, relies on natural parsing of incoming flows of signals followed by a combinatorial organisation of the resulting "units".

[redacted]

The "Garden of earthly delights". Oil painting after Hierony'. Wellcome Collection

On the other hand, the *proprioception sensory system*[233] and motor control of skeletal muscles may also depend on *continuity*, since the incoming signals may not(?) be naturally decomposable into "discrete units".)

But our ergo-structure is built from invisible *internal* units that may be grossly dissimilar to the units of incoming flows; one needs truly universal *discretizers* – "meaningful segmentation" algorithms – to discern these.

Naively, *unit* is anything that can be characterized in a few simple words, but ... these words may be of very different kinds depending not only on the intrinsic properties of such a unit, but also on how it is being processed by a particular ergo-system, e.g., a human ergo-brain.

We follow the lead of natural languages that make units of everything, *qualities, states, actions, processes*, ... by what is called *nominalization*:

everything deserving a name becomes a unit.

Example: Non-Textual Syntactic Units.[234] In order to implement this "definition", one needs to to design a program that would "understand" what "deserves" is. This is most essential for understanding languages, that unlike understanding non-linguistic arrays of signals, decisively depends on the formation of units that are *not geometric fragments of texts*. For instance, the groups of superficially dissimilar words, such as

yes, no, maybe; we, us, our; big, large, huge; smelly, tasty, crunchy.

are a kind of "outlines" of such units. Identification of these is an essential aspect of passage from a library to an ergo-dictionary.

[233] This is the perception of motion, of stresses, and of positions of parts of the body.

[234] The word "syntactic" is understood in the present article as "characteristic of languages".

Analogy, Similarity, Equivalence, Equality, Sameness

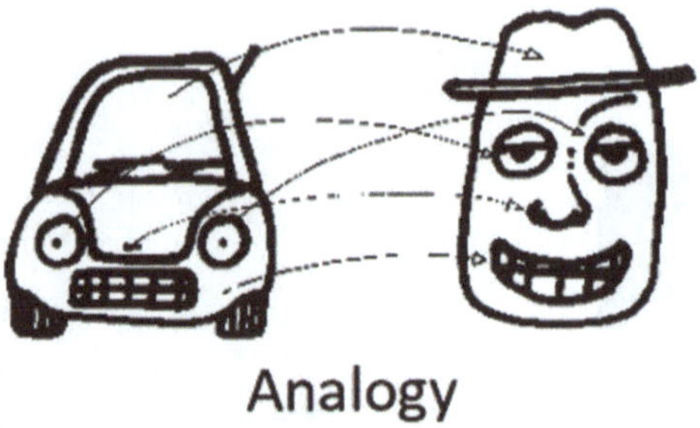

© Clay Crowe,
http://image.wikifoundry.com/image/3/5adc36114eb594a05007ff35681d844d/GW359H220

There are several *similarity relations* between units of languages/images where these relations may differ in kind and in strength.

For example, images may be similar in shape, size, color, subjects they depict, etc., and two sentences may be similar in the kind and style of words they employ, the idea they convey, or in their syntax. The strongest similarities in texts *are letter-wise equalities* of *different* strings.

There is a discrepancy between how the concept of *equality* is treated in mathematics/logic and in natural languages: We happily say

$$2 + 3 \text{ equals } 5$$

but

$$5 \text{ equals } 5$$

appears not very informative even to a logically indoctrinated mathematician – these two "*equal*" are not mutually equal, and the common language has no means to express this inequality. For example,

$$5 \text{ is the same as } 5$$

does not make it look better. But this can be settled if we introduce an ergo-system in the picture, where equalities as well as weaker similarities result from certain *processes* that are qualitatively different from how one arrives at *sameness.*[235]

On Composability of Similarities. Customarily, one defines an *equivalence* as a *symmetric binary relations* on a set[236] S, denoted, say by $s_1 \sim s_2$, that satisfies the transitivity property:

$$s_1 \sim s_2 \ \& \ s_2 \sim s_3 \Rightarrow s_2 \sim s_3.$$

It is more convenient to depict equivalences (and similarities) of signals s in a category-theoretic style by arrows with "names" attached to them, such as

[235]The spirit of this is close to how different levels of "equivalence" are treated in *the n-category theory*.

[236]This definition does not cover equivalences between theories and/or between categories, because these are *are not* relations on *sets*.

$s_1 \overset{f}{\leftrightsquigarrow} s_2$, where one think of such an arrow as an "*implementation of* ~" by some "logical/computational process", e.g., by some co-clustering algorithm.

Then one may compose arrows

$$s_1 \overset{f}{\leftrightsquigarrow} s_2 \overset{g}{\leftrightsquigarrow} s_3 \text{ with the composition denoted } s_1 \overset{f \square g}{\leftrightsquigarrow} s_3.$$

This allows one, for instance, to say that

the composition $f \square g$ of two "strong similarities" f and g is itself a "weak similarity".

Also one can now speak of certain equivalences f and g, e.g., one in color and another one in size, being *incomposable*.

3. Classification, Reduction, Clustering, Compression

Equivalence relations E on a set S go hand in hand with partitions of this S into the corresponding *equivalence classes* that can be conveniently described via the *reduction map* $R = R_E$ from S *onto* a C, such that

$$s_1 \underset{E}{\sim} s_2 \textit{ if and only if } R(s_1) = R(s_2).$$

However, implementations of a *binary* relation $s_1 \underset{E}{\sim} s_2$ and of a *unary* operation $R(s)$ are quite different from a working ergo-system point of view.[237]

It is much harder to record $\approx N^2/2$ bits encoding an equivalence relation on a set S with N units, than $\approx N \log N$ bits needed for defining $R(s)$; *similarities* and *reductions* must be treated separately.

An essential feature of *reductions* from our perspective is *compression of information* and

creation of new units c from the original units s, that are $c = R(s)$.

A more general and less cleanly-defined class of operations is called *clustering*, that is based on similarities that are not sharply defined and are not perfectly *transitive*, unlike what is usually required of "equivalence".

The tautological map $R : s \mapsto c$ associated to a given clustering that assigns to each member s of S the cluster c in S that contains s (this R may not be defined for all s) is still called the *quotient map* or *reduction* from the original set S to the set C of clusters. The reduction that defines co-clustering is an instance of this.

Compression, Morphisms, Functors. Besides the above, there are reductions of quite different type that correspond to "non-local" *compression with a limited loss of information*, where one forgets non-essential information in a text, or in a visual image, while preserving the significant structure/content of it; this is a hallmark of understanding.

---

[237]This is discussed at length in the context of cognitive linguistics by George Lakoff in "*Fire Women and Dangerous Things*" where *classification* is called *categorization*.

It may happen, of course, that a text has little redundancy in it, such as a telephone directory, for instance. Then no significant reduction and no understanding of such text is possible.

In fact, "perfect texts" with no redundancy in them are indistinguishable from random sequences of symbols, but every meaningful text T admits many reductions, depicted by arrows, say $T \overset{r'}{\to} T'$, $T \overset{r''}{\to} T''$, where the bulk of the process of understanding a text consists of a multi-branched random cascade of such reductions.

An example of a significant commonly used reduction is making a *resume* or *summary* of a text. Also giving a *title* is an instance of a reduction – a *terminal reduction*: You cannot reduce it any further without fully degrading its structure.

If we agree/assume/observe that consecutive performance of reductions, say $T_1 \overset{r_{12}}{\to} T_2$ and $T_2 \overset{r_{23}}{\to} T_3$, make a reduction again, denoted $T_1 \overset{r_{13}}{\to} T_3$, also written as *composition*

$$r_{13} = r_{12} \circ r_{23},$$

then reductions between texts can be regarded as *morphisms*, of the **category** (in the mathematical sense) of *texts and reductions* where, strictly speaking the word "*reduction*" suggests these arrows r being *epimorphisms*, i.e., they add *no new information* to texts they act upon.

It may be amusing to encode much (all?) information about a language $\mathcal{L}$ – *syntax, semantics, pragmatics* – in terms of such a category $\mathcal{R} = \mathcal{R}(\mathcal{L})$ of reductions in $\mathcal{L}$, with translations from one language to another, $\mathcal{L}_1 \rightsquigarrow \mathcal{L}_2$, being seen as *functors* between these categories; but it is dangerous to force categories into languages prematurely.

Reduction and Agglomeration of Similarities. There are circular relationships between similarities of different types and/or of different strengths. For instance two signals s_1 and s_2 that have equivalent or just strongly similar reductions may be regarded as weakly similar.

Conversely, if there are "many independent" weak similarity relations between s_1 and s_2 then s_1 and s_2 are strongly similar and possibly equal.

To see what we mean, imagine you have two books, each approximately 200 pages long. Choose any way you want the numbers of pages, say 150 altogether, and count how many times "*the*" appears on each chosen page i of the books 1 and 2.

A similarity of these two "*the*" *contents* N_1 and N_2, such as

$$N_1 - 2 < N_2 < N_1 + 2,$$

for a single pair of pages is not informative, but if this relation holds for all 150 pairs of your chosen pages, you bet that 1 and 2 are copies of the same book.

4. Co-functionality

Some units in a text T or in another kind of flow of signals form relatively tightly knit groups where we say that these units *perform a common function.*

A priori, co-functionality is not a binary relation (albeit it is helpful to assume so when defining *co-clustering*); it can be, however, made binary by giving "names" to these "*functions*" and by regarding functions as new kinds of units.

Then we say that *unit s performs function f* and depict this by a directed edge $s \leftarrow f$. Alternatively, we depict f-co-functional units as being joined by f-colored edges $s_1 \underset{f}{\longleftrightarrow} s_2$.

Connections between Units: their Identification, Nominalisation and Classification

Deciding which units are essentially independent and which have non-trivial *connections/relations* between them is one of the first priorities of an ergo-system that must be accomplished by several algorithms, that can be called *connectors.*

Such connectors, being themselves particular kinds of units, need to be classified by universal algorithms as befits all decent units, where the coarsest classification would separate the following three classes of relations.

- *Similarity.*
- *Cofunctionality.*

(The latter is common not only for pairs but also for triples and possibly quadruples of units that *perform together* certain functions. This "togetherness" is manifested by systematic co-appearance of the corresponding units.)

- *Reductions.*

Warning. Treating relations on an equal footing with initial units opens a pandora's box of self-referentiality in our ergo-system(s). Apparently, this is necessary for the kind of ergo-behaviour we want to achieve but this also inoculates our systems with logical paradoxes. Somehow one has to strike a balance between dumbness and madness of ergo-programs.

After we have developed algorithms for the structural analysis of "incoming flows of signals" along the above guidelines, we shall be able(?) to decide if there is *some unknown "else"* within the human mind crucially involved in the "learning to understand" process that is fundamentally different from formation of units, their classification, and their combinatorial organisation according to their connections and interactions.

The fundamental difficulty we face here appears when we attempt to structuralize not only incoming flows of signals, but also those *created and circulating within* learning system itself, where these "internal flows" are not, at least not apparently, grounded on any structure similar to what underlies "true flows": The linear (temporal or spatial) order between signals.

The data obtained in this regard by neurophysiologists and psychologists do not tell us, at least not directly, how to proceed – we take our cues from what mathematics has to offer.

But while thinking mathematically, we also need to keep in mind possibilities and limitations of the brain, ergo-algorithms must be "broad and shallow": They cannot have many (say, more than 5) *consecutive operations* on each round (unit) of computation (that, roughly, corresponds to what we routinely do on a one-second time scale); yet, allowing several hundred (thousand?) operations running in parallel.[238]

30. Fragmentation, Segmentation, and Formation of Units

Textual units, such as particular fragments of incoming signals,[239] e.g., particular strings of letters such as "*words*",[240] or some distinguished regions in visual fields, such as "*perceived objects*" or "things",[241] lie at the heart of all *combinatorial models of understanding*. Albeit one can hardly give a comprehensive definition of such units, or of *signal-units* in general, these are frequently recognizable by the following property

> *The probability of encountering a textual unit u among a multitude of other signals in the same category as u* (here "category" means *class*) is *significantly greater than the product of probabilities of "disjoint parts of u"*.

For instance the word "probability", which has 11 letters in it, may, a priori, appear only once or twice in a library with a billion books ($<< 26^{11}$ letters) in it.[242]

This does not work quite so nicely for short words: Scrabble dictionaries offer ≈1000 three-letter English words and ≈4000 four-letter words where many of them, e.g., *qat* (an African plant) or (to) *scry* (to practice crystal gazing) appear rarely, but the improbable frequency of such a word may be seen in the appearance of several copies of it in a single volume, or even on the same page.

---

[238] This parallelism is the "technical reason" why our basic mental (ergo)processes are inaccessible to our sequentially structured conscious minds.

[239] We temporarily ignore overlaps between fragments such as "*hard to see*" and "*to see it*" in the unit-phrase "*hard to see it*". ("*Hard to*" makes a perfect "unitary uttering"; yet this is a weaker unit than "*hard to see*".)

[240] A textual unit may be "disconnected", e.g., it may consist of two (more?) strings separated by other strings in a text. This happens, for instance, to separable prefixes in German that are moved to the ends of sentences. Also this **is** *not exceptionally rarely* **seen** in English.

[241] The rigid concept of *object-unit* modifies by classification/reduction and applies to "things" that come in many shapes such as words with flexible morphological forms, the human body, or to something inherently random such as an image of a tree with multiple small branches. When our eye looks at such a tree, our mind, conjecturally, sees (something like) a *branch/shape distribution law* rather than the sample of such a distribution implemented by an individual tree.

[242] The number of different books in the world is estimated at about 100 million.

The abnormal frequency alone, however, does not define units: The string "obabili" appears at least as often as the full "probability"; thus, one has to augment the "definition" of a unit by the following

completeness/maximality condition: If a string s is a unit, then larger strings $s' \supsetneqq s$ are *significantly* less probable than s.

© Anna Kucherova / stock.adobe.com

Segments and Boundaries. Fragmenting texts into units is naturally *coupled* with the process of *segmentation* that is introduction of *division points* that make boundaries of string-units in texts.[243]

The determination when the position d in a string S between two letters may be taken for a division point depends on the strings s "to the left" and "to the right" from d in S, where such a string, say s_{left}, being a unit is an essential indication for d being a division point.

But it may also happen that there are no such clear cut units next to d in S but there is a 20-letter string S' somewhere else in the library that contains isomorphic copies of five-letter strings to the left and to the right from d and such that the corresponding d' is recognizable as a division point in S'. Then we may accept d as a division point in S and to use this for identifying previously unseen units in S.

The coupled *fragmentation + segmentation* is a multistage process, each step of which is a part of the learning transformation $PROG$ on a certain space of pairs $(Frag, Seg)$ that incorporates into the full "understanding space" on the later stages of learning.

This process must comply with the "*please, no numbers*" *principle*: The program $PROG$ we want to implement must function similarly to an infant's brain that, unlike an extraterrestrial scientist, has very limited ability of counting and of manipulating large numbers (e.g., frequencies) as well as small ones (e.g., probabilities).

---

[243] Boundaries of so-called "words" are marked in most written languages by white spaces, but phrases and sentences are pinched between division punctuation signs. But we pretend to be oblivious to this for the moment.

This is achieved by consecutive "internal fragmentation" of the process *PROG* itself into a network of simple processors/directories where, they all, individually, perform (almost identical) "baby operations" with the global result emerging via communication between these processors.

Classification of Words and Partitions into Sentences. Segmentation of texts into strings with more than 2–3 words in them is impossible without preliminary syntactic classification of *basic units* – words and short phrases.[244] But when such a classification is performed the number n of basic units u – this n is about 10^5–10^6 in English – is reduced to much smaller number $\underline{n}$ of classes $\underline{u}$, realistically with $10 \leq \underline{n} \leq 30$. Then a library with N basic units in it would allow one to reconstruct the rule of formation of strings of length about $\log_{\underline{n}} N$. For instance, if we classify with $\underline{n} = 20$, then a modest library with 10^9–10^{10} basic units in it[245] gives an access to 6–8 basic unit long strings, for $\log_{20} 1.3 \cdot 10^9 \approx 7$ that may allow an automatic discrimination between *admissible* and *nonsensical* strings up to, maybe, 12 words in length. Then *generation* of *meaningful strings* becomes a purely mathematical problem.

Gross Contextual Segmentation. In the spoken language, utterings are divided according to when, where, and who is speaking to whom, but texts in written languages are organized into paragraphs, pages, books, topics, libraries, with a similar arrangement of pages on the web.

These partition structures are essential for making a statistical analysis of languages; conversely, texts can be classified/partitioned according to relative frequencies of short-range structural patterns, e.g., basic units, present in them.

31. Presyntactic Morphisms, Syntactic Categories, and Branched Entropy

Deep linguistic structures display some *approximate category-theoretic* features, e.g., *abridgements* may be seen as *semantic epimorphisms*, or as *functors* of a kind rather than mere "morphisms".

Then translations from one language to another come as functors between categories (2-*categories* if abridgements are regarded as functors) of languages, where the category-theoretic formalism should be relaxed to accommodate imprecision and ambiguity of linguistic transformations.

---

[244] The single word in *Bininj Gun-wok* (an aboriginal language in northern Australia)

abanyawoihwarrgahmarneganjginjeng

corresponds to *I cooked the wrong meat for them again*, [13].

[245] There are about 100 basic units on a page, 10^4–10^5 of such units make a book, a 10 000 book library comprises $\approx 3 \cdot 10^9$ units, and the world wide web may contain up to 10^{12} basic units of the English language.

But we shall be concerned at this point with the following more apparent combinatorial category-like structure that is universally seen in all kinds of "flows of signals".

Let a *library* L in, say English, language $\mathcal{L}$ be represented by a collection of tapes with strings s of symbols, e.g., letters or words, written on them, where many different tapes may carry "identical" or better to say *isomorphic* strings with the notation $s_1 \simeq s_2$, with the equality notation $s_1 = s_2$ reserved for the same string in the same location on the same tape.

Let arrows $s_1 \hookrightarrow s_2$ correspond to *presyntactic insertions* between strings, i.e., where such an arrow associates a substring $s_1' \subset s_2$ to s_1, where $s_1' \simeq s_1$.

We assume our strings are relatively short, no more than 10–20 words of length: This is sufficient for describing any "library" since every 10-word-long string *uniquely* (with negligibly rare exceptions) extends (if at all) to longer strings, since the total number of strings in any language is well below $100^{10} << n^{10}$ for n being the number of symbol-words in a language.[246] As for L one might think of something with the number N of words in it in the range 10^6–10^{12}.

The resulting **category** $\mathcal{C}_{\hookrightarrow} = \mathcal{C}_{\hookrightarrow}(L)$ *carries full information about* L.

Justification

[+] *Invariance.* $\mathcal{C}_{\hookrightarrow}$ is invariant under changes of "alphabets" – names of the symbols.

[++] *Universality and Robustness* The categorical description of languages satisfies the most essential ergo-requirement that is UNIVERSALITY.

For instance, *spoken languages* can be similarly described in categorical terms, where, unlike written languages, the arrows must correspond to *approximate* insertion relations between auditory or visual patterns.

In fact, allowing *approximate* presyntactic insertions with *sequence alignments* (with a margin of error 5–10%) in place of syntactic isomorphisms between strings would enhance the *robustness* of categorical descriptions of *written* languages as well.

Features

[*] *Non-locality.* The $\mathcal{C}_{\hookrightarrow}$-description of libraries depends on comparison between strings that may be positioned mutually far away from each other in texts.

[**] *Long-Term Memory.* This comparison between strings, depends on the presence of a structurally organized, albeit in a simple way, *memory* within the learning program.[247]

---

[246] Never mind the saying: "*there are infinitely many possible sentences in a natural language*".

[247] Conceivably, this organisation corresponds to how languages are perceived by their principal learners (1–4 year old children), where the $\mathcal{C}_{\hookrightarrow}$-categorical organisation of memory is the "ground level" of what we call "understanding" of $\mathcal{L}$.

Redundancy and excessive local complexity of $\mathcal{C}_{\hookrightarrow}$

[−] The full category $\mathcal{C}_{\hookrightarrow}(L)$ contains many "insignificant" arrows, e.g., insertions of *single letters* into ten-word sentences and arrows between "non-linguistic" strings, such as "tic stri".

This can be corrected by

allowing only **textual units** *for objects in* $\mathcal{C}_{\hookrightarrow}$

and by

selecting a **representative subdiagram** $\mathcal{D}_{\hookrightarrow} \subset \mathcal{C}_{\hookrightarrow}$.

Such a diagram $\mathcal{D}_{\hookrightarrow}$ (that is a *network* of directed *arrow-edges* between *strings for vertices*) *must generate* (most of?) $\mathcal{C}_{\hookrightarrow}$ as a monoid, and also it must be "small", e.g., being a *minimal* subdiagram generating $\mathcal{C}_{\hookrightarrow}$.

(There is no apparent natural or canonical choice of $\mathcal{D}_{\hookrightarrow} \subset \mathcal{C}_{\hookrightarrow}$, but it may depend on the order in which the learner encounters texts in the library.)

[−+] *Pruning and Structuralizing* $\mathcal{D}_{\hookrightarrow}$. No matter how you choose $\mathcal{D}_{\hookrightarrow}$ it has *too many arrows* issuing from certain (relatively short) strings s, where the number of such arrows grows with the size of a library. Thus, in order to comply with the principles of ergo-logic, our learning algorithms must automatically reorganize $\mathcal{D}_{\hookrightarrow}$ in order to correct for this excessive branching. This is achieved by operations of *reduction*[248] applied to (the sets of) strings and arrows.

Categories $\mathcal{C}^{\hookrightarrow\downarrow}$ and Diagrams $\mathcal{D}^{\hookrightarrow\downarrow}$ of Annotated Texts

If the texts in a library are annotated with tag-strings s' that are written on several levels over original strings s, then the category with "horizontal" arrows $s'_1 \to s'_2$ is augmented by the "vertical" *position arrows* $s'' \downarrow s'$ saying that s'' lies over s', where such "mixed categories" and their representative subdiagrams are denoted by $\mathcal{C}^{\hookrightarrow\downarrow}$ and $\mathcal{D}^{\hookrightarrow\downarrow}$.

The presence of vertical arrows serves two purposes.

[1] Vertical arrows significantly *increase the connectivity* of diagrams since *a bound on the number* of tag-words on the *high levels* of annotations yields the existence of *many* horizontal (syntactic insertion) arrows between strings on these levels that were not present on the lower levels.

[2] And

the notion of a representative diagram $\mathcal{D}^{\hookrightarrow\downarrow}$
is modified in the presence of vertical arrows

by replacing many horizontal arrows issuing from lower-level strings in an annotated text by the corresponding arrows on the higher levels where the "low-level information" is encoded by (inverted) vertical arrows. Thus one (partly) compensates for the excessive branching of $\mathcal{D}_{\hookrightarrow}$.

---

[248]This is also called *clusterisation, classification, categorisation, factorisation.*

Arrow of Time. The category $\mathcal{C}_{\hookrightarrow}$ is unaware of the time direction in linguistic strings, but this direction in (written or sound recorded) strings can be, probably, reconstructed by the predominantly *backward orientations* of self-references in texts.

Structures in Symbols. In our categorical description (the alphabet of) the basic symbols, say letters, carry no internal structure of their own. But in reality letters in alphabets are non-trivially structured in agreement with one of the ergo-logic principles that allows no unstructured set of objects with more than three–four members in it. I am not certain what one should do about it.

Dimension of Vision. Visual signals[249] are customarily recorded on *two-dimensional* backgrounds such as on photographs and/or eye retinas, where the extra dimensions of depth and of time (in moving pictures) carry only auxiliary information. The morphisms $s_1 \hookrightarrow s_2$ here correspond to similarities between visual patterns s_1 and subpatterns in s_2.

But, probably, a significant part of visual perception is *one-dimensional* being implemented/encoded by the neurobiology of *saccadic eye movements*. This suggests unified algorithms for learning to see and for learning to speak.

Syntactic from Presyntactic

Eventually, we isolate strings (sometimes pairs of strings) that serve as *textual units* and also we identify *significant insertions* between them that we call *syntactic insertions.*

Linguistic 2-Spaces $\mathcal{P}_{\hookrightarrow} = \mathcal{P}_{\hookrightarrow}(L)$ and $\mathcal{P}^{\hookrightarrow\downarrow}$

Let us represent strings from a given library L by line segments of lengths equal to the numbers of letters in them. Attach rectangular 2-cells to the disjoint union of all these strings, where these "rectangles" are Cartesian products $s \times [0,1]$, with s being some strings/segments of length ≥ 5 letters each and where the attachment maps are syntactic insertions from the segments $s \times 0$ and $s \times 1$ to some string segments S_0 and S_1, such that the images are *maximal mutually isomorphic* (i.e., composed of the same letters) substrings in S_0 and S_1.[250]

In fact, it is more instructive to use the maps corresponding *not to all* syntactic insertions in the category $\mathcal{C}_{\hookrightarrow} = \mathcal{C}_{\hookrightarrow}(L)$ but only to those from a *minimal* diagram $\mathcal{D}_{\hookrightarrow} \subset \mathcal{C}_{\hookrightarrow}$, that *generates* all morphisms from $\mathcal{C}_{\hookrightarrow}$ on strings of length≥ 5.

Then the resulting two-dimensional cubical (rectangular) polyhedron $\mathcal{P}_{\hookrightarrow} = \mathcal{P}_{\hookrightarrow}(L)$ adequately encodes the library L and if L is sufficiently large, this $\mathcal{P}_{\hookrightarrow}$

[249]There is a demarcation line separating visual structures of *Life* (plants, animals, humans, human artefacts) from those of *non-Life* (stretches of water, rocks, mountains). These two classes of images are, possibly, treated differently by the visual system.

[250]Our ad hoc bound *length* ≥ 5 serves to eliminate/minimise the role of "meaninglessly isomorphic" substrings, (e.g., of individual letters) where the same purpose may be implemented by a natural constraint on strings and gluing maps.

carries all structure knowledge of the corresponding language $\mathcal{L}$ with segmentation into basic units – words and short phrases made visible.

If one deals with the category $\mathcal{C}^{\hookrightarrow\downarrow}$ corresponding to an *annotated* library, or with a subdiagram $\mathcal{D}^{\hookrightarrow\downarrow} \subset \mathcal{C}^{\hookrightarrow\downarrow}$, then one attaches "vertical" rectangles along with "horizontal" ones, where the horizontal rectangles are associated to the arrows $s'_1 \hookrightarrow s'_2$ and the vertical ones to the arrows $s'' \downarrow s'$.

Branching Entropy

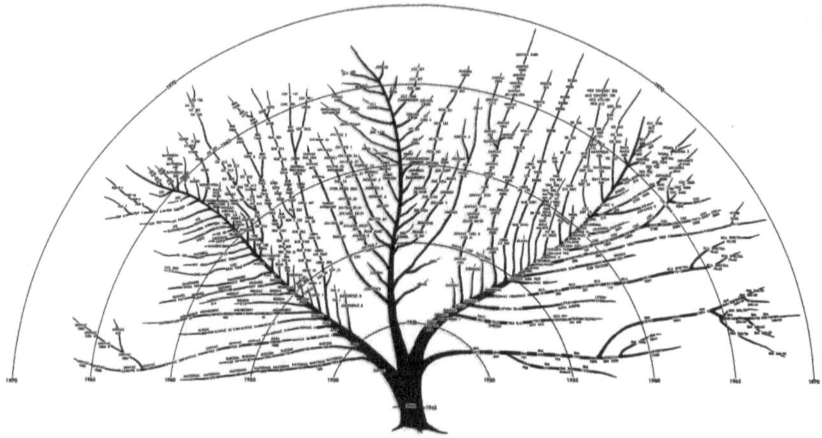

Extensions of a *string-unit*, s, e.g., of a word, by short units t following next after s in a library L define a *probability measure* on these t for

$$p_{\vec{s}}(t) = p_{L,\vec{s}}(t) = N_L(st)/N_L(s),$$

where $N_L(s)$ and $N_L(st)$ denote the numbers of occurrences of the strings s and of st respectively in L.

The collection of numbers $\{p_s(t)\}$ indexed by t serves as an indicator of variability of usage of s in the texts in the library L, where it seems reasonable to use not all t but only a collection T of unit-strings (words) t corresponding to the roughly 10 (it may be something between 3 and 50, I guess, that need be determined experimentally) *largest numbers* among $p_{\vec{s}}(t)$.

The standard invariant of the probability space $\{p_{\vec{s}}(t)\}$ that reflects variability of p and regarded as an invariant of s is the (*one step forward*) *entropy*

$$\overrightarrow{\text{ent}}(s;L) = -\sum_{t\in T} p_{\vec{s}}(t)\log p_{\vec{s}}(t).$$

Similarly, one defines $\overleftarrow{\text{ent}}(s; L)$ via left extensions ts of s as well as the corresponding invariants reflecting relative frequencies of "double extensions" of s that are t_1t_2s, t_1st_2 and st_1t_2.

Such entropies apparently are quite different for the strings "*birds-fly*" and "*pigs-fly*"[251], but "*bats-fly*" will be close to "*birds-fly*" in this respect. (This is more pronounced for extensions not of the strings themselves but of their "syntactic variations".)

32. Similarities and Classifications, Trees and Coordinatizations

Many (most?) linguistic "units", are classes of other units. For instance, *words* are, in reality, *equivalence classes of strings containing these* words, rather than mere "spell-strings". For instance, the two collections of strings

[**bats-eat**]: *bat-with-flapping- ..., bats-from- ..., bats-are-present- ...,*
vampire-bat, bats-catch- ..., inoculation-of-bats,
bat-captured ...

[**bat-hits**]: *training-bat, used-bats-on-sale, made their own bats,*
increase-your-bat- ..., throws-his-bat, ... -bats-per-game,
raised-a-bat ...

represent two different "bat" class-words.[252]

Classifications are often (but not always) achieved by means of *similarity* and/or *equivalence* relations R that, besides *similarity* and *equivalence*, reflect the ideas of

"sameness", "identity", "equality", "isomorphism", "analogy",
"closeness", "resemblance",

where such relations R are regarded as *higher-order units* and are themselves subjects to further classification. For instance, the similarity $\bullet \sim_1 \circ$ is different from $\bullet \sim_2 \blacksquare$ as well as from $\triangle \sim_3 \square$, where $\sim_2$ and $\sim_3$ are similar themselves and dissimilar from $\sim_1$.

Not all similarities lead to what may be called "true classification", partly because the "equivalence axiom" $A \sim A$ is not satisfied in ergo-logic. Indeed you would become mad if you fill your brain with $A \sim A$ for all A in your head.

Also some similarity concepts are applicable only to small groups of objects, such as what brings together

$$\{\textit{sweet, bitter, salty, sour, tangy}\},$$

and that do not extend to a majority of words.

---

[251] The two strings have comparable frequencies on Google.

[252] Non-existence of the string "*bats eat and hit*" shows how far apart the two classes are but ambiguous strings such as "*hit by a flying bat*" effectuate "quasi-poetic bridges" between the two classes.

On the other hand, there are more – about a dozen – different class-words spelled "bat", that are, essentially, subclasses of [**bat-hits**].

Another kind of groups of words having much in common that may or may not be regarded as true classes are those of *morphological word forms* such as

$$\{works, worked, working\} \quad \text{or} \quad \{white, whiteness, whiten\}.$$

On the other hand, traditional parts of speech (*verb, noun, adjective*, ... , etc.) represent typical *classes of words*; also division of words into "common" and "rare" is essential despite being ambiguous.

The two common classification structures are as follows.

1: Classification as a Tree. This may be seen as a sequence of partitions of word units into finer (smaller) classes, where the rule defining each following partition depends on the previous (coarser) one.

A linguistically rather artificial instance of this, is classification/positioning of words in alphabetically organized dictionaries.

A more significant example is where the first partition divides words into the two classes **A** and **B**:

A. Class of content words: {*nouns*, (most) *verbs*, *adjectives*, *adverbs*},

B. Class of function words: {*articles, pronouns, prepositions, etc.*}.

The classes of the second partition are obtained by subdividing words into "parts of speech".

Then the third partition may divide content words according to their "overall meanings", e.g., nouns according to whether they represent "physical objects" or "abstract concepts", etc.

2: Classifications by Coordinates.[253] These are given by several *coordinates* that are functions on objects we try to classify, where definition/determination (but not the value!) of each coordinate does not depend on the rest of the other coordinates.

The classes are formed by assigning particular values to some coordinates.

For instance, one may have the following functions $\mathbf{c}_1$, $\mathbf{c}_2$, $\mathbf{c}_3$, $\mathbf{c}_4$ on phrase-units u.

- $\mathbf{c}_1(u)$ takes values *long*, *medium*, *short*, depending on whether u has at most 4, between 5 and 8, or more than 8 word-units in it.
- $\mathbf{c}_2(u)$ takes values *yes* or *no* depending if u contains a *content verb* in it.
- $\mathbf{c}_3(u)$ assigns the *key word* w in u to u.
- $\mathbf{c}_4(u)$ is the expected age group (3–6, 7–11, 12–...) of a child who is able to understand the phrase u.

In general, coordinatization "imbeds" a potentially large set U of units to the *coordinate space* that is the Cartesian product of several small sets.

Our goal is formulating *universal* classification rules *a priori, applicable to all kinds of strings*, as well as differently-structured signals that would be as good, eventually better, than classifications based on "meaning".

[253] The kind of coordinates we present below are called *features* in the machine learning community.

33. Clustering, Biclustering, and Coclustering

There can be no isolated sign. Moreover, signs require at least two Quasi-minds.[254]

CHARLES SANDERS PEIRCE

Suppose certain pairs of points in a set V of "units" are connected by edges, graphically ∘—•, that represents a certain *resemblance* between these points regarded as vertices of a graph on the vertex set V.

Simple clustering of such a V is a partition of V into subsets, called *clusters*, such that (the numbers of) the connections between members of clusters are stronger (more numerous) than interconnection between different clusters.

The archetypical clustering is division of a graph into its *connected components*, but, in general, there is no mathematical clustering recipe applicable to all graphs – after all, many graphs are non-clusterable.

On the other hand, many "resemblance graphs" in life admit more or less non-ambiguous dissection into clusters.

Simple clustering often applies to sets V with a distance (or distance like) function $d(v_1, v_2)$, e.g., to a subset in the n-space with the *ordinary Euclidean* metric.[255]

Biclustering is more interesting than simple clustering. For instance, imagine a language that admits a simple general definition of *word-unit* and where there is a universal rule for identification of *word boundaries*. (In real life defining what is a word and devising an algorithm for identifying them in a flow of signals is by no means easy.)

Biclustering is a classification of words according to their *functions*: Two words w_1 and w_2 are regarded as *functionally similar* if the other words with which they systematically "cooperate" themselves tend to be similar.

---

[254]Mystifying but inspiring.

[255]This allows one to blissfully use precooked formulas from the book.

The condition

w_1 is similar to w_2 if coworkers of w_1 are often similar to coworkers of w_2

may strike you as being circular. But then you rewrite it as

w_1 is similar to w_2 if coworkers of w_1 are often similar to coworkers of w_2.

Now, it points toward an iterative process (algorithm) that transforms a preliminary clusterization to a more advanced one.

It is often difficult to define and/or to identify *togetherness* of "doing something" for pairs (or larger groups) of words, but it is relatively easy to decide, without any reference to "meaning" or "function" whether two given words, say w_1 and w_2, *often* come *close together*, or, on the contrary they come close relatively rarely.[256]

Both, "*close together*" and "*often*" are variable, where the latter must be adjusted to the former: What is regarded *often* for coming next to each other will be considered rare for the simultaneous presence of these words on the same page in a text.

Granted a specification of "*close*" and "*often*" one arrives at what is called the *co-occurrence graph* G on the set W of words,[257] where w_1 is joined with w_2 by an edge if the two "*often come close together*".[258]

The remarkable fact is that such graphs, if they come from "real life", have huge redundancy in them – they are very far from anything "random".

More specifically, such graphs G typically admit *approximate reductions* to certain much smaller graphs $\underline{G}$.

On terminology

Classification is our preferred word for what is called *categorization* in linguistics and in psychology: The division of "objects" into classes.

A classification achieved by some reduction $G \to \underline{G}$ is called *co-clustering* in linguistics and *bi-clustering* as well as *two-mode clustering* in data mining and in bioinformatics, where one says *clusters* rather than of "classes".[259]

We use the term "coclustering" where functional cooperation may involve more than two units and reserve "biclustering" for the above case of binary cooperation.

---

[256] This pre-assumes that we know what it means to be "*same*" for words positioned at *different* locations in flows of speech or in written texts.

[257] We assume here that words constitute *sets*.

[258] Since we are now firmly on a mathematical ground, ambiguity of "*often come close together*", or vagueness of terminology on a preliminary stage of exposition, in general, poses no danger. Firstly, because it is *not disguised* as being something precise, and secondly because mathematics promotes crystallization of valid ideas into logically solid ones.

This is different from what happens to almost all speculative considerations fulled by "raw intuition" unaided by mathematics.

[259] This kind of analysis, probably, has been used in other branches of science/statistics under different names that makes it hard to find out when and by whom this idea was originally introduced. Not impossibly, this was understood and implicitly used by Aristotle.

Biclustering applies not so much to graphs but rather to *functions* in two variables, $G(u,v)$, where the domains U of u and V of v do not have to be equal. (Graphs are represented by 2-valued *edge/no-edge*, or $\{0,1\}$ for brevity's sake, functions.) Namely,

⋆ *reduction of* $G(u,v)$ to a function $\underline{G} = \underline{G}(\underline{u},\underline{v})$ that is defined on a pair of smaller, often significantly smaller, sets $\underline{U}$ and $\underline{V}$, is a pair of maps from U onto $\underline{U}$ and from V onto $\underline{V}$, say $P^{\downharpoonleft} : U \to \underline{U}$ and $Q^{\downharpoonleft} : V \to \underline{V}$, written as

$$u, v \rightrightarrows \underline{u} = P^{\downharpoonleft}(u), \underline{v} = Q^{\downharpoonleft}(v),$$

such that the composition, sometimes called *superposition*, $G^{\downharpoonleft}(u,v)$, of the functions $\underline{G}$ and $R^{\downharpoonleft}$, that is

$$G^{\downharpoonleft}(u,v) = \underline{G}(\underline{u},\underline{v}) = \underline{G} \circ P^{\downharpoonleft} \& Q^{\downharpoonleft}(u,v) = \underline{G}(P^{\downharpoonleft}(u), Q^{\downharpoonleft}(v))$$

provides a "*good approximation*" to the function $G(u,v)$.

(If $U = V$ and $G(u,v) = G(v,u)$, then one may take $\underline{U} = \underline{V}$ and $P^{\downharpoonleft} = Q^{\downharpoonleft}$. But if $G(u,v) \neq G(v,u)$ then $\underline{U}$ and $P^{\downharpoonleft}$ are not necessarily equal to $\underline{V}$ with $Q^{\downharpoonleft}$ even if $U = V$.)

⋆ *Clusters* (classes) are the subsets in U and in V corresponding to $\underline{u}$ and $\underline{v}$ via $u, v \rightrightarrows \underline{u}, \underline{v}$, i.e., the subsets of those u in U and v in V for which $P^{\downharpoonleft}(u) = \underline{u}$ and $Q^{\downharpoonleft}(v) = \underline{v}$.

"*Approximation*" here is different from the above positional closeness of words. It depends on what kind of function G is and where its values lie. In our examples, besides being a two-valued function it may be *three-valued*, saying whether u and v cooperate *strongly, weakly*, or *not at all*. Also it may be a *number-valued* function with $G(u,v)$ being the relative frequency of co-occurrence of u and v.

In either case, the range of the function, denote this range by I, must be equipped with a *metric* measuring the distances between different values.

If I equals a set of positive numbers, then one takes the absolute value $|i_1 - i_2|$ for such a distance, and if I is an "abstract" two or three point set, one may implement these by numbers, say by $\{0,1\}$ and $\{0,1,2\}$, and use the distance $|i_1 - i_2|$ again.

And then *closeness* may be defined relative to the *Hamming distance* (also called l_1-*distance*)

$$\mathrm{dist}(G, G^{\downharpoonleft}) = \sum_{u \in U, v \in V} |G(u,v) - G^{\downharpoonleft}(u,v))|$$

To see how this works, let

- $U = V$ be a set that is comprised of 100 000 words.
- u and v in W be regarded as "*cofunctional*" if v goes *right after* u.
- "*often*" means "*at least ten times*".

The function G corresponding to this can be seen as a (non-symmetric) $100\,000 \times 100\,000$ matrix with $\{$*often, rare*$\}$ entries. A reliable evaluation of the values $G(u,v)$, that are the entries of this matrix, needs a library of more than $10^{11} = 10 \cdot 10^{10}$ words in it.[260]

But it may happen, and it does often (albeit approximately) happen in "real life", that this huge matrix is (approximately) determined by something much smaller, say by a 300×300 matrix, where you need only $90\,000 < 10^5$ entries to fill in a description of which needs only

$$90\,000 + (2\log_2 300) \cdot 10^5 < 2 \cdot 10^6 \text{ bits}$$

instead of the original 10^{10} bits.

And biclusterisation serves for achieving such simplification by reduction of your big matrix/function G to an $\underline{G}(\underline{u}, \underline{v})$ that is defined on a set $\underline{U} \times \underline{V}$ with $300 \times 300 = 90\,000$ elements in it.

Quasi-Uniqueness of $\underline{G}$ and $P^{\downharpoonleft}\&Q^{\downharpoonleft}$

The existence of a reduction $G \rightsquigarrow \underline{G}$ is, a priori, *extremely unlikely* even if we do not require $G^{\downharpoonleft}(u,v) = \underline{G} \circ P^{\downharpoonleft}\&Q^{\downharpoonleft}(u,v)$ to be an especially fine approximation of $G(u,v)$.

Therefore, even if a certain $G^{\downharpoonleft}(u,v)$ delivers only a rough approximation to $G(u,v)$, the corresponding $\underline{G}(\underline{u},\underline{v})$, $P^{\downharpoonleft} : U \to \underline{U}$ and $Q^{\downharpoonleft} : V \to \underline{V}$ will be (essentially) *unique with an overwhelming probability.*

But if the sets $\underline{U}$ and $\underline{V}$ are small, say consisting of 2–4 elements, then there may be several candidates competing for the roles of $\underline{G}$ and $P^{\downharpoonleft}\&Q^{\downharpoonleft}$ and one has to select the "best ones". A preferred choice here is where the function $\underline{G}$ is the farthest from the most probable one. (This may be sometimes formulated as *minimization of a kind of an entropy.*)

Completion/Extrapolation of the Matrix G

You easily keep in you memory 100 000 words and you (subliminally) remember a few million occurrences of some pairs of them coming close together.[261] But, certainly, this never comes anywhere close to 10 000 000 000 – the number of entries in the matrix G: Most of G in your head is filled with question marks.

On the other hand, a couple of million examples is not so little when it comes to the matrix $\underline{G}$ that has only 90 000 entries, and if you have these, you replace $G(u,v) =?$ by $G(u,v) = G^{\downharpoonleft}(u,v) = \underline{G}(\underline{u},\underline{v})$.

As a result, when necessity arrives, you will not hesitate to accept or to reject as implausible pairs of relatively rare words coming next to each other, such as "intellectually posterior", "hydraulically superior", "superior posterior",

[260] If you check one pair of words per second – eight hours a day, five days a week – it will take more than 10 000 years to go through such a library.

[261] An average book contains about 100 000 words, and there are other sources of words besides books.

"intellectually hydraulically", "corticofugally inhibited", "intellectually candied", etc.[262]

Additive and Probabilistic Biclustering

Let all u in a set U carry *weights*, that are positive numbers denoted $|u|$. Then an *additive reduction* of such a *weighted space* is a map that adds up the weights. Namely, if, say, $P^{\downarrow}$ maps U onto some $\underline{U}$,

$$U \overset{P^{\downarrow}}{\to} \underline{U},$$

then this $\underline{U}$ is endowed with weights $|\underline{u}|$, that are sums of the weights of all u that go to $|\underline{u}|$,

$$|\underline{u}| = \sum_{P^{\downarrow}(u)=\underline{u}} |u|.$$

Similarly if the entries of a matrix (function) $G = G(u,v)$ are positive numbers, then the *additive reduction* $\underline{G}$ of G under

$$U \overset{P^{\downarrow}}{\to} \underline{U}, \quad \text{and} \quad V \overset{Q^{\downarrow}}{\to} \underline{V}$$

is where the weights of the entries from G add up to the corresponding entries in $\underline{G}$,

$$|G(\underline{u},\underline{v})| = \sum_{\substack{P^{\downarrow}(u)=\underline{v},\\ Q^{\downarrow}(v)=\underline{v}}} |G(u,v)|.$$

Normalization and Probability. A positive weight function on a set U, $v \mapsto |v|$ is called *normalized* if these weights add up to one,

$$\sum_{u \in U} |u| = 1.$$

Then these weights $p_u = |u|$ are interpreted as "*probabilities of the events* u".

In general, e.g., when the weights represent frequencies of occurrence of u, one normalizes by setting

$$|u|_{\text{prob}} = \frac{|u|_{\text{freq}}}{\sum_{u \in U} |u|_{\text{freq}}},$$

thus turning $(U, |\ldots|_{\text{freq}})$ to a *probability space* $(U, |\ldots|_{\text{prob}})$.

Clusterisation of Letters: **vowels & consonants**

Additive probabilistic clustering is practical for small sets W, e.g., where A equals the set of 26 letters in the English alphabet, where the roughest such clustering divides the alphabet A into two classes, denote them $\circ$ and $\bullet$, according to relative frequencies of pairs of letters in English.

[262] In truth, you need several coclustering mechanisms working in parallel, e.g., to reject offhandedly "clusterizations scries" and the like.

The weights $|a|$ and $G(a,b)$ we use here are normalized frequencies of a and ab in a given text and we search for the (mathematically most natural) reduction, where the four weights

$$\begin{matrix} |\underline{G}(\circ,\circ)| & |\underline{G}(\circ,\bullet)| \\ |\underline{G}(\bullet,\circ)| & |\underline{G}(\bullet,\bullet)| \end{matrix}$$

of the entries of the 2×2 matrix $\underline{G}$ have *minimal relative entropy* with respect to the matrix of the products of the weights in $\underline{A}=\{\circ,\bullet\}$ that are

$$\begin{matrix} |\circ|\cdot|\circ| & |\circ|\cdot|\bullet| \\ |\bullet|\cdot|\circ| & |\bullet|\cdot|\bullet|. \end{matrix}$$

Most likely – this is certainly known, but I did not check it – this minimal entropy division of A into the classes $A_\circ$ and $A_\bullet$ coincides for the most part with the division of letters into vowels and consonants.

Graphically, G (approximately) "reduces" to the two vertex graph $\circ$—$\bullet$ by dividing the vertex set A into two classes/clusters

$$A = \textbf{vowels} \,\&\, \textbf{consonants}$$

$$\overset{\textbf{vo}}{\circ} \text{—} \overset{\textbf{co}}{\bullet}.$$

Observe that this partition of A does not depend on any a priori knowledge of the "nature" of letters, but only on the relative frequencies of letters and pairs; the idea of *meaning* we attribute to these classes is not the source, but *a product* of the process of mathematical structuralisation where multiple biclusterisation plays an essential role.[263]

Geometrization of G

To be specific, let a function $G(u,v)$ in two variables take values in the set $\{0,1,2\}$ (standing for *nothing*, *something*, *much*) and let

$$U \overset{\mathcal{G}}{\to} \{0,1,2\}^V \text{ defined as } u \overset{\mathcal{G}}{\mapsto} f(v) = g_u(v) = G(u,v)$$

be the tautological map from the domain U of u to the space of $\{0,1,2\}$-valued functions $f(v)$ on the domain V of v, that is the Cartesian product space of copies of $\{0,1,2\}$ indexed by v from V,

$$\{0,1,2\}^V = \underbrace{\{0,1,2\}\times\{0,1,2\}\times\cdots\times\{0,1,2\}}_{V}.$$

For instance, let U be the set of 100 000 words and V a particular subset of say, 30–100 words selected by some preliminary mathematically defined process. For instance, these may be

- 100 most frequent words,

[263] A phonetically more accurate clustering needs tracking *triples* (quadruples?) of letters that will allow distinguishing certain pairs, such as "th" and "wh", for instance.

or, more interestingly,

★ 100 most frequent words from some class obtained by another biclustering algorithm,

such as

- ○ a representative group of function words,
- ∗ 100 most common verbs
- ✤ a list of 30 common four-legged animals
- ☞ a list of 30 common professions.

The space $\{0,1,2\}^V$ comes with many distance-like functions where a preferred one is the Hamming distance

$$\mathrm{dist}_V(f_1(v), f_2(v)) = \sum_{v \in V} |f_1(v) - f_2(v)|,$$

that passes to U via $\mathcal{G}$.

Then (bi)clustering of U according to possible coworkers v of u may be achieved by simple clustering relative to such a "distance" on U.

Reciprocal Clustering

If you choose a subset V of words in U at random, then there will be no preferred clustering of U for the distance coming from $dist_V$. On the other hand some exceptional V would lead to "clean" clusterings of U. Such special V are tight knit groups of similar words, such as, for instance: {*the*, *a*} or

{**Red**, **Green**, **Yellow**, **Orange**, **Purple**, **Pink**, **Brown**, **Black**, **Gray**, **White**}.

Combinatorial Clustering: Why Pigs do not Fly

Let G be a graph on the vertex set V. Then the vertices v of this graph can be classified/clustered according to the combinatorics of subgraphs comprised of vertices and edges in the vicinity of v, where the simplest characteristic of such "vicinity" is the *valency of* v, that is, the number of edges attached to v.

Thus, for instance, one may first divide V into two parts V_{small}, where this valency is small, and V_{large}, where it is large, and then subdivide further according to the values of *pairs of numbers* of edges from v to V_{small} and to V_{large}.

Pigs, Pigeons, and Sparrows. "Birds fly" on 14 000 000 Google pages, "*Pigs fly*" on 3 000 000 pages, "*pigeons fly*" on 1 000 000 pages, and "sparrows fly" on 500 000 pages.

What distinguishes pigs from birds is not the sheer numbers of occurrences of sentences with "pig&fly" or "bird&fly" in them, but the numbers of combinatorial structures displayed by these sentences: There are by far less types of sentences with the words "*pig*" and "*fly*" in them, than of those with "*pigeon*" and "*sparrow*" replacing "*pig*".

This is similar to the use of prepositions in English, e.g., *under* and *in*, that may be accompanied by *different* kinds of nouns and/or verbs; yet the geometries/combinatorics of their vicinities in the "*network of short English sentences*" look, nevertheless, quite *similar* that brings all(?) prepositions to the same cluster.

On Biclustering Algorithms

Detection of clusters of "natural units" u in a U, e.g., of words, may not need the full knowledge of $G(u,v)$ at all (u,v) but only for v taken from special *small subsets* in V. For instance, "*the*" divides other words into two groups according to their systematic occurrence just before or just after "*the*".

This does not work for general G, and biclustering of sets, say, of cardinalities 100 000, associated to reductions $U \to \underline{U}$ and $V \to \underline{V}$ to sets $\underline{U}$ and $\underline{V}$ of cardinalities of order 300 seems, generically speaking, computationally unfeasible.

On the other hand there is a variety of heuristic algorithms that work pretty well for functions G coming from "life".

Words in Contexts: Biclustering and Triclustering

Biclustering may be applied to the function $G(w,x)$ where w are words and the variable x represents a context, e.g., a book from some collection X.

The natural function G encodes (frequent) presence/absence of a w in x and biclusterisation serves to classify books by topics according to their "key words", but the words themselves become classified by topics they are frequently used in, such as: *chemistry of plants, animal foods*, etc.

Structurally more informative classification, e.g., with an organisation of classes as *trees* with several branches, may be achieved with tri-clusterisation for $G(w_1,w_2,x)$ recording pairs of words (w_1,w_2) that appear in the same book x.

In general, however, it is unclear how to proceed with tri-clustering, partly because there is no convincing counterpart to the above "geometrization" $\mathcal{G}$ of G. On the other hand, most(?) multiple interactions appear as "combinations" of binary ones and multi-clustering reduces to several biclusterings.

Conclusion

Co-clustering is neither the final product of building a structure from "flows of words" nor is it an "atomic unit" of such a structure but rather a large molecule with simple, yet non-trivial, internal architecture where this molecule, in turn, serves as a building block for more elaborate syntactic structures.

The simplicity of this "mathematical molecule" makes it quite versatile: One can modify it in many ways and adjust it to building a variety of different global structures.

For instance: Reductions of $U\&V$ that lead to (approximate) clusterisations of U and/or V are (not always) composable and combinatorics of systems of (not quite) commutative diagrams of reductions represents an interesting "higher-order structure" in $U\&V$.

And besides mere classification a more subtle structure of a language may be extracted from a "distance" on V induced by the above $\mathcal{G}$ from a space of functions on some auxiliary set V, where the essential properties of such a "distance" are encoded by the (not-quite) *category* of approximate partial isometries of U with respect to this "distance".

One may continue indefinitely along these lines but one has to stop somewhere. Wings of imagination supplied by the power of mathematics can bring you beyond whatever can be reached by a more pedestrian kind of thinking. But if you fly too high in the sky of math you may miss your destination down on Earth.

References

[1] Bhaskar Acharya, The Ecology of the Dhole or Asiatic Wild Dog (Cuon alpinus) in Pench Tiger Reserve, Madhya Pradesh.
Dissertation submitted to Saurashtra University, Rajkot, Gujarat, for the award of the Degree of Doctor of Philosophy in Wildlife Science. Submitted by Balebail Bhaskar Acharya under the supervision of Dr. A.J.T. Johnsingh. Wildlife Institute of India, Dehradun, January 2007
https://www.researchgate.net/profile/Bhaskar_Acharya/publication/230582544_The_Ecology_of_the_Dhole_or_Asiatic_Wild_Dog_Cuon_alpinus_In_Pench_Tiger_Reserve_Madhya_Pradesh/links/09e41502baab122de7000000/The-Ecology-of-the-Dhole-or-Asiatic-Wild-Dog-Cuon-alpinus-In-Pench-Tiger-Reserve-Madhya-Pradesh.pdf

[2] When Brain Damage Unlocks The Genius Within. Brain damage has unleashed extraordinary talents in a small group of otherwise ordinary individuals. Will science find a way for everyone to tap their inner virtuoso? By Adam Piore, Popular Science, February 19, 2013 https://www.popsci.com/science/article/2013-02/when-brain-damage-unlocks-genius-within

[3] E. Anderesen, A. Dunlea, The impact of input: language acquisition in the visually impaired, First Language, 13 (1993) pp. 23–49.
http://fla.sagepub.com/content/13/37/23.full.pdf

[4] Inge Auerbacher, Finding Dr. Schatz: *The Discovery of Streptomycin and A Life it Saved*, iUniverse, 2006

[5] William H. Calvin, "A Stone's Throw and its Launch Window: Timing Precision and its Implications for Language and Hominid Brains," Journal of Theoretical Biology 104, 121–135 (1983). http://WilliamCalvin.com/1980s/1983JTheoretBiol.htm.

[6] Kevin Brown, *Penicillin Man: Alexander Fleming and the Antibiotic Revolution*, Sutton Publishing Ltd, 2005

[7] Lennard Bickel, *Howard Florey: The Man Who Made Penicillin* (Australian Lives series), Sun Books, 1983.

[8] *Cephalosporins and Penicillins: Chemistry and Biology*, edited by Edwin H. Flynn, Elsevier, 1972, http://www.sciencedirect.com/science/book/9780122614507

[9] J.-P. Cheng, *The Origin of Probability and The Problem of Points*, sites.math.rutgers.edu/~cherlin/History/Papers2000/cheng.html

[10] Oliver Sacks, *A Bolt from the Blue.* The New Yorker Jul, 23, 2007, https://www.newyorker.com/magazine/2007/07/23/a-bolt-from-the-blue

[11] Heather L. Van Epps, *René Dubos: Unearthing Antibiotics*, J. Exp. Med. 2006 Feb. 20; 203(2):259. https://www.ncbi.nlm.nih.gov/pmc/articles/PMC2118194/

[12] Michael Dunnill, *The Plato of Praed Street: The Life and Times of Almroth Wright*, 2000, Royal Society of Medicine Press

[13] Nicholas Evans, Stephen Levinson, *The Myth of Language Universals.* Behavioral and Brain Sciences (2009) 32, 429–492, https://www.cambridge.org/core/services/aop-cambridge-core/content/view/S0140525X0999094X

[14] First probabilists: Galileo, Cardano, Fermat, Pascal, Huygens, http://bricol.net/downloads/misc/HistProb/02.First_Probabilists.pdf

[15] Simon E. Fisher, Gary F. Marcus, The Eloquent Ape: Genes, Brains and the Evolution of Language, Nature Reviews Genetics 7, 9–20 2006. http://www.psych.nyu.edu/gary/marcusArticles/Fisher%20Marcus%202006.pdf

[16] A Hald, A *History of Probability and Statistics and Their Applications Before* 1750, Wiley Series in Probability and Statistics, 1990.

[17] In memoriam, H.W. Florey: An Episode. J Gen Microbiol. 1970 Jun;61(3):289–299. https://www.ncbi.nlm.nih.gov/pubmed/4922763/

[18] R. Johnson and S. Liddell, *Toward a Phonetic Representation of Signs: Sequentiality and Contrast*, Sign Language Studies Volume 11, Number 2, (2010) pp. 241–274. Resume is accessible: https://eric.ed.gov/?id=EJ935239

[19] F. Lawvere and S. Schanuel, *Conceptual Mathematics*, Cambridge University Press, 1997.

[20] *The Handbook of Linguistics*, Mark Aronoff, Janie Rees-Miller editors, Blackwell Publishing, 2002–2007.

[21] G. Macfarlan, *Alexander Fleming, The Man and the Myth*, Oxford University Press, 1985.

[22] G. Macfarlan, *Howard Florey, The Making of a Great Scientist*, Oxford University Press, 1979.

[23] David Bellhouse, *Decoding Cardano's Liber de Ludo Aleae.* Historia Mathematica, Volume 32, Issue 2, May 2005, pp. 180–202, https://doi.org/10.1016/j.hm.2004.04.001

[24] Carol L. Moberg, Zanvil A. Cohn, Launching the Antibiotic Era: Personal Accounts of the Discovery and Use of the First Antibiotics.

[25] Oudeyer, P., Kaplan, F., Hafner, V.V.: Intrinsic Motivation Systems for Autonomous Mental Development. IEEE Transactions on Evolutionary Computation 11:1, (2007), http://www.pyoudeyer.com/ims.pdf

[26] Jason Padgett, *Struck by Genius: How a Brain Injury Made Me a Mathematical Marvel.* Houghton Mifflin Harcourt 2014.

[27] Sources in the History of Probability and Statistics, www.cs.xu.edu/math/Sources/

[28] Robert D. Potter, The race to get enough penicillin (1944), http://clickamericana.com/eras/1940s/the-race-to-get-enough-penicillin-1944

[29] J. Schmidthuber, Formal Theory of Fun and Intrinsic Motivation and Creativity. IEEE Transactions on Autonomous Mental Development, Volume 2, Issue 3, 2010, pp. 230–247, http://ieeexplore.ieee.org/document/5508364/

[30] Nicole Kresge, Robert D. Simoni, and Robert L. Hill, *Selman Waksman: the Father of Antibiotics*, J. Biol. Chem. 142, pp. 519–528, http://www.jbc.org/content/279/48/e7

[31] Selman Waksman and Antibiotics. National Historic Chemical Landmark Dedicated May 24, 2005, at Rutgers The State University of New Jersey. Commemorative Booklet (PDF) http://www.acs.org/content/acs/en/education/whatischemistry/landmarks/selmanwaksman.html

[32] G. Shafer and V. Vovk, The Sources of Kolmogorov's Grundbegriffe, Statist. Sci. Volume 21, Number 1 (2006), 70–98.

[33] *Teaching and Learning: International Best Practice*, edited by McInerney, D.M. and Liem, A.D., Information Age Pub. 2008

[34] Darold A. Treffert, Savant Syndrome: An Extraordinary Condition A Synopsis: Past, Present, Future, Philos Trans. R. Soc. Lond. B Biol. Sci. 2009 May 27; 364(1522):1351–1357. http://www.ncbi.nlm.nih.gov/pmc/articles/PMC2677584/

[35] *Extraordinary People: Understanding Savant Syndrome*, by Darold A. Treffert, Backinprint.com, 2006.

[36] Selman Waksman, *My Life with the Microbes.* Scientific Book Club 1958.

[37] Who Saved the Most Lives in History – Science Heroes News, https://science-heroes.com/index.php?optioncom_content&view=article&id=258&Itemid=232

[38] Wolpert, Daniel M., Diedrichsen, Jörn; Flanagan, J. Randall "Principles of senso rimotor learning", Nature Reviews Neuroscience 12, 739–751, doi:10.1038/nrn3112, http://www.nature.com/nrn/journal/v12/n12/abs/nrn3112.html

[39] Qin Zhua, Geoffrey P. Binghamb, Human readiness to throw, Evolution and Human Behavior 32 (2011) 288–293.

MIX
Papier aus verantwortungsvollen Quellen
Paper from responsible sources
FSC® C105338

If you have any concerns about our products,
you can contact us on
ProductSafety@springernature.com

In case Publisher is established outside the EU,
the EU authorized representative is:
Springer Nature Customer Service Center GmbH
Europaplatz 3, 69115 Heidelberg, Germany

Printed by Libri Plureos GmbH
in Hamburg, Germany

TEXT BOOK OF

Criminology Penology and Victimology

TEXT BOOK OF

Criminology Penology and Victimology

Rabindra K. Mohanty, Ph.D
Satyajit Mohanty, IPS

Himalaya Publishing House

MUMBAI • NEW DELHI • NAGPUR • BENGALURU • HYDERABAD • CHENNAI • PUNE • LUCKNOW • AHMEDABAD • ERNAKULAM • BHUBANESWAR • INDORE • KOLKATA • GUWAHATI

First Edition: 2012

Published by : Mrs. Meena Pandey for **Himalaya Publishing House Pvt. Ltd.**, "Ramdoot", Dr. Bhalerao Marg, Girgaon, **Mumbai - 400 004.** Phone: 022-23860170/23863863, Fax: 022-23877178 **E-mail: himpub@vsnl.com; Website: www.himpub.com**

Branch Offices :

New Delhi : "Pooja Apartments", 4-B, Murari Lal Street, Ansari Road, Darya Ganj, New Delhi - 110 002. Phone: 011-23270392, 23278631; Fax: 011-23256286

Nagpur : Kundanlal Chandak Industrial Estate, Ghat Road, Nagpur - 440 018. Phone: 0712-2738731, 3296733; Telefax: 0712-2721215

Bengaluru : No. 16/1 (Old 12/1), 1st Floor, Next to Hotel Highlands, Madhava Nagar, Race Course Road, Bengaluru - 560 001. Phone: 080-32919385; Telefax: 080-22286611

Hyderabad : No. 3-4-184, Lingampally, Besides Raghavendra Swamy Matham, Kachiguda, Hyderabad - 500 027. Phone: 040-27560041, 27550139; Mobile: 09848130433

Chennai : No. 8/2, Modley 2nd Street, Ground Floor, T. Nagar, Chennai - 600 017. Phone: 044-28144004/28144005.

Pune : First Floor, "Laksha" Apartment, No. 527, Mehunpura, Shaniwarpeth (Near Prabhat Theatre), Pune - 411 030. Phone: 020-24496323/24496333

Lucknow : Jai Baba Bhavan, Church Road, Near Manas Complex and Dr. Awasthi Clinic, Aliganj, Lucknow - 226 024 (U.P.). Phone: 0522-2339329, 4068914; Mobile: 09305302158, 09415349385, 09389593752

Ahmedabad : 114, "SHAIL", 1st Floor, Opp. Madhu Sudan House, C.G. Road, Navrang Pura, Ahmedabad - 380 009. Phone: 079-26560126; Mobile: 09327324149, 09377088847

Ernakulam : 39/104 A, Lakshmi Apartment, Karikkamuri Cross Rd., Ernakulam, Cochin - 622011, Kerala. Phone: 0484-2378012, 2378016; Mobile: 09344199799

Bhubaneswar : 5 Station Square, Bhubaneswar - 751 001 (Odisha). Phone: 0674-2532129, Mobile: 09338746007

Indore : Kesardeep Avenue Extension, 73, Narayan Bagh, Flat No. 302, IIIrd Floor, Near Humpty Dumpty School, Indore - 452 007 (M.P.). Mobile: 09301386468

Kolkata : 108/4, Beliaghata Main Road, Near ID Hospital, Opp. SBI Bank, Kolkata - 700 010, Phone: 033-32449649, Mobile: 09910440956

Guwahati : House No. 15, Behind Pragjyotish College, Near Sharma Printing Press, P.O. Bharalumukh, Guwahati - 781009, (Assam). Mobile: 09883055590, 09883055536

DTP by : Riyaz, Cuttack, Odisha

Printed at : M/s. Seven Hills Printers, Hyderabad, On behalf on HPH.

Partha Nath Mukherji

Professor Emeritus,
Institute of Social Sciences, New Delhi.
Former Director (Vice-Chancellor) of the
Tata Institute of Social Sciences, Mumbai.
Former President, Indian Sociological Society.

FOREWORD

My own familiarity with criminology is at best pedestrian. As an area of sociological interest 'crime' made its appearance in sociology through concerns with social deviance, anomie and the like. Somehow in our country it remained on the margins of the discipline. The Indian Sociological Society does have a Research Committee on *Sociology of Crime and Deviance* indicating the importance that the profession still accords to the advancement of knowledge on crime and criminality. As Mohantys perceptively point out criminology has outgrown its historical linkages with sociology, without losing the thread of continuity, and has since developed concepts and methods of thinking about crime and criminal behavior that are uniquely its own.

My own renewal of interest in criminology took place during my stint at the Tata Institute of Social Sciences (TISS), Mumbai, which perhaps has the oldest centre for the study of criminology. It was there that I first became familiar with the newly emerging field of victimology in the mid-nineties. I sincerely felt that while some excellent social work initiatives were taking place, theoretical, conceptual and methodological concerns, in comparison, were lagging behind.

None will disagree that the fast breeding crime and criminality of various hues in the country and the world at large are the most disturbing symptoms of anomie in our times. It has endangered human security to an extent that is becoming perilous for normal existence, varying in intensity from one region to the other. Rapid advances in information technology have exacerbated the trend. From drug and human trafficking, to trafficking in kidneys, to kidnappings, to sexual abuse, to cyber crimes and corruption, and a host of other crimes raise their hydra heads. At this juncture a multi-disciplinary emergency research is the urgent need of the hour. Knowledge from research and training in criminology can ill-afford to be outpaced by the devastating impact spiraling crime.

I am delighted that at this critical juncture a text book with an elaborate comprehensive coverage is now being made available. Text book writing is, contrary to popular perception, an extremely demanding back-breaking exercise. Eighteen chapters in criminology, six in penology and two in victimology deal with concepts and theories; criminology of collective deviance, including organised and white collar crimes and problems of recidivism; socially problematic crimes relating to gambling, alcoholism, prostitution, drug addiction and cyber crimes; with theories in penology pertaining to deterrence, retribution, restoration, reformation, rehabilitation, probation and parole.

Social sciences today admit of contestations over concepts and theories. Sometimes fundamental questions are raised on definitions of whether certain kinds of behavior defined as deviant or criminal, really merit such a tag. Equally, we are loaded with the Durkheimian proposition that crime is a social fact in society, that society will define certain acts as criminal so as to maintain order. This epistemological dialectic is an inevitable part of sociology of knowledge. I hope and expect that this textbook will prove its worth not only by equipping readers with the basic concepts and theories, but by stimulating discussion and debate that will not end in sterility, but lead to appropriate application for the benefit of society.

I congratulate Sri Satyajit Mohanty a conscientious officer of the Indian Police Service and Dr. Rabindra Kumar Mohanty, a professional sociologist of considerable experience and reputation for having jointly produced this textbook. Singly, this task perhaps would not have been possible.

Date: 01.12.2011

Partha Nath Mukherji

PREFACE

Criminology is an advanced, theoretical field of study. It can be defined as the study of crime, the causes of crime (etiology), the meaning of crime in terms of law and community reaction to crime. Not too long ago, criminology separated from its mother discipline, sociology, and although there are some historical continuities, it has since developed concepts and methods of thinking about crime and criminal behavior that are uniquely its own. It has expanded its scope and subject matter by including the study of the treatment of crime and criminal justice popularly known as penology. Study of victim's facilitation and victim's right has been as important as the crime perpetration. This branch of knowledge under criminology is given the name of victimology.

Study of crime, its treatment and the victim's rights are useful to different groups for different things. On a political level, it provides a measure for the success or failure of governance system and of certain criminal justice policies. On an operational level, it provides important information on how best to allocate state resources. Functioning under the constraints and handicaps of an outmoded system, police performance has undoubtedly fallen short of public expectation. From a civil society perspective, it provides indicators against which to measure criminal justice performance and therefore an important tool to hold accountability of the various role-players in the criminal justice system. At the academic front it may generate a lot of research culminating and terminating in diagnosis and prognosis.

Increasing crime, rising population, growing pressure of living accommodation, particularly in urban areas, violent outbursts in the wake of demonstrations and agitations arising from labour disputes, agrarian unrest by the rural populace, problems and difficulties of students, vulnerabilities of marginalized, politicization of criminals and criminalization of politics, the cult of extremists, enforcement of economic and social legislation and umpteen others have all added new dimensions to the study of criminology and tended to generate research much more urgently than ever before. Changes over time in the levels and patterns of crime have significant consequences that affect not only the criminal justice system but also other critical policy sectors.

Teaching criminology to the undergraduates and post-graduates on the one hand and to the officers of law enforcement on the other, have been nonetheless challenging. A comprehensive text book covering criminology, penology and victimology is yet scarce. Having a handy and comprehensive text book at hand for students is as important as listening to the lectures in at times dry class rooms. At the same time they need a book with all in one, as in the present book criminology, penology and victimology. Besides, this is a subject which addresses students belonging not only to undergradutes and postgraduates in universities and institutes of sociology, criminology, law and social work, but also the law enforcement officers trained at police, judicial and correctional academy. Yet again, criminology attracts learners from diverse academic and socio- cultural background whose learning and skill development needs vary accordingly.

Keeping the aforesaid objective in view this book entitled a text book of criminology, penology and victimology adequately show cases the necessary concepts, theories, methods, acts and rules, case laws, case studies from the perspectives of India and Overseas and most specifically the guidelines for law enforcements. Detailed references given at the end of each Chapter may help the researchers in their academic pursuits.

The book has well knit twentysix chapters. The chapter-1 goes on to discuss the history, nature, subject matter, importance and the inter-disciplinary concerns of criminology. Chapter 2 to 6 are devoted to the exclusive discussion of the concept and theories: sociological, psychological, economic and biophysical. The etiology of crime has been succinctly analyzed in Chapter-7. The themes relating to individual deviance and collective deviance have been dealt under Chapters 8 & 9 respectively. Organised crime, white collar crime and recidivism have been discussed in separate chapters from 10 to 12. Crime trends pertaining to some old and some new crime problems like gambling, prostitution, write drug addiction, alcoholism and cyber crime have been well dealt under Chapters 13 to 18.

Concepts of penology, prison system prison conditions and prison reforms have been covered under Chapters 19 & 20. Chapters 21 to 23 are devoted towards the discussion of theories of penology pertaining to deterrence, retribution, restoration, reformation, rehabilitation, probation and parole. Chapter-24 discusses the much needed concept of social defense and the approaches to the promotion of social defense. The last two chapters are devoted to the discussion of often neglected area in the text books such as victimology (Chapter-25) and victims of terrorism (Chapter-26).

In giving this book its desired shape, we have liberally borrowed from several published sources. Wikipedia resources and individual articles accessed through Google search engine have been of great help. Due recognisation has been given to all wherever they have been referred to. Unintended exclusion is humbly accepted as a pardonable mistake. In accomplishing this book our debts are many.

We are indebted to Professor Partha Nath Mukherji, Former Director, Tata Institute of Social Sciences, Mumbai and ex-President of Indian Sociological Society for very kindly writing a foreword of this book. Our special thanks are due to Sri V. Thiagarajan, IPS, ADGP, Training, Odisha and Sri A. K. Sarangi, IPS, IGP, Training, Odisha and Director, BPSPA for their encouragement and support.

Our gratitude goes to Sri Akshaya Deo, Managing Director, Odisha Criminal Reports for his valued suggestions during discussions. We appreciate the help rendered by the faculties of BPSPA, Prof. A.B. Mishra, Prof. P.K. Nandi, Sri G.B. Pati, OPS, Sri A.K. Choudhury, OPS, Sri R. K. Paikaray, OPS and Sri GPSS Nanda, Inspector of Police. The encouragements rendered by Sri Soumitra K Nanda, Steno to Director BPSPA and Sri Bigyan Mohanty, our Computer faculty are unforgettable.

We are grateful to Himalaya Publishing House, Mumbai for having accepted our proposal for publishing this important book and for having brought this out in shortest possible time. The hard work put by the type setters and proof readers of Himalaya Publishing, despite their busy commitments is praiseworthy.

This is the first edition of the book and there are possibilities of omissions and commissions. Any constructive suggestion given by the students and faculties and the officers in law enforcement shall be profoundly appreciated.

Satyajit Mohanty, IPS

Rabindra Kumar Mohanty, Ph.D

CONTENTS

Chapter-1

Introducing Criminology 1-28

• Introduction • Nature of Criminology • Subject Matter of Criminology • Criminology and Other Social Sciences • Importance of Criminology • History of Criminology • References

Chapter-2

Modern Concepts 29-38

• Introduction • Modern Concepts & Terminologies • Crime and Morality • Characteristics of Crime • Types of Crime

Chapter-3

Economic Theories 39-54

• Introduction to Modern Theories • Becker's Economic Analysis of Crime • Richard Quinney: Reality of Crime • William Bonger: Criminality and Economic Conditions • Bonger in Indian Context • References

Chapter-4

Psychological Theories 55-70

• Introduction • Sigmund Freud: Psychoanalytic Theory • Jean Piaget & Lawrence Kohlberg: Cognitive Development Theory • Lawrence Kohlberg: Development of Moral Reasoning • Hans J. Eysenck: Learning Theory • Gabriel Trade: The Imitation of Deviance • Albert Bandura: Behavior Modeling • Maslow's Theory of Motivation & Needs • Hirschi and Hindelang: Theory of Intelligence and Crime • References

Chapter-5

Bio-Physical Theories 71-80

• Introduction • Phrenology of Franz Joseph Gall • Physiognomy of Johan Casper Lavater • Hereditary Theory of Richard Dugdale • Anthropometry of Cesare Lombroso • Somatotyping Theory of William Sheldon • Evaluation of Biological Explanations • References

Chapter-6

Sociological Theories 81-113

• Durkheim: Crime as a Normal Phenomenon • Differential Association Theory of Edwin H.Sutherland • Enrico Ferri and Social Defense • Natural

Crime - Garofalo • The Ecological School • Merton's Paradigm of Deviant Behaviour • Labeling Theory of Crime • The Social Disorganization Theory of Crime • Critical Theories of Crime • The Broken Windows Theory: George L. Kelling • The Future of Sociological Crime Theories • References

Chapter-7

Etiology of Crime 114-140

• Introduction • Social Causes of Crime • Economic Causes • Physico-Psychological Causes • Physical • Psychological Causes of Crime • Political Causes • References

Chapter-8

Individual Deviance: Juvenile Justice 141-162

• Introduction • History of Juvenile Justice Legislation • Defining a Child • Provisions under JJ Act 2000 & 2006 • Do's and Don'ts for Police in Juvenile Justice • Child in Need of Care and Protection: Guideline for Police • JCL and Adult Criminal: Distinctions • References

Chapter-9

Collective Deviance 163-174

• Introduction • Criminology of Collective Deviance • Approaches to Collective Deviance • Recapitulation • References

Chapter-10

Organized Crime 175-193

• Introduction • Origin and Conceptual Background • Defining Organized Crime • Characteristics of Organised Crime • Types of Organised Crime • Responses to Organised Crime • Control Strategies • References

Chapter-11

White Collar Crime 194-213

• Introducing the Concept • Historical Background of White-Collar Crimes • Occupational Crime and Street Crime • Types of White Collar Crime • Indian Scenario • Criminal Liability of the Corporations • Prosecution • Control Strategies • An Overview • References

Chapter-12

Recidivism 214-222

• Introduction • Types of Recidivists • Characteristic Causes • Recidivism in India Law • Some Case Studies • Summary of Causes in Case Studies • References

Chapter-13

Crime Trends 223-233

• Introduction • Why Study Crime Trends • Crime Trends in India: NCRB Data • References

Chapter-14

Gambling 234-254

• The Concept of Gambling • Economic, Legal and Psychological Connotations • Types of Gambling • History of Gambling • Legalized Gambling • Economic Impacts of Gambling • What is Not "Gambling" • Gambling and Crime • Is Gambling an Addiction? • Causes of Gambling Addiction • The Risk Factors of Gambling Addiction • Types of Gamblers • The Phases of Gambling Addiction • Helping Problem Gamblers • References

Chapter-15

Drug Addiction 255-268

• Concept of Drug Addiction • Causes of Drug Addiction • Commonly Abused Types of Drugs • Effects of Drug Addiction • Drug Addiction In Relation to Crime • Indian Scenario • References

Chapter-16

Alcoholism 269-283

• The Concept of Alcoholism • History and Etymology • Symptoms of Long Term Alcohol Misuse • Social Effects: Drug-Related Crime • Causes of Alcoholism • Disease of Alcoholism: Stages of Growth • The Recent Trends • Indian Alcohol Consumption Trend (2005) • Alcoholic Treatments • Prevention and Policy Prescription • References

Chapter-17

Cyber Crime 284-303

• Introduction • Defining and Measuring Cyber Crime • Scope of Cyber Crime • Theories of Classification of Cyber Crime • Types of Cyber Crime • Cyber Crime Trends • Indian Scenario (NCRB 2009 Report) • Tamil Nadu Landmark Cyber Case • Policy Implications • An Overview • References

Chapter-18

Prostitution 304-333

• Meaning of Prostitution • History of Prostitution • Causes of Prostitution • Types of Prostitution • Prostitution - The Consequent Dangers • Combating Prostitution • The Immoral Trafficking Prevention Act, 1956 ("ITPA"), • Enforcement of Laws • Recent Trends • Feminism, Prostitution and Legal Issues • Prevention • References

Chapter-19

Penology: Concept & Dimensions 334-354

• Concept of Penology • Punishment • Reasons of Punishment • History of Punishment • Types of Punishment • Judicial Discretion in Sentencing Offenders • Sentencing Power of the Courts • Habitual Offenders and Enhanced Sentence • Capital Punishment Controversy in India • References

Chapter-20

Prison Condition in India 355-372

• Introduction • Defining Prison • Number of Prisons in Indian States/UTS as on 2008 • History of Prison Reforms in India • Prison Conditions in India • Role of NGOs • Prison Offences • References

Chapter-21

Theories of Penology: Deterrence Theory 373-383

• The Concept of Deterrence • History of Deterrence • General and Specific Deterrence • Absolute and Restrictive Deterrence • The Role of the Police in Deterrence • A Critique of Deterrence • Deterrence Theory as a Military Strategy • Criticism of Deterrence Theory as a Military Strategy • References

Chapter-22

Theories of Penology: Theories of Retributive & Restorative Justice 384-390

• Theory of Retributive Justice • History • Subtypes • Criticism • The Restorative Justice • History • Recidivism • Practices •References

Chapter-23

Theories of Penology: Reformatory & Rehabilitative Theory of Punishment 391-419

• Reformatory Theory of Punishment • Probation • Parole • Furlough • The Concept of Pardon in India • The Theory of Rehabilitation • References

Chapter-24

Social Defence 420-439

• Introduction • Conceptual Dichotomies • History of Social Defence • Characteristics of Social Defence • Crime Trends and Social Defence • Victims Needing Special Attention • Social Defence and Police • Promotion of Social Defence • References

Chapter-25

Victimology 440-464

• Introduction • History • The Study of Victimology • Victimology Theory • Typology of Victims • Victims Needing Special Attention • Importance of Studying Victimology • Victim Trauma • Victim Expectations • Victim Compensation • Victim's Basic Rights • Victim and Criminal Justice • The Indian Scenario • References

Chapter-26

Victims of Terrorism 465-481

• Terrorism: One Meaning Several Terms & One Term Several Meanings • Victim of Terrorism: Concept and Dimensions • Theoretical Discourse • Methodological Issues • Praxiological Considerations • An Overview • References

CHAPTER 1

Introducing Criminology

Learning Objectives :

- **Acquainting the Learners with the History and Development of knowledge of criminology.**
- **To know the scientific character of criminology.**
- **To grasp the subject matter of modern criminology against Pre-classical and classical notions.**
- **To comprehend the relationship of criminology with other Social Sciences.**
- **To understand the importance of studying criminology as a profession.**

Introduction

Criminology is the social science approach to the study of crime as an individual and social phenomenon. It has been derived from Latin word *crîmen*, meaning "accusation"; and Greek word *logos* which means science or scientific study. Hence etymologically criminology means scientific study of crime in its entirety. Criminological research areas include the incidence and forms of crime as well as its causes and consequences. They also include social and governmental regulations and reactions to crime. Criminology is an interdisciplinary field in the behavioural sciences, drawing especially on the research of sociologists and psychologists, as well as on writings in law. In 1885, Italian law professor Raffaele Garofalo coined the term in Italian, called *criminologia* meaning criminology. The French anthropologist, Paul Topinard used it for the first time in French '*criminologie*' around the same time. Criminology is that branch of social science, which deals with the study of crime in an individual and society.

Criminology focuses on the behaviour that violates the criminal law and seeks explanations for that behaviour. The study of the origin of laws that define certain behaviour as criminal is a primary focus of the sociology of law, although a number of sociologists include in criminology the study of how certain behaviour has come to be defined as criminal. As important as it is to know why laws are passed to criminalize certain behaviour, such knowledge does not explain why certain people violate the law whereas others do not. It is useful to understand the origin of the law of theft, but it is also important to know why some people steal and others do not, and why some of those use force against their victims in robberies whereas use stealth.

Criminology is the body of knowledge regarding delinquency and crime as social phenomena. Certain acts that are considered undesirable are defined by the political society as crimes. In spite of this definition some people persist in the behaviour and thus commit crimes; the political society reacts by punishment, treatment, or prevention. This sequence of interactions is the subject matter of criminology. The objective of criminology is the development of a body of general and verified principles and of other types of knowledge regarding this process of law, crime and reaction to crime. This knowledge will contribute to the development of other sciences, and through these other social sciences will contribute to an understanding of social behaviour. In addition, criminology is concerned with the immediate application of knowledge to programs of social order and crime control. This concern with practical programs is justified, in part, as experimentation which may be valuable because of its immediate results but at any rate will be valuable in the long run because of the increased knowledge which results from it.

According to Edwin Sutherland and Donald Cressey: Criminology is the body of knowledge regarding crime as a social phenomenon. It includes within its scope the processes of making laws, of breaking laws, and of reacting toward the breaking of laws. The objective of criminology is the development of a body of general and verified principles and of other types of knowledge regarding this process of law, crime, and treatment. The definition includes the most important areas of interest to criminologists: (1) the development of criminal law and its use to define crime, (2) the cause of law violation, and (3) the methods used to control criminal behaviour.

Criminology in the broadest sense covers the whole of criminal science. In a narrower sense, it refers to the part of criminal science which empirically describes criminal behaviour and explores individual and social factors associated with crime and criminals. Criminology is aimed principally at elucidating the connection between crime and the personal characteristics of the offender or his environment, with special reference to the origin of the offence (etiology, genesis). As an object of research, however, the genesis of criminal behaviour cannot be separated from the behaviour itself. Therefore, criminology must also contain a descriptive part, criminography. Furthermore, criminology encompasses prognosis or the science of prediction. Victimology, which is researched into the relation between offender and the victim, is also included under criminology.

Crime is ever changing, so it becomes difficult for criminologists to define it. What is not legal here, may be legal in another state or country. While what isn't banned now, might be banned tomorrow. Murder is wrong, and a crime (homicide) but in certain situations and under certain instances, like war, it isn't a crime, instead is a hero's action. Geographical crime is relative to where you live. It changes from where you live. In America, if white person kills a black person, the sentence is not as serious as that of a black man killing a white man. Law is not static. Criminology includes sentencing, punishment, laws and so on. Another way to look at it is society builds the individual, who in turns builds the criminal; thus it can be reverted back to society.

Nature of Criminology

Criminology is an advanced, theoretical field of study. It can be defined as the study of crime, the causes of crime (etiology), the meaning of crime in terms of law, and community reaction to crime. Not too long ago, criminology separated from its mother discipline, sociology, and although there are some historical continuities, it has since developed habits and methods of thinking about crime and criminal behaviour that are uniquely its own.

1. Criminology draws on the range of human and social science disciplines. The subject is evolving in its theoretical and methodological development, reflecting the rapid social changes it tries to capture and the increasing cross-fertilisation of ideas and methods between the human sciences. In its modern forms, it is characterised by robust debates over how to :

- Conceptualise and explain its subject matter.
- Operationalise its theories in conducting research.
- Inform debates over crime control policy, the scope of human rights, the links between criminal and social justice, and the expanding knowledge bases of the crime prevention, security, policing and justice-related professions.
- Develop and enhance its methodological and technical expertise.
- Manage the sensitive ethical issues arising from empirical research.

2. Criminology is both a theoretical and an empirical discipline. Criminology is the science which studies crime rates, individual and group reasons for committing crime, and community or societal reactions to crime. This is known as theoretical criminology. Theory is a complex subject in its own right. Criminological theory is no exception; it also tends to be complex. Similarly there is another branch of knowledge called practical criminology. Applied criminology is the art of creating typologies, classifications, predictions, and especially profiles of criminal offenders, their personalities and behaviour patterns. Criminologist is a person who studies criminology; not to be confused with a "criminalist" who reconstructs a crime scene or works with crime scene evidence for forensic purposes.

3. At the heart of criminology are theoretical debates about a wide range of perspectives. Criminologists believe in one of three perspectives: the consensus view, the conflict view, or the interactionist view. The consensus view dictates that crime is illegal behaviour defined by the existing criminal law, which reflects the values and moral of a majority of citizens. The conflict view states that crime is behaviour created so that economically powerful individuals can retain their control over society. The interactionist view portrays criminal behaviour as a relativistic, constantly changing concept that reflects society's current moral values. According to interactionist view, criminal behaviour

is behaviour so labeled by those in power; criminals are people society chooses to label as deviants. Research methods used by criminologists are surveys, longitudinal studies, record studies, experiments, and observations. Questions about ethical standards arise when information-gathering methods appear biased or exclusionary. These types of issues may cause serious consequences because their findings can have a significant impact on individuals and groups. Criminology emphasises the importance both of theoretical work and of a firm evidence base for its theories. It also engages in formal and critical evaluation of crime prevention, security and crime control policies, as well as of other responses to crime and deviance. However, in furthering these values, it needs to nurture a lively debate and dialogue between a range of theoretical and methodological perspectives, triangulating quantitative and qualitative data. It must guard against attempts to foreclose this dialogue with the premature creation of theoretical or methodological protocols favouring particular sub-discipline fields, whether endorsed by state officials, by the mass media, or by fashions of academic thought.

4. Criminology seeks to generalise on the basis of evidence. It is therefore neither purely deductive nor purely descriptive; theorisation needs both to guide the collection of data and to be grounded in evidence. Similarly, interpretation of data has to be guided by theorisation. Criminologists get opportunities to acquire capacities of thinking in both abstract and concrete terms and to relate the two. Bodies of evidence are often consistent with alternative interpretations embodied in rival theoretical perspectives. There is a need to weigh alternative interpretations in terms of consistency with evidence, logic, fit with other findings, and breadth of explanatory power. Empirically, criminology is concerned with field based observation of :

- Processes of criminalisation and victimisation.
- The causes and organisation of crime and deviance.
- Processes of preventing and managing crime and victimisation.
- Official and unofficial responses to crime, deviance and social harm.
- Representations of crime, offenders, victims and agents and agencies of control.

5. Criminology as an empirical discipline implies that the scientific method should be used in studying criminology. Criminologists use objective research methods to pose research questions (hypotheses), gather data, create theories, and test their validity. They also use every method of established social science inquiry, including analysis of existing records, experimental designs, surveys, historical analysis, and content analysis.

6. Criminology has its own policy orientation. Given its strong policy orientation and close relationship with the criminal justice professions, many of criminology's most significant theoretical advances have been made through empirical studies. Criminology also contributes to and benefits from continuous theoretical debates within the social sciences. The vitality of the discipline also requires a continuous interchange between theory and analytic and evaluative research, and attention to increasingly salient ethical debates about crime, security, and human rights at international, national, regional and local levels.

7. Criminology is an interdisciplinary science because it involves two or more academic fields. Criminologists have been trained in diverse fields. Many of its practitioners were originally trained as

sociologists, psychologists, economists, political scientists, historians, and natural scientists. Criminology has a rich history with roots in the utilitarian philosophy of Beccaria, the biological positivism of Lombroso, the social theory of Durkheim, and the political philosophy of Marx. In the late 1960s, criminal justice programs were created to examine and improve the U.S. system of justice. Criminology and criminal justice are mutually dedicated to understanding the nature and control of criminal behaviour. The study of deviant behaviour also overlaps with criminology because many "deviant" acts are violations of criminal law. The criminological enterprise is comprised of subareas, including criminal statistics, the sociology of law, theory construction, criminal behaviour systems, penology, and victimology. Criminal justice refers to the study of agencies of social control that handle criminal offenders. Criminal justice scholars engage in describing, analyzing, and explaining the operations of the agencies of justice, specifically the police departments, courts, and correctional facilities. They seek more effective methods of crime control and offender rehabilitation. There are some overlapping areas of concern. Criminal justice experts cannot begin to design effective programs of crime prevention or rehabilitation without understanding the nature and cause of crime. They require accurate criminal statistics and data to test the effectiveness of crime control and prevention programs. There are some overlapping areas of concern between criminology and deviance. Deviance refers to the study of behaviour that departs from social norms. Included within the broad spectrum of deviant acts are behaviours ranging from violent crimes to joining nudist colony. Not all crimes are deviant or unusual acts and not all deviant acts are illegal. Under what circumstances do deviant behaviours become crimes ? For example, when does sexually oriented material cross the line from merely suggestive to obscene and therefore illegal. If an illegal act becomes a norm, should society reevaluate its criminal status? For example, there is still debate over the legalization and/or decriminalization of abortion, recreational drug use, possession of handguns, and assisted suicide.

8. Criminology is a contested and often contentious discipline which is very likely to reflect current social, political and public disputes. Therefore learners should be provided with opportunities to develop awareness of their own values and those of their cultural and political environment, and an appreciation of how alternative values impact upon rival interpretations of evidence. Criminology is intrinsically a reflexive discipline, involving an understanding of contested values in the constitution and application of criminological knowledge.

Subjectmatter of Criminology

Criminology, like all academic disciplines, is constantly changing. As such, the importance attached to different classical and contemporary theories will continually change. The constant emergence of new theories generates new areas of criminological enquiry. Such new areas of enquiry may also be generated by changing political and social concerns, or by changes within another discipline, such as philosophy. In spite of this constant production of new knowledge, however, the broad outlines of the subject area remain relatively constant. The philosophical or methodological focus of criminology is on three themes : ontology, epistemology and behaviourism.

1. Ontology : What is the nature of particular crime? Do certain things really exist, or do we make up the concept in our own minds. Each person has a different view, based on their own thinking. Realists believe that certain things do exist, and therefore crime is real, and so is evil, and it is up to us to find and research it. Nominalists on the other hand, do not believe in anything except what we make up in our own minds. We create the concept, and then act as though it were real, but in reality, it is in name only. We search people out and label them, just like witch hunts (age, person, sex, character, looks etc.).

2. Epistemology : How do we find out about things? How do we know what we know? What counts as knowledge? Positivism must have objective knowledge, hard facts. It is the belief there is a real world out there, and through technology and techniques we can find out positive knowledge about crime. Anti-Positivists do not believe that anything can show them truth, for example, a computer statistics read out is not real proof, they desire empathy, but scientists do not like empathy for it can not be recorded.

3. Behaviourism : What motivates Human beings ? Determinists believe things are determined by forces out of ones control, for example, a criminal could not help themselves, it was brought on by an internal malfunction. They liken human behaviour to an item (physical object - like a set of billiard balls on a table, one reacts against another) that can be studied. Voluntarism is all about free will; actions that are brought about by choice. This relates to the criminal by deciding if a criminal chooses to act as they do, or are they forced to by outside forces?

The subjectmatter of criminology consists of six principal divisions, as follows: 1. the sociology of law, 2. Criminal etiology, 3. Penology and Justice System, 4. Prison welfare, 5. Behavioural System and 6. Victimology.

1. Sociology of Law : Sociology of law is an attempt to systematically analyse the conditions under which criminal laws develop and also relates to an explanation of variations in the policies and procedures used in the administration of criminal justice. Sociology of law is a diverse field of study which examines the interaction of law with other aspects of society, such as the effect of legal institutions, doctrines, and practices on other social phenomena and vice versa. Some of its areas of inquiry include the social development of legal institutions, the social construction of legal issues, and the relation of law to social change. Sociology of law also intersects with the fields of jurisprudence, economic analysis of law and more specialized subject such as criminology.

2. Criminal Etiology : Criminal etiology is essentially an attempt at scientific analysis of the causes of crime or the factors responsible for crime. Crime situations are rarely clear-cut, single-source events. There are as many causes of crime as there are theorists. This scientific search for an etiology of crime has contributed an impressive literature to the subject.

3. Penology and Justice Administration : Penology and justice system are concerned with the control of crime. Penology (from the Latin *poena*, "punishment") is a section of criminology that deals with the philosophy and practice of various societies in their attempts to repress criminal activities, and satisfy public opinion via an appropriate treatment regime for persons accused or convicted of criminal offenses. Penology is concerned with the effectiveness of those social processes devised and adopted for the prevention of crime, via the repression or inhibition of criminal intent via the fear of punishment. The study of penology therefore deals with the treatment of prisoners and the subsequent rehabilitation of convicted criminals. It also encompasses aspects of probation (rehabilitation of offenders in the community) as well as penitentiary science relating to the secure detention and retraining of offenders committed to secure institutions. Criminal justice refers to the agencies of social control that handle criminal offenders. It deals with description, analysis, and explanation of the behaviour of the agencies of justice. The criminal justice response to crime begins, when a crime is reported to the police, or when the police discover a crime has been committed. The police + The courts + Corrections = The *criminal justice system.*

4. Prison Welfare : A prison or correctional facility is a place in which individuals are physically confined or interned and are usually deprived of a range of personal freedoms. Prisoning is a legal penalty that may be imposed by the state for the commission of a crime. A jail is considered synonymous with prison. Criminologists venture to study the place where convicts or under trials are housed including the enjoyment of rights by and the facilities for life and living extended to them.

5. Behavioural System : Criminology is the explanation of criminal behaviour, as well as the behaviour of police, attorneys, prosecutors, judges, correctional personnel, victims, and other actors in the criminal justice system. Criminal behaviour involves study of crime types and patterns. Crime is basically an adaptive behaviour. Crime is a compromise, representing for the individual the most satisfactory method of adjustment to inner conflicts which he or she cannot express otherwise. Criminal is basically antisocial- is incapable of significant loyalty to individuals, groups, or social values. Criminal exhibits behaviour that brings him or her into conflict with society. Criminal behaviour may be purposeful for the individual in so far as it addresses certain felt needs. Criminal does not feel empathy with others. Criminal is unable to imagine how others think and feel about his behaviour.

6. Victimology : Victimology is a branch of criminology that studies victims, survivors and their relationships with the people who victimized them. It scientifically examines the degrees of suffering caused by offenders. Victimology researchers investigate the physical, emotional and financial impact of crime.Today's victimology combines criminology with several other disciplines, including psychology, social work, sociology, economics, law and political science. The most commonly studied areas of victimization are murder, rape, spousal abuse, elder abuse, child abuse and kidnapping. As the field expands identity theft, terror attacks, hate crimes and disability discrimination are also being researched.

On the whole, criminology includes knowledge and understanding of the following substantive areas.

1. The development of criminology as a distinct area of study and inquiry; its interdisciplinary nature; alternative theoretical approaches within criminology; contemporary debates about the content and scope of criminology.
2. How crime, deviance and victimisation are socially and legally constructed; the different sources of information about crime and victimisation, how they are produced, including their location in particular legal, political, social and ideological frameworks, and how they can be interpreted; trends in crime; different forms of crime and their social organisation; different theoretical approaches to the study, analysis and explanation of deviance; relationships between crime and social change and the impact of globalisation; and social divisions such as age, gender, social class, race and ethnicity.
3. The social and historical development of public policing; the organisation and powers of the police in different locations; functions, methods and strategies of policing; the practice and

implications of particular policing strategies; the structure and culture of police work; policing diversity; new and emergent forms of private and state policing; the governance and accountability of policing in specific contexts, and issues of human rights and civil liberties in relation to policing.

4. The development, role, organisation and governance of efforts to reduce and prevent crime and harm, and to ensure personal and public safety and security in different locations; the role of non-governmental agencies; the effectiveness of such measures and human rights issues in relation to preventive and pre-emptive measures.
5. The social and historical development of the main institutions involved in crime control in different locations; the philosophy and politics of criminal justice and modes of punishment; the use of discretion in relation to justice processes including issues of discrimination and diversity; governance of criminal and youth justice and other crime control processes; the development of penal and alternative policies in different locations and their relationship to social change; the main forms of sentence and alternatives and the governance, roles and structure of the agencies involved, and offenders' experiences of adjudication and sentence.
6. Representations of victimisation, crime and deviance, and of the main agents and institutions which respond to crime and deviance, as found in the mass media, in official reports and in public opinion.
7. How to develop a reflective approach and a critical awareness of the values of local cultures and local politics, and of the student's own values, biography, and social identity, and how to bring these skills to bear in an informed response to crime and victimisation; and awareness of how political and cultural values, including their own, have an impact on responses to and rival interpretations of safety and security, crime control, policing, criminal and youth justice, sentencing, and alternative responses to offending.
8. How to make ethically sound judgments in relation to research carried out in the field.

Criminology and Other Social Sciences

Criminology can be seen as a rendezvous discipline, a subject at which social scientific disciplines interact. Criminologists approach the distinctions between the different social sciences in dialogic mode, not seeing the differences as barriers but rather as opportunities for debate and development. Such debates may usefully occur about the meaning of concepts which have a strategic place in more than one discipline. Currently the concept of globalisation provides such a scope for debate. It is in this dialogic spirit that some distinctions between criminology and what might be described as both user and contributor disciplines are set out below.

Criminology and Sociology

As a field of study, criminology is held together by a substantive concern about crime. This concern typically forms only a small part of a sociology degree programme. Moreover, while it would be fair to say that there has been a close relationship between the development of criminological and social theory, it is important to recognise that criminologists have also been influenced by theoretical developments in a range of disciplines, of which sociology is only one. Nonetheless, the close relationship between criminology and sociology continues and contemporary criminologists share with sociologists and other social scientists theoretical concerns in relation to social stratification and social division, governance, risk and globalisation, for example, to the development of which several social science disciplines have contributed.

Criminology is also distinctive in that it has developed, and continues to develop, its own theoretical concepts and theoretical traditions. One example is the sub-field of victimology, developed within criminology. The recent growth of interest in 'cultural criminology' also moves beyond the customary concerns of sociology. Similarly recent criminological analyses of genocide and genocide trials use concepts and debates derived from law, politics, and philosophy as well as sociology, but the substantive concerns and the specific integration of these concepts and debates is uniquely criminological.

In general, the substantive areas of concern of criminology, criminal justice studies and police studies are therefore now too many to be included within a sociology degree, given the numerous other requirements of that discipline.

Criminology and Law

The degree of overlap in subjectmatter between criminology and law programmes is very slight. Law learners need to learn rules and their diverse (as between the European Union and the UK) modes of interpretation, as well as legal philosophy. They also study the structure and operation of the legal system and the criminal justice process. Criminologists learn rules and interpretive rules, if at all, only later in their careers in relation to a particular research field. Nor do they necessarily learn jurisprudence, although they are now expected to know about human rights in relation to policing and the criminal justice system. The only substantive areas of overlap lie in the study of police powers and practice, and court processes.

It is also the case that the subject benchmark statement for law says little about empirical research methods beyond retrieval of legal information. Criminology learners, on the other hand, might well be expected to demonstrate some knowledge of the philosophical foundations of different research methods and some proficiency in the use and evaluation of empirical research using qualitative and quantitative methods.

Criminology and Political Science

The areas of common concern between criminology and politics/political science are relatively few in this country, although the greater congruence in the United States, where political scientists have concerned themselves with the operation of state agencies such as criminal courts and regulatory agencies, indicates a potential overlap. More recent work, for example, research on genocide and in relation to state crime, suggests the possibility of a closer relationship between the two disciplines as the processes and effects of globalisation are increasingly addressed by social scientists.

Criminology and Penology

Penology is the branch of knowledge which draws heavily from criminology. Penology is the study of penalties (from the Greek penalty), although in its broadest sense it is also concerned with the consequences and merits of attempting to deal with various kinds of conduct by criminal prohibition ('criminalizing'). It includes not only the study of penal codes of law, but also the investigation of ways in which penal codes are applied by courts in practice, and the manner in which each type of penal measure is applied. For example, even when a penal code appears to oblige courts to pronounce a sentence (such as imprisonment for life in the case of murder), there are ways of avoiding this (such as convicting the offender of a less serious charge of homicide); most penal systems provide legal devices by which a sentence of imprisonment can be terminated before its nominal end. Penologists are interested in all such expedients, and in the criteria which are used by courts, administrators and other personnel to make distinctions between offenders, whether for such purposes or for other reasons. Other reasons may include the belief that certain types of offenders are more likely than others to respond to certain regimes, or that some prisoners are so dangerous that they must be given special sentences, detained longer than is normal for the offence, or given freedom only under specially strict conditions. If cause and effects are interrelated then similar is the relationship between criminology and penology.

Criminology and Social Policy

Social policy and criminology have in common a precarious existence as applied disciplines which means that there may be strong external pressures to underplay the theoretical work on which they depend. Given the common focus on looking at social problems/harms/needs and their alleviation or control (with crime a paradigmatic example of a 'social problem'), it is surprising to note that the two disciplines have had only limited forms of cross-fertilisation in recent decades.

Recent publications indicate that the institutional focus of social policy is on the 'production, organisation and consumption of welfare' (not including freedom from crime) while 'issues in social policy' only mentions crime briefly with regard to 'young people'. In terms of the service-based issues confronted by social policy analysts, crime control appears peripheral to this discipline. Moreover, the transnational and global dimensions of contemporary crime and justice are not addressed by social policy, which remains largely pegged to the activities of state and regional governments. In contrast to social policy, criminology is a fast growing discipline in both teaching and research terms internationally in the post-welfare state era.

The strongest intellectual connection between the two disciplines lies in the shared emphasis on the links between poverty/inequality/marginalisation/social exclusion and criminalisation. A recent example of different approaches to a shared area of concern has emerged in relation to juveniles, marginalized and the violence against the women and aged in families and care homes. Thus there is a definite connection between Criminological Theory and Social Policy

(a) Theories of crime causation imply that changing the conditions, the theory holds responsible for causing crime can reduce it and even prevent it.

(b) Policy: A course of action designed to solve some problem that has been selected from among alternative courses of action.

(c) Every theory has policy implications deducible from its primary assumptions and propositions.

(d) A good theory should offer useful practical recommendations.

Criminology and Forensic Psychology

In terms of subjectmatter, both criminology and forensic psychology study the nature of and reasons for violence and other forms of offending, issues related to the working of the criminal justice system, issues relating to victimisation, and issues relating to punishment. Despite this overlap there are clear differences. Criminologists are concerned with social censure and the cultural and political processes whereby certain kinds of behaviour are criminalised (and others not); these questions are not addressed within forensic psychology; criminologists are also concerned with the social harms resulting from, for example, corporate crime, state crime, or living in a community perceived to be unsafe. In fact, 'forensic psychology focuses more on the individual within a given context, while criminology focuses on the situation (including individuals)'. These differences in the object of analysis arise from differences in theory. Criminology tends to deploy concepts of class, gender, race, ethnicity, the state, the person, the polity and culture which have been developed within sociology, cultural studies, feminist theory and political science. In addition to its own concepts, these social scientific concepts are intrinsic to criminological analyses, whether theoretical or practice oriented. Critical analysis and deployment of these concepts is not intrinsic to psychology. Criminological theory itself develops such concepts further, viz. the example of 'responsibilisation' referred to above as a new way of theorising the relationship between the central, state and the voluntary sector or citizenry.

Importance of Criminology

Criminology has a long theoretical tradition, many parts of which have provided specialised concepts which continue to shape the discipline: Becker's key concept of 'deviance' (1963) is a case in point. Even earlier, Edwin Sutherland (1940) provided the basis for theoretical self criticism by pointing out that theory can obscure as well as enlighten, that the theory (as well as the practice) of his day rendered the crimes of corporations invisible. This history is constitutive of criminology as a discipline, what a lawyer would call 'trite knowledge', that which hardly needs to be spoken.

This unique theoretical tradition continues to be developed, as evidenced by the recent growth of interest in international networks and the widespread use of concepts developed within the discipline of criminology such as 'responsibilisation' (displacing responsibility for crime prevention onto communal organisations), 'transgressive' criminology (which challenges established disciplinary boundaries) or 'status offences ' (activities attracting legal sanctions against women or children for actions which would not be sanctioned for an adult male or, in some cases, any adult). Another example is the continuing development of the concept of state crime, from the active perpetration of harm to culpable failure to protect. In criminal justice and policing programmes the idea of community safety as a public good has emerged.

It is of increasing importance that not only academic criminologists but also practitioners have the capacity constantly to appraise and evaluate research evidence. In terms of methodology, criminology, while deploying the full range of accepted social scientific methods, has also developed methods which are appropriate to its own objects of analysis. The most familiar of these are the survey of self-reported crime and the crime victim survey. The latter technique has been elaborated and supplemented to explore more complex harms such as violence against women and racist harassment and repeat victimisation. Filming and closed-circuit television cameras have also been used to construct both qualitative and quantitative criminological data, most recently on the raced practices of private security personnel and on police treatment of suspects in custody. These significant technical developments are supported by methodological and conceptual advances: concepts such as 'occidentalism' and 'glocalisation' increasingly reshape both fieldwork and analytic strategies.

Study of criminology provides scope to develop a range of skills that will enable the learner to work autonomously in subsequent employment. Criminology attracts learners from diverse academic and social backgrounds. Their learning and skills development needs vary accordingly. The study of criminology enables learners to develop a number of cognitive abilities and skills. These may be acquired in a range of teaching and learning situations, so that learners will be able to become competent in :

- generating and evaluating evidence
- appreciating the complexity and diversity of the ways in which crime is constituted, represented and dealt with
- assessing the merits of competing theories relevant to crime, victimisation and responses to crime and deviance
- assessing the merits and diversity of objectives of competing responses to crime and deviance, including the protection of human rights
- gathering, retrieving and synthesising data and information
- making ethical judgments about published research
- making reasoned arguments
- interpreting quantitative and qualitative evidence and texts
- developing the ability to reflect in critical and constructive ways on their own learning.

The range of subject-specific abilities that learners would normally be expected to develop through the study of criminology include :

- the ability to identify criminological problems, formulate questions and investigate them
- competence in using criminological theory and concepts to understand crime, victimisation and responses to crime and deviance
- competence in using criminological theory to elucidate representations of crime, victimisation, and responses to these, as presented in the mass media and official reports

- competence in explaining complex social problems in terms of criminological theory
- the capacity to analyse, assess and communicate empirical information about crime, victimisation, responses to crime and deviance, and representations of crime
- the ability to identify human rights issues in responses to crime and deviance
- the ability to recognise a range of ethical problems associated with research and to take action in accordance with the guidelines of ethical practice developed by the Society and cognate professional bodies
- the ability to identify and deploy a range of research strategies including qualitative and quantitative methods and the use of published data sources and to select and apply appropriate strategies for specific research problems
- the ability to present the philosophical and methodological background to the research of others and to one's own research.

Criminology provides some of the technical skills to the learners as well which are generic to all social sciences. These include :

- written and oral communication skills, including the clear presentation of research procedures, academic debates, and the student's own arguments
- skills of time planning and management
- the ability to work productively in a group
- the capacity to present data and evidence in an appropriate format for a variety of audiences
- the ability to formulate researchable problems within a general area of concern
- the ability to evaluate evidence of diverse kinds and to draw appropriate conclusions
- research design and data collection skills in relation to crime, victimisation, and responses to crime and deviance, including knowledge of survey, experimental, and case study design; the identification of an appropriate sampling method; structured, semi-structured and indepth interviews; ethnography; evaluation methods; and the critical use of published data sources
- the ability to analyse data including indexing and retrieval of qualitative data, and an understanding of basic statistics (sampling, measures of significance and knowledge of the relevant software)
- reading skills: the ability to identify the most important arguments or evidence in a text and to record and/or represent these
- bibliographic and referencing skills: the identification of relevant published and web-based materials in relation to a particular topic
- computing skills in relation both to text and the presentation of basic research data.

The importance of studying criminology in terms of intended objectives and threshold achievements is schematically presented below.

TABLE NO. 1.1

Studying Criminology : Intended Objectives and Threshold Achievements

Intended Objective	Threshold Achievement
1. An understanding of the key concepts and theoretical approaches that have developed and are developing in relation to crime, victimisation and responses to crime and deviance.	1. Able to describe a range of key concepts and theoretical approaches within criminology.
2. An awareness of how crime and victimisation are constructed in the media and by agents and practices of crime control.	2. Able to identify political and social processes of victimisation and criminalisation.
3. Able to understand and demonstrate the relationship of social class, gender, age, race, ethnicity and other salient aspects of diversity in relation to crime, victimisation and responses to these phenomena.	3. Able to recognise patterns of social diversity and social inequality in relation to crime, victimisation and responses to these phenomena.
4. An understanding of the social and historical development of policing, of the changing values governing police work including human rights, of the structure and culture of police work in different locations, of policing diversity, and of new and emergent forms of private and state policing.	4. Able to recognise different police cultures, historical and contemporary trends in police work, and the implications of changes in the values governing police work and police practice in a diverse society.
5. An understanding of the social and historical development of punishment including courts and hearings for adults and young people; the governance and values of the relevant institutions; the theory and practice of sentencing; prison and community based penalties; and the place of human rights in these processes.	5. Able to recognise values and processes that underpin developments in youth and criminal justice and the practices of agencies which administer sentencing and alternatives.
6. An understanding of the organisation and governance of efforts to prevent harm and ensure personal safety, and of human rights issues in relation to these.	6. Able to identify and describe efforts to prevent harm and ensure personal safety and to describe value problems.

Contd...

Intended Objective	Threshold Achievement
7. An understanding of the nature and appropriate use of research strategies and methods in relation to issues of crime, victimisation, and responses to crime and deviance.	7. Able to identify a range of different research strategies and methods; able to identify an appropriate strategy for specific research problems.
8. An understanding of the value of comparative analysis.	8. Able to recognise and illustrate the use of different approaches to comparison in relation to crime, victimisation, and responses to crime and deviance.
9. An understanding of complex social problems and the relationships between them in terms of criminological theory and empirical evidence in relation to crime, victimisation, and responses to crime and deviance.	9. Able to recognise and illustrate the relationship between a range of social problems identified by criminological theory and evidence.
10. An understanding of the distinctive character of the discipline of criminology in relation to other forms of understanding, such as other disciplines and everyday explanations.	10. Able to recognise ways in which the discipline of criminology can be distinguished from other forms of understanding.

In summary, criminology has emerged as a discipline distinct from other, related, social sciences in terms of its subjectmatter, its theoretical concerns and development, its methodologies and techniques.

Aspects of the conceptual framework are shared with some of the cognate disciplines: the 'grand theories' with sociology, social policy and politics; theories of justice and of human rights with law; somewhat less is shared with psychology: some quantitative methodologies and some substantive issues which none the less may be differently conceived. Criminology has developed a distinct theoretical tradition, which has enabled it to expand its substantive concerns in an exponential manner, resulting in a plethora of optional courses, any one of which could itself develop into a new core unit. Above all, it is largely this explosion of imaginative growth that cannot be contained within the established disciplines.

In addition, and increasingly, professional criminologists and graduates are being called upon to advise and inform the work of crime control agencies: from preventing youth offending to advising the prison service about deaths in custody; from the role of the police in community safety teams to the structure and functioning of the people trade; from how to count family violence to how to prevent it; from institutional racism to the management of diversity. Criminology must develop in its own way to meet these challenges of the twenty-first century. The new problems demand new theories and new methods, and teaching must reflect this. In summary, not only does criminology not share a sufficiency of the conceptual framework of any other discipline, but the movement is centrifugal. This rapidly growing separation needs the particular kind of light touch steering which only subject specific benchmarks can provide.

History of Criminology

Crime is as old as Society. Till the emergence of Criminology as a scientific discipline there were varieties of explanation given to understand crime and criminal behaviour. Criminology as a new branch of knowledge arose in the mid-18th century, as social philosophers (The Classical School) gave thought to crime and concepts of law. The father of classical criminology is generally considered to be Cesare Bonesana Marchese di Beccaria (1738-94) for his book- *On Crimes and Punishment* (1764). His theory grew against the pre-classical notion of criminality.

It is true that modem criminology owes its origin to the European criminologists such as Beccaria, Ferri, Garafalo, Tarde and others but this does not mean that the knowledge of criminal science was completely unknown to ancient India. The Indian epics and other authoritative sources such as Manusmriti, Nyaya Mimansa and Kautilya's Arthshastra contain exhaustive references to crime and criminals which clearly indicate that a well defined criminal policy was in vogue in early period of Hindu society. The most striking feature of the penal law of ancient India was that it used religion and morality as the basis of determining what was criminal and what was not. People in ancient India showed greater respect for religion, morality, ethical values and law; the social solidarity of the community kept people conscious about their duties towards their fellowmen. The occurrence of crime was therefore, a rare phenomenon. Moreover, the fear of caste expulsion and social disapprobation acted as an effective deterrent to keep persons away from criminality. It was far more humiliating and disgraceful than actual corporal punishment.

Later, with the socio-political changes due to advent of Moghul and English rule in this country, the Indian society witnessed a radical change. Due to the impact of western culture the past traditions and ethical values of life have lost all their significance and there has been a drift into excessive materialism which has created an atmosphere conducive to multiplicity of crimes. India's criminal policy, penal laws and procedural laws are all modelled on British criminal justice system. It is therefore, difficult to link up the ancient penal laws of India with the present law of crime and procedure.

To many people interested in the roots and origin of Criminology, Lombroso is considered to be the father of the modern science, meaning that he explored it and examined it from many angles in order to expose it. It was used by a law professor in Italy, Raffaele Garofalo, who actually used the Italian term for the word, criminologia. There are some who believe that it is possible that the study of bumps on an individual's skull, referred to as Phrenology, is a better mark of representation when it comes to the scientific origins of the term. Lombroso is considered to be the father of modern criminology because of the many advances that he attempted to make in the field. If not for him, criminology might not be as developed as it is today.

Over time, many different theories were composed as to what caused criminal behaviour. These theories have their roots in biology, psychology, and sociology, to name a few. Over time, by using these foundations for new theories, different schools of thought began to emerge. Through these different schools of thought, the theoretical science of the criminal element could be rationalized. Individuals often debate in order to attempt to define the reason for crime, so that it may then be eliminated by solving the problem of what causes crime. By identifying and responding to the motivator for crime and deviant behaviour, more effective measures of social control could be put into action. Currently, we are able to take care of the symptoms of crime. For example, we have police officers who catch criminals. When found guilty by a panel of their own peers, the criminal will go to jail. That takes the criminal out of society. However, there are new criminals emerging every day. We have learned to treat the symptoms, the individual criminals, but we have not yet been able to determine the instigating factor in these developments.

Throughout the history of criminology, this is the main goal of the science to study the symptoms of crime. Since it is a theoretical science, all of the theories that correspond to the topic cannot immediately be proven as accurate or inaccurate, merely theoretical. However, the continuous effort of individuals to idealize the workings of the individual and society have been transformed and re-defined since the beginning use of the term. If the history of criminology shows us anything, it is that the human nature to establish right from wrong and the differences between individuals distinguishing the difference is virtually a timeless theme, at least since its birth. By continuously researching and documenting the variables and changes within society and crime, people who are interested in criminology can line up the theories with the established facts. After this is done, it is possible to sit down and analyze the coinciding factors in order to determine the best course of action when it comes to evolving society so as to limit crime and intolerance as much as possible. Following table explains the thematic issues and theoretical explanations in history about the emergence of criminology as a science.

TABLE NO. 1.2

Thematic Issues and Theoretical Explanations in History

THEORY	THEME
Demonology/ Theology (5,000 BC-1692 AD)	Demonic Influence/ God's will
Classical School (1764—)	Free will/reason/hedonism
Positive School (1860—)	Determinism/beyond individual control
Imitation (1843-1905)	Mind on mind crowd influences
Psychoanalysis (1895-present)	Subconscious guilt/defense mechanisms
Sociology (1908-present)	Social/environmental factors
Ecology (1927-present)	Relation of person with environment
Culture Conflict (1938-1980)	Conflict of customs from "old" country
Differential Association (1939-present)	Learning from bad companions
Anomie (1938-present)	State of normlessness/goal-means gap
Social Defense (1947-1971)	Soft targets/absence of crime prevention

Contd...

THEORY	THEME
White-collar Crime (1945-present)	Cutting corners/bordering on illegal
Control Theory (1961-present)	Weak social bonds/natural predispositions
Strain Theory (1954-present)	Anger, relative deprivation, inequality, Stress, Hassles, Interpersonal Relations
Subcultures (1955-present)	Criminal values as normal within group
Labeling Theory (1963-1976)	Self-fulfilling prophecies/name-calling
The "New Criminology" (1973-1983)	Ruling class oppression
Conflict Criminology (1969-present)	Structural barriers to class interests
Critical Criminology (1973-present)	Segmented group formations
Radical Criminology (1976-present)	In articulation of theory/praxis
Left Realism (1984-present)	Working class prey on one another
Feminism (1980-present)	Patriarchal power structures
Low Self Control Theory (1993-present)	Impulsiveness, Sensation-seeking

Taking the above table into account and without recourse to any disciplinary inclination of Criminology, the history of explanations of crime can safely be divided under three major heads :

1. Pre-Classical Explanations
2. Classical Explanations
3. Modern Criminology

1. Pre-Classical Explanations

The earliest form of punishment was private revenge, in which the victim or the victim's kin retaliated for injury and the community did not interfere. The problem was that private revenge often escalated into blood feuds that could continue for many years until one or the other family was completely wiped out. The loss of life and property became so great that the communities slowly started to impose trials and official penalties on offenders in order to restrict private vengeance. For many centuries, this community trial and punishment largely was carried out in the context of religion. Criminal acts were said to be affronts to the gods, who might express their anger through plagues, earthquakes, or other desolation. Punishment proportionate to the wrongdoing was said to lessen the gods' anger. For example, the lex talionis ("an eye for an eye and a tooth for a tooth"), as found in the Bible, prescribed this correspondence between crime and punishment. Properly read as "no more than an eye for an eye" it also significantly limited the excesses of private revenge in an attempt to reduce the consequences of the blood feuds.

While these religious and spiritual approaches to crime and punishment dominated early thinking, naturalistic approaches also go back to antiquity. For example, Plato (429?–347 B.C.) argued that the basis of law was the prevailing social morality rather than the laws of the gods. Thus, every action against that morality constituted a crime. In his Republic and Laws he delineated four types of offenses : (1) against religion (theft within a temple, impiety, or disrespect); (2) against the state (treason); (3) against persons (poisoning, use of drugs, witchcraft, sorcery, infliction of injury);

and (4) against private property (killing a thief caught stealing at night was not punishable). Plato also made various other arguments : that crime was the product of a faulty education, that the severity of punishment should be determined by the degree of culpability, that criminals are sick individuals who must be cured, and that if they cannot be cured they must be eliminated.

In Aristotle's (384–322 B.C.) view, humans were a synthesis of a body and a soul, endowed with intelligence, emotion, and desire. In his Nicomachean Ethics, Aristotle defined crime as the act of free will, stimulated by desire. Thus he argued that children, idiots, the mentally ill, and individuals in a state of ecstasy should not be held responsible for criminal actions.

According to Aristotle, societal responses to crime could be preventive or repressive. Preventive responses could be : (1) eugenic (some children should be nurtured and educated while others should be abandoned and left to die because of some deformity); (2) demographic (the number of births should be limited, and unnecessary pregnancies should be terminated); and (3) deterrent (punishment should be designed to intimidate the offender and deter the onlookers). Repressive responses originally were limited to allowing private revenge, but later were extended to include such measures as banishment and turning the offender over to the victim's family.

Rome was the source of the world's most powerful early legal influences. The Law of the Twelve Tables (Leges Duodecim Tabularum, more informally simply Duodecim Tabulae) was the ancient legislation that stood at the foundation of Roman law. The Law of the Twelve Tables formed the centerpiece of the constitution of the Roman Republic and the core of the mos maiorum (custom of the ancestors).

According to traditional, semi-legendary historical accounts preserved in Livy, during the earliest period of the Republic the laws were kept secret by the pontifices and other representatives of the patrician class, and were enforced with untoward severity, especially against the plebeian class. A plebeian named Terentilius proposed in 462 BC that an official legal code should be published, so that plebeians could not be surprised and would know the law.

Like most other early codes of law, they combine strict and rigorous penalties with equally strict and rigorous procedural forms. In most of the surviving quotations from these texts, the original table that held them is not given. Scholars have guessed at where surviving fragments belong by comparing them with the few known attributions and records; many of which do not include the original lines, but paraphrases. It cannot be known with any certainty from what survives that the originals ever were organized this way, or even if they ever were organized by subject at all.

Out of 12 (twelve) tables eighth table detailed the crimes and their punishments and hence was important.

TABLE VIII (Torts) Ancient Roman Law

Qui malum carmen incantassit...

Those who have incanted an "evil song"... (interpreted by Cicero as slander, in City of God 2.9)

Si membrum rupsit, ni cum eo pacit, talio and esto.

If one has maimed another and does not buy his peace, there be retaliation in kind.

Manu fustive si os fregit libero, CCC, si servo, CL poenam subito si iniuriam faxsit, viginti quinque poenae sunto.

Someone who breaks another's bone by hand or club must pay 300 sesterces; for a slave, 150; if he has done simple harm against another, 25.

Qui fruges excantassit... neve alienam segetem pellexeris

Someone who kills crops with a spell, or another's corn...

Patronus si clienti fraudem fecerit, sacer esto.

If a patron defrauds his client, let him be outlawed.

Qui se sierit testarier libripensve fuerit, ni testimonium fatiatur, inprobus intestabilisque esto.

If one has been called to witness, or hold the scales, unless he gives his testimony, let him be dishonoured and incapable of further testimony.

Si telum manu fugit magis quam iecit, arietem subicito.

If an object flies unaimed from your hand rather than aimed [and causes injury], you will owe a ram.

Intentional homicide, setting fire to a dwelling or harvested crop, treason, and parricide were all were punished by death. The intentional infliction of injury was punished by a fine or by the infliction of a comparable injury if the fine was not paid. Punishment for theft generally was compensation equal to double the value of the stolen goods, although a thief caught in the act could be killed. If the thief was a free man, he could be given to his victim as a slave. Death sentences were also imposed on judges or arbitrators caught taking bribes and on witnesses giving false testimony. However, the sentences could only be carried out with the consent of the whole assembly of citizens, and citizens of Rome were rarely put to death. After the second century A.D., exile and banishment became common punishments. The institution of slavery decisively influenced the evolution of the penal system in Rome because the very severe sanctions devised for slaves were later extended to the entire population, with the exception of a limited number of privileged and wealthy citizens. When the population of Rome reached one million, during the second century A.D., permanent tribunals were established, composed of thirty or more jurors presided over by a praetor. At first the jurors had to be of the senatorial class, but gentlemen, wealthy citizens, and soldiers later became eligible. These tribunals were empowered to deal with cases of treason, homicide, adultery, corruption, and kidnapping, and there was no appeal from their decisions.

2. Classical Explanations

The Classical "School" of Criminology is a broad label for a group of thinkers of crime and punishment in the 18th and early 19th centuries. Its most prominent members, Cesare Beccaria and Jeremy Bentham, shared the idea that criminal behaviour could be understood and controlled as an outcome of a "human na'ure" shared by all of us. The pair argued that human problems, namely on crime, ought to be tackled by the application of reason, rather than tradition, religion or bogus superstition. Human beings were believed to be hedonistic, acting in terms of their own self-interest, but rational, capable of considering which course of action was *really* in their self-interest. A well-ordered state, therefore, would construct laws and punishments in such a way that people would understand peaceful and non-criminal actions to be in their self-interest—through strategies of punishment based on deterrence. The classical school of criminology seeks to assess the nature, the causes and the extent of crime and further monitor the response to it.

Cesare Beccaria and other members of the Classical School of Criminology believed that criminal behaviour could be minimized using the basics of human nature. The school was based on hedonism, the idea that human beings act in their own self-interests. They believed that rational people enter into a social contract in which they realize that having a peaceful society would be most beneficial to themselves. The school sought to reduce crime through reform to the criminal punishment system, which they felt tended to be cruel and excessive without reason as well as an ineffective deterrent. It mainly grew as a reaction against the pre classical notion of criminality as can be seen from the table below.

TABLE NO. 1.3

Pre-classical Notion and Classical Notion: A Comparison

Pre-classical Notion	**Classical Notion**
Sin	**Crime**
Religion	**Law**
Arbitrary Justice	**Proportionate**
Inhuman Treatment	**Deterrence**
Actor	**Action and Intention**
Cure	**Prevention**
No Scope to confront Accusers	**Confront Accusers**
Discretionary power of Judges	**No such power of Judges**

Though there is some kind of semblance between sin and crime, the two radically differ in their content, scope and consequences. The concept of sin emanates from religion whereas crime is a legal proposition. Sin results in violation of rules of religion or morality while crime involves breach of law. A sinner is punished by God but a criminal is punished by the State. There is no direct injury or harm in case of a sin but a crime necessarily involves some kind of direct injury. The remedy for a sin is penance whereas a person who commits a crime is subjected to a term of sentence by the law court.

The Classical School of Criminology posits that punishment can be a deterrent for crime, so long as the punishment is proportional, fits the crime, and is carried out promptly. The most effective deterrent for criminal behaviour would be swift punishment rather than long trials. They felt that people who broke the law were irrational and did not know how to act in their best self-interests or society's. Members of the school contended that punishments needed to be consistently enacted for specific crimes with no special circumstances in order to demonstrate to people that criminal activity will not benefit them because there are definite consequences.

A major part of the criminal punishment reform that the Classical School of Criminology fought for was fair and equal treatment of accused offenders. Prior to the school's fight for reform, judges could punish criminals at their own wills regardless of the severity of the crime, which may have led some to view the criminal punishment system as tyrannical. Cesare Beccaria and other members fought for punishments for specific crimes to be set by legislature and not to allow judges unbridled power. They felt that if judges could only apply legislatively sanctioned punishments, trials would be quick and criminals would receive their punishments faster.

The idea behind the Classical School of Criminology's fight for swift trials and clearly defined punishments was that criminals were more likely to be deterred if they knew what type of punishment they would receive and how quickly. Members of the school believed that preventing crime was actually more important than punishing it, but by having a clear punishment system in place, criminals would use reasoning to deduce that crime would not be in their best self-interests. The Classical School of Criminology was accepted by European rulers in the late eighteenth century and is considered to have influenced the Western justice system. The views of Beccaria and Bentham are summarized below.

A. Beccaria- Crime and Punishment (1764)

1. The law should place as few restrictions on individual freedom - the law should not enforce moral virtue -excessive prohibition increases rather than decreases crime.
2. Specific guaranties of defences against the administration of justice - rights of the accused to be scrupulously protected & importance of the presumption of innocence - the state must proof its case.
3. The criminal law should be clear & certain - offences & punishments should be defined in advance - retrospective laws, judicial discretion or interpretation is bad. Clear knowledge of the law safeguards rights & deters offenders more effectively than any system of uncertain terror.
4. As per 3 written code of law represents the social contract. Laws should be used to maintain the social contract.
5. The justification of punishment is retribution not revenge & to protect individual rights. All people should be treated equally. The use of torture to fain confessions should be abolished.
6. Punishment must be proportionate & strictly limited - it should not exceed the harm done. The evil inflicted on an offender should exceed any advantage gained from the crime "*beyond this is superfluous and tyrannical*".
7. Penalties should correspond with the nature of the offence & its extent in society; e.g. fines for theft, physical punishment & labour for violence. Savage penalties replaced by punishments applicable to a society that values liberty above else - hence the rise of imprisonment. (Tarde's doctrine that in every epoch the main punishment reflected the most valued goods).
8. Punishment should be certain & speedy. Punishment should be prompt and effective. Deterrence was more effective if it was certain, fast & mild. Punishment should be based on the pleasure/ pain principle. Punishment should be based on the act, not on the actor. The punishment should be determined by the crime.
9. Exemplary punishment was prohibited nor should punishment be imposed to reform beyond pain that is proportional to the harm done. Nor varied to suit the personality or circumstances of the crime.

10. The potential offender is a reasoning independent individual who calculates & thus all are equally responsible and all deserve the same punishment for the same crime.
11. Prevention of crime "*it is better to prevent crimes than to punish them. This is the chief aim of every good system of legislation*".
12. Only legislators should create laws. Judges should impose punishment only in accordance with the law. Judges should not have discretion about the laws.

B. Jeremy Bentham's- Principles of Morals and Legislation (1789)

1. Utilitarianism is based on the principle of utility, which prescribed "the greatest happiness for the greatest number."
2. Utilitarianism assumes that human actions are calculated in accordance with their likelihood of bringing happiness (pleasure) or unhappiness (pain).
3. Bentham proposed the "felicific calculus".
4. Any human action at all should be judged moral or immoral by its effect on the happiness of the community. Rational behaviour is behaviour that is consistent with logic.
5. Hedonism: A doctrine with the central tenet that the achievement of pleasure or happiness is the main goal of life. Hedonism is a method by which individuals are assumed to logically weigh the anticipated benefits of a given course of action against its possible costs.
6. Free will enables human beings to purposely and deliberately choose to follow a calculated course of action. If crime is to be deterred, punishment (pain) must exceed the pleasures gained from the fruits of crime.
7. Bentham may have been both the last of the old classical criminologists and the first of the positivist criminologists.

Problems of Classicism

Classical explanations is criticisized for several of its flaws :

- Too legalistic, juristic & deductive reasoning of the classicist.
- Over-reliant on punishment.
- Does not recognize stigmatization.
- Over-emphasizes the act rather than the circumstances.
- Reliance on free will and equality in punishment — political concepts.
- Positivism's rejection of the way responsibility in the law was regulated & interpreted.
- It did not distinguish between :
  - Book view and Field view
  - Insanity and Sanity
  - Child and Juvenile and Adult criminal
  - Actor, Action and Intention.
  - First Offender and Habitual Offender
  - Guilt was less important to society than the dangerousness of the offender.

Criminological authors C. Coleman and C. Norris point out that the assumption that all people are equal before the law and able to make equally rational choices is flawed. For example, young children, the mentally ill, as well as those who are unable to form logical, free willed, rational choices are unable to make equally rational choices in a way similar to the demotic man or woman.

Criticisms exist as aforementioned, however, eminent criminologist Jock Young highlights that classicism as a theory has the largest history of any criminological theory and still continues to be a source of many conceptions and ideas today.

3. Modern Criminology

More accurately today, criminology is the study of crime as a social trend—its overall origins, its various manifestations and its impact upon society as a whole. That makes it more a form of sociology than a law enforcement tool. But the trends it studies have a huge impact on the way the police do their jobs, the way society treats its criminals, and the way a given community goes about maintaining law and order. Modern criminology is usually traced to Cesare Lombroso, a nineteenth century Italian doctor who treated criminal patients in prison. He developed a positivist theory of criminology which stated that some criminals are biologically predisposed towards crime. This ran counter to earlier thoughts on criminology, which held that people are basically decent and driven to crime only through extenuating circumstances, such as poverty. Modern criminology turned radically different from classical school as can be seen from the table below :

TABLE NO. 1.4

Classical School and Modern Criminology Distinguished

Parameter	Classical	Modern Positivist
Historical Period	18th-century Enlightenment, early period of Industrial Revolution	19th-century Age of Reason, mid–Industrial Revolution
Leading Figures	Cesare Becarria, Jeremy Bentham	Cesare Lombroso, Raffael Garofalo, Enrico Ferri
Purpose of School	To reform and humanize the legal and penal systems	To apply the scientific method to the study of crime and criminality
Image of Human Nature	Humans are hedonistic, rational, and have free will. Our behaviour is motivated by maximizing pleasure and minimizing pain.	Human behaviour is determined by psychological, biological, or social forces that constrain our rationality and free will.
Image of Criminals	Criminals are essentially the same as noncriminals. They commit crimes after calculating costs and benefits.	Criminals are different from noncriminals. They commit crimes because they are inferior in some way.
Definition of Crime	Strictly legal; crime is whatever the law says that it is.	Based on universal human abhorrence; crime should be limited to inherently evil (mala in se) acts.
Core concepts Major themes	Punishment Deterrence Social control Social contract, sure and proportionate.	Treatment Indeterminacy rehabilitation Treatment & rehabilitation Social scientist as judges.

Lombroso was revelatory because he recognized distinctions in motivations, which inferred that methods other than punishment could be used to reduce crime. Modern criminology is a beautiful amalgam of French Thought, Chicago School, and the Strain Theory.

French sociologist Emile Durkheim posited a later theory of criminology that was more deterministic than Lombroso. He believed people were the products of society, rather than the other way round, and felt that crime was a natural and even beneficial necessity (he believed that it help spur greater reforms in law and social justice). His countryman Alexandre Lacassagne helped bridge the gap between the "nature/nurture" debate by citing both hereditary and environmental factors in creating criminals.

The Chicago School was catalyzed by Clifford Shaw and Henry McKay in the 1940s. They drew upon the concentric zone model of urban growth described by Robert E. Park and Ernest W. Burgess some 15 years earlier. It stated that crime remained worse in "high transition" parts of the city, where people move often and poverty facilitated breakdown of standard social contracts. As neighborhoods go downhill, the more affluent depart for safer areas, which increases the rate of descent and fosters a culture of crime. This social ecological model of criminology continues to hold a great deal of influence today.

Strain theory arose from sociologist Robert K. Merton, who worked at about the same time the Chicago School was fermenting. He noted the dichotomy between the "American Dream" of prosperity through hard work and the inequities people actually experienced in real life. The resulting strain leads some to resort to crime in order to bridge the gap, as well as joining rebellious social groups, such as street gangs or vagrants.

Modern criminology incorporates different aspects of these theories as well as numerous smaller ones. The primary purpose is to identify the causes of crime and thus to seek amenable solutions (either through crime prevention, punishment rehabilitation, or alleviation of the social stresses which lead to crime). Most criminologists don't believe that crime will be erased in our lifetime or even many lifetimes (those who follow Durkheim's thinking may claim that society would suffer without at least a certain amount of crime). But by better understanding it as a social phenomenon, they can help reduce its negative impact, while helping law enforcement in the process.

Legacy of the Modern Positivists

1. Individual behaviour a crucial element in crime. A few criminals are responsible for most crime. Crimes of the powerful, corporations or of the State are under-represented & neglected.
2. Most known criminals are: male, young, minority status, lower class & unskilled or unemployed. Most crimes involves offences against property while violence is rare. The role of family, school, peers and community are crucial factors in crime causation.
3. Provoked & stimulated many disciplines to examine the problems of crime: biology, anthropology, medicine, psychiatry, psychology etc.
4. Constitutional factors lead to an emphasis on social factors which stimulated the growth of the social sciences, criminology, social work etc.

5. International bodies formed & helped develop penology & the treatment of criminals & the administration of justice.
6. A practical influence on penal codes especially children, juveniles, mentally ill & other issues e.g. reliability of witnesses, victim behaviour & police & forensic science.
7. Emphasis on accurate information about the prevalence & variations of crime. It suggests to be systematic about building on the ideas of the theory which can be coherent and logical.
8. A wider definition of crime is preferred because legal definitions are narrow & do not reflect true consensual values or are limited to acts against the law rather than "deviance".
9. Inconsistencies & contradictions are addressed. Inborn individual factors were the primary causes of criminal behaviour & the criminologists were positivists because they used experimental & inductive methods of the natural & social sciences instead of the juristic & deductive reasoning of the classicist.
10. Explanations need to be comprehensive so as to be able to explain as many different crimes and aspects of offending. It can be testable & verifiable by the usual methods of investigation. Inferences need to be predictive & provide practical applications. Positivism's rejection of free will was based on criticism of the way responsibility in the law was regulated & interpreted especially definitions of insanity. Guilt was less important to society than the dangerousness of the offender.

Sutherland's Theory of Modern Criminology

Edwin H. Sutherland was an American sociologist and criminologist, often called the "Doin of Criminology" for his pioneering work in this area. He is considered one of the most influential criminologists of the twentieth century, probably best known for defining "differential association"—a general theory of crime that explains how deviants come to learn the motivations and the technical knowledge for criminal activity—and his definition of the "white-collar crime."

Sutherland was the author of the leading textbook, Criminology, published in 1924. In it, he rejected the early twentieth century theories about criminal behaviour that held that criminals were products of feeble-mindedness. In his third edition of the book (1939) he proposed the theory of "differential association," which explained how criminals come to learn the motivations and the technical knowledge for criminal activity.

Sutherland believed that criminal behaviour was learned through social interaction with others. He rejected the notions of criminal behaviour being caused by psychopathological or economic factors. He developed his own view of it, which he expressed in his "differential association" theory. In short, it says that a person becomes delinquent when in his surrounding the "definitions" that favor violation of law exceed the "definitions" that favor law. In other words, when a person is exposed to criminal behaviour in his near environment more than he is exposed to prosocial behaviour, he is likely to engage in criminal behaviour himself.

Sutherland's theory of modern criminology can be summarized as below.

1. Criminal behaviour is learned;
2. Criminal behaviour is learned from other people in the process of communication;

3. Learning criminal behaviour means learning motivation, rationalization, attitudes and techniques to commit the crime;
4. Particular motives and drives for criminal act is learned from definitions of the legal codes;
5. One engages in criminal behaviour because of an excess of definitions favorable to violation of law over definitions unfavorable to violation of law;
6. Differential association differ in intensity, frequency, duration, and priority;
7. The process of learning criminal behaviour is the same as learning any other type of behaviour;
8. Criminal behaviour is an expression of general needs and values, but it cannot be explained by those needs and values alone.

Sutherland believed that direct communication was essential in learning the criminal behaviour. Impersonal communication, such as movies or newspapers, had little influence in committing criminal behaviour. He often collaborated in his research with criminals, studying their lives and observing their behaviour. Such, discussions with the con man, Broadway Jones, led to publishing of The Professional Thief in 1937. The professional thief is a person who steals professionally. In the book, he claimed that not everyone can become a professional thief. Rather one must be accepted into a group of professional thieves where he would further learn about the profession.

Sutherland coined the phrase white-collar criminal in a speech to the American Sociological Association on December 27, 1939. He claimed that "conventional generalizations about crime and criminality are invalid because they explain only the crime of the lower classes" (White-Collar Crime, 1949), and that official statistics should include white-collar crimes. Sutherland defined a white-collar crime as "a crime committed by a person of respectability and high social status in the course of his occupation" (Ibid). He believed that white-collar crime was among the most dangerous types of crime in the modern world.

In recognition of his life and work, in 1960 the American Society of Criminology established the annual award that carries Sutherland's name. It is given to scholars for their contributions to theory or research in criminology.

References

Ahuja Ram, (2000) *Criminology,* Rawat Publications.

Barak, Gregg (ed.). (1998). Integrative Criminology, Aldershot: Ashgate/Dartmouth.

Beirne, Piers. (1994). The Origins and Growth of Criminology: Essays on Intellectual History, 1760-1945 Aldershot, Hants, England: Dartmouth.

Carrabine, Eamonn. (2004). Criminology: A Sociological Introduction. London: Routledge.

Gottfredson, M., T. Hirschi (1990). A General Theory of Crime. Stanford University Press.

Hale Chris et al, (2005), Criminology, Oxford University Press, New York.

Hallsworth, S. (2005), Street Crime, Cullompton: Willan Publishing.

Katz, J. (1988), The Seductions of Crime, New York: Basic Books.

Mehta Rohinton, (1999) Crime and Criminology — A socio legal Analysis of the phenomenon of Crime,

Snow White, First edition.

Merton, Robert (1957). Social Theory and Social Structure. Free Press.

Mike Maguire, Morgan Rod and Reiver Robert, (2007), The Oxford Handbook of Criminology, Oxford University Press, 14th ed.,at 741-742.

Paranjape, N.V. (2008) Criminology and Penology, 13th Edn. Central Law Publication, Allahabad.

Ponnian, M. (2003) Criminology and Penology, Pioneer Books, Third edition.

Presdee, M. (2000), Cultural Criminology and the Carnival of Crime, London: Routledge

Reiner, R. (2007) Law and Order, Cambridge: Polity Press.

Siddique Ahmad, (2003) *Criminology — Problems perspectives,* Eastern Book Co., Fourth edition.

Sirohi J. P., (2000) Criminology *and Criminal Administration,* Allahabad Law Agency.

Srivastava S. S., (2002) Criminology *and Criminal Administration,* Central Law Agency, Second edition.

Sutherland, Edwin H, (1929), Crime and the conflict process. Whittier, Calif: California Bureau of Juvenile Research.

Sutherland, Edwin H, (1973), On Analyzing Crime. Chicago: University of Chicago Press.

Sutherland, Edwin H, (1985) (original 1949). White-collar Crime. Yale University Press.

Sutherland, Edwin H. (1988) (original 1937). The Professional Thief. Chicago: University of Chicago Press.

Sylvester, Sawyer F, (1972), The Heritage of Modern Criminology. Cambridge, MA: Schenkman Pub. Co.

■■■

CHAPTER 2 Modern Concepts of Criminology

Learning Objectives :

- ❒ **Acquainting the Learners with the modern concepts and terminologies of Criminology.**
- ❒ **To define crime, *mensrea* and *actus reus.***
- ❒ **To grasp the legal and social connotations of crime.**
- ❒ **To comprehend the relationship between crime and morality.**
- ❒ **To understand the Characteristics of crime.**
- ❒ **To know about the types of crime.**

Introduction

Modern criminology ran counter to earlier thoughts on criminology, which held that people are basically decent and driven to crime only through extenuating circumstances. The pioneers include, Cesare Lombroso, Raffaele Garofalo, Enrico Ferri, Emile Durkheim, Edwin Sutherland, RK Merton and umpteen others. Modern criminology turned radically different from classical school and we find today several schools of thought under it such as Sociological Theories and other non-sociological theories such as Psychological Theories, Bio-physical Theories and Economic Theories. Modern criminology incorporates different aspects of these theories as well as numerous smaller concepts. The primary purpose is to identify the causes of crime and thus to seek amenable solutions (either through crime prevention, punishment rehabilitation, or alleviation of the social stresses which lead to crime). Most criminologists don't believe that crime will be erased in our lifetime or even many lifetimes (those who follow Durkheim's thinking may claim that society would suffer without at least a certain amount of crime). But by better understanding it as a social phenomenon, they can help reduce its negative impact, while helping law enforcement in the process.

MODERN CONCEPTS

The starting point of the discussion, like any other area of study, is to define and delimit the concepts and terminologies. This has to be done, at least to some extent, despite the fact that it is difficult to appreciate a definition without knowing something about the subject in the first instance and definitions should be attempted towards the end rather than in the beginning of the discussion of the subject. The most convenient starting point for understanding the nature and scope of criminology is obviously the concept of crime itself since the subject concerns itself with the study of crime and criminals from various perspectives.

Defining Crime

A crime free society with order, peace and non-violence is a myth which has never been realized in the history of mankind. Generally speaking, almost all societies have certain norms, beliefs, customs and traditions which are implicitly accepted by its members as conducive to their well being and healthy development. Infringement of these cherished norms and customs is condemned as anti-social behaviour. Thus many writers have defined 'crime' as an anti-social, immoral or sinful behaviour. However, according to the legal definition, 'crime' is any form of conduct which is declared to be socially harmful in a State and as such forbidden by law under pain of some punishment. Therefore, Tappan (1960:80) has defined crime as, "an intentional act or omission in violation of criminal law committed without any defence or justification and penalised by the law as felony or misdemeanour."

According to Kenny, "crimes are wrongs whose sanction is punitive, and is in no way remissible by any private person, but is remissible by the Crown alone, if remissible at all". But this definition has evoked criticism on the ground that there are indeed a number of compoundable offences that are remissible by the consent of the parties. Expressing his views on definition of crime, Roscoe Pound commented that "a final definition of crime is impossible, because law is a living and changing thing, which may at one time be based on sovereign will and at another time on juristic science, which may at one time be uniform, and at another time give much room for judicial discretion, which may at one time be more specific in its prescription and at another time much more general."

Cross & Jones define crime as a legal wrong the remedy for which is punishment of the offender at the instance of the State. John Gillin (1971:6) defines crime as an act that has been shown to be actually harmful to the society or that is believed to be socially harmful by a group of people that has power to enforce its beliefs and that places such act upon the ban of positive penalties. Thus he considers crime as an offence against the law of the land.

According to Blackstone (1905:5) a crime is an act committed or omitted, in violation of a public law either forbidding or commanding it. He, however, realised at a later stage that this definition may prove to be misleading because it limits the scope of crime to violations of a 'public law' which normally covers political offences such as offences against the State. Therefore, he modified his definition of crime and stated, "a crime is a violation of the 'public rights and duties' due to the whole community." Stephen, the editor of Blackstone's commentaries further modified the above definition and said, "a crime is a violation of a right, considered in reference to the evil tendency of such violation as regards the community at large." Thus both, Blackstone and Stephen stress that crimes are breaches of 'those laws which injure the community'. Stephen further added that "crime is an act which is both forbidden by law and revolting to the moral sentiments of the society."

Rejecting this juridical concept of crime the well known Italian criminologist Raffeale Garofalo preferred sociological definition of crime and stated that crime is an act which offends the basic sentiments of 'pity' and 'probity'. Yet another view about crime is to treat it as an anti-social behaviour which is injurious to society. Supporting this contention Sutherland characterises crime as a symptom of social disorganization. The tendency of modem sociological penologists is, therefore, to treat crime as a social phenomenon which receives disapprobation of the society. In the words of Donald Taft, "crime is a social injury and an expression of subjective opinion varying in time and place". Halsbury defines crime as an unlawful act which is an offence against the public and the perpetrator of that act is liable to legal punishment.

From the foregoing definitions, it may be said that a crime is a wrong to society involving the breach of a legal wrong which has criminal consequences attached to it i.e. prosecution by the State in the criminal court and the possibility of punishment being imposed on the wrongdoer. It is significant to note that though the legal definition of crime has been criticised because of its relativity and variable content yet Halsbury's definition is perhaps the most acceptable one as compared with other definitions because of its elaborate and specific nature and element of certainty. Further, it also provides for the machinery and procedure to determine the violations and to identify the offenders. Accordingly the legal and social dimensions of crime run on parallel lines.

Legal and Social Definitions Controversies

Crime being an unlawful act, the obvious way of defining it is in legal terms, in order to distinguish it from sin, religious and moral wrongs. The basic premise lies there in which the pitfalls resulting from individual or group opinions are avoided, to give, as far as possible, a scientific and precise character to criminology. Legal connotation looks at crime as "an act or omission in violation of criminal law", since any defence or justification is to be found within the criminal law, and there is no question of violating the criminal law if some defence or justification is available for a particular act or omission in certain circumstances. Further, it is not necessary for an act or omission to be intentional in order to be a crime; it could be made punishable on the basis of knowledge, recklessness or negligence or even without any reference to the mental element of the wrongdoer i.e. based on the concept of strict responsibility.

Yet there is another school of thought, which considers the legal definition to be inadequate and unsuitable for the purpose of criminology. It is called the social definition. Crime accordingly is defined as "an act which the group (social) regards as sufficiently menacing to its fundamental interests, to justify formal reaction to restrain the violator" (Johnson, 1974:11). Raffeale Garofalo, one of the three leading exponents of the Italian school of criminology, rejected the 'juridical' conception of crime which according to him fails in that it both includes and excludes behaviour properly encompassed in a "sociological notion of crime". For his sociological purpose, he formulated his theory of "natural crime". By "natural crime" he meant acts which offend the basic moral sentiments of pity (revulsion against the voluntary infliction of suffering on others) and probity (respect for property rights of others) (Allen,1964:67).

The exponents of each one of the above schools accuse the other of being unscientific in approach. The legal definition has been criticised on the ground that whether any act or omission is recognized as a crime at a given time in a society depends upon values which are relative and not on any intrinsic worth of the act or omission and that makes the study of crime unscientific. As aptly commented upon by an advocate of the social approach, the categories set up by criminal law are of a "fortuitous nature" and do not arise intrinsically from the nature of the subject-matter the scientists attempt to analyse (Sellin, 1938:21). This charge of variation in the legal attitude towards various acts is countered by the supporters of the legalistic approach by pointing out that not only do the legal norms vary due to various circumstances but also all the social norms which are essentially relative and impermanent. It is pointed out that criminal law not only gives precise definitions of forbidden acts but also has the machinery and procedure to determine the violations and, therefore, is able to identify the offenders, which is not possible in cases where certain conduct is branded as criminal in social terms irrespective of prevailing legal notions. This certainly is the advantage in the legal definition over the social one despite the various inherent weaknesses of criminal law processes like non-reporting, non-prosecution of many offenders, lack of or hostile witness, the possibility of false conviction, greater possibility of failure to convict all guilty persons and of innumerable cases pending disposal. The best exposition of the case for legal definition has been made by Paul W. Tappan in the article *"Who is the Criminal?"* in the following words:

"The validity of this contention (based on social definition) must depend, of course, upon what the nature of the subject-matter is. These scholars suggest that, as a part of the general study of human behaviour, criminology should concern itself broadly with all anti-social conduct, behaviour injurious to society. We take it that anti-social conduct is essentially any sort of behaviour which violates some social interest. What are these social interests which are weighty enough to merit the concern of the sociologist, to bear the odium? What shall constitute a violation of them? Particularly where, as is so commonly true in our complicated and unintegrated society, these interests are themselves in conflict? Roscoe Pound's suggestive classification of the social interests served by law is valuable in a juristic framework, but it solves no problem for the sociologist who seeks to depart from legal standards in search of all manner of anti-social behaviour. However desirable may be the concept of socially injurious conduct for purposes of general or abstract description, it does not define what is injurious. It sets no standard. It does not discriminate cases, but merely invites the subjective value-judgment of the investigator (1947:96-102).

It is because of the 'confusion' caused by the social definition that the use of the expression 'white-collar crime' by Professor Sutherland irks Tappan. He clinches the issue in favour of the legal definition by observing that convicted criminals represent the closest possible approximation to those who have in fact violated the law even if this group may not be complete or fully representative of all those who have committed crime. Further, the criminal law establishes substantive norms of behaviour, standards more clear-cut, specific and detailed than the norms in any other category of social control.

Actus reus and Mens rea

The terms actus reus and mens rea developed in English Law, are derived from the principle stated by *Edward Coke*, namely, actus non facit reum nisi mens sit rea, which means: "an act does not make a person guilty unless (their) mind is also guilty"; hence, the general test of guilt is one that requires proof of fault, culpability or blameworthiness both in behaviour and mind. Actus reus, sometimes called the external element or the objective element of a crime, is the Latin term for the "guilty act" which, when proved beyond a reasonable doubt in combination with the mens rea, "guilty mind", produces criminal liability in the common law-based criminal law jurisdictions of Canada, Australia, India, Pakistan, New Zealand, England, Ireland and the United States. In the United States, some crimes also require proof of an attendant circumstance.

In order for an actus reus to be committed there has to have been an act. An act consists of commission, omission or possession. Various common law jurisdictions define act differently but generally, an act is a "bodily movement whether voluntary or involuntary". In Robinson v. California, 370 U.S. 660 (1962), the U.S. Supreme Court ruled that a California law making it illegal to be a drug addict was unconstitutional because the mere status of being a drug addict was not an act and thus not criminal.

Omission : Omission involves a failure to engage in a necessary bodily movement resulting in injury. As with commission acts, omission acts can be reasoned causally using the but for approach. But for not having acted, the injury would not have occurred. The Model Penal Code in US specifically outlines specifications for criminal omissions: the omission is expressly made sufficient by the law defining the offense; or a duty to perform the omitted act is otherwise imposed by law. So if legislation specifically criminalizes an omission through statute; or a duty that would normally be expected was omitted and caused injury, an actus reus has occurred.

Possession : Possession holds a special place in that it has been criminalized but under common law does not constitute an act. Some countries like the United States have avoided the common law conclusion in Regina v. Dugdale by legally defining possession as a voluntary act. As a voluntary act, it fulfils the requirements to establish actus reus.

In this respect, the role of automatism is highly relevant in providing a positive explanation of the need to demonstrate the voluntariness of the behaviour for liability. This is supported by the English case Hill v Baxter, which held that the act must be voluntary for the defendant to be guilty. Voluntariness is one of the key points in establishing if an actus reus has been committed. A person suffering from a seizure who stabs somebody trying to help them has not committed an actus reus because it was an involuntary act. Definitions of a voluntary or involuntary act have varied over time but legal scholars have over time developed tests. Oliver Wendell Holmes, in his 1881 book The Common Law explained that "A spasm is not an act. The contraction of the muscles must be willed." Indeed, the Model Penal Code in US, which is utilized by many U.S. states in constructing their penal codes, specifically describes what are considered involuntary acts and thus not criminal : (1) a reflex or convulsion; (2) a bodily movement during unconsciousness or sleep; (3) conduct during hypnosis or resulting from hypnotic suggestion; (4) a bodily movement that otherwise is not a product of the effort or the determination of the actor, either conscious or habitual.

Voluntariness does not exclude omission because it is implicit in omission that the actor voluntarily chose to not perform a bodily movement and thus caused an injury. The purposeful, reckless, or negligent absence of an action is considered a voluntary action and completes the voluntary requirement for actus reus.

Thus, in jurisdictions with due process, there must be an actus reus accompanied by some level of mens rea to constitute the crime with which the defendant is charged. The doctrine had been discussed elaborately by Will. J. in Tolsoll's case (1869) 23 QBD 168. The Criminal Law does not usually apply to a person who has acted with the absence of mental fault; this is the general rule. The exception is strict liability crimes. In the civil law, it is not usually necessary to prove a subjective mental element to establish liability, say for breach of contract or a tort, although if intentionally committed, this may increase the measure of damages payable to compensate the plaintiff as well as the scope of liability. Quite simply therefore mens rea refers to the mental element of the offence that accompanies the actus reus. In some jurisdictions the terms mens rea and actus reus have been superseded by alternative terminology. In Australia for example, the elements of all federal offences are now designated as "fault elements" (mens rea) and "physical element" (actus reus). This terminology was adopted in order to replace the obscurity of the Latin terms with simple and accurate phrasing.

The traditional common law definitions and the modern definitions approach the crime from different angles. In the traditional common law approach, the definition includes : (1) actus reus : unlawful killing of a human being and (2) mens rea : malice aforethought.

Modern Law approaches the analysis somewhat differently. Homicide is a crime in that it forbids any "intentional" or "knowing" conduct that results in the death of another human being. "Intentional" in this sense means the actor possessed a "purpose" or "desire" that his or her objective (i.e. death of another human being) be achieved. "Knowing" means that the actor was aware or practically certain that the death would result. Thus, the actus reus and mens rea of homicide in a modern criminal statute can be considered as follows : (1) actus reus : Any conduct resulting in the death of another individual and (2) mens rea : The conduct resulting in the death was done intentionally or knowingly.

In the modern offense element approach, the attendant circumstances tend to take over for the traditional mens rea, indicating the level of culpability as well as other circumstances, i.e. the crime of theft of government property would include as an attendant circumstance that the property belong to the government.

Criminologists use words/phrases a certain way to indicate relationships between causes (independent variables) and effects (dependent variables). Here are few such examples.

1. "varies with" — this means things fluctuate together; as one thing goes up, the other thing goes down; usually used to describe a possible inverse relationship but also used to describe a direct relationship.
2. "where..." — while not technically a verb, this word usually indicates a feedback relationship, where things go up or down in response to one another. Often, but not always, the case involves an important Z factor which moderates, distorts, or confounds the relationship. Relationals like "varies", "fluctuates", "predominates", "associated with", and "overrepresented by" are usually found when the theorist is dealing with socio-demographic variables, like age, race, or social class.
3. "seems to be" — this wishy-washy language usually means that the theorist suspects a weak relationship, probably way less than 50%.
4. "tends" — this might mean, but not always, that there are important Z factors which are antecedent, intervening, or contingent; that is, that come before, in the middle, or after an X and Y relationship. Or, it may be a cojoint relationship.

5. "is conducive to" — this usually means that the cause is a mysterious, unknown onstruct; typically found in highly abstract theories involving words like anomie, relative deprivation, norms, or controls. In some cases, it refers to a confounding or contextual relationship.

Crime and Morality

The word 'crime' is derived from the Latin word '*krimos*' which means 'to accuse'. It covers those acts which are against social order and deserve disapprobation and condemnation of society. Linking crime with morality, Garofalo (1914: 59), observed that "crime is an immoral and harmful act that is regarded as criminal by public opinion because it is an injury to so much of the moral sense as is possessed by a community.

The perception of crime as an immoral act had its roots in the medieval legal system when laws were mostly based on religious precepts and the State was subordinated to the Temporal power. The Penal Code was considered as a body of ethical rules making all immoral acts punishable. But with the change in time and advance of legal science, the social norms too have radically changed. Therefore, now a crime is defined as an act forbidden and punishable by law and it is immaterial whether such an act is moral or immoral from the ethical point of view. No doubt, most of the immoral acts which were traditionally considered as crime are treated as crime even today, but there are a number of conducts which though immoral are not considered as crime. For example, ingratitude, callous disregard for sufferings of others, hard-heartedness etc. are not regarded as crime though they are against morality. It would be pertinent to quote the observations made by the authors of the Indian Penal Code in this regard :–

"Many things which are not punishable are morally worse than many things which are punishable. The man who treats a generous benefactor with gross ingratitude and insolence deserves more severe reprehension than the man who aims a blow in passion, or breaks a window in a frolic; yet we have punishment for assault and mischief, and none for ingratitude. The rich man who refuses a mouthful of rice to save a fellow-creature from death may be far worse than the starving wretch who snatches and devours the rice; yet we punish the latter for theft, and we do not punish the former for hard-heartedness. .."

It would, therefore, be seen that if the social expediency along with so other factors which makes an act a 'crime' it is not material whether that act is moral or immoral.

Characteristics of Crime

There are certain characteristics of a crime which make an unlawful act or omission punishable under the law of the land. The characteristics of a crime are discussed as below :

1. External consequences : Crimes always have a harmful impact on society may it be social, personal, emotional or mental.

2. Act (Actus Reus) : There could be an act or omission to constitute a crime. Intention or mens-rea alone shall not constitute a crime unless it is followed by some external or overt act. Generally, omitting to do something will not amount to actus reus of an offence. The criminal law usually punishes individuals for positive conduct and not for inaction. There are, however, some notable exceptions. For example, a police officer may have a duty to act to prevent an assault and if he does not, he will be liable to be punished under the law.

3. Mens-rea or guilty mind : Mens-rea is one of the essential ingredients of a crime. It may, however be direct or implied. The implied mens-rea is otherwise termed as constructive mens-rea. Mens rea implies that there must be a state of mind with respect to an actus reus, that is, an intention to act in the proscribed fashion.

4. Prohibited act : The act should be prohibited or forbidden under the existing penal law. An act, howsoever immoral, shall not be an offence unless it is prohibited by law of the land.

5. Punishment : The act in order to constitute a crime should not only be prohibited by the law but should also be punishable by the State. The punishment is usually set out in terms of a maximum and the actual punishment in any particular case is left to the discretion of the judge. Both, the defence and the prosecution have a right to appeal against the quantum of sentence.

6. Society and Time Specificity : Crime and social policy are inter-related and the concept of crime and punishment depends largely on the social values, accepted norms and behavioural patterns of a particular society at a given time.

7. Permissibility : Like the society, crimes are also a varying content changing with the changes in social structure. What is crime today may become a permissible conduct tomorrow and vice versa. For example, abortion which was considered to be a heinous crime because of the immorality involved in it is no longer an offence under certain circumstances consequent to the enactment of law legalising abortion. Even prenatal determination of the sex of a child has been prohibited and made punishable with a view to prevent parents aborting a girl-child.

8. Relativity of Crime : Crime is a relative term, therefore, what is wrongful (crime) at one place may not be necessarily so in another place. Thus adultery is a criminal offence in India but in England it is merely a civil wrong redressible by payment of compensation. Again, in India consuming liquor is an offence in many States under their respective prohibition laws but it is not so in wet areas where there is no licensing on sale of liquor. This relativity of crime obviously reflects upon the varying social reactions to human conduct in different places.

9. Index of social progress : The moral tune of the society can easily be gauged from the law of crimes enforced in that society at a given time. This in other words means that criminal law is an index of social progress of a given society. The emergence of law of crime and criminological knowledge has been through a definite process of evolution corresponding to different phases of social evolution. The modem complexities of human life have contributed to the rising incidence of crime. But there is nothing to be purturbed about this rising trend in criminality. In fact, it is a myth to think of a crimeless society. Modem criminologists have even gone to the extent of reckoning increase in crime as a symptom of social progress.

10. With the passage of time, the emphasis has shifted from 'crime' to 'criminal' : The modem view regarding penal policy favours individualisation of the offender through clinical treatment methods. This has led to the emergence of reformatory era in the field of penology thus rendering the earlier deterrent, retributive and retaliatory methods completely obsolete and outdated.

Classification of Crimes

The existence of crime in a society is a challenge to its members due to its deleterious effect on the ordered social growth. In fact, it leads to a colossal waste of human energy and an enormous economic loss. Therefore, with the advance in the field of criminology and behavioural sciences, efforts are being constantly made to work out a commonly acceptable classification of crimes and criminals for providing a rational basis of punishment for various categories of offenders.

There are a variety of crimes such as violent personal crimes, occasional property crimes, occupational crimes, political crimes, public order crimes, conventional crimes, organised crimes, professional crimes, white-collar crimes, sexual crimes, crimes against property, person, decency, public order etc. Broadly speaking, these may be categorised into three main heads, namely, (i) offences falling under Code of Criminal Procedure; (ii) offences under Indian Penal Code; and (iii) offences under local or special laws or enactments.

Some writers have preferred to classify crimes into legal, political, economic, social and miscellaneous crimes.

1. Legal crimes : Legal crimes can be termed as traditional crimes such as theft, Robbery, dacoity, rape, hurt and rioting etc.

2. Political offences : Political offences are those which are motivated politically or committed in violation of the election laws or norms set out for the politicians in course of their political activities.

3. Economic crimes : Economic crimes include white-collar offences such as tax evasion, smuggling, prostitution, gambling, foreign exchange violations, offences under the MRTP (Amendment) Act, 1991 etc.

4. Social crimes : Social crimes are those which are committed under social legislation such as The Child Marriage Restraint Act, 1978; Protection of Civil Rights Act, 1955; Immoral Traffic (Prevention) Act, 1956; Indecent Representation of Women (Prohibition) Act, 1986; Commission of Sati (Prevention) Act, 1987; The Dowry Prohibition Act, 1961 as amended in 1983 and 1986; Juvenile Justice (Care and Protection of Children) Act, 2000 as amended in 2006; Scheduled Castes & Scheduled Tribes (Prevention of Atrocities) Act, 1989 to name a few.

5. Crime without a Victim : There are certain offences which though punishable under the law, do not have any direct harm on others. Such offences may be termed as victimless crime. For example, drunkenness and related offences, sale and use of prohibited substances, vagrancy, begging, soliciting, bestiality etc. do not cause any harm to any person and hence they may be categorised as victimless crime. It may be stated that legality or illegality of such crimes depends mostly upon the morality and economic interests of the community. It is mainly for this reason that many of these activities are decriminalised and taken out from the purview of criminal law. Thus, drunkenness and homosexuality are decriminalised in U.K. and many other European countries when not committed in a public place. The justification advanced for decriminalisation of these offences is that use of non-criminal methods such as social service programmes etc. to help drug addicts or drug-traffickers or homosexuals would perhaps be more helpful than the use of criminal law to control their behaviour.

6. Cyber crime : More recently, a new specie of crime known as cyber crime has emerged as a result of development of computer science and information technology during last quarter of the 20th century. These are crimes either computer generated crime or where computer software is itself a target of crime. A special Act called the Information Technology Act, 2000 has been enacted to tackle these offences.

7. Miscellaneous crimes : All other remaining crimes which are committed under local or special Acts are termed as miscellaneous crimes, for example, offences under the Prevention of Food

Adulteration Act, 1954; Drugs Act, 1940; Consumer's Protection Act, 1986; Prevention of Illicit Traffic in Narcotic Drugs & Psycho-tropic Substances Act, 1988 etc.

8. Classification of offences under I.P.C : Under the Indian Penal Code (IPC), various offences have been classified into fifteen broad categories on statistical basis. They are :

TABLE NO. 2.1

SL. NO.	TYPE	EXAMPLES	CHAPTER IPC
1.	Offence Against Person	Hurt, Murder, Rape, Kidnapping, etc.	XVI
2.	Offences against property	Theft extortion , Robbey, Dacoity, Cheating etc.	XVII
3.	Offences relating to documents	Forgery of documents, currency, notes, property	XVIII
4.	Offences affecting mental order	Insult, intimidation and annoyance	XXII
5.	Offences against public tranquility	Unlawful Assembly, Riotous, aftray	VIII
6.	Offences against state	Conspiring Attempting and waging war	VI
7.	Offences relating to public servants	False personality, training incorrect records disobeying direction etc.	IX, X
8.	Offence relating to coin preparing stamps	Govt. stamp counterfit coins	XII
9.	Offences affecting public health, safety, morals dowry	Nuisance of all sorts	XIV
10.	Offence against religion	Insulting religion etc	XV
11.	Offences relating to marriage	Adultery, False Marriage etc	XX
12.	Offence relating to cruelty to married women	Dowry torture, Domatic violence	XXA
13.	Offences relating to defamation	Using defamatory words, printings and engraving defamation matter	XX
14.	Offence relating to weight and measures	Keeping to using false weight and measures	XIII
15.	Offences relating to election	False presentation briber, Booth capturing etc.	IXA

This classification seems to be more rational and elaborate from the point of view of administration of criminal law and penal justice.

■■■

CHAPTER 3

Economic Theories

Learning Objectives :

- ❒ **Acquainting the Learners with the Becker's, Quinney's and Bonger's economic theories of crime.**
- ❒ **To understand Bonger's text in Indian context.**
- ❒ **To grasp the critical appreciation of economic theories of crime.**

Introduction

In recent years, there seems to have been a transformation of criminological views regarding somewhat sceptical question of criminal accountability. Modern critics attack the traditional criminological view on the ground that their search for characteristic differences between the class of criminals and the class of non-criminals rests upon erroneous assumption. This false dichotomy has been based on a misconceived characterisation of criminals as 'criminal type'. As Michael Phillipson aptly observes that to take crime out of its social context and to try to explain it as a product of physical characteristics or mental deficiencies is a myth. He summarises his criticism of traditional criminology by suggesting that it contains four false assumptions, namely :-

1. that there are universal causes of crime;
2. that the human population can be divided into two groups, criminals and non-criminals;
3. that crime can be located by the study of individual criminals; and
4. that the official statistics are indices of trends in crime.

The proponents of modern criminology attempt to explain criminality in terms of social conflict. Engels (1971) pointed out that resentment among the deprived class of society due to their exploitation and demoralisation was one of the reasons for growing criminality. Therefore, there was need to change the whole of the social and economic structure of society. Thus modern criminology attributes societal reasons for general criminality and suggests a pragmatic approach to the resolution of the problem.

The advocates of modern criminology firmly believe that distinction between criminals and non-criminals is the direct outcome of a mistaken notion of labeling certain individual offenders as 'criminal types'. Modern criminologists prefer to identify the criminal with a particular social type who has been a victim of well known inequalities between social classes, private wealth, private property, social power, and life chances. Thus there is nothing like 'criminal type' as suggested by traditional criminologists. The modern criminologists have succeeded in substituting the traditional belief regarding crime causation by social deviance as a cause of criminal behaviour.

It may be stated that modern criminologists lay greater stress on multiple causation theory because they consider crime as a social phenomenon, the political society reacting through punishment, treatment or preventive measures as the sequence of interaction is the ultimate object of criminology. They suggest that social control mechanism must be very selective in the legal norms for enforcement in their disposition of law violators. They advocate restructuring the existing social arrangements to eliminate crime from society. The current theory of Indian criminal jurisprudence is based on seven fundamental notions, viz., the principle of legality, mens rea, conduct, consequence of mens rea and conduct, harm, causation and punishment. With the change in time, the criminal law has radically changed and the concept of criminal liability therefore, faces new problems.

It may well be said that changing nature, extent and types of crime can not be explained by a single all embracing theory, no matter how systematic it may be to account for all dimensions. Modern theories are discussed under four major heads, viz. economic theories, psychological theories, biophysical theories, sociological theories.

ECONOMIC THEORIES

Becker's Economic Analysis of Crime

In 1968, Gary Becker provided the modern, mathematical formulation of what is now generally considered to be the economic model of crime. Becker was awarded the Nobel Memorial Prize in Economic Sciences in 1992 for his Economic Analysis of Crime and punishment as one of his thrust areas of specialization. Becker posits that criminals, indeed all individuals, are rational maximizers of expected utility. On the one side, criminal acts can generate benefits for perpetrators—for example, the utility they might derive from hurting someone they are angry with or the resources obtained through theft. On the cost side, society imposes sanctions, generally with uncertain probability, on those who commit crimes. The Beckerian rational calculator weighs the expected costs against the expected benefits, and if the first are lower than the second, the person commits the crime. From this very simple model, Becker emphasized that society could reduce the amount of crime by either lowering the benefits that criminals garnered from their illegal activities or by raising the costs that would befall them (or both).

Becker's analysis includes various forms of crime. Becker argued that crime is not an aberration outside the scope of rational analysis, but rather the predictable outcome of opportunities for gain. The economic analysis of crime starts with one simple assumption: Criminals are rational. A mugger is a mugger for the same reason one is a professor-because that profession makes him better off, by his own standards, than any other alternative available to him. Here, as elsewhere in economics, the assumption of rationality does not imply that muggers (or professors) calculate the costs and benefits of available alternatives to seventeen decimal places-merely that they tend to choose the one that best achieves their objectives. The assumption of rationality applies not only to criminals but to everyone included in the analysis. Judges, policemen, legislators and potential victims are all, like criminals, rational individuals pursuing their own ends as best they can. Economic theory gives us no reason to assume that criminals are less rational than judges, or judges less self-interested than criminals.

If muggers are rational, we do not have to make mugging impossible in order to prevent it, merely unprofitable. If the benefits of a profession decrease or its costs increase, fewer people will enter it-whether the profession is plumbing or burglary. If little old ladies start carrying pistols in their purses, so that one mugging in ten puts the mugger in the hospital or the morgue, the number of muggers will decrease drastically-not because they have all been shot but because most will have switched to safer ways of making a living. If mugging becomes sufficiently unprofitable, nobody will do it.

When the analysis moves from descriptive to prescriptive, it seems natural to apply a similar approach to ourselves. In designing institutions to control crime, our concern is not sin but cost. Eliminating all murders, even all muggings, would no doubt be a fine thing-but whether we ought to try to do it depends on how much it costs. If reducing the annual murder rate from ten to zero requires us to turn half the population into police, judges, and prison guards, it is probably not worth doing.

While Becker acknowledged that many people operate under a high moral and ethical constraint, criminals rationally see that the benefits of their crime outweigh the cost such as the probability of apprehension, conviction, and punishment, as well as their current set of opportunities. From the public policy perspective, since the cost of increasing the fine is trivial in comparison to the cost of increasing surveillance, one can conclude that the best policy is to maximize the fine and minimize surveillance. However, this conclusion has limits, not the least of which include ethical considerations.

One of the main differences between this theory and Jeremy Bentham's rational choice theory, which had been critisized in criminology, is that if Bentham considered it possible to annihilate crime completely (through the panopticon), Becker's theory acknowledged that a society could not eradicate crime beneath a certain level. For example, if 25% of a supermarket's products were stolen, it would be very easy to reduce this rate to 15%, quite easy to reduce it until 5%, difficult to reduce it under 3%, and nearly impossible to reduce it to zero (a feat that would cost the supermarket so much in surveillance, etc. that it would outweigh the benefits even if it were possible).

He argues that a decision to engage in illegal activity is the outcome of an individualistic calculus; benefits and costs (both monetary and non-monetary) are weighed up, and the individual makes a decision which reflects the expected balance of them. One way to conceptualise decisions of this kind is as a rather special kind of investment activity. Many of the crucial decision variables-probability of apprehension and conviction, likely punishment, alternative earnings possibilities in legitimate occupations - are empirically observable, and hence their effect on observed crime rates can in principle be tested. As usual Becker's contribution has mainly been to analyse and suggest possibilities for hypothesis testing, but his graduate students and other interested economists have been quick to pick up the challenge. A good deal of evidence has been accumulated to support the plausibility of Becker's contention that criminal behaviour responds to changes in costs and benefits.

Unusually for Becker, the argument is couched throughout in normative terms. The model of criminal behaviour put forward is devised to be used in conjunction with cost functions for law enforcement and a simple social welfare function in order to generate conclusions about the optimal levels of policy variables such as the extent of enforcement, type of punishment and perhaps even what should be a crime. Becker is not, however, arguing for major policy changes. Given the behavioural responses to legal and illegal incentives which he discerns, and given the costs and benefits of enforcement and punishment programmes, he suspects that the authorities, at least in the USA, get things roughly right-perhaps a surprising conclusion, given his scepticism of the efficacy of government action in other spheres.

One could reduce the benefits of crime in a number of ways. As Becker emphasized in his Nobel lecture, one approach might be to enhance education or job training so that the opportunity cost of crime would be higher, and individuals would find options in the legitimate sector to be more attractive than in the criminal one. More targeted approaches could also be used to reduce the benefits of particular criminal choices (Becker, 1993: 389-392). For example, one could reduce the benefits of theft by making car radios inoperable if taken from the car without having a security code.

Becker also stressed that policymakers could use two tools in an effort to raise the costs of crime: increasing the probability of apprehension and increasing the severity of the punishment for those who are caught. Some have believed that Becker is the intellectual champion behind the policy of increasing incarceration, but this is only partially correct. Becker's basic model actually favored monetary fines over incarceration since the former, being mere transfers, are socially costless: one reduces the utility of the miscreant by transferring his wealth to the state. Unlike fines, locking up inmates is obviously socially costly: the miscreant suffers a loss of utility, but the state ends up expending resources rather than collecting them. In addition, convicts are frequently unable to work productively during their time in prison, so the diminished work output also represents a social cost. Note then that Becker's basic model focused on pure deterrence rather than incapacitation; fines can deter but certainly do not incapacitate.

Becker's insight is important and suggests that society should look for opportunities to use fines when appropriate. Unfortunately, when we are talking about street crime, Becker's basic model has the extreme shortcoming of depending on a sanction—a monetary fine—that is of little value for the large percentage of the offending population that is essentially judgment proof (or close enough to make large fines ineffectual). This shortcoming was further compounded by Becker's emphasis that the combination of low probability of apprehension and high sanction was optimal because it reduces the social costs that accompany high rates of arrest and prosecution.

Becker's model was brilliant and would be unassailable if the model fully captured the calculus of potential criminals and the factors influencing the decision to commit crime, and if criminals were fully rational, well informed, sufficiently solvent to be responsive to high monetary sanctions, and risk neutral. Unfortunately, none of these conditions are true. Thus, a number of lessons that emerged from Becker's work—rely on fines as much as possible, and reduce the likelihood and increase the severity of punishment—are probably bad advice for setting good criminal justice policy for the type of street crime that Americans have been most worried about over the last four decades. As Cesare Beccaria had written centuries ago, higher likelihood of apprehension and less extreme sanctions probably work best to dampen criminal propensities. The more impulsive the criminal, the more Beccaria's insight is likely to be important and the less valuable Becker's recommendation. I should stress that Becker did not overlook this point; he is far too sharp a mind for that, and he specifically noted that if potential criminals are risk loving instead of risk averse one would want to increase the likelihood of apprehension.

It is somewhat unfair to blame the massive increase in incarceration on Becker's views about punishment, which are actually far more nuanced than some have realized. Indeed, his emphasis on monetary sanctions was part of his belief that the criminal justice system relied too heavily on incarceration as a punitive sanction. Becker pointed out that if criminals are risk neutral, they might perceive equal expected punishment from a regime that caught a high percentage of criminals and gave them all a relatively light sentence and one that caught a low percentage but really punished these few very harshly. If equal levels of deterrence could be obtained from these two regimes, Becker explained that we should opt for the low probability of catching the criminals since this will lead to lower costs of processing them through the criminal justice system. Becker correctly stated that risk-averse criminals would actually be more deterred by the low probability, high-sanction regime, but that risk-loving criminals would respond more to a Beccarian rather than a Beckerian approach (Levitt and Miles, 2005: 4-5).

Thus, the three messages to emerge from Becker's work were: (1) we need to raise the expected costs of criminal behaviour if we are to reduce crime; (2) the most efficient way to achieve these higher expected costs is through fines (where fines are possible) and through greater resort to incarceration (or executions, if potential murderers perceive a death sentence to be worse than a life sentence) when fines are not feasible (obviously, those without resources cannot pay fines and thus will not be deterred); and (3) increasing penalties for the relatively few criminals that are caught achieves the goal of deterrence at lower costs since fewer defendants would need to be caught and processed through the criminal justice system. But moving from Becker's theory into a public policy arena, where it was recognized that reliance on fines was not a practically feasible deterrent to the street crime that most Americans feared, the Beckerian model has not generated unvarnished success. Becker's work, though nuanced, provided a strong intellectual stimulant to the massive increase in incarceration, which was effectuated through more frequent and longer impositions of terms of imprisonment as well as through the war on drugs. There is certainly some reason to fear that the mass incarceration policy, however appropriate the initial increases were, may have gone too far. Moreover, the war on drugs, while successful in some respects, has not followed Beckerian predictions since the effort to raise the price of drugs by tough criminal enforcement has in fact failed. Indeed, over the last 20 years, Caulkins and Reuter (2006) assert that the inflation-adjusted price of cocaine has fallen by more than two-thirds.

At the same time, Becker also weighed in as a supporter of the death penalty on the grounds that it acted as a "substantial deterrent" of murder. Specifically, he claimed to support the use of capital punishment "for persons convicted of murder because, and only because ... it deters murders" (Becker, 2006: 1). Many knowledgeable commentators were highly skeptical of the view that the death penalty, as implemented in the United States, could serve as much of a deterrent. For example, as early as 1967, sociologist Thorsten Sellin, through a comparison of homicide rates in contiguous states, created doubt about the existence of a deterrent effect. Basically, he tried to choose comparable states that differed only in their reliance on the death penalty and then used their similar patterns of homicide to suggest that the death penalty had no impact on crime (Sellin, 1967). Sellin's work had the advantage of simplicity and transparency but the defect of not being able to statistically control for all of the factors—for example, changes in the rate of incarceration or in the numbers of police—that might influence homicide rates. Others challenged the deterrence hypothesis by questioning its sociological effects on communities. For example, William Bowers and Glenn Pierce (1980) argued that the death penalty would make a statement about the acceptability of deadly retributions, leading to a "brutalization effect" on society. Consider whether the recent execution of Saddam Hussein in Iraq may have stimulated more killings than it deterred.

Unfortunately, some of the followers of Becker applied their views about the economic model of crime too monolithically. "Raise the price and you will get less of it," while certainly true if other things remained equal, became a mantra that often obscured some of the other things that did not remain equal. While others followed the view championed by Beccaria that criminals were most likely to be deterred by high certainty of punishment (thereby allowing more lenient sentences), Becker's model was used to champion the notion that greater deterrence would be found in infrequent resort to harsh penalties. In this way, a woodenly applied version of Becker's model, which many came to associate with the economic model of crime, seemed to champion, among other things, the administration of the US death penalty apparatus.

One other important development in the last two decades is the American justice system's resort to sentencing guidelines designed to reduce the variability in sentencing. Interestingly, this is one area where the law has moved in a direction counter to the thrust of the economic model of punishment that Becker advanced. He predicted that if criminal sentences are kept the same in terms of harshness but become more uniformly applied, risk-neutral criminals would be unaffected. Furthermore, risk-averse criminals actually feel less deterrent pressure: punishments applied with greater certainty are less onerous than the more variable prospect of punishment.

The great public works project of our time—prison construction—has been an essential feature of the massive growth in America's prison population. In 2003, about 2 million Americans were imprisoned, with roughly 600 out of 100,000 behind bars, a figure that is far higher than analogous rates in other developed countries. To place the growth in incarceration in historical context, consider that between 1933 and 1973, "incarceration in the United States varied within a narrow band of roughly 100 to 120 prisoners per 100,000 population" (Donohue, 2005: 48). To achieve today's large prison population, the fraction of those imprisoned grew by an annual average of 5 percent starting in the mid-seventies. At times, the debate over the level of incarceration in the United States has seemed to divide into polar camps: those opposing the post-1970s increase or those fully endorsing it (and even calling for further prison expansions).

There seem to be two main weaknesses to Becker's arguments. The first is the assumption of social homogeneity implicit in the notion of a social welfare function, when it is widely held (not least among economists) that some groups of the population have greater political power than others, leading to legislation and enforcement patterns which reflect the influence of sectional interests. Secondly, it is difficult not to feel that Becker's enthusiasm for the economic approach does tend at times to run away with him. Although differences in incomes and assets, alternative earnings possibilities, probabilities of conviction and so forth are much more important in determining behaviour than they are often given credit for, there are surely variations in attitudes and degrees of honesty which affect the propensity to commit crimes even among individuals facing similar economic circumstances. While Becker would accept this, by implication he regards them as not particularly significant, possibly assuming that such variations in 'tastes' are randomly distributed.

Some of the strongest critics have questioned whether increased incarceration really lowers crime, pointing out that the deterrent impact of incarceration (as opposed to the admitted incapacitative benefits) has not been established and arguing that there are negative consequences on those incarcerated that may enhance subsequent levels of criminal misconduct. This concern about possible adverse effects on crime rates from increasingly punitive sanctions is buttressed in a recent study by Chen and Shapiro (2005), which uses a regression-discontinuity analysis to study the effects of prison conditions and sentence length on recidivism rates. Specifically, they exploit the fact that prisoners in their sample are assigned to prisons based on a security-level score, allowing them to study prisoners whose scores lie close to the cutoff points. For example, the cutoff between a lower and a higher security prison provides a natural experiment to assess the effects of prison conditions, as inmates on the dividing line are relatively equal in their initial criminality characteristics but are exposed to substantially different prison conditions.

Nevertheless, Becker's achievement is substantial. This one paper has stimulated a mushrooming literature on the economics of crime which it would be foolish for criminologists to ignore. The debate about deterrence, for instance, has been raised to a new level of sophistication by the introduction of econometric techniques by followers of Becker.

Richard Quinney : Reality of Crime

Richard Quinney's book "The Social Reality of Crime" published in 1970 shifted the discussion of criminality from police and courts to issues of social justice. He held that crime was not a static set of rules violations but a structure defined by wealth, power and the individual drive for freedom. Quinney's book reflected the cultural upheaval of the late 1960s with its emphasis on the worth of the individual and its suspicion of the rich and powerful. Until Quinney's book, the predominant issue about crime and criminals in the United States was punishment vs. rehabilitation. Crime was seen as a sign of pathology.

Richard Quinney sought to study the problem of crime and criminality with focus on the conflicts of groups in the context of the various institutions, political, economic, religious, kinship, educational and public. In particular, his thinking was moulded by the protest movement against the Vietnam war around 1970. The diverse groups, according to Quinney, were linked to the major institutional concerns but the interest structure was characterised by the unequal distribution of power. He rejected the pluralist notion that law represents the compromising of diverse interests; rather law supports some interests at the expense of others (*Carey*, 1978:133-36).

Quinney held the wealthy and powerful decided the rules, making crime a social justice issue. He attributed to humans a desire for freedom that made them innately resistant to the structures set by society, turning them into potential criminals. Quinney's thesis was in part intended to counter the growing power of B.F. Skinner in psychology and more broadly in society. Skinner's behaviourism saw human action as an inevitable function of outside stimuli instead of the result of innate or internal drives. Quinney begins this piece with differentiating "static and dynamic interpretations of society": He contends that the static view, taken by most sociologists, focuses on the social structure as static and consensual. Quinney's own view is dynamic. By that he means that his theoretical perspective is based on emphasizing that social relations are built over time, and that they do not generally come of one-trial learning.

Quinney is offering us a continuum as a tool for analysis, the static - dynamic continuum like the positivist - postmodern continuum, the conflict - consensus continuum, the individual - society continuum, the individual as passive - active continuum. Another facet of this dynamic interpretation Quinney advocates is his philosophy that man is active, not passive. Quinney insists that man is active, has choices, makes choices, and is not nearly as controllable as behaviourism suggests. Quinney relates this active nature of man to a tendency to struggle against control: "It is thus against something that the self can emerge." As for example, the adolescent needs to rebel to establish that sense of self as independent. Quinney sees this social construction of both reality and identity as interdependent. The story of the white-footed mice shows that even creatures exhibit a need to control their own environment. The problem with using rats and psychology sophomores for research is that they never told us some of the rats just don't run, some of the psychology sophomores are devious and don't play fairly, Deviants, undoubtedly.

Quinney says that crime is created. By that, he is referring to the social definition of crime, to the fact that the system of government we have created for ourselves was and is constructed by those who have titled authority and power. In creating the structure they protect their own privilege. Those who are rewarded with titled places in the structure also protect those privileges for they aspire to their share in them. Those who are hired into the agencies and institutions of the system gain higher authority and title, as well as some form of sharing in the profits, by following the agenda of the system. It's a little like everyone owning a "piece of the rock." The owner class may have more than the others, but each is a stake holder to a small extent.

Since the definitions of crime are enacted by those in power, this ever widening gap in power as well as income, puts even more power in the hands of the defining group. Not only that, but the criminal laws and regulations have been in effect now for a long enough time that they seem to just be there as a normative consensus. The original source of the definitions of deviance and crime are long forgotten. We no longer think to look to unstated assumptions, for the injustices. The laws have come to feel "right."

Quinney further reminds us that the definition of criminal accompanies more accurately our reaction to the behaviour than does the behaviour itself. If a student is late turning in his paper, but he is always thorough, and on time, that lateness may be overlooked. But if a student, who is often sloppy, not so accurate in his answers, and often late, turns in his paper late, we are more likely to react to that as one more offense on that student's part. Each student engaged in the same behaviour. Yet our reaction labelled one student's behaviour more negative and more egregious than the other's.

Quinney laid down six propositions regarding crime and operation of criminal law based on his earlier theories.

Proposition 1 (Definition of Crime): Crime is a definition of human conduct that is created by authorised agents in a politically organised society.

Proposition 2 (Formulation of Criminal Definitions): Criminal definitions describe behaviours that conflict with the interest of the segments of society that have the power to shape public policy.

Proposition 3 (Application of Criminal Definitions): Criminal definitions are applied by the segments of society that have the power to shape the enforcement and administration of criminal law.

(Propositions 2 and 3 are clearly based on Void's group interest theory (Void and Bernard, 1986:277-80) and proposition 4 linked to Sutherland's differential association theory. Between Void and Quinney there is, however, the difference that while Quinney included all criminal behaviour in his proposition, Void had specifically excluded "impulsive, irrational acts of a criminal nature that are quite unrelated to any battle between different inter-groups in organized society").

Proposition 4 (Development of Behaviour Patterns in Relation to Criminal Definitions): Behaviour patterns are structured in segmentally organised society in relation to criminal definitions and, within this context, persons engage in actions that have relative probabilities of being defined as criminal.

Proposition 5 (Construction of Criminal Conceptions): Conceptions of crime are constructed and diffused in the segments of society by various means of communication.

Proposition 6 (The Social Reality of Crime): The social reality of crime is constructed by the formulation and application of criminal definitions, the development of behaviour patterns related to criminal definitions, and the construction of criminal conceptions.

Some of the aspects in the propositions may be elaborated as follows :

- Law in operation is an aspect of politics. Thus crime is not a matter of individual pathology, "but is a judgment made by some about the actions and characteristics of others".
- In the creation and interpretation of laws, the values of some necessarily taken into account may negate the values held by others. The relative power and influence of the segments to which individuals may belong would determine the situation.
- The *politics of reality* is that people grant those in power the authority to carry out the actions that best promote the interests of the authorities.

Quinney has often been typecast as a Marxist, but his philosophy was influenced by religious sources as well, including a stay in a monastery and interaction with Buddhist philosophers. But Quinney's view of crime created a revolution in criminology. Academia was quick to pick up on his theories, and college courses began to look harder at social causes of crime instead of just the pathology of the criminal. Excerpts from his books are routinely included in college course books, as are discussions of his ideas.

William Bonger : Criminality and Economic Conditions

William Bonger (1916) has tried to show the influence of the economic system and of these consequences upon the social instincts of man. Mere egoistic tendency does not by itself make a man a criminal. It is possible for the environment to create a great egoist, but this does not imply that the egoist will necessarily become criminal. For example, a man who is enriched by the exploitation of children may nevertheless remain all his life an honest man from the legal point of view. He does not think of stealing, because he has a surer and more lucrative means of getting wealth, although he lacks the moral sense which would prevent him from committing a crime if the thought of it occurred to him. As a consequence of the present environment, man has become very egoistic and hence more capable of crime, than if the environment had developed the germs of altruism.

The present economic system is based upon exchange. A society based upon exchange isolates the individuals by weakening the bond that unites them. When it is a question of exchange the two parties interested think only of their advantage even to the detriment of the other party. In the second place the possibility of exchange arouses in a man the thought of the possibility of converting the surplus of his labour into things which increase his well-being in place of giving the benefit of it to those who are derived of the necessaries of life. Hence the possibility of exchange gives birth to cupidity.

The exchange called simple circulation of commodities is practiced by all men as consumers, and by the workers besides as vendors of their labour power. However, the influence of this simple circulation of commodities is weak compared with that exercised by capitalist exchange. It is only the exchange of the surplus of labour, by the producer, for other commodities, and hence is for him a secondary matter. As a result he does not exchange with a view to profit (though he tries to make himself as advantageous a trade as possible), but to get things which he cannot produce himself.

Capitalistic exchange, on the other hand, has another aim – that of making profit. A merchant, for example, does not buy goods for his own use, but to sell them to advantage. He will, then, always try, on the other hand, to buy the best commodities as cheaply as possible, by depreciating them as much as he can; on the other hand, to make the purchaser pay as high a price as possible, by exaggerating the value of his wares. By the nature of the mode of production itself the merchant is therefore forced to make war upon two sides, must maintain his own interests against the interests of those with whom he does business. If he does not injure too greatly the interests of those from whom he buys, and those to whom he sells, it is for the simple reason that these would otherwise do business with those of his competitors who do not find their interest in fleecing their customers. Wherever competition is eliminated for whatever cause the tactics of the merchant are shown in their true light; he thinks only of his own advantage even to the detriment of those with whom he does business. 'No commerce without trickery' is a proverbial expression (among consumers), and with the ancients Mercury, the god of commerce, was also the god of thieves. This is true, that the merchant and the thief are alike in taking account exclusively of their own interest to the detriment of those with whom they have to do.

The fact that in our present society production does not take place generally to provide for the needs of men, but for many other reasons, has important effects upon the character of those who possess the means of production. Production is carried on for profit exclusively; if greater profits can be made by stopping production it will be stopped – this is the point of view of the capitalists. The consumers, on the other hand, see in production the means of creating what man has need of. The world likes to be deceived, and does not care to recognize the fact that the producer has only his profit in view. The latter encourages this notion and poses as a disinterested person. If he reduces the price of his wares, he claims to do it in the interest of the public, and takes care not to admit it is for the purpose of increasing his own profits. This is the falsity that belongs inevitably to capitalism.

In general this characteristic of capitalism has no importance for the morality of the consumer, who is merely duped, but it is far otherwise with the press, which is almost entirely in the hands of the capitalists. The press, which ought to be a guide for the masses, and is so in some few cases, in the main is in the hands of capitalists, who use it only as a means of making money. In place of being edited by men who, by their ability and firmness, are capable of enlightening the public, newspapers are carried on by persons who see in their calling only a livelihood, and consider only the proprietor of the sheet. In great part the press is the opposite of what it ought to be; it represents the interests of those who pay for advertisements or for articles; it increases the ignorance and the prejudices of the crowd; in a word, it poisons public opinion.

Besides this general influence upon the public the press has further a special place in the etiology of crime, from the fact that most newspapers, in order to satisfy the morbid curiosity of the public, relate all great crimes in extenso, give portraits of the victims, etc., and are often one of the causes of new crimes, by arousing the imitative instinct to be found in man.

As we have seen above the merchant capitalist makes war in two directions; his interests are against those of the man who sells to him, and of the man who buys from, him. This is also true of the industrial capitalist. He buys raw materials and sells what he produces. But to arrive at his product he must buy labour, and this purchase is 'sui generis'.

Thus it becomes plain that economic conditions occupy a much more important place in the etiology of crime than most authors have given them. The present economic system and its consequences weaken the social feelings. The economic interests of men are necessarily found to be in opposition. This is a trait that capitalism. But its principal characteristic is that the means of production are in the hands of a few, and most men are altogether deprived of them. Consequently, persons who do not possess the means of production are forced to sell their labour to those that do, and these, in consequence of their economic preponderance, force them to make the exchange for the mere necessaries of life, and to work as much as their strength permits.

This state of things especially stifles men's social instincts; it develops, on the part of those with power, the spirit of domination, and of insensibility to the ills of others, while it awakens jealously and servility on the part of those who depend upon them. Further the contrary interests of those who have property, and the idle and luxurious life of some of them, also contribute to the weakening of the social instincts.

The material condition, and consequently the intellectual condition, of the proletariat are also a reason why the moral plane of that class is not high. The work of children brings them into contact with persons to associate with whom is fatal to their morals. Long working hours and monotonous labour brutalize those who are forced into them; bad housing conditions contribute also to debase the moral sense, as do the uncertainty of existence, and finally absolute poverty, the frequent consequence of sickness and unemployment. Ignorance and lack of training of any kind also contribute their quota. Most demoralizing of all is the status of the lower proletariat.

The economic position of woman contributes also to the weakening of the social instincts. The present organization of the family has great importance as regards criminality. It charges the legitimate parents with the care of the education of the child; the community concerns itself with the matter very little. It follows that a great number of children are brought up by persons who are totally incapable of doing it properly. As regards to the children of the proletariat, there can be no question of the education properly so called, on account of the lack of means and the forced absence of one or both of the parents. The school tends to remedy this state of things, but the results do not go far enough. The harmful consequence of the present organisation of the family make themselves felt especially in the case of the children of the lower proletariat, orphans, and illegitimate children. For these the community does but little, though their need of adequate help is the greatest.

Prostitution, alcoholism, and militarism, which result, in the last analysis, from the present social order, are phenomena that have demoralising consequences. As to the difference kinds of crime, the very important group of economic criminality finds its origin on the one side in the absolute poverty and the cupidity brought about by the present economic environment, and on the other in the moral abandonment and bad education of the children of the poorer classes. Then, professional criminals are principally recruited from the class of occasional criminals, who, finding themselves rejected everywhere after their liberation, fall lower and lower.

Upon the basis of what has gone before, we have a right to say that the part played by economic conditions in criminality is preponderant, even decisive. This conclusion is of the highest importance for the prevention of crime. If it were principally the consequence of innate human qualities (ativism, for example), the pessimistic conclusion that crime is a phenomenon inseparably bound up in the social life would be well founded. But the facts show that it is rather the optimistic conclusion that we must draw, that where crime is the consequence of economical and social conditions, we can combat it by changing those conditions.

However important crime may be as a social phenomenon, however terrible may be the injuries and the evil that it brings upon humanity, the development of society will not depend upon the question as to what are the conditions which could restrain crime or make it disappear, if possible; the evolution of society will proceed independently of this question.

Thus while establishing a co-relationship between economic conditions and crime, WA. Bonger drew the following conclusions :

1. He prepared a statistical data and demonstrated that almost 79 per cent of the criminals belong to non-profitable class. Thus he tried to establish a co-relationship between poverty and delinquency. In his doctoral thesis entitled Criminality and Economic Conditions, Dr. Bonger made a detailed study of the economic literature of whole Europe and concluded that crimes relating to property such as theft, stealing, robbery, dacoit, house-breaking etc. record an abnormal increase during the periods of depression when the prices are high.
2. Bonger further observed that the influence of economic conditions on delinquency is essentially due to the capitalistic economy which breeds disparity and leads to unequal distribution of wealth. The capitalist resort to hoarding and monopolistic trends thus creating artificial scarcity and consequent rise in prices. This in turn stops production which ultimately leads to unemployment of labour, as a result of which offences such as alcoholism, vagrancy, beggary, assault, violence, etc. record an upward trend.
3. In an economic system based on capitalism, economic cycles of inflation and deflation are frequent. Inflation gives rise to bankruptcy and insolvency with the result the persons affected thereby are forced to lead an anti-social life and some of them may even resort to criminality.
4. Another peculiar feature of capitalistic economy is the competitive tendency among entrepreneurs. Efficiency, low-production cost and better quality of products are some of the admirable results of competitive economy. But when these efforts fail to meet the competition, unlawful devices such as violation of laws relating to trade marks, copyright, patents etc., are committed by the manufacturers. This gives rise to increase in crime rate.
5. There is yet another danger of the capitalistic economy which contributes to enormous increase in crimes. The employment of children and women furnishes soothing ground for criminality despite effective legislative restriction! banning their improper utilisation in industrial establishments. It has been rightly observed that employment of children as labour is in itself a potential cause for crimes because a child who earns his wages does not know how to spend it usefully. Consequently, he is apt to spend his money on undesirable items such as smoking, gambling, drinking, staking, womanizing and so on, which ultimately drag him into the criminal world.

Bonger in Indian Context

The issue relating to right of children to be protected from economic exploitation and their utilisation in hazardous work has engaged the attention of world community during the preceding decades. The

International Labour Organisation has vehemently opposed child labour in view of the provisions of Article 18 of the United Nations Convention on the Rights of the Child, 1989 which prohibits employment of children for works which are harmful to their health or physical, mental, spiritual moral or social development. The Government of India, however, perceives child labour as a necessary evil, a concomitant of poverty which cannot be done away with unless poverty itself is eradicated from society. The Child Labour (Prohibition & Regulation) Act, 1986, prohibits engagement of children in certain hazardous employment. Thus, it does not completely ban child labour but only seeks to 'protect' working children from exploitation. Section 14(1) and (2) of the Act provides for penalty for contravention of the provisions of the Act. The implementation and enforcement of the provisions of the Act was far from satisfactory. Therefore, the Government came out with another law in 2006 prohibiting employment of children below the age of 14 years in hotels, shops, bakeries or as domestic servants etc. It must, however, be noted that child labour is rooted in the socio-economic conditions of the people and therefore it cannot be wiped off unless poverty itself is completely eradicated.

The employment of women also has a demoralising effect on children. With the outdoor occupational activities of mothers, the children are not properly looked after. The lack of parental care and control over children in homes may detract them from righteous path and they are likely to fall into bad company of delinquents out of sheer frustration and want of proper attention towards them. That apart, greed for money often induces women to agree to immoral acts. Particularly, in the context of Indian society, the condition of working women is deplorable because of the lack of adequate protection to her from social dangers while she is at work. Commenting on this point Gillin (1971:6) has rightly observed that while lack of employment seems to be hazardous for adult males the employment of women and children is associated with an increase in criminality.

It must, however, be reiterated that crime and poverty are endemic to any society, whatever be its form-whether it is capitalistic or socialistic and whatever may be the extent of its material development. Just as poverty cannot be wiped out so also crimes cannot be wiped off though they can be reduced and controlled with all out effort towards material development and prosperity of a nation.

The relevance of poverty to crime is sufficiently highlighted though some of the Judicial decisions wherein the accused were compelled to commit gruesome murder under pressure of extreme poverty. Thus, in Maragatham (AIR 1961 Mad 498) the accused were husband and wife who were starving for about ten days without any food or work for their subsistance. Therefore, they decided to put an end to their lives along with their one and a half month old female infant. They tied themselves together with a rope and jumped into a well. They were, however, rescued but unfortunately the infant was drowned. They were convicted of attempt to murder of their infant child and committing suicide under Section 307 read with Sections 34 and 309 of the Indian Penal Code.

In yet another case of Shreerangyee v. State of Madras (1973: 1 MLJ 205) the accused was a hardworking but unfortunate woman who was deserted by her husband. She had five children but was unable to support them for want of adequate earning. Her financial position further worsened due to her youngest child's severe illness and doctor's demand for money for the treatment. Shreerangyee tried in vain to raise the sum. Having exhausted all the legitimate means to earn a living, she, in exasperation, killed all her five children by drowning them and finally jumped into the well. She was, however, rescued and convicted, under Section 302 IPC for killing her children. The Court in this case ruled out poverty as an excuse for the murder of innocent children and attempt to suicide.

The cases referred to above, clearly indicate that social perspective of justice had been completely lost sight of in the rigidity of legal process. Though in some of the subsequent cases (Hussainara v. State of Bihar, AIR 1979 SC 1369 ; Bandhua Mukti Morcha v. Union of India, AIR 1984 SC 802) the courts have evinced deeper concern for the poorer sections of the society, but there is urgent need for restructuring of the existing judicial mechanism so as to make it more humane and re-assuring that the poor and weak are not forced to resort to criminal acts out of sheer disgust and desperation.

Judiciary's deeper concern for the poorer sections of the society who suffer more within our legal system than others, was once again reflected in the case of Bavadas Bowri v. State of Assam (1982) Cr. L.J. 213 (Gau.). In this case, the appellant, who was an indigent and disabled man belonging to a backward class was convicted of murder under Section 302 of I.P.C. and was sentenced to imprisonment for life. In the exercise of his right of private defence, he used a pen-knife against strong adversary who was assaulting him with a bamboo stick. It was for this reason that the appellant was forced to fight for his life and the thrust given in these circumstances had caused the death of the assailant.

The High Court of Gauhati accepted the right of private defence of the appellant and observed that the entire case had been conducted sluggishly. "Poor quality of justice dispensed to the poor is a common feature of the judicial administration. Justice Lahiri, inter alia, observed, "a public prosecutor should have the strength not to disown the poor he must exercise power of withdrawal under Section 321 of Cr.P.c. if he finds that the charges are not genuine. The primary duty and conduct of judiciary is to do justice within the four corners of law..." In Joyannathan v. State's the "involvement of the appellant (a small farmer of Perambur) entangled in a village quarrel landed him in a police case (1983) Cr. L.J. 1748 (T.N.). Police took over a year to frame a charge-sheet during which the appellant's life became a hell as he had to part with his two acres of land and milk business in order to. Meet frequent visits to Magistrate's Court and also to support his wife and three helpless children. It was only after the revision petition was filed by the brother of the appellant in High Court of Madras, that his revision was allowed. Moved by the plight of the poor appellant, the High Court ordered to quash all the criminal cases where the police had kept the FIR's unduly pending over six months.

The effect of modernisation and excessive materialism has changed the very concept of crime. Therefore, there is greater influx of socio-economic crimes in the present time. They include tax evasion, hoarding, black-marketing, violation of FERA/ MRTP Act, financial scams/ adulteration, etc. The cyber crimes have added new dimensions to white-collar criminality in the computer age of 21st Century. The reformative measures have failed to tackle these non-traditional offences effectively and social legislations have not been able to prevent these crimes due to their ineffective enforcement. It is, therefore, necessary that with the changing patterns of criminal behaviour, more stringent laws should be enacted to bring socio-economic crimes under control. Despite COFEPOSN and FERA regulations in force for several years in India, there has not been any significant change in the crime index relating to smuggling and foreign exchange violations which are adversely affecting the Indian economy. The criminal law enforcement agencies should, therefore, initiate drastic measures to curb this menace.

Bonger's Predictions and Prescriptions

What is the direction that society will take under these continual modifications ? This is not the place to deal fully on this subject. In fact, the facts indicate quite clearly what the direction will be. The productivity of labour has increased to an unheard of degree, and will assuredly increase in the future. The concentration of the means of production into the hands of a few progresses continually; in many branches it has reached such a degree that the fundamental principle of the present economic system, competition, is excluded, and been replaced by monopoly. On the other hand the working class is becoming more and more organized, and the opinion is very generally held among working men that the causes of material and intellectual poverty can be eliminated only by having the means of production held in common.

With material poverty there would disappear also that intellectual poverty which weighs so heavily of the proletariat; culture would no longer be the privilege of some, but a possession common to all. The consequence of this upon criminality would be very important, for even in our present society with its numerous conflicts, the members of the propertied classes, who have often but a veneer of civilization, arc almost never guilty of crimes of vengeance. There is the more reason to admit and in a society where interests were not imposed and where civilization was universal, these crimes would be no longer present, especially since alcoholism also proceeds in large part from the intellectual poverty of the poorer classes. And what is true of crimes of vengeance, is equally true of sexual crimes in so far as they have the same etiology.

A large part of the economic criminality (and also prostitution to a certain extent) has its origin in the cupidity excited by the present economic environment. In a society based upon the community of the means of production, great contrasts of fortune would, like commercial capital, be lacking, and thus cupidity would find no food. These crimes will not totally disappear so long as there has not been a redistribution of property according to the maxim, 'to each according to his needs', something that will probably be realized, but not in the immediate future. The changes in the position of woman which are taking place in our present society, will lead, under this future mode of production, to her economic independence, and consequently to her social independence as well. It is accordingly probable that the criminality of woman will increase in comparison with that of man during the transition period. But the final result will be the disappearance of the harmful effects of the economic and social preponderance of man.

As to the education of children under these new conditions it is difficult to be definite. However, it is certain that the community will concern itself seriously with their welfare. It will see to it that the children whose parents cannot or will not be responsible for them, are well cared for. By acting in this way it will remove one of the most important causes of crime. There is no doubt that the community will exercise a strict control over the education of children; it cannot be affirmed, however, that the time will come when the children of a number of parents will be brought up together by capable persons; this will depend principally upon the intensity that the social sentiments may attain.

As soon as the interests of all are no longer opposed to each other, as they are in our present society, there will no longer be a question either of politics ('a fortiori' of political crimes) or of militarism. Such a society will not only remove the causes which now make men egoistic, but will awaken, on the contrary, a strong feeling of altruism. In such a society there can be no question of crime properly so called.

References

Allen Francis (1964) Borderland of Criminal Justice,Univ of Chicago Press p.67

Becker, Gary S "Nobel Lecture (1993) The Economic Way of Looking at Behaviour." Journal of Political Economy 101:3, pp. 385-409.

Becker, Gary S 2006 "On the Economics of Capital Punishment." The Economists' Voice 3:3 (March) <www.bepress.com/ev>.

Becker, Gary S., (1968) "Crime and Punishment: An Economic Approach." Journal of Political Economy 76:2: 169-217.

Blackstone William, (1905) Commentaries, Vol. IV, Oxford, at the Clarendon Press p. 5.

Bonger, W.A., (1916) Criminality And Economic Conditions, translated by Henry P. Horton (Boston, Little Brown & Co.).

Bowers, William J., and Glenn L. Pierce, (1980) "Deterrence or Brutalization: What Is the Effect of Executions?" Crime and Delinquency 26:3: 453-484.

Carey, James T, (1978). An *Introduction to Criminology*. Englewood Cliffs, NJ: Prentice-Hall, pp. 133-36.

Caulkins, Johathan P., and Peter Reuter, (2006) "Illicit Drug Markers and Economic Irregularities." Socio-Economic Planning Sciences 40:1 (March): 1-14.

Chen, Keith, and Jesse Shapiro, (2005) "Does Prison Harden Inmates? A Discontinuity-based Approach." Executive summary available at <http://www.som.yale.edu/faculty/keith.chen/papers/ exesum.pdf>.

Donohue, John J., and Justin Wolfers, (2005) "Uses and Abuses of Empirical Evidence in the Death Penalty Debate." Stanford Law Review. 58: 791-846.

Dubber, Markus D. (2002). Criminal Law: Model Penal Code. Foundation Press.

Elizabeth A. Martin, ed (2003). Oxford Dictionary of Law. Oxford: Oxford University Press.

Garofalo Raffaele (1914) Criminology, Kissinger Publishing p. 59.

Gillin J.L., (1971) Criminology & Penology, Greenwood Press 1971 3rd Ed., p. 6.

Holland, Winnifred H. (1982-1983). "Automatism and Criminal Responsibility". Crim. L.Q. 25: 106.

Johnson, Elmer Hubert, (1974) Crime Correction and Society, Dorsey Press, USA p.11.

Levitt, Steven, and Thomas J. Miles, (2005) "The Empirical Study of Criminal Punishment." The Handbook of Law and Economics. Eds. Mitchell Polinsky and Steven Shavell.

Quinney Richard, (1970) The Social Reality of Crime, Little Brown & Company, pp. 15-23.

Sellin Thorsten, (1938) Culture conflict and Crime (New York: Social Science Research Council) p.21.

Sellin Thorsten, (1975) "The Conflict of Conduct Norms", from (Eds. Farell and Swiger Social Deviance, p. 310.

Sellin, Thorsten, (1967) "Homicides in Retentionist and Abolitionist States." Capital Punishment. Ed. Thorsten Sellin. New York: Harper and Row.

Tappan Paul W, (1960) Crime, Justice & Correction, New York, McGrew Hill p. 80.

Tappan Paul W, (1947) "Who is the Criminal?" American Sociological Review, Vol. 12 No. 1, pp. 96-102.

Void George B. and Thomas J. Bernard, (1986) Theoretical Criminology, Oxford University Press , New York 3rd Edn., pp. 277-80.

■■■

CHAPTER 4 Psychological Theories

Learning Objectives :

- ❒ **Acquainting the Learners with the Freud's, Piaget's, Kohlberg's and Exsenck's psychological theories of crime.**
- ❒ **To understand Tarde's theory on imitation of deviance and Bandura's behaviour modelling.**
- ❒ **To comprehend Maslow's theory motivation and needs.**
- ❒ **To grasp the interrelationships between intelligence and crime.**

Introduction

Criminology when defined from psychological perspective, in a broad way relate to the explanation of criminal behaviour, as well as the behaviour of police, attorneys, prosecutors, judges, correctional personnel, victims, and other actors in the criminal justice system. Criminals have distinct psychological features and are either exceptionally calm and composed or overtly aggressive and temperamental. Some criminals are extraordinarily intelligent whereas many criminals are of average and low intelligence. Psychological Theories of Crime look at criminality from individual perspective and take crime as an adaptive behaviour. Crime is a compromise, representing for the individual the most satisfactory method of adjustment to inner conflicts which he or she cannot express otherwise. Criminal is basically antisocial– is incapable of significant loyalty to individuals, groups, or social values. Criminal exhibits behaviour that brings him or her into conflict with society. Criminal behaviour may be purposeful for the individual insofar as it addresses certain felt needs. Criminal does not feel empathy with others and is unable to imagine how others think and feel. Psychological theories of crime begin with the view that individual differences in behaviour may make some people more predisposed to committing criminal acts. These differences may arise from personality characteristics, biological factors, or social interactions.

1. PSYCHOANALYTIC THEORY : SIGMUND FREUD

Psychoanalytic theories of crime causation are associated with the work of Sigmund Freud who believed that people who had unresolved deep-seated problems were **psychopaths**. Persons characterized by no sense of guilt, no subjective conscience, and no sense of right and wrong. They have difficulty in forming relationships with other people; they cannot empathize with other people. They are also called sociopaths or antisocial personalities. **Criminal behaviour is *maladaptive,* or the product of inadequacies inherent in the offender's personality. Something that happened in the past affected the way a person is today.** The actions and behaviour of an adult are understood in terms of childhood development. Behaviour and unconscious are related and criminality is a psychological conflict

According to Sigmund Freud, Criminal behaviour comes from 3 possible causes :

- Overbearing Conscience : A conscience so overbearing it arouses feelings of guilt. One is guilty even before committing the crime.
- Weak Conscience : A conscience so weak it cannot control its impulses.
- Immediate Gratification : Immediate satisfaction of the impulses of a person.

Human Personality Contains a Three Part Structure: The Id, Ego and Superego.

The Id is the primitive part and it is present at birth. It is the part of a persons personality where a baby recognizes he/she is hungry. The Id represents all unconscious biological drives a person has: Hunger, Sex drive, Life Sustaining Necessities. It is the seat of all irrational, antisocial and instinctual behaviour. The Ego is the conscious or rational part of the mind understood through reality principle. The Ego is the part of a persons conscious mind that recognizes social order. The Ego is the Reality Principle, it takes into account what is practical and conventional. The Ego develops when a child finds that some things cannot be immediately gratified. The Superego relates to the moral judgment of the mind and passes judgments on behaviour.

Id prompts, Ego acts and Superego prevents. Ego mediates between Id and Superego. If Id predominates, Ego becomes a non-conformist deviant. Id operates on the Pleasure Principle (instinctual urges) and employs Primary Process Thinking (immediate gratification). If Superego dominates over Id, Ego becomes a conformist based on internalized morals/values with a sense of right and wrong. Superego suppresses instinctual drives of ID (through guilt and shame) and serves as the moral conscience.

All humans have natural drives and urges repressed in the unconscious. Furthermore, all humans have criminal tendencies. Through the process of socialization, however, these tendencies are curbed by the development of inner controls that are learned through childhood experience. Freud hypothesized that the most common element that contributed to criminal behaviour was faulty identification by a child with her or his parents. The improperly socialized child may develop a personality disturbance that causes her or him to direct antisocial impulses inward or outward. The child who directs them outward becomes a criminal, and the child that directs them inward becomes a neurotic.

2. COGNITIVE DEVELOPMENT THEORY: JEAN PIAGET & LAWRENCE KOHLBERG

Developmental theories are usually attempting to explain the development of offending from childhood to adulthood, and hence based on longitudinal studies that follow up individuals over time. The emphasis of such theories is on continuity rather than discontinuity from childhood to adulthood. A common assumption is that the ordering of individuals on an underlying construct such as criminal potential is relatively constant over time. According to this approach, criminal behaviour results from the way in which people organize their thoughts about morality and the law.

Much of modern cognitive theory, including its relationship to socialization, stems from the work of the Swiss psychologist, **Jean Piaget**. In the 1920s Piaget observed children reasoning and understanding differently, depending on their age. He proposed that all children progress through a series of cognitive stages of development, just as they progress through a series of physical stages of development. According to Piaget, the rate at which children pass through these cognitive stages may vary, but they eventually pass through all of them in the same order.

Piaget introduced several other important concepts. According to Piaget, cognitive development occurs from two processes: adaptation and equilibrium. **Adaptation** involves the child's changing to meet situational demands. Adaptation involves two sub-processes: assimilation and accommodation. **Assimilation** is the application of previous concepts to new concepts. An example is the child who refers to a whale as a "fish." **Accommodation** is the altering of previous concepts in the face of new information. An example is the child who discovers that some creatures living in the ocean are not fish, and then correctly refers to a whale as a "mammal." **Equilibrium** is the search for "balance" between self and the world, and involves the matching of the child's adaptive functioning to situational demands. Equilibrium keeps the infant moving along the developmental pathway, allowing him or her to make increasingly effective adaptations.

Jean Piaget's ideas on mental and moral development in children focus on four stages of development :

1. Sensorimotor: birth to 2 years.; learn through senses. The child learns by doing: looking, touching, and sucking. The child also has a primitive understanding of cause-and-effect relationships. Object permanence appears around 9 months.
2. Preoperational: 2-7 years.; learning language, drawing other skills. The child uses language and symbols, including letters and numbers. Egocentrism is also evident. Conservation marks the end of the preoperational stage and the beginning of concrete operations.
3. Concrete operations: 7-11 years.; logical thinking and problem solving. The child demonstrates conservation, reversibility, serial ordering, and a mature understanding of cause-and-effect relationships. Thinking at this stage is still concrete.
4. Formal operations: 11-15 years; abstract ideas. The individual demonstrates abstract thinking, including logic, deductive reasoning, comparison, and classification.

Moral development of individuals is closely related to conforming to certain norms and standards of society, and control of antisocial or negative behaviour is absolutely necessary in preventing delinquent behaviour. However, the stages of moral development should be understood to delineate the positive or negative impact of non conformity. After complete moral development, individuals may choose to not conform to social standards and try to move beyond norms. This is not always negative and would be associated with novelty and creativity. However, when moral development is hindered at an early stage of life when children ignore moral values and ethics of society for their own selfish pursuits and there is a complete lack of external control, the moral development is truncated and lead to criminal behaviour in later life. A restraining factor in crime is rational thinking and if individuals can rationalize, reflect and understand the seriousness of crimes, they would be deterred from engaging in crimes and would thus in turn aid their own moral and personal development. According to psychoanalysis we all have criminal/aggressive tendencies but a criminal is the one who directs the basic urges of aggression towards other people and the external world. Moral development of children would be dependent on several factors and according to psychologist Lawrence Kohlberg the first stage of moral reasoning is of obedience and avoiding punishment, the second stage is keeping up to expectations of family members and the final stage is moving beyond social conventions. Criminals are people who have incomplete moral development as their behaviour may by motivated by a desire to break the law and challenge obedience and punishment, and this is largely different from mature nonconformity that is at times constructive. A criminal is thus recognized by his destructive and negative personality.

3. Lawrence Kohlberg : Development of Moral Reasoning

In 1958, Lawrence Kohlberg, a developmental psychologist, formulated a theory concerning the development of moral reasoning. He modified and elaborated on Piaget's ideas about moral thinking. He posited that there are three levels of moral reasoning, each consisting of two stages. During middle childhood, children are at the first level of moral development. At this level, the Pre-conventional/ Pre-formal level, moral reasoning is based on obedience and avoiding punishment. In the pre-formal stage, which is seen exclusively in infants and young children (until about age 7) and occasionally in adults or older children, is a basic level of moral reasoning ? That is, there is a punishment/reward system of what's right and wrong. The person who is in the pre-conventional phase reasons like this: "If I got punished for it, or I expect to be punished for it, it's wrong.If I get or expect to be rewarded for it, it's right."

The second level, the Conventional/Formal level of moral development, is reached at the end of middle childhood. The moral reasoning of individuals at this level is based on the expectations that their family and significant others have for them. In the formal reasoning phase, which is seen children beginning around age 7, and is still seen in many adults, is a good/bad system. That is, someone may reason, "I obey the laws because it's a good thing to do, it's the right thing to do." They may reason, "I don't steal because it's bad, and it's wrong. This level of moral reasoning focuses a lot on what society says about a given action. Stealing is wrong and bad because society says it is. Society also says that we should follow laws."

Kohlberg found that the transition to the third level, the Post-conventional/ Post-formal level of moral development, usually occurs during early adulthood. At this level, individuals are able to go beyond social conventions. They value the laws of the social system; however, they are open to acting as agents of change to improve the existing law and order. People who do not progress through the stages may become arrested in their moral development, and consequently become delinquents. The latest stage, post-formal, is seen only in a small percentage of adults, and doesn't even develop until adolescence (and isn't complete until, likely, a person's mid-20's).This is a "what's truly right/wrong" system. That is, the person who reasons this way doesn't follow what they've been told to do or what society says to do, necessarily. They take time to think about what THEY, individually, believe is right and wrong, and incorporate many of their own experiences and beliefs. Civil disobedience began because of this type of thinking.

Kohlberg's theory is criticized as inconsistent for different crimes. Convicted criminals who had offended for no financial gain (assault, murder, sex offences) showed more mature moral judgement than those who offended for money (robbery, burglary, theft, fraud). Research has focused on the offender's beliefs and attitudes (content), which can be contrasted with the offender's actions (process). One can argue eloquently and convincingly about social/moral issues yet have a personal set of values which are entirely self-serving, hedonistic or anti-social. Several well-known experiments have shown that people will behave in ways which they believe or know to be wrong, being influenced by the present situation rather than their individual disposition to behave morally. Most people don't reason entirely at one level. Some people who reason at the post-formal level on serious matters may choose not to commit crime because they don't want to be punished (pre-formal). Someone may ordinarily choose to do what society dictates is right (formal), but question a particular new law (post-formal).

4. LEARNING THEORY : HANS J. EYSENCK

Learning theory is based upon the principles of behavioural psychology. Behavioural psychology posits that a person's behaviour is learned and maintained by its consequences, or reward value. These consequences may be in three forms of reinforcement: external reinforcement, vicarious reinforcement, and self-reinforcement. External reinforcement occurs as a direct result of their behaviour (e.g. money, social status, and goods). Vicarious reinforcement occurs by observing the behaviour of others being either reinforced or punished (e.g. observing others who are being reinforced as a result of their behaviour). This is imitation. Imitation refers to the process by which people learn how to act by observing others. Delinquent behaviour arises whenever non- secure attachments are created. Behaviour is determined by environmental consequences which it produces for the individual concerned. Major determinants of behaviour are found in the environment surrounding the individual. Self-reinforcement refers to ones' sense of pride, or as meeting of standards in ones' own behaviour.

According to social learning theory, juveniles learn to engage in crime in the same way they learn to engage in conforming behaviour: through association with or exposure to others. Primary or intimate groups like the family and peer group have an especially large impact on what we learn. In fact, association with delinquent friends is the best predictor of delinquency other than prior delinquency. However, one does not have to be in direct contact with others to learn from them; for example, one may learn to engage in violence from observation of others in the media.

Individuals may teach others to engage in crime through the reinforcements and punishments they provide for behaviour. Crime is more likely to occur when it (a) is frequently reinforced and infrequently punished; (b) results in large amounts of reinforcement (e.g., a lot of money, social approval, or pleasure) and little punishment; and (c) is more likely to be reinforced than alternative behaviours.

According to social learning theory, some individuals are in environments where crime is more likely to be reinforced (and less likely to be punished). Sometimes this reinforcement is deliberate. For example, the parents of aggressive children often deliberately encourage and reinforce aggressive behaviour outside the home. Or the adolescent's friends may reinforce drug use. At other times, the reinforcement for crime is less deliberate. For example, an embarrassed parent may give her screaming child a candy bar in the checkout line of a supermarket. Without intending to do so, the parent has just reinforced the child's aggressive behaviour.

According to social learning theory, criminal behaviour is acquired through observational learning. The learning takes place in three contexts, the family, prevalent subculture, and the social environment. The reinforcement for criminal behaviour comes from both the internal and external sources. The reinforcement can be in the form of tangible rewards of the criminal activity itself (i.e., money), or from social rewards (i.e., increase in peer status). Criminal behaviour is maintained through a complex schedule of reinforcement and punishment throughout the life of the individual. This is similar to the differential reinforcement concept of factors influencing criminal behaviour. According to the theory, if criminal behaviour has been reinforced in the past, it is expected that it will be reinforced in the future.

According to social learning theory, criminal behaviour is acquired through observational learning. The learning takes place in three contexts, the family, prevalent subculture, and the social environment. The reinforcement for criminal behaviour comes from both the internal and external sources. The reinforcement can be in the form of tangible rewards of the criminal activity itself (i.e., money), or from social rewards (i.e., increase in peer status).

According to learning theorists, deviant behaviour can be eliminated or modified by taking away the reward value of the behaviour. Hans J. Eysenck, a psychologist that related principles of behavioural psychology to biology, postulated that by way of classical conditioning, operant conditioning, and modeling people learn moral preferences. Classical conditioning refers to the learning process that occurs as a result of pairing a reliable stimulus with a response. Eysenck believes, for example, that over time a child who is consistently punished for inappropriate behaviour will develop an unpleasant physiological and emotional response whenever they consider committing the inappropriate behaviour. The anxiety and guilt that arise from this conditioning process result in the development of a conscience. He hypothesizes, however, that there is wide variability among people in their physiological processes, which either increase or decrease their susceptibility to conditioning and adequate socialization.

Eysenck further believes that crime and deviance can be explained through inherent character traits and different personality traits which predispose individuals to crime. Thus criminality is an inherited characteristics like personality and intelligence. Eysenck believed that there are two basic personalities, the introvert and the extrovert, in which people are born with. Introverts are shy and quiet whereas extroverts are confident and outgoing. He believed that the extroverts are more likely to be criminals because they were more likely to get into situations where they might get into trouble, 'he craves excitement, takes chances, often sticks his neck out, acts on the spur of the moment and is generally an impulsive individual'. They are also harder to be socialized into the norms, values and laws prevalent and social control has less hold upon them. People with an extrovert personality are more likely to break the law as they crave attention, take chances and act on the spur of the moment. Introverts are the opposite of extroverts and are therefore less likely to be criminal.

Eysenck's original research found two main dimensions of temperament: neuroticism and extraversion-introversion.

Neuroticism

Neuroticism is the name Eysenck gave to a dimension that ranges from normal, fairly calm and collected people to one's that tend to be quite "nervous." His research showed that these nervous people tended to suffer more frequently from a variety of "nervous disorders" we call neuroses, hence the name of the dimension. But he was not saying that people who score high on the neuroticism scale are necessarily neurotics — only that they are more susceptible to neurotic problems. Eysenck was convinced that, since everyone in his data-pool fit somewhere on this dimension of normality-to-neuroticism, this was a true temperament, i.e. that this was a genetically-based, physiologically-supported dimension of personality. He therefore went to the physiological research to find possible explanations.

The most obvious place to look was at the **sympathetic nervous system**. This is a part of the autonomic nervous system that functions separately from the central nervous system and controls much of our emotional responsiveness to emergency situations. For example, when signals from the brain tell it to do so, the sympathetic nervous systems instructs the liver to release sugar for energy, causes the digestive system to slow down, opens up the pupils, raises the hairs on your body (goosebumps), and tells the adrenal glands to release more adrenalin (epinephrine). The adrenalin in turn alters many of the body's functions and prepares the muscles for action. The traditional way of describing the function of the sympathetic nervous system is to say that it prepares us for "fight or flight."

Eysenck hypothesized that some people have a more responsive sympathetic nervous system than others. Some people remain very calm during emergencies; some people feel considerable fear or other emotions; and some are terrified by even very minor incidents. He suggested that this latter group had a problem of sympathetic hyperactivity, which made them prime candidates for the various neurotic disorders. Perhaps the most "archetypal" neurotic symptom is the **panic attack**. Eysenck explained panic attacks as something like the positive feedback you get when you place a microphone too close to a speaker: The small sounds entering the mike get amplified and come out of the speaker, and go into the mike, get amplified again, and come out of the speaker again, and so on, round and round, until you get the famous squeal that we all loved to produce when we were kids.

Extraversion-introversion

His second dimension is extraversion-introversion. By this he means something very similar to what Jung meant by the same terms, and something very similar to our common-sense understanding of them: Shy, quiet people "versus" out-going, even loud people. This dimension, too, is found in everyone, but the physiological explanation is a bit more complex.

Eysenck hypothesized that extraversion-introversion is a matter of the balance of "inhibition" and "excitation" in the brain itself. These are ideas that Pavlov came up with to explain some of the differences he found in the reactions of his various dogs to stress. **Excitation** is the brain waking itself up, getting into an alert, learning state. **Inhibition** is the brain calming itself down, either in the usual sense of relaxing and going to sleep, or in the sense of protecting itself in the case of overwhelming stimulation.

Someone who is extraverted, he hypothesized, has good, strong inhibition: When confronted by traumatic stimulation — such as a car crash — the extravert's brain inhibits itself, which means that it becomes "numb," you might say, to the trauma, and therefore will remember very little of what happened. After the car crash, the extravert might feel as if he had "blanked out" during the event, and may ask others to fill them in on what happened. Because they don't feel the full mental impact of the crash, they may be ready to go back to driving the very next day.

The introvert, on the other hand, has poor or weak inhibition: When trauma, such as the car crash, hits them, their brains don't protect them fast enough, don't in any way shut down. Instead, they are highly alert and learn well, and so remember everything that happened. They might even report that they saw the whole crash "in slow motion!" They are very unlikely to want to drive anytime soon after the crash, and may even stop driving altogether.

Now, how does this lead to shyness or a love of parties? Well, imagine the extravert and the introvert both getting drunk, taking off their clothes, and dancing buck naked on a restaurant table. The next morning, the extravert will ask you what happened (and where are his clothes). When you tell him, he'll laugh and start making arrangements to have another party. The introvert, on the other hand, will remember every mortifying moment of his humiliation, and may never come out of his room again. (I'm very introverted, and again I can vouch to a lot of this experientially! Perhaps some of you extraverts can tell me if he describes your experiences well, too — assuming, of course, that you can remember you experiences!)

One of the things that Eysenck discovered was that violent criminals tend to be non-neuroticistic extraverts. This makes common sense, if you think about it: It is hard to imagine somebody who is painfully shy and who remembers their experiences and learns from them holding up a Seven-Eleven! It is even harder to imagine someone given to panic attacks doing so. But please understand that there are many kinds of crime besides the violent kind that introverts and neurotics might engage in !

Neuroticism and Extraversion-introversion

Another thing Eysenck looked into was the interaction of the two dimensions and what that might mean in regard to various psychological problems. He found, for example, that people with phobias and obsessive-compulsive disorder tended to be quite introverted, whereas people with conversion disorders (e.g. hysterical paralysis) or dissociative disorders (e.g. amnesia) tended to be more extraverted. Eysenck came to recognize that, highly neuroticistic people over-respond to fearful stimuli; If they are introverts, they will learn to avoid the situations that cause panic very quickly and very thoroughly, even to the point of becoming panicky at small symbols of those situations — they will develop phobias. Other introverts will learn (quickly and thoroughly) particular behaviours that hold off their panic — such as checking things many times over or washing their hands again and again. Highly neuroticistic extraverts, on the other hand, are good at ignoring and forgetting the things that overwhelm them. They engage in the classic defense mechanisms, such as denial and repression. They can conveniently forget a painful weekend, for example, or even "forget" their ability to feel and use their legs.

Psychoticism

Eysenck came to recognize that, although he was using large populations for his research, there were some populations he was not tapping. He began to take his studies into the mental institutions of England. When these masses of data were factor analyzed, a third significant factor began to emerge, which he labeled psychoticism. Like neuroticism, high psychoticism does not mean you are psychotic or doomed to become so — only that you exhibit some qualities commonly found among psychotics, and that you may be more susceptible, given certain environments, to becoming psychotic. The kinds of qualities found in high psychoticistic people include a certain recklessness, a disregard for common sense or conventions, and a degree of inappropriate emotional expression. It is the dimension that separates those people who end up institutions from the rest of humanity.

5. Gabriel Tarde and the Imitation of Deviance

One of the earliest formulations of a learning perspective on deviance is discovered in the writings of the French social theorist Gabriel Tarde (1843-1904). Tarde believed that people learn from one another through the process of imitation. He wrote several short volumes on his family and the town in which he lived, as well as editing and republishing the papers of outstanding family members. While in school, Tarde retained a permanent distaste for socially imposed discipline whenever it limited individual freedom. Tarde published a series of articles criticizing the "Italian school" represented by Lombroso, Garofalo, Ferri and emphasized the preponderance of social factors—especially the socialization and imitation—behind crime. (Tarde, 1969a: 2-5)

Tarde directed attention to the social processes whereby forms of behaviour and ways of thinking and feeling are passed on from group to group and person to person. His was a theory of "imitation and suggestion." The origins of deviance were pictured as very similar to the origins of fads and fashions. Each was a socially learned acquisition, governed by what Tarde referred to as the "three laws of imitation." These included (1) the law of close contact, (2) the law of imitation of superiors by inferiors, and (3) the law of insertion.

These three laws of imitation describe why people engage in crime. First, individuals in close intimate contact with one another imitate each other's behaviour (Tarde, 1969: 30). By the law of close contact, Tarde meant simply that people have a greater tendency to imitate the fashions and customs of those with whom they have the most contact. If someone were regularly surrounded by people involved in a world of deviant behaviour or lifestyles, they would be more likely to imitate these people than they would others with whom they had little association. Direct contact with deviance was believed to foster more deviance. Tarde theorized that there was short-term behaviour (fashion) and long-term behaviour (custom). He suggested that, as population became denser, behaviour would be oriented more toward fashion than toward custom.

The functions of the higher senses are more transmissible through imitation than those of the lower. We are much more likely to copy someone who is looking at or listening to something than someone who is smelling a flower or tasting a dish (Tarde, 1903a: 195). Imitation, contrary to what we might infer from certain appearances, proceeds from the inner to the outer man (Tarde, 1903: 199).

As we look at indirect contact, we think of a world in which much of our contact with people, their actions, and their beliefs are mediated by mass communications. Tarde's writing anticipated such a world of indirect imitation. He believed that the media played a central role in the proliferation of such nineteenth-century "epidemics of deviance" as the rise in mutilations of women, the practice of women disfiguring the faces of male lovers, and the rash of "Jack the Ripper"-type murders became evident. In Tarde's own words, "infectious epidemics spread with air or wind; epidemics of crime follow the telegraph."

Tarde's second law of imitation spreads from the top down; consequently, youngsters imitate older individuals, paupers imitate the rich, peasants imitate royalty, and so on. Crime among young, poor or low-status people is really their effort to imitate wealthy, older, high-status people. This law suggests perhaps people follow the model of high-status in hopes their imitative behaviour will procure some of the rewards associated with being of a "superior" class. In any event, Tarde's ideas have a particular relevance in our own age of visibly "high-class" deviance. Does post-Watergate knowledge of the deviance of "superior" persons, such as high governmental officials and corporate executives, increase the likelihood of deviance by us all ?" Tarde's law of imitation of superiors suggests that possibility.

Tarde's third law is the law of insertion: new acts and behaviours are superimposed on old ones and subsequently either reinforce or discourage previous customs. This law refers to the power inherent in newness or novelty; new fashions were said to replace old "customs." For example, drug taking may be a popular fad among college students who previously used alcohol. However, students may find that a combination of both substances provides even greater stimulation, causing the use of both drugs and alcohol to increase. Another example would be a new criminal custom developing that eliminates an older one – truck hijacking replacing train robbing. When two mutually exclusive ways of doing something come into conflict, Tarde believed the newer one would ordinarily win out. The replacement of the knife by the gun as a weapon of deviant destruction was also cited as an example of this process.

From its early inception in Tarde's three laws of imitation, the learning perspective, has exerted an enormous impact on the study of deviance and social control. It is the product of learning in the world in a particular way, learning with and from others about how to define, feel, and act within a world which we create together.

As we examine social learning more, we see a lot of theories integrated, which originated to some extent from Tarde's imitation theory. Social psychologists suggest that drug abuse patterns may result from the observation of parental drug use. Parental drug abuse begins to have a damaging effect on children as young as two years old, especially when parents manifest drug-related personality problems, children imitate their behaviour. Children whose parents abuse drugs are more likely to have persistent abuse problems than the children of nonabusers, because one is more exposed intimately than the other.

Adolescents respond to peer group influences more readily than adults because of the crucial role peer relationships play in identity formation. Youth's greater desire for acceptance and approval renders them more susceptible to peer influences as they adjust their behaviour and attitudes to conform to those of their contemporaries. Significantly, young people "commit crimes, as they live their lives, in groups." We are led to copy from others everything that seems to us a new means for attaining our old ends, or satisfying our old wants, or a new expression of our old ideas; and we do this at the same time that we begin to adopt innovations which awaken new ideas and new ends in us.

6. Albert Bandura : Behaviour Modeling

Social learning is the branch of behaviour theory most relevant to criminology. Social learning theorists view violence as something learned through a process called behaviour modeling. In modern society, aggressive acts are usually modeled after three principle sources. The most prominent models are family members. Albert Bandura, a social learning theorist, reports that studies of family life show that children who use aggressive tactics have parents who use similar behaviours when dealing with others. A second influence on the social learning of violence is provided by environmental experiences. People who reside in areas in which violence is a daily occurrence are more likely to act violently than those who dwell in low-crime areas where norms stress conventional behaviour. A third source of behaviour modeling is provided by the mass media. Films and television shows commonly depict violence graphically. Moreover, violence is often portrayed as an acceptable behaviour, especially for heroes who never have to face legal consequences for their actions.

Bandura first presented the principles of social learning theory in 1963. The study demonstrated that modeling is one of the most effective ways to teach children ways of behaving and their consequences. These theories are relevant to studies that have shown a link between movie portrayals of behaviour and the behaviour of juvenile viewers.

Social learning theorists argue that people are not actually born with the ability to act violently but that they learn to be aggressive through their life experiences. These experiences include personally observing others acting aggressively to achieve some goal or watching people being rewarded for violent acts on television or in movies. People learn to act aggressively when, as children, they model their behaviour after the violent acts of adults. Later in life, these violent behaviour patterns persist in social relationships. The boy who sees his father repeatedly strike his mother with impunity is the one most likely to grow up to become a battering parent and husband. Bandura's social learning theory when applied to effects of mass media, is an important concept. It was the "backbone" of subsequent research, studying the impact of television violence on children. This is a contemporary look at what Tarde spoke about, but because this technology was not available during his lifetime, he spoke mainly of verbal communication via telegraphs and newspapers. The conclusion from Bandura's observational learning research relates to mass communication, particularly the "effects" of film and television on youth. Bandura spoke about modeling, or observational learning when he performed the Bobo Doll Experiment. He made a film of a young woman, beating up a Bobo Doll, and showed it to kindergarteners. The kids imitated the young woman's actions. The research proves children will imitate and learn behaviour performed by symbolic models on television.

In integrating Tarde's imitation theory, Edwin Sutherland put forth a few propositions, which just need to be mentioned because of its relevance to imitation. Edwin Sutherland spoke of differential association. He hypothesized that "any person can be trained to adopt and follow". Sutherland, in summation, felt that criminal behaviour is learned, and learned in interaction with other persons in a process of communication. He also proposed that the learning part of criminal behaviour occurs within intimate personal groups. This parallels Tarde's second law of imitation, close contact. (Sutherland and Cressey, 1994a: 192). However in this Sutherland felt the process of learning criminal behaviour by association with criminal and anticriminal patterns involves all of the mechanisms that are involved in any other learning. Negatively, this means that the learning of criminal behaviour is not restricted to the process of imitation. A person who is seduced, for instance, learns criminal behaviour by association, but this process would not ordinarily be described as imitation. (Sutherland, 1994: 194).

Sutherland suggested that the distinction between lawbreakers and the law-abiding lies not in their personal fiber but in the content of what they have learned. Those with the good fortune of growing up in a conventional neighborhood will learn to play baseball and to attend church services; those with the misfortune of growing up in a slum will learn to rob drunks and to roam the streets looking to do mischief.

Now, after examining the theory of imitation and its relevant integrated theories, we need to look at some criticisms of these works. Learning theorists fail to account for the origin of criminal definitions; How did the first "teachers" learn criminal techniques and definitions. Learning theories also imply that people systematically learn techniques that allow them to be active and successful criminals, but they fail to adequately explain spontaneous and wanton acts of violence and damage and other expressive crimes that appear to have little utility or purpose, i.e. a random shooting. Little evidence exists that people learn the techniques that enable them to become criminals before they actually commit criminal acts (Siegal, 1998: 207)

Tarde's three laws are rather loose and have been criticized for being overly simplistic and for neglecting a host of other physical, psychological, social, political, and economic factors related to deviance. Some of the dynamics of these laws were never specifically laid out. Why, for instance, was newness more attractive than established custom ? Are we more likely to accept new forms of doing things if they do old things better, for example, alcohol to crack, 45 magnums to machine guns ? Tarde was not clear about such issues. Nonetheless, his ideas about the imitative origins of deviance opened the door for an interpretation of deviance as learned behaviour. Tarde rejected the biological theories as well as explanations, which viewed society as independent of the activities of its members. He planted the theoretical seeds of a perspective, which later came to fruition in Edwin Sutherland's theory of a differential association. Note the importance placed upon associative imitation in the following excerpt from Tarde's Penal Philosophy:

Many have adapted and refined Tarde's work. His research on deviance and how it is manifested has caused many contemporaries to take notice and embark on new avenues that emit from it. Tarde's three laws, close contact, superiors and inferiors, and insertion all have been expanded upon in today's contemporary criminological research.

7. MASLOW'S THEORY OF MOTIVATION & NEEDS :

Maslow believed that human beings are motivated by five basic levels of needs and that people choose crime because they cannot (or will not) satisfy their needs legally. He views crime as one of several adaptations to the helplessness caused by oppression.

According to Maslow human needs can be structured and measured. Maslow's theory belongs to the category of Need theories of motivation. Need theories are based on different hierarchies of needs, dividing needs according to their importance. Maslow's hierarchy of needs places physiological needs on the lowest stage.

Maslow's hierarchy of needs consist of four layers. He based his hierarchy on human needs and these four human needs are: physiological needs (includes hunger, etc), safety needs, belonging and love needs, esteem needs and self-actualization needs. Hierarchy is constructed in a form of the pyramid and upper needs cannot be satisfied until the needs from the lower levels are not satisfied. The physiological needs and the consummatory behaviour involved with them serve as channels for all sorts of other needs as well. For example, when the person is hungry or in danger he is very unlikely to worry about his esteem and self-actualization.

If all the needs are unsatisfied, and the organism is then dominated by the physiological needs, all other needs may become simply non-existent or be pushed into the background. If the physiological needs are relatively well gratified, there then emerges a new set of needs, which we may categorize roughly as the safety needs. The tendency to have some religion or world-philosophy, social security, science, protection of marginalized, aged women and children that organizes the universe and the men in it into some sort of satisfactorily coherent, meaningful whole is also in part motivated by safety-seeking. Then there will emerge the love and affection and belongingness needs. Now the person will feel keenly, as never before, the absence of friends, or relations, or a wife, or children. He will hunger for affectionate relations with people in general, namely, for a place in his group, and he will strive with great intensity to achieve this goal. He will want to attain such a place more than anything else in the world and may even forget that once, when he was hungry, he sneered at love. Satisfaction of the self-esteem need leads to feelings of self-confidence, worth, strength, capability and adequacy of being useful and necessary in the world. But thwarting of these needs produces feelings of inferiority, of weakness and of helplessness. These feelings in turn give rise to either basic discouragement or else compensatory or neurotic trends. Further even if all these needs are satisfied, we may still often (if not always) expect that a new discontent and restlessness will soon develop, unless the individual is doing what he is fitted for. A musician must make music, an artist must paint, a poet must write, if he is to be ultimately happy. What a man can be, he must be. This need is called self-actualization.

Criminal justice agency deals with the second level as it guarantees safety for people. People who do not feel safe cannot pass to other levels, such as belonging needs and esteem needs. The concept of justice also occupies and important place in Maslow's theory. Some specialists place justice to the first category as one of basic needs.

Motivation is an important part of the work of police. In order to make the citizens be satisfied with the work of criminal justice agency it is important to reach police satisfaction. In order to reach satisfaction of physiological needs police workers should get adequate salary which would enable them to satisfy these needs. If the needs of the first level of the hierarchy are not satisfied people become centered on them only. Need for food and sleep becomes extremely important and they cannot concentrate on other needs. As soon as primary of physiological needs are satisfied, the person can pass to another level of hierarchy. Psychological needs and needs for belonging also should be satisfied. If they are satisfied it is possible to pass to the category of esteem and appreciation. These factors are very important because they may be distinguishing features which make the work of criminal justice agency different from any other kind of jobs. Traditionally, the work in this sphere was considered prestigious and respected one. If this tendency is saved, the forth level of Maslow's hierarchy will be involved in the work of criminal justice agency. For this purpose the workers of this agency should know that their work is appreciated in the society. Job satisfaction is very important criteria of fulfilling psychological needs. It is the task of contemporary society to provide necessary estimation and appreciation of criminal justice agency. Only in this way its workers will be able to provide high level service and correspond to the demands of the society.

Maslow believes that we should study and cultivate peak experiences as a way of providing a route to achieve personal growth, integration, and fulfillment. Peak experiences are unifying, and ego-transcending, bringing a sense of purpose to the individual and a sense of integration. Individuals most likely to have peak experiences are self-actualized, mature, healthy, and self-fulfilled. All individuals are capable of peak experiences. Those who do not have them somehow depress or deny them. Maslow originally found the occurrence of peak experiences in individuals who were self-actualized, but later found that peak experiences happened to non-actualizers as well but not as often.

Maslow is criticized for not having answered the following questions:

Why can't people satisfy their basic needs legally, or why do they choose not to ?

Why don't societies ensure that basic needs can be satisfied legally so that the choice to satisfy them illegally makes no sense ?

Why does society oppress many people, and why aren't more effective measures taken to greatly reduce that oppression ?

Certain other basic problems are (a) the problem of values in any definitive motivation theory, (b) the relation between appetites, desires, needs and what is 'good' for the organism, (c) the etiology of the basic needs and their possible derivation in early childhood, (d) redefinition of motivational concepts, i. e., drive, desire, wish, need, goal, (e) implication of our theory for hedonistic theory, (f) the nature of the uncompleted act, of success and failure, and of aspiration-level, (g) the role of association, habit and conditioning, (h) relation to the [p. 396] theory of inter-personal relations, (i) implications for psychotherapy, (j) implication for theory of society, (k) the theory of selfishness, (l) the relation between needs and cultural patterns. These questions must be considered as motivation theory attempts to become definitive.

8. THEORY OF INTELLIGENCE AND CRIME: HIRSCHI AND HINDELANG

"Intelligence" generally consists of different abilities, such as the ability to reason, solve problems, think abstractly, learn and understand new material, adapt to novel situations quickly, grasp complex relationships, and profit from past experience. In other words, intelligence is "catching on," "making sense" of things, or "figuring out" what to do. The study of intelligence in criminological research has ebbed and flowed considerably during the past century. In the first quarter of the 1900s, hundreds of studies categorized criminal offenders as "feebleminded" and "mentally deficient." Fifty studies conducted from 1910 to 1914 identified an average of 51 percent of institutionalized delinquents as feebleminded (Sutherland). In 1931, however, E. H. Sutherland challenged this prevailing view. He compared the IQ scores of adult offenders to those of army draftees—representative of the general population—and the two groups had nearly identical IQ levels. He concluded that intelligence was not a "generally important cause of delinquency" (p. 362). This rejection of IQ was widely accepted in the criminological literature through the mid-1970s. In 1977, Hirschi and Hindelang (1977:586) reviewed a half-dozen well-known empirical studies and concluded that IQ predicted delinquency as strong, if not more strongly, than race and social class—two variables prominently featured in criminological theory. This revisionist perspective stimulated greater interest in IQ and crime over the next two decades. Herrnstein and Murray (1994) published their highly controversial book The Bell Curve in which they argued, among other things, that racial differences in crime rates resulted from racial differences in intelligence. This book has received widespread negative reaction and has possibly created a general backlash against studies of intelligence and crime.

A final argument against IQ holds that even if all people commit crime with equal frequency, less intelligent people would be less able to evade detection and would be arrested more often. This detection hypothesis has received some empirical support in that IQ scores tend to correlate more strongly with officially recorded crime than self-reported crime. However, most studies still find a significant correlation between IQ and self-reported crime, which is not easily explained by differential police detection.

In contrast to the above spurious arguments, some explanations emphasize IQ as a cause of crime. The earliest causal explanation, popular during the early 1900s, portrayed criminals as so "feebleminded" and "mentally deficient" that they could neither distinguish right from wrong nor resist criminal impulses. This feeblemindedness hypothesis, however, lost favor long ago as it became clear that few criminals are actually mentally deficient and most recognize, though may not follow, behavioural norms.

A more recent, and more compelling, causal explanation emphasizes the importance of intelligence—especially verbal intelligence—during childhood socialization. The socialization of children involves constant verbal communication and comprehension of abstract symbols; therefore, children with poor verbal and cognitive skills have greater difficulty completing the socialization process, which puts them at risk of undercontrolled, antisocial behaviour. Empirical studies overall have supported this developmental hypothesis , and it fits with the especially strong correlation between verbal IQ and crime.

The 2009 Handbook of Crime Correlates stated that reviews have found that around eight IQ points, or 0.5 SD, separate criminals from the general population, especially for persistent serious offenders. It has been suggested that this simply reflects that "only dumb ones get caught" but there is similarly a negative relation between IQ and self-reported offending. That children with conduct disorder have lower IQ than their peers "strongly argue" against the theory (Lee et. al. 2009).

A final causal explanation links IQ to crime through school performance. Less intelligent students do less well in school, which results in academic frustration. This frustration, in turn, weakens their attachment and commitment to schooling, and a weakened bond to school, as per social control theory, allows for more criminal behaviour (Hirschi and Hindelang,1977:586). This school-performance hypothesis has received strong support from empirical studies, and it is probably the most widely accepted explanation of the IQ-crime correlation.

One last approach to IQ and crime deserves mention even though few criminological studies have examined it. Rather than low IQ increasing criminal behaviour, criminal behaviour might decrease IQ. Many facets of a criminal lifestyle can impair cognitive abilities, including physical injuries, especially head traumas, drug use, and withdrawing from school.

Turning to the future of IQ-crime studies, two questions stand out. Will criminologists continue to study intelligence and crime, and if so, which research questions should be pursued ?

Starting with the second question, one issue for future studies involves society's response to intelligence. Menard and Morse hypothesized that school teachers and administrators negatively label low-IQ students thus increasing their risk of criminal behaviour. In addition to testing this hypothesis, studies should examine other societal reactions to IQ as well. Stories abound of classmates stigmatizing bright students as "brains" and "geeks," especially in schools with overall low scholastic achievement. Bright students might avoid these negative labels by cutting back on schoolwork and acting out antisocially. Indeed, peers' labels of high-IQ students may cause more harm than officials' labels of low-IQ students.

Another issue for study involves the nature of IQ's effect on criminal behaviour. Up until now, causal arguments have assumed that low IQ increases criminal behaviour; however, it is possible that in various ways high IQ actually increases criminal behaviour. For example, more-intelligent individuals may feel greater confidence of committing crimes without getting caught, which, as per deterrence theory, should lead to more criminal behaviour. More-intelligent individuals might also have more opportunities for some crimes, such as white-collar crime.

References

Bandura, Albert (1973). Aggression: A social learning analysis. Engle-wood Cliffs, NJ: Prentice Hall.

Cole, M. & Cole S. R. (1993). The development of children. New York: W.H. Freeman and Company.

Eysenck, H.J. (1964). Crime and Personality. Boston: Houghton Mifflin.

Eysenck, H.J., & Gudjonsson, G.H. (1989) The causes and cures of criminality. Contemporary Psychology, 36, 575-577.

Freud, S. (1961). The Complete Works of Sigmund Freud (Vol.19). London: Hogarth.

Herrnstein, Richard J., and Murray, Charles. 1994 *The Bell Curve: Intelligence and Class Structure in American Life.* New York: The Free Press.

Hirschi T and M Hindelang (1977) The Meaning of the IQ-Delinquency Relationship, American Sociological Review, Vol. 42, No. 3, Aug pp. 571–587.

Kohlberg, L. (1976). Moral stages and moralization: The cognitive-developmental approach to socialization In J. Lickona, Moral development behaviour: Theory, research, and social issues. New York: Harper & Row.

Lee Ellis, Kevin M. Beaver, John Wright; (2009) Handbook of Crime Correlates; Academic Press

Siegal, Larry J. (1998), Criminology: Theories, Patterns and Typologies, Sixth Edition. p.145.

Sutherland, Edwin H., and Cressey, Donald R. (1994). Theories of Deviance. p.192.

Tarde, Gabriel (1969). On Communication & Social Influence. p.2-5,30.

Tarde, Gabriel. (1903). The Laws of Imitation, pp.195,199.

Wilson, J.Q. & Herrnstein, R. (1985). Crime and Human Nature. New York: Simon and Schuster.

Wrightsman, L.S., Nietzel, M.T., & Fortune, W.H. (1994). Psychology and the Legal System. Belmont:Brooks Cole Publishing Company.

■■■

CHAPTER 5 Bio-Physical Theories

Learning Objectives :

- ☐ **Acquainting the Learners with the bio-physical theories of crime.**
- ☐ **To comprehend the interrelationships between phrenology, physiognomy, hereditary, anthropometry, somatotyping and crime.**
- ☐ **To make the critical appreciation of biological theories of crime.**

Introduction

Biological theories of crime causation are based on the belief that criminals are physiologically different from noncriminal. It is also called the physical theory of crime. For its scientific and experimental character it is designated as biological positivism. The cause of crime is biological inferiority or physical abnormalities. According to biological theories, a criminal's innate physiological makeup produces certain physical or genetic characteristics that distinguish criminals from noncriminal. Biology and Psychology tend to focus on the question of individual, who is more likely than others to commit crime. Physical criminology & phrenology became the modern approach to the study of crime and Lombroso's famous study of the abnormalities of Italian prisoners & conscripts became the first widely known example.

Believably prior to Lombroso, there were some biological theories of crime such as Phrenology of Franz Joseph Gall, Physiognomy of Johan Casper Lavater, and Hereditary Theory of Richard Dugdale. Lombroso's theory came to be known as Anthropometry. Then has come William Sheldon's Somatotype Theories.

1. Phrenology of Franz Joseph Gall

It was Aristotle who originally put forward the idea that the brain is the organ of the mind. Phrenology is the study of the shape of a person's skull, especially the natural bumps on it, in order to determine his character. Phrenologists elaborated the idea and assumed that the exterior of the skull conformed to its interior and therefore to the shape of the brain. Further, different functions of the mind were assumed to be associated with different parts of the brain. Hence, phrenologists attempted to study the external shape of the skull, especially the bumps on it, to find out how the mind functioned.

Franz Joseph Gall (1758-1828), an eminent European anatomist, is generally given credit for the systematic development of the study of phrenology. In 1791 he started publishing materials on the relations between head conformations and the personal characteristics of individuals. Gall listed twenty-six special faculties of the brain and stated that they were responsible for the behaviour of a person. Acquisitiveness, amativeness, amorousness, benevolence, combativeness, conjugality, constructiveness, cautiousness, ideality and self-esteem are some of the faculties described by Gall. To him, the various faculties were grouped into three regions or compartments: (1) the 'lower' or active propensities, (2) moral sentiments and (3) intellectual faculties. The lower propensities, such as amativeness, acquisitiveness, combativeness and secretiveness, were stated to be responsible for criminal behaviour.

Gall argued that the lower propensities could be held in check by the moral sentiments or the intellectual faculties. They further argued that the various faculties or 'organs' of the mind get strengthened by exercise and enfeebled by disuse, just as any other organ of the body, 50 criminal tendencies in any person could be removed by strengthening the organs of the desirable faculties through careful training in right living and, at the same time, inhibiting the use of lower propensities. Gall called his theory "craniology" and gave it status in the intellectual circles of Europe and America. It was named "phrenology" in the year 1815 by Thomas Ignatius Forster, a naturalist from England.

Phrenology disappeared as a scientific discipline by the middle of nineteenth century because no one was able to substantiate with verifiable data its conceptions of physiological 'organs' of the mind. It is now very well established that the brain is not divided into compartments each of which can be related exclusively to a certain aspect of a person's character. Human behaviour is too complex to be explained by such a simple principle.

2. Physiognomy of Johan Casper Lavater

Physiognomy is the art of making judgments about people's character from the appearance of their face. This art has been in practice in different parts of the world since ancient times. Ancient Greeks and Romans believed in "physiognomy" (physical features can reveal a person's natural disposition). "Do not trust beardless men and bearded women". Johan Casper Lavater (1741-1801) is recognized as the first person to have done exhaustive work on physiognomy. He published a four volume work on the subject entitled Physiognomical Fragments, in the year 1775. 'Face reading' is a full time occupation for some tribal communities from Bhadrachalam area in Andhra Pradesh even today. However, many of the observations made by Lavater, such as beardlessness in men is an unfavorable trait, were based on popular beliefs and did not have any scientific basis. However, his study gave impetus to a better organized and logically more impressive field of study called Phrenology.

3. Hereditary Theory of Richard Dugdale

Explanations of human behaviour in terms of heredity go far back in antiquity and are based on the commonsense observation that children tend to resemble their parents in physical as well as mental characteristics. It is now well established that genes are the transmitters of hereditary characteristics. Several studies have attempted to determine if criminality is hereditary by studying family trees, statistics, identical and fraternal twins and adopted children. The genes are located on chromosomes which are complex structures in the nuclei of plant and animal cells. Every cell of the normal human contains 23 pairs of chromosomes. At the time of conception, an ovum and a sperm, each of which normally contains 23 chromosomes, unite to form a single cell with 23 pairs of chromosomes.

The early attempts to prove a genetic link with criminality usually involved study of the family trees of known criminals. The most famous of such studies was conducted by Richard Dugdale which was published in a book entitled The Jukes: A Study in Crime, Pauperism and Heredity(1877). Dugdale found six members of a family he called 'jukes', in a country jai. He then traced the genealogy of the family back over 200 years and found that there was history of criminality, prostitution and pauperism in the family. He came to the conclusion that heredity was the reason for the criminality in the family. It is apparent that Dugdale's study was based on unreliable and incomplete information. The high proportion of criminals in the Jukes family does not prove any link between criminality and genetic factors. Such a close family similarity might be explained by similar environmental factors acting on all members of that family.

Twin Studies

As pointed out above, in order to precisely study the effect of heredity on criminality, environmental factors have to be controlled and eliminated. However, it is very difficult to control all the environmental factors in any study. Another way to study the relationship is by controlling the hereditary factor and finding out to what extent people with the same hereditary disposition differ with regard to their criminality. This is done through the study of twins.

There are two types of twins-monozygotic or identical twins and dyzygotic at fraternal twins. Identical twins are the product of a single fertilized egg and are genetically identical. Fraternal twins are born as a result of simultaneous fertilization of two eggs with two sperms. So they are no more similar than ordinary brothers and sisters. Because of the identical genetic inheritance, any difference in the behaviour of identical twins would have to be attributed to the environmental factors. By the same reasoning, similarities of behaviour between identical twins could be attributed to their identical inheritance. Of course, the similarities could be because of similarities in training or similar environmental conditions. However, any general tendency to greater similarity of behaviour in case of identical twins sets up a strong presumption that the similarity is due to influence of heredity.

One of the famous twin studies was conducted by German physiologist Johanne Lange (Lange, 1929). His study revealed that when one of the identical twins had a record of imprisonment, the other twin had also been imprisoned in 77 percent of the cases. The percentage in respect of fraternal twins and ordinary brothers of nearest age was found to be 17 and 8 respectively. Some other similar studies also showed greater similarity of criminal behaviour among identical than among fraternal twins. However, these earlier researches suffered from two basic research problems. Firstly, in most of the studies at least one of the twins was drawn either from prison or from a psychiatric clinic, giving an unrepresentative cross section of the twin population. Secondly, the tests used for determining whether the twins were identical or fraternal were not scientific. Often it was done by a visual comparison of the physical similarities.

Two twin studies conducted by Christiansen (1968 and 1974) and Dalgaard and Kringlen (1976) tried to overcome the deficiencies of Lange's study. Christiansen drew his sample of twins from the official "Twins Register" of Denmark. Information was collected on all the twins who were born between 1881 and 1901 and lived up to the age of at least 15 years. There were about 6000 such pairs. He then used the official "Penal Register" to determine whether either or both twins had criminal record. He found a concordance rate of 35.8 percent in respect of identical male twins, i.e. when one of the identical male twins had a criminal record, the other twin also had a criminal record in 35.8 percent of cases (24 out of a total of 67 cases). The concordance rate in respect of male fraternal twins was only 12.3 per cent (14 out of the 114 cases). The differences were even more marked in respect of females - 21.4 percent in respect of identical twins but only 4.3 percent in respect of fraternal twins.

Dalgaard and Kringlen studied 139 pairs of Norwegian male twins. They found a criminal concordance rate of 25.8 percent for identical twins as compared to only 14.9 percent for fraternal twins. However, when the twins were grouped according to their mutual psychological closeness, all differences between identical and fraternal twins disappeared. So they concluded that higher concordance rate for identical twin might be due to the similarity of their upbringing and was not due to hereditary factors. Cloninger and Gottesman (1987) did not agree with these conclusions. They argued Dalgaard and Kringlen should have grouped the twins according to early home environment, rather than psychological closeness to rule out the possibility of similarity in upbringing. According to them, the above study did not reveal any difference in concordance between twins brought up in very close environmental conditions and those brought up in different environments, which in fact proved that environment was not the cause of criminality.

Adoption Studies

Hutchings and Mednick studied the correlation between the criminality of the biological parent and that of the adoptee. High correlation would indicate more probability of criminal characteristics being inherited. Hutchings and Mednick studied all non familial male adoptions in Copenhagen in which the adoptee had been born between 1927 and 1941. They found that boys with criminal biological fathers were more likely to be criminal than those with law-abiding fathers. They also found that those with criminal adoptive fathers were also more likely to be criminal than those with law-abiding fathers. However, the effect of criminality of the adoptive father was found to be not as great as the effect of criminality of biological father. They further found that there was an interactive effect between the criminality of the biological and the adoptive father. When both the biological father and the adoptive father were criminal, the effect upon the criminality was quite marked. In such cases 36.2 per cent of the adoptee was found to be criminal, as compared to 11.5 per cent when only adoptive father was involved in criminal offence and 22 per cent when only biological father was a criminal.

These findings indicate how heredity and environment interact with each other in determining the behaviour of a person. Heredity may predispose an individual to criminality but that individual is likely to exhibit criminal behaviour if the environmental influences are favorable to such behaviour.

4. Anthropometry of Cesare Lombroso

The attempt to derive character traits by measuring the human body in criminology is known as anthropometry. These theories are also known as body type theories that is, the diagnosing of the individual's constitution and behaviour by the shape of their body, as attempts are made to establish some relation between physical characteristics and criminal predisposition. Body type theorists argue that there is a high degree of correspondence between the physical appearance of the body and the temperament of the mind. Anthropometrical approaches to crimininality include Lombroso's measurement of atavistic stigmata, and Sheldon's measurement of general physique, or "somatotype." Atavism refers to Lombroso's theory that while most individuals evolve; some devolves, becoming primitive or "atavistic". These evolutionary "throwbacks" are "born criminals," the most violent criminals in society. Born criminals could be identified through their atavistic stigmata. Stigmata, as a term of medicine refers to the physical marks and characteristics that suggest an individual is abnormal. For Lombroso, 'atavistic stigmata' were those physical characteristics that suggested an individual to be atavistic. Such stigmata included abnormal skull sizes, hawk-like noses, large jaws and cheekbones, and fleshy lips. Somatotyping is a methodology related to the derivation of behavioural types from particular forms of the body. Somatotyping was applied to criminology by William Sheldon.

Cesare Lombroso's Criminal Man (1876) is considered the first book devoted solely to the body type causes of criminality. He, an Italian prison doctor working in the late 19th century and sometimes regarded as the "father" of criminology, was one of the largest contributors to biological positivism and founder of the Italian school of criminology. Lombroso took a scientific approach, insisting on empirical evidence and clinical experimentation, for studying crime. Considered as the founder of criminal anthropology, he suggested that physiological traits such as the measurements of one's cheek bones or hairline, or a cleft palate, considered to be throwbacks to Neanderthal man, were indicative of "atavistic" criminal tendencies. This approach, influenced by the earlier theory of phrenology and by Charles Darwin and his theory of evolution, has been superseded. Enrico Ferri, a student of Lombroso, believed that social as well as biological factors played a role, and held the view that criminals should not be held responsible when factors causing their criminality were beyond their control.

He proposed that criminals were biological throwbacks to an earlier evolutionary stage, a reversion to an earlier and more primitive being who was both mentally and physically inferior. In other words, the criminal reflected our lower and more ape-like ancestors. Lombroso used the term 'atavistic', a term borrowed from Darwin, to describe such people. He held that atavism was indicated by some physical characteristics in people. The physical characteristics mentioned by him included the following :

(a) peculiar size or shape of head; (b) asymmetry of the face; (c) peculiarities of the eyes; (d) enlarged jaw and cheekbones; (e) unusually large or small ears; (f) peculiarities of nose-twisted, upturned, flattened or beak-like; (g) protruding and fleshy lips; (h) abnormal dentition; (i) receding chin; (j) abundance of wrinkles or hair; (k) excessive length of arms; (l) extra fingers or toes; (m) retreating forehead; (n) asymmetry of the brain. Lombroso saw criminals as suffering from an illness, this illness cannot be beaten or cured and people are therefore allocated criminality at birth, there is nothing the criminal can do about this. He suggested that if a person's arm span (that's both arms measured together) is longer than a person's height then that person would be pre-disposed to crime.

From time to time, the list was changed, but it was always made up of similar types of physical anomaly. Any person having five or more anomalies was classified by Lombroso as a born criminal. The born criminals, according to Lombroso, portrayed certain other peculiarities in addition to the physical characteristics. Sensory peculiarities, fictional peculiarities (greater agility, more ambidexterity etc.,), tendency to express ideas pictorially, lack of moral sense and excessive idleness were some such features listed by Lombroso.

Lombroso developed his ideas during the course of his work as a doctor in the Italian army. In those days there was a popular belief in Italy that people from Southern Italy, including Sicily, were lazy, incapable, barbaric and also criminal. Like many others, Lombroso was concerned about the problems of soldiers who came from Southern Italy. One day, while performing postmortem on a thief from Southern Italy, he noticed a depression on the occipital part of the skull. He named it 'median occipital fossa', because of its location in the middle of the occiput as in inferior animals, especially rodents. On discovering such a characteristic feature of inferior animals in the skull of the thief, Lombroso had a flash of an insight about the nature of the criminals. In his own words:

"I seemed to see all of a sudden, lighted up as a vast plain under a flaming sky, the problem of the nature of the criminal- an atavistic being who reproduces in his person the ferocious instincts of primitive humanity and the inferior animals."

Initially Lombroso claimed that all criminals were born criminals and that biological factors were responsible for criminality. His thinking, however, changed over the years. In his later writings, he mentioned that factors such as climate, rainfall, the price of grain, sex and marriage customs, banking practices, the structure of government, church organization, the state of religious beliefs and education, have an effect on criminality. Lombroso ultimately divided criminals into the following categories :

Born criminals : The criminals in whom atavistic qualities are disclosed. They were born or atavistic or the natural criminal. According to Lombroso, they constituted about one-third of the total number of offenders.

Insane criminals : Insane criminal: Not born criminal, but rather become criminals as a result of an alteration of their brain - mental disease of organic kind. In this category Lombroso included idiots, imbeciles, paranoiacs, sufferers from melancholia, and those afflicted with paralysis, dementia, alcoholism, epilepsy or hysteria.

Criminals by passion : This category is distinguished by the high level of impetuousness and ferocity. This type of criminal, according to Lombroso, commits a crime because he or she is compelled to do so by a pure spirit of altruism. Two examples of this type of criminal are the wife who kills her unfaithful husband and the brother who kills the man who raped his sister. Lombroso stated that criminal by passion is more likely to be female than male.

Occasional criminals : This group was divided in to three distinct sub-groups: pseudo-criminals, Criminaloids and habitual criminals. Pseudo-criminals comprise those who commit crime involuntarily, either due 10 perverse laws or in defense of the person or the family. Criminaloids represent a large general class without special physical or recognizable mental disorders, but whose mental and emotional make up are such that under certain circumstances they indulge in vicious and criminal behaviour. The most important element in their criminality was the environment or the opportunity to commit crime. Habitual criminals were defined by Lombroso as individuals who regarded systematic violation of the law in the light of an ordinary trade or occupation and committed their offences with indifference. These criminals turn to crime for reasons of poor education and training at an early age, or are drawn into crime by close association with criminals.

Lombroso's theory consisted of the following propositions :

1. Criminals are, by birth, a distinct type.
2. That type can be recognized by physical characteristics, or stigmata, such as enormous jaws, high cheekbones, and insensitivity to pain.
3. The criminal type is clearly distinguished in a person with more than five stigmata, perhaps exists in a person with three to five stigmata, and does not necessarily exist in a person with fewer than three stigmata.
4. Physical stigmata do not cause crime; they only indicate an individual who is predisposed to crime. Such a person is either an **atavist** or a result of degeneration.
5. Because of their personal nature, such persons cannot desist from crime unless they experience very favorable lives.
6. Asymmetry of the face or head, large monkey-like ears, large lips, twisted nose, excessive cheek bones, long arms, excessive skin wrinkles.
7. The male with five or more of these physical anomalies is marked as a born criminal.
8. Female criminals are also born criminals, but they may be identified with as few as three anomalies.
9. Tattoos were significant to Lombroso. Most of the "born criminals" had them. Obscene nature of their depictions and messages were reflected in their nature and behaviour. Tattoos stood as evidence of both insensitive to physical pain and immorality.

Criticisms of Lombroso

- Charles Goring, who in 1913 published a book entitled *The English Convict: A Statistical Study* concluded that there is no such thing as a physical criminal type. Goring's book was the first to adopt the position that criminality is probably the result of the interaction of a variety of hereditary and environmental factors at a time when theorists thought in terms of either/or, and a bridge between early biological positivism and the less deterministic psychological and psychiatric schools.
- Lombroso's work was based in an Italian prison. His sample was therefore biased.
- The prisoners were usually from a poor background, their diets would have been bad and this could explain their deformities.
- Moreover the fact that a number of prisoners had these deformities could cause them to be social outcasts and therefore commit crimes because of this not because of their actual physical deformities.

5. Somatotyping Theory of William Sheldon

A further systematic study linking of body types with delinquency was done by William Sheldon in 1949. The following are the main assumptions of his theory.

A person's temperament or constitutional make-up can be known from the shape of his body. The shapes of the body as well as the temperament are strongly influenced by conditions at birth and they remain constant throughout the life of the person. The human embryo is made up of three different tissue layers, namely, endoderm (inner layer), mesoderm (middle layer), and ectoderm (outer layer). The endoderm gives rise to the digestive organ. The bones, muscles and the tendons of the motor-organ system develop from the mesoderm, the connecting tissue of the nervous system and skin develop from the ectoderm. Basing his ideas on these facts, Sheldon constructed a corresponding physical and mental typology, as under :

(a) Endomorphic : Persons of this body type have a tendency to put on fat. They have soft roundness in various regions of their body with short tapering limbs, small bones and soft and smart skin. They are said to possess 'viscerotonic' temperament which is characterized by love for comfort and extroversion. They tend to be friendly and sociable.

(b) Mesomorphic : These persons have large trunk, heavy chest, large wrists and heavy hands. They are hard, muscular and athletic. Their personality is strong and assertive, with a tendency to be aggressive. Sheldon termed such temperament as 'somotonic.'

(c) Ectomorphic : Persons of this body type have lean, fragile and delicate body with small skeleton and weak muscles. They are weak and thin. They suffer from chronic fatigue, insomnia, allergies and skin troubles. They tend to be hypersensitive, shy and introverted. The temperament of such people was termed as 'cerebrotonic'.

Figure 5.1 Body Construction and Temperament

Sheldon maintained that each person possesses the characteristics of the three types to a greater or lesser degree. He used numbers 1 to 7, to indicate the extent to which the characteristics of the three types were present in a given individual. Thus, according to Sheldon, no one could be characterized purely as one somatotype and every person was more of one type than another. For example, an ectomorph might be 2-2-7, an endomorph 6-2-3, and so on.

Sheldon studied 200 males held in a rehabilitation home in Boston as well as 200 college students to test his theory. He found that the average somatotype of the delinquent boys from the rehabilitation center was 3.5-4.6-2.7 and that of the college students 3.2-3.8-3.4. As he felt that the difference between the two groups with regard to mesomorphy and ectomorphy was significant, he concluded that most delinquents tended towards mesomorphy and lacked in ectomorphy. Sheldon did not consider that delinquents are more likely to be mesomorphic because, for example, mesomorphs are more likely to be selected for gang membership.

The association between mesomorphy and delinquency was also found in a study by the Gluecks (1950). They Compared 500 persistent delinquents with 500 non-delinquent control group. When photographs of the boys were mixed together and then visually assessed for the predominant body type, 60 percent of the delinquents were found to be mesomorphs as opposed to only 31 percent of the non-delinquents. Their study also revealed that mesomorphs who became delinquents were characterized by feelings of inadequacy and emotional instability whereas the non-delinquent mesomorphs did not generally have these traits. The Gluecks also found that delinquency in mesomorphs was associated with three social-cultural factors- careless household routine, lack of family group recreations, and meagerness of recreational facilities at home. In view of the same they concluded that delinquency was probably a combination of biological, environmental and psychological factors. The main criticism against the Gluecks study is that the method of somatotyping involved only visual assessment and not precise measurements.

Another early researcher in this area was Ernest Kretschmer of Germany. Kretschmer identified three body types: the 'leptosome' or 'asthenic' type was tall and thin; the 'athletic' type had well developed muscles and the 'pyknic' type was short and fat. The body type, which did not come under any of these types was called 'dysplastic' or 'mixed.' These body types were associated with different types of crimes as under.

TABLE 5.1

Body Type and Crime

Types of Crime	Body type likely to be involved in the crime
Violent Crime	Athletic
Petty thievery and fraud	Asthenic
Deception and fraud	Pyknic
Offences against decency and Morality as well as violent crimes	Dysplastic

Kretschmer's study did not compare his results with non criminal population. He also did not present his statistical analysis.

Evaluation of Biological Explanations

- Most of the prison population is made up of lower working class males. Reasons for committing such crimes can be linked to their class rather than their physical deformities.
- Those with extra chromosomes/physical deformities etc have an abnormal appearance and this may cause individuals to behave oddly as it can prevent them from having a normal life.

- Crime is too diverse for biological explanations to account for all behaviour.
- There are methodological problems in research studies. Not everyone who breaks the law ends up in prison. Biological theorists studied largely the very poor people whose physical development had been affected by poverty, poor nutrition. This type of theorising neglects the idea that there is a "grey area" of criminality - people who commit crimes but who are not caught and therefore not imprisoned.
- Biological Explanations cannot easily account for group rate differences.
- Social policy implications become doubtful as we cannot change biology.

References

Christiansen, K. O. (1968). Threshold of tolerance in various population groups illustrated by results from Danish criminological twin study. In A. V S. de Reuck & R. Porter (Eds.), Ciba Foundation Symposium on the Mentally Abnormal Offender (pp. 107-116). London: Churchill.

Christiansen, K. O. (1974). Seriousness of criminality and concordance among Danish twins. In R. Hood (Ed.), Crime, criminology, and public policy (pp. 63-77). London: Heinemann.

Cloninger, C. R., & Gottesman, I.I. (1987). Genetic and environmental factors in antisocial behaviour disorders. In S. A. Mednick, T. E. Moffitt, & S. A. Stack (Eds.), The causes of crime: New biological approaches (pp. 92-109). New York: Cambridge University Press.

Dalgaard, O. S., & Kringlen, E. (1976). A Norwegian twin study of criminality. British Journal of Criminology, 16, 213-232.

Ernest Kretschmer, (1931) Physique and Character (International Library of Psychology) (1931), Routledge.

Gall Franz Joseph (1833) "The Anatomy and Physiology of the Nervous System, "The American Journal of the Medical Sciences" Southern Society for Clinical Investigation.

Glueck, S. & Glueck, E, (1950) Unraveling Juvenile Delinquency. Cambridge, Mass.: Harvard University Press.

Hutchings, B., & Mednick, S. A. (1971). Criminality in adoptees and their adoptive and biological parents: A pilot study. In S. A. Mednick & K. O. Christiansen (Eds.), Biosocialbases of criminal behaviour (pp. 127-141). New York: Gardner.

Lange, Johanne, (1929) Crime and Destiny, Zwillingen, Leipzig.

Lavater, John Casper (1783) Essays in Physiognomy (Translated) , London: Ward, Lock & Co.

Lombroso Cesare (1876) Criminal Man (Translated Reprint) Duke University Press, USA 2006.

Richard Dugdale, (1877)The Jukes: A Study in Crime, Pauperism, Disease and Heredity. G.P. Putnam's Sons, New York.

William Sheldon, (1949) The Varieties of Human Physique: An Introduction to Constitutional Psychology. New York: Harper.

■■■

CHAPTER 6

Sociological Theories

Learning Objectives :

- **Acquainting the Learners with the theories of crime propounded by Durkheim, Sutherland, and Merton's theory of crime.**
- **To understand other theories such as the Ecological School, Social Disorganization, Critical theories and the Broken window Hypothesis on crime.**
- **To comprehend the future of Sociological theories of crime.**
- **To make a critical appreciation of Sociological theories of crime.**

Introduction

Crime is a very complex social issue to be dealt with. The central thesis of the sociological school is that criminality has taken two principal forms. First, sociologists have attempted to relate variations in crime rates to variations in social disorganization, including the variations in large institutional systems such as the processes of mobility, culture conflict, competition, and stratification; political, religious and economic ideologies; population density and composition; and the distribution of wealth, income, and employment. This kind of analysis has fallen into disfavour in recent years, principally because criminologists have become aware of the extreme unreliability of crime statistics. Because, variations in crime rates may represent mere differences in statistical procedures, rather than real variations in the frequency of crime. Second, sociologists have attempted to define the processes by which a person becomes a criminal. These analyses are related to general theories of social learning and have utilized such concepts as imitation, attitudes, values, differential association, and some extent, compensation and frustration-aggression. Further, some sociologists have analysed how lack of control by various social institutions such as the family and the school results in delinquent and criminal behaviour.

A greater interplay is now found between the Psychological-Psychiatric School and the Sociological School. The Sociological School recognizes that psychogenic traits must be taken into account in the explanation of criminal behaviour, and members of the Psychological School have recognized the importance of social and cultural conditions. But there is disagreement as to the extent to which 'personality' and 'culture' should be emphasized in criminological theories, largely because there is no consensus as to the specific manner in which personality and culture interact to produce specific forms of criminal behaviour. Some writers in both schools have made classifications of criminals, with the conception that one class is due to personality and another due to culture. Lindesmith and Dunhan (1941: 307-314) have suggested that criminals constitute a continuum with personality and culture as two variables; personality as a cause is zero at one extreme and 100 per cent at the other, with culture ranging between the same extremes in the reverse order. Thus, at one extreme would be the individualized type, represented by the psychotic, whose 'crime' is an expression of individual values and not supported by culture; at the other extreme is the 'social criminal' represented by the professional thief, whose objectives are approved by the general society and whose methods are approved only within his smaller variant culture. This conception of the relation between personality and culture seems to form a basis for merging the Psychological and Sociological approaches to an explanation of criminal behaviour. But the problem of how, or whether, basic personality traits and culture combine in the less extreme cases remains as a point of controversy. However, the insights from sociological theories have resulted in reformulation of theories in criminology.

1. Durkheim: Crime as a Normal Phenomenon

Emile Durkheim was a French sociologist who contributed significantly to criminological thought. His theories were in part a reaction to the classical assumptions that humans were free and rational in a contractual society. He was a positivist. However, unlike Cesare Lombroso, who had biological focus, Durkheim focused on society and its organization to explain human behaviour. He also maintained that "social facts" such as customs, law, morality and religious beliefs were quantifiable and measurable things on the basis of which scientific analysis could be conducted.

Crime is the Cause and Effect of Social Change

In his "The Division of Labour in Society' (1893), Durkheim viewed the societies as developing from the more primitive "mechanical" form into the more advanced "organic" form. In the mechanical form there is little division of labour. In small-scale societies (rural, pre-industrial, societies), social organisation was seen to be based upon close, personal, norms and values. Durkheim argued that since norms and values tend to be the social glue that binds people together in groups, the combination of informal social controls, restricted geographic mobility and the like, served as the basis for social order. In large-scale societies based on organic characterized by division of labour, the legal system develops as a functional necessity to regulate relationships that cannot be adequately maintained by the operation of informal norms (mainly because instrumental nature of most relationships mean that moral ties are much weaker). In effect, people would no-longer know what was morally right or morally wrong and so would be unsure as to where the boundaries of acceptable behaviour lay. Durkheim argued that the basis of social order in complex, industrial, societies was to be organic solidarity. Organic solidarity would produce a collective conscience which would be the underpinning of social order.

The collective conscience represents the will of society as a whole. Legal rules represent a form of objective expression of the collective conscience. In effect, laws develop out of (and are underpinned by) the collective conscience of a society. Durkheim argues that people are shaped by their social experiences (they experience society as a moral force bearing down upon them) if the collective conscience is weakened, the moral ties that bind people together are also weakened.

Durkheim departed fundamentally from the conventional path. While most criminologists treated crime as a pathological phenomenon and sought psychological causes in the mind of the criminal, Durkheim saw crime as normal in terms of its occurrence, and even as having positive social functions in terms of its consequences. Crime was normal in that no society could enforce total conformity to its injunctions, and if society could, it would be so repressive as to leave no leeway for the social contributions of individuals.

Deviance from the norms of society is necessary if society is to remain flexible and open to change and new adaptations. "Where crime exists, collective sentiments are sufficiently flexible to take on a new form, and crime sometimes helps to determine the form they will take. How many times, indeed, it is only an anticipation of future morality—a step toward what will be." But in addition to such direct consequences of crime, Durkheim identified indirect functions that are no less important. A criminal act, Durkheim reasoned, elicits negative sanctions in the community by arousing collective sentiments against the infringement of the norm. Hence it has the unanticipated consequence of strengthening normative consensus in the common weal. "Crime brings together upright consciences and concentrates them."

As societies develop and grow, the moral ties which bind people are weakened. They cannot be continually reinforced by close, personal, contact. As societies become more-complex in the multitude of social relationships, a mechanism to regulate relationships has to be developed - a legal system. Legal systems develop to codify moral behaviour and this process lays the groundwork for our understanding of the functions of both law and crime. The codification of morals into laws requires public displays that such boundaries exist. Boundary marking displays are made by the mass media and the Courts. Crime is publicized by the mass media. The Courts present the rituals of the legal system which serve to set it apart from everyday behaviour.

An important point to remember is that, for Durkheim, what constitutes a crime is something culturally relative rather than universal. A crime is a crime because it offends values, not because it is something fundamentally wrong. By punishing criminals, society reaffirms it own values. If crimes were not committed, then the values of society would become blurred. If there is no punishment, then there would be no way of reestablishing the values that the crime offends.

Non-conformity to social norms occurs when social structure is under pressure (or strain). Rapid technological change (Industrial Revolution) changes the way institutions relate to one another. Change produces new norms, new values. Those socialised to the "old" norms take time to adapt- there is a "time lag" people have to be re-socialised in order to develop new norms / values that fit with structural changes. "Normative confusion" anomie is likely to occur. Thus, strain at the institutional (structural) level of society leads to a weakening of the Collective Conscience and, as a consequence, increased levels of crime / deviance. In effect, people are born into an existing society and are taught the rules and relationships governing social behaviour. Social behaviour, therefore, is explained by changes in the social structures that produce consequent changes in human behaviour.

Crime and Law

Law and crime function to mark the boundaries of acceptable and unacceptable behaviour in any large-scale society. That is, they represent broad social guide-lines for people's behaviour. Laws are a public form of boundary marking. They are written-down and applicable to everyone. Criminal behaviour was seen to be the way in which legal boundaries were tested. Because laws were necessarily slow to change once adopted, changes in people's behaviour over time would fail to be reflected in changes in the law, unless a mechanism existed to accommodate change. Criminal behaviour, in effect, represents the dynamic force whereby people are made to reassess the applicability of various laws. Where laws are out-of-step with general behaviour, they need to be changed.

According to Durkheim, law plays an important role in maintaining social solidarity in each of these two types of societies, but in different ways. In mechanical societies, its main function is to enforce the uniformity and the status quo. It is thus oriented toward repressing any deviation from the established norms of the society. In organic societies, the main function of law is to integrate the diverse parts of the society. In mechanical societies the various social institutions such as family, church and school, play a vital role in providing informal pressure to conforms to the norms. On the other hand, formal regulatory agencies are given the responsibility of ensuring compliance to law, in organic societies.'

Crime as a normal phenomenon

Durkheim viewed crime as a normal occurrence, and said that it is impossible to have a society totally devoid of crime. In seeking to argue his case, Durkheim points to the fact that crime seems to be omnipresent across time and space. All societies, bar none, have experienced crime. Durkheim argued that every society had crime therefore it must serve some purpose. Thus, crime seems to represent something of a condition of normality. The dividing line between behaviours considered criminal and those considered morally reprehensible is not clearly marked in any society. If there is decrease in behaviours designated as criminal, then there may be a tendency to move behaviours previously designated as morally reprehensible into the criminal category.

Functions of Crime

Emile Durkheim was a functionalist. He believed if something exists in society it must have a purpose for existing. It must serve a function. Since crime existed it is necessary to explain the functions it performed. Elaborating his concept of 'normalcy of crime' in his book 'The Rules of Sociological Method' (1995), Durkheim argued that crime is not only normal but also functional in several ways.

Firstly, it is necessary to maintain social solidarity in the society. In every society there are a large number of people which obeys the laws of the land and refrains from committing any type of crime. These people, who fulfill the demands of the collective conscience, feel a sense of moral superiority in identifying themselves with the good and the righteous, and opposing the morally inferior criminals. The sense of moral superiority of the majority of population helps in fostering social solidarity. The punishment of the criminals further reinforces the sense of superiority and strengthens the social solidarity.

The second functional role of crime is that it helps in social progress. A primary source of social change is individual or group deviation from social norm. Even though such deviants are frequently declared as law-breakers or criminals, they may inspire others to follow suit and in the process social change may occur.

Thirdly, another functional aspect of crime is that it serves as a warning bell and draws the attention of the officials and policy makers to serious problems in the society. The crime rates in any area over a period of time can be studied and appropriate preventive measures can be taken before any problem assumes serious proportions.

Finally, crime has an integrating function. The public alarm and outrage at criminal acts serve to draw the law-abiding closer together. This increases social solidarity (the communal feelings that people have for one another). The bonds between people were seen to be continually reinforced by criminal behaviour.

Although Durkheim was clear about the functional significance of crime, he was also aware of the fact that too much crime in a society would create problems. A too high a level of crime could create problems, given the distinction that can be made between acceptable / unacceptable forms of behaviour and the idea of values consensus produced through socialization ? In this respect, Durkheim saw that too a high level of criminal behaviour would be potentially disruptive (or dysfunctional) because it would seriously weaken the moral order in society

Anomie in Organic Societies

The state of inadequate regulation was called "anomie" by Durkheim. In an organic society, the function of law is to regulate the interactions of the various parts of the society. If the regulation is inadequate, an unhealthy level of criminality is likely to arise. Anomie occurs when traditional norms of behaviour were undermined without being replaced by new norms. In the absence, people experience feelings of anxiety, aimlessness, purposelessness and, disorientation.

As society becomes increasingly pluralistic in function, the rules of behaviour, including laws, increase in abundance, making it difficult for people to remember the rules. Similarly, the definition of particular crimes may become unclear and difficult to define in a highly complex and heterogeneous society. Anomie was also the result of a lack of societal norms or regulations over people's desires and aspirations. Anomie is associated with the frustration resulting from the difficulty of achieving goals or success, in a society beset by normative complexity and a breakdown of traditional social institutions.

To Durkheim, anomie was a very dangerous phenomenon, when people no-longer believe in their obligations, they revert to self-interest. Thus, high levels of criminal behaviour weaken the collective conscience and produce anomie. Since human beings cannot live in a state of true anomie for long, social collapse occurs and then the establishment of some new form of moral order takes place. Not only poverty but also prosperity are the results of anomie or normlessness.

Suicide

Suicide, one of the outcome of anomie, was the subject of Durkheim's most famous research. In his book 'Le Suicide' (1897) Durkheim concluded that suicide rate tended to increase sharply both in periods of economic decline and economic growth. Durkheim also found 'social facts', such as religious beliefs and affiliation, family and community cohesion, to be directly related to rate of suicide in a society. Durkheim related suicide rates with two social conditions: social integration and social regulation. High levels of integration can result in Altruism. On the other hand, low levels of integration may result in egoism. High and low levels of social regulation produce fatalism and anomism respectively.

Durkheim categorized suicides into four types. Sacrifice of life made by a person in the interest of the community or out of loyalty to one's group would be characterized as "Altruistic". As the society becomes increasingly urban and modernized, such altruistic sentiments decrease. The second type of suicide is "Egoistic" suicide. Egoism is the opposite of altruism. Durkheim saw egoism primarily as a social, rather than a personal, phenomenon. In an egoistic society, individual activities take precedence over communal obligations and persons get detached from various community support groups. Under such circumstances, any one failing in his egoistic pursuits might resort to suicide. If the society disallows individual expression or pursuit of goals through excessive regulation, people may feel choked by the oppressive discipline and commit suicide called "Fatalistic" suicide. Then is the suicide related to anomie. "Anomic" suicides occur when the goals and norms of community are in a state of confusion and when there is ineffective regulation of the needs of the individuals.

Durkheim's work was criticized on the following grounds.

- One of the major criticisms of Durkheim's general work in relation to crime has been the idea that he ignores the way in which power is a significant variable in relation to the way in laws are created and maintained in any society.
- Durkheim's assumption that pre modern societies had strong social controls and little crime, was not correct. It is more likely that these societies had little social control and a great deal of crime especially violent crime.
- Crime rates did not increase over the period of the French revolution or during the industrial revolution, periods depicted by Durkheim as being a state of upheaval and anomie.
- In contrast to Durkheim's argument that punishments would become less violent in organic societies, it was found that more developed societies were characterized by severe punishments.

2. Differential Association Theory of Edwin H. Sutherland

Sutherland was an American sociologist who attempted to formulate a general theory of criminal behaviour. The theory emerged gradually over an extended period of time in several editions of his book on criminology. A systematic and formal presentation of the theory appeared only in the third edition of the book, published in 1939, almost 18 years after his first ideas on criminological theory began to take shape.

The theory was outlined by Sutherland in the form of a series of propositions. There were originally seven propositions. In the fourth edition of the book, which appeared in 1947, the list was revised to nine. The following are his propositions.

(a) Criminal behaviour is learned.

(b) Criminal behaviour is learned in interaction with other persons in a process of communication.

(c) The principal part of the learning of criminal behaviour occurs within intimate personal groups.

(d) When criminal behaviour is learned, the learning includes (a) techniques of committing the crime, which are sometimes very complicated, sometimes very simple; (b) the specific direction of motives, drives, rationalizations, and attitudes.

(e) 'The specific directions of the motives and drives are learned from definitions of the legal codes as favorable or unfavorable. In some societies an individual is surrounded by persons who invariably define the legal codes as rules to be observed, while in others he is surrounded by /persons whose definitions are favorable to the violation of legal codes.

(f) A person becomes delinquent because of an excess of definitions favorable to violation of law over definitions unfavorable to violation of law. This is the principle of differential association.

(g) Differential associations may vary in frequency, duration, priority, and intensity. This means that associations with criminal behaviour and also associations with anticriminal behaviour vary in those respects..

(h) The process of learning criminal behaviour by association with criminal and anti-criminal patterns involves all of the mechanisms that are involved in any other learning.

(i) While criminal behaviour is an expression of general needs and values, it is not explained by those general needs and values since non-criminal behaviour is an expression of the same needs and values.

The differential association theory is basically a theory of learning. Fundamentally it asserts that criminal behaviour is learnt by association. The theory bears a great resemblance to Tarde's theory of "imitation". However, instead of imitation, Sutherland's theory was based upon a wider idea involving all the normal mechanisms of the learning process. According to Sutherland, the learning of criminal behaviour takes place in small informal group settings, through direct and personal contact with other group members. The communication through which the interaction takes place is verbal in many ways but also includes "the communication of gestures". Impersonal agencies of communication, such as movies and newspapers play a relatively unimportant part in the genesis of criminal behaviour. When a person learns criminal behaviour, he learns not only the techniques of committing the crime (How the crime is committed) but also the motives, drives, values and attitudes for it committal (Why it is committed). Finally, whether a person takes part in criminal activities depends upon the amount of contact with criminal activities or with those who support those criminal activities. When persons become Criminal, they do so because of contacts with criminal patterns and also because of isolation from anti-criminal patterns. Associations, which are neutral so far as crime is concerned, have a little or no effect on the genesis of criminal behaviour.

The delinquent activity may also be affected by the frequency, duration, priority and intensity of the definitions either for law-abiding or law-breaking behaviour. "Frequency" and "Duration" as modalities of associations are obvious and need no explanation. "Priority" is assumed to be important in the sense that behaviour, whether lawful or unlawful, developed in early childhood may persist throughout life. "Intensity" has to do with such things as the prestige of the source of criminal or anti-criminal pattern and the emotional reactions related both the source and the content of information.

There are three key concepts in Sutherland's theory : Differential association, culture conflict and differential social organization. Differential association refers to the acquisitioned process by which some persons, through association with anti--criminal behaviour patterns, acquire an 'excess of definitions' that are conducive to violation of law, and commit delinquent/criminal acts. 'Culture conflict' refers to the differences in the attitudes, values and behavioural patterns among different groups in a society. Finally, differential social organization refers to the existence of divergent associations and groups in any society, organized around different interests and different purposes. Initially, Sutherland used the term 'social disorganization', to explain the presence of culture conflict in a society. However, in the final version of his theory, the term 'differential social organization' was introduced in place of 'social disorganization'. Similarly, the term 'culture conflict' was replaced with the word 'normative conflict', by Sutherland's co-author of the later editions of the book, Donald R. Cressey, after Sutherland's death. Normative conflict refers to the situation in which different social groups hold different opinions about appropriate ways to behave in specific situations and circumstances. The substitution of the term was done, according to Cressey, to clarify the confusion over the interpretation of the term 'culture conflict'.

Sutherland's theory is based to a considerable extent on the "symbolic interactionism" theory of George Herbert Mead. According to Mead, human beings act toward things on the basis of the meanings that the things have for them. The meanings of such things are derived from the social interaction one has with one's fellows. Further, human beings generalize the meanings they have derived from particular situations, and develop a relatively set way of looking at things. As pointed out by Void, Bernard and Snipes (1998), the basic elements of Sutherland's theory (a) the 'content' of what is learned and (b) the process by which the learning takes place, are derived from Mead's theory. The content component (which includes cognitive elements such as techniques, motives, attitudes, rationalizations etc.) is quite similar to the cognitive factor "meanings" in Mead's theory. The second element, process of learning, directly follows from Mead's views on how "meanings of things" are derived. Sutherland's work can be seen as a reaction to the biological and psychological determinism of the positivists. It was an attempt to integrate the various atomistic, individual explanations of criminal behaviour that were dominant at the turn of the century with the situational learning theories.

The following are the main criticisms against the Differential Association theory:

- The theory cannot be used to explain all criminality. For example, it cannot explain the criminal behaviour of irrational, impulsive, opportunist or passionate criminals.
- It does not explain the criminality of those who have never been subjected to criminals or to people who hold criminal ideas. Key concepts in the theory such as frequency, duration, priority, intensity and 'excess of definitions' are not defined precisely and cannot be reduced to quantitative elements. As such, it is impossible to empirically test the theory.
- The theory does not explain fully why one person has one set of associations and another person has a different set, although the two may come from similar backgrounds.
- The theory does not adequately incorporate personality factors.

3. Enrico Ferri and Social Defense

Ferri, the second among the three major Italian positivists, is closer to the contemporary line of thinking, because he insisted that crime "is the effect of multiple causes". He took into account a large number of physical and social factors also, apart from anthropological factors, while explaining criminal behaviour. The anthropological factors included sex, age, race, organic constitution, and anomalies. Physical factors were represented by climate, geographic location, temperature, fertility of the soil, meteoric conditions, etc. Finally, economic and civil status, profession, social rank, density of population, customs, religion and education made up the social factors.

Enrico Ferri was instrumental in formulating the concept of "social defense" as a justification for punishment. The theory of punishment asserts that its purpose is not to deter or to rehabilitate but to defend society from criminal predation. Ferri's concern was social preservation and not the nature of criminal behaviour. Given the assumptions of biological positivism, the only reasonable rationale for punishing offenders is to incapacitate them for as long as possible so that they no longer posed a threat to the peace and security of society. This theory of punishment provides us with an example of how anthropological assumptions drive policies for dealing with crime and criminals. He was, however, an ardent proponent of measures to prevent crime among "occasional criminals" through social reform, and of efforts to rehabilitate them.

Ferri listed six types of criminal :

1. Born / instinctive: - propensity for crime from birth
2. Insane: - affected by mental disease
3. Passional: - crimes of passion or emotion
4. Occasional : - product of family / social conditions
5. Habitual: - criminal by habit, product of social environment
6. Involuntary criminals, who commit crime because of negligence, imprudence or lack of foresight.

The first two categories are more or less similar to the born and insane criminals of Lombroso. Two types of passional criminals were identified by Ferri -those who suffer from prolonged and chronic mental state and those who suffer from explosive and emotional outburst. The fourth category, occasional criminals, according to Ferri, represented the vast majority of those who commit crimes as a result of 'family and social milieu'. Two more types - habitual criminal and involuntary criminal - were later on added to the original four. Ferri argued that commission of crime becomes an acquired habit for the habitual criminal because of a variety of reasons such as growing up in an environment represented by poverty, poor education and bad companions, and subsequent acquisition of criminal habits through contact with criminals when placed in jail. Finally, involuntary criminals, according to Ferri, are those criminals who commit crime not because of any malicious intention but because of negligence, imprudence or lack of foresight. Ferri was a positivist with a socialistic orientation. He was of the view that the state is the principal instrument through which better conditions are to be attained. He suggested a large number of preventive measures to prevent crime as well as for social betterment. These included free trade, abolition of monopolies, inexpensive dwellings for workmen, public savings banks, better street lighting, birth control, freedom of marriage and divorce, state control of manufacture of weapons and provision for public recreation.

4. Natural Crime- Garofalo

Like Lombroso and Ferri, Garofalo (1914) rejected the doctrine of free will and supported the position that crime can be understood only where it is studied by scientific methods. He attempted to formulate a sociological definition of crime and is perhaps best known for his efforts to formulate a "natural" definition of crime. Classical thinkers accepted the legal definition of crime uncritically; crime is what the law says it is. This appeared to be rather arbitrary and "unscientific" to Garofalo (like the British-American system of linear measurement) who wanted to anchor the definition of crime in something natural (like tying linear measurement to the circumference of the earth, as in the metric system). Garofalo felt that definitions of crime should be anchored in human nature, by which he meant that a given act would be considered a crime if it were universally condemned, and it would be universally condemned if it offended the natural altruistic sentiments of probity (integrity, honesty or respect for the property rights of others) and pity (compassion, sympathy or revulsion against the voluntary infliction of suffering on others).

Garofalo made a distinction between "natural crime" and "police crime," to explain his view, "police crimes" are those behaviours which do not offend the altruistic sentiments but are nonetheless treated as criminal acts by law. Natural crimes are evil in themselves (*mala in se*), whereas other kinds of crimes (*mala prohibita*) are wrong only because they have been defined as such by the law. As can be inferred from the above, Garofalo focussed on psychic anomalies rather than on physiological anomalies. He, however, gave a biological explanation for the anomalies and argued that altruistic sentiments had an organic basis and were not the result of environmental or economic factors.

Garofalo developed four categories of criminals, each meriting different forms of punishment viz. (1) "extreme"-murderers, (2) "impulsive"-violent criminals, (3) "professional"-criminals deficit in probity, and (4) "endemic"-lascivious criminals. The first type, murderers, were identified with those people who completed lacked in altruism. Violent criminals were subdivided into two categories. Those who committed crimes that are specific to a geographical region were termed as endemic criminals and those who committed crimes under the influence of passion or as a result of external causes designed to arouse anger (such as alcohol, high temperature etc.,) were called passional criminals. The third type of criminals, according to Garofalo, committed offences against property due to lack of respect for the property of others. Sexual impulse was stated to be the reason for the commission of crime by the lascivious criminals.

Garofalo, a magistrate by profession, rejected the classical principle that punishment should fit the crime, arguing instead that it should fit the criminal. As a good positivist, he believed that criminals have little control over their actions. This repudiation of free will (and, therefore, of moral responsibility) and fitting the punishment to the offender would eventually lead to sentencing aimed at the humane and liberal goals of treatment and rehabilitation. So he was naturally concerned with penal policies and reforms in criminal procedure. He outlined a detailed theory of punishment based on Darwinian principle of adaptation. He argued that society, like any natural body, must eliminate all those who fail to adapt to civilized life and turn to criminality. He suggested three means of elimination : (1) death, (2) partial elimination including life imprisonment and transportation, and (3) enforced reparation. While death penalty was recommended for those with permanent psychological anomaly which rendered them incapable of any social life, imprisonment or transportation was recommended for the young offenders who could be reformed. The third type of punishment in the form of imposition of fine or forcible collection of money was suggested for those who committed crimes under exceptional circumstances and were not likely to commit them again. He favoured death penalty as the most efficient means of eliminating criminals as well as their progeny from the society.

For Garofalo, however, the only question to be considered at sentencing was the danger the offender posed to society, which was to be judged by an offender's "peculiarities." By "peculiarities," Garofalo was not referring to Lombrosian stigmata, but rather to those particular characteristics that place offenders at risk for criminal behaviour. Society could only be defended from extreme criminals by swiftly executing them, regardless of the crime for which they are being punished. Here Garofalo departed from Lombroso and Ferri, both of who were against the death penalty, although Lombroso gradually came to accept it for born criminals and for those who committed particularly heinous crimes. Impulsive criminals, a category which included alcoholics and the insane, were to be imprisoned. Professional criminals were psychologically normal individuals who utilize the hedonistic calculus before committing their crimes, and thus require "elimination," either by life imprisonment or transportation to a penal colony overseas. "Endemic crimes", by which Garofalo meant crimes peculiar to a given location or region (*mala prohibita*), could best be controlled by changes in the law, not by imposing harsh punishments on offenders.

5. The Ecological School

Durkheim's argument that rapid social change was associated with increases in crime due to the breakdown of social controls was used by members of the Department of Sociology at the University of Chicago in the 1920s to pinpoint the environmental factors associated with crime and to determine the relationship among those factors. However, instead of focusing on rapid change in entire societies, they focused on rapid change in neighborhood. Their procedure involved correlating the characteristics of each neighborhood with the crime rates of that neighborhood. Since the research was based on an image of human communities taken from plant ecology, it became known as the Chicago School of Human Ecology (also referred to as the Chicago School).

The Chicago School was primarily the brainchild of Robert Park and Ernest Burgess. Park saw the city not just as a set of buildings in a particular geographic location, but as a living ecological environment, or as a kind of 'super organism". By this, he meant that the people and the institutions were so intimately bound together that they tended to interact as a whole. As with any ecological community, there were 'natural areas' where the inhabitants were mainly of a particular type or types: racial or ethnic communities; immigrant communities; similar income and occupational groupings, etc. Within each community there were symbiotic relationships: the grocer needed the customers and vice versa, whilst the inhabitants of different geographical areas needed each other to demand and supply work, etc.

Apart from the concept of 'natural areas' and 'symbiosis', Park also took another concept from plant ecology. This involved the process by which the balance of nature in a given area might change. A new species may invade the area, come to I dominate it, and drive out other forms of life. Ecologists call this process "invasion, dominance, and succession". This can also be seen in human societies. The processes of invasion, dominance, and succession were further explored by Park's associate, Ernest Burgess. He pointed out that cities-do not merely grow at their edges but have a tendency to expand radially from their center in patterns of concentric circles, each moving gradually outward. Burgess described these concentric circles as 'zones.' They saw cities as consisting of five zones (CBD - Central Business District, transition, workingman, residential, and commuter), each gradually invading and dominating the adjacent zones, with an overall growth outward. The "natural" process was one of invasion, dominance, and succession; much like the way a new species of plant takes control of an ecosystem.

Zone I is the central business district. The major banks, large department stores, expensive shops, and the main administrative buildings of the city are usually located in this Zone. Zone II generally is the oldest section of the city, and it is continually involved in a process of invasion, dominance, and succession by business and industry that are expanding from Zone I. Houses in this area are usually in a dilapidated condition which are unlikely to be renovated because they may be demolished any time to make way for incoming business and industry. Since this is the least desirable residential section of the city, it is usually occupied by the poorest people, including the most recent immigrants to the city. Zone III is the zone of relatively modest homes and apartments, occupied by workers and their families who have escaped the deteriorating conditions in Zone II. The final Zone within the city itself is Zone IV, the residential areas or more and the satellite- expensive houses. Beyond the city limits are the suburban areas and the satellite cities, which constitute Zone V, the commuter zone. Each of the five zones grow and thus gradually move outward into the territory occupied by the next Zone. Their "zonal hypothesis" was that delinquency is greatest in the zone of transition.

The above theories of Park and Burgess were used as a basis for a Study of the juvenile delinquency in Chicago by Clifford R. Shaw and Henry D. McKay. They recorded the number of juveniles who appeared before the juvenile court, mapped out the areas from which they came, and then calculated the areas with the highest delinquency rates, defined as the number of juvenile crimes per hundred juveniles in the area. They therefore measured official levels of juvenile crime and recorded the area where the delinquent lived, and not the area where the crime was committed. By this measurement, the neighborhoods with the highest delinquency rates were those in the centre of the city, i.e. within or immediately adjacent to areas of heavy industry or commerce. These neighborhoods had a large number of condemned buildings.

They were sparsely populated and' that population was further decreasing as the land was being taken over by industry. Crime rates decreased as the neighborhoods got further from the centre: on the zonal maps, Zone I or the "loop" had the highest crime rate, and the rate decreased as the Zones moved outwards away from the center. Shaw and McKay also found that the crime rate of a particular area over time remained constant, despite vast changes in the inhabitants of that area. The central zones retained their high crime rate despite shifts of almost all the population of the areas and irrespective of the changes in the ethnic origin of the people who lived there. At the same time there was no increase in delinquency rates in the areas into which the older immigrant communities moved. It was further found that within similar areas, each ethnic group, had a delinquency rate that was proportional to the overall area. Each group produced delinquency rates that ranged from the lowest to the highest in the city, depending on the type of area surveyed. From all this they concluded that the delinquency rate was more a result of environmental conditions than of racial or ethnic characteristics. According to Shaw and McKay, high delinquency rates in the central zones are due to a variety of social problems, which are directly traceable to the rapid shift in populations. The formal social organizations that existed in the neighborhood tend to disintegrate as the original population retreats. Because the neighborhood is in transition, the residents no longer identify with it and do not care as much about its appearance or reputation. There is a marked decrease in 'neighborliness''. Because new people are continuously moving into the area, residents no longer know their own neighbors or their neighbors' children. Thus children who are out of their parents' sight may be under almost no control. The high mobility of the residents also results in high turn over in the school, which is disruptive both to learning and to discipline. The area also tends to be a battle ground between invading and retreating cultures. This can generate a great deal of conflict in the community, which tends to be manifested in individual and gang conflicts between the youth of the two cultures.

Although other areas also periodically undergo this process, areas in Zone II are continually invaded both by the industry and by successive waves of immigrants from rural areas. These new immigrants already have many problems associated with their adjustment to the new culture, in addition, the neighborhood into which the immigrants move is in a chronic state of 'social disorganization'. As a result, the delinquency among their youth tends to increase. McKay and Shaw concluded that the following factors contributed to delinquency in delinquency areas:-.

- The parents and neighbors frequently approved of delinquent behaviour, so that the child grew up in a social world in which delinquency was an accepted and appropriate form of conduct.
- The neighborhoods included many opportunities for delinquent activities such as availability of junk dealers, professional fences and willing residents for purchase of stolen goods; and existence of dilapidated buildings for junking.
- Delinquent activities in these areas began at an early age as a part of play activities of the street in which there was a continuity of tradition of transmission of such criminal techniques as jack rolling and shoplifting, from older boys to younger boys.

Because Shaw felt that juvenile delinquency- was generated by social disorganization in interstitial areas (so called because they are in between the organized natural areas), he did not believe that treatment of individual delinquents would have much effect in reducing overall delinquency rates. He thought that the answer to the problem could be found only in the development of programs, which sought to effect changes in the conditions of life in specific local communities. With that view he launched the Chicago Area Project in 1932. Under the project, seventy-two neighborhood centres were established in six areas of Chicago. These centres had two primary functions. First, they were to coordinate such community resources as churches, schools, labour unions, clubs, and other groups in addressing and resolving community problems.

Second, they were to sponsor a variety of activity programs including handicraft workshops, community programs, and recreation. Control of these centres rested with the local residents. Though the Chicago Area Project operated continuously for a period of twenty-five years and was effective in achieving many admirable goals such as providing access to the juveniles to educational and occupational opportunities, increasing involvement of juveniles in recreational activities and increasing interagency cooperation in addressing community problems.

Shaw and McKay developed a theory called Cultural Transmission Theory. There were three ways in which "criminal traditions" got embedded into communities.

- *residential instability/mobility* — population turnover where many residents move frequently or are looking to move (nobody invests in community they are going to leave anyway)
- *racial/ethnic heterogeneity* — different races, cultures, and language create the perception of insurmountable barriers (diversity not happening as a good thing) with people isolating themselves and avoiding meaningful interaction or common ground to share
- *poverty* — inadequate resources, primarily due to a low tax base, which prevent a community from being able to deal with local problems (significant portions of the population are focused on survival in an area they would much prefer to leave as soon as possible)

Shaw and McKay were also concerned about the three D's of poverty: Disease, Deterioration, and Demoralization. They never said that poverty causes crime. Poverty, by itself, is never a cause of crime; it only facilitates crime by deprivation of adequate resources to deal with crime. They only said that "poverty areas" tended to have high rates of residential mobility and racial heterogeneity that made it difficult for communities in those areas to avoid becoming socially disorganized. These two population variables (mobility or turnover which impedes informal structures of social control, and heterogeneity which obstructs the quest to work together on common problems) have become the primary causal variables for social disorganization theories. The causal chain is sometimes called the "gradient tendency" for the mechanistic way in which successive declines in the effectiveness of neighborhoods leads to higher and higher rates of delinquency. One of the best ways to defeat the gradient tendency is to stop sending so many community residents to prison, then back to the community, and then back to prison again.

Although there is probably much disagreement over it, the following are some of the important assumptions about the ecological school.

1. Social Ecology attempts to link the structure and organization of any human community to interactions with its localized environment. **Ecological theory** is a type of sociological approach that emphasizes demographics (the characteristics of population groups) and geographics (the mapped location of such groups relative to one another) and that sees the social disorganization that characterizes delinquency areas as a major cause of criminality and victimization. This is otherwise known as the Criminology of Place.
2. The Criminology of Place, also called Environmental Criminology, is an emerging perspective within the contemporary body of criminological theory that builds upon the contributions of routine activities theory and situational crime prevention, as well as the ecological approaches.
3. Crime and delinquency are caused primarily by social factors (environmental determinism). This theory is about place, not people (contextual, not compositional).
4. Official statistics are no less appreciated, but fieldwork is better (acceptance of official arrest data).
5. The city is a perfect natural laboratory (cities like Chicago reflect society as a whole).
6. Components of social structure are unstable (conflict, anomie, social disorganization).
7. Instabilities and their effects are worse for the lower classes (lower class crime focus).
8. Human nature is basically good (social ability thesis) but subject to vulnerability and inability to resist temptation.

9. It emphasizes the importance of geographic location and architectural features as they are associated with the prevalence of victimization.
10. The greatest contribution the ecological school made to criminological literature can be found in the claim that society, in the form of the community, wields a major influence on human behaviour.
11. Theories of social disorganization are often associated with the perspective of social ecology and with the Chicago School of criminology, which developed during the 1920s and 1930s Chicago School of criminology is an ecological approach to explaining crime that examined how social disorganization contributes to social pathology.
12. Chicago School demonstrated the tendency for criminal activity to be associated with urban transition zones, which, because of the turmoil or social disorganization that characterized them, were typified by lower property values, impoverished lifestyles, and a general lack of privacy.
13. The facts speak imperfectly for themselves, but better if fitted into theory (positivism). In criminology, social disorganization is usually treated as both perspective and theory, while ecology is an approach or "school." The ecological school refers to a group of professors associated with the Department of Sociology at the University of Chicago from 1920 to 1932.

The ecological school has been criticized on the following grounds:-

- The study relied on official police and court records to measure delinquency in a given area. Subsequent studies based on the records of both criminal justice and social welfare agencies as well as self report surveys found that delinquency is fairly evenly distributed among all social classes.
- Cities may not develop in 'concentric circles' as proposed by Burgess. In many cities the industrial areas are located in the outskirts and not in the centre of the city.
- Their claim that different racial and ethnic groups have similar rates of delinquency in similar neighborhood has been found to be incorrect. Subsequent studies found that Japanese and other Oriental families had markedly lower crime rates than other groups in all areas.

6. Merton's Paradigm of Deviant Behaviour

Robert Merton, an American sociologist, practiced a functional approach to deviant behaviour within society. As a functionalist, Merton believed that social institutions act to further society and ultimately keep stability within the social structure. It is this framework from which Merton tried to understand how deviance shaped society (and vice versa). Deviance can be described as, the characteristics of behaviours of individuals which violate group norms (including cultural mores and moral standards). The deviance, in order to be properly labeled, must induce a negative response from the group. The second piece of this definition is actually the most important in that; without the negative reaction, the behaviour would not be considered deviant.

Because deviance is found naturally within society, Merton believed that it was society itself which causes it. Further, Merton believed that when societal norms, or socially accepted goals (such as the American Dream), place pressure on the individual to conform they force the individual to either work within the structure society has produced, or instead, become members of a deviant subculture.

Merton forwarded two distinct theories which he did not always clearly distinguish. First, there is presented a theory of anomie, referring to a de-institutionalization of norms that occurs when there is a disjunction between the emphasis on cultural goals and institutional means (Merton 1938:673, 1968:189). Second, Merton also presents a strain theory of deviant behaviour which holds that people are more likely to pursue illegitimate means to attaining culturally prescribed goals when they are blocked from accessing the institutionalized means to these goals (Merton 1938:679, 1968:211). Merton's theory on deviance relates to the analysis of the relationship between culture, structure and anomie. Merton defines culture as an "organized set of normative values governing behaviour which is common to members of a designated society or group." Social structures are the "organized set of social relationships in which members of the society or group are variously implicated." Anomie, state of normlessness, then occurs when there is "an acute disjunction between the cultural norms and goals and the socially structured capacities of members of the group to act in accord with them." Merton's concept of anomie refers to "a specific imbalance where cultural goals are overemphasized at the expense of institutionalized means". Merton describes anomie in terms of a particular type of disjunction of means and goals: when society stresses the achievement of cultural goals but does not as equally emphasize the prescribed means to those goals a de-institutionalization of means occurs and "the integration of society becomes tenuous and anomie ensues" (Merton 1938:674). In his theory, Merton links anomie with deviance and argues that the discontinuity between culture and structure have the dysfunctional consequence of leading to deviance within society. Merton believes that all subscribe to the American Dream, but the ways in which people go about obtaining the Dream are not the same because not everyone has the same opportunities and advantages as the next person. This leads to a considerable amount of deviance. This theory is commonly used in the study of criminology (specifically the strain theory).

Typology of Deviance

Merton also created a typology of the different options individuals have when deciding whether or not deviance from norms is an option. Merton identified five different ways or modes of adaptation in which an individual can respond to this problem of anomie. The five adaptations are Conformity, Innovation, Ritualism, Retreatism and Rebellion. Merton clarified that the modes of adaptation refer to behaviour, not personality, and the same person may use different modes of adaptation in different circumstances. The relationship of each of these types of adaptations to cultural goals and institutionalized means is shown in the following table.

TABLE 6.1

Merton's Paradigm of Deviant Behaviour

Attitude to Goals	Attitude to Means	Modes of adaptation
accept	accept	Conformity
accept	reject	Innovation
reject	accept	Ritualism
reject	reject	Retreatism
reject/accept	reject/accept	Rebellion

Conformists : Conformity entails acceptance of both cultural goals and institutionalized means. Persons with this adaptation strive to achieve wealth and other culturally valued goals through approved methods and will continue to do so whether they succeed or not. The stability of the society can be attributed to the fact that most people fall into this category. Conformists are the group which chooses to both accept the goals of society and accept the standard means by which to attain their goals. For example, in order to have a happy life by social standards, they gain a college education, work in a 9 to 5 job, and eventually complete their goal.

Innovators : In innovation mode of adaptation there is acceptance of cultural goals but rejection of means for achieving the goals. Persons with this adaptation try to acquire wealth by methods, which may be innovative but illegitimate. The methods include criminal activities such as theft, burglary, robbery, gambling, prostitution, drug peddling and different forms of white-collar crime. While conformity is the attaining of societal goals by socially accepted means, innovation is the attaining of those goals in unaccepted ways. Innovators look to fulfill the goals of society but instead of going about it through established channels, as conformists do, they will find other means to reach their goals. Innovators find and create their own way to go about obtaining what they want, and a majority of the time, these new ways are considered to be socially unaccepted and deviant. Merton saw innovation as most common among the lower class Americans because of their vulnerability to unemployment and reduced opportunities to achieve the culturally aspired goal (i.e. acquisition of wealth) through legitimate means. Examples may include business moguls such as Steve Jobs, or groups such as the Mafia. Both examples took alternate routes to reaching the socially constructed goal.

Ritualists : Ritualism represents abandonment of the cultural goals with a continued and compulsive adherence to the institutionalized means. According to Merton, this is the adaptation of the persons who do not wish to find themselves in any trouble for not following the established norms. Bureaucrats who blindly follow the rules and procedures without bothering about the ultimate purpose, for which such procedures were laid down, come under this category.

Retreatists : In retreatism mode of adaptation, individual rejects both goals and the means of achieving them. The individuals who fall into this category include the vagrants, vagabonds, tramps, drunkards, drug addicts and psychotics. According to Merton, this adaptation does not necessarily arise from a lack of commitment to the culture. It may also occur when an individual is strongly committed to both the goals and means but finds that there is no possibility of achieving success. Retreaters want to find a way to escape from everything and therefore reject the goals and the means and are seen as deviant. The retreatist will stop engaging in society altogether and instead might live in solidarity or in communes with likeminded individuals. By far, the best example of this is the Hippie of the early 1960s. Innovation and ritualism are the pure cases of anomie as Merton defined it because in both cases there is a discontinuity between goals and means.

Rebellion : Rebellion is the last of the possible adaptations to the problem of anomie. In this mode of adaptation, individual not only rejects the existing goals and means but also attempts to substitute new goals in place of the conventional ones. Rebels reject both aspects but instead of leaving society, they try to change it. One example of such a new goal is the establishment of a socialist society. The movement of social and political activism among college students is a good example of this group.

Merton thus saw anomie as a contradiction between the culture of the society (which places a disproportionate emphasis on the achievement of goal and accumulation of wealth) on one hand and the social structure (which limits the possibilities to achieve this goal through the use of institutionalized means) on the other. Unlike Durkheim, Merton considered anomie to be an endemic condition but as one which affected only certain sections of the society. Merton argued that legitimate opportunities to achieve wealth through institutionalized means were concentrated in higher classes. The absence of such opportunities gives rise to strain and results in criminality in the lower classes. Merton further stated that his theory of anomie is intended to explain certain types of deviant behaviour customarily described as criminal or delinquent and not all behaviours prohibited by the criminal law. Ritualism is the acceptance of the means but the forfeit of the goals. Ritualists continue to subscribe to the means, but they have rejected the overall goal; they are not viewed as deviant. In comparison to the conformist, the ritualist will reject the goal but still go about the standard means of doing so. A ritualist may go through the motions of going to college and working 9 to 5, but will not attempt to fulfill the goal of earning excess amounts of money.

Merton as a functionalist further says that deviants often help society change and stay fluid. It is in this way that deviant behaviours might be less of a hindrance and actually have a more progressive influence on society. Merton's attempt to reconcile deviant behaviours with society at large is the single greatest reason why his Strain Theory has lasted as a viable paradigm to this day.

However, research has found even in some groups even if these conditions exists they may not commit crime depending on their personality profile, expose to delinquent behaviour, family support systems, values and beliefs male or female and other demographic factors like age, religion and ethnic background will mediate and affect the likelihood of crime. That is crime is a complex phenomenon and strain theory alone will not explain all types of crime and in all groups and in all countries. Some theories or combination of theories are more applicable to have an insight of the issue crime and crime prevention than base policies on one crime theory alone. People also can change over time because of new experiences and life events crime is a dynamic phenomenon not a static one.

7. Labeling Theory of Crime

The labeling theory emphasizes the criminalization process as the cause of some crime- the way people and actions are defined as criminal. The labeling theory argues that once a person commits a first criminal act and gets processed in the system, they are labeled negatively as a criminal. The label becomes a self-fulfilling prophecy. Crimes are not inherently evil or immoral acts, but defined that way. Crime has no meaning unless people react to it. Crimes are illegal because society defines them that way. It is also known as social reaction theory. Labeling theory (synonymous to "identifying against") holds that deviance is not inherent to an act, but instead focuses on the identifying tendency of majorities to negatively label minorities or those seen as deviant from norms. The theory is concerned with how the self-identity and behaviour of individuals may be determined or influenced by the terms used to describe or classify them, and is associated with the concept of a self-fulfilling prophecy and stereotyping. Labeling theory is also closely related to interactionism and social construction.

The labeling theory was propounded by Tannenbaum in 1938 who believed that tagging, defining, identifying, segregating and describing criminals by labeling them under different heads was helpful in treatment of offenders. Frank Tannenbaum is considered the grandfather of labeling theory. His Crime and Community (1938), describing the social interaction involved in crime, is considered a pivotal foundation of modern criminology. While the criminal differs little or not at all from others in the original impulse to first commit a crime, social interaction accounts for continued acts that develop a pattern of interest to sociologists. What concerned Tannenbaum in particular was the effect the label would have on the boy's self-concept. The boy is likely to be overwhelmed by the community reaction and begin to think of himself as the type of person who would do such things. Over time the label becomes a self-fulfilling prophecy. The juvenile becomes bad because he is defined as bad by others. This process, which Tannenbaum called the "dramatization of evil," plays a greater role in the making of a deviant than perhaps any other experience. The reactions of others to deviance, no matter how compassionate the intent, may trap the deviant individual in a role from which he (or she) cannot escape. The deviant individual, now "tagged," lives in a different world. In the eyes of others, the person who steals is a thief, the person who kills is a murderer. "One of the central tenets of the theory is to encourage the end of labeling process. In the words of Frank Tannenbaum, "the way out is through a refusal to dramatize the evil", the justice system attempts to do this through diversion programs. The growth of the theory and its current application, both practical and theoretical, provide a solid foundation for continued popularity."

Thereafter Lemert (1951) preferred to label offenders as persons with primary deviance and secondary deviance. The former do not see themselves as deviant whereas the latter accept their deviant status. According to Lemert primary deviance arises out of biological, psychological and or sociological reasons while the secondary deviance is caused by social reaction to primary deviation. The primary deviance is the experience connected to the overt behaviour, say drug addiction and its practical demands and consequences. Secondary deviation is the role created to deal with society's condemnation of the behaviour. Therefore societal reaction has a direct bearing on these two deviant types. Being a 'criminal' becomes a matter of status for these deviants and because of this self-image, there is a constant pressure on them to behave as a deviant.

While it was Lemert who introduced the key concepts of labeling theory, it was Howard Becker who became their champion. He first began describing the process of how a person adopts a deviant role in a study of dance musicians, with whom he once worked. He later studied the identity formation of marijuana smokers. This study was the basis of his Outsiders published in 1963. This work became the manifesto of the labeling theory movement among sociologists. In his opening, Becker writes :

"...social groups create deviance by making rules whose infraction creates deviance, and by applying those roles to particular people and labeling them as outsiders. From this point of view, deviance is not a quality of the act the person commits, but rather a consequence of the application by other of rules and sanctions to an 'offender.' The deviant is one to whom that label has been successfully applied; deviant behaviour is behaviour that people so label (1963:9)." While society uses the stigmatic label to justify its condemnation, the deviant actor uses it to justify his actions. He wrote: "To put a complex argument in a few words: instead of the deviant motives leading to the deviant behaviour, it is the other way around, the deviant behaviour in time produces the deviant motivation (1963:26)."

Howard Becker developed his theory of labeling (also known as social reaction theory) on the assumption that people are likely to engage in rule-breaking behaviour as essentially different from the members of rule-making or rule-abiding society. The law-breakers see themselves morally at odds with those who are law-abiding. Becker labels these law-breakers as "outsider" and holds that they accept the label attached to them and they begin to view themselves as different from "mainstream" of society. The deviants first indulge in primary deviance intentionally or unintentionally and gradually step into the arena of secondary deviance. A deviant who is denied means to carry out daily routines, turns to illegitimate means to make a living. The affiliation of the labeled deviant with an organised group of criminals provides him moral support and he joins the organised crime thus learning new forms of deviance through differential associations.

Becker recognises four types of citizens according to their behaviour in society labeling them differently. The numbers of society that are rule-abiding and free of labels are described as conforming citizens, while those who are labeled without breaking a law are termed as falsely accused. Those citizens who exhibit law-breaking behaviour are labeled as pure deviants, while those that break law yet avoid labeling are called secret deviants.

The modern labeling theory, however, recognise that 'societies' create crime by enacting laws and therefore the substantive nature of law should be the object of study. Sometimes these are called criminalization theories and they have some resemblance to societal reaction to crime. The labeling theory focuses on the official reaction to crime and makes a rather counterintuitive argument regarding the causes of crime. According to labeling theory, official efforts to control crime often have the effect of increasing crime. Individuals who are arrested, prosecuted, and punished are labeled as criminals. Others then view and treat these people as criminals, and this increases the likelihood of subsequent crime for several reasons. Labeled individuals may have trouble obtaining legitimate employment, which increases their level of strain and reduces their stake in conformity. Labeled individuals may find that conventional people are reluctant to associate with them, and they may associate with other criminals as a result. This reduces their bond with conventional others and fosters the social learning of crime. Finally, labeled individuals may eventually come to view themselves as criminals and act in accord with this self-concept.

Labeling theory was quite popular in the 1960s and early 1970s, but then fell into decline-partly as a result of the mixed results of empirical research. Some studies found that being officially labeled a criminal (e.g., arrested or convicted) increased subsequent crime, while other studies did not. Recent theoretical work, however, has revised the theory to take account of past problems. More attention is now being devoted to informal labeling, such as labeling by parents, peers, and teachers. Informal labeling is said to have a greater effect on subsequent crime than official labeling. Ross Matsueda (1982) discusses the reasons why individuals may be informally labeled as delinquents, noting that such labeling is not simply a function of official labeling (e.g., arrest). Informal labeling is also influenced by the individual's delinquent behaviour and by their position in society-with powerless individuals being more likely to be labeled (e.g., urban, minority, lower-class, adolescents). Matsueda also argues that informal labels affect individuals' subsequent level of crime by affecting their perceptions of how others see them. If they believe that others see them as delinquents and trouble-makers, they are more likely to act in accord with this perception and engage in delinquency. Data provide some support for these arguments.

John Braithwaite (1989) extends labeling theory by arguing that labeling increases crime in some circumstances and reduces it in others. Labeling increases subsequent crime when no effort is made to reintegrate the offender back into conventional society; that is, when offenders are rejected or informally labeled on a long-term basis. But labeling reduces subsequent crime when efforts are made to reintegrate punished offenders back into conventional society. In particular, labeling reduces crime when offenders are made to feel a sense of shame or guilt for what they have done, but are eventually forgiven and reintegrated into conventional groups-like family and conventional peer groups. Such reintegration may occur "through words or gestures of forgiveness or ceremonies to decertify the offender as deviant" (pp. 100-101). Braithwaite calls this process "reintegrative shaming." Reintegrative shaming is said to be more likely in certain types of social settings, for example, where individuals are closely attached to their parents, neighbors, and others. Such shaming is also more likely in "communitarian" societies, which place great stress on trust and the mutual obligation to help one another (e.g., Japan versus the United States). Braithwaite's theory has not yet been well tested, but it helps make sense of the mixed results of past research on labeling theory.

Bruce G., and Phelan (1999:361-67) discussed two types of labeling called hard and soft labeling in the context of mental illness. People who believe in hard labeling believe that mental illness does not exist. It is merely deviance from the norms of society that cause people to believe in mental illness. Thus, mental illnesses are socially constructed illnesses and psychotic disorders do not exist. People who believe in soft labeling believe that mental illnesses do, in fact, exist. Unlike the supporters of hard labeling, soft labeling supporters believe that mental illnesses are not socially constructed.

Becker's immensely popular views were also subjected to a barrage of criticism, most of it blaming him for neglecting the influence of other biological, genetic effects and personal responsibility. It must be stated that Becker's theory of labeling though popular has encountered several criticisms. Many sociologists and criminologists view this theory as a purely theoretical approach to crime and criminals without any empirical basis. The distinction between primary and secondary deviance seems to be futile as it is influenced by changing variables. Some criminologists view labeling theory as declining in its importance due to lack of empirical support.

Some sociologists have called the labeling theory as "the shoe fits, wear it" theory. It suggests that society makes deviance by passing laws infringement of which constitutes deviance and the law-breakers are .labeled as 'deviants'. Deviance, according to this theory is not the act that a person commits, but rather a consequence of the application by others, of rules, laws and sanctions to the "offender". A deviant is one to whom people so label.

Critics say that labeling theory is especially crucial to understanding juvenile delinquency because it is during the time of adolescence that juvenile's self-identities are formed. Labeling theory also helps to explain the long-term consequences of a deviant label on a person's social status. Thus if a juvenile is labeled as deviant or delinquent, then his self-identity may develop as such and he will be more prone to become a hardened criminal. Because of his/her negative self-concept he or she is likely to choose a crime carrier and associate with other professional or organised groups of criminals.

8. The Social Disorganization Theory of Crime

The leading sociological theories focus on the immediate social environment, like the family, peer group, and school. And they are most concerned with explaining why some individuals are more likely to engage in crime than others. Much recent theoretical work, however, has also focused on the larger social environment, especially the community and the total society. This work usually attempts to explain why some groups-like communities and societies-have higher crime rates than other groups. In doing so, however, this work draws heavily on the central ideas of control, social learning, and strain theories.

Theories of social disorganization are perspectives on crime and deviance that sees society as a kind of organism and crime and deviance as a kind of disease or social pathology. Social Pathology refers to those human actions which run contrary to the ideals of residential stability, property ownership, sobriety, thrift, habituation to work, small business enterprise, sexual discretion, family solidarity, neighborliness, and discipline of will. The tem social pathology refers simply to behaviour not in keeping with the prevalent norms and values of the social group. Social disorganization as a condition is said to exist when a group is faced with social change, uneven development of culture, maladaptiveness, disharmony, conflict, and lack of consensus.

Social disorganization theory seeks to explain community differences in crime rates (Sampson and Groves,1989 ; Bursik and Grasmick, 1992). The theory identifies the characteristics of communities with high crime rates and draws on social control theory to explain why these characteristics contribute to crime.

Crime is said to be more likely in communities that are economically deprived, large in size, high in multiunit housing like apartments, high in residential mobility (people frequently move into and out of the community), and high in family disruption (high rates of divorce, single-parent families). These factors are said to reduce the ability or willingness of community residents to exercise effective social control, that is, to exercise direct control, provide young people with a stake in conformity, and socialize young people so that they condemn delinquency and develop self-control.

The residents of high crime communities often lack the skills and resources to effectively assist others. They are poor and many are single parents struggling with family responsibilities. As such, they often face problems in socializing their children against crime and providing them with a stake in conformity, like the skills to do well in school or the connections to secure a good job. These residents are also less likely to have close ties to their neighbors and to care about their community. They typically do not own their own homes, which lowers their investment in the community. They may hope to move to a more desirable community as soon as they are able, which also lowers their investment in the community. And they often do not know their neighbors well, since people frequently move into and out of the community. As a consequence, they are less likely to intervene in neighborhood affairs-like monitoring the behaviour of neighborhood residents and sanctioning crime. Finally, these residents are less likely to form or support community organizations, including educational, religious, and recreational organizations. This is partly a consequence of their limited resources and lower attachment to the community. This further reduces control, since these organizations help exercise direct control, provide people with a stake in conformity, and socialize people. Also, these organizations help secure resources from the larger society, like better schools and police protection. Recent data provide some support for these arguments.

Social disorganization theorists and other criminologists, such as Hagan, John (1994), point out that the number of communities with characteristics conducive to crime-particularly high concentrations of poor people-have increased since the 1960s. These communities exist primarily in inner city areas and they are populated largely by members of minority groups (due to the effects of discrimination). Such communities have increased for several reasons. First, there has been a dramatic decline in manufacturing jobs in central city areas, partly due to the relocation of factories to suburban areas and overseas. Also, the wages in manufacturing jobs have become less competitive, due to factors like foreign competition, the increase in the size of the work force, and the decline in unions. Second, the increase in very poor communities is due to the migration of many working- and middle-class African Americans to more affluent communities, leaving the poor behind. This migration was stimulated by a reduction in discriminatory housing and employment practices. Third, certain government policies-like the placement of public housing projects in inner-city communities and the reduction of certain social services-have contributed to the increased concentration of poverty.

9. Critical Theories of Crime

Critical theories also try to explain group differences in crime rates in terms of the larger social environment; some focus on class differences, some on gender differences, and some on societal differences in crime. Several versions of critical theory exist, but all explain crime in terms of group differences in power. Here is discussed only Marxist and Feminist theories.

Marxist Theories : Human nature is such that people make history or "man fighting back". Marx said "Men make their own history, but they do not make it just as they please, they do not make it under circumstances chosen by themselves, but under circumstances directly encountered, given and transmitted from the past. The tradition of all the dead generations weighs like a nightmare on the brain of the living."

Marxism rejects the dualism of determinism & free will. Free will is not added at the end of chain of pre-determined circumstances (strain) or by the impact of labeling & free will (as in labeling). There is no pre-social human essence or reason existing outside of the social context or the historical period. The transition to capitalism from feudalism requires the discipline of wage labour in order to create capital. But the initial coercion of labour in capital formation is replaced by the silent compulsion of the "free market". Consensus is not merely an illusion but arises out of the contradictory nature/reality of capitalism the so called "double freedom" of the worker in capitalism.

Marxist theories address the issue that who makes the rules and why laws are the products of the ruling classes hegemonic [when the social order is seen as natural by the ruled] means of establishing legitimacy. Marxist theories take various forms & vary the emphasis they place on ideological & economic control — between base and super-structure is crucial.

Marxist theories argue that those who own the means of production (e.g., factories, businesses) have the greatest power. This group-the capitalist class-uses its power for its own advantage.

Capitalists work for the passage of laws that criminalize and severely sanction the "street" crimes of lower-class persons, but ignore or mildly sanction the harmful actions of business and industry (e.g., pollution, unsafe working conditions). And capitalists act to increase their profits; for example, they resist improvements in working conditions and they attempt to hold down the wages of workers. This is not to say that the capitalist class is perfectly unified or that the government always acts on its behalf. Most Marxists acknowledge that disputes sometimes arise within the capitalist class and that the government sometimes makes concessions to workers in an effort to protect the long-term interests of capitalists (Chambliss, and Mankoff, 1976).

Marxists explain crime in several ways. Some draw on strain theory, arguing that workers and unemployed people engage in crime because they are not able to achieve their economic goals through legitimate channels. Also, Marxists argue that crime is a response to the poor living conditions experienced by workers and the unemployed. Some draw on control theory, arguing that crime results from the fact that many workers and the unemployed have little stake in society and are alienated from governmental and business institutions. And some draw on social learning theory, arguing that capitalist societies encourage the unrestrained pursuit of money. Marxist theories, then, attempt to explain both class and societal differences in crime.

Sellin, Thorsten (1938) applied both Marxism and Conflict Theory to an examination of the cultural diversity of modern industrial society. In a homogeneous society, norms or codes of behaviour will emerge and become laws where enforcement is necessary to preserve the unitary culture. But where separate cultures diverge from the mainstream, those minority groups will establish their own norms. Socialization will therefore be to the subgroup and to the mainstream norms. When laws are enacted, they will represent the norms, values and interests of the dominant cultural or ethnic group which may produce Border Culture Conflict. When the two cultures interact and one seeks to extend its influence into the other, each side is likely to react protectively. If the balance of power is relatively equal, an accommodation will usually be reached. But if the distribution of power is unequal, the everyday behaviour of the minority group may be defined as deviant. The more diversified and heterogeneous a society becomes, the greater the probability of more frequent conflict as subgroups who live by their own rules break the rules of other groups.

Feminist Theories : Feminist theories focus on gender differences in power as a source of crime. These theories address two issues: why are males more involved in most forms of crime than females, and why do females engage in crime. Most theories of crime were developed with males in mind; feminists argue that the causes of female crime differ somewhat from the causes of male crime. Thus, Sociology of Crime and Deviance tends to focus mainly on men. Although that it is true that approximately 80% of offenders are men, there is the other 20% who are women that are simply ignored in many sociological theories. Heidensohn (1996), has criticized the male dominance of the subject and has suggested four reasons behind it :

1. Male dominance of offenders : As the majority of offenders are men for many sociologists it would therefore make sense to focus on the majority rather than the minority of women offenders.

2. Male Dominanation of Sociology : According to Heidensohn sociology topics of investigation reflect a male view and interests. As it was usually the case that the majority of academics were men.

3. *Vicarious identification :* What interests males is what's studied, and applied to crime their interest lies in the lives of the marginal and the exciting.

4. *Sociological Theorising :* Male sociologists constructed their theories without ever thinking about how they could be applied to women. Most traditional theories are 'gender blind', which in effect means they ignore the specific viewpoint of women.

Feminist Debates and Criminology :

Though male sociologists have largely ignored female offending; feminist writers have sought to include some criminology analyses within their approaches.

Liberal Feminism : This approach to feminism is based on the idea that by bringing women onto the agenda and by demonstrating how women have been ignored in research, then there will be a greater understanding of female deviance. In particular, new theories can be developed which will cover both men and women.

Radical Feminism : Radical feminists argue that the only way to understand crime is to see it through a female perspective – and research should be based on the assumption that all men are prepared to commit crimes against women if given the chance. Women should construct their own unique approaches to explaining crime and deviance and this should incorporate the threat from men.

Socialist Feminist : This approach stresses that the position of men and women in general and with reference to crime, can only be understood by locating males and females within the context of societies that are divided by both sexism and capitalism.

Postmodern Feminism : Smart and Cain argued that the very concerns criminology have (burglary, street crime etc) are actually a reflection of male concerns and therefore women should be looking beyond these to the things that are most harmful to them. In other words they should look at how harm comes to women in the widest sense possible and not just accept the (male) boundaries of criminology.

Postmodernism and Transgression : In response to the need for a feminist version of criminology Carol Smart introduced the idea of transgressive criminology. She suggested that criminology itself as a discipline was tied to male questions and concerns and that it could never offer answers to feminist questions. Instead of looking at how feminism affects criminology, she said that they should look at how criminology can affect feminists. This can be done through looking at a range of activities both illegal and legal that harm women and to then look at how they came about and how they can be changed. This idea led to feminists (and sympathetic male sociologists) looking more closely at the way women stayed in at night for fear of becoming victims, at domestic violence and how women were treated by the law in issues of rape and harassment (where they form the overwhelming bulk of the victims).

Male Roles, Postmodernity and Masculinity Smarts idea of transgression linked to the growing importance of postmodern analysis. Some sociologists began to go beyond the traditional confines and to revisit the issue of why most crime is male crime.

Gender differences in crime are said to be due largely to gender differences in social learning and control. Females are socialized to be passive, subservient, and focused on the needs of others. Further, females are more closely supervised than males, partly because fathers and husbands desire to protect their "property" from other males. Related to this, females are more closely tied to the household and to child-rearing tasks, which limits their opportunities to engage in many crimes.

Some females, of course, do engage in crime. Feminist theories argue that the causes of their crime differ somewhat from those of male crime, although female crime is largely explained in terms of strain theory. Meda Chesney-Lind and others argue that much female crime stems from the fact that juvenile females are often sexually abused by family members (Chesney-Lind, 1997). This high rate of sexual abuse is fostered by the power of males over females, the sexualization of females-especially young females-and a system that often fails to sanction sexual abuse. Abused females frequently run away, but they have difficulty surviving on the street. They are labeled as delinquents, making it difficult for them to obtain legitimate work. Juvenile justice officials, in fact, often arrest such females and return them to the families where they were abused. Further, these females are frequently abused and exploited by men on the street. As a consequence, they often turn to crimes like prostitution and theft to survive. Theorists have pointed to still other types of strain to explain female crime, like the financial and other difficulties experienced by women trying to raise families without financial support from fathers. The rapid increase in female-headed families in recent decades, in fact, has been used to explain the increase in rates of female property crime. It is also argued that some female crime stems from frustration over the constricted roles available to females in our society.

Some Criticisms of Feminist Theories of Crime

- Some feminist theories are accused of hypocrisy as they are said to over focus on gender in their explanations of crime and deviance.
- Some feminist theories are criticised for not understating the issues of class and ethnicity.
- Some feminists are accused of an over focus on victimisation and of simplifying the causes of offending.

10. Integrated Theories of Crime

Several theorists have attempted to combine certain of the above theories in an effort to create integrated theories of crime. The most prominent of these integrations are those of Terence P. Thornberry and Delbert S. Elliott and associates (Thornberry,1996). Elliott's theory states that strain and labeling reduce social control. For example, school failure and negative labeling may threaten one's emotional bond to conventional others and investment in conventional society. Low social control, in turn, increases the likelihood of association with delinquent peers, which promotes the social learning of crime. Thornberry attempts to integrate control and social learning theories. Like Elliott, he argues that low control at home and at school promotes association with delinquent peers and the adoption of beliefs favorable to delinquency. Thornberry, however, also argues that most of the causes of crime have reciprocal effects on one another. For example, low attachment to parents increases the likelihood of association with delinquent peers, and association with delinquent peers reduces attachment to parents. Likewise, delinquency affects many of its causes: for example, it reduces attachment to parents and increases association with delinquent peers (an argument compatible with labeling theory). Further, Thornberry argues that the causes of crime vary over the life course.

For example, parents have a much stronger effect on delinquency among younger than older adolescents. Factors like work, marriage, college, and the military, however, are more important among older adolescents.

Thus criminals are exposed to several lots of factors which predispose them to criminal action. Integrated theories of crime can be summed up in terms of the following chart.

FIGURE 6.1
MULTIPLE FACTORS AND THE DECISION TO COMMIT A CRIME

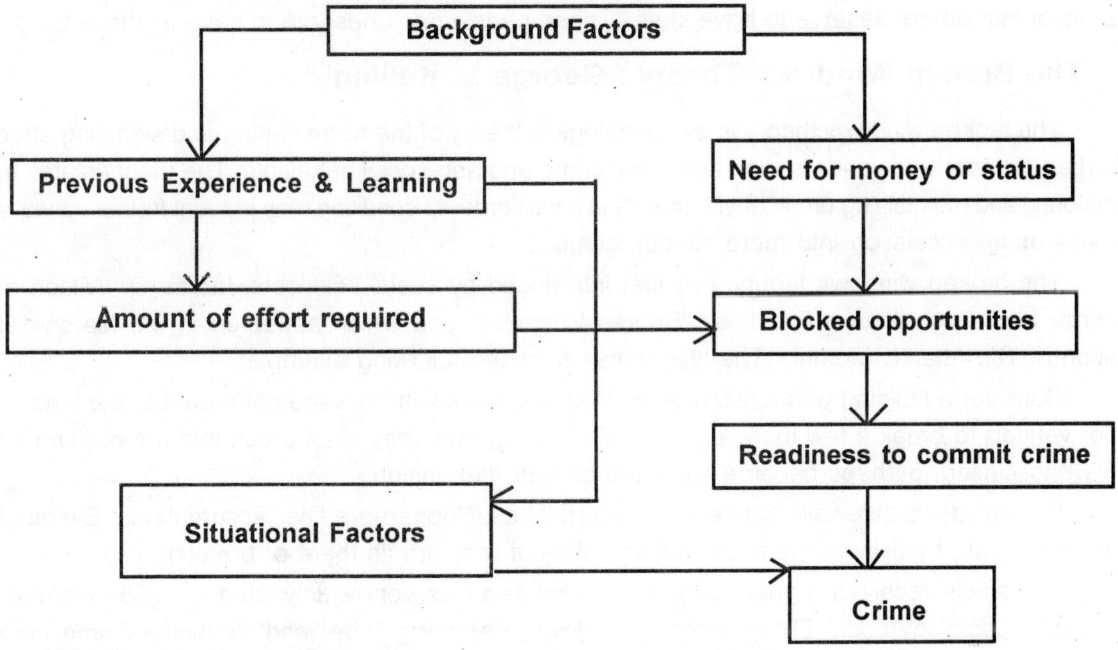

This chart shows that free will is assumed, but there are certain background and situational factors that might predispose someone toward crime. *Background factors* would include psychological characteristics, like how intelligent the person is; social factors like family background; and demographic factors, like what kind of neighborhood the criminal comes from. *Situational factors* would include persuasion by friends, arguments with spouse, or whether the person has consumed alcohol or drugs, along with a whole host of other possible situational, or urge-inducing, factors. Motive is listed as *need for money or status*, but in this classical view of crime, there is really no need to think about motive. All that is required are some generalized needs: the desire to get ahead, to show off, excitement, etc.

Previous learning and experience refers to direct or vicarious experience with crime. The offender would be comparing the target to others he has been successful with in the past, or if he is new at it, he would be comparing the target with what others have been successful with in the past. This factor also refers to the criminal's self-perception of his/her own skills, ability to elude law enforcement, and get rid of the stuff afterwards. *Blocked opportunities* is a concept from strain theory and would probably not be used by a rational choice theorist, but it is the criminal's assessment of what legitimate avenues are available for satisfying needs. The decision to be made is whether the same amount of money, for example, can be made by work, gambling, borrowing, or avenues other than crime. The *amount of effort required* fits into this as the amount of time spent considering and evaluating whether the rewards (and costs) of crime outweigh alternative avenues for satisfying the same needs.

Readiness to commit crime is best understood as the offender "psyching" up. Again, strain theory offers a good perspective on this as withdrawal of legitimacy or belief in an unjust world, but "neutralization" captures the essence of the idea. The offender engages in self-talk, maximizing the belief that he deserves to rob the house ("That house and everything in it is mine"), and minimizing the belief that others deserve to have stuff ("Those people don't deserve to have all those things").

11. The Broken Windows Theory : George L. Kelling

The broken windows theory is a criminological theory of the norm setting and signaling effects of urban disorder and vandalism on additional crime and anti-social behaviour. The theory states that monitoring and maintaining urban environments in a well-ordered condition may prevent further vandalism as well as an escalation into more serious crime.

The broken windows theory was first introduced by social scientists James Q. Wilson and George L. Kelling, in an article titled "Broken Windows" and which appeared in the March 1982 edition of The Atlantic Monthly. The title comes from the following example:

Consider a building with a few broken windows. If the windows are not repaired, the tendency is for vandals to break a few more windows. Eventually, they may even break into the building, and if it's unoccupied, perhaps become squatters or light fires inside.

Or consider a sidewalk. Some litter accumulates. Soon, more litter accumulates. Eventually, people even start leaving bags of trash from take-out restaurants there or breaking into cars.

The article received a great deal of attention and was very widely cited. A 1996 criminology and urban sociology book, Fixing Broken Windows: Restoring Order and Reducing Crime in Our Communities by George L. Kelling and Catharine Coles, is based on the article but develops the argument in greater detail. It discusses the theory in relation to crime and strategies to contain or eliminate crime from urban neighborhoods.

A successful strategy for preventing vandalism, say the book's authors, is to fix the problems when they are small. Repair the broken windows within a short time, say, a day or a week, and the tendency is that vandals are much less likely to break more windows or do further damage. Clean up the sidewalk every day, and the tendency is for litter not to accumulate (or for the rate of littering to be much less). Problems do not escalate and thus respectable residents do not flee a neighborhood. The theory thus makes two major claims: that further petty crime and low-level anti-social behaviour will be deterred, and that major crime will, as a result, be prevented. Criticism of the theory has tended to focus only on the latter claim.

Theoretical Explanation

The reason why the state of the urban environment may affect crime may be described as due to three factors: Social norms and conformity, The presence or lack of monitoring, and Social signaling and signal crime.

A major factor in determining individual behaviour is social norms, internalized rules about the appropriate way to act in a certain situation. Humans constantly monitor other people and their environment in order to determine what the correct norms for the given situation are. They also monitor others to make sure that the others act in an acceptable way. In other words, people do as others do and the group makes sure that the rules are followed. But when there are no people around, as is often the case in an anonymous urban environment, the monitoring of or by others does not work. In such an environment, criminals are much more likely to get away with robberies, thefts and vandalism. When there are no or few people around, individuals are forced to look for other clues—called signals—as to what the social norms allow them to do and how great the risk of getting caught is. An ordered and clean environment sends the signal that this is a place which is monitored, people here conform to the common norms of non-criminal behaviour. A disordered environment which is littered, vandalized and not maintained sends the opposite signal: this is a place where people do as they please and where they get away with that, without being detected. As people tend to act the way they think others act, they are more likely to act "disorderly" in the disordered environment. Broken Windows Thesis is based on the following assumptions:

- Broken windows thesis emphasizes that physical deterioration and an increase in unrepaired buildings lead to increased concerns for personal safety among area residents.
- Offenders from other neighborhoods are then increasingly attracted by the area's perceived vulnerability.
- Physical evidence of disorder, left unchecked, leads to crime by driving residents indoors and sending a message to would-be offenders that a neighborhood is out of control.

Application of the Broken Windows Theory

In 2005, Harvard University and Suffolk University researchers worked with local police to identify 34 "crime hot spots" in Lowell, Massachusetts. In half of the spots, authorities cleared trash, fixed streetlights, enforced building codes, discouraged loiterers, made more misdemeanor arrests, and expanded mental health services and aid for the homeless. In the other half, there was no change to routine police service. The areas that received additional attention experienced a 20% reduction in calls to the police. The study concluded that cleaning up the physical environment is more effective than misdemeanor arrests, and that increasing social services had no effect (Johnson, Carolyn Y, 2005).

Other side effects of better monitoring and cleaned up streets may well be desired by governments or housing agencies, as well as the population of a neighborhood: broken windows can count as an indicator of low real estate value, and may deter investors. Fixing windows is therefore also a step of real estate development, which may lead, desired or not, to gentrification.

In the education realm, the broken windows theory is used to promote order in classrooms and school cultures. The belief is that students are signaled by disorder or rule-breaking and that they, in turn, imitate the disorder. Several school movements—most notably with the charter school movement—encourage strict paternalistic practices to enforce student discipline. Such practices include language codes (governing slang, curse words, or speaking out of turn), classroom etiquette (sitting up straight, tracking the speaker), personal dress (uniforms, little or no jewelry), and behavioural codes (walking in lines, specified bathroom times). Several schools have made remarkable strides in educational gains with this philosophy such as American Indian Public Charter School.

Criticism of the theory

According to most criminologists who speak of a broader "backlash", the broken windows theory is not theoretically sound. The term "backlash" refers to the wave of "tough-on-crime" policies in the U.S. beginning in the 1980s. They claim that the "broken windows theory" closely relates correlation with causality, a reasoning which is prone to fallacy. David Thacher, assistant professor of public policy and urban planning at the University of Michigan, stated in a 2004 paper that:

"Social science has not been kind to the broken windows theory. A number of scholars reanalyzed the initial studies that appeared to support it ... Others pressed forward with new, more sophisticated studies of the relationship between disorder and crime. The most prominent among them concluded that the relationship between disorder and serious crime is modest, and even that relationship is largely an artifact of more fundamental social forces".

It has also been argued that rates of major crimes also dropped in many other U.S. cities during the 1990s, both those that had adopted "zero-tolerance" policies and those that had not. In the Winter 2006 edition of the University of Chicago Law Review, Bernard Harcourt and Jens Ludwig looked at the later Department of Housing and Urban Development program that re-housed inner-city project tenants in New York into more orderly neighborhoods (Harcourt and Ludwig, 2006). The broken windows theory would suggest that these tenants would commit less crime once moved, due to the more stable conditions on the streets. Harcourt and Ludwig found instead that the tenants continued to commit crime at the same rate.

In a 2007 study called "Reefer Madness" in the journal Criminology and Public Policy, Harcourt and Ludwig found further evidence confirming that mean reversion fully explained the changes in crime rates in the different precincts in New York during the 1990s. Further alternative explanations that have been put forward include the waning of the crack epidemic, unrelated growth in the prison population due to Rockefeller drug laws, and that the number of males aged 16–24 was dropping regardless due to demographic changes (Levitt and Stephen, 2005).

A low-level intervention of police in neighborhoods has been considered problematic. Accordingly, Gary Stewart writes that "The central drawback of the approaches advanced by Wilson, Kelling, and Kennedy rests in their shared blindness to the potentially harmful impact of broad police discretion on minority communities" (Gary Stewart,1998, pp. 2249–2279). This was seen by the authors, who worried that people would be arrested "for the 'crime' of being undesirable". According to Stewart, arguments for low-level police intervention, including the broken windows hypothesis, often act "as cover for racist behaviour"(Ibid).

12. The Future of Sociological Crime Theories

There is the old saying that you can take the criminal out of a bad environment, but you can't take the bad environment out of the criminal. Negative influences of social environment predispose some people to lives of crime, and that such negative influences may remain active even when people's circumstances change. Sociological theory relates to a perspective that focuses on the nature of the power relationships that exist between social groups and on the influences that various social phenomena bring to bear on the types of behaviours that tend to characterize groups of people. Crime is the product of hostility, marginalization and class struggle. The perspective emphasizes the nature of existing power relationships between social groups.

Social structure is the pattern of social organization and the interrelationships among institutions characteristic of a society. Crime is the result of an individual's location within the structure of society.

Social structure theories explain crime by reference to the economic and social arrangements (or structure) of society. These theorists see the various formal and informal arrangements between social groups as the root causes of crime and deviance. They highlight those arrangements within society that contribute to low socio-economic status of identifiable groups as significant causes of crime. They view members of socially and economically disadvantaged groups as being more likely to commit crime, and they see economic and social disenfranchisement as fundamental causes of crime.

Social process refers to the ongoing interaction between and among social institutions, individuals and groups. Social life and social process are inextricably interwoven. Social life is the ongoing (typically) structured interaction that occurs between persons in a society, including socialization and social behaviour in general. Crime is the end product of various social processes, especially inappropriate socialization and social learning.

The sociological views on crime causation go beyond traditional criminological theories of strain, cultural deviance, social control, differential association and social learning to present emerging and integrated models of the origins of crime, including antisocial peer socialization, social development, interactional theory, behaviour genetics, and community determinants in the context of Social structure, institutions and processes.

Sociologists continue to refine existing theories and develop new theories of crime, including integrated theories of crime e.g., Charles Tittle's control balance theory (*Tittle,*1995). Sociologists, however, are coming to recognize that it is not possible to explain crime solely in terms of the immediate social environment. As a consequence, they are devoting more attention to the larger social environment, which affects the immediate social environment. And they are devoting more attention to the situations in which people find themselves, which affect whether predisposed individuals will engage in crime. Further, sociologists are coming to recognize that they need to take account of the factors considered in biological, psychological, and other theories of crime. Most notably, they must take account of individual traits like intelligence, impulsivity, and irritability. These traits influence how individuals respond to their social environment. An irritable individual, for example, is more likely to respond to strain with crime. These traits also shape the individual's social environment. Irritable individuals, for example, are more likely to elicit hostile reactions from others and select themselves into social environments that are conducive to crime, like bad jobs and marriages. (At the same time, the social environment influences the development of individual traits and the ways in which individuals with particular traits behave.)

Further, sociologists are increasingly recognizing that their theories may require modification if they are to explain crime in different groups and among different types of offenders. As indicated above, theories may have to be modified to explain female versus male crime. And theories may have to be modified to explain crime across the life course. For example, the factors that explain why young adolescents start committing crime likely differ somewhat from those that explain why some older adolescents continue to commit crimes and others stop. Much recent attention, in fact, has been devoted to the explanation of crime across the life course, as described in the text by Vold, Bernard, and Snipes (1998). Also, theories will have to be modified to explain crime among different types of offenders. Some offenders, for example, limit their offending to the adolescent years. Others offend at high rates across the life course.

Sociological theories, then, will become more complex, taking account of individual traits, the immediate social environment, the larger social environment, and situational factors. And modified versions of such theories will be developed to explain crime in different groups and among different types of offenders.

References

Ahuja Ram, (2000) *Criminology,* Rawat Publications.

Becker, Howard. Outsiders. (1963), New York, Free Press, (1997 Reprint).

Bernard E. Harcourt, (2001)"Illusion of Order: The False Promise of Broken Windows Policing", Harvard.

Braithwaite J., (1989) Crime, Shame and Reintegration, Cambridge University Press.

Bruce G., and Jo C. Phelan, (1999), "The Labelling Theory of Mental Disorder (II): The Consequences of Labeling." A Handbook for the Study of Mental Health: Social Contexts, Theories, and Systems. Cambridge: Cambridge UP, 361-76.

Bursik, Robert and Harold Grasmick (1992) Neighborhoods and Crime: The Dimensions of Effective Community Control, New Lexington Press.

Chambliss, William J. & Mankoff, Milton (eds.) (1976) Whose Law ? What Order ? A Conflict Approach to Criminology. New York: John Wiley.

Chesney-Lind, M. , (1997). The female offender: Girls, women and crime. Thousand Oaks: Sage Publications.

Featherstone, Richard, and Mathieu Deflem, (2003) "Anomie and Strain: Context and Consequences of Merton's Two Theories." Sociological Inquiry 73(4):471-489.

Featherstone, Richard, and Mathieu Deflem, (2003) "Anomie and Strain: Context and Consequences of Merton's Two Theories." Sociological Inquiry 73(4):471-489.

Garofalo Raffaele, (1914) Criminology, Kessinger Publishing, LLC (Reprint 2008).

Gary Stewart, (1998) Black Codes and Broken Windows: The Legacy of Racial Hegemony in Anti-Gang Civil Injunctions. The Yale Law Journal, Vol. 107, No. 7, pp. 2249–2279.

George Kelling and Catherine Coles, (1996). Fixing Broken Windows: Restoring Order and Reducing Crime in Our Communities Simon & Schuster; 1st Touchstone Ed.

Hagan, John (1994) Crime and Disrepute, Northwestern University Press.

Hale Chris et al, (2005), Criminology, Oxford University Press, New York.

Harcourt, Bernard E.; Ludwig, Jens (2006). "Broken Windows: New Evidence from New York City and a Five-City Social Experiment". University of Chicago Law Review 73.

Heidensohn, Frances, (1996) Women and Crime, Palgrave Macmillan; 2nd Edition.

James Q. Wilson and George L. Kelling (1982) "BROKEN WINDOWS: The police and neighborhood safety" , Atlantic Monthly.

Johnson, Carolyn Y(2005) "Breakthrough on 'broken windows'". The Boston Globe, Counterspin.

Keizer, K; Lindenberg, S; Steg, L (2008). "The Spreading of Disorder". Science 322 (5908): 1681–5.

Lemert, E. M. (1951) Social Pathology. New York: Mcgraw-Hill. pp. 75-76.

Levitt, Steven D.; Stephen J. Dubner (2005). Freakonomics: A Rogue Economist Explores the Hidden Side of Everything. New York, NY: HarperCollins.

Lindesmith, Alfred R., and *Dunham*, H. Warren, (1941) "Some Principles of Criminal Typology." Social Forces 19, pp.307–314.

Martin, R., Mutchnick, R. & Austin, W. Timothy. (1990). Criminological Thought:Pioneers Past and Present. New York: Macmillen Publishing Company.

Matsueda, Ross L, (1982) "Testing Control Theory and Differential Association: A Causal Modeling Approach." American Sociological Review 47:489 504.

Mead, G. H. (1934) Mind, Self, and Society. Chicago: University of Chicago Press.

Mehta Rohinton, (1999) Crime and Criminology — A socio legal Analysis of the phenomenon of Crime, Snow White, First edition.

Merton, Robert K. (1938) "Social Structure and Anomie". American Sociological Review 3 (5): 672–682.

Merton, Robert K. (1968) Social Theory and Social Structure,New York, Free Press.

Mike Maguire, Morgan Rod and Reiver Robert, (2007), The Oxford Handbook of Criminology, Oxford University Press, 14th ed., pp. 741-742

Paranjape, N.V. (2008) Criminology and Penology, 13th Edn. Central Law Publication, Allahabad.

Pfohl, S. (1994). Images of Deviance and Social Control: A Sociological History (2nd ed). New York: McGraw-Hill, Inc.

Ponnian, M. (2003) Criminology and Penology, Pioneer Books, Third edition.

Ritzer, George (2007). Sociological Theory (7 ed.). McGraw-Hill Higher Education.

Sampson,Robert J, W Byron Groves, (1989)Community Structure and Crime: Testing Social-Disorganization Theory, American Journal of Sociology, Volume: 94, Issue: 4, pp. 774-802.

Sellin, Thorsten. (1938) Culture Conflict and Crime. New York: Social Science Research Council.

Siddique Ahmad, (2003) *Criminology — Problems perspectives,* Eastern Book Co., Fourth edition.

Sirohi J. P., (2000) Criminology *and Criminal Administration,* Allahabad Law Agency.

Srivastava S. S., (2002) Criminology *and Criminal Administration,* Central Law Agency, Second edition.

Sumner, C. 1994. The Sociology of Deviance: An Obituary. New York: Continuum Publishing.

Sutherland. (1939), Principles of Criminology. *J.B. Lippincott Company* (1974 Reprint).

Tannenbaum, F. (1938) Crime and Community. London and New York: Columbia University Press

Tepperman, L., & Curtis, J (2006). Principles of Sociology: Canadian Perspectives, p. 117. Oxford University Press, Canada.

Thacher, David, (2004) "Order Maintenance Reconsidered": Moving Beyond Strong Causal Reasoning", Vol-94 No-2, Journal of Criminal Law and Criminology pp.381-414

Thornberry, Terence P. (1996) Empirical support for interactional theory: a review of the literature in Delinquency and Crime edited by J. David Hawkins, Cambridge Studies in Criminology Series.

Tittle, Charles R. (1995) Control *Balance*: Toward a General *Theory* of Deviance, Westview Press.

Void, G.B., T.J. Bernard and J.B. Snipes (1998) Theoretical criminology, fourth edition. New York: Oxford University Press.

■■■

CHAPTER 7

Etiology of Crime

Learning Objectives :

- **Acquainting the Learners with the criminal intention and bodily structure and circumstantial factors leading to crime perpetration.**
- **To comprehend the factors responsible for crime : Social, Economic, Physical, psychological and Political crime.**
- **To make a critical appreciation of etiology of crime.**

Introduction

The reasons why people commit crime willfully has been a matter of great concern. Study of causes of crime refers to a branch of knowledge under criminology called etiology of crime. There are as many causes of crime as there are explanations or theorists. This scientific search for an etiology of crime has contributed an impressive literature to the subject. Criminology has now grown into a major subject of inquiry that is characterized by an inter-disciplinary approach. Explanations of criminal behaviour draw concepts from a vast array of social and physical sciences. The reasons why people are motivated to commit crime range from an Ecological perspective (Shaw & McKay, 1942) to Differential Association (Sutherland, 1924), Cultural Conflict (Sellin, 1938), Sub-Cultural influences (Cohen, 1955; Matza, 1969), Anomie and Strain (Durkheim, 1951; Merton, 1968), Lack of Opportunities (Cloward & Ohlin, 1960) to Lack of Self Control (Gottfredson & Hirschi, 1990).

The questions that animate this chapter concern the challenges that are posed for criminology by the economic, cultural, and political transformations that have marked late twentieth-century social life. Rate, nature, trends and dimensions of crime have changed drastically over a period of time. The restructuring of social and economic relations, the fluidity of social process, the speed of technological change, and the remarkable cultural heterogeneity that constitute 'late modernity' pose intellectual challenges for criminology that are difficult and sometimes discomfiting but which are ultimately too insistent to ignore in the context trends of change and public policy. It would also be to depart from the canons of clarity, perspicacity and relevance that worthwhile criminological work has always observed. Criminologists, particularly those who draw upon a sociological tradition, have always sought to ground their analyses in a nuanced sense of the world as it is, and as it is becoming, not least because the phenomena of crime and disorder have so regularly been traced to the effects of social upheaval and dislocation.

Crime situations are rarely clear-cut, single-source events. Clearly, the nature of the society, its culture and history are significant variables in understanding criminal behaviour. It follows that the nature of criminality in India must be explained as a consequence of its peculiar socio-cultural, economic, psychological and political conditions. Hence are discussed below the causes/explanations of crime.

SOCIAL CAUSES OF CRIME :

Crime is an offense against the law of the land. It varies in character and degree on account of the act itself and also on account of the law. A mild offense against the law is called a misdemeanor. A serious offense is called a crime. The only difference between a crime and a misdemeanor is in the gravity of the offense, and, since the estimate of the seriousness of an act varies from place to place, these are not the same in different communities. In early society, when natural justice prevailed and each man settled his own difficulties with his fellows, crime in a legal sense was unknown. Cruelty, savagery, and bestiality might have existed, but they did not become criminal until the judgment of society pronounced them so in formal law. Even after society began to recognize certain acts as criminal, they were treated solely as offenses against the person involved and not as against society at large. But now every criminal act is considered an offense against society. In a sociological sense a serious offense against society may be a social crime, even though the law has not been passed defining such act as criminal.

1. Family and Crime

Family is the miniature society. It is well said that "The *family is the first great training school in either behaviour or misbehaviour*". The family is the root cause for several offences in the society. Broken home converts the children into delinquents and housewives into prostitutes and other criminals.The causes of crime relating to family are discussed below :

(i) Child Abuse : Criminologists research on child abuse and found several child abuse cases, out of them some were died and many were affected with brain damage. The situation is being continued in entire world without any exception. The parents quarrel with each other, and show their anger on the children. The children become preys of the angers of their elders. Broken houses, family disputes convert the children into delinquents, and cause truancy. Some of poor parents send their children to work near rich people. Some of such rich people harass the poor children on trifling causes. They pay very meager salary to them. They do not provide proper food, cloth and shelter. Some of the industrialists employ children in their industries and factories at lowest rates of wages. These children are barely exploited physically and economically which forces them to commit crime.

(ii) Spouse Abuse : The women are praised in words. But they are ill-treated from the ancient period up to this day throughout the World. The criminologists found that husbands assault their wives very cruelly. Majority of the husbands in the world use violence against their wives. Assaults, viz. kicking, punching, stabbing, burning, choking, etc. on wives are common throughout the globe.

Some of the husbands do not provide them the basic amenities. Some of the husbands do not trust their wives and they are always suspicious on their chastity. Even though, they do debauchery, they want their wives in chastity. It is a crime, according to the penal laws, where a husband forcibly does sex with his wife without her consent. However, generally no wife dares to complain against her husband on his sadistic and sexual perversion. In India, the greedy husbands squeeze and tease their wives demanding to bring dowry, additional dowry from their parents' houses. If the wives do not consent, or fail in bringing money from their parents' houses, the husbands become more atrocious and seriously attack on their wives, and some times brutally kill them. The disputes between the wife and husband greatly affect on their children. Children are psychologically depressed. They start delinquency, truancy, cigarette smoking, small thefts, telling lies, going to pictures in the school times, etc. Gradually, they become juvenile delinquents.

(iii) Large Families : Children reared by large families with "criminalistic" values would accept a criminal lifestyle as normal. Children neglected by their families would be more strongly influenced by nonfamilial associates, some of whom might be pro-criminal. Some of the parents do not adopt family planning operation. They continue to procreate the child after child. The number of children increase more than what the parents can afford for their life and living. The father becomes weak to earn money sufficient to feed his family. The grown up children seek the alternative ways of earning. They gradually tend towards to delinquency at childhood and spoil their future.

(iv) Negligence by Parents : Families are seen to be the primary socializing agents, parents are expected to provide warmth and protection as well as guidance. Conversely, absence of affection and inadequate discipline have been seen as sources of crime. In some of the families, both the parents happen to be employees. They cannot concentrate on their children. Education should begin at home, with enlightened teaching in family relations, the sanctity of marriage, proper child-rearing and respect for authority. Often, parents ignore and neglect the proper training and teaching of their children. They are too busy pursuing their own interests to pay attention to the needs of their children. Too often, therefore, children are just "raised," not reared. They grow up in a permissive atmosphere of abandon and are not properly taught respect for authority. They are left on their own, without proper guidance, and encouragement. The problem is compounded when divorce rips homes asunder. The only way to establish a proper life is to ensure right education at home on issues that initially affect life and relationships. In the absence of such monitoring and watching, their children are attracted to bad friends, and are habituated to vices, such as, truancy (school absconding), cigarette smoking, debauching, gambling, etc. .

(v) Dowry Deaths : Although crimes within the family typically go unrecorded, violence between husband and wife accounts for a significant proportion of recorded criminal assaults and homicides. In India, dowry deaths have become very common. It is very heinous crime known to the world. The greedy husbands and their parents squeeze, tease and harass their wives and their parents-in-laws and demand for more and more dowry. If the wives fail to bring more money; husbands start ill-treatment, and then kill them brutally. This is very shameful to our society. It is punishable under Section 304-B of I.P.C.

(vi) Son Preference : In India, majority of the couples like to have a son, to perform *Sradha Karma* and *Pinda Karma* after their death. There is the fear in the Hindus that the person who have no sons goes to hell. Hence majority of the couples like to procreate a son. For this purpose, they get the sex determination test done at the ladies at the time of pregnancy and consequent. If the tests give the result that the child in fetus is female, the couple prepare to go for abortion. Sex selective abortion is a crime.

2. Defective Education

There exists a positive correlation between illiteracy and criminality. The reason for worldwide unrest, unhappy marriages, the spiralling divorce rate, juvenile delinquency, crime and other mounting social problems is a lack of the right kind of education. All these conditions are a direct result of the missing dimension in education. Similarly curriculum based examination with fixed timing, distance from school, lack of vocational training and uncertain future lead the youth to abstain from school and engage in anti-social activities. A child with poor academic performance may indicate something wrong at school. Perhaps he does not like school due to external influences; i.e. bullying, difficult teachers, taunting by other students, or peer pressure. Modern education in India is very defective. It does not design the character of students in the right directions. Instead it has lead to an increase in selfishness and disorderliness in the students. Offenders have poor academic achievement, having fail grades or are truants or drop outs. They tend not to participate in school activities. Suspensions/ expulsions from educational institutions have their negative effects as well.

Besides this, even after completing his education an individual does not become capable of earning his livelihood and many well educated people remain unemployed for years. Late employment leads to late marriage. All these causes tend to aggravate criminal activity like theft, alcoholism, prostitution, terrorism etc.

3. Cinema, Newspapers and Media

Crime films shape the ideas about heroism and they keep neophytes in touch with their own histories by reinterpreting famous crimes of the past, the causes of crime, and criminal responsibility. Films are based on real events, though details may be altered for the purposes of storytelling. The popularity of TV brought about the emergence of TV series featuring detectives, investigators, special agents, lawyers, and, the police. Cinema has helped in the increase of crime in India. According to Blumen and Hansen, cinemas indirectly influence the male criminals. The cinema arouses criminal tendencies in men by teaching crime techniques, by exhibiting many kinds of crimes, by stimulating day dreams about crimes. In much the same way it directs or urges by setting alight the sexual desire, the desire for exhibition, love-making, etc. Once socialized into illegal activities, people may continue them. Of course, not everyone who watches a Rambo movie or its television equivalent becomes a criminal. The "glorification" of violence on television has little effect on most folks, but it has a powerful effect on those who are social misfits. It dehumanises them and becomes a self-fulfilling process. Repeated exposure to violence may, through a process of disinhibition, make violence seem acceptable, even desirable.

Newspapers increase criminal activity by publishing methods of crime through news, by printing many news items relating to crime, by making crime a general subject, by printing news of major crimes and showing their advantages, by circulating the names of criminals, by highlighting causes which tend to increase crime in the name of studies of crimes and by warning and alerting criminals by publishing police methods.

In today's society, murder, rape, suicide, gang warfare, riots, drive-by shootings and other graphic depictions of violence are no further away than the nearest remote control. All of these images and more can be viewed from the comfort of one's own home on television. Exposure to graphic violence does bear an important and consistent relationship to aggression. Imitation includes more than simply applying a crime technique the criminal learned by watching television. Fictional treatments of crime can inspire and empower potential criminals. John Hinckley drew inspiration and encouragement in his attempt to shoot President Reagan from the dozens of times he watched Taxi Driver, a movie about an assassin who stalks a presidential candidate and wins a young woman's affection. The man who murdered twenty-two people in Luby's Cafeteria in Killeen, Texas, in October 1991 was found with a ticket to the film The Fisher King in his pocket. The link between media and violence becomes more than theory when you consider the overwhelming correlation cases.

4. Defects of the Marriage and Dowry System

Two links between crime and age of marriage have been forged in the literature. First, several studies suggest that delinquents marry at younger ages than do nondelinquents. Second, criminality tends to decline at about the time that marriage takes place. Perhaps because of the popular belief that marriage has a settling effect, researchers have sometimes concluded that marriage reduces crime. Yet at least three accounts of the relationship between marriage and crime can be given. Delinquents may marry when they are ready to settle down, delinquents who are less criminally inclined may be more likely to many (with marriage marking no change in motivation), or marriage may produce change.

Defects of the marriage customs and the evils of dowry system have given rise to a series of domestic violence. On the one hand, the dowry system urges the fathers of girls to earn money through illegitimate means for their daughters' marriages, and on the other hand, it leads to suicide by many young girls who cannot bear to see the degraded condition of the parents. In extreme cases, the lack of wealth compels a girl to remain unmarried throughout her life that leads in some cases to prostitution. Failure of parents to give dowry leads to bride burning, bride beating etc. These are the vital conditions relating to marriage that lead to crime.

5. Religious Fundamentalism

Religious Fundamentalism means blind love toward ones own religion with blind bitter feeling towards other religions that leads to communalism. Communalism leads to communal riots and crimes such as looting, murder and so on perpetrated in order to increase the followers of one's own religion and to injure another religions. This is always happening in India. Its best examples are the crimes that occurred during Babari Masjid Demolition and Godhra Carnage in 2002 etc.

But the crimes committed during communal riots in the name of religion, when the country was partitioned in 1947 were unequalled in the history of the world. Helpless women were stripped and taken in procession through the main roads all the time being whipped. Their breasts were cut off, they were violated and mutilated, and their children's hearts were cut out in front of their eyes. The fundamentalist insistence on strict observation of religious laws leads to an accusation of legalism in addition to exclusivism in the interpretation of metaphysical beliefs. The goals of such action are usually a mixture of religious objectives and the politico-economic interests of one's own community as against those of similarly defined other communities.

6. Agencies of Moral Risk

Every crime is legally a wrong, but not every wrong is defined as a crime. Morality can be distinguished from law or from justice according to the way in which the latter is publicly enforced and sanctioned through the power of the state, while the former is regarded as a private matter where wrongs are to the moral discredit of a person but not such as to allow legal recourse for those wronged. The difference between morality and justice comes not from the difference between actions and consequences (as between morality and energetic ethics) but from the difference between motives and actions. Agencies or institutions of moral risk are frequently mentioned as an important factor in the causation of crime. Amusements through cheap dance halls, gambling parlours and dens of vice are considered to be responsible for the demoralization of young children. Many a times the criminals visit these places and pick up their companions from there or teach such practices to others and as a result all of them are involved in committing crime to some extent. Indeed, studies of thinking prior to offending show that the criminal is not concerned with moral issues, but rather with the likelihood of being successful.

7. Factions and Feuds

The hostility shown towards strangers and other intruders is reflected in violence between and within local groups. Among the excitable males with their somewhat over-developed self-esteem, violence in response to real or imagined insults is widespread. A feud, or vendetta, is an extended argument between two groups of people, usually started as the result of an insult, violence, or even murder. Today the term is more popularly associated with celebrities and sports rivalries, but historical blood feuds were fairly commonplace, and there were even rules and laws—like dueling—that were set up in order to help resolve them. Some feuds managed to escalate beyond individual groups and families, and in a few extreme cases, like the War of the Roses, they even led to large-scale conflicts in which thousands were killed. Factions and feuds in villages are a common feature in this country. The unfortunate legacy of the caste system which tended to create narrow channels of thought and a tendency to think and act in terms of little groups is the main factor in the creation of tension. There are also a number of instances where blood feuds have been carried on between two branches of a single village. In villages so affected, unending criminal and civil litigation is the heritage of their residents.

8. Superstition

With superstition ruling the roost in the tribal hinterland of Odisha, villagers young and old continue to be tortured and killed on the suspicion of practising witchcraft. A strong belief in spells and sacrifice to obtain material wealth is a frequent cause of crime. Many a times murders are committed in the belief that human sacrifice would reveal hidden treasure. Thus, superstition continues to exist in considerable measure among the illiterate and ignorant masses and among schedule tribes and aborigines and even among more advanced communities in varying degree.

9. Cultural Conflict

Cultural conflict refers to mental conflict within an individual (or group of individuals) between two cultures, both of which are partially accepted and which provide certain contradictory standards and offering loyalties. The Chicago School of Sociology Researchers collected quantitative data in regards to how and where young offenders lived. Most of the data pointed towards young males living in areas of disrepair with low rents, high unemployment rates and a high turnover of the population, all of this combined with a lack of policing and, no apparent, governmental intervention (poor amenities and public services) created a hotbed for gangs and sub-cultures to emerge (these areas are badged "zones of transition"). It stated that "street gangs provided the main community for young offenders" this could be because a street gang possibly provides attainable goals or a moral handy-cap which young males can identify with (opposed to normal).

According to Thorsten Sellin both criminality and non-criminality are attributable to the kinds of conduct norms that have been experienced. But learning of divergent conduct norms presupposes the existence of a society in which the conduct norms of one group are in conflict with the conduct norms of another. People form criminal groups with separate values and norms from those of mainstream society in order to attain the goals that they desire. Conflicting cultural norms just simply give more occasions for crime and deviance to occur. Thus a condition of culture conflict underlies a condition of high crime rates.

10. Social Stratification

It refers to a relatively permanent ranking of status and roles in a social system in terms of differential privileges, prestige, influence and peculiar social stratification involves inequality, arising either from the actual functions performed by the persons involved or from the superior power and control of resources possessed by certain individuals or groups, or both. Caste based prejudice, Racism and discrimination make people crime prone due to in-egalitarian hierarchy.

It has been assorted that crime varies in amount by social strata. The individuals in the lower strata are suspected of displaying greater risk and liabilities owing to their sociological positions for getting involved in crime and for being acted upon officially.

A social structure theory of crime causation has been developed by R.K. Menton as part of a more general theory of deviance. The notion is that high crime rats are reflection of a situation in which a society places great emphasis upon the goal of individual 'success' while effectively blocking for some parts of the population, the paths to achievement of that goal. In the study of criminology crisis of poverty is as much a problem as the crisis of prosperity.

11. Differential Association

This is a theory developed by Edwin H. Sutherland holding that criminal behaviour is learnt in social interaction. The hypothesis maintains that criminal patterns of behaviour are learnt in interaction with criminals in primary group situations (such as the family, among school friends) and in relative isolation from anti-criminal social norms. The hypothesis of differential association is that criminal behaviour is learnt in association with those who define such behaviour favorable and in isolation from those who define it unfavorably, and a person in appropriate situation engages in such criminal behaviour. Generally delinquents/criminals have delinquent/criminal peers. This is also called "hanging around with the wrong crowd."

12. Fashion & Crime

In certain occasions fashions tend to crime. "Miss World", "Miss Country", "Miss Town" etc. beauty competitions lead unhealthy atmosphere. They lead agitations, disturbances in the city, state, etc. clashes occur between the supporters and opposing groups of such competitions. They lead to chaos. The youth drive the vehicles fashionably and speedily and cause the motor accidents. Lotteries are fashionable crimes.

13. Neighborhoods

Social Disorganization Theory links (directly) high crime rates to neighborhood ecological characteristics suggesting that youths from disadvantaged neighborhoods were participants in a subculture in which delinquency was approved behaviour. Criminal behaviour is acquired in a social and cultural settings through a process of interaction. Percentage of units in multi-unit housing structures in urban areas is a strong predictor of violent crime as it leads to anonymity.

ECONOMIC CAUSES :

There is no widely accepted definition of economic crime, and it is impossible to enumerate briefly the various definitions, theories, and offenses included in this category. There are three theoretical traditions that explore the aspects of economic crime: offender motivations, economic outcomes, and economic processes.

The first tradition refers to economic crimes as illegal acts in which offenders' principal motivation appears to be economic gain. Here, an economic crime is conceived of as any offense in which individuals or collectivities of people purposively act in an illegal manner in order to gain financial returns (e.g., robbery, drug selling, tax evasion, computer crime, and abuses of economic aid). Although conceptually appealing, this tradition has several drawbacks. For example, it assumes that offenders' motivations are readily observable or knowable from the criminal act itself. Although the motive behind robberies may appear to be the desire for property, perpetrators' primary motivation may be different (e.g., thrill seeking or racial hatred). Some crimes have multiple motives and economic gain may be a secondary goal. Furthermore, offenders themselves are not always conscious of their motives and they may be unable to distinguish between the reasons that precipitated their actions and the rationalizations or justifications that follow them.

A second tradition avoids difficulties associated with trying to infer motives and focuses on illegal acts that successfully provide offenders with an economic return. However, excluding attempted crimes from analysis limits our understanding; successfully completed offenses may differ in important ways from those that are failures. A variation of this tradition defines economic crime as offenses for which victims incur an economic cost. Typical victims include individuals, groups, or organizations against which the act was directed; however, a much wider group of victims may have been indirectly affected by such crimes. This occurs in cases in which a criminal act subverts or undermines the commercial effectiveness of normative business practices and the negative consequences extend beyond those at whom the specific immediate harm was intended (e.g., computer hacking, insider trading in stock market transactions). This definition addresses a common oversight in criminology—ignoring or under-representing victim issues—nonetheless, it is too narrow in some respects and too broad in others. For example, it excludes "victimless" crimes that have economic implications (e.g., prostitution) and includes any offense for which victims experience a cost (e.g., an assault that results in medical expenses or loss of wages).

A third tradition contends that the processes that lead to criminal behaviour are the same as those that guide consumer behaviour in the marketplace. This approach informs most theoretical work on crime offered by economists since the late 1960s. Its most cogent statement is found in Gary Becker's neoclassical or "economic" approach to explaining crime (1968). The remainder of this entry describes this approach and discusses its advantages and weaknesses; reviews other social science perspectives that address some of the shortcomings of the neoclassical approach; and summarizes recent directions in the study of economic crime.

The roots of crime are diverse and a discipline like economics, predicated on rational behaviour, may be at something of a disadvantage in explaining a phenomenon largely viewed as irrational. The foray by economists into this area is relatively recent, dating back to Gary Becker's pathbreaking contribution in 1968. As part of a larger model designed to explore optimal criminal justice policy, he developed the "supply of offense" function, which indicates the factors affecting the number of crimes a rational individual commits. Since then there has been much progress in both expanding on this important relationship and utilizing it for more theoretically grounded analyses of criminal behaviour.

A recent survey suggests that three general issues are of central concern in the economics of crime literature: the effects of incentives on criminal behaviour, how decisions interact in a market-setting, and the use of cost-benefit analysis to assess alternative policies to reduce crime (see Freeman, 1999). In this entry we will focus on the role of incentives on criminal behaviour.

Crime is a major activity for young males. Crime is like basketball; it's a young man's game. As one researcher has observed: "Actual rates of illegal behaviour soar so high during adolescence that participation in delinquency appears to be a normal part of teen life."

Some social scientists argue that crime is closely related to work, education, and poverty and that truancy, youth unemployment, and crime are by-products or even measures of social exclusion. "Blue-collar" criminals often have limited education and possess limited labor market skills. These characteristics partly explain the poor employment records and low legitimate earnings of most criminals. These sort of issues originally led economists to examine the relationship between wages and unemployment rates on crime. More recently economists have also considered the benefits and costs of educational programs to reduce crime.

A related question concerns the impact of sanctions. For example, does increased imprisonment lower the crime rate ? How does the deterrent effect of formal sanctions arise ? Although criminologists have been tackling such issues for many years, it is only recently that economists have entered the arena of controversy. This is not surprising given the high levels of crime and the associated allocation of public and private resources toward crime prevention. The expenditure on the criminal justice system (police, prisons, prosecution/defense, and courts) is a significant proportion of government budgets. In addition, firms and households are spending increasingly more on private security. Amidst the above controversies and issues following are the some of agreed economic causes of crime.

1. Poverty

Poverty is a major cause of crime, since a hungry man can do anything to appease his hunger. One comes across instances where a criminal commits a crime immediately on his release from prison so that he may be imprisoned again where he is assured of a square meal. Thefts increase in number when the prices go up, and decrease when prices fall. In this way, a close relation is found to exist between the number of crimes and the prices of food grains in India. One major cause of prostitution is poverty. One even comes across news of suicides and murders of families due to poverty. The best example of crime committed due to poverty is the violence of Maoists in the tribal pockets of Odisha, Andhra Pradesh and Chattisgarh, the so called Red Corridor.

2. Unemployment

Unemployment, much like poverty, is a major cause of crime in India. Many young men commit suicides and get involved in thefts and other such crimes when they are frustrated by continued unemployment. Among the people who are seen creating disturbances, causing violence, etc., the majority are the unemployed. Most gangsters do not work and gangsterism is their only occupation. Unemployment increases the relative attractiveness of large and less violent gangs engaging more in property crime. Domestic violence tends to rise as unemployed men are at home, household finances come under strain, and there is greater tension within families.

3. Industrialization

Industrialization is also an important cause of crime in India. It has destroyed the family life of millions of labourers. They work very hard for eight to ten hours in the day during which they have suffered many deficiencies and for this they have to live in towns far away from their families. Continued existence of this life leads them to indulge in drinking alcohol and accompanying prostitutes and this encourages other crimes. In India, the most prospering centers of prostitution are important industrial towns which, at the same time, foster crimes such as murder, assault, disturbance, robbery, etc.

4. Urbanization

In cities the society fails to exercise control over the individual who loss his identity in the mob after having committed a crime. Density of population in urban neighborhoods has been considered a factor in the production of delinequency and crime, the thought being that the incidence is greater in the thickly populated areas than in the sparsely populated areas. Mr. Burnt, in his study of delinquency in London, found that 21 percent of his delinquency cases, as compared to 16 percent of non-delinquents, came from overcrowded homes. Gambling dens and wine shops that provide means of spurious recreation in urban areas also encourage crime.

5. Smart Money and Technology

Smart Cards and Cyber Money help increase of money laundering activities and provide new opportunities for money transfer that are difficult if not impossible for government authorities to monitor and detect. Money Laundering efforts remain a holistic and comprehensive strategies against drug smugglers and other transnational criminal organizations. As these groups find ways to circumvent new restrictions and display the kind of ingenuity that is difficult to counter. Not only is the money laundering process integral to the functioning of criminal organizations, but it is also an indicator to their success. Money Laundering and its consequence provide the juncture between what is termed as the "upperworld" and the "underworld". This is one of the major ways in which criminal organizations penetrate the licit economy with the help of so called reputed members of the society. To allow money laundering to go unchallenged, therefore, would have a corrosive impact on the integrity of financial institutions.

6. Prosperity

The rise in crime is a function of rising affluence and not of penury alone. The rich are just as much addicted to crime as the poor. Occupational crime differs from common street crime in that it is committed by people in a position to do so. The progress of civilization may not destroy crime. In spite of unprecedented prosperity, America leads the world in crime as well as in divorce. Prosperity brought with it easier access to alcohol and drugs; greater disparities of wealth, a boom in consumer goods that excite envy and are easily stolen; a culture that treats theft more leniently than it once did; a marked decline in the likelihood of imprisonment for a crime, rising corruption and changing moral standards. Millions have made the acquisition of material wealth their goal in life. Husbands and wives — in their never-ending desire for more material goods — have gone head over heels in debt! This means endless arguments and marital strife and that too in front of the children. Parents have tried to provide their youngsters with more material possessions than they had as children. But this has not brought them closer. Instead, the generation gap is becoming wider. The dictionary defines prosperity as: The state of being successful, thriving or wealthy. Prosperity is not something you have, it's something you are. By definition, it's a state of being rather than in amassing of what you can buy, or store, or covet or lose. To be prosperous is about merit, not material. To be successful is to enjoy personal power, not purchasing power. And to be a wealthy is to be healthy and whole; to have a presence about you, and to own your own brilliance, your own joy and to be fully autonomous, fully free to pursue the happiness you seek.

7. Consumerism

Now people judge themselves and others by what we and they consume. The advertisers know this and boost consumerism. It's a world driven by competition for consumer goods and paid-for experiences, of hi-tech and high-end shopping signals that have become the means by which men keep score with each other. As the sociologist Zygmunt Bauman points out, to be a successful consumer defines what it is to be "normal". Therefore to be "abnormal" is to be a failed consumer. The lot of the failed consumer is miserable. This new poor may be better off in absolute terms than the poor of previous generations. The failed consumer suffers not just from exclusion from normal society but isolation. The final ignominy of today's poor is that they don't want to overthrow the rich to create a new order, they just want to be like them. So strong is this urge that the failed consumer will lie, cheat and steal to "earn" the trappings of success. In the world of the "me generation", people become calculating rather than law-abiding in their overwhelming desire to be normal. This is crime driven by the rampant egoism of turbo-consumerism, where enough is never enough. And precisely because of its competitive nature, consumer-driven crime cannot be switched off through tougher laws.

8. Indebtedness

Indebtedness means the state, of being in debt, without regard to the ability or inability of the party to pay the same. But for indebtedness, there must be an actual liability at the time, either to pay then or at a future time. Slavery in early society explains the nature and consequences of indebtedness. Slavery refers to the condition of an individual who works for another individual against his or her will as a result of force, coercion, or imprisonment, regardless of whether the individual is paid for the labor. Compulsory labor under torturous conditions for the satisfaction of debts has led them to indulge in violent struggles. Compelling the performance of labor or personal services in lieu of debt is also a crime as it happens.

In today's world, hunger, deprivation and poverty increase the problem of indebtedness which culminate and terminate in various forms of crime including suicide. Farmer's suicide is a case in view. Unable to meet any of the commitments like crop failure due to natural calamities and devastations, payment to the landlords and paying off bank loans, they kill themselves in despair. The reasons for suicide again are loss of crop and apprehensions over loan repayment, mainly from private sources. Similarly bankcruptcy, Crimes of fraud, concealment, and misrepresentation continue to victimize all sorts of groups and individuals in society.

9. Occupations

Some occupations are especially lucrative and crime prone. Occupational crime occurs when crimes are committed to promote personal interests, say, by altering records and overcharging, or by the cheating of clients by professionals. Occupational crime is crime committed by individuals in the course of their employment such as crimes committed by professionals such as physicians, engineers, teachers, police, lawyers and clerks in offices in the course of their practices. It seems that every occupational category generates a considerable number of criminals and the higher the prestige of the occupation, the more their criminal activities cost the general public. It is different from other crime as the function of opportunities but others look for opportunities to commit crime. Organizational or corporate crime occurs when corporate executives commit criminal acts to benefit their company by overcharging or price fixing, false advertising, etc.

10. Jal, Jungle and Jameen Issues

Jal, Jungle and Jameen mean so much to the tribal and marginalized. The present atmosphere of violence over jal, jungle and jameen is due to the continuity of process of marginalization in spite of several plans and programs. Politicians put forth Jal (water), Jameen (land) and Jungle (forest) as their main poll plank generating factions feuds of which anti-socials take advantage. Maoists champion the cause of the marginalized and remain mysterious warriors who emerge and vanish like the jungle mist. Critics of present day justice system argue that until India ends impunity for genocidal killings, it will continue to be a nation ruled by men, and not the law. The issue of tribal land alienation was correlated with that of displacement, and the judicial system was used to get an order declaring unconstitutional the transfer of land from a tribal to a non-tribal through the medium of the state (in its land acquisition capacity). Violence, murder and law and order situations occur due to land alienation, inadequate rehabilitation and resettlement policies clearing of pavements and eviction of hawkers. Eviction of Hawkers, and of dwellers, is an important issue, and the demands include classification not only of shelter, land and monetary compensation, but of livelihood too.

PHYSICO-PSYCHOLOGICAL CAUSES :

Crime, like all human behaviour is a complicated issue with many facets and there are different causes and effects of each crime committed. The different psychological approaches to crime place the emphasis on different factors, ranging from biology, personality and unconscious conflict to social learning.

(A) PHYSICAL

Some people are prone to be criminals due to their genetic make-up. Genetically inherited characteristics predispose people towards deviance. Scholars associate physical characteristics with personality characteristics and the criminality in them. A choleric person has a "hot" temperament, is irritable and easily roused to anger. A melancholic person is sensitive and tends to feel sad. A phlegmatic person is not easily aroused to excitement and lacks emotion expression. A sanguine person is confident and optimistic. To them bodily constitution indicates whether a person is "born criminal". "Born criminal" violates the laws. "Born criminal" is an "atavist" (throwback to an earlier stage of human evolution).

Lombroso (1876), the Italian Doctor found a link between physical characteristics and criminality. He is often referred to as the founding father of criminology. He believed that delinquents are throwbacks to primitive man and can be identified through several key characteristics: Large jaw, High cheek bones, Large ears, Extra nipples, toes and fingers, Insensitivity to pain, High arched eyebrows, Love of orgies and Irresistible craving for pain.

1. Physical Deformities & Inferiority : Physical defects also make a man criminal. Lombroso suggested that if a person's arm span (that's both arms measured together!!) is longer than a person's height then that person would be pre-disposed to crime. He found that individuals with a mesomorphic build were more aggressive and more prone to criminality. Due to deformities he lags behind his fellow beings in many activities and it is in order to remove this difference he turns to crime. Criminals have usually been found to be extremely ugly. On examining prisoners in the Parkhurst prison, Charles Goring found most of them to be inferior, from the physical viewpoint, to the common individual. Mental defects are just as much causative of crime as are physical defects. Feeble-mindedness is also found to be a cause of crime.

Criminal behaviours have also been associated with hormone abnormalities, especially those involving: (1) Testosterone (a male sex hormone) and (2) Progesterone and estrogen (female sex hormones). Administering estrogen to male sex offenders has been found to reduce their sexual drives.

2. Hereditary Factor : Heredity, according to some criminologists is one main cause of crime. Lineage and descent have some influence upon criminality. Goring's findings viewed that crime must be explained by the heredity of weak-mindedness as well as by the inheritance of a predisposition. Early sexual maturity, mental defects and nervous instability etc., are abnormalities, an individual inherits, which help in making him a criminal. Introduction of some types of teratogens, such as illegal drugs, alcohol, and nicotine, may represent substances that, regardless of when they are introduced, could potentially be harmful to the exposed fetus. Much attention has recently been paid to the association between maternal smoking during pregnancy and negative behavioural outcomes among exposed fetuses. These negative outcomes include impulsivity and attention problems. Prenatal nicotine exposure has also been associated with criminal offending. Maternal smoking during pregnancy is linked to criminal violence in exposed offspring. Chemicals contained in cigarette smoke may alter fetal brain neurochemistry. Moreover, exposure to cigarette smoke prenatally may increase risk for asphyxia.

More recently attempts have been made to demonstrate the biological inheritance of criminality by studying cases of criminal twins. Among these the most notable investigation is that of Mr. Lange in Bavaria. In the cases of monozygotic twins who were supposed to be identical and to have derived from the same ovum, it was found that out of 13 sets, ten sets were found to be both criminal. In the cases of dizygotic twins, which were supposed to be non-identical and to have developed from separately germinated ova, out of 17 pairs, only two of them had prison record. Thus, it was concluded that inherited tendencies play a part in the causation of crime.

If we do not inherit the acquired faculties and habits of our parents, it is unfortunately too true that we inherit their diseases and the connection between disease and crime is a fact which cannot be denied. In many cases it is perfectly true that people suffering from disease or physical degeneracy do not become criminals, in most cases they do not; at the same time a larger proportion of such people fall into a lawless life than is the case with people who are free from inherited infirmities. The undoubted tendency of physical infirmity is to disturb the temper, to weaken the will, and generally to disorganize the mental equilibrium. Such a tendency, when it becomes very pronounced, leads its unhappy possessor to perpetrate offenses against his fellow-men, or, in other words, to commit crime.

3. Constitutional Factor : Lombroso contended that there was, among various kinds of criminals, a born criminal type which was characterized by irregularities of body and mind. "Some of the many anatomical and mental anomalies of the born criminals were : asymmetry of skull; low retreating forehead; strongly arched brows; large outstanding ears; sparse beard; tattooing; left-handedness; relative insensibility to pain; acuteness of vision; obtuseness of the sense; violent temper; extreme vanity; improvidence; and so forth.

The criminals possess a certain type of physique and character. They display definite types of body built as well as characteristic and associated physic disposition. The basic types of body built are the aesthetic, athletic, pyknic and certain mixed and unclassified types.

4. Glandular Factor : Hormones are chemical messengers produced by the endocrine glands, the brain, gastrointestinal organs, sex organs, the kidney, the heart, the pineal gland, the skin and the hair. They exert a strong influence on behaviour, principally by inducing brain events that prompt people to behave in certain ways to environmental stimuli. Hormones not only influence reactive or conscious behaviour after environmental cues, but they also influence anticipatory or unconscious behaviour (by knowing what the body needs beforehand, hormones make a person think they want something before they see it). The key concept of endocrinology is central motive state. In psychology, the equivalent concept is desire. Some simple examples are thirst, hunger, and sex, and more complicated pathways or "circuits" exist for various other states of excitation, arousal, or quiescence. Hormonal reactions also don't occur in a vacuum. Desires have an appetitive (searching) phase and a consummatory (fulfilling) phase. Although individuals by themselves can probably generate the appetitive phase of a hormonal reaction, they need the hormonal reactions of others (pheromones) or additional environmental stimuli to generate the consummatory phase.

The functioning of the glands is made a basic determiner of temperament, intelligence, character and personality as well as of physical and mental pathology. Glandular disturbances could account for many disharmonic, defective, unstable and inadequate qualities found in the suspected constitutional inferiority of criminals. The intelligent higher type criminals and defective and deficient constitution of criminals are all explained by abnormal or deficient functioning of the glands. Following table shows the effects of Hormone released on individual behaviour.

TABLE 7.1

Effect of Hormone on Individual Behaviour

Endocrine gland	Hormone released	Behaviour influenced
Pituitary (anterior lobe)	Growth hormone, Prolactin, ACTH, LH, TSH, FSH	Food preference, brooding, parenting, quiescence.
Pituitary (posterior lobe)	Arginine vosopressin, Oxytocin, Endorphins, Enkephalins	Hunger, attachment behaviour, sexual behaviour, pain, aggression
Pineal	Melatonin	Sleepiness
Thyroid	Thyroxine, Calcitonin	Nervousness
Parathyroid	Parathyroid hormone	Sickness, immune system
Heart	Atrial natriuretic factor	Water/sodium ingestion
Adrenal (cortex area)	Androgen, Glucocorticoids, Mineralocorticoids	Aggression, dominance, genital size
Adrenal (medulla area)	Epinephrine, Norepinephrine	Sodium homeostasis
Kidney	Renin	Sodium hunger
Skin	Vitamin D	Mood, melancholy
Liver/lung	Preangiostensin, Angiotensin	Thirst, sodium hunger
Pancreas	Insulin, Glucagon	Food satiety
Stomach/intestines	Cholecystokinin, Peptides, Bombesin, Somatostatin	Food satiety
Gonads (ovary)	Estrogen, Progesterone	Female sex drive, aggression, territoriality
Gonads (testis)	Testosterone, Substance P	Male sex drive, pain
Lymphocytes	Cytokines	Sleepiness

5. Physical development : Factors concerning physical development may also sometimes be causes of crime. As a general rule it can be said that it is between the ages of 17 to 24 that the greatest number of crimes are committed. In old age rape and other sexual crimes increase in number. Murderous crimes are usually committed by young men. Embezzlement, fraud, vagrancy, alcoholic excesses and murder, etc. are peculiar to middle age. Girls who nature early is more prone to becoming involved in clandestine affairs. In adolescence the aggressive tendency grows.

6. Neurotransmitters : Neurotransmitters are chemicals that allow for the transmission of electrical impulses within the brain and are the basis for the brain's processing of information. Many studies have indicated that the levels of three neurotransmitters, namely, serotonin, dopamine, and nor epinephrine have an influence on behaviour. Genetic defects in two neurotransmitters, dopamine and serotonin, have been identified in violent individuals and certain drug abusers. Although neurotransmitter levels are initially determined by genetics, they can be manipulated with drugs, such as lithium carbonate (for serotonin), reserpine (for nor epinephrine) and various antipsychotic drugs (for dopamine). Changes in diet can increase the level of these transmitters. On the other hand, stressful conditions can reduce serotonin levels considerably resulting in increased tendency to engage in violent and antisocial behaviour. Alcoholism can also affect neurotransmitter levels.

7. Nervous System : The outer portion of the brain, the cerebral cortex, is divided into four lobes: Frontal temporal, parietal, and occipital. The frontal and the temporal lobes are involved with goal-directed behaviour, impulses, and emotions. Hence it is believed that these two lobes are responsible for aggression and violent behaviour. Various studies on the brain, using brain imaging techniques (such as computerized tomography, magnetic resonance imaging, positron emission tomography and single photon emission tomography), revealed that persons with damaged frontal and temporal lobes may commit violent and sexual offences respectively. Studies relating to measurement of brain activity of criminals using electroencephalograph (EEG) have also revealed that habitual violent offenders are characterized by abnormal brain wave patterns.

There is evidence to indicate that in addition to the brain and the central nervous system, the autonomic nervous system (ANS) also plays some role in antisocial behaviour. The ANS controls many of the body's involuntary functions such as blood pressure, heart and intestinal activity, and hormone level. When an individual faces a danger situation (a 'fight or flight' situation), the ANS prepares the body for such situation by increasing the heart rate, diverting the blood from the stomach to the muscles, dilating the pupils, and increasing the respiratory rate. Lie detectors measure these functions. Eysenck gave an explanation for criminal behaviour, based on the functioning of autonomic nervous system. His theory has been discussed separately.

8. The limbic system : Violent criminal behaviour has also been linked to disorders in other parts of the brain. Recent evidence suggests that chronic violent offenders have much higher levels of brain disorder than the general population. The limbic system is a structure surrounding the brain stem that is believed to moderate expressions of violence and controls the life functions of heartbeat, breathing, and sleep also.

Critique of biological explanations

Some of the criticisms against biological explanations of criminality are given below :

1. Theory overlooks the bright and handsome criminals.
2. Theory ignores those who are ugly and live lives of productive and cooperative labor.
3. Theory does not look at the variations over time in crime rates. Since genes change very slowly, there should be a steady rate of crime over the centuries. That is not the case. Crime rates vary dramatically.
4. Biological theorists were studying the very poor - people whose physical development had been affected by poverty and poor nutrition.
5. Not everyone who breaks the law ends up in prison.
6. This type of theorising neglects the idea that there is a "grey area" of criminality - people who commit crimes but who are not caught and therefore not imprisoned.

(B) PSYCHOLOGICAL CAUSES OF CRIME

1. Mental Deficiency : Crime and deviance are linked to maternal deprivation. Mentally deficient and lack of intelligence can be crime prone. "Intelligence" generally consists of different abilities, such as the ability to reason, solve problems, think abstractly, learn and understand new material, adapt to novel situations quickly, grasp complex relationships, and profit from past experience. In other words, intelligence is "catching on," "making sense" of things, or "figuring out" what to do. While mental deficiency is one of the many earmarks of the so-called constitutional criminal, sub normality and feeble-mindedness have been considered important causes of crime in their own right. Charles Goring concluded that weak-mindedness was an outstanding factor in the production of crime. Relying on the results of psychometric testing of juvenile reformatory inmates, he claimed that low-grade mentality was the greatest single cause of crime. He concluded that "at least 50 percent of all criminals are mentally defective."

2. Extrovert Character : Crime and deviance can be explained through inherent character traits. People with an extrovert personality are more likely to break the law as they crave attention, take chances and act on the spur of the moment. Personality can be seen as a spectrum ranging from extrovert to introverts. Introverts are the opposite of extroverts and are therefore less likely to be criminal. Extroverts tend to be more naturally active, expressive, and interested in many things. Introverts tend to be more reserved, private, cautious, and interested in fewer interactions, but with greater depth and focus.

3. Mental Disorders : Mental disorders play a very important part in the etiology of crime. Crimes result from abnormal, dysfunctional, or inappropriate mental processes within the personality. Dr. East who studied about 234,000 persons admitted to prisons in England from March 1923 to March 1926 found that only about 2,000 i.e. 1.5 percent were dealt with under the Lunacy and Mental Deficiency Act and similar were the results shown by the surveys made by the National Committee for Mental Hygiene (1931). Dunham who worked in the research division of the mental hospitals in Illinois, also came to the conclusion that the criminal insane comprised a very insignificant part of the total institutionalized insane.

4. Unconscious Emotional Difficulties : Freud's approach explains criminal behaviour in terms of unconscious emotional difficulties or motivations which frequently manisfest themselves in faulty character development and which makes it difficult for individuals to conform to reasonable expectations. According to him, the mind is composed of Id, Ego and Super Ego. The Id represents all the basic and instinctive urges which are not socially acceptable and are generally repressed by Ego into the unconscious. The Super Ego on the other had represents all that is moral. The Ego strikes a balance between the demand of the Id and Super Ego. In the case of the criminals the unconscious desires find expression in antisocial acts due to the malfunctioning of the Ego.

5. Motivation to Satisfy Wishes : Criminal behaviour may be purposeful for the individual in so far as it addresses certain felt needs. Maslow believed that human beings are motivated by five basic levels of needs and that people choose crime because they cannot (or will not) satisfy their needs legally. W.I Thomas says human motive could be reduced to four basic types of wishes. They are experience, response, recognition and security. This system of motivation is based on organic needs arising from the body and the external social situations. Individuals turn directly to delinquency in order to get their wishes satisfied or resort to delinquency as a result of blockages in the satisfaction of wishes in normal channels. Thus it can be said that unsatisfactory fulfillment of wishes and associated emotional distress is expected to constitute an important course of crime.

6. Feeling of being Treated Unjustly : Many psychological elements especially the feeling of justice or of being unjustly treated, may originate the idea with it an internal process in which the personality is directly involved. This process is called cirminogenesis. It develops a criminogenic attitude which does not necessarily lead to the commission of the offence unless other stages passed through. Many honest persons remain honest although their criminogenic attitude is at least internally manifested. Thus there is a distinction and eventually a passing from criminogenic process to actual criminal behaviour, the letter manifesting itself in the commission of the offence. People don't commit crimes merely because they are deprived, but because of a lack of connections between their goals and the means to attaining these goals. Therefore, they employ illegitimate means of criminal acts because the legitimate means are not available to them.

7. Contradictory Individual and Social Self : Man becomes abnormal when the social harmony is disturbed or altered by morbid process and when there is a physio-psychic unbalanced conditions and conflicts between the individual and the social self. All these mean that the genesis of criminal bahaviour, is the result of contradiction between the individual and social self, or between individual and social interests. More specifically, every time the individual self prevails over the social self and as a result anti-social actions take place.

Freudian psychologists propose that a person can only lead a stable life if the Id, Ego and Super Ego are well balanced. If the Super Ego is weak and the Id is strong, a person will seek physical pleasure, undeterred by the threat of punishment dictated by parents and society. The failure to develop a strong enough Super Ego (which allows us to behave in a positive way) is attributed to absent or poor parenting. Normal law-abiding citizens who have control over their Id acknowledge the threat of punishment and this provides sufficient deterrent from antisocial behaviour.

8. Personality Disorders : Psychopath and personality disorders may end in criminal activities or anti-social behaviour. A psychopath is the person who does not feel empathy with others. It is unable to imagine how others think and feel and is able to inflict pain and engage in cruelty without empathy for the victim. The psychotic offender experiences irrational ideas and distorted perceptions. Anti-social personality is basically unsocialized and exhibits behaviour that brings him or her into conflict with society. It is incapable of significant loyalty to individuals, groups, or social values. Criminal behaviour is maladaptive, or the product of inadequacies inherent in the offender's personality.

9. Imitations : People learn how to act by observing others which is called Imitation. Delinquent behaviour arises whenever non- secure attachments are created. Low self-control is the premier individual-level cause of crime. Social learning theory outlines three requirements for people to learn and model behaviour include attention: retention (remembering what one observed), reproduction (ability to reproduce the behaviour), and motivation (good reason) to want to adopt the behaviour. One method by which violent entertainment may promote criminal violence is simply imitation. Children in particular are very susceptible to mimicking what they see in the media. Surveys of young American male violent felons found that many had imitated crime techniques they watched on television programmes.

10. Schizophrenia : Schizophrenia literaly means "split mind". It relates to a psychiatric disease relating to a group of severe psychotic disorders characterized by disorganized and delusional thinking, disturbed perceptions, inappropriate emotions and actions. The subtypes of Schizophrenia are (1) *Paranoid* : relates to preoccupation with delusions or hallucinations. Delusions means false beliefs, often of persecution or grandeur, that may accompany psychotic disorders. Hallucinations relates to false sensory experiences such as seeing something without any external visual stimulus. (2) *Disorganized* : It is linked to disorganized speech or behaviour, or flat or inappropriate emotion. (3) *Catatonic* : Immobility (or excessive, purposeless movement), extreme negativism, and/or parrotlike repeating of another's speech or movements.

POLITICAL CAUSES :

From its inception criminology has been embedded in politics. Despite frequent claims to scientific objectivity, criminological inquiry has been defined and sustained by political concerns. Affinities between political orientations and explanations of crime have often been noted, and debates over theoretical differences have typically included references to such affinities. Indeed, pointing out the ideological assumptions and implications of theories has been a standard element in assessments of their worth—with or without regard for research findings. If politics is the authoritative allocation of values, then crime and politics are inextricably linked. The symbols of crime, law, and order shape political rhetoric and public expectation. The symbols of crime affect the nature of political promise and expectation.

In the eighteenth and early nineteenth centuries the idea of studying crime and criminals was closely associated with that of making governance more effective. European intellectuals saw the arbitrariness and cruelties of despotic rule as threats to social order. Their views were crystallized in 1764 by Cesare Beccaria, who forcefully and concisely argued that punishment of offenders should be "public, necessary, the minimum possible under the circumstances, [and] proportionate to the crime". By the 1830s the movement to rationalize governmental social control through law promoted statistical studies of the "dangerous classes" (ultimately leading to Cesare Lombroso's search for "born criminals") and the mapping of associations between crime and various indicators of moral deficiency.

Until the 1960s, disagreements among criminologists centered almost entirely on how best to measure and explain the characteristics of people who violate the law, or who were statistically likely to do so. It was generally assumed that the goal of criminology is to learn what pathologies, individual and/or environmental, cause criminal behaviour. That assumption was challenged by a growing number of "conflict" criminologists who argued (1) that criminality is defined by a lawmaking process influenced mainly by the more powerful classes in society, and (2) that the prime directive of law enforcement is to protect the interests of the higher classes, so that (3) the lower classes are more likely both to commit the kinds of acts legally defined as crimes (while the often much more harmful behaviours of the higher classes are not so defined) and to be labeled as criminals regardless of their behaviour.

Now-a-days many criminals go scot free with prepared legal advice because there is opportunity to prove truth to be false and falsehood to be true. This encourages them to engage in further criminal activity. On the other hand, the innocents, who are prosecuted and convicted in their place, also become criminals as a reaction. The ill treatment to which the prison inmates are subjected also hardens the softest of criminals. Crimes are further encouraged by the inefficiency, immorality and corruption of the police. People connected with political groups also unobtrusively assist criminals and make use of them to inflict injury upon and to defame members of opposing groups. Targeting those with "politically incorrect" motives undermines the principle of objective law that undergirds our legal system's protection of rights.

Criminalization of Politics

Criminalisation of politics has become an issue of grave concern among the Indian intelligentsia. And though the top leaders of all political parties agree that those with criminal record should be debarred from contesting elections, the number of such people is only increasing. The criminalization of politics continues to be a big concern, with an increase in the number of MPs with criminal records from 128 in 14th Lok Sabha to 150 in 15th Lok Sabha. It is the evidence to the undemocratic and autocratic selection and nomination of candidates by political parties. In order to ensure the winability of candidates, parties ignored principles to give preference to money power and muscle power.

Criminalization of politics can be visualized in two different lenses. In narrow sense it refers to the direct entry and interference of Criminals into State legislatures and parliament of India, while on wider sense it refers to interference of criminals into politics either directly or indirectly like financing any candidate, providing anti-social manpower, booth capturing, contract killing of rival candidates, providing muscle power services, as well as campaigning or canvassing for any candidate contesting elections. The competitive use of anti-social forces for the mobilization of party funds, for management of elections, organizing meetings and conference and even recruiting workers at lower levels from among anti-social elements has increased many folds. Approximately all political parties from national to regional taking services of criminals to win election. Earlier criminals used to lend outside support but now they themselves have entered in the electoral arena and have become not only members of the house and even have become ministers. Thus we have a new phenomenon in Indian politics called "Tainted ministers." In 2004, about one in five MPs had a criminal record, including some with charges of heinous crimes such as murder, rape, dacoity and kidnapping and not just violation of Section 144, or something similar. Around 20% of the members of the current Lok Sabha have criminal cases pending against them.

The corruption has spread its tentacles far & wide, there is corruption from womb to tomb, from maternity hospital to grave yard. The injustices mated out, the atrocities perpetrated by public servants are worse than Britishers. The biggest confounding factor in the political environment of business is criminalization of politics: people with criminal backgrounds becoming politicians and elected representatives. Let us discuss the components of criminalization of politics here below.

1. Muscle Power : The influence of muscle power in Indian politics has been a fact of life for a long time. As early as in 1977, the National Police Commission headed by Dharam Vira observed: "The manner in which different political parties have functioned, particularly on the eve of periodic election, involves the free use of musclemen and 'Dadas' to influence the attitude and conduct of sizable sections of the electorate. The Panchayat elections, like other elections in the recent past, have demonstrated once again that there can be no sanity in India as long as politics continues to be based on caste and Gangsterism. The common people who constitute the voters are in most cases too reluctant to take measures that would curtail the criminal activities. Once the political aspect joins the criminal elements the nexus becomes extremely dangerous. Many of politicians chose muscle power to gain vote bank in the country, and they apply the assumption that, if we are unable to bring faith in the community then we can generate fear or threat to get the power in the form of election.

2. Money Power : The elections to Parliament and State Legislatures are very expensive and it is a widely accepted fact that huge election expenditure is the root cause for corruption in India. A candidate has to spend lakhs of rupees to get elected and even if he gets elected, the total salary he gets during his tenure as an MP/MLA will be meagre compared to his election expenses. Opportunities are looked into in order to bridge the gap between the income and expenses publicly through donations and secretly through otherwise. The expenditure estimation for an election estimated as ₹ 5 per voter as election expenditure, and the expenses run to crores. Some politicians have money and muscle, so they win and help others in their party to win as well.

3. Ticket Distribution & Winability : The most important factor, which determines the ticket distribution, is the "winability" of the candidate. Hence more and more people with money and muscle power are getting tickets from the political parties. Another factor that has played an important role is that the criminal elements think that they can escape punishment by becoming a member of legislative bodies in the states or at the centre.

4. Vicious Circle : As goes the old adage – "There ain't no such thing as a free lunch". A person with criminal background entering politics will demand his pound of flesh from the party in power to which he has extended support. Apart from seeking protection from the criminal cases going against him, he will also want a share from the developmental schemes going on in his areas. Thus the government funds meant for the poor go into the pockets of these people.

5. Vote Bank : The political parties and independent candidates have astronomical expenditure for vote buying and other illegitimate purposes through these criminals or so called goondas. A politician's link with them constituency provides a congenial climate to political crime. Those who do not know why they ought to vote comprise the majority of voters of this country and as such are maneuverable and purchasable. Most of them are individually timid and collectively coward. To gain their support is easier for the unscrupulous than the conscientious.

We have long witnessed criminals being wooed by political parties and given cabinet posts because their muscle and money power fetches crucial votes. Elections are won and lost on swings of just 1% of the vote, so parties cynically woo every possible vote bank, including those headed by accused robbers and murderers. Legal delays ensure that the accused will die of old age before being convicted, so parties virtuously insist that these chaps must be regarded as innocent till proved guilty.

6. Corruption : Corruption has always been a close companion of crime. In every election all parties without exception put up candidates with a criminal background. Even though some of us whine about the decision taken by the parties, the general trend is that these candidates are elected to office. By acting in such a manner we fail to realize that the greatest power that democracy arms the people is to vote incompetent people out of power.

Independence has taken place through a two-stage process. The first stage was the corrupting of the institutions and the second stage was the institutionalization of corruption. As we look at the corruption scene today, we find that we have reached this stage because the corrupting of the institutions in turn has finally led to the institutionalization of corruption. The failure to deal with corruption has bred contempt for the law. When there is contempt for the law and this is combined with the criminalization of politics, corruption flourishes. India is ranked 66 out of 85 in the Corruption Perception Index 1998 by the German non-government organization Transparency International based in Berlin. This means that 65 countries were perceived to be less corrupt than India and 19 were perceived to be more corrupt.

7. Loopholes in The Functioning of Election Commission : The Election Commission must take adequate measures to break the nexus between the criminals and the politicians. The forms prescribed by the Election Commission for candidates disclosing their convictions, cases pending in courts and so on in their nomination papers is a step in the right direction if it applied properly. Too much should not be expected, however, from these disclosures. They would only inform people of the candidate's history and qualifications, but not prohibit them from casting their votes, regardless, in favor of a criminal.

For the past several general elections there has existed a gulf between the Election Commission and the voter. Common people hardly come to know the rules made by the commission. Bridging this gap is essential not only for rooting out undesirable elements from politics but also for the survival of our democratic polity. This is an incremental process, the rate of success of which is directly proportional to the increase in literacy rate in India. The electorate have made certain wrong choices in the past, but in the future national interest should guide them in making intelligent choices.

8. Denial of Justice and Rule of Law : Criminalization is a fact of Indian electoral politics today. The voters, political parties and the law and order machinery of the state are all equally responsible for this. There is very little faith in India in the efficacy of the democratic process in actually delivering good governance. This extends to accepting criminalization of politics as a fact of life. Toothless laws against convicted criminals standing for elections further encourage this process. Under current law, only people who have been convicted at least on two counts be debarred from becoming candidates. This leaves the field open for charge sheeted criminals, many of whom are habitual offenders or history-sheeters. It is mystifying indeed why a person should be convicted on two counts to be disqualified from fighting elections. The real problem lies in the definitions. Thus, unless a person has been convicted, he is not a criminal. Mere charge-sheets and pending cases do not suffice as bars to being nominated to fight an election. So the law has to be changed accordingly.

9. Lack of Control of Local Bodies : Local bodies like caste panchayats play a vital role in controlling crime. Previously the caste panchayats in the village kept control over the behaviour and conduct of the members of the caste. Today, these social organizations no longer exist. If they exist, they have been deprived of all their authority. This situation leads to prevalence of crime.

10. Safety Valve in Law : A banal remark has been in frequent use in Indian politics since last few years that "innocent until proved guilty." This has become a boon for the criminal-politicians because investigations can always be manipulated and notorious judicial delays ensure that even the worst criminals are not convicted for decades.

11. Politics of Morality : Lack of ethics or values in Indian politics, generally all major or minor political parties in India used to play blame game, instead of finding a collective solution of the problem. Let me tell you the significance of ethics in politics of India by mentioning "Omar Abdullah episode" happened recently in J&K legislative assembly. On 28th July 2009, Jammu and Kashmir Chief Minister Omar Abdullah resigned on moral grounds from the post after a senior PDP leader alleged that he was involved in the Srinagar sex scandal. There was high drama in the Assembly when Abdullah said he would step down and not return till the allegations were proved wrong. Members of his party, National Conference, surrounded him and tried to physically stop him from leaving. Congress in-charge of state Prithviraj Chavan had tried to stop Abdullah and asked him not to resign. Most noticeable thing during this whole episode was reactions which, came from political lobby of India. Many senior and prominent politicians in Indian politics declared Omar Abdullah an emotional fool and immature fallow because he resigned on the civil or moral grounds. It clearly shows that resigning on moral grounds in India considered as an act of foolishness. Does it mean that Railway Minister [Lal Bahadur Shastri] during Nehru regime, who gave his resignation on moral grounds after a rail accident, was an emotional fool ? We all know the moral of story.

12. Coalition Politics : Politics of coalition is an emerging trend in India. Coalition politics is the sharing of power in which different interests are democratically accommodated. A perception persists that coalitions are inherently unstable and undermine democracy. Actually, the type of government depends on the type of election result and a fractured verdict by the electorate can only result in a coalition. And a verdict given by the electorate cannot be visualised as any violation of the electorate's mandate. It cannot be treated as undemocratic. Unfortunately coalition politics has generated several criminal processes like abduction, horse trading, conspiracy and brutalities.

13. Politics of Factionalism : The lack of homogeneity in the Indian population causes division between different sections of the people based on religion, region, language, caste and race. This has led to the rise of political parties with agendas catering to one or a mix of these groups. Some parties openly profess their focus on a particular group, for example, the Dravida Munnetra Kazhagam's focus on the dravid population, and the Shiv Sena's pro-Marathi agenda. Some other parties claim to be universal in nature, but tend to draw support from particular sections of the population, for example, the Rashtriya Janata Dal (translated as National People's Party) has a vote bank among the Yadav and Muslim population of Bihar and the All India Trinamool Congress does not have any significant support outside West Bengal. The Bharatiya Janata Party, the party with the second largest number of MPs in the 15th Lok Sabha, has an image of being pro-Hindu. Such support from particular sections of the population affects the agenda and policies of such parties, and refute their claims of being universal representatives. The Congress may be viewed as the most secular party with a national agenda, however, it also practices vote bank politics to gain the support of minorities, especially Muslims, through appeasement and pseudo-secularist strategies. The narrow focus and vote bank politics of most parties, even in the central government and central legislature, sidelines national issues such as economic welfare and national security. Moreover, internal security is also threatened as incidences of political parties instigating and leading violence between two opposing groups of people is a frequent occurrence.

14. Inadequate Rules on Party Funding : The inadequacy of the rules and regulations relating to political party funding (including provisions allowing for anonymous donations) comes out starkly as perhaps the prime reason for colossal amounts of money going into the party exchequer and allowing extensive opportunities for non-transparent and illegal financing, which cannot be traced or sanctioned. In this context, an RTI filed with the Election Commission of India by Common Cause, disclosed startling information. Many parties including two national parties have not been filing their contribution reports before the Election Commission, in accordance with the provisions of Section 29C of Representation of the People Act, 1951. According to Section 8(3) of the Representation of People Act (RPA), 1951 "A person convicted of any offence and sentenced to imprisonment for not less than two years [other than any offence referred to in sub-section (1) or sub-section (2)] shall be disqualified from the date of such conviction and shall continue to be disqualified for a further period of six years since his release." For certain offences listed under sub-section (1) of sec (8), conviction per se acts as a disqualification while for certain other offences given in sub sections (2) and (3) the quantum of sentence affects the conviction. This however did not prevent these political parties from wrongly claiming income tax exemption under Section 13A of the Income Tax Act. It is shocking that the Income Tax Department has been routinely granting tax exemption even to those political parties who are not entitled to it under the law, despite clear Supreme Court directions to the contrary.

LEGAL CONCERNS :

1. Vohra Committee

Series of bombs blasts that shook Bombay on 13 March 1993, had involved the collaboration of a diffuse network of criminal gangs, police and customs officials as well as their political patrons, a commission were institutes to investigate the so-called nexus. The report by N.N.Vohra found such deep involvement of politicians with organised crime all over India. Vohra observed "the various crime syndicate/mafia organisations have developed significant muscle and money power and established linkage with governmental functionaries, political leaders and other to be able to operate with impunity. As highlighted by the Vohra Committee, our elections involve a lot of black money and it is this use of black money in elections which has also brought about the criminalization of politics. After all, the story of the Hawala scam started by the police stumbling to the Jain diaries in their effort to trace the money received by the Kashmir militants. The scam brought out the linkage between the corrupt businessmen, politicians, bureaucracy and the criminals. The 1993 Bombay blasts which took away the life of 300 people was made possible because RDX could be smuggled by allegedly bribing a customs official with ₹ 20 lakhs. Some 15 years ago Vohra committee submitted its report to curb criminalization of politics but the fact is that no application in this way is being made. This was mentioned in the petition submitted by the Speaker of Lok Sabha and President of India on 16th May that- "The subject of criminalisation of politics is one that concerns the entire nation closely. It is deeply disturbing that on the one hand, our polity is tolerant of 'fake encounters' (summary executions) of alleged criminals and terrorists, while our highest representative body – Indian Parliament – harbours people caught red-handed in acts of human trafficking, and convicted on charges of abduction and suspected murder."

2. Supreme Court's Judgment

The Supreme Court judgment of May 2, 2002 (Union of India Vrs Democratic Reforms and Others) mandated that candidates disclose their criminal antecedents, if any, as also their financial and educational background. The Election Commission had proposed amendment of statutory rules and the format of nomination papers, to give effect to this judgment of the Supreme Court. The Apex Court judgment to check corruption among public servant is a welcome step. No law should provide protective shield to the corrupt public official and the court has rightly held that no prior sanction of competent authority would be required to prosecute them. With this order, 93 MPs and 10 ministers in Manmohan Singh's ministry are under the scanner on various criminal charges. This is appalling. It is ironical that the executive and legislatures who make and implement policies and guidelines for the development are themselves acting as stumbling block in the development of the nation. The role of Supreme Court becomes very important here. The Apex Court as custodian of constitution should take all necessary steps to strengthen democracy in the country. The legislature and executive have been complaining about the Supreme Court's intervention on their domain, but it becomes imperative in such kind of unwanted situation. The Supreme Court of India upheld a PIL which made it mandatory for everyone seeking public office to disclose their criminal, financial and educational history. It was a way to ensure that the voters knew the important details about their "honourable" leaders, and steamed them were indeed.

Some of the parties would be able to draw advantage from the Supreme Court order because they have had less opportunity to indulge in crime and corruption. They would have a greater chance of watching candidates of other parties squirm and suffer in agony. Some others might be happy because they already have efficient watchdog systems and batteries of lawyers in place that would permit them to file counter-affidavits and challenge nominations of opposing candidates within hours of their being filed.

3. Right to Information Act and Criminalization of Politics

The Court held that the right to information - the right to know antecedents, including the criminal past, or assets of candidates - was a fundamental right under Article 19(1) (a) of the Constitution and that the information was fundamental for survival of democracy. In its Judgment of May 2, 2002, it directed the Election Commission to call for information on affidavit from each candidate seeking election to Parliament or the State Legislature as a necessary part of the nomination papers on: Whether the candidate has been convicted / acquitted / discharged of any criminal offence in the past - if any, whether the candidate was accused in any pending case of any offence punishable with imprisonment for two years or more, and in which charge was framed or cognizance taken by the court of law. If so, requires the details thereof; the assets (immovable, movable, bank balance, etc.) of a candidate and of his/her spouse and that of the dependents; liabilities, if any, particularly of any overdue of any public financial institution or Government dues; educational qualifications of the candidate. The Right to Information Act 2005 is a historical Act that makes Government officials liable for punishment if they fail to respond to people within a stipulated timeframe. Many public servants are leading luxurious lifestyles, beyond the legal sources of their income. Many public servants are filing false affidavits about their annual income, wealth details to Election Commission of India / Vigilance Commission / other authorities, as the case may be. These authorities are not properly verifying these affidavits. Many scams, scandals are coming to light day in & day out, politicians are accusing each other of involvement in scams. Whereas, the said authorities are keeping mum, as if those affidavits filed by tainted public servants are true. The tainted public servants are not even providing full, right information to public as per RTI Act, lest the truth come out.

This seen is very normal now a day that some public servants, caught red-handed during luxurious spending, they easily say that it is at their political party's expense or their well wisher's expense. However no entries are found in the account books of said parties to that respect. The law forbids public servants from accepting gifts, hospitality, favors beyond the value of rupees one hundred (Rs. 100), as it may be a form of bribe. But one may ask all these under RTI. Right to know is an inherent attribute of every person. Right to know differs only in one sense with right to information. Right to know is a natural right and right to information is a provision given by government to its people. Natural rights do not have any value legally until they are legally considered. Hence right to know as such implied in the freedom of speech and expression which is a legally considered right must have to be given a special value. Right to information as such will bring transparency of the government activities and allow the people to find remedies for those things by which they suffered.

4. Concerns of Election Commission

The Election Commission has been underlining the issue of increasing criminalization in politics since 1998. In his letter to the Prime Minister, the Chief Election Commissioner Mr. T.S. Krishnamurthy articulated the Commission's concerns on the subject. "…… There have been several instances of persons charged with serious and heinous crimes like murder, rape, dacoity, etc. contesting election, pending their trial, and even getting elected in a large number of cases. This leads to a very undesirable and embarrassing situation of lawbreakers becoming lawmakers and moving around under police protection. The Commission had proposed that the law should be amended to provide that any person who is accused of an offence punishable by imprisonment for five years or more should be disqualified from contesting election even when trial is pending, provided charges have been framed against him by the competent court. Such a step would go a long way in cleansing the political establishment from the influence of criminal elements and protecting the sanctity of the Legislative Houses………"

Former Chief Election Commissioner Sri B.B Tandon also suggested that political parties evolve a code of ethics among themselves and not give ticket to such criminal elements. "It is only the will-power of the political parties which could help Indian democracy get rid of this menace."

5. The Asian Human Rights Commission (AHRC) Report

The Election Commission's views have been forcefully reiterated by the Asian Human Rights Commission. In its report published in April 2007, the AHRC noted that "Criminalization of politics in India is a growing problem, despite legal attempts to address it. In 2003 a law was introduced to prohibit the election of criminals or alleged criminals to state or central legislature. However, persons with criminal or alleged criminal backgrounds continue to hold seats in parliaments all over the country. In fact, current statistics show that some 23.2 percent of members of parliaments in India have criminal investigations or cases pending against them. The state election in Uttar Pradesh in April 2007 is a case in point. An estimated 35 percent of candidates have trials and cases pending against them. The charges include rape, armed robbery and murder. The primary reason for criminal elements gaining political power is the failure of rule of law. Criminals protect the illegitimate interests of politicians and in turn obtain protection from them and their parties. This mutually beneficial relationship works against the establishment of the rule of law. As a consequence, the criminal justice system is increasingly corrupted and manipulated. This decrepit state of affairs is essential to the continued criminal hold on parliaments across the country, as reasonably well-functioning police, prosecutors and courts would be a death blow to those in control."

References

Becker, G., (1968), Crime and punishment: an economic analysis, Journal of Political Economy, Vol. 76, pp. 169-217.

Cloward, R. & Ohlin, L. (1960). Delinquency and Opportunity. NY: Free Press.

Cohen, A. (1955). Delinquent Boys. NY: Free Press.

Durkheim, E. (1951). Suicide. NY: Free Press.

Fiorentini, G. and Peltzman, S., (ed)(1995), Introduction, in The Economics of Organised Crime, Cambridge University Press and CEPR.

Fishbein, D. (1990). "Biological perspectives in criminology" Criminology 28(1): 27-72.

Gottfredson Michael R. and Travis Hirschi (1990), A General Theory of Crime, Stanford University Press.

Matza David (1969) Becoming Deviant, London Prentice Hall.

Merton R K, (1968) Social Theory and Social Structure, NY: Free Press.

Rafter, N. (2008). The criminal brain: Understanding biological theories of crime. NY: NYU Press.

Scerbo, A. & A. Raine. (1992). "Neurotransmitters and antisocial behaviour" in A. Raine (ed.) The Psychopathy of Crime.

Sellin, T. (1938). Culture conflict and crime. New York: Social Science Research Council.

Shaw, Clifford R. & McKay, Henry D. (1942). Juvenile Delinquency in Urban Areas. Chicago: University of Chicago Press.

Sutherland, Edwin H. (1924) 'Principles of Criminology' Chicago: University of Chicago Press.

Wax, Emily (2008). "With Indian Politics, the Bad Gets Worse"Washington Times. http://www.washingtonpost.com/wp-dyn/content/article/2008/07/23/AR2008072303390.html.

■■■

CHAPTER 8

Individual Deviance : Juvenile Justice

Learning Objectives :

- **To Understand the complexities in defining a child.**
- **To comprehend the history of juvenile justice legislation.**
- **To be acquainted with the Provisions of juvenile justice legislation.**
- **To make clear the distinctions between juvenile and adult criminal.**

Introduction

Children are born as children not as delinquent or destitute. A child is a part of the society in which he lives. Due to his immaturity, he is easily motivated by what he sees around him. It is his environment and social context that provokes his actions. Juvenile legislation attempts to cure his illness by treating the juvenile without doing anything to treat the causes of the illness. It is naïve to believe that poverty, unemployment, inequalities and changing values will not impact children existing in its midst, and that they will grow unaffected. The juvenile justice system as it is currently envisaged, at its best, can only help the child to cope and tolerate the maladjusted and dysfunctional society. If the state genuinely has the interest of children at heart, it should not only look after its children, but also take measures to improve the situation faced by their family and other support structures.

The focus of juvenile legislation is on the juvenile's reformation and rehabilitation so that he also may have a chance to opportunities enjoyed by other children. But there is a contrasting view that loudly states that juvenile offenders are committing violent crimes from which society should be protected, and that the juvenile justice system is mollycoddling them. It is apprehended that this latter strain of thinking will gather momentum and pressure will be created to treat juvenile offenders on par with adult offenders or toughen juvenile legislation, especially in respect of serious offences.

Working with juveniles in conflict with law is not an easy task. The majority of juveniles within the juvenile justice system are without families and homes; they have migrated to a different region where they are earning a livelihood. They perceive the juvenile justice system as a hindrance. They abhor this "protection" as they have been on their own since many years, with nobody to depend on or advise them. They have been making their independent decisions, and are "little adults". The juvenile justice system aims at converting these "little adults" into children.

HISTORY OF JUVENILE LEGISLATION

International developments that led to special emphasis on the need for a specific law on juvenile justice amongst many countries, including India were:

- The Declaration on the Rights of Children adopted by the Erstwhile league of Nations in 1924 and then the United Nations in 1959;
- Setting up of a Special Working Group in the 70's within the UN Commission of Human Rights for drafting a Convention on Child Rights;
- The United Nations Standard Minimum Rules for Administration of Juvenile Justice, 1985, also known as the Beijing rules;
- The United Nations Convention on the Rights of the Child, adopted by the General Assembly in 1989.

From the early 20th century, the different Indian States had enacted their own Children Acts. The Madras Children Act 1920 was the first Children Act to be enacted, closely followed by Bengal and Bombay in 1922 and 1924, respectively. Though the Bombay Children Act was enacted 4 years after the Madras Children Act, it was the first Children Act to become functional. In February 1924, a voluntary state-aided agency, viz., the Children's Aid Society (CAS), was formed to implement the provisions of the Bombay Children Act within the municipal corporation limits of Bombay. CAS established institutions for the care and protection of children, and even today manages these institutions.

The State's Children Acts brought within its ambit two categories of children, viz., (i) youthful offenders, and (ii) destitute and neglected children. Both these categories of children were to be handled by the Juvenile Courts. During this period throughout the world, children were dealt with under the "welfarism" mode. The well-being of the child was at centre stage for both these categories of children and adjudication of guilt was not stressed, hence Probation Officers played an important role and legal representation was unheard of.

The National Children's Act came into existence in 1960. After independence, greater emphasis was given to protection of children and the National Children's Act was revised in 1986 to be called the Juvenile Justice Act, 1986. This was the first specific law on juvenile justice in India. Unlike other countries, this law covered both children who are alleged to have committed an offence as well as those who are neglected or exploited and need care and protection.

The Juvenile Justice Act, 1986 was repealed and a new law was formulated in 2000, which is called the Juvenile Justice (Care and Protection of Children) Act, 2000. This new law marked the beginning of a shift away from custodial measures and detention or institutional care to alternative non-institutional forms of care such as adoption, quality care and rehabilitation through a child-centered juvenile justice system and special protection from persons in whose care/control/charge they are. Reasons for change in the 1986 law are as below.

- The 1986 law was based on the welfare model, had focus on custodial measures and institutionalisation of children.
- It was unable to fulfill its objective of protecting and promoting the rights of children on many counts.
- Legal anomalies in the 1986 law had been ignored for too long.
- Administration of juvenile justice too required streamlining and strengthening.
- India ratified the United Nations Convention on the Rights of the Child (UNCRC), 1989 on December 11, 1992. This demanded that child-related laws be brought in conformity with the UNCRC. It also demanded a child-friendly system of justice delivery, with a clear focus on alternative measures of care and rehabilitation rather than detention.

In 2006, there was an amendment to the 2000 law to remove some ambiguities and incorporate various judgments of the Supreme Court on critical issues. Important elements of the amended law include determination of juvenility on the basis of

- the date on which the offence was allegedly committed,
- prescription of a maximum sentence of three years,
- prescription of penalty on the media for disclosure of identity of a child,
- inclusion of definition of adoption to ensure life long protection of rights of adopted children,
- periodic review of pending cases, and
- speedy justice.

Since juvenile justice is a State subject also, the States have always been required to frame their own rules, based on a set of Model Rules framed by the Centre. Subsequent to the 2006 amendment in the law, the Central Model Rules on Juvenile Justice dated 2001 were repealed and replaced by the Juvenile Justice (Care and Protection of Children) Rules, 2007. The Model Rules 2007 too have regarded the Principle of Best Interest as being fundamental for the application, interpretation and implementation of JJA 2000, and is to be of primary consideration while administrating juvenile justice.

DEFINING A CHILD

A juvenile is basically a child. The concept of child is shrouded in mystery due to lack of consensus over minimum age and upper age. There is no uniform definition of a child in India. In society, children get defined differently in different contexts. The constitutional and legal conception of a child tends to vary depending upon the purpose of the law which defines it. The UN Committee on the Rights of the Child constantly emphasizes the desirability of setting the minimum age as high as possible. It has, in particular, criticized countries where the age is set at 10 or below.

In Oxford Dictionary, a 'child' is defined as 'a young human being below the age of full physical development'. In societal context, for parents, their children always remain children. For police, only the very young, i.e. children below 10 years, get seen as children, while the older children, especially adolescents are seen as young persons with a mind of an adult and capable of planning their own life, including planning crime. When it comes to marriage, suddenly children look like they have grown up and are ready to get married. Incidence of child marriages in India is very high.

The Constitution of India under Article 21 A says "State shall provide free and compulsory education to all children of the age of six to fourteen years" read with Article 45. Under Article 24, "No child below the age of fourteen years shall be employed to work in any factory or mine or engaged in any other hazardous employment." Article 326 stipulates "...every person who is a citizen of India and who is not less than eighteen years of age... shall be entitled to be registered as a voter..."

Sec. 82 of The Indian Penal Code, 1860 says "Nothing is an offence which is done by a child under seven years of age." Upper limit is further specified under Sec. 83 clarifying "Nothing is an offence which is done by a child above seven years of age and under twelve, who has not attained sufficient maturity of understanding to judge of the nature and consequences of his conduct on that occasion."

The Indian Contract Act, 1872, The Indian Majority Act, 1875, The Mines Act, 1952, The Motor Vehicles Act, 1988 and The Juvenile Justice (Care and Protection of Children) Act, 2000, as amended in 2006 recognize the child to be of below 18 years of age. But the Prohibition of Child Marriage Act, 2006 spells out "Child means a person who, if a male has not completed 21 years of age and, if a female, has not completed 18 years of age."

There are no clear International guidelines on the minimum age of criminal responsibility (ACR). There is a wide variation in minimum ACR levels around the world which shows the range to be from 8 to 18 years. The Guidelines for Periodic Reports of the Committee on the Rights of the Child in Article 40, request information on 'the establishment of a minimum age below which children shall be presumed not to have the capacity to infringe the penal law'.

The Beijing Rules state, in Rule 4: 'In those legal systems recognizing the concept of the age of criminal responsibility for juveniles, the beginning of that age shall not be fixed at too low a level, bearing in mind the facts of emotional, mental and intellectual maturity.'

Moreover, if children are not defined legally and treated as legal persons, social relations governing them and matters concerning their rights will be left to varied subjective interpretations, thus keeping them out of the ambit of justice. Societal norms on childhood/maturity may and often do go against the best interests of the child, and a legal definition helps to curb and change such norms over time. Efforts should therefore be made to agree on a reasonable lowest age limit that is applicable internationally. The only child-friendly law in India that conforms to the UN definition of a child as a person below 18 years is the Juvenile Justice (Care and Protection of Children) Act, 2000 as amended in 2006.

PROVISIONS UNDER JJ ACT 2000 AND 2006

1. Statement of Objects

Statement of objects and reasons of The Juvenile Justice (JJ) Act is as under :

1. The justice system as available for adults is not considered suitable for a juvenile. It overrides all other existing legislations in matters involving detention, prosecution, penalty or sentence to be imposed on children alleged to have committed an offence.

2. Need for large involvement of informal systems specially the family, the voluntary organisation and the community.
3. Juvenile justice system to appreciate the developmental needs in comparison to criminal justice system.
4. To bring the juvenile law in conformity with the UN convention on the Rights of the child. It recognises the vulnerability of children to harm of all kinds, including their induction into crime. It lays down the minimum standards of protection to be followed for care, protection, detention and rehabilitation of children.
5. To prescribe uniform age of eighteen years for both boys and girls. It provides separate adjudicating machinery, distinct from that of adults, for handling matters concerning children.
6. Speedy disposal within a time limit of four months. It stresses on the need for child-friendly legal procedures and seeks to lay them down as far as possible.
7. To create special juvenile police units with humane approach. It identifies and establishes the need for rehabilitation of children who come in contact with law either as children in conflict with law or as children in need of care and protection.

Accordingly, the JJ Act seeks to provide care/protection to children by :

- Catering to a child's development needs
- Adopting a child-friendly approach
- Keeping the child's "best interest"
- Keeping restoration/rehabilitation of the child in view
- Greater intervention of NGOs & local authorities
- Special emphasis on rehabilitation and social reintegration of the children
- Adoption of children simplified
- Registration of all child care institutions within six months of the amendments coming into force
- Setting up special juvenile police units
- Setting up Juvenile Justice Boards - JJBs and Child Welfare Committees – CWCs -in every district.

2. Two Categories of Children

It recognizes two categories of children that are in conflict with law and those requiring "care and protection". The law seeks to promote a child-friendly juvenile justice system in India. The first example of this lies in the use of different and child-sensitive terminology. A child alleged to have committed an offence is known as **"a child in conflict with law"** and not a criminal/alleged criminal, or juvenile delinquent. A juvenile can be **'apprehended'** and not arrested, sent to justice 'home' not jail. When a child in conflict with law is apprehended by police, such child is supposed to be produced before the **'Juvenile Justice Board'** and not Court. A child in conflict with law or a juvenile is subjected to an **"inquiry"** and not a trial.

Under the 1986 Act the classification of delinquent juveniles and neglected juveniles was meant to separate the two categories of children with the Juvenile Welfare Board and the Juvenile Home meant for the neglected juvenile and the Juvenile Court and Special Home meant for the delinquent juvenile.

Thus the argument went that often children who had committed serious offences were kept in the same institution as children whose only crime was that they were neglected children as per the Act. Keeping this argument in mind, the state has now ensured a complete separation between the two categories as now juveniles in conflict with the law are kept in the observation home and children in need of care and protection are sent directly to the juvenile home.

(a) Child in need of care and protection means a child who is Homeless, Having incapacitate parent, Abandoned, missing or run away, Victim of natural calamities, Victim of armed conflict, Victim of civil commotion, Victim of sexual abuse, Mentally or physically challenged, Mentally or physically ill, Suffering with terminal diseases, AIDS/HIV affected, Victims of trafficking, Survivors of Drug Abuse and Children working in restaurant, *dhaba*, home and used for entertainment purposes.

(b) Juvenile in conflict with law is a child below 18 yrs who is alleged to have committed an offence. Section-3 of the Act states that when an enquiry has been initiated against a juvenile in conflict with law, the inquiry may be continued and orders may be made in respect of such person as if such person continues to be a juvenile.

There appeared a controversy. Whether the day of appearance of the child before the competent authority to be the deciding factor for determining age or the day of the alleged offence was committed ? Till 2002 the views of the Apex Court on separate judgments (Amit Das vs State of Bihar-2000(2) Crime 2008, Rajinder Singh vs State of Chhatisgarh, 2002 S.C.) was that the age of the juvenile when first appeared before the competent authority is relevant. This judgment of Amit Das vs State of Bihar as deserved was widely critiqued. It was criticised as it diverted from a well-settled principle of law thereby depriving young persons of the beneficial provisions of juvenile legislation. Many felt that the judgment had failed to interpret the law in its correct spirit. Moreover, it did not consider a 1982 three-Judge Bench decision of the Supreme Court that had unambiguously held that the relevant date was the date of offence. In *Umesh Chandra's* case (1982, 2 SCC 202)the Supreme Court had held :

"As regards the general applicability of the Act, we are clearly of the view that the relevant date for the applicability of the Act is the date on which the offence takes place. Children Act was enacted to protect young children from the consequences of their criminal acts on the footing that their mind at that age could not be said to be mature for imputing mens rea as in the case of an adult. This being the intendment of the Act, a clear finding has to be recorded that the relevant date for applicability of the Act is the date on which the offence takes place...We are clearly of the view that the relevant date for applicability of the Act so far as age of the accused, who claims to be a child, is concerned, is the date of the occurrence and not the date of the trial."

Ultimately a five-Judge Bench settled this issue in *Pratap Singh vs. State of Jharkhand & Ors.* [(2005) 3 SCC 551;] reverting back to the seasoned findings that had been incorrectly overturned in *Amit Das'* judgment. The Apex Court in *Pratap Singh's* case was faced with a query as to "Whether the date of occurrence will be the reckoning date for determining the age of the alleged offender as juvenile offender or the date when he is produced in the court / competent authority." All five Judges unanimously opined, "The reckoning date for the determination of the age of the juvenile is the date of the offence and not the date when he is produced before the authority or in the court."

After *Pratap Singh's case*, the legislature through the 2006 amendment removed any doubt by setting out in the definition itself that "juvenile in conflict with law" means a juvenile who is alleged to have committed an offence and has not completed eighteenth year of age as on the date of commission of such offence13. In case of continuous offence, i.e., an offence committed over a period of time, juvenility is to be determined on the date of commencement of the offence and if the juvenile thereafter crosses 18 years, he is still to be dealt with under juvenile legislation irrespective of when the FIR is registered.

Thus the controversies were brought to rest by amending the JJ Act in 2006 brought into effect from 22-8-2006 and incorporating proviso 2(l) which reads – "juvenile in conflict with law means a juvenile who is alleged to have committed an offence and has not completed eighteen year of age as on date of commission of such offence."

3. Determination of Age

Age is the most important factor to register a FIR. The following guidelines have to be kept in mind while dealing with cases involving a juvenile :

In case the age of the child in conflict with law is not known or not clear, efforts should be made to ascertain the age of the child before proceeding legally. In case of doubt, the benefit of doubt shall be in favor of the child in conflict with law for treating him/her as a juvenile. In order to collect the age proof of juvenile in conflict with law, efforts should be made to collect his school certificate, municipal certificate, etc. In case of non-availability of any document, age should be confirmed by Medical Board constituted at Govt. Hospital and not by ossification report.

In every case, the age of the juvenile shall be determined within a period of 30 days from the date of making of the application before the Committee and the age determination inquiry shall be conducted by the Court of the Board by seeking evidence by obtaining :

(a) (i) The matriculation or equivalent certificates, if available. High School Certificate shall prevail over other certificates (Janardhan Pandey v. State of U.P. 1997 cl Crimes 201)

(ii) The date of birth certificate from the school (other than a play/open school). Delhi Police Guideline says that the proof of age can not be based on ration card, Janampatri etc. Age mentioned in the certificate of Open School cannot be depended upon in heinous cases.

(iii) The birth certificate given by a corporation or a municipal authority or a panchayat. Where Birth certificate is on record, medical opinion uncalled for (Dail Singh vs Punjab 1994 Crime 861). Birth Certificate etc. shall prevail over medical opinion (Anita v. Atal Bihari 1993 Cr L.J. 549).

(b) In the absence of (a), the age of juvenile is to be determined from a duly constituted medical board at Govt. Hospital. While dealing with question of determination of the age of the accused for the purpose of finding out whether he is a juvenile, a hyper-technical approach should not be adopted. And if two views may be possible on the said evidence, the court should lean in favor of holding the accused to be a juvenile in borderline cases. (Rajinder Chandra vs Chhatisgarh; Crl Appeal No.113 of 2002 S.C.).

4. Adjudicating Authority for Child in Conflict with the Law

The law pertaining to what are now called children in conflict with the law has undergone a few changes. The adjudicating authority has been redesignated as the Juvenile Justice Board and the composition has changed from an adjudicating authority which was a Magistrate with a panel of two social workers to assist her as prescribed under the old law to a Bench which is composed of two social workers and one Magistrate. The shift in composition of the Board can bring about a shift in the line of inquiry from intention to motive.

(a) JJB

The chief purpose for distinct handling of a juvenile's case is that such case requires a socio-legal approach as reformation and rehabilitation, and not punishment is the goal. Under JJA 2000, Juvenile Justice Board is the "competent authority" in relation to juveniles in conflict with law. The constitution of the JJB reflects this objective of juvenile legislation. The JJB has to tread a fine path; juveniles are culpable for their criminal acts, but they should not be penalised for such action, instead the aim should be to persuade them away from the enticements of a life of crime. The 2000 Act has given equal importance to the Magistrate and the social workers; they jointly constitute the competent authority to deal with juvenile cases.

The JJB consists of a Metropolitan Magistrate or a Judicial Magistrate of the first class in a non-metropolitan area, and two social workers one of whom at least should be a woman. The Magistrate and the social workers are to function as a Bench, i.e., together, but their roles are distinct. The Magistrate plays an important role in deciding whether the juvenile has committed an offence or not. When the JJB is satisfied that an offence has been committed, then the social workers play an important role in deciding what should be done for the comprehensive rehabilitation of the juvenile, keeping in view the circumstances in which the offence was committed. It has been rightly put by Barry C. Feld (1999) that the Magistrate takes care of the deed and the social workers of the needs of the juvenile.

The JJB is to have fixed place, days and timings of its sittings. The frequency of its sittings will depend upon the pendency of cases before a particular JJB. Expeditious completion of an inquiry by JJB is vital so that the juvenile's life is not unnecessarily disrupted for a long period, and his rehabilitation process starts at the earliest. Prolonged incarceration pending an inquiry causes trauma to the juvenile, which can be easily avoided.

The law recognizing the importance of speedy inquiry has mandated the JJB to complete an inquiry within 4 months from the date of its commencement, and if the same is not possible due to the special circumstances of a case, the JJB is required to extend the stipulated period for completion of inquiry by a well reasoned order. In case the Board is not sitting, the juvenile in conflict with law shall be produced before the single member of the Board as per the provision laid down under sub-section (2) of Section 5 of the Act. When a juvenile is produced before an individual member of the Board and an order is obtained, such order shall need ratification by the Board in its next meeting.

(b) The Child Welfare Committee (CWC)

It is the sole authority to deal with matters concerning children in need of care and protection. A Committee has to be constituted for each district or group of districts, and consists of a chairperson and 4 other persons one of whom at least should be a woman.

The committee has the final authority to dispose of cases for the care, protection, treatment, development and rehabilitation of the children as well as to provide for their basic needs and human rights. A child rescued from hazardous occupation, brothel, abusive family or other such exploitative situation must be produced before the CWC who will conduct an inquiry to ensure optimum rehabilitation with minimal damage to the child.

Children in need of care and protection are sent directly to the children's home by the Child Welfare Committee that passes necessary orders for their rehabilitation, restoration and social re-integration.

(c) Juvenile Justice Homes

Juvenile Justice Homes or special institutions mandated to be set up under the Juvenile Justice Act 2000. The homes for JCL include observation homes and special homes. The homes for CNCP shall include children's homes and shelter homes. Juvenile Justice Homes must be located in a congenial child-friendly environment in natural surroundings with adequate open space for children to engage in outdoor activities. The location and size of an institution at a particular city or town will depend on the caseload of juveniles committed by the Juvenile Justice Board and children referred by the Child Welfare Committee. Every institution should be well connected by various modes of transportation and be within the vicinity of schools and technical and other vocational training institutions so that they are easily accessible to the juveniles/children in care.

The homes for JCL and for CNCP shall function from separate premises. The accommodation in each institution shall be separate observation homes for girls and boys. Classification and segregation of JCL and CNCP must be made according to their age group preferably 7-11 years, 12-16 years and 16-18 years, giving due consideration to physical and mental status and degree of the offence committed and the type of need.

(i) Shelter Homes : The JJ Act empowers State Governments to recognize reputed and capable voluntary organizations and provide them assistance to set up and administer Shelter Homes for children in need of care and protection. These homes function as drop-in-centers cum night-shelter for children in need of urgent support who are brought to such homes.

(ii) Observation Homes : As provided for under the JJ Act, 2000, the State Government is empowered to establish and maintain either by itself or under an agreement with voluntary organisations, observation homes in every district or group of districts for the temporary reception of juveniles in conflict with law during the pendency of any inquiry regarding them under the Act. When not released on bail, he shall be kept in observation home until he can be brought before the board.

(iii) Children's Homes : The JJ Act, 2000 empowers the State Government either by itself or in association with the voluntary organisations to set up Children's Homes in every district or group of districts for the reception of the child in need of care and protection during the pendency of any inquiry and subsequently for their care, treatment, education, training, development and rehabilitation.

(iv) Special Homes : The State Governments are empowered to establish and maintain either by itself or under an agreement with voluntary organizations, special homes in every district or group of districts for the reception and rehabilitation of juveniles in conflict with law.

5. Special Juvenile Police Unit

JJA 2000 envisages the setting-up of the SJPU in every district and city, and the designation of at least one police officer attached to a police station as "the juvenile or the child welfare officer". SJPUs are situated at police stations, headed by a senior level police officer, and its members are the Child Welfare Officers (also called "Designated Police Officers") attached to different police stations within that zone.

Section 63 of the JJ Act deals with SJPU which states that in order to enable the police officers who frequently or exclusively deal with juveniles or are primarily engaged in the prevention of juvenile crime or handling of the juveniles or children to perform their functions more effectively, they shall be specially instructed and trained. In every police station at least one officer with aptitude and appropriate training and orientation may be designated as JWO who will handle the juvenile or child in co-ordination with the police. SJPU be created in every district and city to coordinate and to upgrade the police treatment of juveniles and the children.

SJPU coordinates and functions as watch-dog for providing legal protection against all kinds of cruelty, abuse and exploitation of child or juvenile. It takes cognizance of adult perpetrators of crimes against children and sees to it that they are without delay apprehended and booked under the appropriate provisions of the law and for this purpose the district level units shall maintain liaison with other units of police station. SJPU is mandated to seek assistance from voluntary organizations, punchayats RWA in identifying juveniles in Conflict with Law as well as reporting cases of violence against children, child neglect and child abuse.

In case of a minor offence, the endeavour is to "divert" the juvenile with the permission of the JJB. Diversion is a process by which a juvenile offender, in appropriate cases, does not enter the juvenile justice system, and hence, is not compulsorily made to face an inquiry before the JJB. A list of Members of JJB and Child Welfare Committees (CWC) and the list of children home must be available under each SJPU.

Juvenile Welfare Officer : Every Police Station shall designate Juvenile Welfare Officers in order to enable the police officers who frequently or exclusively deal with juveniles or are primarily engaged in the prevention of juvenile crime or handling of the juveniles or children under this Act to perform their functions more effectively, they shall be specially instructed and trained. Only such officers shall be designated who have aptitude to act as Juvenile Welfare Officers. The list of Juvenile Welfare Officers will be sent to concerned Juvenile Justice Board by District SPs. "Special juvenile police unit, of which all the police officers designated as above, to handle juveniles or children will be members, may be created in every district and city to co-ordinate and to upgrade the police treatment of the juveniles and the children." It will be the sole responsibility of the SHO/Inspector/ Incharge of the unit to ensure that the complete staff is briefed and is aware of the procedure to be followed on apprehension of a juvenile in conflict with law.

6. Legal Protections for Child in Conflict with the Law

Whether the juvenile commits a bailable or non-bailable offence, the child shall be released on bail with or without surety. The only grounds on which the juvenile can be detained is if there is reasonable ground for believing that the release is likely to bring him into association with any known criminal or expose him to moral, physical or psychological danger or that his release would defeat the ends of justice. Further detention can only be in observation home and not in prison or police station. The bail provisions for young offenders are far more liberal than those applicable under the criminal procedure code (Sec 12).

- In terms of the orders which the Board may pass regarding the juvenile, there is a discretionary power to send the child home after admonition or advice, order the juvenile to perform community service, release the child on probation of good conduct etc (Sec 15).
- The arrest, detention or imprisonment of a child shall be in conformity with the law and shall be used only as a measure of the last resort and for the shortest appropriate period of time.
- The Act also clearly prohibits sentencing a child to death or life imprisonment or committing to prison in default of payment of fine or furnishing of security (Sec16).
- For a juvenile sentenced under the Act, there is a clear mandate such juvenile shall not suffer any disqualification attaching to a conviction of an offence under the law. Further after a reasonable period of time the Board is enjoined to remove the records (Sec19).
- The Act also stipulates that there shall be no joint trial of a juvenile with a person who is not a juvenile (Sec 18).
- The Act also protects the privacy of the child by mandating that no report by newspaper etc shall disclose the name, address or school or other particulars calculated to lead to the identification of the juvenile nor shall any picture of such juvenile be published (Sec 21).

Principle of Privacy and Confidentiality : No report in any newspaper, magazine or visual media on any inquiry against a juvenile in conflict with law under this act shall be made disclosing the name, address or school or other particulars calculated to lead the identification of the juvenile nor shall any picture of any such juvenile be published. As per Section 21 of the Juvenile Justice (Care & Protection of Children) Act 2000, any person who contravenes this provision shall be liable to a penalty which may extent to twenty five thousand rupees. In addition, the law clearly prohibits maintenance of records of children in conflict with law and disqualification of a person from employment on grounds of any record of crime committed by such person as a juvenile.

Expansion of category of child in need of care and protection : The category of children in need of care and protection has been expanded to include victims of armed conflict, natural calamity, civil commotion, child who is found vulnerable and likely to be inducted into drug abuse etc. Restoration is taken as option for child in need of care and protection.

Being poor, idle and disorderly cannot be a crime. It is often easy to catch hold of poor, young, idle and disorderly children and band them as criminals. We must not forget that many of us as adults too all fall in this category. While adults may find their way out, for children it is not easy to be heard and defended. Children therefore must not be charged on such biases. Behavioural problems of children, such as vagrancy, truancy, runaways and other such acts must therefore be decriminalized.

Children in conflict with law are not criminals. They are victims of their own socio-economic conditions, who more often than not have been denied their right to education, health, shelter, care and protection. They are alleged as /accused of/ recognised as having infringed the penal law.

The innovation the law makes with respect to children in need of care and protection is the conceptualization of restoration of the child as being the focal point, with restoration being conceptualized as restoration to parents, adopted parents or foster parents. (Sec-39). This being the crux, the law then outline four options for children in juvenile homes and special homes which include adoption, foster care, sponsorship and after care.

In this respect the moment the individual enters the institution he or she is classified by the Classification Committee which looks into age, physical and mental status, length of stay period, degree of delinquency and his character...(Rule-21). After the child is classified then the child is subjected to the daily routine prescribed by the rules. Each institution shall have a well regulated daily routine for the inmates which should be displayed and should provide among other aspects, for regulated disciplined life, physical exercise, educational classes, vocational training, organized recreation and games, moral education, group activities, prayer and community singing. (Rule 14)

What is to be noted is that this programme of 'reintegration" is to be carried out with respect to the child in conflict with the law for a period of not less than 2 years with respect to a child who is over 17 and below 18 and for other juveniles till he ceases to be a juvenile. For children in need of care and protection, the child would continue in the children's home if the other measures conceptualized under the Act, like foster care, adoption, sponsorship and after care are not suitable.

7. Bail of Juvenile

When a juvenile in conflict with law for a bailable or a non-bailable offence is apprehended, or detained, or appears, or is brought before a Board, such juvenile shall be released on bail with or without surety but he shall not be so released if there appears reasonable grounds for believing that the release is likely to bring him in association with any known criminal or expose him to moral, physical or psychological danger or that his release would defeat the ends of justice. While replying to bail application, it should not be mentioned that he is hardened or notorious but specific grounds should be mentioned. Similarly, extension of judicial custody remand can not be based on the ground of pending investigation of the case but on the ground that there exists reasonable ground for exposing him to moral, physical or psychological danger in accordance with Rule No. 12.

When such a delinquent having been arrested is not released on bail, the officer causing the apprehension shall cause him to be kept only in an Observation Home for such period during the pendency of the inquiry against him, as may be specified in the order.

8. Process in case of Cognizable Offences

According to Rule 11 (Juvenile Justice Care & Protection of Children Rules – 2007) in dealing with cases of juveniles in conflict with law, matters involving simple offences, the police or the Juvenile Welfare Officer or the Child Welfare Officer from the nearest Police Station shall record information regarding the offence alleged to have been committed by the juvenile in the Daily Diary followed by a report containing social background of the juvenile and circumstances of apprehension and the alleged offence and forward it to the Board before the first hearing.

In case a FIR has already been registered for the punishment of an offence prescribed below seven years and after investigation it came to notice that offence was committed by a juvenile then police has to file a final report, social report, DD entry etc. before the Juvenile Justice Board.

The social report of juveniles in conflict with law shall be prepared on the basis of background, source of income, education, social status etc. of their parents/guardians. Peer group background of the juvenile in conflict with law should be mentioned in the social report. The proforma for preparing Social Report is available in the model rules 2007. This report shall be attached with the records containing details of circumstances of apprehension and the alleged offence.

a) No FIR can be registered if the juvenile is involved in an offence where imprisonment is upto seven years.

(b) If a FIR has already been registered before it is known that a juvenile is involved, the same can be investigated.

(c) FIR can be registered where juvenile is involved along with adult persons can be investigated.

(d) FIR can be registered where there is no age proof but it seems to be a borderline case.

(e) In cases where, after determining the age, the accused is found to be adult, immediately an FIR is to be registered and investigation is to be taken up.

9. Period for Completion of Investigation of Cases

Where a complaint is filed or First Information Report is lodged against a child below the age of 18 years for an offence punishable with imprisonment of 7 years and more, the investigation shall be completed within a period of 3 months from the date of filing of the complaint or lodging of the First Information Report and if the investigation is not completed within this time, the case against the child shall be treated as closed. Investigation is completed only when the charge-sheet/final report is filed in the court concerned. Procedure for completing the investigation has been discussed in Section-167 of the Cr.P.C., which is as follows :

(a) 90 days, where the investigation relates to an offence punishable with death, imprisonment for life or imprisonment for a term or not less than 10 years;

(b) 60 days, where the investigation relates to any other offence.

The charge-sheet with respect to a juvenile should be filed within 90 days where the juvenile is on bail and as per provision of Section-167 of Cr.P.C. where the juvenile is remanded to Observation Home.

10. Offences Committed against Children

Apart from setting out the legal and administrative processes in matters involving children in conflict with law as well as those in need of care and protection, the Juvenile Justice Act of 2000 recognizes certain special offences committed against children by those having the charge of the child or in whose custody/control a child has been placed. This can include parents, school, institutions where children live (including adoption agencies, drop-in-centres, short stay homes or shelter homes etc.) The 'Special Offences' are listed under Sections 23 to 26 of the Juvenile Justice Act of 2000. They relate to the following: Section 23 deals with cruelty against children. Section 24 penalizes use and abuse of children for begging. Section 25 deals with using children for drugs and substance abuse. Section 26 penalizes employment of child labor. Offences covered under Section 24, 25 and 26 are cognizable. The juvenile justice law is silent on whether they are bailable or non-bailable. Schedule I, Part II of Cr.PC deals with classification of offences against laws other than IPC. It very clearly states those offences punishable with imprisonment of 3 years or more but less than 7 years are cognizable and non-bailable and triable by a Magistrate of First Class. Therefore, going by Schedule I Part II of the Cr.PC, with the exception of the offence of cruelty (Section 23 of JJ Act), all other special offences u/s 24, 25 and 26 of JJ Act are to be treated as non-bailable. Indicators of selected forms of abuse of or offenses against children are as below.

TABLE 8.1

Forms of Child Abuse

Physical Abuse	Severe forms	Emotional Abuse	Girl Child Neglect
Sexual Abuse Emotional Abuse Girl Child Neglect Slapping/ kicking Beating with stave/stick Pushing Shaking	Sexual assault- Making the child fondle private parts Making the child exhibit private body parts. Exhibiting private body parts to a child Photographing a child in the nude Other forms : Forcible kissing Sexual advances during travel situations Sexual advances during marriage situations Exposure- children forced to view private body parts Exposure- Children forced to view pornographic materials	Humiliation is the lowering of the self esteem of child by harsh treatment, ignoring, shouting or speaking rudely, name calling and use of abusive language Comparison between siblings and with other children	Lack of the attention to girls as compared to brothers Less share of food in the family Sibling care by the girl child Gender discrimination

(a) Whoever having actual charge of or control over the juvenile, assaults, abandons, exposes or willfully neglects the juvenile or causes or procures him to be assaulted, abandoned, exposed or neglected in any manner likely to cause such juvenile unnecessary mental or physical suffering (Sec. 23 of Juvenile Justice (Care and Protection of Children) Act, 2000).

(b) Whoever employs or uses any juvenile or child for the purpose of causes any juvenile to beg (Section 24 of the Act).

(c) Gives or causes to be given to any juvenile, intoxicating liquor in public place or any narcotic drug or psychotropic substance, except upon the order of a duly qualified medical practitioner or in case of sickness (Sec. 25 of the Act).

(d) Ostensibly procures a juvenile for the purpose of any employment and withholds the earnings of the juvenile or uses such earnings for his own purposes (Section 26 of the Act).

(e) Whenever it comes to the notice of a police officer that a person employs or uses a juvenile for the purpose of begging or causes any juvenile to beg, he shall register a case u/s 24 of the Juvenile Justice (Care and Protection of Children) Act, 2000 which is a cognizable offence. The officers should carefully question all juveniles found begging and trace out the persons who are using these juveniles for the aforesaid purposes.

Punishment for cruelty to juvenile or child : If any person having the actual charge of, or control over, a juvenile or a child, assaults, abandons, exposes or willfully neglects the juvenile or causes or procures him to be assaulted, abandoned, exposed or neglected in any manner likely to cause such juvenile or the child unnecessary mental or physical suffering, he shall be punishable with imprisonment up to 6 months, or fine, or both.

DO'S AND DON'TS FOR POLICE IN JUVENILE JUSTICE

As per central rule 11 (2) The Juvenile Justice (Care and Protection of Children) 2007 when a juvenile in conflict with law is apprehended by the police, the juvenile shall be placed under the charge of the Juvenile or Child Welfare Officer of the police station, who shall produce the juvenile before the Board within twenty-four hours, as per sub section (1) of section 10 of the Act. Where such juvenile or Child Welfare Officer has not been designated, as per provisions laid down under sub-section (2) of Section 63 of the Act, or is not available for some official reasons, the police officer who apprehended the juvenile shall produce him before the Board.

Do's

The basic steps to be followed are :

1. Any child apprehended by the police should be handed over to the Special Juvenile Police Unit (SJPU) or the Designated Police Officer (known as the Juvenile Welfare Officer) as soon as possible.
2. Police officer to inform parent/ guardian and the Probation Officer as soon as possible that the juvenile has been arrested.
3. Record the child's version of the incident. Investigate the child's version. If it comes to notice that the child has come into the conflict situation due to an adult, action has to be initiated against such adult person.
4. If found appropriate and necessary, the child may be apprehended and in that case an Apprehension Memo will have to be duly filled.
5. If apprehension does not seem to be necessary, the child may be released on an undertaking signed by his/her parent/guardian.
6. Get the child medically examined.
7. Personal search of the child may be conducted using proper Format.
8. Extra care should be taken to create an impression in the mind of juvenile that he is not under interrogation as a thief or as a liar but that an effort is being made to find out the truth in order to help him.

9. Display of sympathy and understanding so as to give the impression that the police officer is mainly interested in the welfare or good of the juvenile.
10. Enquiries from the juvenile should be in privacy and in a comfortable environment.
11. It should be ensured that there is no distraction when the juvenile is being questioned.
12. The parents and guardians may be allowed to be present during the questioning of juvenile.
13. Care should be taken not to embarrass the juvenile. Girl juvenile should be handled by Women Officers only.
14. Statement made by a juvenile should be verified from independent sources as there is a possibility of exaggeration and distortion of facts.
15. Physical needs of the juvenile, like hunger or thirst or the need to go to toilet should be properly attended to.
16. Produce the child before the Juvenile Justice Board. If the Board is not sitting, the child should be produced before any individual member of the Board.
17. Prepare and submit a Social Background Report. NGOs/Resident Welfare Associations and Civil Societies shall be associated in the investigation of Juveniles in conflict with law to know the social background details of the juvenile. Similarly, help of a psychologist shall be taken in every case of juvenile in conflict with law before producing final report to the Juvenile Justice Board.
18. Prepare and submit a Final Report/Police Investigation Report to the Board.

Don'ts

While the rights of a juvenile are clearly specified in the principles, objectives and the content of the JJ Act, even in the Cr.PC certain rights of children are recognized vis-à-vis police. These are:

As far as possible, the juvenile should not be called to the police station but should be questioned either at his own home or some other premises which does not give the impression of a Police Station. No bullying should be done during questioning. The officer should not lose temper and make false promises. In no case should the child be kept in police station after the codal formalities have been completed. No juvenile/child shall be handcuffed or fettered (R-42).

Section 160 of the Criminal Procedure Code prohibits the detention of males under the age of 15 years or females of any age for the purpose of investigation or questioning by the police.

Arbitrary detention is illegal. Section 50, 56 and 57 of the Code of Criminal Procedure Code mandate that no person can be detained in custody without knowing the grounds of arrest and that a detainee must be presented before the magistrate within twenty four hours of his arrest.

There can be no application under Section 167 Cr.PC for detention in police custody. No confessions can be taken from a juvenile in the form of a statement signed by a child in front of the police under Section 161 of Cr.PC and such statements cannot be used in evidence against the juvenile.

No proceeding shall be instituted under Chapter VIII of Cr.P.C. (e.g. proceedings u/s 107, 109, 110 CrPC) against a juvenile.

No proceeding shall be instituted against escaped juvenile (S-22). Police Officer to take charge of a juvenile escaped from a special home or an observation home and send him back from where escaped without any criminal proceeding.

No juvenile shall be charged with, or tried for any offence, together with a person who is not a juvenile and who is accused of an offence under Section 223 of the Code of Criminal Procedure, 1973 (2 of 1974) or any other law for the time being in force. The Board taking cognizance of that offence shall direct separate trials of the juvenile and the other persons.

CHILD IN NEED OF CARE AND PROTECTION: GUIDELINE FOR POLICE

The best guideline for police in handling a child in need of care and protection has been prepared by Delhi Police under the guidance of Delhi High Court. Accordingly whenever a juvenile in need of care and protection is found in the jurisdiction of any police station, the following action should be taken :

1. A Police Officer dealing CNCP should strictly follow the provisions of Juvenile Justice (Care and Protection of Children) Act, 2000, Amendment Act 2006 and Rules 2007. He / She should also be conversant with the provisions of The Child Labour (Prohibition and Regulation) Act, 1986 and The Bound Labour System (Abolition) Act, 1976, and rules framed under these Act. He/ She should also make sincere effort to acquire knowledge of other acts/laws/notifications which recognize and protect rights of children.
2. A Police Officer while dealing with CNCP should always be in plain clothes.
3. All basic amenities and food requirements of child should be taken care of by the police officer during the period child remains in his/her charge.
4. A police officer should be fully aware of the Child Welfare Committee (s), its place and days sittings as well as of eh names, addresses and phone numbers of its individual members.
5. A police officer dealing with the CNCP should have with him/her readily available list of various governmental and non governmental organizations (NGO's) working with children in the area, so that necessary emotional and legal support can be provided to CNCP through these organizations. A copy each of said list should also be kept with the SHO and Duty Officer of the concerned police station.
6. A police officer dealing with CNCP should have a list of shelter homes / children homes/fit institution/ child helplines, recognized and certified under the JJ Act (as amended upto date), with him.
7. A list of government hospitals, with pediatric unit, should also be available with the police officer dealing with CNCP, so that necessary medical aid can be provided to a child in custody of such officer, in case need arises.

Procedure to be adopted in case of different categories of children in need of care and protection of law.

(A) Found Child

(This category of children includes street children, child beggars, missing/lost children, homeless children/children with nobody to take care of and runaway children)

(a) Make a detailed DD entry in this regard with full name and mobile number of the concerned police officers.

(b) Try to find out parent(s)/guardian of the child. If possible, a child hailing from NCT of Delhi, should be taken for spot visit to the area where he resides to trace out his parent (s)/ relatives/guardians. A assistance can also be taken from RWA's of the area.

(c) A wireless message with particulars of the child should sent immediately in all the police stations. The record of police station (s) in vicinity of area where child is living should be meticulously checked for missing reports, if any.

(d) Hue and cry notice should be issued and published at the earliest.

(e) Photographs of the child found should be taken for his/her identity.

(f) Particulars of the child must be relayed/telecast on electronic media to seek public help in his / her identification.

(g) The information should be passed on immediately to District Missing Persons Squad as per procedure laid down and efforts should be made to link it with procedure followed by crime branch for missing children.

(h) In case of a child, who hails from state other than Delhi, if the child is able to give some identifiable address then a wireless message should be sent immediately to the concerned SSP/ Police station.

(i) Form of the found child should be filled up.

(j) If it is not possible to find out parent / guardian of child, the police officer concerned should produce the child before the concerned Child Welfare Committee within 24 hours, excluding the journey time.

(k) In case the Child Welfare Committee is not sitting, the child shall be taken to the residence of individual member of the committee, and if none is available then to an appropriate institution of children, registered and certified under the JJAct (as amended upto date). The same institution will produce the child before CWC, with 24 hours, as per Model rule 27 sub rule 6.

(l) If a child in charge of a police officer needs any medical aid; same should be made available to him/her without any delay.

(m) MLC of a child should be prepared only if any physical or sexual abuse or sickness is suspected or stated by child. Any medical or gynecological examination is not a pre-requisite for production of child before CWC as per Model rule 27 sub rule 8.

(n) During the period of the custody of the child, the concerned police officer should take care the food and other basic amenities that are required by the found child.

(o) A Police Officer dealing with children should be fully aware of Child Welfare Committee(s) and the names, addresses, phone numbers etc. of the members of the committee(s).

(p) All expenses incurred for handling CNCP, their care, food and transport etc. shall be claimed under the head of Investigation Expenses'.

(q) In all cases where adults are seen exploiting or abusing CNCP, immediate intervention and protections should be provided to the victim child and intervention of a social worker of the SJPU or a NGO should be sought.

(r) While dealing with any CNCP, the concerned police officer immediately should get in thought with an NGO working for children so that necessary emotional and legal support can be provided to children.

(B) Child Labor (children under 14 years of age)

a) Contact and coordinate with concerned Child line, Deputy Commissioner of Labor and Child Welfare Committee(s).

(b) FIR should be registered under section 14, 16 of The Child Labor (Prohibition and Regulation) Act 1986; section 23,26 of The Juvenile Justice Act, 2000 (as amended upto date); Section 16/17,18 and 19 of The Bonded Labor System (Abolition) Act 1976 as per circumstances.

c) Get the child medically examined and have his/her MLC prepared.

(d) Produce the child before the concerned CWC.

(e) In case provisions of the Bonded Labor System (Abolition) Act, are attracted the should be presented before the concerned SDM to acquire relevant certification of bonded labor. This would include those children who provide labor or service but are not paid remuneration or are paid remuneration which is less than the minimum wage.

(C) Abused Child

(This category of children includes all children who are subjected to any form of physical / emotional/mental harm and includes cases of sexual abuse of children)

(a) On a complaint of a cognizable offence involving a child victim being made, concerned police officer shall record the complaint promptly and accurately.

(b) The investigation of the case shall be made over to an officer not below the rank of Sub-Inspector, preferably a lady officer, sensitized by imparting appropriate training to deal with child victims of sexual crime.

(c) The statement of the victim shall be recorded verbatim.

(d) The officer recording the statement of the child victim should not be in police uniform.

(e) The statement of the child victim shall be recorded at the residence of the victim or any other place where victim can make a statement freely without fear.

(f) The statement should be recorded promptly without any loss of time. As far as possible the statement of the victim child should be recorded in question answer form. When the victim child is of a tender age and is unable to describe/give details of the offensive act, any gesture made by the child, keeping in mind the child's age and capacity, should also be taken note of by the police officer concerned.

(g) The parents of the child or any other person in whom the child reposes trust and confidence will be allowed to remain present.

(h) The investigating officer to ensure that at no point the child victim should come in contact with the accused.

(i) The child victim shall not be kept in the police station overnight on any pretext, whatsoever, including medical examination.

(j) The investigating officer recording the statement of the child victim shall ensure that the victim is made comfortable before proceeding to record the statement and that the statement carries accurate narration of the incident covering all relevant aspects of the case.

(k) In the event the Investigating Officer should so feel the necessity, he may take the assistance of a counselor/psychologist/psychiatrist.

(l) The investigating officer shall ensure that the child victim is medically examined at the earliest preferably within twenty four hours (in accordance with Section 154A Cr. P.C) at the nearest government hospital or hospital recognized by the government.

(m) The investigating officer shall ensure that the investigating team visits the site of the crime at the earliest to secure and collect all incriminating evidence available, under his personal supervision.

(n) The investigating officer shall promptly refer for forensic examination clothing and articles necessary to be examined, to the forensic laboratory which shall deal with such cases on priority basis to make its report available at an early date.

(o) The investigation of the cases involving sexually abused child may be investigated on priority basis and completed preferably within ninety days of the registration of the case. The investigation shall be periodically supervised by senior officer (s).

(p) The investigating officer shall ensure that the identity of the child victim is protected from publicity.

(D) Abandoned Child of any age up to 18 years

(a) Detailed DD entry with full name and mobile number of concerned police officer should be recorded.

(b) The child should be medically examined.

(c) The child be hospitalized if the doctor recommends.

(d) Concerned CWC should be informed with a copy of DD entry and medical report.

(e) After discharge by the hospital the child shall be produced before CWC along with all relevant original papers.

(f) In case of abandoned infant, the child shall be placed in an adoption agency/children home, recognized and certified under JJ Act, 2000(as amended up to date), or pediatric unit of a government hospital followed by production of the child before Child Welfare Committee within 24 hours. (Updated list be circulated at regular intervals).

(E) Surrendered Child

(a) The parent (s) who wants to surrender a child should be produced before concerned CWC.

(b) The rest of the responsibility is that of the CWC. Police Officer's responsibility gets over once the parent (s) and the child is produced before CWC.

JCL and Adult Criminal : Distinctions

The underlying rationales of the juvenile justice system are that youth are developmentally different from adults and that their behaviour is malleable. Rehabilitation and treatment, in addition to community protection, are considered to be primary and viable goals. On the other hand in adult criminal justice system, rehabilitation is not considered a primary goal, which operates under the assumption that criminal sanctions should be proportional to the offense. Deterrence is seen as a successful outcome of punishment. In juvenile justice system limitations are placed on public access to juvenile records because of the belief that juvenile offenders can be successfully rehabilitated, and to avoid their unnecessary stigmatization. Court proceedings may be confidential to protect privacy. In adult criminal justice system open public access to criminal records is required, and all court proceedings are open to the public. The juvenile justice system follows a psychological casework approach, taking into account a detailed assessment of the youth's history in order to meet his or her specific needs. The juvenile offender faces a hearing, rather than a trial, which incorporates his social history as well as legal factors. Law enforcement has the option of preventative detention — detaining a youth for his own protection or the community's protection. Defendants in the criminal justice system are put on trial, which is based largely on legal facts and have the right to apply for bond or bail. A juvenile offender is judged "delinquent" rather than "guilty." Because of the individualized nature of the juvenile justice system, sentencing varies and may cover a wide range of community-based and residential options. The disposition is based on the individual's offense history and the severity of the offense, and includes a significant rehabilitation component. The disposition can be for an unspecified period of time; the court can send a youth to a certain facility or program until it is determined he is rehabilitated, or until he reaches the age of majority. The disposition may also include a restitution component and can be directed at people other than the offender, for example his parents. An adult defendant is found "innocent" or "guilty." The offender is sentenced to a specified period of time which is determined by the severity of the offense, as well as the defendant's criminal history. Following table gives a schematic outline of the distinctions between a JCL and an adult offender.

TABLE 8.2

Distinctions between a JCL and an Adult Offender

Parameter	JCL	Adult Criminal
Age	Below 18 years	Adult
Attitude	Sympathy	Hate
Punishment	Cooperative/suggestive	Punishment/Deterrence
History	Modern	Age-old
Strategy	Spontaneous/Unplanned	Planned/mens rea
Post phase	Special Home	Jail
Police Role	Counseling	Prosecution
Police Uniform	Detection without Uniform	Arrest in Uniform
Trial	Juvenile Board	Judge Court
Handcuff/ Hazzat	NO	YES
Escort from Relations	Required	Not Required
SBR	Required	Not Required
Bail Procedure	Child friendly	Complex
Civil Society Help	NGO Help Sought	Not Required

References

1. Alston, Philip, (1994) The Best Interests of the Child, Clarendon Press, Oxford.
2. Baxi , Upendra, (1982) The Crisis of the Indian Legal System, Vikas Publishing House, New Delhi.
3. Delhi Police (2009) Guidelines for Police Officers of the Special Juvenile Police Unit, First Edition, New Delhi.
4. Feld Barry C. (1999) Bad Kids: Race and the Transformation of the Juvenile Court, Oxford University Press.
5. Goonesekere, Savitri, (1998) Children, Law and Justice, Sage Publications, New Delhi.
6. Mears, David, (2000) Assessing the Effectiveness of Juvenile Justice Reforms: A closer look at the criteria and the impacts on diverse stakeholders, Law and Policy, vol. 22. No. 2.
7. Morill, Calvin et.al., (2000) 'Telling Tales in School: youth culture and conflict narratives' ,Law & Society Review, vol. 34 No.3.
8. Sabnis, M.S, (1996) Juvenile Justice and Juvenile Correction: Pride and Prudence, Somaiya Publications Pvt. Ltd. (Bombay). p.31.

■■■

CHAPTER 9 Collective Deviance

Learning Objectives :

- ❑ **Acquainting the Learners with the criminology of collective deviance.**
- ❑ **To comprehend the statistical, absolutist, legalistic, reactive, group evaluation and synthetic approaches to collective deviance.**

Introduction

The magnitude of social harm inflicted by collective or organizational or group deviance is often far greater than individual acts of deviance. The economic cost of corporate crime is estimated to be about twenty times the cost of street crime. Serial killers may claim a dozen victims, but genocidal state campaigns take millions of lives. Collective deviance discourse illustrates the importance of power and morality in defining deviance. Often the meaning of collective deviance is contested. Rule breakers rationalize their behaviour and often consider it morally acceptable. It is only when there is a shift in the balance of power that the behaviour becomes identified and deviant and sanctioned. The behaviours and rationalizations of organized crime are not much different from that of corporate crime. Terrorist attacks against a nation and its citizens are sometimes quite similar to terrorist attacks by a nation against it's or another nation's citizens. In some cases, there are convergences between various forms of collective deviance. Organized crime figures and networks may intermingle with corporate ones, as well as with terrorist organizations, naxal outfits and gang wars. The same terrorists may operate outside a government at one time and with it at another. A course on organizational deviance will show that there is often a fine line between an organized crime lord and an inventive captain of white-collar crime industry. This chapter focuses on the nature of collective deviance, organized crime, white-collar criminality and recidivism.

Criminology of Collective Deviance

The term "deviance" usually refers to some behaviour that is inconsistent with standards of acceptable conduct prevailing in a given social group, although the term has also been used to designate personal conditions, ideas, or statuses that are stigmatized or disreputable. Collective deviance is a form of categorical thinking, acting, and reacting that constitutes a typification of any and all persons to whom the label applies. Here, deviance is a quality possessed not by an individual but by an entire group or collectivity. To a large extent, criminology and studies of deviance have developed along separate tracks although they show much overlap. Criminologists have typically limited themselves to issues about legality, crime, or crime-related phenomena. Students of deviance, on the other hand, have studied crime as well as a wider range of behaviours or conditions that are deviant by one or another of the definitions reviewed but are not necessarily illegal, such as suicide, alcoholism, homosexuality, mentally disordered behaviours, stuttering, and even such behaviours as public nose picking or flatulence, sectarian religious behaviours, and body mutilation. Hence, it is difficult to distinguish criminology clearly from studies of deviance (Bader et al.1966).

Many criminologists concede that illegal acts are not fundamentally different from legal but deviant acts, except by the fact of illegality itself, which is largely an arbitrary designation by legal functionaries. At the same time, students of deviance readily acknowledge that many deviant acts are also illegal and they find data about crime especially useful because it is more systematic than most data concerning legal forms of deviance. Recognizing this overlap is obvious among those deviance scholars who employ a legalistic definition of deviance, but almost every comprehensive treatment of deviant behaviour, regardless of the definition used, includes a subsection on criminal acts that are also deviant. Furthermore, both camps have raised similar questions and have come to share a common set of theories for explaining the phenomena in their domains. Among other issues, criminologists as well as students of deviance want to explain why the acts they study are deviant or criminal; they want to describe and explain the distribution, frequency, prevalence, and change in the occurrence of various criminal or deviant acts; they want to explain why and how criminal or deviant acts are committed; they want to explain how social groups manage and respond to crime and deviance and how people who are accused or guilty of crime or deviance respond to being accused or managed; and they want to understand how criminal or deviant phenomena affect and are affected by other aspects of social life.

Because of the overlap between crime and deviance, some scholars regard distinctions between criminology and deviance studies as false and counterproductive, and they have called for a merger of the subject-matters. Since all definitions of deviance, except the legalistic one, portray deviant behaviour as a more inclusive concept, merger might imply subsuming criminology under the umbrella of deviance studies. Under that conceptualization, criminal behaviour would be treated as a special case of deviant behaviour—that which is prohibited in law, thereby meriting the possibility of officially imposed sanctions that legal forms of deviance escape. On the other hand, some contend that criminology has already preempted deviance studies so that deviance as a separate subject matter no longer exists or matters (Sumner, 1994), and still others contend that the two fields should clearly differentiate themselves by allocating legally related phenomena exclusively to criminology, leaving other forms of disapproved behaviour to deviance studies (Bader et al.1966).

There are two main intellectual barriers to merging criminology with studies of deviance. First, some criminal behaviour in some places (such as gambling) is not deviant, at least by most definitions of deviance, so would be subject to criminological study but not to study by students of deviance. The political processes that produce laws can result in behaviours being declared illegal although the conduct is not deviant by any definition except a legalistic one. How those political processes unfold and how enforcement might fare when illegality does not match social reality are of great importance to some criminologists. In addition, some behaviours are deviant at the time they become illegal but later become non-deviant without the law having been changed. Therefore, those intellectually opposed to merger might note that subsuming criminology under deviance studies would exclude some of the more interesting aspects of criminology.

A second intellectual obstacle to conceiving of crime and deviance as one subject matter hinges on disagreements about the nature of human behaviour. Some criminologists and some students of deviance contend that all forms of behaviour, including specific acts or types of crime or deviance, are more or less distinct, requiring unique explanations. By observing a plethora of differences, such as whether committed by males or females, blacks or whites, young or old, in some circumstances rather than others, whether legal or not, whether regarded as especially serious or not, whether violent or simply immoral, whether committed with planning or spontaneously, and the like, some conclude that specific behaviours have few similarities that would justify their being explained by the same theories. This orientation has generated a large number of explanations of specific behaviours, such as predatory crime (Braithwaite, 1989), common juvenile misdeeds (Hagan, 1989), embezzlement (Cressey,1953), mental illness (Scheff,1984), homicide by females (Ogle et al. 1995), and many others.

Challengers, however, contend that various deviant behaviours are only superficially different because similar underlying causal processes operate in most if not all forms of deviance. They often use the analogy of focusing so closely on individual trees that the forest is overlooked. In addition, those who advocate general theory contend that it is necessary in order to achieve scientific goals of explanation and prediction based on assumptions about unity in nature; and that general theory is more parsimonious because specific accounts already overlap, often in unrecognized ways. The generalist orientation has led to a number of theories that aim to account for wide ranges of deviance in a variety of circumstances. These general accounts usually take one of two forms.

One form of general theorizing assumes a universal causal process that generates different forms of deviance under different conditions. The theories attempt to identify that causal process and specify the conditions under which it produces one form of deviance rather than another (Akers, 1985; Gottfredson and Hirschi, 1990; Agnew, 1992; Tittle, 1980). A second approach to general theory assumes that different causal processes operate at different times and under different conditions. The theories merge several causal processes by specifying when or why one or another will come into play to produce a given form of deviance at a specific point in time or in a given circumstance (Braithwaite, 1989; Elliott et al. 1985; Thornberry, 1987).

The generalist orientation clearly implies that crime and deviance are one entity. However, no general theory has yet been generally accepted as better than specific, focused accounts. Therefore, debates concerning the intellectual benefits of general theory and the relative advantages of differentiating criminology and studies of deviance will continue.

The main barriers to conceptualizing deviance and crime as a single entity are not intellectual, however—they are ideological and practical. Criminologists claim a more central role in addressing issues of acute public concern and they position themselves more favorably to receive government funds for research and to offer useful advice for controlling crime. Indeed, criminology has traditionally been identified with practical concerns. In addition, because they regard many forms of deviant behaviour that are not illegal as unworthy of serious attention, some criminologists resist identity as students of deviance. Finally, by focusing on illegal conduct and limiting themselves to modern, politically, and geographically demarcated societies, criminologists avoid some of the problems encountered by students of deviance who struggle to identify group boundaries, to measure opinions of group members, or to document disapproval of various behaviours. Students of deviance, on the other hand, often resist the idea that they are also criminologists, maintaining that criminological study is too narrowly focused on behaviours arbitrarily designated by simple acts of legislative bodies. By so limiting itself, criminology ignores a large domain of phenomena crucial for understanding human behaviour and social organization. Some students of deviance also regard criminology as a handmaiden of government, committed to preventing acts that are contrary to the interests of powerful groups who influence the content and enforcement of the criminal law.

Perhaps the major practical barrier to criminology and studies of deviant behaviour becoming one subject-matter is the tendency of criminologists to orient themselves concretely while students of deviance are relativistically oriented. Assuming that concepts of serious illegality are largely invariant across time and place (Wellford,1975), most criminologists have focused on explanations of criminal behaviour or its distribution, with only a minority showing concern with explaining variations in the law. Students of deviance, however, have traditionally emphasized variations in social attributions of deviance and have mainly tried to understand why something is or is not deviant in a given place and time, with only a minority being concerned with explaining behaviour itself.

Approaches to Collective Deviance

Social scientists disagree, about a precise definition of deviance because they use different approaches in trying to determine exactly what the standards of conduct or the acceptable statuses and conditions are in a given group (Gibbs, 1981). At least five ways of conceptualizing deviance are used.

(a) The Statistical Approach : One way of defining standards of conduct and deviance from them is to observe how people in a particular group actually behave (Wilkins, 1964). Accordingly, if a large proportion of people in a group smoke cigarettes, smoking is "normal" while failure to smoke would be atypical, or deviant. With a "statistical" perspective, sky diving, eating snails, and murder are all deviant since they are all unusual. On the other hand, highway speeding, "fudging" on one's income tax, and pilfering small items from employers are all conforming behaviours since they are currently committed by a fairly large proportion, if not a majority, of the population.

A statistical approach rests on a common observation: many people assume that something that "everybody does" cannot be in violation of conduct standards. Indeed, a frequent justification (or rationalization) for conduct that is being threatened with sanctions is to claim that everybody else or at least most others do it. And, many people decide what is appropriate by watching what others do.

Even though a statistical approach appears to correspond with the everyday thinking of many laypersons, it is not widely used by social scientists. Scholars have found that statistical patterns only superficially reflect how social groups formulate standards of conduct. Most people in the United States, for example, would feel uncomfortable classifying behaviours like church tithing, abstinence from all alcohol use, and maintenance of premarital sexual virginity (all atypical behaviours) as "deviant." There is an underlying agreement that even though they are unusual, these behaviours are somehow "good," acceptable, even desirable, and that they are to be encouraged as ways of behaving. Similarly, most people feel discomfort in thinking of adultery, lying to one's spouse or sweetheart, or pilfering from an employer as appropriate, even though they are frequently committed by a large proportion of the populace. Yet, most can easily endorse the inappropriateness of acts that are atypical in a negative way—more evil, unacceptable, or undesirable than the average. Therefore, most students of deviance contend that despite some tendency for people to refer to statistical guidelines, normative standards mainly revolve around notions of rightness and wrongness or with what people think others "ought" to do.

(b) The Absolutist Approach : A second approach applies ideal conduct standards set down by a social scientist (or group of social scientists) to all groups and individuals under study. A social scientist decides what is good, useful, or just, and then measures deviations from those evaluative criteria. For example, some theorists (functionalists) view societies as interdependent mechanisms; all parts that work together for maintenance of the society are regarded as essential and in that sense "good" or non-deviant. But a society may contain dysfunctional (dangerous or destructive) elements (Gross,1959), which are regarded as deviant. Most who use this approach assume that societies will usually condone inherently good behaviours and condemn those that are inherently bad. Indeed, it has been argued that contemporary societies exist because throughout evolutionary history they practiced and condoned useful behaviour while avoiding and condemning dangerous behaviour. Presumably, social groups that failed to do this did not survive the ravages of time (Parsons,1951).

Functionalists assume that an investigator can, through logic and research, actually determine what is good for a society. For example, incest is thought to be dysfunctional (Davis,1950; Murphy,1980) because if widely practiced, it could lead to biological deterioration of the population, destruction of orderly social relations, and disruption of the mechanisms for efficient child rearing. According to some, therefore, incest is inherently and obviously deviant because it is socially dangerous. Most members of any existing society presumably will disapprove of incest and will refrain from practicing it because only those societies that in the past developed and enforced social rules prohibiting incest would have survived to be represented in the contemporary world. Similar arguments can be made for murder, rape, assault, homosexuality, child abuse, mental illness, and other behaviours.

Other absolutist thinkers employ a different rationale. Radical, Marxist, and humanist scholars often maintain that a sensitive informed researcher can apply absolute moral standards to behaviour in any given society or specific situation to decide whether various activities are unjust or evil (deviant) (Schwendinger and Schwendinger, 1983). Some believe that any exploitation of one person or a category of persons for the benefit of another, or any conduct that threatens the dignity and quality of life for specific people or humanity as a whole, is inherently evil, and thereby deviant (Simon, 1999). Others contend that any behaviour causing people to suffer or violating any person's right to self-actualization and freedom is inherently immoral, or deviant (Platt). Marxist scholars, for instance, point up the exploitative nature of economic relations in capitalistic societies and regard this inherent exploitation, along with the selfish and insensitive acts it breeds, as deviant or "criminal" because it corrupts human qualities (Bonger,1916; Quinney, 1970). Similarly, humanists regard racial discrimination as deviant because it deprives a whole group of people of equal rights and human dignity.

Absolutist approaches to deviance are not widely used because of their subjectivity. Trained (sensitized), careful observers disagree about what is good or bad for society, what is contrary to human dignity, and what is fair or unfair. And what one observer believes to be functional for a society another may find dysfunctional. It has even been argued that a certain amount of deviance itself may benefit society. Dealing with deviance may help a group differentiate its members, crystalize the norms so group members will know how to behave, provide a means for tension to be reduced, keep the mechanisms of social control in good working order so that they will be efficient in true emergencies, and generate cohesion as the members of a group unite in opposition to deviance (Cohen,1966; Dentler and Erikson,1959). Since there is no reason to believe that the values or insights of social scientists are in any way superior, more desirable, or defensible than those of anyone else, or that the values of any particular social scientist are any more justifiable than those of another social scientist, an absolutist approach in the study of deviance cannot be used in a consistent and meaningful way.

(c) The Legalistic Approach : A third way of identifying conduct standards and deviant behaviour is to simply use illegality as the criterion of whether a given activity is in violation of behavioural norms. Accordingly, if the law prohibits an act, it is deviant; and if the law requires an act, failure to perform it is deviant. If the law is silent about, or permits, an act, then that act is considered consistent with conduct standards, or conforming.

The rationale for a legal criterion of deviance differs depending on how the law is viewed. Some contend that the law expresses collective sentiment indicating that particular activities are dangerous or threatening enough to require efforts at control (Tittle,1994). Although recognizing that law making is a political process, which often reflects conflicting interests and the clash of power, some nevertheless believe that, in the main, law reflects popular sentiment as well as efforts to promote the public good.

Others see the law as an instrument by which the powerful maintain their elite positions and protect their privileges, but these scholars accept illegality as the appropriate criterion of behavioural standards because they believe conformity and deviance are inherently whatever people powerful enough to impose their own views say they are. Accordingly, norms (or conduct standards) are "definitions of the situation" imposed upon a social group by power elites (McCaghy, 1976; Quinney, 1970).

Still others view law as a combination of popular sentiment and elite desires. They contend that some law expresses consensus among the population (such as laws prohibiting assault, murder, child abuse) while other laws reflect the desires of special interests (such as laws prohibiting importation of competitive products, requiring licenses to provide certain services, or denying the right of laborers to strike). For these scholars, whether law is collectively or particularistically oriented is irrelevant; in both instances, law expresses coercive potential—a key element of behavioural norms. Thus, the conduct rules that matter are those that can be enforced. Since the law reflects behavioural rules backed up by power of enforcement, it necessarily encompasses conduct norms—at least those worth social scientists' study.

A legalistic approach is straightforward and usually easily applied (since a scholar need only refer to the codification of laws), and it hinges on an extremely important element of social life—the exercise of political power. Moreover, a related body of inquiry—criminology—almost exclusively uses a legalistic approach to define its subject-matter. Yet the legalistic conceptualization is not generally used for the larger study of deviance, of which crime may be a part. For one thing, not all societies have a clearly defined body of written statutes that can be identified as law. Primitive societies, for example, have deviant behaviour but no formally written law that defines it as such (Malinowski, 1926; Hoebel, 1968). Using a legalistic approach in non-literate societies assumes the resolution of a prior definitional problem—what is law?

In addition, in any society many of the activities that are illegal nevertheless appear to be normatively acceptable. For instance, despite legal prohibitions on the sale of tobacco products to minors, in many places such products are easily available to minors in vending machines and minors are officially allowed to smoke in designated areas by many schools. Moreover, it is rare for the police to arrest anyone for selling tobacco to minors; and most people may not regard such sale as bad, dangerous, or abnormal (although public opinion about this appears to be undergoing a change). Thus, laws are often out of synch with actual behaviour and public sentiment, sometimes because society changes without the laws being changed or repealed and sometimes because laws result from political action by interest groups whose agendas may not correspond with views of the general public.

By contrast, many activities that appear to be inconsistent with general conduct standards are not illegal. It is not a crime to lie to one's spouse or sweetheart although evidence suggests that a large proportion of the people disapprove of it (Tittle,1980). Similarly in some places it is not illegal to operate a topless bar, yet such establishments frequently meet with scorn, protests, and sometimes violent opposition by neighbors. Further, legal statutes rarely prohibit eating human flesh although it is clearly outside the bounds of acceptable conduct in contemporary societies.

Finally, despite the importance of the legal realm, conduct norms are not limited to that realm but instead are ubiquitous at all levels of social life from the interpersonal to the societal. A legalistic approach, therefore, narrowly focuses attention on one tier of a multi-tiered system.

(d) The Reactive Approach : A fourth way of defining deviance is by social reaction (what people do about behaviour or a condition). According to this approach, when social reaction to some behaviour is condemnatory, punitive, or simply disapproving, it indicates that the behaviour is in violation of behavioural standards prevailing in that group and is therefore deviant. One variation of the reactive approach emphasizes the "typical" reaction to a class of behaviours. Another stresses social reaction to particular instances of behaviour while assuming that this particular reaction implies nothing about the deviance of the entire class of behaviours of which the particular case is an instance. Such basic differences are complicated by questions concerning which part of the social system must react negatively to qualify something as deviant. Some emphasize negative reactions by official agents and functionaries, but others accord more importance to informal reactions by a collective social audience.

Probably the best-known reactive approach to deviance is that embodied in the "labeling perspective" (Tannenbaum,1938; Becker,1963;Gove,1980). Labeling theorists do not agree about their focus, and they are sometimes ambiguous in presentation. Nevertheless, most scholars agree that the predominant concern of the labeling perspective is legal reaction to specific acts, particularly the reaction of agents of the criminal justice system (such as police) to particular instances of behaviour disapproved by power holders, whose interests are embodied in the legal codes (Gove, 1980). Accordingly, some labeling proponents recognize no categories or classes of deviant behaviour. To them, murder, rape, child abuse, or smoking marijuana are not necessarily deviant. Rather, specific acts of murder, rape, child abuse, or marijuana use may or may not be deviant depending upon whether officials arrest the perpetrator and label him/her as deviant. When labeling occurs, the particular act of murder, rape, child abuse, or marijuana use is deviant; otherwise, it is not.

Some labeling theorists are more restrictive in what they define as deviant. They maintain that an act is not deviant unless and until a collective social audience has accepted the label of deviant for the act and/or the perpetrator. It is said that a label attached by officials must "stick"; that is, it must be recognized by a social audience and serve as the vehicle through which the group attributes bad character to the actor, or attributes badness to the act (Becker, 1963; Kitsuse,1962).

Another variant of the reactionist approach accepts the generally deviant nature of some categories of behaviour. Accordingly, if a social collectivity or its chief representatives typically react negatively to some behaviour or typically attach a stigma to those who are caught, that kind of behaviour is deviant even if a particular perpetrator escapes being labeled. For instance, if a social group usually shows its condemnation of some class of behaviours by punishing specific acts of that class or if it usually attributes bad character to those who commit such acts, then theorists would assume that in that social group those behaviours are deviant. Using this approach, apprehension, punishment, or group attribution of bad character for a specific act may be treated as problematic but such actions are not essential for a given act to be deviant.

Still another variation of the reactive/labeling approach does not limit itself to behaviour but includes as deviance, statuses, states of being, or physical conditions (Lemert, 1951). According to this approach, "deviance" inheres in disreputable or pejorative statuses, styles of life, or physical attributes; therefore, deviance is a condition not a behaviour or set of behaviours. Accordingly, race, gender, poverty, and criminality have all been identified as stigmatized "deviant" statuses while physical handicaps, unattractive appearances, speech impediments, and small statures have been designated as stigmatized conditions qualifying as deviant. Although most approaches to deviance raise issues about how and why particular acts or categories of acts come to be deviant and why individuals come to commit acts that are deviant or that are likely to be regarded as deviant, this particular approach poses different questions—how stigma comes about and how people who are targets of stigma react to, or manage, the deviant identity associated with it.

The reactive approach has been widely used, and during the 1960s and 1970s was the dominant orientation. Even now many people associate the study of deviance exclusively with the reactive approach. This popularity stemmed partly from the key idea that deviant acts or conditions are those disapproved, either by officials or by group members, or in one version of this approach, are considered important enough by key functionaries of the social system to warrant attention. The reactive approach was also popular because of the appeal of the larger labeling argument, from which many of the reactive conceptualizations sprang. The notion that deviant behaviour and stigmatized conditions are highly problematic intrigued many. Moreover, the contentions of the labeling school that deviance is a creation of the very forces intending to do something about it and that designating acts as deviant represents unequal power in action, with large consequences for those labeled and for society, meshed with the ideological bent of many social scientists of the time.

In due course, however, numerous problems caused the reactive approach to lose its dominance. Among other concerns, many behaviours that appear to be in violation of conduct standards are not necessarily the focus of negative reactions, especially not official reactions. For instance, adultery in the United States is almost never dealt with by the police or the courts, and it rarely results in a deviant label for the actor. Even when there are citizen complaints, the police normally refuse to make arrests. Yet surveys show that most people believe adultery to be wrong, bad, or inappropriate, and it is thought by many social scientists to have crucial implications for a major social institution—the family.

Furthermore, the police frequently arrest people, and courts sometimes impose severe penalties, for behaviours that are not widely disapproved and do not seem to have much impact on society (such as marijuana use during the 1960s). Third, the narrower reactive definitions created unusual conceptual inconsistencies. If murder is regarded as deviant only when discovered and reacted against by official agents, then it must also be regarded as conformity if the perpetrator escapes punishment—regardless of the number of group members who may disapprove of it, how socially dangerous it might be, or how many people do it. Moreover, according to this kind of reactive definition, deviance can only be studied on a case-by-case basis after the fact, thereby making generalization or consideration and explanation of categories of deviance impossible.

Finally, practical application of some reactive definitions has proven difficult. Empirically ascertaining when labeling has occurred (is an arrest a labeling act, or must one be convicted?) or when a label sticks (how do we know the boundaries of a social audience, how many must accept the label, what constitutes stigma ?) renders some versions of the reactive approach almost completely unworkable (Gove, 1980). Furthermore, those reactive definitions that focus on disreputable statuses and conditions raise limited questions leading to repetitious research about the importance of power in the stigmatization process and about the way various stigmatized people manage their identities. In addition, evidence and logical critiques have eroded the luster of labeling theory, which had earlier energized reactive approaches (Gibbs, 1966; Gove, 1980; Hagan, 1973; Mankoff, 1971; Wellford, 1975).

(e) The Group Evaluation Approach : A fifth method of identifying conduct standards and deviance is by the beliefs or opinions of group members. Accordingly, deviant behaviour is that regarded as unacceptable, inappropriate, or morally wrong in the opinion of the members of a group.

One problem is deciding how many people in a group must believe some behaviour to be unacceptable for it to qualify as deviance, although it is generally assumed that a significant consensus exists among group members about rightness or wrongness and about how people ought to behave. If this assumption is correct, social disapproval indicates that some behaviour is outside acceptable standards of conduct. And such disapproval would suggest deviance regardless of the typicality or prevalence of the behaviour, whether anything is actually done about the offense, or whether the shared concepts about rightness or wrongness grew out of common experiences or out of careful socialization by particular interest groups with an investment in promoting specific ideas.

A group evaluational definition of deviance has never been the dominant approach although it is conceptually clear, has intuitive meaning, permits deviance to be treated meaningfully as a continuous variable expressing the extent to which different behaviours are disapproved, and provokes many sociologically relevant questions—such as why some people do disapproved things, why and how social processes sometimes produce shared opinions of disapproval, why general opinions sometimes change, and why social processes sometimes come to activate collective responses to behaviour but not at other times. Perhaps the major reason for under use of the group evaluational approach is its assumption that all group members' opinions are of equal value. Research shows that the opinions of some are more important than the opinions of others because some people can implement their opinions with coercive action, and some are more influential in persuading others to their points of view. In addition, some question the usefulness of a conceptualization of deviance that treats as a violation of conduct standards some acts about which nothing is done and for which no sanctions are brought to bear. Furthermore, efficient application of this approach requires survey data—public opinion information about perceptions of conduct. Such information is rarely available for all the acts that one might want to consider as potentially deviant, and in some societies no survey data about any behaviour are available. Finally, this approach assumes that the boundaries of groups are clear enough to permit scholars to ascertain the thoughts of most people within those boundaries about the appropriateness of various behaviours. In reality, however, group boundaries are often ambiguous, particularly in a heterogeneous society with overlapping subcultures and diverse group identities.

(f) The Synthetic Approach : Some scholars merge different definitions of deviance in an effort to produce more effective, though more complicated, "integrated" approaches. For example, one synthetic definition (Tittle and Paternoster, 2000) combines the reactive and the group evaluation approaches because they pose especially salient questions about social behaviour—do most people believe it is wrong and are there usually negative sanctions associated with the behaviour ? Deviance is defined as any type of behaviour that the majority of a given group regards as unacceptable or that typically evokes a collective response of a negative type. In this definition, "unacceptable" means the behaviour is disapproved, or regarded as wrong, inappropriate, bad, or abnormal—in short, behaviour that a group evaluates negatively. "Majority" means that over half of the people in a specified, bounded group regard the behaviour as unacceptable. Although this is an arbitrary cutoff point, it is used because the main objective is to array behaviour on a continuum from very non-deviant (hardly anybody considers it unacceptable) to very deviant (almost everybody in the group considers it unacceptable), with any particular act falling somewhere on the continuum. Collective response of a negative type implies that the majority of the specified group typically does something to express its displeasure, or the officials who possess coercive power over the members of the group typically respond to the behaviour in a way that expresses negative evaluation. Whether synthetic, integrated conceptualizations become widely accepted remains to be seen.

Recapitulation

Deviance generally implies pejorative departures from social standards of one kind or another. However, there are many conceptualizations of deviance and how it is conceptualized affects what and how deviance is studied. One key unifying theme is the relativity of deviance. Because much deviance is also illegal, there is great over-lap between criminology and deviance studies. Some contend that the two areas of study should be regarded as one, but there are intellectual and practical barriers to such a merger. A related issue concerns whether various forms of deviance can be explained with large general theories or requires special ad hoc accounts. Much work remains to be done before the key questions about collective deviant behaviour can be answered.

References

Agnew, R, (1992) "Foundation for a General Strain Theory of Crime and Delinquency." Criminology 30: 47–87.

Akers, R. L, (1985)Deviant Behaviour : A Social Learning Approach. 3d ed. Belmont, Calif.: Wadsworth.

Bader, CB, (1996) Paul J.; and Desmond, S, "Reclaiming Deviance as a Unique Course from Criminology." Teaching Sociology 24: 316–320.

Becker, H. S, (1963)Outsiders : Studies in the Sociology of Deviance. New York: Free Press.

Bonger, W A, (1916)Criminality and Economic Conditions. Translated by H. P. Horton. Boston: Little, Brown, and Company.

Braithwaite, J, (1989)Crime, Shame, and Reintegration. New York: Cambridge University Press.

Cohen, A K (1966)Deviance and Control. Englewood Cliffs, N.J.: Prentice-Hall.

Cressey, D R, (1953)Other People's Money. New York: Free Press.

Davis, K, (1950) Human Society. New York: Macmillan.

Dentler, R A., and Erikson, K T. (1959)"The Functions of Deviance in Groups." Social Problems 7: 98–107.

Elliott, D S, Huizinga, D; and Ageton, S S. (1985) Explaining Delinquency and Drug Use. Newbury Park, Calif.: Sage.

Erikson, K T. (1966) Wayward Puritans: A Study in the Sociology of Deviance. New York: Wiley.

Gibbs, J P. (1966)"Conceptions of Deviant Behaviour: The Old and the New." Pacific Sociological Review 9: 9–14.

Gottfredson, M R., and Hirschi, T (1990)A General Theory of Crime. Stanford, Calif.: Stanford University.

Gove, W R., ed. (1980) The Labelling of Deviance: Evaluating a Perspective. 2d ed. Beverly Hills, Calif.: Sage.

Gross, L, ed. (1959) Symposium on Sociological Theory. White Plains, N.Y.: Row-Peterson.

Hagan, J, (1973)"Labelling and Deviance: A Case Study in the 'Sociology of the Interesting." Social Problems 20:447–458.

Hoebel. E. Adamson. (1968) The Law of Primitive Man. New York: Atheneum.

Kitsuse, John I. (1962) "Societal Reactions to Deviant Behaviour: Problems of Theory and Method." Social Problems 9: 247–257.

Lemert, E M. (1951)Social Pathology. New York: McGraw-Hill.

Malinowski, B, (1926)Crime and Custom in Savage Society. London: Routledge and Kegan Paul.

Mankoff, M. (1971)"Societal Reaction and Career Deviance: A Critical Analysis." The Sociological Quarterly 12: 204–218.

Mccaghy, C H. (1976) Deviant Behaviour: Crime, Conflict, and Interest Groups. New York: Macmillan.

Murphy, Robert F (1980) The Dialectics of Social Life: Alarms and Excursions in Anthropological Theory. New York: Columbia University Press.

Ogle, R, Maier-Katkin, D, and Bernard, T S. (1995) "A Theory of Homicidal Behaviour among Women." Criminology 33:173–193.

Parsons, T, (1951)The Social System. New York: Free Press.

Quinney, R, (1970) The Social Reality of Crime. Boston: Little, Brown.

Scheff, T J, (1984)Being Mentally Ill: A Sociological Theory. 2d ed. New York: Aldine de Gruyter.

Schwendinger, H, And Schwendinger, J (1983) Rape and Inequality. Beverly Hills, Calif.: Sage.

Simon, D R. (1999)Elite Deviance. 6th ed. Boston: Allyn and Bacon.

Sumner, C (1994)The Sociology of Deviance: An Obituary. New York: Continuum.

Tannenbaum, F (1938)Crime and the Community. Boston: Ginn.

Thornberry, T P (1987) "Toward an Interactional Theory of Delinquency." Criminology 25:863–892.

Tittle, C R. (1980)Sanctions and Social Deviance: The Question of Deterrence. New York: Praeger.

Tittle, C R., and Paternoster, R (2000) Social Deviance and Crime: An Organizational and Theoretical Approach. Los Angeles, Calif.: Roxbury.

Wellford, C F (1975) "Labeling Theory and Criminology: An Assessment." Social Problems 22:335–347.

Wilkins, L T (1964) Social Deviance, Social Policy, Action, and Research. Englewood Cliffs, N.J.: Prentice-Hall.

■■■

CHAPTER 10

Organized Crime

Learning Objectives :

- **To Understand the meaning, characteristics and the origin of organized crime.**
- **To comprehend the types of organized crime.**
- **To discuss Vohra Committee Report.**
- **To be acquainted with the control strategies of organized crime.**
- **To grasp the response of law enforcement to organized crime.**

Introduction

Organized crime refers to any continuing unlawful activity by an individual, singly or jointly, as member of an organised crime syndicate, by use of violence or threat of violence or intimidation or coercion or other unlawful means, with the objective of gaining pecuniary benefits, or gaining undue economic or other advantage for himself or for any other person. Organized crime or criminal organizations are groups or operations run by criminals, most commonly for the purpose of generating a monetary profit. Organised crime is the product of a self-perpetuating criminal conspiracy involving the ruthless exploitation of the social, political and economic institutions of the society. The phenomenon thrives on muscle power and corruption; fear caused by the muscle power and corruption and by the illegal money generated by the organised crime itself. Very appropriately organised criminality has been described as a parasite like the tapeworm, organised crime being to the society what a tapeworm is to the alimentary canal in human beings. Just as each segment of the tapeworm parasite is an independent, viable entity and is capable of breeding independently of the parent, a similar process is discernible in the working of the criminal organisation at the various hierarchical levels. Organized crime as a concept is to some extent synonymous with certain historical organization such as MAFIA and CAMORRA.

MAFIA and CAMORRA are the old Italian families working in USA. Its activities spread by intimidation, bribery and they controlled Naples. They had a quasi-police organization to counter opposition. In 1880 efforts were made to crush them and culminated in 1911 murder trial at which numerous members were imprisoned or hanged. The word 'Mafia' is of Sicillian origin and it is difficult to assign any precise meaning to it. In the USA it was originally used to refer to organised criminals migrating from Sicily and mainland Italy. The Mafia (also known as Cosa Nostra) is a criminal syndicate that emerged in the mid 19th century in Sicily. In the United States, "the Mafia" generally refers to the Italian American Mafia. Other organizations described as mafias include the Russian Mafia, the Irish Mob, the Chinese Triads, the Japanese Yakuza, the Neapolitan Camorra, Calabrian 'Ndrangheta, and the Apulian Sacra Corona Unita. There are also a number of localized mafia organizations around the world bearing no link to any specific ethnic background. Since1830, it is a loose association of criminal groups that share a common organizational structure and code of conduct. Each group, known as a "family", "clan", or "cosca", claims sovereignty over a territory in which it operates its rackets – usually a town or village or a neighborhood (borgata) of a larger city. Offshoots of the Mafia emerged in the United States during the late 19th century following waves of Italian emigration as well as in Canada and Australia. The Camorra is a Mafia-type criminal organization, or secret society, originating in the region of Campania and its capital Naples in Italy. It is one of the oldest and largest criminal organizations in Italy, going back to the 18th century. In 1922 Mussolini took over and Camorra was crushed.

Now the word Mafia is used, almost universally, for all organised criminality. It was mainly bootlegging. Legendary figures like Al Capone headed Chicago mafia after Torrio was seriously wounded in an assassination attempt and stepped down in 1925. Throughout his reign he ran the streets of Chicago with his mob. Capone had city aldermen, mayors, legislators, governors, congressmen, and over half the Chicago police force on his payroll. The film "God Father" gives a deep view of such gang. But there are many rich Americans with background of organized crime. John Jacob Aster made millions by capturing fur trade from Red Indians with unparallel brutality. He used the money for construction of houses and estate business and collected huge rent by force. He became richest man of America in 1930s. Henry Ford propounded a philosophical explanation for the association of business and crime as "There is something sacred about big business. Anything which is economically right is morally right". He employed Harry Benett as "Director Services" who employed a gang of criminals and ex-convicts. The workers in Ford plant were terrorized by them and any Union activities were ruthlessly crushed. The Organized Crime Control Act (U.S., 1970) defines organized crime as "The unlawful activities of .... a highly organized, disciplined association...".

Some criminal organizations, such as terrorist organizations, are politically motivated. Gangs sometimes become "disciplined" enough to be considered "organized". An organized gang or criminal set can also be referred to as a mob. Some criminal organizations, such as terrorist organizations, are politically motivated. They are known as Violent non-state actor (VNSA). The act of engaging in criminal activity as a structured group is referred to in the United States as racketeering. In the U.S., organized crime is often prosecuted federally under the Racketeer Influenced and Corrupt Organizations Act (RICO), Statute (18 U.S.C. Part -I).

Origin and Conceptual Background

"If we take a global rather than strictly domestic view, it becomes evident even crime of the organized kind has a long if not necessarily noble heritage. The word 'thug' dates to early 13th-century India, when Thugs, or gangs of criminals, roamed from town to town, looting and pillaging. Smuggling and drug-trafficking rings are as old as the hills in Asia and Africa, and extant criminal organizations in Italy and Japan trace their histories back several centuries..." (Sullivan:2002).

"Piracy and banditry were to the pre-industrial world what organized crime is to modern society" (Lunde : 2004). Today, crime is thought of as an urban phenomenon, but for most of human history it was the rural world that was crime-ridden. Pirates, highwaymen and bandits attacked trade routes and roads, at times severely disrupting commerce, raising costs, insurance rates and prices to the consumer. As Lunde states, "Barbarian conquerors, whether Vandals, Goths, Norsemen, Turks or Mongols are not normally thought of as organized crime groups, yet they share many features associated with successful criminal organizations. They were for the most part non-ideological, predominantly ethnically based, used violence and intimidation, and adhered to their own codes of law" (Ibid).

Although medieval feudal lords were not usually engaged in what moderns would consider "criminal activities" (except for irregular robber barons, self-enthroned Viking adventurers, and mercenary "free company" leaders), their hierarchical courts, monopoly of violence, extension of protection to their serfs in exchange for labor and a percentage of harvests and durability are structurally similar to classic organized crime groups like the Mafia. In the modern world, it is difficult to distinguish some corrupt and lawless governments from organized crime gangs. These regimes, characteristic of some of the newly independent states of the former Soviet Union, use the state apparatus to control organized crime for their own ends.

Organized crime most typically flourishes when a central government and civil society is disorganized, weak, absent or untrusted. This may occur in a society facing periods of political, economic or social turmoil or transition, such as a change of government or a period of rapid economic development, particularly if the society lacks strong and established institutions and the rule of law. The dissolution of the Soviet Union and the Revolutions of 1989 in Eastern Europe that saw the downfall of the Communist Bloc created a breeding ground for organized criminal organizations.

The newest growth sectors for organized crime are identity theft and online extortion. These activities are troubling because they discourage consumers from using the Internet for e-commerce. E-commerce was supposed to level the playing ground between small and large businesses, but the growth of online organized crime is leading to the opposite effect; large businesses are able to afford more bandwidth (to resist denial-of-service attacks) and superior security. Furthermore, organized crime using the Internet is much harder to trace down for the police (even though they increasingly deploy cybercops) since most police forces and law enforcement agencies operate within a local or national jurisdiction while the Internet makes it easier for criminal organizations to operate across such boundaries without detection.

In the past criminal organizations have naturally limited themselves by their need to expand. This has put them in competition with each other. This competition, often leading to violence, uses valuable resources such as manpower, equipment and finances. In the United States, the Irish Mob boss of the Winter Hill Gang (in the 1980s) turned informant for the Federal Bureau of Investigation (FBI). He used this position to eliminate competition and consolidate power within the city of Boston which led to the imprisonment of several senior organized crime figures including Gennaro Angiulo, underboss of the Patriarca crime family. Infighting sometimes occurs within an organization, such as the Castellamarese war of 1930–31 and the Boston Irish Mob Wars of the 1960s and 1970s.

Today criminal organizations are increasingly working together, realizing that it is better to work in cooperation rather than in competition with each other. This has led to the rise of global criminal organizations such as Mara Salvatrucha and the 18th Street gang. The Italian-American Mafia in the U.S. have had links with organized crime groups in Italy such as the Camorra, the 'Ndrangheta, Sacra Corona Unita, and Sicilian Mafia. The Cosa Nostra has also been known to work with the Irish Mob (John Gotti of the Gambino family and James Coonan of the Westies are known to have worked together, with the Westies operating as a contract hit squad for the Gambino family after they helped Coonan come to power), the Japanese Yakuza and the Russian Mafia. The United Nations Office on Drugs and Crime estimated that organized crime groups held $322 billion in assets in 2005.

This rise in cooperation between criminal organizations has meant that law enforcement agencies are increasingly having to work together. The FBI operates an organized crime section from its headquarters in Washington, D.C. and is known to work with other national (e.g Russian Federal Security Service (FSB), and the Royal Canadian Mounted Police), federal (e.g., Bureau of Alcohol, Tobacco, Firearms, and Explosives, Drug Enforcement Administration, United States Marshals Service, and the United States Coast Guard), state (e.g., Massachusetts State Police Special Investigation Unit and the New York State Police Bureau of Criminal Investigation) and city (e.g., New York City Police Department Organized Crime Unit and the Los Angeles Police Department Special Operations Division) law enforcement agencies.

In India earliest organized plunderers and thugs were Chengis Khan, Mahmood Ghazni, Mohammad Ghori and Sultan Mamood. Present day legendary organized criminals are Vardarajan Mudaliar, Dawood Ibrahim, Rajan Nair, Abdul Latif of Bombay Latif gang and Abdul Wahab of Gujurat. In Bihar, Thakur gang, Suraj Deo Singh, Yadav gangs are today dominating. Kidnapping and ransom are a very lucrative business in Bihar. In U.P. it is called "Goonda Tax" and in Bihar "Rangdari Tax". Gangsters like Brijesh Tiwari, Tribhuvan gang of Varanasi, Muktar Ansari, Babloo Srivastav, Abhay Singh are prominent gangsters. In Odisha Pratap Swain gang and Nilu Patjoshi gang of Ganjam, Hyder and Tito of Kendrapara, Mangaraj group in Bhubaneswar, Padia, Kalia of Cuttack are famous.

Vohra Committee Report

In 1993, the Government of India appointed a committee under the chairmanship of N.N. Vohra, Cabinet Secretary, to take stock of all available information about the crime syndicates/mafia organisations which had developed links with and were being protected by government functionaries and political personalities. This was apparently done in the aftermath of the serial Mumbai bomb blasts, occurring shortly after the demolition of the Babri mosque in Ayodhya, for which certain dons of the underworld; both within and outside India, were believed to be responsible. The committee examined the reports of the Secretary, Research and Analysis Wing (RAW) and directors of the Central Bureau of Investigation (CBI) and the Intelligence Bureau (IB) for an appraisal of the situation and to suggest various measures to curb the joint criminality of the mafia, politicians and bureaucrats. The submissions of the directors of the RAW, CBI and IB revealed the ground realities, which have been incorporated in the final report of the committee.

Director CBI has reported that Crime Syndicates all over India 'have become a law unto themselves. Even in the smaller towns and rural areas, musclemen have become the order of the day. Hired assassins have become a part of these organisations. The nexus between the criminal gangs, Police, Bureaucracy and Politicians has come out clearly in various parts of the country. The existing criminal justice system, which was essentially designed to deal with individual offences / crimes, is unable to deal with the activities of the Mafia; the provisions of law in regard to economic offences are weak; there are insurmountable legal difficulties in attaching / confiscation of the property acquired through Mafia activities.'

The Director of IB issued the "warning signals of sinister linkage between the underworld politicians and bureaucracy" and recommended immediate attention to :

(i) Identification of the nexus between the criminals/mafias and anti-national elements on the one hand and bureaucrats, politicians and other sensitively-located individuals on the other.

(ii) Identification of the nature and dimensions of these linkages and the modus operandi of their operations.

(iii) Assessment of the impact of these linkages on various institutions viz. the electoral, political, economic, law and order and administrative apparatus.

(iv) Nexus, if any, between the domestic linkages with foreign intelligence.

(v) Necessary action to show effective action to counteract/neutralize the mafia activities.

(vi) Political and legal constraints in dealing with the covert/illegal functioning of the linkages.

The Directors submitted that there has been a rapid spread and growth of criminal gangs, armed Senas, drug Mafias, Smuggling gangs, Drug peddlers and economic lobbies in the country which have, over the years, developed an extensive network of contacts with bureaucrats / government functionaries at the local levels, politicians, media persons and strategically located individuals in the non-state sector. Some of these syndicates also have international linkages, including the foreign intelligence agencies.

It was further observed that the linkages developed by crime syndicates get generally confirmed when pressure is mounted on the concerned agencies not to take action against the offenders or to go slow in the cases against them. Such pressures are mounted either immediately after a raid is conducted or at the time when prosecution is about to be initiated. Pressures are also exerted whenever corrupt and undesirable officers are shifted from sensitive assignments.

The startling revelations by the report confirm the seriousness of the situation. The report was tabled in the Parliament in August 1995. In view of the gravity of the matter, the Supreme Court directed the government to appoint a high level committee to examine the allegations referred to in the Vohra Committee Report in March 1997. Under the direction of the court, the government constituted a three-member committee with the same N.N.Vohra as Chairman, former Cabinet

Secretary B.G.Deshmukh and Central Vigilance Commissioner S.V.Giri as members in September 1997. Hence the government completed the legal obligation but the real attitude of the government can be judged from the following observation of the then Home Minister. 'There is nothing we can do to end the unholy nexus between criminals and politicians and bureaucrats. All we can do is to pass legislations, which can not be implemented. I see no end to the problems in the country. Especially as there are criminals in every section of society.'

The Report reads "the various agencies presently in the field take care to essentially focus on their respective charter of duties, dealing with the infringement of laws relating to their organisations and consciously putting aside any information on linkages which they may come across" (Para 11.1)

Defining Organized Crime

Howard Abadinsky says "Organized crime is a nonideological enterprise involving a number of persons in close social interaction, organized on a hierarchical basis, with at least three levels/ ranks, for the purpose of securing profit and power by engaging in illegal and legal activities. Positions in the hierarchy and positions involving functional specialization may be assigned on the basis of kinship or friendship, or rationally assigned according to skill. The positions are not dependent on the individuals occupying them at any particular time. Permanency is assumed by the members who strive to keep the enterprise integral and active in pursuit of its goals. It eschews competition and strives for monopoly on an industry or territorial basis. There is a willingness to use violence and/or bribery to achieve ends or to maintain discipline. Membership is restricted, although nonmembers may be involved on a contingency basis. There are explicit rules, oral or written, which are enforced by sanctions that include murder." (Abadinsky, 1990: 6).

To Block and Chambliss "Organized crime is a term that refers to those illegal activities connected with the management and coordination of racketeering (organized extortion) and the vices - particularly illegal drugs, illegal gambling, usury, and prostitution." (Block and Chambliss, 1981: 12).

Patrick Ryan, goes on to argue "Most people define organized crime as the activities of the Mafia, which they think of as some sort of mysterious, quasi-mystical confederation of Italians organized by families and dedicated to such pursuits as murder, extortion, bookmaking and gambling, and racketeering. Organized crime once meant the Mafia, but today it is much more. Organized crime is large-scale criminal activity, operated by big organizations with complex hierarchies. Although the activities of organized crime remain fairly similar, the people involved have changed radically. No longer is organized crime the work of one ethnic group. No longer does organized crime operate more or less locally. Like legitimate big business, organized crime has gone international." (Ryan, 1995, 1).

Michael Woodiwiss, says "My understanding of the phrase 'organized crime' is literal and short - organized crime is systematic criminal activity for money or power. Organized crime, understood in this way, is as old as the first systems of law and government and as international as trade. Piracy, banditry, forgery, fraud, and trading in stolen or illegal goods and services are all ancient occupations that often involved the active participation of landowners, merchants, and government officials. More recently, corporate business interests, and public and private professionals, have also been actively involved in organized crime. While organized crime has never been exclusive to anyone race, ethnic group, class, profession, or gender, many studies have shown that those with power, influence, and respectability in local, regional, national, or international society have tended to organize crime more successfully and securely than those without. Myth and misrepresentation have, however, distorted this historical record." (Woodiwiss 2001: 3).

Under Maharashtra Control of Organised Crime Act, 1999 "organised crime" means any continuing unlawful activity by an individual, singly or jointly, either as a member of an organised crime syndicate or on behalf of such syndicate, by use of violence or threat of violence or intimidation or coercion, or other unlawful means, with the objective of gaining pecuniary benefits, or gaining undue economic or other advantage for himself or any person or promoting insurgency (Section 1, e)

Characteristics of Organised Crime

Organised crime has many peculiar characteristics which distinguish it from ordinary crime. The distinction between ordinary and professional criminals on the one hand and the organised criminals on the other hand are well on record in criminology. The ordinary criminal typically works alone or with a partner or two, or as part of a larger group temporarily assembled for a specific job. The professional often works alone but is frequently a member of a more or less cohesive team or 'troupe'- many confidence men, and most pickpockets work in highly-organised, quasi-permanent gangs. Most of these groups, however, represent the association of independent criminals on equal terms, on a share-and-share-alike basis. The organised criminal, by contrast, is usually the temporary hireling, the steady employee, or the administrator of a criminal bureaucracy. Organised crime is not only 'more' organised than ordinary or professional crime, but differently organised. The combinations used by ordinary and professional criminals tend to be voluntary, cooperative, and non-hierarchical. Given below are some of the characteristics of organised crime :

1. Team work : It involves association of a group of criminals which is relatively permanent. In organised crime there is a group of persons of considerable size which engages itself in continuous crime over a long, usually indefinite period of time.

2. Hierarchical structure : It has a structure with grades of authority from the lowest to the highest, involving a system of specifically defined relationships with mutual obligations and privileges.

3. Planning : It involves advance arrangements for committing crimes, minimising risks and insuring safety and protection. Functioning of organised criminals has been likened to the working of corporations and big business houses. There is division of labor, delegation of duties and responsibility and specialisation of functions. Like any modem business, organised crimes also involve careful planning, risk insurance and have expansive and monopolistic tendencies.

4. Centralised authority : It functions on the basis of centralised control and authority which is vested either in the hands of one individual or a few members. The organisation is generally highly centralised; the authority is vested in one or just a few members of the group. Such Individual is called a DON. Criminals are elected to the position of DON after experiencing several stages of growth from street smart to predatory stage to protection money to power stage, finally terminating in the position of DON.

5. Reserved fund : In order for a criminal organization to prosper, some degree of support is required from the society in which it lives. Thus, it is often necessary to corrupt some of its respected members, most commonly achieved through bribery, blackmail, and the establishment of symbiotic relationships with legitimate businesses. It maintains a reserve fund from profits which serves as capital for criminal enterprises, seeking help of the police, lawyers and even politicians, and for providing security to the arrested members and their families.

6. Specialisation : Some groups specialise in just one crime while some others may be simultaneously engaged in multiple crimes. Those groups which are engaged in multiple crimes are more powerful and influential. Criminal organisations adopt measures to protect the group and to guard against the prevention of their activities. To this end, arrangements are made with doctors, lawyers, policemen, judges, politicians and government officials. The mechanics used by them are methodical selection of targets, fair amount of groundwork, target survey, dry runs, arrangement of spotters, separate channel for weapons and finally organizing crime.

7. Division of labor : Organised crime involves delegation of duties and responsibilities and specialisation of functions. They have strategic strong internal laws with right person having specialized skills is fitted in right work.

8. Violence : It depends upon use of force and violence to commit crimes and to maintain internal discipline and restrain external competition. It has a tendency to dominate, through political clout or corruption, the law enforcement agencies.

9. Monopoly : It has expansive and monopolistic tendencies. Initially, organised criminal gangs operate in a limited area, and are engaged in a limited type of crime with a limited number of persons, but, gradually they expand into a wider range of activities extended over large geographical areas, involving a large number of carefully selected criminals. In whatever area they operate, they secure monopoly in their criminal enterprises. They do not hesitate to use violence or threats of violence to eliminate competition.

10. Protective measures : It arranges permanent protection against interference from law-enforcement authorities and other agencies of government. The protective measures include contacts with policemen, lawyers, doctors, politicians, judges, and influential persons in society. Giving money in cash or in gifts, providing help in elections, threatening their competitors, and arranging their (influential persons') foreign trips are some of the methods used by these gangs to secure protection and avoid arrest and conviction.

11. Conduct norms : It frames rules of conduct, policies of administratioc, and methods of operation for members and for the operarion of crime. This helps in maintaining discipline, efficiency, loyalty, obedience, and mutual confidence. Penalties are imposed violating rules.

12. Victimization : Organized crime often victimize businesses through the use of extortion or theft and fraud activities like hijacking cargo trucks, robbing goods, committing bankruptcy fraud (also known as "bust-out"), insurance fraud or stock fraud (inside trading). Organized crime groups also victimize individuals by car theft (either for dismantling at "Chop shops" or for export), burglary, credit card fraud, and stock fraud ("pump and dump" scam). Some organized crime groups defraud national, state, or local governments by bid-rigging public projects, counterfeiting money, smuggling or manufacturing untaxed alcohol (bootlegging) or cigarettes (buttlegging), and providing immigrant workers to avoid taxes. Organized crime groups seek out corrupt public officials in executive, law enforcement, and judicial roles so that their activities can avoid, or at least receive early warnings about, investigation and prosecution.

13. Multifarious Illegal activities : Organized crime groups also provide a range of illegal services and goods, such as loansharking of money at very high interest rates, bookmaking and gambling, prostitution, drug trafficking, gunrunning, providing murder for hire, illegal dumping of toxic waste, people smuggling and trafficking in human beings. Organized crime groups also do a range of business and labor racketeering activities, such as casino skimming, Insider trading, setting up monopolies in industries such as garbage collecting, construction and cement pouring, bid rigging, getting "no-show" and "no-work" jobs, using non-union labor and pocketing the wage difference, money laundering, political corruption, and bullying.

14. Networking : Criminal gangs tend to work together in mutual professional interest. This has led to greater networking and the rise of global criminal organizations. This rise in cooperation between criminal organizations has meant that law enforcement agencies increasingly have to work together. Criminal organizations may function both inside and outside of prison. Corrupt political leaders may have links to existing organized crime groups, either domestic or international, or else may simply exercise power in a manner that duplicates the functioning and purpose of organized crime. In addition to what is considered traditional organized crime involving direct crimes of fraud swindles, scams and racketeering motivated for the accumulation of monetary gain, there is also well networked non-traditional organized crime which is engaged in for political or ideological gain or acceptance. Such crime groups are often labeled terrorist organizations and include such groups as Al-Qaeda, the Jewish Defense League, Irish Republican Army, Taliban, Maoists and Naxal outfits to name a few. The characteristic of these groups is the fight for power and domination. This very characteristic is the most dangerous for the national security system.

Types of Organised Crime

Organised crime can be classified into the following seven major categories :

1. Organised Gang Criminality : Bank robbery, hijacking, murder, kidnapping, automobile and jewel thefts are some of the common examples of organised gang criminality. In the USA, "they live violently and expect violence"; the situation is not very different in other countries these days. This kind of criminality reached maximum height in the USA during the two world wars but was controlled to a great extent subsequently.

India too has witnessed the growth of organised gang criminality during the last few years, particularly murders and kidnappings for ransom. Uttar Pradesh and Bihar, especially the latter, are the two States which are conspicuous in this regard. Dhanbad, a major coal-mining area, and West Champaran are the two prominent places as regards the mafia operators in Bihar. The most disturbing aspect is the social and political acceptance of these criminals as indicated by the patronage and protection bestowed on them by politicians and policemen. One of the criminals engaged in kidnappings for ransom in Bihar was killed in the house of a close relative of a political leader, while another was arrested from the house of a Member of the Legislative Assembly (Rajgopal, 1988:130). In Karachi, the biggest city of Pakistan, where conditions in terms of lawlessness, terrorism and political criminality are at least as bad as in India, even worse, the problem of kidnappings for ransom has attained incredible dimensions.

Lawlessness has reached such a level making it appear that the Government has lost the will to enforce the laws. The law enforcement agencies including courts are reluctant to take cognizance of the activities even when gangsters fire at advocate/terrorize witnesses/police in the precincts of the Court of law. Judges are transferred and replaced for a favorable decision. In Pakistan Chief Justice of Supreme Court could be removed. Governors do not allow so called majority party to form Government.

In this state of affairs, when even the existing laws (Section 110 of the Criminal Procedure Code) are not being applied, the fate of new legislation is obvious. An example is the enactment of the Gangsters Act by the U.P. Government in 1986 to meet the challenge of the Mafia in the State. As could be expected, the law failed miserably to make any dent in Mafia criminality. The sordid picture is presented in an empirical study regarding the working of the Act. "A study of its (Gangsters Act) implementation in 15 districts reveals that so far not even a single person has been convicted although thousands of persons were booked under the Act. In eighty per cent of cases even a charge could not be framed in the court. Where the charges were framed, the acquittal rate has been hundred per cent" (Rajgopal, 1988:141-42).

"Organised gang criminality, despite being dangerous and often extremely violent, does not constitute so much threat to the society as is posed by racketeering and syndicated crime. It strikes in the open and can be attacked and brought under control by courageous, alert, and vigorous police action. Racketeering and syndicated crime, on the other hand are 'insidious, subterranean, and undermining'(Caldwell,1965:77).

2. Racketeering : Extorting money from legitimate or illegitimate business amounts to what is known as racketeering. This is achieved by intimidation or actual force. It is in this particular context that the description of organised criminals as parasites is most appropriate. Those in illegal business like prostitution, gambling and drug trafficking are especially vulnerable to racketeering since they cannot approach the police and other authorities for protection. In fact, money is demanded and paid as consideration for the 'protection' given by a criminal group to the operators of illegitimate business.

By capturing the unions, organised criminals are in a position to exploit both employers and employees in business and industry; unions providing the criminal leverage and facade to Mafia leaders. The labour unions of Bharat Coking Coal Ltd., situated in the famous colliery town of Dhanbad, are known for their domination by the organised crime which leads to various abuses in relation to orders, allocation and transport of various raw materials coming out of the collieries. Members of the Mafia also indulge in pilferage, smuggling and illegal mining of coal. Inter-union rivalries are common resulting in violent clashes between various groups. Murder and mayhem are not unusual phenomena in such circumstances.

Construction work through contracts is another area where organised criminals have taken their share by employing political clout and muscle power. In many States in India, it is the Mafia which often prevails regarding terms and conditions and the eventual award of the government and public-sector tenders. It is not, therefore, surprising that police protection has been often sought by the engineers facing threats of violence and police force has to be deployed sometimes in connection with the allocation of work to contractors (Siddique, 2005:438).

3. Syndicate Crime : This is the other significant area of organised crime, and consists of the supply of illegal goods and services. Smuggling, bootlegging, gambling, prostitution and foreign exchange violations are examples of such crimes. The leaders of syndicate crime possess all the visible signs of respectability in such society and ordinarily they operate without resorting to violence. There are front men at various levels executing the works assigned to them. The so called 'bosses' generally function behind the scenes from their well-appointed headquarters and that make them somewhat immune from the enforcement authorities. Giving rise to parallel economy, black money and violence, this variety of organised crime has the greatest potentiality of damaging the social, economic, political and moral fabric of the society. The worst aspect of the problem is that the money generated through illegal means is invested in legitimate business enterprises.

4. Smuggling : It is a highly-prevalent criminal activity in developing countries like India where the official position is to encourage local products and industries and to conserve maximum foreign exchange. Smuggling is the direct result of the import and export policy of the government of a country. Besides the craze for foreign goods, due to their real or supposed superior quality, the gap between the prices of the indigenous and foreign products is also responsible for illegal imports. Further, the politicians and enforcement officials sharing the fruits of the illegitimate enterprises develop a vested interest in the continuation of smuggling operations.

India has vast seashore and a large border area along with neighboring countries and the most vulnerable areas are Gujarat, Maharashtra, the terai area of Uttar Pradesh, Punjab and the eastern region. The main items smuggled into the country are synthetic fabrics, watches, electronic goods, gold and narcotic drugs. Items smuggled out of the country also include foreign and Indian currency and animal skins. The magnitude of the problem can be gauged by the following comment appearing in a newspaper. "Smuggling has assumed such dangerous proportions as to threaten the nation's social and political fabric. It is widely believed that smuggling equals India's official trade. It cannot be seriously contested that among the economic crimes smuggling has had the most deleterious effect in certain States as Maharashtra, Gujarat, Tamil Nadu, Punjab, Bihar and Bengal" (Times of India, dated July 19, 1987).

Among the items smuggled into the country, gold tops the list in terms of supply and demand. Traditionally, India has always been the 'bottomless pit for gold consumption'. Over the years a new dimension has been added because gold is used in exchange for arms, drugs and other smuggled goods. In recent years the Government has intensified efforts to prevent gold smuggling. The Coastal Guards and the Customs Department have been provided with better class of weapons, speedboats, vehicles and motorcycles for patrolling, special metal detectors, night-vision binoculars, closed-circuit TVs, walkie-talkies and semi-automatic guns. The scheme for rewarding informers as well as officers responsible for detection was further liberalised in 1985 to provide payment of 20 per cent of the value of contraband seized.

Trafficking in drugs is perhaps the most sinister of all the smuggling activities. Not only the economy and health problems of extreme gravity are involved, it has political and related aspects of far-reaching consequences. In India, trafficking in drugs and terrorism has come to be intimately connected with each other. Terrorists get money for much-needed arms and ammunition by working hand in glove with smugglers. Violence by terrorists is only a part of the violence unleashed by the smuggling of drugs. The inter-gang rivalry, of which Mumbai provides a typical example, generates violence of extreme severity leading to killings and counter-killings. According to report of the U.N. conference held in 1987 in Vienna, 80 per cent of the heroin found in the streets of the U.K. and Western Europe passed through Bombay. A number of factors have contributed to this development; some of them are given below (Siddique, 2005:440-41):

- While the cultivation of opium has been significantly reduced in India, the cultivation in Pakistan went up by about 25 per cent during the last few years. The situation in Afghanistan and the exodus of refugees into Pakistan also resulted in an unprecedented glut of drugs, a good part of which found its way into India.
- For Pakistanis it became safer to push drugs through India instead of carrying them directly to Western Europe in view of the greater vigilance on Pakistanis entering those countries. The period of the early eighties also synchronised with the beginning of terrorism in Punjab. Around the same time, Chittagong in Bangladesh became an important place to receive drugs from the 'Golden Triangle'; drugs reaching the eastern part of India being the next logical step.
- Silver prices in India were much lower than elsewhere before 1983 and this led to a great deal of illegal export of silver from the country. The situation, however, hanged after the prices of silver crashed in the international market in 1983 leading to the replacement of silver export by illegal export of narcotics.

5. Identity Theft and Online Extortion : The newest growth sectors for organized crime are identity theft and online extortion. These activities are troubling because they discourage consumers from using the Internet for e-commerce. E-commerce was supposed to level the playing ground between small and large businesses, but the growth of online organized crime is leading to the opposite effect; large businesses are able to afford more bandwidth (to resist denial-of-service attacks) and superior security. Furthermore, organized crime using the Internet is much harder to trace down for the police (even though they increasingly deploy cybercops) since police forces and law enforcement agencies in general operate on a national level while the Internet makes it even more simple for criminal organizations to cross boundaries and even to operate completely remotely.

Once viewed as a local or at most a regional threat, organized crime has become a highly sophisticated transnational affair. In the new global age, borders have opened up, trade barriers have fallen and information speeds around the world at the touch of a button. Business is booming -and so is transnational organized e-crime. E-crime generally refers to a criminal activity where a computer or computer network is the source, tool, target, or place of a crime. Despite the inevitable references to 'computers' or 'online activity', e-Crime nevertheless encompasses a whole raft of 'traditional' crimes - such as fraud, theft, blackmail, forgery and embezzlement. E-crime is notoriously difficult to detect and punish because of its sheer technical complexity and because unseen attackers can strike victims from hundreds or even thousands of miles away. Due to the nature of e-Crime, and its ability to evolve with technology, new threats are emerging with an alarming degree of regularity and simply maintaining a better appreciation of the risks we face could have a significant effect on your ability to respond to them if the need arises.

6. Trafficking of Women & Children : Lacking much of the paperwork that is common to legitimate organizations, criminal organizations can usually evolve and reorganize much more quickly when the need arises. They are quick to capitalize on newly-opened markets, and quick to rebuild themselves under another guise when caught by authorities. This is especially true of organized groups that engage in human trafficking. Trafficking of women and girls is the third largest illegal trade in the world after trading in arms and drugs. It generates billions of dollars annually. The dictionary definition of Trafficking is "an illegal trade in a commodity", in this case, that commodity is women and children. UN Protocol to Prevent, Suppress and Punish Trafficking in Persons Especially Women and Children, 2000 defines trafficking as "the recruitment, transportation, transfer, harbouring or receipt of persons, by means of the threat or use of force or other forms of coercion, of abduction, of fraud, of deception, of the abuse of power or of a position of vulnerability or of the giving or receiving of payments or benefits to achieve the consent of a person having control over another person, for the purpose of exploitation." The victims are taken from one place by deception, fraud, force, using one's power, exploiting someone's vulnerability to another place through recruitment, transportation, buying, selling with or without payment for forced labor, services, domestic labor, sexual exploitation etc.

Techniques used by Traffickers

- Seek help of local person or contacts to approach the victim's family as well wishers.
- Make contacts with unsuspecting women & girls at bus stand/railway station.
- Make friends with young girls/children while at play/school.
- Approach poverty stricken family to part with or sell their children to meet their debt.
- Luring women & girls with false promises of well paid jobs or marriage.
- Use false documents/passports to trafficked people across borders.
- Kidnapping/abduction.
- Rape or other forms of sexual abuse.

7. Corruption : Corruption is any course of action or failure to act by individuals or organizations, public or private, in violation of law or trust for profit or gain. Transparency International (TI) has chosen a clear and focused definition of the term: Corruption is operationally defined as the abuse of entrusted power for private gain. TI further differentiates between "according to rule" corruption and "against the rule" corruption. Facilitation payments, where a bribe is paid to receive preferential treatment for something that the bribe receiver is required to do by law, constitute the former. The latter, on the other hand, is a bribe paid to obtain services the bribe receiver is prohibited from providing. Corruption undermines political, social and economic stability and damages trust in institutions and authorities. It also fuels transnational crime. Terrorists and organized criminals are aided in their illegal activities by the complicity of corrupt public officials. Corruption is of particular concern for the world's police and judicial systems, as corruption in one country can compromise an entire international investigation. Political and bureaucratic corruption in India are major concerns. A 2005 study conducted by Transparency International in India found that more than 45% of Indians had first-hand experience of paying bribes or influence peddling to get jobs done in public offices successfully.

Corruption and organised crime today are inextricably intertwined. Corrupt practices evolve with time, growing from spontaneous incidents to organized groups and can even be franchised in a company, government or industry. The heuristic linkage between the two is found at five levels.

1st level is called strategic bribery. The bribe is offered or solicited for a single act. For example: to obtain a passport, licenses or to get information designed to enable criminal advantage. The organized crime groups obtains an advantage like false identities and documents, advance information about police activity, manipulation of official records, disappearing evidence or access to jury identities. At the 2nd level called collusion, it becomes continuous and perennial acts of bribery ("On payroll"). This encourages gradual strengthening of nexus which ensures a flow of information and protection from police intrusion into criminal activities. They obtain constant access to confidential information allowing them to maintain patterns of illegal activity remaining "one step ahead" of law enforcement agencies. Next at the 3rd level of compromise, it starts infiltration into Government agencies. This infiltration is sporadic within lower ranking official positions. The organized crime group members and friends gain access to law enforcement agencies, judiciary, prosecutor's office and other operations for criminal advantage. They either apply for job vacancies with the Government or the organized crime groups and/or buy the job by bribing or blackmailing officials to place its members and friends in the government post. This infiltration can encompass entire branches or higher ranking officers in Law Enforcement, Prosecutor's Office and other sensitive government offices. At the next level it gets organized and starts infiltration into higher level of the Government. Accordingly there occurs the placement of corrupted officers into Government office. The organized crime groups use bribery and blackmail to support officers previously corrupted by their organization to higher ranks with broader access to information and service. This "support" is repaid to them with increased protection and access to more useful secret Government information and service. Bribery, blackmail and coercion are used to control the chief decision-maker of a branch of government. The organized crime can continue operating with very little risk of discovery or successful prosecution. The fifth final stage called franchise is the infiltration into the Political Arena. The organized crime groups may operate participating in political campaigns (giving money, media support and others) or buying votes for candidates and corrupting the democratic election process or lobbying other politicians for support using bribery and blackmail or exploiting organized crime group members' family links or creating "debts" for politicians to "repay" later by using blackmail and extortion.

Organized corruption represent an ever increasing danger for the State and the society, since organised crime invests them on one side in completely legal business - money laundry, while on the other side they represent an enormous corruption potential. U.S. Attorney General John D. Ashcroft said "In the real world of limited resources, we know that we can only detect, investigate and prosecute a small percentage of those officials who are corrupt...I remain convinced that there is no more important area in the fight against corruption than the challenge for us within the law enforcement and justice sectors to keep our own houses clean."

Responses to Organised Crime

There are plenty of problems that appear in the field of the fight against organised crime. The sanctions and the punishments are not adapted to the new situation. Because of the great mobility of the capital throughout the geographically unlimited areas, new techniques have appeared for the money laundry. The strategy of the criminal associations is advancing rapidly while the co-operation between the police forces and between other services is too slow. The organised crime is not the problem of the police alone; it is essential to begin the suppression of organised crime on the bases of new tactics and strategies, with the co-operation of all parts of the society and through a carefully prepared programme of national safety, as well as through international co-operation. We therefore need a coordination of valid legal regulations in the area of criminal justice, penal procedure legislation, administrative legislation and judicial and constitutional legislation.

The laws are unstable and different, especially regarding special methods and means. New evidential procedures are complicated and are not coordinated between the States. On the other hand new forms of organised crime appear each year, the territories that such groups control, are increasing, the power of the organisations increases, especially with their investing in completely legitimate businesses. If the police introduce new methods in the fight against organised crime, the criminals immediately begin with the use of new, even more sophisticated methods for the acquisition and spreading of their powers. It can be seen that the traditional forms of fight against crime are not adequate for the modern forms of organised crime and that it would be necessary to start a joint action of all institutions in a society, which could in any way contribute in the fight against organised crime. We specifically think here of stronger bonds amongst the public, the police and the Courts. So far three forms of fight against organised crime have proven to be successful: the control of the drug market, the control of money circulation and the international co-operation.

Dealing with the issue of remedies against organised crime, one aspect to be kept in mind is that the problem is to be tackled by methods different from those employed in ordinary crimes. This was aptly put by an Attorney-General of the USA thus: "Innovative measures are necessary for the effective prosecution of organised crime cases because the traditional law-enforcement process is by and large designed for the control of individuals not for the control of organisations." Two approaches are possible in dealing with the criminal organizations responsible for organised crimes (Siddique,2005):

1. **The Law Enforcement Perspective :** According to this approach, organised crime is the creation of certain distinct groups in the society and the task is to deal with them by applying legal sanctions. What is required, therefore, is to enact adequate and appropriate laws and their efficient implementation. This appears to be the basis of the approach of the President's Commission on Law Enforcement and Criminal Justice in the USA. The Commission identified the source of organised crime in the USA in "a society that seeks to operate outside the control of American people and their working government." According to the Commission, this society unites thousands of criminals who work together in complex organisational structures and are subject to laws and rules that are more rigidly enforced than those of legitimate government. The Commission found that organised crime in America was controlled by a national alliance of at least Tightly-knit gangs or "families" whose members are of mainland Italian or Sicilian descent. The Commission identified the organisation as 'La Cosa Nostra' for which the term 'Mafia' was employed by the Kefauver Committee (the Senate Special Committee to investigate Organised Crime in Inter-States Commerce, 1950-51).

As regards the law-enforcement approach, besides the Penal Code which provides some basic principles relating to conspiracy, attempts, abetment and frauds, there are quite a few other enactments specifically designed to deal with organised criminality. The Customs Act, Conservation of Foreign Ex-change and Prevention of Smuggling Activities Act (COFEPOSA), Smugglers and Foreign Exchange Manupulators' (Forfeiture of Property) Act and Foreign Exchange Management Act, 1999 are some of these enactments. Apparently these laws have not hit their targets very effectively; the main hurdles being political interference and corruption of enforcement officials. The real task is to insulate the enforcement agencies from political influences as far as possible. To achieve this, the creation of a central task force constituted out of personnel from the police, income tax, customs and any other appropriate agency has been suggested. By its nature, organised crime is not ordinarily confined to a single State in a given situation and quite often investigations cut across various States. As of now, the Centre has no powers to investigate into the crimes committed in States. Even the Central Bureau of Investigation can operate in a State only if consent is given by the State. Viewed from all these angles a central task force having all-India jurisdiction may be highly desirable.

The problem of implementation of a couple of penal provisions in the Indian enactments referred to above may be briefly considered. Since organised crimes are committed for economic gain, the punishment should relate to offsetting the ill-gotten gains and confiscation of property of convicted criminals is, therefore, an appropriate sanction. The principle of confiscation of property is strengthened in the U.K. by the Drug Traffic Offences Act which contains a rebuttable presumption that all of the assets at the offender's disposal were the proceeds of crime. There are, however, some snags which make the effective application of the sanction somewhat difficult. Confiscation of property under the Smugglers and Foreign Exchange Manipulators' (Forfeiture of Property) Act is not permissible until a valid detention is made under the COFEPOSA; the legality of detention is invariably contested and quite often it involves prolonged judicial proceedings. This may provide an opportunity to alleged criminals to manipulate the property in such a way so as to make it immune from subsequent penal action. It is, therefore, absolutely necessary to devise methods and procedures to cut short the length of judicial proceedings relating to detention under the COFEPOSA or the National Security Act.

Under Section 31 of the Drug Act passed in 1985, very severe punishments are permissible in case of offenders having previous convictions to their credit. Though a sound policy in itself, again the implementation is not possible unless all-India records are carefully maintained by a central agency of convictions under the Act all through the country. The principle of confiscation of property is strengthened in the U.K. by the Drug Traffic Offences Act which contains a rebuttable presumption that all of the assets at the offender's disposal were the proceeds of crime.

National Investigation Agency Act of 2008

Some of the salient features of NIA are –

- The Bill for constitution of the National Investigation Agency (NIA) to be applicable to whole of India, citizens of India, outside India and persons on ships and aircraft registered in India.
- Officers of the NIA to have all powers, privileges and liabilities which the police officers have in connection with investigation of any offence.
- The police officer in charge of a police station on receipt of the report of the offence shall forward it to the state government which in turn will send it to the Centre.
- If the Centre feels the offence is terror related, it shall direct the NIA for investigation.
- Provision for transfer of investigation and trial of offences to state government with Centre's prior approval.
- NIA may investigate other offences connected with terror-related offences.

- A state government shall extend all assistance to NIA for investigation of terror-related offences.
- Provisions of the Act with regard to investigation shall not affect powers of the state government to investigate and prosecute any terror crime or other offences.
- The Centre shall constitute special courts for trial of terror-related offences.
- Special Courts may sit at any place for any of its proceedings.
- For speedy and fair trial, the Supreme Court may transfer any case pending with the special court to another special court in the same state or any other state, and the High Court may transfer such cases to any other special court within the state.
- Offences punishable with imprisonment for less than three years may be tried summarily.
- Bill for federal agency tabled in Lok Sabha.
- Special court to have all powers of the court of sessions under CrPC for trial of any offence under the Act.
- Proceedings to be held 'in camera' if special court deems it necessary.
- Trial to be held on day-to-day basis on all working days and to have precedence over the trial of other offences.
- State governments empowered to constitute one or more special courts.
- No appeal shall be entertained after the expiry of 90 days. Highlights of Unlawful Activities (Prevention) Amendment Bill, 2008.
- Use of bombs, dynamite, poisons or noxious gases, biological radioactive nuclear substances are terror act.
- Aiding, abetting or committing a terrorist act shall be punishable with imprisonment up to ten years.
- Funding terror activities, organising training camps and recruiting persons for committing terror acts shall be punishable with at least five years' imprisonment.
- Detention of accused up to 180 days if investigation not completed.
- No bail shall be granted if accused is not an Indian citizen and has entered the country unauthorisedly.
- No accused, if in custody, to be released on bail or on his own bond.
- Court shall presume, unless contrary is shown, that accused has committed offence.

2. The Social and Economic Perspective : According to some of the sociologists who studied the problem of organised crime in the USA, the phenomenon is not an alien conspiracy but is a part of the social functioning or dysfunctioning and market economy. What is needed for the eradication of the malady are effective economic strategies, e.g., sound policies regarding production and distribution, reducing the spread between profits and costs, or breaking up organised crime monopolies by creating legitimate markets through decriminalisation or legalisation.

With regard to tackling organized crime certain proactive measures and control strategies need to be developed.

Proactive Steps

- Assessment of potential activities/ regions – illicit liquor, gambling, flesh trade, trade rivalries, civil disputes, etc
- Effective legal action
- Police action to target the big fish
- Tackling the support systems – politicians, officials, lawyers
- Surveillance
- Documentation
- Training
- Cleaning the jail
- Gaining Public Confidence
- Preventing Potential Recruits
- Internal vigilance against subversion

CONTROL STRATEGIES

Criminal Intelligence :

- Optimum use of interrogation opportunities
- Contacts, observers, sources etc.
- Technical surveillance – call data analysis, monitoring of communications
- Databank-compilation as a continuous process
- Partnership with other agencies

Investigations :

- beyond the case in question
- into linkages, networking etc.
- including identification, seizure, forfeiture of property
- Institutionalizing Co-ordination mechanisms at the State level
- Co-operation and mutual assistance between districts and states

Suggestions :

- Not enough to have laws, Laws must be just and implementable.
- Need to create STF
- To collect intelligence against the gangs
- To act against gangs in co-ordination with District police

- To take action against inter districts gangs
- Willing Officers and men with specialized Training, latest equipments and other resources
- Special allowance and other amenities including adequate insurance coverage
- In camera trial; long distance trials
- Proper reward to personnel and officers
- More fast track courts to be created
- Presiding Officer for the Special Court needs to be specially chosen

References

Abadinsky, Howard (1994), Organized Crime, 4th ed., Chicago.

Albanese JS (2008). "Risk Assessment in Organized Crime Developing a Market and Product-Based Model to Determine Threat Levels". Journal of Contemporary Criminal Justice 24 (3).

Block, Alan A., Chambliss, William J. (1981), Organizing Crime, New York/Oxford.

Bovenkerk F (2003). "Organized crime and ethnic reputation manipulation". Crime, Law, and Social Change 39 (1).

Caldwell Robert G., (1965) Criminology New. York: Ronald Press, p. 77.

Hagan FE (1983). "The Organized Crime Continuum: A Further Specification of a New Conceptual Model". Criminal Justice Review 8 (52).

Katyal, K.K. (2002) "A mighty challenge, a feeble response", in The Hindu (on Vohra Committee).

Levi, Michael (1998), Perspectives on 'Organised Crime': An Overview, The Howard Journal, 37(4), 335-345.

Lunde, Paul (2004) Organized Crime, DK Adult Publisher.

Nordstrom, Carol (2007). Global Outlaws: Crime, Money and Power in the Contemporary World. Berkeley: University of California Press.

Rajgopal PR (1988) Violence and response: a critique of the Indian criminal justice system, New Delhi, Uppal Pub. House.

Ryan, Patrick J. (1995), Organized Crime: A Reference Handbook, Santa Barbara, CA: ABC-CLIO.

Schloenhart, A (1999). "Organized crime and the business of migrant trafficking". Crime, Law and Social Change 32 (3).

Siddique Ahmad, (2005) Criminology — Problems perspectives, Eastern Book Co., Fourth edition.

Smith DC(1978). "Organized Crime and Entrepreneurship", International Journal of Criminology and Penology 6 (2).

Sullivan, Robert, ed (2002) Mobsters and Gangsters: Organized Crime in America, from Al Capone to Tony Soprano. New York: Life Books.

Woodiwiss, Michael (2001), Organized Crime and American Power, Toronto: University of Toronto Press.

■■■

CHAPTER 11 White-Collar Crime

Learning Objectives :

- ❒ **To Understand the meaning, characteristics and the origin of white collar crime.**
- ❒ **To comprehend the types of white collar crime.**
- ❒ **To discuss the Indian Situation and trends.**
- ❒ **To be acquainted with the control strategies of white collar crime.**
- ❒ **To grasp the role of law enforcement in prosecution of white collar crime.**
- ❒ **To grasp the criminal liability of corporations.**

Introduction

The phrase "white-collar crime" or 'incorporated governance' was coined in 1939 during a speech given by Edwin Sutherland to the joint meeting of the American Sociological Society and the American Economic Association at Philadelphia. The whole address was aimed to shatter the conventional and stereotyped images of the criminals as grown and brought up on the dark side of a town, and the belief that the epicenter of the overall Crime problem was that of the lower Strata of the society. Sutherland defined the term as "crime committed by a person of respectability and high social status in the course of his occupation."

Subsequently he modified his earlier definition of white-collar crime as "crimes committed by a person of the upper socio-economic class who violates the criminal law in the course of his occupational activities and Professional activities" (1949) and in his work he challenged the traditional image of the criminals and the predominant etiological theories of crimes of his days. The white-collar criminals, he identified were often middle aged men of respectability and high Social Status and his definition of white collate Crimes established Status, Occupation and Organization as Central features. He was of the opinion that white-collar criminals were often found in the affluent neighborhoods, and they were all well respected in the community. Sutherland opined that the conceptions and notions of crime in his days were not so satisfactory or rather to say were "misleading and incorrect" which were mainly based on the "biased samples" of criminals and their criminal behaviours.

Prior to Sutherland, Scholars like W. A. Bonger (1916) EA Ross (1907) Sinclair (1906) and Steffens (1903), laid emphasis on the misdeeds by businessmen and elites. Even before these scholars Edwin C. Hill (1924) laid emphasis on the criminal behaviour of the elites in the American Congress in 1872 in his paper 'Criminal Capitalists' but the work of Sutherland was more pioneering compared to others.

The definition given by Sutherland is too broad as he does not specifically mention what he means by a person of respectability, nor does he describe the scope and ambit of the term 'occupation'. Coleman and Moynihan (1996) pointed out that the lack of definite criteria for determining who are the 'persons of respectability and status' has made Sutherland's definition of crime the most controversial. The term 'high social status' as used in the definition also leads to discrepancy since the term is a relative one and therefore not specific. As mentioned earlier, the 'occupation' term used in his definition, leads to ambiguity as he does not categorise what occupation should come within its ambit. Another criticism put up against Sutherland's definition is that white-collar crimes, in the general sense of term, also includes those violations which are not committed in the course of occupation or profession and at the same time, white-collar crimes are not strictly committed by the members of the higher socio-economic strata. For example, the white-collar crime of tax-evasion is not only committed by persons belonging to the higher strata but also by persons belonging to middle or even lower strata of the society.

However, it was not only Sutherland, but many others who preferred to define white-collar crimes on the status basis. Hartung (1950:25-34) defines "a white-collar offense as a violation of law regulating business, which is committed for a firm by the firm or its agents in the conduct of its business." Moreover there are some who argue that differentials of power and influence are the keys to identify, satisfactorily frame, and unravel fundamental questions about crime and crime control. For example, investigations of sexual harassment have shown there the offender has always been of a superior status than that of the victim. Therefore the principal reason propounded behind the commission of a white-collar crime is that of greed and arrogance and not of necessity or any compulsion. White-collar crimes throw light on the diverse criminal behaviour. The status-based definition of white-collar crime draws on and reaffirms criminological ties to macro level analyses of social and political processes.

However, there are others who refuse to give much emphasis on offender's social position and repute as the only basis of classifying white-collar crimes. They advocate that collaring should be done not of the criminals but of the crimes. The concept of abuse of trust, for example, can be considered to be a white-collar crime. Abuse of trust means exploitation of fiduciary position of the agent responsible for the safe custody, discretion, information or for property rights. Here since the trust crime is being committed by citizens wearing both blue-collars and white-collars. Similar instance can be done as that of adulteration of food, where this type of offence is committed not only by the upper class people but also by middle class businessmen involved in the food business. Therefore, one should not always peep into the offender's wardrobe and his official position but consider the modus operandi of his misdeeds and the ways he exploits. Whatever the merits of the position, they hold strong that in white-collar crimes the offenders are not always powerful and the victims are ordinary citizens.

Therefore, Sutherland's restriction of white-collar crime to acts committed by men/women in the course of their operation, excludes extra vocational crime. Whereas some other views refused to restrict themselves within the circumference of the elite class and tried to go deep into it. However, whatever definitions have been offered, it failed to acquire general acceptance as it did not seem to be adequate.

Hugh Barlow (1995) was of the opinion that white-collar crime is not only committed by the people of high social status in their occupational capacity but also is committed by the people of lower strata. Criminologists were of the opinion that Sutherland himself creates lots of confusion regarding the concept of white-collar crime.

There is an enormous overlap between criminal and civil law in terms of their rules of liability or moral judgments, they imply about behaviours and purposes. The argument is that the distinction between white-collar offenses and criminal offenses is primarily procedural rather than behavioural. Therefore to conclude it can be said that while the legal treatment of some acts as crimes and others as civil violations (torts and regulatory violations) is itself worthy of sociological study, this difference in the definition of illegal behaviour should not be accepted uncritically, as it has been in the past. The absence of a precise definition of white-collar crime has plagued criminologists to analyze and interpret the concept of white-collar crime in their own manner. Subsequently it was very much advent that for some scholars, it was really very hard to accept 'Position' as primordial factor. For that very reason today there exists profound disagreement over the precise definition of white-collar crime. Definitional issues pertaining to white-collar crime have reference to three main indicators: the type of offense, the type of offender and the organizational culture.

(a) Federal Bureau of Investigation has adopted the narrow approach, defining white-collar crime as "those illegal acts which are characterized by deceit, concealment, or violation of trust and which are not dependent upon the application or threat of physical force or violence" (1989:3). The type of offense, include property crime, economic crime, and other corporate crimes like environmental and health and safety law violations.

(b) By the type of offender, it takes into account the social class or high socio-economic status, the occupation of positions of trust or profession, or academic qualification, researching the motivations for criminal behaviour, e.g. greed or fear of loss of face if economic difficulties become obvious. Necessity is not always the motive behind the commission of a crime. Thus, the only way that one crime differs from another is in "the backgrounds and characteristics of its perpetrators; the poor and disreputable fodder routinely encountered in police stations and in studies of street crime are seldom in evidence here. Most if not all white-collar offenders by contrast are distinguished by lives of privilege, much of it with origins in class inequality.

(c) By organizational culture rather than the offender or offence which overlaps with organized crime, Appelbaum and Chambliss (1997:117) offer a twofold definition: (a) Occupational crime which occurs when crimes are committed to promote personal interests, say, by altering records and overcharging, or by the cheating of clients by professionals and (b) Organizational or corporate crime which occurs when corporate executives commit criminal acts to benefit their company by overcharging or price fixing and false advertising.

Despite of its fundamental awkwardness, the term "white-collar crime" has now been deeply embedded within our legal, moral, and social science vocabularies. The term persists and proliferates not so much in spite of its lack of definitional precision, but because of it speakers attribute to it those meanings that correspond to their own particular analytical or ideological concerns.

Historical Background of White-Collar Crime

The earliest documented case of white-collar crime law dates back to 15th century England. There has been a case popularly known as the Carrier's case of 1473, where the agent was entrusted to transport wool and he attempted to steal some of it for him. Therefore, the Star Chamber and Exchequer Chamber of the English Court of Law adopted the 'breaking bulk' doctrine as it constituted the crime of larceny. The term breaking bulk means taking anything out of a package or parcel and thereby destroying its entirety. It was thus important in connection with the subject of bailment, involving as it did the distinction that where a bailee received possession of goods in a box or package, and then sold them as a whole, he was guilty only of a breach of trust, but if he "broke bulk", the doctrine of breaking bulk would apply and he would be guilty of felony, i.e. liable for confiscation of a convicted person's land and goods.

However, the growth of industrial capitalism in the eighteenth century ushered a new history of crime and criminality. The very base of industrial capitalism is based on coercion and robbery. Now, before we discuss the topic let us understand the meaning of capitalism. In the words of V.L. Lenin (1964), "Capitalism is the name given to the social system under which the land, factories, implements, etc. belong to a small number of landed proprietors and capitalists, while the mass of a people possess no property or very little property and are compelled to hire themselves out as workers. The land owners and factory owners hire workers and make them produce wires of different kind which they sell on the market. The factory owners furthermore, pay the workers only such a wage as provides a bare subsistence for them and their families, while everything that the worker produces over and above these amount goes to the factory owner's pocket as his profit."

Therefore, it can be derived from the definition is that, since there has been nothing to the major mass of the society except their bare subsistence, they were left with no better option than to sell their labor-power to survive. Whereas, the exploiting class, generally termed as the bourgeoisies, took the advantage of the situation and made more and more wealth out of the labors of the working class. Consequently, the wealth remained concentrated in the hands of only certain class of people, where most of them formed a part of the exploited class left with no freedom for work but to sell their labor-power.

The process of emergence of these conditioned was termed by Karl Marx as 'primitive accumulation' while in the words of Adam Smith, it was 'previous accumulation.' Therefore, the Dutch Marxist, William Bonger contended that criminal attitude develops among the working class under capitalism due to conditions of misery and at the same time the criminal attitude develops among the bourgeoisie from the avarice fostered when capitalism strives. Therefore, it is to be understood that industrial capitalism brought about a grave change in the mind and attitude of the people. Accumulating money at the cost of others became the sole goal of the individual who were in a position to dominate over the will of the others. With this began the cat fight of grabbing money by hook or by crook.

Britain, among the other European countries were concerned, played the most nasty game in acquiring wealth giving up all the ethics and morals they ever had. They made wealth by looting their colonies and adopting the devastating barbaric method in doing it. According to the Parliamentary Reports, it has been found that the British East India Company, only within their first ten years of colonial rule in India extorted wealth from the Indians in the name of 'gifts' which worth round six million pound sterling. Therefore, there can be no denial of the fact the very structure of the industrial capitalism is built of crimes and criminality.

Therefore, the birth of white-collar crimes took place with the Industrial Revolution in Western industrial societies which led a vast impact all throughout the globe. The developing countries, which at that time were the colonies of the great European Nations, were the worst victims of these crimes so committed against them. New companies rose into power, they were able to squelch competitors and then implement monopolistic policies without fear of being outsold by other companies. The public had to pay outrageously exorbitant prices for the same thing that was available previously at a much cheaper rate. The reason behind the exorbitant rise in the price rate was nothing but the corporate greed.

The greed of making money by the people of the upper class was not confined within the boundaries of Europe. The Industrial revolution threw a global impact whereby worldwide people indulged in making money. These gave rise to resentment and political movements rallied for laws so as to prevent monopolistic practices. It succeeded in United States of America in 1890, when Congress passed the Sherman Antitrust Act which took the initiative to make the monopolistic trade illegal. Other industrialized countries like Great Britain had a history of penalties involving white-collar crime by that time, but it was not as sweeping as the Sherman Act. Some nations implemented a smattering of these laws, known as competition or antitrust law, but did not have a strong binding force for a long time.

But more anti-white-collar crime sentiment rose in the late 19th century and early 20th century in the United States as a result of a group of journalists known as muckrakers who more often than not started exposing corruption both in the public and private sectors. They laid much focus on the prevalent stock fraud, insurance fraud and underhanded practices of monopolistic companies that had grabbed under the Sherman Act. The muckrakers' exposes gave rise to public resentment and thereby called for in some reform. By 1914, Congress attempted to solidify and strengthen the sentiment laid down by the Sherman Act, which was used against labor unions, with the Clayton Antitrust Act. This Act was much stricter and went much further than the Sherman Act in making particular monopolistic practices illegal.

In addition to these anti-trust laws, there have been certain other fiscal laws which were passed prior to Sutherland's address at the American Sociological Society. After the economic depression prevailing at that time, people went to great lengths to rebuild their financial security, and it is theorized that this led many hard workers who felt they were underpaid to take advantage of their positions. Despite the great lengths, Sutherland went to describe exactly what he categorized as white-collar crime. Though he tried to bring about clarity of the subject, it suffered from certain ambiguities as discussed later.

Much of Sutherland's work was focused at separating and defining the differences in blue-collar street crimes such as murder, burglary, theft, assault, rape, arson, vandalism, so on and so forth which are often blamed on psychological, associational, and structural factors. Instead, he evaluated that white-collar criminals are opportunists, who over time look for the opportunity to take advantage of their circumstances to accumulate financial gain. They are educated, intelligent, affluent, confident individuals, who were qualified enough to get a job which allows them the unmonitored access to often large sums of money. Many also use their intelligence to con their victims into believing and trusting in their credentials.

Occupational Crime and Street Crime

Although there has been some debate as to what qualifies as a white-collar crime, the term today generally encompasses a variety of nonviolent crimes usually committed in commercial situations for financial gain. White-collar crime therefore overlaps with corporate crime because the opportunity for fraud, bribery, insider trading, embezzlement, computer crime, and forgery is more available to white-collar employees.

In 1979, U.S. Department of Justice official Herbert Edlehertz described white-collar crime as an illegal act or series of illegal acts committed by non-physical means and by concealment or guile, to obtain money or property, or to obtain business or personal advantage. The FBI defines it very simply as "lying, cheating, and stealing" and it is classified by the Department of Justice in USA under the Administration Improvement Act (AIA) of 1979 as "non violent illegal activities which principally involve traditional notions of deceit, deception, concealment, manipulation, breach of trust, subterfuge or illegal circumvention." The term today generally encompasses a variety of nonviolent crimes usually committed in administrative, occupational and commercial situations for material gain. White-collar Crime is the crime in the Suite associated with occupations. They are the people who wear white-collars to work steal and kill, by using guilt and deceit. More money is stolen and more people die every year as the result of scams and willful illegal corporate activity than as a result of the activities of street criminals.

Occupational crime is the crime committed by individuals in the course of their employment. Every occupational category generates a considerable number of criminals and the higher the prestige of the occupation, the more their criminal activities cost the general public. Occupational crime is different from other crime in the sense that it is the function of opportunities but others look for opportunities to commit crime. Occupational crime occurs when crimes are committed to promote personal interests, say, by altering records and overcharging, or by the cheating of clients by professionals.

The most glaring example and common form of occupational crime is employee theft and vandalism. This might range from draining company funds by creative bookkeeping or sophisticated computer techniques to stealing small items of little value from the office. In the same manner vandalism by disgruntled employees may run the gamut from scrambling important computer data to scratching graffiti on newly painted walls. Just as we all pay for street crime through the portion of our taxes that goes to support criminal justice system, we all pay for employee crime because victimized companies merely pass on their loses to their customers via higher prices. Employee crime may also lead to business going bankrupt and employees losing the jobs this business provided.

Occupational crime differs from common street crime in that it is committed by people in a position to do so. The motives of occupational criminals are the same as those of street criminals. Chronic white-collar offenders were similar to street criminals in their patterns of prior deviance. Studies reveal that high status white-collar criminals are about 10 years older than the average street criminal, and males are overrepresented. Following Table presents the differences between white-collar and street criminality.

TABLE 11.1

White-Collar Crime and Street Crime

White-collar Crime	**Street Crime**
Committed by people in a position	Open & Independent of Status
Lower ethical and moral standards	*mens rea*
Function of opportunities, prey comes to them	Look for opportunities to commit crime
More costly to society	Less costly
Non-violent guile	Violent
Prosecution difficult	Prosecution easy
Fraud, Tax Law Violations & Tax Fraud, Embezzlement	Robbery, Burglary, Theft

Not only are there characteristic differences between the two but also are there the differences in treatment meted out to them. The empirical data clearly demonstrate a double standard between white-collar crimes and so-called street crimes. There are a number of reasons to explain why white-collar criminals are not more rigorously pursued. By virtue of their relative affluence, those accused as white-collar offenders are able to afford the fees of the best lawyers, and may have friends among senior ranks of the political elite (Cronyism), the judiciary and the law enforcement agencies. Cronyism is partiality to long-standing friends, especially by appointing them to positions of authority, regardless of their qualifications. Hence, cronyism is contrary in practice and principle to meritocracy. Cronyism exists when the appointer and the beneficiary are in social contact; often, the appointer is inadequate to hold his or her own job or position of authority and for this reason the appointer appoints individuals who will not try to weaken him or her, or express views contrary to those of the appointer. Politically, "cronyism" is derogatorily used. These connections often not only ensure favorable treatment on an individual basis, but also enable laws to be drafted or resource allocations to be shifted to ensure that such crimes are not defined or enforced too strictly. It is a fact that virtually no police effort goes into fighting white-collar crime, and the enforcement of many corporate crimes is put into the hands of government investigation agencies which can act only as watchdogs and point the finger when an abuse is discovered. Different public policies are at work and there are differences in the level of public interest, case complexity, and a lack of white-collar related literature, all of which has a significant effect on the way white-collar offenders are sentenced, punished, and perceived by the public.

Another reason for differential treatment might be the fact that criminal penalties tend to be more reflected to the degree of physical force or violence involved than to the amount of monetary loss, all other things being equal. Because white-collar crimes are committed by those with opportunities that do not require violence, they are far less likely to garner more severe criminal penalties. For example, someone who mugs a victim on the street by threatening to knife them, and steals their wallet, might very likely be punished with a more severe sentence than an inside trader who cheats shareholders out of a million dollars.

Blue-collar Crime

The types of crime committed are a function of the opportunities available to the potential offender. Thus, those employed in relatively unskilled environments and living in inner-city areas have fewer "situations" to exploit (Clarke: 1997) than those who work in "situations" where large financial transactions occur and live in areas where there is relative prosperity. Newman and Clarke (2003) applied the Situational Crime Prevention strategy to e-crime where the opportunities can be more evenly distributed between the classes. Blue-collar crime tends to be more obvious and attract more active police attention (e.g. for crimes such as vandalism or shoplifting which protect property interests), whereas white-collar employees can intermingle legitimate and criminal behaviour and be less obvious when committing the crime. Thus, blue-collar crime will more often use physical force whereas in the corporate world, the identification of a victim is less obvious and the issue of reporting is complicated by a culture of commercial confidentiality to protect shareholder value. It is estimated that a great deal of white-collar crime is undetected or, if detected, it is not reported. In the truest sense, the terms white and blue-collar crime refers to police slang for an arrest of a suspect, or collar. Blue-collar crimes are those that involve local police (known for wearing blue, or, "Men in Blue") and white-collar crimes are those involving Federal agents, such as FBI (who typically wear suits and ties with white shirts).

State-corporate Crime

Because the negotiation of agreements between a state and a corporation will be at a relatively senior level on both sides, this is almost exclusive a white-collar "situation" which offers the opportunity for crime. Organizational or corporate crime occurs when corporate executives commit criminal acts to benefit their company by overcharging or price fixing, false advertising to name a few. Newcomers entering such corporate environments are socialized into the prevailing way of doing things. The literature indicates that those who engage in corporate crime tend to have an external locus of control. The lower the stage of moral development, the more likely the person will engage in corporate wrongdoings. It appears that people who choose business careers tend to have lower ethical and moral standards than people who choose other legitimate careers. Efforts to differentiate between offenders and non-offenders in the white-collar world have focused largely on locus of control, moral reasoning, and Machiavellianism. Machiavellianism is the unprincipled and uncaring intelligent manipulation of others for personal gain.

Corporate Crime is criminal activity on behalf of a business organization. During much of American history the primary legal stance relating to activity of business was decidedly laissez-faire (leave it alone to do as it will). Under the influence of capitalist philosophy American Courts adopted the view that governments should not interfere with businesses by attempting to regulate it. For a long period of history victims of defective and dangerous products could not sue corporations for damages because the guiding principle was caveat emperor (let the buyer beware), and unhealthy and dangerous working conditions in mines, mills and factories were excused under the freedom of contract clause of the constitution.

Although attitudes in the 20th century changed considerably, crimes of frauds, concealment, and misrepresentation continued to victimize all sorts of groups and individuals in society. Crimes such as these are actions committed during the course of fulfilling the name of corporate profits and growth.

The first major scandal of the first decade of 21st century is known as The Enron Scandal. It is that Enron Corporation in US, which has been called 'one of the most intricate pieces of financial chicanery in history' and for investors in its stock and its employees. The Enron scandal did tremendous damage to the company created a crisis of investor's confidence the links of which has been seen since the great depression.

Enron was a $100 billion corporate empire that had more then 200,000 employees in 40 countries and controlled about one quarter of all trading in natural gas and electricity in the United Status. The company poured millions of dollars into political campaign and lobbyists arguing of further deregulation of energy markets. The Enron hierarchy expected the never-ending innovation and growth from its executive to feed this monster enterprise. In response executives created imaginary markets, 'paper partnerships' and phantom growth that enabled them to report profits that did not exist and to hide debts hat did. Executives were accomplished at cooking the financial books in many ingenious ways, which kept Enron's stock prices rising and thus their own compensation. Because mush of the competition received by Enron's executives was stock based, they had major incentives to make the company look as good as possible to investors by reporting high profits.

TYPES OF WHITE-COLLAR CRIME

White-Collar Crimes may be divided into Occupational Crime and Organizational Crime but in common parlance there exist 25 popular types of White-Collar Crimes as :—

1. **Bank Fraud :** To engage in an act or pattern of activity where the purpose is to defraud a bank of funds.
2. **Blackmail :** A demand for money under threat to do bodily harm, to injure property or to expose secrets.
3. **Bribery :** When money, goods, services or any information is offered with intent to influence the actions, opinions and decisions of the taker, constitutes bribery. One may be charged with bribery whether you offer the bribe or accept it.
4. **Cellular Phone Fraud :** Unauthorized use or tampering or manipulating cellular phone services. This can be accomplished by either use of a stolen phone or where an actor signs up for service under false identification or where the actor clones a valid electronic serial number (ESN) by using an ESN reader and reprograms another cellular phone with a valid ESN number.

5. **Embezzlement :** When a person who has been entrusted with the money or property, appropriates it for his or her own use and benefit..
6. **Counterfeiting :** Occurs when someone copies or imitates an item without having been authorized to do so and passes the copy off for the genuine or original item. Counterfeiting is most often associated with money however can also be associated with designer clothing, handbags and watches.
7. **Forgery :** When a person passes false or worthless instruments such as cheque or counterfeit security with intent to defraud or injure the recipient.
8. **Tax-Evasion :** Frequently used by the middle — class to have extra-unaccounted money.
9. **Adulteration :** Adulteration of foods and drugs.
10. **Professional crime :** Professional occupational crime is the one committed by professional practitioners such as physicians, engineers, teachers and lawyers in the course of their practices.
11. **Computer fraud :** Where computer hackers steal information sources contained on computers such as: bank information, credit cards, and proprietary information.
12. **Credit Card Fraud :** The unauthorized use of a credit card to obtain goods of value.
13. **Extortion :** Occurs when one person illegally obtains property from another by actual or threatened force, fear, or violence, or under cover of official right.
14. **Health Care Fraud :** Where an unlicensed health care provider provides services under the guise of being licensed and obtains monetary benefit for the service.
15. **Insider Trading :** When a person uses inside, confidential, or advance information to trade in shares of publicly held corporations.
16. **Insurance Fraud :** To engage in an act or pattern of activity wherein one obtains proceeds from an insurance company through deception.
17. **Investment Schemes :** Where an unsuspecting victim is contacted by the actor who promises to provide a large return on a small investment.
18. **Kickback :** Occurs when a person who sells an item pays back a portion of the purchase price to the buyer.
19. **Larceny/Theft :** When a person wrongfully takes another person's money or property with the intent to appropriate, convert or steal it.
20. **Money Laundering :** The investment or transfer of money from racketeering, drug transactions or other embezzlement schemes so that it appears that its original source either cannot be traced or is legitimate.
21. **Racketeering :** The operation of an illegal business for personal profit.
22. **Securities Fraud :** The act of artificially inflating the price of stocks by brokers so that buyers can purchase a stock on the rise.
23. **Telemarketing Fraud :** Actors operate out of boiler rooms and place telephone calls to residences and corporations where the actor requests a donation to an alleged charitable organization or where the actor requests money up front or a credit card number up front, and does not use the donation for the stated purpose.

24. **Welfare Fraud :** To engage in an act or acts where the purpose is to obtain benefits (i.e. Public Assistance, Food Stamps, or Medicaid) from the State or Federal Government.

25. **Weights and Measures :** The act of placing an item for sale at one price yet charging a higher price at the time of sale or short weighing an item when the label reflects a higher weight.

Thus the most common white-collar offenses include: antitrust violations, computer and internet fraud, credit card fraud, phone and telemarketing fraud, bankruptcy fraud, healthcare fraud, environmental law violations, insurance fraud, mail fraud, government fraud, tax evasion, financial fraud, securities fraud, insider trading, bribery, kickbacks, counterfeiting, public corruption, money laundering, embezzlement, economic espionage and trade secret theft.

Indian Scenario

White-collar crimes are to be considered as a global phenomenon to which India is no exception. If there is an industry in which India has surpassed the developed west then it is the field of white-collar crimes. White-collar crimes in India is not in total based on the theory propounded by Prof. Sutherland but on the concept as well by Prof. Hugh Barlow as "Crimes committed by the people of lower strata in their occupational status." As discussed earlier, white-collar crimes emerged in India with the advent of the British colonisation during the period of industrial capitalism. Prior to that, instances of men working with the District treasury embezzling with the money kept under his safe custody or bribing practised among the officials were found. Therefore, the white-collar crimes were confined to this limit. Thus, the people indulging in the white-collar crimes then can to said to be mere 'grass eaters' the people in the modern times have reached the stage of 'meat-eaters'.

Thus the white-collar crimes become phenomenon to be reckoned with industrial revolution. Modern industrial capitalist economy which evolved with time became complex in nature as it developed a growing commercial nexus among insurance, banking, stocks and related corporate matters. This in turn, gave rise to critical legal intricacies relating to property rights and other legal matters which paved the way for the birth of a new class of professionals of advocates who in the name of providing justice started abetting in the wrong and thereby pursued their own narrow interest. A large number of advocates evolved who forget the pious oath of serving the society and start looking for the legal loopholes and concentrate mainly in helping out the rich entrepreneurs to grow richer. They make extensive study to try out ways for maximum tax evasion for these rich corporate personalities as well as for themselves. The white-collar crimes committed by these legal practitioners are not only confines in sorting out illegal methods of tax-evasion. There are very frequent instances of unscrupulous and unethical practises like that of fabricating false evidence, engaging professional witnesses, thereby violating ethical standards of legal profession and dilatory tactics in collusion with the ministerial staff of the courts.

Next are the instances of white-collar crimes committed in the Indian society by the lawyers. Many lawyers also work on a free-for-service basis, thus generating the same temptations to increase their incomes by fraudulent means. Frauds perpetrated by lawyers can include major embezzlement of client's funds, bribery of witnesses and judges, persuading clients to pursue fraudulent or frivolous lawsuits, billing clients for hours not worked, filling unnecessary motions and complicating a simple legal matter to keep clients on hook. They in the name of interpreting the laws often act as the protective shield for the goons having or not-having any political colour and allow them to go free whereas they should have been subject to deterrence. It is the most unfortunate situation at the same time devastating, because here the crimes are committed by those individuals who are being given by the State the responsibility to ensure justice.

In India, the white-collar crimes are so widespread that it does not confine itself in the legal arena. Similar unfortunate instances can be drawn from other professions too, like that of medical practitioners, engineers, educationalists, businessmen, politicians and the list goes on. The medical practitioners are often found involved in issuance of false certificates, carrying out illegal abortions, selling out sample drugs and medicine, even in some cases adulterated drugs and medicines to the patients. Dilatory tactics are often adopted by them in providing treatment to their patients with a mens rea to extract huge amount of money, no matter the person has good practice. Some of the notorious instances are like that of Nithari case, where the medical professionals put up before the society the optimum level of brutal character they can reach for the crave of making money. Misleading and fake advertisement claiming absolute cure is also one of the frequent malpractices being carried out in the medical profession. Frauds committed by physicians include such practices as filing insurance claims for tests or procedures not performed; performing expensive and unnecessary operations; steering patients to laboratories; pharmacies or some facility in which the doctor has some financial stakes; referring the patients to other doctors in return for kickbacks; and prescribing and charging for brand name drugs and substituting generic drugs. The problem lies in the fact is, they often escape punishment, since they cannot be said to have violated the letter of law, but, by violating the spirit of law, they commit crimes which are truly anti-social and creates enormous damage to the public health and safely at large.

Speaking of the engineers' role in having their role to play in white-collar crimes, we often find instances of underhand dealing with contractors, suppliers, passing of sub-standard works and maintenance of bogus reports of the labor work. They financially earn more for their low grade works from the contractors, than they can earn for the genuine work. Therefore, many of them, out of the greed of earning more and more, play dangerously with thousands of lives of the individuals.

When the matter comes to the white-collar crimes educational institutions do come in the league to operate with impunity. A nastier role is played by the private institutions who are least bothered in providing the education, but only concentrate of making business at the cost of the children's future. Even rackets operate in these institutions for procuring students to appear in the examinations on the basis of manipulated eligibility certificates, thereby damaging the standard of education in India. When it comes to the Governmental institutions, the teachers and staffs of the institutions are often found to be involved in unscrupulous practices, since they can hardly make fortune from the inadequate salary they receive from the government. Teachers often drag the students for taking private tuitions and even go to the extent of blackmailing them of ruining their future, if they deny doing so.

These are only a handful of instances of white-collar crimes practised in day to day life by certain professionals in the course of their profession. The major role in committing white-collar crimes are played by the business tycoons and politicians, whose greed and wants multiply with the more they acquire. In India, whenever any major scandal comes to the media focus, a thorough investigation always finds an unlawful involvement of political parties in it. So far as the businessmen are concerned, their acts of white-collar crimes go beyond count. They are termed as the corporate criminals who more often than not are involved in illegal contracts, combination and conspiracies of trade restraints, unfair labor practices, selling of adulterated foods and drugs, bribing of public officials so on and so forth. They take advantage of the corporate veil and indulge in a number of crimes. The recent Satyam scam case is one of the worth-mentioning illustrations, where it was seen how an individual, hiding himself in the veil of incorporation, indulge in defrauding crores of money.

The white-collar crimes are often master minded and are carried out in a planned manner by technocrats in the form of scams, frauds, etc. facilitated by technological advancements. In these types of offences not only the individual is being with pecuniary loss but also, such offences like peddling drugs and narcotic substances, counterfeiting of currency, financial scams are some of the crimes which evoke serious concern and impact to the national security and governance. Violation of foreign exchange regulations and import and export laws are frequently resorted to for the sake of huge profits.

Another major white-collar crime very common among the individuals, no matter he/she belongs to the middle or upper strata of the society is that of evading taxes. The complexity of the taxation laws provided a number of loopholes through which many individuals tried to escape. A tax-evasion has known no professional or class boundary in our country. Be it an engineer, doctor, advocate, a business tycoon or a simple small industry trader- all has learnt the trick of evading taxes. The main difficulty posed before the Income Tax Department is to acquire true information of the real and exact income of these professionals. It is often alleged that only an insignificant amount of their total income is posed a income before the Income Tax Department and the rest therefore goes into the circulation as black money. The frequent modifications in the tax-laws of the country has been able to add very little to put a check on this continuing menace which is throwing a great negative impact in the Governmental revenue and thereby the growth of the country.

White-collar crimes are also committed by the people of lower social strata in their occupational capacity (by Prof. Barlow) as adulteration of milk by the milk man, selling adulterated food by the shopkeeper, selling expired medicine, taking out few kilos of gas from the cylinder and so on.

Gradually white-collar crimes acquired an established place within the society and with the introduction of famous license-quota-permit Raj in the earlier seventies it took a long strive forward to the increasing trend of white-collar criminality by giving tremendous power to the Trio as — the politicians, the bureaucrats and the businessmen. These unholy three put their jaws in our administrative system in such a manner that it nurtured the total system of quasi corrupt Indian society to a complete corrupted one.

Traders in India were not so far behind in creating an irreparable damage to the society at large. They have mainly engaged themselves in hoarding, profiteering and black marketing of the essential commodities. Moreover sometimes for monetary gains they used to engage themselves in adulteration of foods which may turn dangerous to a person's life. The Report of the Vivian Bose Commission of Inquiry (1956) highlighted how the industries indulge into white-collar crimes such as fraud, falsification of accounts, tampering of records for personal gains, tax evasion, so on and so forth.

To comment specifically on Indian traders for Black Marketing of the essential commodities, profiteering and Hoarding, Monopolies Inquiry Commission, (MIC) which submitted its report in 1965, gave a graphic account of that in the following words as: "There is hardly anybody in India who has not been a victim of the practice of hoarding, cornering and profiteering. Whenever there is a slight shortage — even temporary — in any consumer goods for which the demand is urgent and inelastic, almost every trader — it is perhaps unnecessary to use the qualification 'almost' — conceals his stock and blindly tells the customers that he has not got the commodity in stock...". Based on the MIC recommendations, India enacted its first competition law called the Monopolies and Restrictive Trade Practices Act, 1969 (MRTP Act). The MRTP Act aimed at achieving twin objectives of preventing concentration of economic wealth by private enterprises, and ensuring non-prejudicial public interest in economic activities within India. The MRTP Act provided for a setting up of a Commission to implement the provisions of the law.

The Santhanam Committee Report in its finding (GOI,1964: 251-253) presented a clear picturesque description that in India the white-collar crimes so defined, are mostly rampant among the businessmen, industrialists as well as corrupt officials and private practitioners. Highlighting the magnitude of the crimes in India, the Commission on "Prevention of Corruption" in its report quoted,

"The advancement of technological and scientific development is contributing to the emergence of 'mass society' with a large rank of file and a small controlling elite, encouraging growth of monopolies, the rise of the managerial class and intricate institutional mechanisms. Strict adherence to high standard of ethical behaviour is necessary for the even and honest functioning of the new social, political and economic processes. The inability of all sections to appreciate this need in full results in the emergence and growth of white-collar and economic crimes renders enforcement of all laws, themselves not sufficiently deterrent, more-difficult. Tax evasion and avoidance, share-pushing, malpractices in the share market and administration of companies, monopolistic control, usury, under-voicing or over-voicing, hoarding, profiteering, substandard performance of contracts of constructions and supply, evasion of economic laws, bribery and corruption, election offences and malpractices are some examples of white-collar crime."

It was on the basis of the report of this Commission that a new chapter was suggested to include into the Indian Penal Code, 1860 by making amendments to the Act. The motive behind such suggestion was to bring white-collar crimes under the purview of the criminal law and at the same time broadly classify the offences that can come under the purview of white-collar crimes. The matter was referred to the Law Commission by the Government of India for considering it. However, the Law Commission disagreed with the proposal and observed that such offences which are put under the banner of white-collar crimes are to be left to be dealt with by special and self-contained enactments which act as supplement to the basic criminal law. The reason behind that was, as discussed earlier, there have been a lot of debates regarding criminalising acts constituting white-collar crimes.

According to V. R. Krishna Iyer, J. "economic offences often are subtle murders practised on the community or sabotage of the national economy." So it may be termed as the "White-collar Economic Offences". These economic offences can devastate an entire community rather than robbing a lone victim. Their impact can last for years, stealing crucial services or a lifetime's savings through crimes invisible to their victims. Similarly, illegal accumulation of foreign exchange through just one type of fraud or through under invoicing of exports and over invoicing of imports — is calculated to be estimated between ₹40-50 crores every year.

Two instances of embezzlement and fraud as provided by a report made by Vivian Bose Commission are — Notorious Dalmia-Jain and Mundhra case in which loss amounted were estimated at ₹ 3.5 crore. While dealing with the investigation of Mundhra case, Mr. M. C. Chagla made following observations as: "Mundhra is a flamboyant personality and a financial adventurer whose only ambition is to build up an industrial empire by dubious means."

In spite of the fact that a large number of economic offences have been unearthed in our country in the last five years as Securities Scam, Hawala Scam, Urea Scam, Sugar Scam, Banking Scam, Tele-Communication Scam, Fodder Scam, Stamp scam etc. and an effort from the government of appointing numerous Commissions as Vivian Bose Commission, Bakshi Tek Chand Committee, Parliamentary Committee on the jeep scandal, Railway corruption Inquiry Commission, Sadasivam Committee of enquiry, S. R. Ray Commission of Inquiry, S. R. Das Commission, M.C. Chagla Committee and many more in which thousands of crores of rupees were involved, surprisingly no offenders have been convicted so far.

Are there really any remedies to curb these malpractices at the advent of our Welfare state ? Why those obligations imposed on the state to achieve the status of welfare state are not implemented properly ? Restrictions imposed on the ownership and the distribution of the national wealth run from the following provisions of our Indian Constitution that "The state shall in particular direct its policy towards securing that the ownership and control of the material resources of the community are so distributed as best to sub serve the common goods; that the operation of the economic system does not result in the concentration of wealth and means of production to the common detriment." This philosophy forced our law makers to pass certain legislations, violations of which may lead to tremendous fillip to white-collar criminals or occupational related crime in India which include Essential Commodities Act, 1955; Industrial (Development and Regulation) Act, 1957; Import and Export (Control) Act, 1947; Companies Act, 1956; Foreign Exchange (Regulation) Act, 1973; Central Excises and Salt Act, 1944; Income-tax Act, 1961; Customs Act, 1962; The Conservation of Foreign Exchange and Prevention of Smuggling Activities Act, 1974 as amended in Act 52 of 1993; The Smugglers & Foreign Exchange Manipulators (Forfeiture of Property) Act, 1976; The Consumer Protection Act, 1986; Narcotic Drugs and Psychotropic Substances Act, 1985; Monopolies and Restrictive Trade Practice Act, 1992 are enacted to prevent these offences which causes irreparable damage to the public health.

Specifically, Section 4 of the Prevention of Corruption Act, 1947, where there is a presumption for instance, "that money received other than legal remuneration by a public servant is an illegal gratification." They laid more emphasis on the workings of the public servant as they may use their occupational capacity to exploit the mass in general.

Subsequently on the report submitted by Santhanam Committee certain legislations earlier enacted were amended as - Anti-Corruption Laws (Amendment) Act, 1964; Foreign Exchange (Amendment) Act, 1964; Prevention of Food Adulteration Act, 1954; Wealth Tax (Amendment) Act, 1964 and more powers have been conferred on the investigating officers and on the Magistrates for conducting the proceedings of the summary trials.

Criminal Liability of the Corporations

There are a number of corporate crimes that have come into light now days. One of the major havoc that is created in present times is because of mysterious disappearance of corporations. The CBI probe against Win Chaddha and Hindujas in the Bofors Scandal and FEERA enquiries against Delhi firms Jyotsna Holding Pvt. Ltd. Which involved 6.5 crores of illegal commission for a contract of pipes with ONGC in 1988 was a glaring illustration. The ever biggest multi-thousand crore security scam by Harshad Mehta in 1991-92 involving persons of high positions in banks, including SBI further puts up the glimpse of massive corruption, fraud and mismanagement in the banks and nexus between bankers and underworld brokers. Again another scam of stock market share in 2001, master-minded Ketan Parikh, is another illustration of manipulation of black market. The 2G Spectrum Scam of 2011 has left the worst example. In Indian as well as in international perspective the whole legal framework to hold the corporations and multi national companies accountable is systematically being dismantled, even as corporations and other agents of globalization dictate policies of nations. The corporate sector enjoys far more rights than the common people. With the onset of the new trade regime, national laws are being changed to empower corporations with the right to hire and fire at will, to get the first right over natural and community resources.

But now the time has come to put a check over their arbitrary acts. There has been a debate as to whether a Corporation can be held criminally liable. The explanation can be given with the help of the following theories :

Nominalist Theory of Corporate Personality : According to this theory, corporations as nothing more than collectives of individuals. In this an individual first commits the offence; the responsibility of that individual is then imputed to the corporation.

Realist Theory : According to these theory corporations have an existence, which is to some extent independent of the existence of its members. Here, the responsibility of corporation is primarily. In my opinion realist theory looks more convincing and practically applicable.

One of the argument which sustains the presumption that corporate can be held criminally liable is that in many cases it is the corporation itself, through its policies or practices, that has done wrong and prosecution and punishment should be directed at the real wrongdoer. In many cases there is no individual who, alone, has committed a crime.

It is the conjunction of the practices of several individuals, all-acting in compliance with a company's sloppy or non-existent procedures that has caused the harm. Alternatively, in many cases companies have complex structures with responsibility buried at many different layers within the corporate hierarchy making it difficult, if not impossible, to determine where the true fault lies.

Whether a company being not a physical entity can be made subject to sentence of imprisonment ? There is no controversy when fine is only punishment given under any statute. There is also no lies when statute entrusts the court with discretion to inflict fine or imprisonment, as in this case court shall inflict only fine on company. Because a company being a Juristic person cannot obviously be sentenced to imprisonment as cannot suffer imprisonment. Judicial controversy lies in that situation when statute prescribes mandatory imprisonment with fine as a punishment for an offence. In 2003 Supreme Court in Assistant Commissioner, Assessment-II, Banglore & Ors. v. Velliappa Textiles Ltd & Anr [JT 2003(suppl. 2) SC 99; (2003) 11 SCC 405] took the view that since an artificial person like a company could not be physically punished to a term of imprisonment, such a section, which makes it mandatory to impose minimum term of imprisonment, cannot apply to the case of artificial person.

However, Supreme Court in 2005 in Standard Charted Bank v. Directorate of Enforcement [*JT 2005 (5) SC 267; (2005) 4 SCC 50*] in majority decision expressly overruled the Velliapa Textiles case *(quoted supra)* on this issue. K.J Balkrishanan J. in majority opinion held "We hold that there is no immunity to the companies from prosecution merely because the prosecution is in respect of offences for which punishment prescribed is mandatory imprisonment. We overrule the views expressed by the majority in Velliappa Textiles on this point." In every case in which the offence is only punishable with imprisonment or imprisonment and fine and the offender is the company or other body corporate or an association of individuals, it shall be competent to the court to sentence such offender to fine only.

PROSECUTION

Many white-collar crimes are especially difficult to prosecute because the perpetrators use sophisticated means to conceal their activities through a series of complex transactions. The rich gets richer and the poor gets prison. Although typically the government charges individuals white-collar crimes, the government has the power to sanction corporations as well for these offenses. The penalties for white-collar offenses include fines, home detention, community confinement, paying the cost of prosecution, forfeitures, restitution, supervised release, and imprisonment. However, sanctions can be lessened if the defendant takes responsibility for the crime and assists the authorities in their investigation. Any defenses available to non-white-collar defendants in criminal court are also available to those accused of white-collar crimes. A common refrain of individuals or organizations facing white-collar criminal charges is the defense of entrapment. For instance, in United States v. Williams, 705 F.2d 603 (2nd Cir. 1983), one of the cases arising from "Operation Abscam," Senator Harrison Williams attempted unsuccessfully to argue that the government induced him into accepting a bribe.

If we have specific legislations to trace out white-collar criminality then why these offenders go unpunished ? Main reasons for which these white-collar criminals or occupational criminals go unpunished are :

- The perpetrators use sophisticated means to conceal their activities through a series of complex transactions. Nexus and Political interference hinder the process;
- Many of the more serious white-collar crimes are incredibly complex and require thousands of person hours and lots of money to unravel, thus making them difficult to equate with street crime in terms of being able to neatly discover, tabulate, and report them. Difficult to detect unless otherwise trapped;
- White-collar criminals, because of the segmented application of criminal and civil law, are administratively segregated from conventional criminals;
- Great wealth does confer a certain degree of immunity from prosecution and/or conviction;
- False identities and documents;
- Advance information about police activity;
- Manipulation of official records;

- Disappearing evidence;
- Access to jury identities;
- Corporate Individuals commit crime which go in the name of Corporation and a company being a Juristic person cannot obviously be sentenced to imprisonment;
- Legislators and the law implementers belong to the same group or class to which these occupational criminals belong;
- Less police effort; and
- Favorable laws.

Control Strategies

As has been studied so far, white-collar crimes are not consequences of any drive of necessity. It is simply greed for multiplying wealth and to retain so called meaningless status in the society by pursuing rivalry with the competitors. These people indulging themselves in these activities are to be considered to the leeches, sucking blood of the society and in turn creating grave hindrances to the progress of the society. Since their acts rarely affect an individual directly, these are often ignored of by a common mass and thereby no single individual takes the pain of prosecuting against the offender. But it must not be forgotten that these crimes and criminals are slowly and steadily spreading their wings towards the society at large and much wrong has already done by now. Therefore, there is an immense need to curb it making no waste of time; otherwise it might totally paralyze and ultimate ruin the society.

The biggest hindrance that lays in the prevention of white-collar crimes is the lack of public awareness. Since the individuals involved in these acts, do not generally affect a single individual, an urge of eradicating them, do not usher. Moreover, white-collar crimes are yet not considered to be as grave as that of traditional crimes, so to it's certain characteristics as mentioned earlier. Though they are likely to take shape of heinous crimes in times to come, yet there lie certain obstacles in considering them at par with traditional crimes. The acts coming under the purview of white-collar crimes cannot be termed as crimes in the strict sense of term. Therefore, apart from having the stringent penal or regulatory laws for deterring the persons involved in them, there is a first and foremost need of raising public awareness regarding the issues. The government has a comprehensive role to play in doing so through media or by initiating task forces who would act as watch dogs. Proper and methodical string action should be used randomly so as to catch the individuals involved in the white-collar crimes red-handed.

At this present juncture what we need is the strengthening of our enforcement agencies such as Central Bureau of Investigation, The Enforcement Directorate, The Directorate of Revenue Intelligence, The Income-tax Department and The Customs Department. Concentration and distribution of national wealth must be done in a proper manner. Speedy trial should be arranged by appointing more Judges. Central Vigilance Commission must keep a constant vigil on the workings of the top ranking officers. General public must not avoid being engaged themselves in the prosecution of the white-collar criminals as the offence in general is directed towards them. Lastly if they are traced and proved guilty then Deterrent Theory of punishment is an apt one. A study of literature available on white-collar crime reveals the following measures in dealing with white-collar crime.

- Strengthen the socio-economic status of public officers with fair and competitive salaries.
- Implement strong penalties for breaches of public duties.
- Encourage "whistle-blowers."
- Train and assign Integrity Officers to corruption-prone Govt operations.
- Reduce individual discretion.
- Transparency in the hiring process including public notification of potential candidates.
- Strengthening promotional testing and strict secrecy of test questions. Vetting process thorough background investigations and polygraph test (updated financial statements, family ties, etc.)
- Public disclosure of assets.
- The State have to increase the financial support of the political parties, during political campaigns.
- Strengthening the Laws of Media concentrations.
- Greater collaboration with civil society.
- Right to Information mechanism be scrupulously followed and be strengthened.

An Overview

Therefore, to conclude, it can be inferred that whatever reasons of justifications are there for the methods used for the control of white-collar crimes, the ambivalence of the social response to this sort is so related to wider social factors which have both objective and subjective dimensions. As has been so mentioned a more subjective source of ambivalence in the social response to white-collar crimes is the assumption that there is less public concern about these behaviours so termed as white-collar crimes, and therefore there is a less support for severe sanctions than in the cases of crimes which are traditional street crimes. But even if there was greater public ambivalence towards white-collar crimes in comparison to the traditional crimes, writers such as Box has regarded this as a further challenge 'to sensitize people to not seeing processes in which they are victimised disasters or accidents.'

A full fledged Organised Crime Syndicate is like a mighty tree – cutting it down is extremely difficult. So it is prudent not to allow it to grow. Therefore, the motto should always be that prevention is better than cure. Since the acts involved defrauding public faith and belief, public as a whole mass should come forward to protect the whole society from these greedy people who are destroying the ethics and morality of the society slowly and slowly for their sole aim of pursuing narrow self interest.

References

Appelbaum, Richard P. & Chambliss, William J. (1997). Sociology: A Brief Introduction. New York: Longman.

Barnett, Cynthia. (2006) The Measurement of White-Collar Crime Using Uniform Crime Reporting (UCR) Data.

Bonger, W.A. (1916) Criminality And Economic Conditions, translated by Henry P. Horton (Boston, Little Brown & Co.).

Braithwaite John, (1985) White-Collar Crime, Annual Review of Sociology, vol. 11.

Bureau of Justice Statistics, (1981) United States Department of Justice, Dictionary of Criminal Justice Data Terminology (2d ed) p. 215.

Clarke, Ronald (ed), (1997). Situational Crime Prevention: Successful Case Studies (2nd edition). New York: Criminal Justice Press.

Coleman, Clive & Jenny Moynihan, (1996), Understanding Criminal Data, Open University Press, Buckingham, pp. 8-10

Cressey Donald R., (1950), The Criminal Violation of Financial Trust, American Sociological Review, XV, p. 741.

Edwin C. Hill (1924) The Iron Horse, New York: Grosset & Dunlap.

Government of India (1964) Report of the Commission on Prevention of Corruption, New Delhi, p. 13.

Green, Stuart P. (2006). Lying, Cheating, and Stealing: A Moral Theory of White-Collar Crime. Oxford: Oxford University Press.

Hartung Frank B, (1950) "White-Collar Offenses in the Wholesale Meat Market Industry in Detroit," American Sociological Review 56 (1950), pp. 25-34.

Hugh Barlow (1995) "White-Collar Crimes and Other Crimes of Deception: Connecting Policy to Theory" in his (Ed)Crime and Public Policy: Putting Theory to Work, Boulder, Colo. : WestviewPress.

Leap, Terry L. (2007) Dishonest Dollars: The Dynamics of White-Collar Crime. Ithaca: Cornell University Press.

Lenin, V. I. (1964), Collected Works, vol. 4, Moscow, p.311.

Manheim Harmann, (1965) Comparative Criminology Part II, International Library of Sociology, p. 470.

Mishra Ganesh and Panday Braj Kumar, (1998) White-Collar Crimes, Gyan Publishing House, p. 33.

Newman, Graeme R. & Clarke, Ronald V. (2003). Superhighway Robbery: Preventing E-commerce Crime. Portland, Willan Publishing.

Ratanlal & Dhirajlal, (2006) 'The Indian Penal Code', 30th edition, Wadhwa.

Reiman, J. (1998). The Rich get Richer and the Poor get Prison. Boston: Allyn & Bacon.

Ross, Edward A. (1907) Sin and Society: An Analysis of Latter-Day Iniquity. New York: Houghton Mifflin.

Salinger M. Lawrence, (2005) Encyclopedia of White-Collar and Corporate Crime, A-I, Vol. No.-I, Sage.

Shover, Neal & Wright, John Paul (eds.) (2000). Crimes of Privilege: Readings in White-Collar Crime. Oxford: Oxford University Press.

Simon, D. & Hagan, F. (1999). White-Collar Deviance. Boston: Allyn & Bacon

Sinclair, Upton (1906) The Jungle (The Corruption of the American Meatpacking Industry), Tucson, Arizona.

Steffens Lincoln (1903) The Shame of the Cities, McClure's Magazine XXIV (1).

Sutherland, Edwin H. (1949), White-Collar Crime, New York: Dryden Press, p. 25.

Vittal, N. (1999) 'White-Collar Crimes- (Talk delivered in the 'DST Programme on Forensic Laboratories on 02.05.99)', Central Vigilance Commissioner, New Delhi.

Weisburd David, Elin Waring and Ellen F. Chay (2001) White-Collar Crime and Criminal Careers, Cambridge University Press.

■■■

CHAPTER 12

Recidivism

Learning Objectives :

- ❑ **To Understand the meaning, characteristics and the types of recidivism.**
- ❑ **To comprehend the concept of recidivism in Indian Law.**
- ❑ **To discuss some case studies to locate causal dimensions.**

Introduction

Recidivism is normally seen in the career patterns of criminals. It is commonly noticed that the earlier a person gets into the criminal justice system the greater are the chances that he will continue in it. There are some who become expert in breaking safes, in burglary and take pride in their workmanship. Such people, whether operating alone or in gangs, often tend to make crimes a career. No amount of punishment or correctional approach can wean them away from crime. Recidivism is also understood as the percentage of former prisoners who are rearrested.

The term recidivism is most frequently used in conjunction with substance abuse and criminal behaviour. For example, scientific literature may refer to the recidivism of sexual offenders, meaning the frequency with which they are detected or apprehended committing additional sexual crimes after being released from prison for similar crimes. Statistics of recidivists are not often reliable as in many instances it is found that the same person keeps appearing in prison and other institutions. If to be counted as recidivism the re-offending requires voluntary disclosure or arrest and conviction, the real recidivism rate may differ substantially from reported rates. As another example, alcoholic recidivism might refer to the proportion of people who, after successful treatment, report having, or are determined to have, returned to the abuse of alcohol.

Etymologically, recidivism, has been taken from Latin word *recidîvus* meaning "recurring", from *re-* "back" + *cadô* "I fall") is the act of a person repeating an undesirable behaviour after they have either experienced negative consequences of that behaviour, or have been treated or trained to extinguish that behaviour. The recidivists tend to be a small but persistent minority with hard-core problems in terms of socialization. A recidivist is also known by other names such as a habitual offender and a serial killer.

A Habitual Offender

A habitual offender is a person who has repeatedly committed the same crime. Various states and jurisdictions may have laws targeting habitual offenders, and specifically providing for enhanced or exemplary punishments or other sanctions. They are designed to counter criminal recidivism by physical incapacitation via imprisonment. The nature, scope and type of habitual offender statutes vary, but generally they apply when a person has been convicted a minimum of twice for various crimes. Some codes may differentiate between classes of crimes (for example, some codes only deal with violent crime) and the length of time between convictions. Usually the sentence is greatly enhanced, in some circumstances it may be substantially more than the maximum sentence for the crime. Habitual offender laws may provide for mandatory sentencing - in which a minimum sentence must be imposed, or may allow judicial discretion in allowing the court to determine a proper sentence. One example of a habitual offender statute is a provision requiring the revocation of a driver's license for a person convicted multiple times of driving under the influence.

The practice of imposing longer prison sentences on repeat offenders than on first-time offenders who commit the same crime is not an innovation.[Franklin E. Zimring, Gordon Hawkins, and Sam Kamin, 2001: 4.] For example, New York has a 'persistent felony offender law that dates back to the late 19th century.

Criticism of Habitual Offender Law

There has been various criticism of Habitual Offender Laws such as judicial discretion and lack of uniformity and incompatibility with the principles of fundamental rights.

Habitual Offender Laws, depending on their scope and discretionary room given to judges can lead to persons being punished quite severely for relatively minor offenses. The discretionary nature of the laws means that they can be applied unevenly. In Australia laws relating to dangerous and Habitual offenders have been criticized as ignoring the principle of certainty in sentencing. Another major concern in Australia is the considerable disparity that exists in the requirements for dangerous offender status and in the available sentences for such offenders across jurisdictions. Age and offense requirements, indeterminate or fixed sentencing provisions, and review procedures are quite different from state to state.

Some unusual scenarios have arisen, particularly in California in the United States— the state punishes shoplifting and similar crimes involving over $500 in property as felony petty theft if the person who committed the crime has a prior conviction for any form of theft, including robbery or burglary. As a result, some defendants have been given sentences of 25 years to life in prison for such crimes as shoplifting golf clubs (Gary Ewing, previous strikes for burglary and robbery with a knife), nine videotapes (Leandro Andrade, received double sentence of 25 year-to-life for two counts of shoplifting), or, along with a violent assault, a slice of pepperoni pizza from a group of children (Jerry Dewayne Williams, four previous non-violent felonies, sentence later reduced to six years). The laws have been challenged on the basis of violating fundamental rights. Similarly Indian Habitual

Offenders Act 1952 has received several criticisms as a no better act than earlier Criminal Tribes Act 1871, which has been discussed under the heading Indian Law later in this chapter.

A Serial Killer

A serial killer is a person who murders three or more people over a period of less than thirty days, with a "cooling off" period between each murder, and whose motivation for killing is largely based on psychological gratification.[Burkhalter Chmelir, Sandra, 2003] Often, a sexual element is involved with the killings. The murders may have been attempted or completed in a similar fashion and the victims may have had something in common; for example, occupation, race, appearance, sex, or age group.

Serial killers are often confused with being mass murderers, who commit multiple murders at one time or spree killers, who commit murders in two or more locations with virtually no break in between. Coinage of the English term serial killer is commonly attributed to former FBI Special Agent Robert Ressler in the 1970s. The concept had been described earlier, e.g. by German police inspector Ernst Gennat coining the same term in 1930. Author Ann Rule postulates in her 2004 book Kiss Me, Kill Me that the English-language credit for coining the term "serial killer" goes to LAPD detective Pierce Brooks, mastermind of the ViCAP system.

Types of Recidivists

It is difficult to classify recidivists. But majority of recidivists can be classified in terms of personality. Others are known through the cause associated with their behaviour.

1. The inadequate dependent repeater : This category is probably the most frequent and numerous of all recidivists. He is frequently arrested for drunken and disorderly behaviour, vagrancy and other minor offences. It is not unusual for many of them to have accumulated more than 100 arrests and convictions on minor charges before reaching the age of 50. Many are alcoholics and some are "protected" by the police as good sources of information about more serious offences by others.

2. The dyssocial or sub-cultural repeater : This is usually a person engaged in a business where arrest is an occupational hazard. Moon-shiners come repeatedly to prisons for making intoxicant drinks upon which tax has not been paid. Prostitutes frequently have a series of arrests. Many people living in this pattern of crime have fewer arrests because they have been protected by the "Organization". Gambling, selling drugs, bootlegging and pimping are a few of the crimes in which these repeaters participate. They are called dyssocial and sub-cultural, because they get along well in their own group, providing goods and service the people will buy, but the practice violates the laws of larger society.

3. The compulsive recidivists : This category begins criminal activities early in life, remains with them throughout, committing the same crime over and over again. The progression begins with (1) the situation in life and in the prison with which the individual cannot cope, (2) failure to solve the *problem* after sincere and diligent effort, followed by, (3) replacement of realistic efforts by substitute, regressive behaviour, falling back on more immature solutions, (4) intensification of the original problem by failure of substitute methods, (5) repeatedly grasping for an answer - any answer - and, finally (6) the compulsive repetition of the one answer he has found, whether it works or not.

4. The impulsive recidivists : The impulsive recidivists may repeat a variety of crimes over and over again. This type tends to function in a manner characteristic of the psychopath, sociopath or anti-social personality disorder. He is impulsive without anxiety, and is willing to do anything without regard for others or for society. His anti-social and a social outlook permit him to commit property and bodily offences.

The recidivist is not able to learn by experience or to develop proper insight. It is obvious that they normally don't respond to a society's system of rewards and punishments. Sometimes correctional processes also go wrong especially in the case of juveniles. There are many types of treatment for different groups. To find out to which group a particular delinquent is to be attached, so that he may respond successfully, is often difficult. Often wrong people are put in wrong groups so that even a "good' youth caught up in the system has every chance that he will remain there.

5. Positive recidivism : Since the word effectively means 'continued rebellion against authority', the ethics and morals of recidivism are dependent on who defines 'undesirable behaviour'. Positive examples of recidivism would include several notable individuals of the French Resistance, who continued exhibiting behaviour not desired by the *de facto* rulers of France at the time, and many activists in the American civil rights movement. To be counted as recidivism, the re-offending requires voluntary disclosure of arrest and conviction, so the real recidivism rate may differ substantially from reported rates. As another example, alcoholic recidivism might refer to the proportion of people who, after successful treatment, report having, or are determined to have, returned to the abuse of alcohol. Similarly the freedom fighters in India and Srilanka and *Jehadis* in Muslims are branded in academic circles as positive recidivists.

Characteristic Causes

Some of the characteristic features as well as the causative factors are discussed below :

Past Experience : Recidivists are known for their past experiences in similar crime. An accused's history of convictions are called antecedents, known colloquially as "previous" in the UK and "priors" in the United States and Australia. Several studies have revealed that released rapists are arrested for another rape. Those who had served time for homicide are rearrested for homicide. Similarly those in prison for possessing or selling stolen property and those in custody for possessing, using, or selling illegal weapons are rearrested for same crime.

Social Misfit : One of the main reasons why they find themselves back in jail is because it is difficult for the individual to fit back in with 'normal' life. They have to reestablish ties with their family, return to high-risk places and secure formal identification; they often have a poor work history and now have a criminal record to deal with. Many prisoners report being anxious about their release; they are excited about how their life will be different "this time" which does not always end up being the case.

The Pressures of Prisons : It has been seen that the earlier a person becomes involved in the criminal justice system, the longer he stays in it. Firstly, the emotional and social problems of the individual impede his successful adjustment to society. Secondly, the prison environment intensifies the problems, dichotomises the individual from social authority, and produces an institutionalized apathy and dependency that reduces the possibility of successful adjustment in free society.

Rejecting the Rejecters : Present society is characterized by a theoretical forced equality where everybody is treated alike. The prisoners' problems are shared with other prisoners, because they have all been rejected by society. The prisoners, in turn, reject their rejecters.

Inmate Codes : Mutual loyalty, affection, respect and common standing in opposition to the forces of their rejecters result in what is commonly known as the "inmates' codes". The inmates are united in their common opposition against their captors. A person who leaves the prison, experiences social rejection, finds support of other former inmates, often returns to his prison attitude, and continues his criminal career.

Technology : Sociologist Roger Roots has suggested that the increasing computerization and accessibility of criminal records is having a negative impact on recidivism rates as technology advances. Prior to the computer revolution, persons with criminal records were often able to relocate and start their lives over with clean slates in new communities. Former criminals rose to become some of America's greatest leaders in law, industry, and politics. This possibility seems to be narrowing as criminal records become electronically stored and accessible.

Psychopathy : Criminal recidivism is highly correlated with psychopathy. The psychopath is defined by an uninhibited gratification in criminal, sexual, or aggressive impulses and the inability to learn from past mistakes. Psychopaths also have a markedly distorted sense of the potential consequences of their actions, not only for others, but also for themselves. They do not, for example, deeply recognize the risk of being caught, disbelieved or injured as a result of their behaviour. Individuals with this disorder gain satisfaction through their antisocial behaviour and lack remorse for their actions.[Hare, 1995]. Findings indicate psychopathic prisoners have a 2.5 time higher probability of being released from jail than undiagnosed ones, even though they are more likely to recidivate. Punishment and behaviour modification techniques do not improve the behaviour of a psychopath. Psychopathic individuals have been regularly observed to become more cunning and better able to hide their behaviour. They are generally considered to be not only incurable but also untreatable.[Harris, 2006]

Effect of Incarceration

The effect of **incarceration** on former prisoners has been a very common topic of discussion for many years. In most cases, it is believed that many prisoners will find themselves right back where they started, in jail. In the United States, 68% percent of males and 58% of females are rearrested, and 53% and 39% respectively are re-incarcerated as reported in 2003.

In recent history, the rate of incarceration in the U.S. has increased dramatically, resulting in prisons being filled to capacity with bad conditions and environment for inmates. In many prisons, crime continues inside the prison walls. Gangs exist and flourish on the inside, often with many key tactical decisions being made by leaders who are in jail.

Many other things need to be taken into consideration as well, such as the individual's circumstances before incarceration, the things that happened while they were incarcerated, and the period after they are released from prison, both immediate and long term.

Recidivism in India Law

The Criminal Tribes Act, 1871 was amended a number of times. After Independence, the leaders and social reformers paid attention to this problem. In 1949, the Central government appointed a committee to study the utility of the existence of this law. The committee viewed that the act was against the spirit of the Indian Constitution. It recommended suitable steps to be taken for amelioration of the pitiable conditions of the Criminal Tribes rather than stigmatising them as criminals. As a result, the Criminal Tribes Act of 1871 was repealed in 1952 and the Habitual Offenders Act was enacted in its place. According to the Habitual Offenders Act, a habitual offender is one who has been a victim of subjective and objective influences and has manifested a set practice in crime, and also presents a danger to society in which she/he lives. The Habitual Offenders are usually hardened criminals whose major part of life has been spent in jails. It is because they commit offences at frequent intervals and are sent back to jail.

National Human Rights Commission, in February, 2000 recommended repeal of the 'Habitual Offenders Act 1952' (Radhakrishna, 2006). Later in March 2007, the UN's anti-discrimination body Committee on the Elimination of Racial Discrimination (CERD), noted that "the so-called denotified and nomadic which are listed for their alleged 'criminal tendencies' under the former Criminal Tribes Act (1871), continue to be stigmatized under the Habitual Offenders Act (1952) (art. 2 (1)), and asked India to repeal the Habitual Offenders Act (1952) and effectively rehabilitate the denotified and nomadic tribes. According to the body, since much of 'Habitual Offenders Act (1952)' is derived from the earlier 'Criminal Tribes Act 1871', it doesn't show a marked departure in its intent, only gives the formed notified tribes a new name i.e. Denotified tribes, hence the stigma continues so does the oppression, as the law is being denounced on two counts, first that "all human beings are born free and equal", and second that it negates a valuable principle of the criminal justice system – innocent until proven guilty.

In 2008, the National Commission for Denotified, Nomadic and Semi-Nomadic Tribes (NCDNSNT) of Ministry of Social Justice and Empowerment recommended that same reservations as available to Scheduled Castes and Scheduled Tribes be extended to around 11 crore people of denotified, nomadic or semi-nomadic tribes in India; the commission further recommended that the provisions of the Scheduled Caste and Scheduled Tribe (Prevention of Atrocities) Act, 1989 be applicable to these tribes as well. Today, many governmental and non-governmental bodies are involved in the betterment of these denotified tribes through various schemes and educational programs [Rao,2008].

Sec.75 IPC : Enhanced punishment for certain offences under Chapter XII or Chapter XVII after previous conviction. Whoever, having been convicted by a Court in India, of an offence punishable under Chapter XII (of offences relating to Coins and Government stamps) or Chapter XVII (of offences against property) of this Code with imprisonment of either description for a term of three years or upwards, is again guilty of any offence punishable under either of those Chapters with like imprisonment for the like term, shall be subjected far-every such subsequent offence to imprisonment for life, or to imprisonment of either description for a term which may extend to ten years.

Section 75 IPC deals with persons with previous conviction. The previous sentence need not necessarily be in force when the subsequent offence is committed. Pratap vs. State of U.P., 1973, S.C.C. (Cri) 496: (1973) 3 see 690.

Some Studies :

1. Minnesota Department of Corrections

Many studies have shown a correlation between prisoners attending rehabilitation programs while incarcerated and their likelihood of recidivism. Most have no significant results, although, some studies have shown a positive correlation. The findings that have shown significant results are normally boot camp experiments that have aftercare programs for at least four months.

The Minnesota Department of Corrections did a study on criminals who are in prison to see if rehabilitation during incarceration correlates with recidivism and/or saved the state money. They used the Minnesota's Challenge Incarceration Program (CIP) which consisted of three phases. The first was a six month institutional phase followed by two aftercare phases, each lasting at least six months, for a total of about eighteen months. The first phase was the "boot camp" phase. Here inmates had daily schedules sixteen hours long where they participated in activities and showed discipline. Some activities in phase one included physical training, manual labor, skills training, drug therapy, and transition planning. The second and third phases were called "community phases." In phase two the participants are on intensive supervised release (ISR). ISR includes being in contact with your supervisor on a daily basis, being a full-time employee, keeping curfew, passing random drug and alcohol tests, and doing community service while continuing to participate completely in the program. The final phase is phase three. During this phase one is still on ISR and has to remain in the community while maintaining a full-time job. They have to continue with community service and their participation in the program. Once phase three is complete participants have "graduated" CIP. They are then put on supervision until the end of their sentence. Inmates who drop out or fail to complete the program are sent back to prison to serve the rest of their sentence.

Information was gathered through a quasi experimental design. This compared the recidivism rates of the CIP participants with a control group. The findings of the study have shown that the CIP program did not significantly reduce the chances of recidivism. However, CIP did increase the amount of time before rearrest. Moreover, CIP early release graduates lower the costs for the state by millions every year.

This study has shown that the CIP program significantly increases the amount of time after a prisoner is let out to commit another crime. It also saves millions of dollars every year for the state.

2. Jefferson County, Kentucky

A study was done by Robert Stanz in Jefferson County, Kentucky which discussed an alternative to jail time. The alternative was considered "home incarceration", where the defendant would complete his or her time at home instead of in jail. According to the text, "Results show that the majority of offenders do successfully complete the program, but that a majority are also re-arrested within 5 years of completion". In doing this, they added to the rate of recidivism.

In doing a study on the results of this program, Stanz considered age, race, neighborhood, and several other aspects. Most of the defendants who fell under the recidivism category included those who were younger, those who were sentenced for multiple charges, those accruing fewer technical violations, males, and those of African-American descent.

3. Methadone Maintenance Therapy (MMT)

A study was done on the recidivism rate of inmates going through MMT (Methadone Maintenance Therapy). What this therapy does is instead of putting heroin addicts through withdrawal, they would be administered small doses of methadone, which was hoping to aid in the recovery of heroin addicts. The study was done with 589 inmates who were booked between November 22, 2005, which was the start of the MMT program, and October 31, 2006. The outcome of the study was that there was no significant effect to the recidivism rate. "There was no statistically significant effect of receiving methadone in the jail or dosage on subsequent recidivism risks". Some of the recidivism rate factors included age, gender, final dosage received (mg), race and previous bookings days in jail. The study explained that aiding in the withdrawal process of heroin addicts has no adverse side effects on the recidivism rate of those individuals. This is viewed as important to the authors because it shows that individuals are able to help out in the withdrawal period without necessarily causing them to slip back into addiction (or at least be any more prone to being rearrested than they were before).

4. United States

Two studies were done to attempt to provide a "national" recidivism rate for the U.S. One was done in 1983 which included 108,580 State prisoners from 11 different states. The other study was done in 1994 on 272,111 prisoners from 15 states. Both studies represent two-thirds of the overall prisoners released in their corresponding years.

This is an image which portrays the percent of parolees returning to prison in each state in 2006. This study was done by Matt Kelley. According to this study, in 2006, there was more recidivism in the southern states, particularly in the Midwestern region. However, for the majority, the data is spread out throughout the regions.

Summary of Causes in Case Studies

It has been suggested that the cause suggest that for many decades, correctional observers did not give priority to the reality that offenders who reenter society face a varied assortment of daunting challenges that predictably lead to high recidivism rates. Various researchers also noted that prisoners are stripped of civil rights and are reluctantly absorbed into communities which leads to their further alienation and isolation. A final factor when looking at the causes of Recidivism is the difficulty of a released offender when faced with finding a job. Owners of business will seem hesitant when hiring a convicted felon.

References

Bailey, Kristen (2009) "The Causes of Recidivism in the Criminal Justice System and Why It Is Worth the Cost to Address Them." Nashville Bar Journal.

Burkhalter Chmelir, Sandra (2003). "Serial Killers". in Robert Kastenbaum. Macmillan Encyclopedia of Death and Dying. New York: Macmillan.

Duwe, G., & Kerschner, D, (2008). "Removing a Nail From the Coffin." Crime & Delinquency, 54.

Franklin E. Zimring, Gordon Hawkins, and Sam Kamin (2001), *Punishment and Democracy: Three Strikes and You're Out in California* (New York: Oxford University Press, 4.

Hare, Robert D (1995), Psychopaths: New Trends in Research. The Harvard Mental Health Letter, September.

Henslin, James (2008). "Social Problems: A Down-To-Earth Approach."

Jiler, James (2006) "Doing Time in the Garden: Life Lessons Through Prison Horticulture." New Village Press.

Jill S. Levenson, John W. Morin (2000) [Treating Nonoffending Parents in Child Sexual Abuse Cases] p. 7 SAGE.

Marvin Zuckerman (1991) Psychobiology of personality Cambridge University Press, p. 390.

McMillan, Garnett P, 2008, "The effect of a jail methadone maintenance therapy (MMT) program on inmate recidivism", Addiction, 103:2017-2023.

Radhakrishna, Meena (2006) "Dishonoured by history", The Hindu, (Special issue with the Sunday Magazine).

Rao, Raghvendra, (2008), Panel favours reservation for nomadic tribes, Indian Express, New Delhi.

Roger Roots, (2004) "When the Past is a Prison: The Hardening Plight of the American Ex-Convict," 1 Justice Policy Journal 3.

Schechter, Harold (2003). The Serial Killer Files: The Who, What, Where, How, and Why of the World's Most Terrifying Murderers. Ballantine Books.

Stanz, Robert, (2000), "Predictors of Success and Recidivism in a Home Incarceration Program" Prison Journal, 80:326-345.

Visher, Christy A, (2003). "Transitions From Prison To Community: Understanding Individual Pathways". The Urban Institute, Justice Policy Center, District of Columbia Washington, 2003.

■■■

CHAPTER 13 Crime Trends

Learning Objectives :

- ❐ **To Understand the importance of studying crime trends in criminology.**
- ❐ **To take stock of crime trends in India as per NCRB Data.**

Introduction

Study of crime trends is useful to different groups for different things. On a political level, it provides a measure for the success or failure of governance system and of certain criminal justice policies. On an operational level, it provides important information on how best to allocate state resources. From a civil society perspective, it provides indicators against which to measure criminal justice performance and therefore an important tool to hold accountability to the various role-players in the criminal justice system. At the academic front it may generate a lot of research terminating in diagnosis and prognosis.

Descriptive information and explanatory research on crime trends across the nation that are not only accurate but also timely are pressing needs in the nation's crime-control efforts. Without useful and reliable information, national and local policy makers fly blind when formulating and evaluating the effectiveness of policy interventions or they may generate ad hoc outcome indicators for each new policy assessment. Comparable evaluation capabilities do not exist in the policy area of criminal justice (Rosenfeld, 2006). The local police department is perhaps the last remaining complex organization in global and local society without the capacity to anticipate and plan for changes in its environment, yet it is assigned virtually all of the responsibility for responding to crime increases and more than its share of the blame when the responses prove inadequate.

Why Study Crime Trends

There are substantial misperceptions of crime, both of the number of incidences and trends in crime, among the general population. As a consequence, public policymakers should exercise additional care when framing important law enforcement decisions taken against the background of the public's potential misperceptions of crime. There can be and indeed often is a discrepancy between the public's perception of the likelihood of crime victimisation and the actual risk of victimisation. This discrepancy is apparent in the public's concern regarding a perceived increase in crime amidst declining crime rates (Cohen 2000).The public perceptions of crime rates and crime trends often do not match with police statistics of recorded crime or surveys on victimisation rates (Rex & Tonry 2002; Roberts & Indermaur 2009). The precinct level statistics should also be used by community safety and security practitioners to develop, implement and measure the success of social crime prevention initiatives. It is critical that the crime statistics are credible and reliable else national, provincial, local and community level decision-makers and crime prevention/reduction practitioners may plan and act inappropriately to crime problems.

Changes over time in the levels and patterns of crime have significant consequences that affect not only the criminal justice system but also other critical policy sectors. Yet compared with such areas as health status, housing, and employment, the nation lacks timely information and comprehensive research on crime trends. Crime and violence are a development issue. The high rates of crime and violence in any region have both direct effects on human welfare in the short-run and longer run effects on economic growth and social development. There is no one "ideal" approach (Rosenfeld et.al Ed., 2008) to anlyse the issue. The common denominator is that successful interventions are evidence-based, starting with a clear diagnostic about types of violence and risk factors, and ending with a careful evaluation of the intervention's impact which will inform future actions. A lesson from recent research is the value of distinguishing short-from long-run changes in crime rates and in the factors hypothesized to explain those changes. A final and related insight from the recent research on crime trends distinguishes sources of change in crime rates that apply across local areas from those that may be specific to particular jurisdictions. Some factors, such as economy-wide changes in unemployment or wage rates, potentially affect crime conditions across the nation, whereas others, such as gang conflicts or local changes in policing strategies, influence crime rates in some places but not others.

Against this backdrop, it is important to be able to articulate some of the good reasons that make sense to analyze the crime trends. Some good reasons are listed here.

- Analyze crime to take advantage of the abundance of information existing in law enforcement agencies, the criminal justice system, and the public domain.
- Analyze crime to maximize the use of limited law enforcement resources.
- Analyze crime to inform law enforcers about general and specific Crime trends, patterns, and series in an ongoing, timely manner.
- Analyze crime to have an objective means to assess crime problems locally, regionally, statewide, nationally, and globally within and between law enforcement agencies.
- Analyze crime to be proactive in detecting and preventing crime.
- Analyze crime to meet the law enforcement needs of a changing society.

Thus recent crime trends represent a major social phenomenon and a fundamental research issue for which criminological researchers should provide concrete answers. We should and can do better than "a lot of things mattered," but doing so will require a more vibrant research agenda that focuses on modeling recent crime trends in a much more comprehensive and systematic way. The goal here is to make a modest contribution to this effort by significantly broadening the scope of analysis to incorporate not only the factors often considered (e.g., police force size, drug use and market activity, age structure, incarceration) in the literature, but also various others that have been highlighted in theory but only occasionally considered in empirical studies (e.g., immigration, wages, alcohol consumption, levels of domesticity, and youthful cohort "quality"). This chapter focuses on some important old and new crime problems such as Gambling, Drug Addiction, Alchoholism, Cyber Crime and Prostitution. The chapter concludes with a discussion of Crime Trends in India as per latest reports of National Crime Records Bureau (NCRB), New Delhi, for intuition and introspection by crime analysts and the agencies of law enforcement.

Crime Trends in India : NCRB Data

Crime records play a vital role in the scheme of police working for prevention and detection of crime. The Indian Police, over the years, have sought to improve the efficiency of the crime records systems to discharge their responsibilities with greater efficiency and effectiveness. Though, police is a State subject under the Indian constitution, the Central Government through the Ministry of Home Affairs has been assisting and aiding the States in the modernization of the State Police Forces with the financial aid and through the constitution of Study Groups, Committees and formation of central organizations to help States fight the menace of crime more effectively. Some of the important milestones towards this directions are :

INDIAN POLICE COMMISSION - 1902. This Commission, for the first time, introduced standard police forms and registers for information documentation at the level of Police Station and District which were uniformly adopted throughout the country.

CENTRAL FINGER PRINT BUREAU, CALCUTTA - 1956. This Bureau used to maintain finger print slips of convicted criminals of all the States to facilitate exchange of information on inter-state criminals on the basis of finger print records.

INTER-STATE CRIME RECORDS DIVISION IN CBI - 1964. Apart from maintaining conviction records of criminals in the CFPB, a new set up was made responsible for collection and dissemination of information on selected crime and criminals having inter-state ramifications.

DIG, CID Conference - 1970. A sub committee was set up by this Conference to look into the feasibility of computerization of crime records in view of the growing volume of information collected at various centers. The sub committee recommended computerization of crime records with detailed modalities thereof.

HOME MINISTRY'S SCHEME FOR MODERNISATION OF STATE POLICE FORCES - 1970. The Ministry of Home Affairs formulated a ₹ 100 crores scheme for rendering financial aid to the

States for modernization of police forces. Under this scheme ₹ 10 crores were to be disbursed to all the States annually on pro-rata basis, of which 50% was a grant and 50% loan recoverable in easy installments. Modernisation of crime records through computerization based on the recommendations of the DIsG CID Conference was included in this scheme in the year 1975-76. This scheme was extended by another 10 years period.

SUB COMMITTEE ON COMPUTERISATION OF FINGERPRINTS - 1972. The sub-committee analysed the process of fingerprint classification, evolved a new system of classification so as to admit of computerization of fingerprint records and made recommendations for adoption of this strategy for computerization of fingerprints.

DIRECTORATE OF COORDINATION POLICE COMPUTERS (DCPC) - 1976. In order to implement the modernisation scheme in respect of crime and fingerprint records in the States, this organization was set up to introduce computer based systems in the States and assist the States in procurement and installation of computers and implementation of Crime-Criminals Information System.

STUDY GROUP ON COMPUTERISATION OF CRIME RECORDS - 1976. The study group was set up to make recommendations for appropriate strategies and modalities of computerizations of crime records keeping in view the latest technological developments in the field of Electronic Data Processing and telecommunications.

NATIONAL POLICE COMMISSION - 1977. The Commission was set up to look into all aspects of policing in the country and make recommendations with a view to bring about the requirea organizational, procedural and cultural changes in the State Police to bring it in tune with the national aspirations of a democratic and welfare State.

COMMITTEE ON CRIME RECORDS - 1978. The Committee was set up to review the existing crime records and associated procedures and to recommend changes with a view to enable the police forces to meet the present day requirements of crime and criminal information.

Accepting the recommendations of the National Police Commission - 1977, the Ministry of Home Affairs constituted a Task Force in 1985 to work out the modalities for setting up of the National Crime Records Bureau (NCRB). The Government accepted the recommendations of the Task Force and constituted the NCRB with headquarters at New Delhi in January, 1986. As per the government Resolution dated 11.3.1986 the following objectives were set for the NCRB :-

I. To function as a clearing house of information on crime and criminals including those operating at National and International levels so as to assists the investigators, and others in linking crimes to their perpetrators;

II. To store, coordinate and disseminate information on inter-state and international criminals from and to respective States, national investigating agencies, courts and prosecutors in India without having to refer to the Police Station records;

III. To collect and process crime statistics at the National level;

IV. To receive from and supply data to penal and correctional agencies for their tasks of rehabilitation of criminals, their remand, parole, premature release etc.

V. To coordinate, guide and assist the functioning of the State Crime Records Bureaux:

VI. To provide training facilities to personnel of the Crime Records Bureaux; and

VII. To evaluate, develop and modernise Crime Records Bureaux.

VIII. To execute and develop computer based systems for the Central Police Organisations - and also cater to their data processing and training needs for computerization.

IX. To function as the National storehouse of fingerprint (FP) records of convicted persons including FP records of foreign criminals.

X To help trace inter state criminals by fingerprint search.

XI. To advise Central and State Governments on matters related to fingerprints and footprints, and to conduct training courses for fingerprint experts.

As a first step towards streamlining of the system of crime-criminal information management at the Centre, the Ministry of Home Affairs merged the following crime records establishments of various Central Police Organisations with the NCRB :

1. Directorate of Coordination Police Computers (MHA)
2. Inter-State Criminals Data set up of the Central Bureau of Investigation.
3. Crime Statistics set up of the Bureau of Police Research and Development'
4. Central Finger Print Bureau, Calcutta of the Central Bureau of Investigation.

Further on the recommendations of the Director, NCRB, the Government sanctioned 316 posts of various levels and approved in principle procurement of Computer Systems at a cost of ₹ 2 crores approx. for the NCRB.

The NCRB really started work according to its new charter with the posting of a Director along with a token supporting staff of 5 personnel in January, 1986. While the NCRB was taking shape with the merger of various crime record establishments between February 1986 to April, 1988, a comprehensive proposal of the NCRB for manpower and equipment was finally approved by the Ministry in August, 1987. NCRB has a total sanctioned strength of 435, as on date.

Recently the Bureau has been sanctioned 27 additional posts for the implementation of Crime and Criminal Tracking Network & System (CCTNS). The Crime and Criminal Tracking Network & Systems (CCTNS) is a plan scheme conceived in the light of experience of an earlier scheme namely - Common Integrated Police Application (CIPA). The NCRB, at present, functions through (i) Crime Records Branch, (ii) Central Finger Print Bureau (iii) Statistical Branch, (iv) System Development Branch, (v) State Implementation Branch,(vi) System Maintenance Branch (vii) Training Branch and (viii) Data Centre and Technical Branch.

Crime in India Reports : NCRB

Crime in India is the oldest and the most valuable publication brought out by NCRB. The data for the report is collected by State Crime Records Bureaux (SCRBx) from the District Crime Records Bureaux (DCRBx) and sent to NCRB at the end of the year under reference. Data from mega-cities (cities having population of 10 lakh or more as per the latest census) is also collected separately. District-wise data on some IPC heads is collected and published separately. The first edition of 'Crime in India' pertains to the year 1953 and the latest edition of the report pertains to the year 2009. The report contains comprehensive information on (i) cases registered and their disposal and (ii) persons arrested and their disposal under major heads of Indian Penal Codes and Special and Local Laws. Further, age-group-wise and sex-wise details of persons arrested under these crime-heads are also available in the report. The report contains a chapter on crime against some vulnerable sections of the society - women, children, Scheduled Castes / Scheduled Tribes. The above crime data is presented in graphic / chart format, table format and map format.

Salient Fields : An entire chapter has been devoted to police strength, expenditure and infrastructure. Similarly, information on police casualties and police firing & casualties is given in separate chapters to ensure adequate focus to these issues. Information on complaints against police personnel and custodial crimes has also been given in separate chapters. Continuous attempts have been made to keep pace with the requirements of the time. Thus, information on cyber crimes and human trafficking has been presented in different chapters. Data on economic crimes, recidivism and crime in railways has also been given in separate chapters. Information on kidnapping and abduction is presented in respect of total population, women & girls and children separately.

Data Generates Research and Decision Making : The report is the only, and most comprehensive, databank available with the Government of India on the subject. The data contained in the report is used by policy makers, NGOs, researchers and public at large. Keeping in view the extensive, and increasing, dependence of various stakeholders on the information contained in the report, NCRB have, on their own initiatives, digitised all the editions of the report from 1953 to 2008 and made them available on the website.

TABLE-13.1

Incidence of Cognizable Crimes (IPC) Under Different Crime Heads During 1953 to 2004

Sl. No.	Year	Murder	Attempt to commit Murder	C.H. not amounting to murder	Rape	Kidnapping & Abduction	Dacoity	Preparation & Assembly for dacoity	Robbery	Burglary (House-Breaking)	Theft	Riots
(1)	(2)	(3)	(4)	(5)	(6)	(7)	(8)	(9)	(10)	(11)	(12)	(13)
1	1953	9802	*	*	*	5261	5579	*	8407	147379	256567	20529
2	1954	9765	*	*	*	5514	5395	*	7600	132457	223866	22777
3	1955	9700	*	*	*	5529	4779	*	6710	121744	212028	23609
4	1956	10025	*	*	*	5905	5397	*	7618	134556	236214	24700
5	1957	10419	*	*	*	5821	5560	*	7408	129632	233239	23750
6	1958	10661	*	*	*	6043	4658	*	7120	124695	236103	24942
7	1959	10712	*	*	*	6549	3774	*	6267	118205	233052	26987
8	1960	10910	*	*	*	6024	3981	*	6263	114540	228842	26890
9	1961	11188	*	*	*	6698	4213	*	6428	122605	232868	27199
10	1962	11586	*	*	*	7119	4890	*	7551	134324	252453	29096
11	1963	10754	*	*	*	6924	4997	*	7694	137025	242487	28114
12	1964	11748	*	*	*	8050	5287	*	8336	153862	273676	32693
13	1965	12310	*	*	*	7927	4955	*	8067	142015	273702	32940
14	1966	12631	*	*	*	7854	4817	*	8585	150180	298701	34696
15	1967	13398	*	*	*	8192	6300	*	10252	173575	339861	42447
16	1968	13849	*	*	*	8830	6384	*	10194	156206	315546	45801
17	1969	14732	*	*	*	8464	6049	*	9922	145429	300140	55796
18	1970	15708	*	*	*	10111	9837	*	16958	166339	337211	68331
19	1971	16180	*	2357	2487	9647	11193	*	18402	165807	335204	64114
20	1972	15475	*	2196	2605	9402	10411	*	17054	167062	346382	65781
21	1973	17072	*	2408	2919	10223	10627	*	18857	181433	379412	73388
22	1974	18649	*	2514	2962	10543	13697	*	22286	199878	436918	80547
23	1975	17563	*	2502	3376	11139	12506	*	21656	192854	421891	67241
24	1976	16673	*	2584	3893	11250	10910	*	17974	168655	365138	63675
25	1977	18376	*	2615	4058	12240	12599	*	22725	193622	432046	80449
26	1978	19314	*	2728	4558	13616	13195	*	22923	183991	437187	96488
27	1979	20349	*	3008	4300	13125	14028	*	22860	168574	430483	98896
28	1980	22149	*	3032	5023	13595	15194	*	23493	166695	431036	106957
29	1981	22727	*	3272	5409	13833	14626	*	22996	157540	422059	110361
30	1982	23339	*	3427	5427	13341	12700	*	21938	142726	375240	106511
31	1983	25112	*	3793	6019	13842	12382	*	21310	139103	353536	108101
32	1984	25786	*	4093	6740	15141	12301	*	23204	136272	330669	101460
33	1985	25970	*	3995	7289	16051	11254	*	22501	130354	330554	99757
34	1986	27269	*	4195	7952	15667	10444	*	22395	128946	323533	94197
35	1987	28513	*	3721	8559	15251	10036	*	22917	125466	310575	90789
36	1988	28771	20689	3755	9099	15771	9306	931	21611	124304	319848	94587
37	1989	31222	23748	4100	9752	17318	9896	1015	22480	129020	341240	98943
38	1990	35045	27095	4281	10068	18474	11089	1286	25440	131331	353191	102846
39	1991	39174	29778	4243	10410	20079	10831	1393	26428	132087	362928	105309
40	1992	40105	31202	4621	11708	20518	11308	1297	26444	127281	350582	104749
41	1993	38240	29725	3890	12218	19830	9357	1102	24354	123020	320434	93838
42	1994	38577	30020	3946	13208	20983	9271	946	23933	121536	303564	94344
43	1995	37464	29571	3830	13754	20426	8335	961	22443	116507	294306	96520
44	1996	37671	29597	3728	14846	20848	8035	888	22705	115097	284985	92831
45	1997	37543	29322	3869	15330	21898	7867	1049	22141	113319	277077	91812
46	1998	38584	30577	3681	15151	23520	8091	1266	23603	119260	287967	90767
47	1999	37170	29628	3912	15468	23236	7079	1501	21332	111296	271907	80838
48	2000	37399	30743	3773	16496	22871	6825	1505	20926	105391	258588	80456
49	2001	36202	31523	3367	16075	22487	6154	1614	19901	101182	252803	76222
50	2002	35290	30380	3624	16373	21850	6101	1841	18764	96461	247462	68945
51	2003	32716	25942	4029	15847	19992	5303	2302	17512	92827	245237	57334
52	2004	33608	27890	3935	18233	23327	5311	2340	18458	92490	273045	59971

TABLE-13.1 (Contd.)

Incidence of Cognizable Crimes (IPC) Under Different Crime Heads During 1953 to 2004 (Concluded)

Sl. No.	Year	Criminal Breach of Trust	Cheating	Counterfeiting	Arson	Hurt	Dowry Deaths	Molestation	Sexual Harassment	Cruelty by husband or Relatives	Importation Of Girls	Causing Death by Negligence	Other IPC crimes	Total Cognizable Crimes under IPC
(1)	(2)	(14)	(15)	(16)	(17)	(18)	(19)	(20)	(21)	(22)	(23)	(24)	(23)	(24)
1	1953	*	*	*	*	*	*	*	*	*	*	*	148440	601964
2	1954	15860	9934	815	*	*	*	*	*	*	*	*	122929	556912
3	1955	14644	9461	597	*	*	*	*	*	*	*	*	126435	535236
4	1956	15466	9519	596	*	*	*	*	*	*	*	*	135221	585217
5	1957	15362	9388	629	*	*	*	*	*	*	*	*	140163	581371
6	1958	16017	9503	555	*	*	*	*	*	*	*	*	173887	614184
7	1959	15878	9677	443	*	*	*	*	*	*	*	*	188772	620326
8	1960	15862	9207	554	*	*	*	*	*	*	*	*	183294	606367
9	1961	16895	9511	485	*	*	*	*	*	*	*	*	187561	625651
10	1962	18092	9738	423	*	*	*	*	*	*	*	*	199194	674466
11	1963	18438	8854	289	*	*	*	*	*	*	*	*	193254	658830
12	1964	21180	11670	317	*	*	*	*	*	*	*	*	232194	759013
13	1965	21354	11935	436	*	*	*	*	*	*	*	*	235974	751615
14	1966	22353	11606	655	*	*	*	*	*	*	*	*	242655	794733
15	1967	23950	12701	1413	*	*	*	*	*	*	*	*	249892	881981
16	1968	22408	12524	1425	*	*	*	*	*	*	*	*	268795	861962
17	1969	21118	12001	739	*	*	*	*	*	*	*	*	270777	845167
18	1970	22679	12331	650	*	*	*	*	*	*	*	*	295267	955422
19	1971	20270	11412	641	*	*	*	*	*	*	*	*	299711	952581
20	1972	21004	12646	670	*	*	*	*	*	*	*	*	318886	984773
21	1973	21837	14392	582	*	*	*	*	*	*	*	*	349358	1077181
22	1974	22274	15380	718	*	*	*	*	*	*	*	*	371387	1192277
23	1975	23287	17772	951	*	*	*	*	*	*	*	*	373660	1160520
24	1976	23656	19588	887	*	*	*	*	*	*	*	*	395491	1093897
25	1977	22868	19623	784	*	*	*	*	*	*	*	*	451672	1267004
26	1978	23255	19821	636	*	*	*	*	*	*	*	*	514542	1344968
27	1979	22016	18370	525	*	*	*	*	*	*	*	*	526942	1336168
28	1980	20684	17416	830	*	*	*	*	*	*	*	*	550480	1368529
29	1981	20579	17764	1004	*	*	*	*	*	*	*	*	582265	1385757
30	1982	18259	17471	939	*	*	*	*	*	*	*	*	612586	1353904
31	1983	18514	19767	809	*	*	*	*	*	*	*	*	627578	1349866
32	1984	18428	19077	1118	*	*	*	*	*	*	*	*	664371	1358660
33	1985	18417	21016	1504	*	*	*	*	*	*	*	*	696069	1384731
34	1986	19021	22579	1687	*	*	*	*	*	*	*	*	727950	1405835
35	1987	17847	22115	1653	*	*	*	*	*	*	*	*	749550	1406992
36	1988	17352	22705	1537	*	*	*	*	*	*	*	*	750090	1440356
37	1989	17541	23691	2075	*	*	*	*	*	*	*	*	797803	1529844
38	1990	16552	24466	2576	*	*	*	*	*	*	*	*	840709	1604449
39	1991	17495	27466	4467	*	*	*	*	*	*	*	*	886287	1678375
40	1992	17925	29397	5133	*	*	*	*	*	*	*	*	907071	1689341
41	1993	17039	30079	3728	*	*	*	*	*	*	*	*	903082	1629936
42	1994	16523	31207	2851	*	*	*	*	*	*	*	*	924342	1635251
43	1995	15503	30678	2203	12028	203812	4648	26856	4689	28579	*	*	722583	1695696
44	1996	15368	33823	2796	12425	223977	5513	28939	5671	35246	*	*	714587	1709576
45	1997	15077	35228	2228	12363	228497	6006	30764	5796	36592	*	*	726042	1719820
46	1998	16157	38271	1353	12913	235870	6975	30959	8053	41375	*	*	744422	1778815
47	1999	15454	41403	1347	11218	236313	6699	32311	8858	43823	1	*	763835	1764629
48	2000	14581	41701	2299	10392	240580	6995	32940	11024	45778	64	*	779757	1771084
49	2001	14798	44727	1683	10534	271487	6851	34124	9746	49170	114	57182	701362	1769308
50	2002	14027	46271	1522	11820	265025	6822	33943	10155	49237	76	64044	730297	1780330
51	2003	13432	47478	2055	9365	261444	6208	32939	12325	50703	46	60672	700412	1716120
52	2004	14176	51939	1529	8637	276868	7026	34567	10001	58121	89	69423	741031	1832015

Source : NCRB, New Delhi

TABLE-13.2

Crime Figures at a Glance-2010

SL. No.	CRIME HEADS	CASES REPORTED	% TO TOTAL IPC CRIMES	RATE OF CRIME	CHARGE-SHEETING RATE	CONVI-CTION RATE
A VIOLENT CRIMES						
1	MURDER	33335	1.5	2.8	84.2	36.7
2	ATTEMPT TO COMMIT MURDER	29421	1.3	2.5	89.9	29.9
3	C.H. NOT AMOUNTING MURDER	3782	0.2	0.3	84.9	38.9
4	RAPE	22172	1.0	1.9	94.5	26.6
5	KIDNAPPING & ABDUCTION	38440	1.7	3.2	72.4	27.7
6	DACOITY	4358	0.2	0.4	76.4	21.9
7	PREPARATION & ASSEMBLY FOR DACOITY	2615	0.1	0.2	95.5	25.9
8	ROBBERY	23393	1.1	2.0	70.6	28.3
9	RIOTS	67571	3.0	5.7	90.9	21.7
10	ARSON	8508	0.4	0.7	68.1	19.3
11	DOWRY DEATHS	8391	0.4	0.7	93.2	33.6
TOTAL VIOLENT CRIMES		241986	10.9	20.4	84.6	27.7
B CRIME AGAINST WOMEN						
1	**KIDNAPPING & ABDUCTION OF WOMEN & GIRLS**	29795	1.3	2.5	74.2	28.1
2	MOLESTATION	40613	1.8	3.4	96.7	29.7
3	SEXUAL HARASSMENT	9961	0.4	0.8	86.7	52.0
4	CRUELTY BY HUSBAND AND RELATIVES	94041	4.2	7.9	94.2	19.1
5	IMPORTATION OF GIRLS	36	0.0	0.0	90.6	20.00
TOTAL CRIME AGAINST WOMEN (IPC+SLL)		213585	9.6	18.0	92.0	27.8
C ECONOMIC CRIMES						
1	CRIMINAL BREACH OF TRUST	16678	0.7	1.4	70.2	32.7
2	CHEATING	78999	3.6	6.7	72.1	29.2
3	COUNTERFEITING	2589	0.1	0.2	42.7	37.9
TOTAL ECONOMIC CRIMES		98266	4.4	8.3	70.6	30.3
D PROPERTY CRIMES						
1	BURGLARY	90179	4.1	7.6	43.4	34.5
2	THEFT	330312	14.8	27.9	37.7	37.5
TOTAL PROPERTY CRIMES		420491	18.9	35.5	38.9	36.7
E CRIME AGAINST SCs						
TOTAL CRIME AGAINST SCs		32712	1.5	2.8	90.7	35.0
F CRIME AGAINST STs						
TOTAL CRIME AGAINST STs		5885	0.3	0.5	96.0	25.0
G CRIME AGAINST CHILDREN						
TOTAL CRIME AGAINST CHILDREN		26694	1.2	2.3	83.9	34.6
H COGNIZABLE CRIMES UNDER IPC						
TOTAL COGNIZABLE CRIMES UNDER IPC		2224831		187.6	79.1	40.7
I COGNIZABLE CRIMES UNDER SLL						
TOTAL COGNIZABLE CRIMES UNDER SLL		4525917		381.7	94.7	91.7
J COGNIZABLE CRIMES UNDER IPC + SLL						
TOTAL COGNIZABLE CRIMES UNDER IPC + SLL		6750748		569.3	89.8	81.3

NOTE : A4,A11 ARE PART OF CRIME AGAINST WOMEN ALSO; A6,A7 & A8 ARE PART OF PROPERTY CRIMES ALSO

SOURCE : NCRB New Delhi, Crime In India-2010 P.iii

Figure-13.1 **Crime Clock 2004**

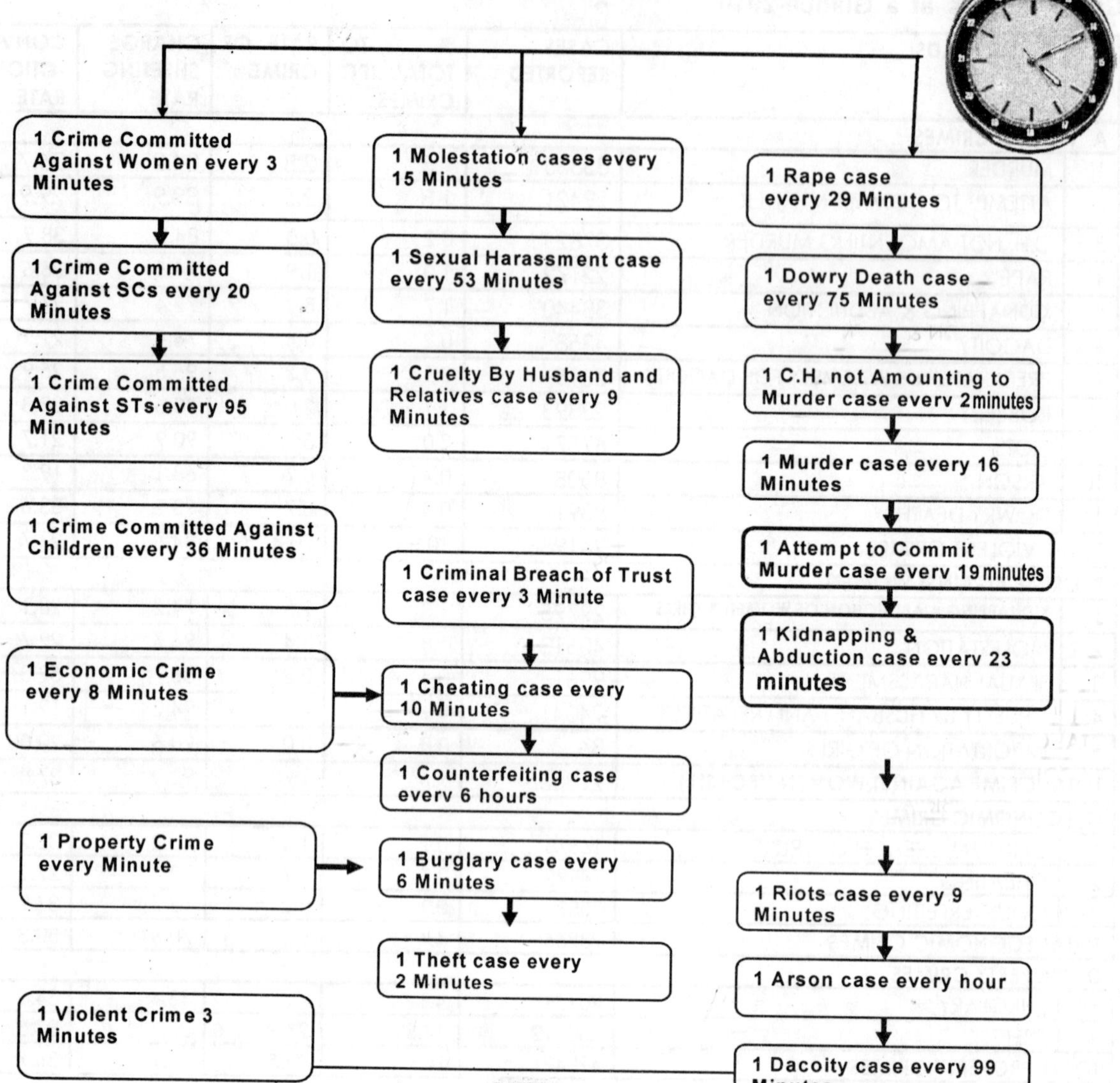

Figure-13.2

Sources : NCRB,New Delhi

TABLE- 13.3

Incidence & Rate of Crime Committed Against Women during 2001-2010

Year	Incidence		Rate Of Total Cognizable Crime		Percentage Contribution to All-India Total	Rank as per Rate of total cognizable crime	Rank as per percentage share
	Orissa	All India	Orissa	All India	Orissa	Orissa	Orissa
2001	5357	1,43,795	14.6	14.0	3.7	14	11
2002	4779	1,43,034	12.8	14.1	3.2	17	14
2003	4946	1,40,601	13.1	13.2	3.5	14	13
2004	5239	1,54,333	13.7	14,2	3.4	13	13
2005	6249	1,55,553	16.1	14.1	4.0	11	10
2006	6825	1,64,765	17.4	14.7	4.1	11	09
2007	7304	1,85,312	18.4	16.3	3.9	12	11
2008	8303	1,95,856	20.7	17.0	4.2	10	09
2009	8120	2,03,804	20.1	17.4	4.0	12	09
2010	8501	2,13,585	20.8	18.0	4.0	10	10

Source: NCRB, New Delhi Crime Statistics 2001-2010, TABLE-5.1.

References

Cohen MA 2000. Measuring the costs and benefits of crime and justice, in LaFree G (ed), *Measurement and analysis of crime and justice*. Criminal Justice 2000(4): 263–315. Washington, DC: National Institute of Justice, US Department of Justice.

Land, Kenneth C., and Patricia L. McCall. (2006). Uncertainty in crime forecasts: A reconsideration. Paper presented at the meeting of the American Sociological Association, Montreal (August).

Levitt, Steven D. (2004). Understanding why crime fell in the 1990s: Four factors that explain the decline and six that do not. Journal of Economic Perspectives, 18, 163-190.

Rex S & Tonry M 2002. *Reform and punishment: The future of sentencing*. Cullompton: Willan Publishing.

Roberts L & Indermaur D 2009. *What Australians think about crime and justice: Results from the 2007 Survey of Social Attitudes*. Research and public policy series no. 101. Canberra: Australian Institute of Criminology.

Rosenfeld, Richard. (2006). Connecting the dots: Crime rates and criminal justice evaluation research. Journal of Experimental Criminology, 2, 309-319.

Rosenfeld, Richard and Arthur S. Goldberger (Ed) (2008), Understanding Crime Trends, National Academic Press, Washington DC.

Zimring, Franklin E. (2007). The great American crime decline. New York: Oxford University Press.

■■■

CHAPTER 14

Gambling

Learning Objectives :

- ❒ **To Understand the legal, psychological and economic connotations of Gambling.**
- ❒ **To comprehend what is not gambling.**
- ❒ **To discuss the history and the types of gambling.**
- ❒ **To be acquainted with the phages of gambling addiction.**
- ❒ **To grasp the causes and risk factors of gambling addiction.**

The Concept of Gambling

From times immemorial, many people from all walks of life have been indulging in this vice of gambling. Gambling is the betting or staking of something of value, with consciousness of risk and hope of gain. The outcomes of gambling games may be determined by chance alone, as in the purely random activity of a tossed pair of dice or of the ball on a roulette wheel, or by physical skill, training, or prowess in athletic contests, or by a combination of strategy and chance. The rules by which gambling games are played sometimes serve to confuse the relationship between the components of the game, which depend on skill and chance, so that some players may be able to manipulate the game to serve their own interests.

A rough estimate of the amount of money legally wagered annually in the world is about $10 trillion (illegal gambling may exceed even this figure). In terms of total turnover, lotteries are the leading form of gambling worldwide. State-licensed or state-operated lotteries expanded rapidly in Europe and the United States during the late 20th century and are widely distributed throughout most of the world. Organized football (soccer) pools can be found in nearly all European countries, several South American countries, Australia, and a few African and Asian countries. Most of these countries also offer either state-organized or state-licensed wagering on other sporting events.

Events or outcomes that are equally probable have an equal chance of occurring in each instance. In games of pure chance, each instance is a completely independent one; that is, each play has the same probability as each of the others of producing a given outcome. Probability statements apply in practice to a long series of events but not to individual ones. The law of large numbers is an expression of the fact that the ratios predicted by probability statements are increasingly accurate as the number of events increases, but the absolute number of outcomes of a particular type departs from expectation with increasing frequency as the number of repetitions increases. It is the ratios that are accurately predictable, not the individual events or precise totals.

Webster Dictionary says: gambling is a practice to stake or risk money or anything of value on the outcome of something involving chance or risk. Encyclopedia Britannica defines gambling as "the betting, staking of something of value with consciousness of risk and hope of gain, on the outcome of a game, a contest, or an uncertain event, the result of which may be determined by chance or accident or have an unexpected result by reason of the bettor's miscalculation." The words gambling, gaming and game are derived from the Anglo -Saxon Gamen, Gamon ("sport" or "play"). Even though many normal activities such as business ventures and military strategies also involve an element of risk and some uncertainty, only those activities where the element of chance is more are generally referred to as gambling.

A.L. Kroeber, an anthropologist tried to discover whether or not gambling was distributed among the cultures of the world in systematic patterns that might reveal a general association between gambling and other cultural traits. He found that aversion to gambling might be more in stricter sects of puritanical religions such as Protestantism, Buddhism and Islam. Another anthropologist named John Roberts classified games into three types: (a) games of physical skill (which might or might not include strategy or chance), (b) games of strategy (without physical skill but perhaps with chance), and (c) games of chance alone. On the basis of the study of different tribes from all over the world, he came to the conclusion that cultures in which games of chance were played usually had religious beliefs in gods or spirits that were benevolent and could be coerced. These games of chance were left to the hands of supernatural power. Games of strategy on the other hand were associated with learning of social roles, mastery of the social system, whereas games of physical skills were associated with mastery of both the self and the environment.

Gambling as a problem has economic, legal and psychological connotations

Gambling has a specific **economic** definition, referring to wagering money or something of material value on an event with an uncertain outcome with the primary intent of winning additional money and/or material goods. Typically, the outcome of the wager is evident within a short period of time. This is illegal participation in games of skill or chance for money and/or other items of value. The voluntary risking of a sum of money on the outcome of a game or other event is gambling.

Legally, all offenses which relate to promoting, permitting, or engaging in gambling are included in the category of gambling. Accordingly, it is an unlawful engaging in playing, operating, or assisting in operating a game of chance for money or some other stake. The legalized gambling will be discussed later

At the **psychological level** gambling is an activity characterised by a balance between winning and losing that is governed by a mixture of skill and chance. *Four models are available in* *literature discussing the socio-psychological dimensions of gambling as a problem: the disease model, the social learning model, the cognitive-behavioural model and the psychodynamic model.*

The Disease/Medical Model : This model proposes that problems with gambling are the result of a disease — a condition that an individual "has" or "does not have." The model states that compulsive gambling is progressive (it will continue to worsen if the individual continues to gamble), and irreversible (a compulsive gambler will not be able to return to non-problem gambling).

Compulsive gamblers are qualitatively different from other gamblers. Compulsive gamblers gradually lose control, and are eventually unable to stop gambling. Compulsive gambling is a progressive condition, and progresses inexorably through a number of stages, beginning with initial success at gambling; then experiencing less success; irrational optimism about winning; psychological distress; chasing losses and possibly engaging in illegal activities to get money to gamble; unsuccessful attempts to cut down or quit gambling; and eventually hitting bottom. Compulsive gambling is a permanent and irreversible condition. The only cure is total abstinence. If the gambler were to resume gambling, all of the symptoms described above would manifest again.

There are two main criticisms of the disease model. The first is that the model does not seem to be able to account for those who may experience problems with gambling from time to time, but who do not progress to the stage of persistent, recurrent compulsive gambling. The second criticism is that the disease model, with its emphasis on internal disease or medical processes, may seem to absolve the compulsive gambler from taking responsibility for the problem. The counter-argument is that even if the compulsive gambler cannot be responsible for having the disease, he or she is responsible for taking action to deal with the disease.

The Social Learning Model : The social learning model views gambling as operant behaviour, subject to reinforcement schedules, which strengthen gambling behaviour. Initially, money was theorized to be the positive reinforcement, but more recent thinking, supported by research, has elevated the importance of the reinforcing properties of physiological arousal in this model.

Interestingly, Skinner (1953) argued that the individual's gambling behaviour is a function of his or her previous reinforcement history. He theorized that initial success with gambling led to an increased likelihood that the gambling behaviour would continue, even if the reinforcement ratio declined, i.e., the individual won less often. Skinner was able to demonstrate this learned pattern in rats and pigeons. However, his hypothesis was not tested on humans. Later, Custer (1982) emphasized the importance of the early big win in the development of pathological gambling.

Social learning theorists still wonder why people affected by problem gambling tend to persist even when there is very little reinforcement, i.e., they lose far more than they win. Two explanations have been suggested. One, reinforcement schedules that pay off only intermittently (like gambling) are known to produce a greater persistence in the behaviour after the reward is stopped, than would a schedule of continuous reward. This is known as the "partial reinforcement extinction effect" (pree). Another aspect of the pree that accounts for persistence is that persistence does actually pay off in a sense, because the gambler will eventually experience a win if he or she continues to gamble. Thus persistence to gamble, even after many losses, is strengthened by the win (Griffiths, 1995).

The second explanation considers the role of arousal. Brown (1987) proposed arousal as a key determinant of problem gambling, thus including physiological concepts in the realm of the social learning model. According to Brown, individuals have different psychophysiological arousal needs, and some learn to use gambling to regulate their arousal needs. Thus, arousal, not winning, may be the primary reinforcer of the gambling behaviour.

Because the social learning model takes the position that behaviour is acquired through a complex interaction of a variety of internal and external factors, it suggests that gambling can be conceptualized as a continuum — from problem-free gambling to varying degrees of problem gambling. Further, the development of gambling problems does not follow set stages (although there may be patterns of gambling), nor are serious gambling problems considered permanent and irreversible. The course that problem gambling can take varies from one individual to another.

Stein (1993) added an interesting perspective to the social learning model from developmental theory, by viewing compulsive gambling as a symptom of cognitive developmental delay in the transition from adolescence to adulthood. The compulsive gambler is seen as having a limited way of understanding his or her environment and developing solutions to his or her problems. As a result, his or her problem-solving strategies — namely, gambling to gain pleasure or diminish pain — are developmentally delayed. The cognitive-behavioural model is a more specific application of the social learning model.

The Cognitive-Behavioural Model : The cognitive-behavioural model is based on the theory that behaviour is initiated, maintained (or discontinued), based on principles of learning (such as imitation, observational learning, schedules of reinforcement), and cognition (how the individual attends to, interprets, and draws conclusions about events that go on around him or her).

Cognitions play an important role in the maintenance of gambling. Gamblers who misinterpret the causes of wins and losses, and develop a belief system that they will be successful at gambling, become more susceptible to problem gambling.

Whether or not the gambler will gamble when faced with an urge (triggered by exposure to a conditioned stimulus), seems to be related to the gambler's coping skills — the individual's ability to control increased arousal, analyze distorted cognitions, delay reinforcement, and apply problem solving skills (Sharpe & Tarrier, 1993).

The Psychodynamic Model : The psychodynamic model proposes that personal problems, like excessive gambling, lie within the psyche are an attempt at self-healing or a strategy of resolving unconscious psychic conflicts. The source of the conflict is beyond voluntary control. Driven by these conflicts, the individual continues to engage in gambling behaviour in an unsuccessful attempt to ease the psychological pain.

Within the psychodynamic model there are several psychodynamically based theories that have been put forward to explain psychopathological behaviour, including excessive gambling. Three main components are : (1) Gambling is an unconscious substitute for pre-genital libidinal/aggressive outlets, (2) Gambling involves an unconscious desire to lose — a wish to be punished in reaction to guilt and (3) Gambling is a medium for continued enactment (but not resolution) of psychological conflict (Griffiths, 1995, p. 11).

Rosenthal (1986) also claimed that the majority of compulsive gamblers were narcissistic. They have feelings of inadequacy that lead to the creation of a fantasy world in which gambling is seen as the solution to their problems. This fantasy allows them to feel important, respected, powerful and independent, and, in fact, many gamblers report such feelings of power and respect while "in action." Rosenthal notes that the narcissistic personality is susceptible to swings of arousal and depression, and that gambling behaviour is an attempt to regulate these swings.

Rosenthal also claimed that gamblers are preoccupied with freedom from responsibility, intimacy, and the constraints of consciousness. Defence mechanisms, the psychodynamic equivalent of coping skills, are believed to be primitive in the compulsive gambler. Defence mechanisms prevent exploration and resolution of the unconscious conflicts, allowing the problem to continue.

Another predominant theme in psychodynamic theory is loss, including physical separation such as death of a loved one, loss of job or status, or divorce; or emotional losses, such as loss of self-esteem, or a sense of security. The theory is that pervasive loss is a central issue for the compulsive gambler, resulting in intolerance of future losses. Gambling is both precipitated by an emotional loss, and also a way to avoid taking any further emotional risks. Compulsive gambling is seen as a fixation or regression to the psychic development of a child who is beginning to perceive the loss of a mother's love. According to this theory, gambling is used as a substitute for the mother's love.

Critics of the psychodynamic model have pointed out that the model is more theoretical than scientifically based, since concepts like "psyche" and the "unconscious" are largely untestable. Much of the "evidence" for this model is based on single case histories and elaborate conjecture (Ferris, Wynne & Single, 1999).

However, the psychodynamic model has made a contribution by advocating that counsellors understand the internal, unseen processes, in addition to focusing on the gambling behaviour itself. Many gamblers may indeed benefit from counselling that explores the issues of loss, emotional intimacy, and the drive to create the fantasy of importance, respect, power and control that is often central to the gambling experience. For some gamblers, stopping gambling may bring these unconscious conflicts into greater awareness, creating an uncomfortable tension that could be a factor in relapse if left untreated.

Types of Gambling

Types of gambling include Casino games, Non-casino games, Fixed-odds gambling, Betting and Internet Gambling.

Casino games : While almost any game can be played for money, and any game typically played for money can also be played just for fun, some games are generally offered in a casino setting. Casino games can be further divided to two: the "beatable" games and the "unbeatable" games. "Beatable" casino games are games that may require a certain amount of skill, experience and knowledge to win the game. Example of this sub-type includes poker, black jack, and sports betting. But, "unbeatable" casino games are games that are purely random and unpredictable. It requires no skill to win, but players are bound to lose in the long run. Examples of this are baccarat, roulette, and pachinko.

Non-casino gambling games : Non- casino gambling games are games that are usually played recreationally and sometimes, without stakes. Examples to this are bingo, dead pool, lotteries, pull-tab games and scratch cards. These games take place outside of casinos. Other non-casino gambling games include :

- Card games, such as Liar's poker, Bridge, Basset, Lansquenet, Piquet, Put, Teen patti
- Coin-tossing games such as Head and Tail, Two-up
- Confidence tricks such as Three-card Monte or the Shell game
- Carnival Games such as The Razzle or Hanky Pank
- Dice-based games, such as Backgammon, Liar's dice, Passe-dix, Hazard, Threes, Pig, or Mexico

Although coin tossing isn't usually played in a casino, it has been known to be an official gambling game in some Australian casinos.

Fixed-odds gambling : Fixed-odds mean that the probability of winning or losing is equal to .5 or 50% each. This kind of gambling is common in sports games like horse racing, basketball, football, boxing, and the like. Except for horse racing, which bettors can bet on the second or third horse, all the other examples require players to bet on any of the two competing teams. Thus fixed-odds gambling frequently occurs at many types of sporting events, and political elections. In addition many bookmakers offer fixed odds on a number of non-sports related outcomes, for example the direction and extent of movement of various financial indices, the winner of television competitions such as Big Brother, and election results. Interactive prediction markets also offer trading on these outcomes, with "shares" of results trading on an open market.

Parimutuel betting : One of the most widespread forms of gambling involves betting on horse or greyhound racing. Wagering may take place through parimutuel pools, or bookmakers may take bets personally. Parimutuel wagers pay off at prices determined by support in the wagering pools, while bookmakers pay off either at the odds offered at the time of accepting the bet; or at the median odds offered by track bookmakers at the time the race started.

Arbitrage betting : Arbitrage betting is a theoretically risk-free betting system in which every outcome of an event is bet upon so that a known profit will be made by the bettor upon completion of the event, regardless of the outcome. Arbitrage betting is a combination of the ancient art of arbitrage trading and gambling, which has been made possible by the large numbers of bookmakers in the marketplace, creating occasional opportunities for arbitrage.

Other types of betting : One can also bet with another person that a statement is true or false, or that a specified event will happen (a "back bet") or will not happen (a "lay bet") within a specified time. This occurs in particular when two people have opposing but strongly-held views on truth or events. Not only do the parties hope to gain from the bet, they place the bet also to demonstrate their certainty about the issue. Some means of determining the issue at stake must exist. Sometimes the amount bet remains nominal, demonstrating the outcome as one of principle rather than of financial importance. Betting exchanges allow consumers to both back and lay at odds of their choice. Similar in some ways to a stock exchange, a better may want to back a horse (hoping it will win) or lay a horse (hoping it will lose, effectively acting as bookmaker).

Internet Gambling : The development of the internet and mobile telephones, within recent years, provides new opportunities for many types of games. Examples include gaming and gambling websites and other "pay to use" sites on the web. It is not difficult for someone who wishes to gamble, without travelling, to find a site where they can bet money (using a credit card) in order to play Black Jack, Roulette, Poker or numerous types of lottery games. These games, which could be considered traditional forms of entertainment have now been given a new genre. Dating websites are abundant on the internet and often require a subscription fee, which can be quite costly. A derivative of such sites also exists on mobile phones.

The broadband Internet connection is popular and is currently used within many homes. Furthermore, many video games are evolving to offer a "network". MMORPGs (Massive Multiplayer Online Role Playing Games) which bring together of a large number of players, can now collect up to several million participants. In general, a supplement is added to the Internet subscription fee, in order to gain access to such games. For network based role-play games, it is also possible to buy a figure who is advanced within the game, and this can be quite costly.

Regarding mobile phones, there has been a significant increase in the number of competitions requiring entry by SMS. These types of competition, which offer a range of prizes by sending premium rate SMS messages, are frequently aimed at young people. People of all ages can also download to their phone, ringtones and wallpaper images of their favourite artist. Again, there is a charge for making downloads. Record publishers have recognised that they have an effective way to regain the money, lost by downloading pirate music from the internet, and they receive an impressive percentage from downloaded ringtones.

These new types of games and other subscriptions and downloads significantly increases the risk of negative consequences (such as loss of money, isolation, withdrawal from everyday life) for the population affected by such offers. Adolescents and young adults, who frequently use new technologies, are the most vulnerable. Many parents are concerned to see their child spend all their pocket money on their mobile phone or subscriptions to diverse gaming websites or chat rooms.

History of Gambling

Gambling is one of mankind's oldest activities, as evidenced by writings and equipment found in tombs and other museums. It was regulated, which as a rule meant severely curtailed, in the laws of ancient China and Rome as well as in the Jewish Talmud and by Islam and Buddhism, and in ancient Egypt inveterate gamblers could be sentenced to forced labor in the quarries. The origin of gambling is considered to be divinatory: by casting marked sticks and other objects and interpreting the outcome, man sought knowledge of the future and the intentions of the gods. From this it was a very short step to betting on the outcome of the throws. Thus, the beginning of gambling and gambling tools can be traced to religious rituals in primitive pre-historic societies. The evolution of these rituals included the sacrifices from the participants to encourage the Fate or the Gods to deliver positive signs and help. The sacrifices were first personal stakes which turned those involved from mere observers into active players of the process. They gambled their stakes on the outcome that depended on Chance only. Eventually the same type of chance-based procedures expanded into every day life. This helped with decision making process like who will get a better piece of a killed prey or get himself a better looking woman. This process ultimately culminated in that evolution toward pure gambling which happened when people decided to gamble with their stakes for material gain only.

Around 7000 B.C. the fertile soil of Mesopotamia – the region between Tigris and Euphrates in modern Iraq – gave humans the first opportunity to change lifestyle from nomadic to a settled one. That led to establishing of a stable society, which was necessary for developing skills, knowledge and tools. The civilization of the human race started from that point in history. No wonder that the earliest gambling tools were discovered among archaeological findings in that part of the world.

Four-sided astragali dating 6000 B.C. and first six-sided dice cubes marked with pips dating 3000 B.C. were found in many archaeological sites. Small bones such as knucklebones (astragali) became the direct predecessors of the modern dice. Astragali have an irregular, not symmetrical shape, which has 4 large sides and two more or less rounded ends. After being thrown, an astragalus will stop on one of its 4 big sides delivering a quasi-random outcome of the throw. Eventually the sides of an astragalus were given value and markings. The lowest value of one was attached to the largest side of a concave shape because of a higher chance that the bone will rest on it. The highest value of 6 was given to the smallest side of a convex shape for the obvious reason. Discovered gaming boards from the same period revealed that the population played board games structurally similar to modern Backgammon. The result of the throw of the dice was used to determine how far a game piece could be moved on the board.

Gambling with dice was very popular in Persia, which covered the territory of modern Iran. Greek historian and biographer Plutarch told the story about a dice game between King Artarxerxes and his mother Parysatis. Sometime before the game the King ordered one of his servants to cut the head of his younger brother Cyrus who had attempted a coup to grab the power but had been defeated in the battle. Parysatis wanted to revenge Cyrus. She offered Artaxerxes to play a friendly dice game. She lost on purpose to entice the King to play for higher stakes. The stakes included the servant who beheaded Cyrus. The Mother Queen proceeded to win and take a revenge on that servant.

Egyptians liked to gamble as much as their Persian and Mesopotamian brothers in gambling. Ancient Egyptian tombs were adorned by sculptural reliefs depicting numerous gambling scenes. Survived hieroglyphic records from old dynasties show that anti-gambling laws were established by 3000-4000 B.C. That means that by that time gambling was already a significant part of the Egyptian life. Plutarch recorded an Egyptian myth about the gods who gambled. Cubical dice dating back to 2000 B.C. have been found in archaeological excavations. Most of the games played by the Egyptians involved 2 dice and a gaming board. The dice were tossed and the game pieces were moved across the board according to the result of the throw. A player who moved his pieces across the whole board first was a winner. Other forms of gambling included "Odds and Evens" games and also guessing games. In guessing games the players had to guess correctly how many fingers were extended by an opponent or how many small objects he had in his hand.

Indian culture adopted gambling from the beginning of Indian civilization, which started 4000 years ago. Great Indian epic Mahabharata dating to 1500 B.C. has the main story, which starts with a dice match. Different members of a royal family competed for the throne. Finally it came down to a dice game. The game was not honest probably thanks to the loaded dice. One of the competitors lost all his royal possessions including his wife. The story, however, has a happy ending. Unlike Mesopotamians who used astragali for dice games, Indians employed the nuts of the "vibhitaka" tree (Schwartz 2006). These nuts have 5 more or less flat sides, which make it possible to use them as dice. Eventually vibhitaka dice were replaced by astragali and cubical dice. Besides dice games

Indians were fond of gambling that involved animals – particularly cock and ram fights. There were many gambling houses for gamblers to assemble and play like there was no tomorrow. Those houses were supervised by an official to assure orderly gambling and collect from the players a percentage for the King.

Chinese gambling tradition is also remarkable. Gambling followed the development of Chinese society every step of the way and by 1000 B.C. it became as inseparable from Chinese culture as wonton soup or a roasted duck. Every big street in Chinese towns had a gambling den. Very popular form of gambling was betting on animal fights and races. Cricket and cock fights and dog and horse races always attracted huge number of gamblers and heavy betting. In the 5th Century B.C. board games, which combined luck and skill grew in popularity. Around 7th Century A.D. Chinese modified western dice into Chinese dominoes. The dominoes included 31 pieces representing 21 numbers (some numbers were repeated). They were made of ivory and other materials and usually had red and black pips. In 12th Century A.D. during the rule of Emperor Seun-Ho Chinese playing cards entered the gambling scene giving another boost and adding another flavor to already rich and diverse gambling life in China.

This evolutionary point of view on gambling origins is accepted by modern history and anthropology. Ancient explanations credited Gods and mythological heroes with the invention of gambling. For ex, the Egyptians thought that god Thoth who was the divine physician, inventor of writing and supreme judge invented gambling. For ancient Greeks gambling also came from Gods. According to Greek Mythology, Zeus, Poseidon and Hades were casting lots to claim parts of the Universe. In result Zeus got the heavens, Poseidon the sea and Hades took possession of the underworld. Later Greeks decided that mythological hero Palamedes created games of chance to entertain Greek troops during the Trojan War. He was also thought to be the inventor of the dice.

The Bible contains many references to the casting of lots to divide property. One well-known instance is the casting of lots by Roman guards (which in all likelihood meant that they threw knucklebones) for the garment of Jesus during the Crucifixion. This is mentioned in all four of the Gospels and has been used for centuries as a warning example by antigambling crusaders. However, in ancient times casting lots was not considered to be gambling in the modern sense but instead was connected with inevitable destiny, or fate. Anthropologists have also pointed to the fact that gambling is more prevalent in societies where there is a widespread belief in gods and spirits whose benevolence may be sought. The casting of lots, not infrequently dice, has been used in many cultures to dispense justice and point out criminals at trials—in Sweden as late as 1803. The Greek word for justice, dike, comes from a word that means "to throw," in the sense of throwing dice.

The science of probability owes its start to games of chance. An Italian mathematician, Gerolamo Cardano is believed to have written the first mathematical analysis of the cast of dice in 1526. Both Galileo and Pascal also undertook analysis of dice playing. However, a detailed mathematical analysis on dice games was provided only in the middle of 17th Century by Christian Huygens.

European history is riddled with edicts, decrees, and encyclicals banning and condemning gambling, which indirectly testify to its popularity in all strata of society. Organized gambling on a larger scale and sanctioned by governments and other authorities in order to raise money began in the 15th century with lotteries—and centuries earlier in China. With the advent of legal gambling houses in the 17th century, mathematicians began to take a serious interest in games with randomizing equipment (such as dice and cards), out of which grew the field of probability theory.

Apart from forerunners in ancient Rome and Greece, organized sanctioned sports betting dates back to the late 18th century. About that time there began a gradual, albeit irregular, shift in the official attitude toward gambling, from considering it a sin to considering it a vice and a human weakness and, finally, to seeing it as a mostly harmless and even entertaining activity. Additionally, the Internet has made many forms of gambling accessible on an unheard-of scale. By the beginning of the 21st century, approximately four out of five people in Western nations gambled at least occasionally. The swelling number of gamblers in the 20th century highlighted the personal and social problem of pathological gambling, in which individuals are unable to control or limit their gambling. During the 1980s and '90s, pathological gambling was recognized by medical authorities in several countries as a cognitive disorder that afflicts slightly more than 1 percent of the population, and various treatment and therapy programs were developed to deal with the problem.

Legalized Gambling

Many forms of legal gambling exist in many countries. Horse racing is permitted in many countries. Similarly, State licensed or State operated lotteries exists in many countries. Of late, Casino gaming has also become very popular in many parts of the world. In some countries Casinos account for a sizable portion of the national income of the country. Slot machines or Poker machines are available in thousands of clubs and restaurants in the Western world as well as some Asian countries. Under US federal law, gambling is legal in the United States, and states are free to regulate or prohibit the practice. Gambling has been legal in Nevada since 1931, forming the backbone of the state's economy, and the city of Las Vegas is perhaps the best known gambling destination in the world. In 1976, gambling was legalized in Atlantic City, New Jersey, and in 1990, it was legalized in Tunica, Mississippi; both of those cities have developed extensive casino and resort areas since then. Since a favorable U.S. Supreme Court decision in 1987, many Native American tribes have built their own casinos on tribal lands as a way to provide revenue for the tribe. Because the tribes are considered sovereign nations, they are often exempt from state laws restricting gambling, and are instead regulated under federal law. Additionally, almost all states have legalized gambling in the form of a state-run lottery and most states allow for limited non-profit organizations to host Bingo nights. The Indian Gaming Regulatory Act (IGRA), enacted in 1988 (US), defined three classes of gambling: Class I, which includes social games for prizes of minimal value and traditional forms of Indian gaming that are played as part of tribal ceremonies or celebrations; Class II, which includes bingo, bingo-like games, and non-banking card games (for example, games such as poker that are played against other players, rather than against the house); and Class III, which includes casino games, slot machines, and banking card games.

In our country, the government's attitude towards gambling has not been consistent. Some forms of gambling such as lotteries and betting on horses are not illegal in India at present. Even with regard to lotteries, the opinion has changed from time to time. At times, only State owned lotteries were permitted and other types of lotteries are banned. Now, there is a proposal to enact a comprehensive legislation with regard to all types of lotteries. Some time ago, there was a move to permit operation of Casinos in places like Goa, which attract large number of foreign tourists, in order to get foreign exchange. However, because of opposition from various sections, the idea was dropped. Goa and Sikkim are the two states which allow casino gambling. There is one casino in Sikkim and 12 in Goa. Other than lotteries, legal gambling in India is limited to betting on horse racing.

Legalized gambling brings other hidden costs, from beefed-up police needs to the expense of treating those addicted to gambling. One frequently overlooked hidden cost involves children and the elderly. According to a recent *Reader's Digest* article: "Today, more kids gamble than are involved with drugs, smoking or drinking" (Mar. 2001, p. 160). While it is illegal for minors to play lotteries, studies show that in states such as Texas and Louisiana about one-third of adolescents buy tickets. Additionally, many casinos run "free" shuttles to retirement homes, conveniently timed to coincide with the arrival of Social Security checks at the beginning of the month, exploiting an often lonely and vulnerable segment of society. Gambling can exert a corrupting influence on government. None of these hidden costs of gambling even begin to address the problem of organized crime. The gambling industry and organized crime have had a long-time love affair reaching back to the early 20th century. Many do not understand why the newly legalized gambling industry is so important to crime syndicates; one key reason is the large amounts of cash generated by narcotics and other illegal activities. How can such vast quantities of cash be accounted for ? One cannot walk into the bank and make large cash deposits day after day without serious questions being asked. Casinos are an ideal front, as illegal cash can be hidden among a casino's profits, difficult to trace. Legalized gambling, particularly land-based casinos and riverboats, thus plays a vital role fronting for organized crime in America.

Economic Impacts of Gambling

There is a debate over whether or not gambling can be good for an economy. Proponents of the view that gambling is harmful use a quote of Nobel Laureate Paul Samuelson.

"(Gambling) involves simply sterile transfers of money or goods between individuals, creating no new money or goods. Although it creates no output, gambling does nevertheless absorb time and resources. When pursued beyond the limits of recreation, where the main purpose after all is to kill time, gambling subtracts from the national income."

Other economists have taken exception to Samuelson's characterization of gambling. They point out that his criticism could be applied to movies or Disneyland. These are products that don't add to the ability of the economy to produce more. But they still have value because they provide satisfaction, or utility in the economist's jargon, to consumers. These economists are more concerned about the costs of banning gambling, that is the ensuing enforcement costs and the incentives to lobby and bribe public officials to allow illegal gambling to occur. Clearly all of the places of gambling have seen an influx of investment and many new jobs being created. Small cities develop into a major city with a low tax burden, a tourist destination and increased migration. Any industry that draws money from outside is helpful. Gambling is often legalized to promote economic development of depressed areas.

What is not "gambling"

Many risk-return choices are sometimes referred to colloquially as "gambling." Whether this terminology is acceptable is a matter of debate, but generally the following activities are not considered gambling :

- Emotional or physical risk-taking, where the risk-return ratio is not quantifiable (e.g., skydiving, campaigning for political office, asking someone for a date, etc.).

- Insurance is a method of shifting risk from one party to another. Insurers use actuarial methods to calculate appropriate premiums, which could be considered similar to calculating gambling odds. However, insurers can set their premiums to obtain a long term positive expected return.
- Situations where the possible return is a secondary reason for the wager/purchase (e.g. buying a raffle ticket to support a charitable cause).
- Investments are also usually not considered gambling, although some investments can involve significant risk. Examples of investments include stocks, bonds and real estate. Starting a business can also be considered a form of investment. Investments are generally not considered gambling when they meet the following criteria :
- Positive expected returns (at least in the long term)
- Economic utility
- Underlying value independent of the risk being undertaken
- Some speculative investment activities are particularly risky, but are still usually considered separately from gambling:
- Securities derivatives, such as options or futures, where the value of the derivative is dependent on the value of the underlying asset at a specific point in time (typically the derivative's associated expiration date)
- Foreign currency exchange (forex) transactions

Gambling and Crime

Gambling is often associated with crime. The relationship is easy to understand. Many types of gambling have been, indeed still are, illegal. Hence, by definition, criminals were the only operators of games. When gambling restrictions were relaxed, criminals were the first to open up legal gambling establishments. A lax regulatory framework may not prevent members of organized crime from openly owning and operating casinos. To some degree, criminals make gambling viable because no one else had their expertise and experience. Because of this history, the concern about organized crime usually is raised whenever legalizing gambling is discussed. Even when New Hampshire began its state lottery in 1964, there was concern that organized crime would take over.

The casino companies sometimes suggest that they are devoid of organized crime influence because they are :

- Dominated by publicly-held companies, many with household names like Hilton and Sheraton.
- Answerable to their shareholders who are thousands of individuals and institutional investors.
- Answerable to the Securities Exchange Commission.
- Indistinguishable from any other business with accountants, attorneys, payroll specialists, auditors, and market researchers.
- Licensed and tightly-regulated by state governments.

Nevertheless, there remains an ever present concern about organized crime. The sheer volume of money, cash in particular that is generated by gambling, makes it a tempting target. Organized crime has been successful infiltrating ancillary businesses such as machine maintenance or those that provide other services. Agencies and authorities sometimes take monetary payoffs to approve an expansion gambling business. The individuals attempting to buy influence are found connected to organized crime families. Researchers state that organized crime is more of a product of illegal or poorly regulated gambling than well-regulated gambling.

Gambling is a natural target for criminals because of the large amounts of cash. Thus proper operations and security are very important. Crime has been an Issue for Card clubs. There have been robberies and assaults of patrons who had left the card clubs after making money. There is dramatic increase in crime in the area where a new club opened with more people coming into the neighborhood or criminals following the money. Some of the crime may be unrelated to the gambling that occurs at the cardclubs. Rather, the club is a suitable place to meet individuals who are willing to buy stolen property, drugs, or cars without registration. Other activities such as loan sharking and credit card/ check fraud and crime planning are done at the clubs. Another crime issue associated with gambling is street crime. The crime rates were significantly higher in the places that allowed gambling.

Those engaged in gambling business have, with few exceptions, been identified with the criminal elements. Gambling house proprietors are normally associated with swindlers, confidence men or notorious bandits including robbers and murders. Gambling houses have flourished to a great extent in localities where lawlessness and corruption have prevailed. The business of gambling has also been characterized by dishonesty. There are various methods in which gambling houses cheat their customers.

Is Gambling an Addiction ?

Addiction refers to habitual psychological and physiological dependence on a substance or practice beyond one's voluntary control. The literal definition of gambling addiction implies that it is a behavioural problem, a psychological problem, a physiological problem and none the less a form of substance abuse. Contrary to what many believe, gambling addiction is not merely a money problem. It is a psychiatric disorder characterized by persistent, recurrent and maladaptive gambling behaviour which disrupts the subject's personal, family and working life. It is a progressive disorder characterized by a continuous or periodic loss of control over gambling; a preoccupation with gambling and with obtaining money with which to gamble; irrational thinking; and a continuation of the behaviour despite adverse consequences. Most people gamble to escape from stress, a painful past, to seek arousal, or to be in "action." Gambling addiction is also known as compulsive gambling, pathological gambling, and problem gambling. The technical definitions of gambling addiction are :

Like alcohol, gambling is classified as a drug. However, alcohol is classified as a depressant while gambling is classified as a stimulant. As a drug, gambling is a psycho-stimulant, but at the same time it also has an anesthetic effect. Gambling has the same psychoactive properties as cocaine and heroin—the same chemicals are affected (adrenaline, dopamine, endorphins) in all three drugs when the addicts are in action. Gamblers do seek "action" through the psychoactive drug-like effects offered by gambling. The same chemicals, or hormones, are released in the brains of gambling addicts and drug addicts when they are in "action." When problem gamblers quit suddenly or they're forced to because they are out of money, they face the same physical withdrawal symptoms as drug addicts. This is the reason that well-educated, intelligent, wealthy people with no addiction background turn to crime to feed their gambling habit after being "hooked."

Many have argued that gambling is much more dangerous than alcohol, cocaine, or heroin. Chronic Alcoholism may take 5 to 20 years to develop, but people can become addicted to gambling on their very first try. Like cocaine and heroin addicts, problem gamblers go through rituals and seek immediate gratification. These rituals might include what shirt to wear, what machines to play, what table to sit at, what team to bet on, what dealer to play with, or what lucky artifact to bring with them when they're in action. Opponents of gambling believe that it is worse than alcohol and drugs.

- There are no visible signs. Compulsive gambling has been called an invisible disease because gambling addicts exhibit few signs. A drug addict might act strange or have bodily marks, an alcoholic might smell and stagger, but a problem gambler exhibits no physical signs of his or her problem.
- It is widely available. There are many places one can gamble at a moment's notice: the phone or computer, community halls (bingo), the workplace (raffles and office pools), corner grocery stores (lottery), fund raisers, shopping malls, social events (poker nights, charity), restaurants and bars. Gambling is everywhere.
- It is legal. Gambling is advertised as a form of entertainment, and in some cases as a lifestyle, giving the public the perception that it is a harmless activity. Gambling is also advertised as a form of charity, prompting people to believe that if they gamble they are contributing to the good of the community.
- It can lead to financial ruin in a short period. A gambler can lose hundreds, thousands, or millions of dollars in less than 30 seconds by placing a single bet. A Gambler's financial losses are unlimited. No other addiction can lead to financial ruin as quick as gambling.

Causes of Gambling Addiction

People gamble for different reasons. Some want to gain a lot of money with little effort. Others, especially problem gamblers, seek thrill and excitement through games of chance, and feel a "high" described as similar to a heroin-induced rush when prolonged tension, awaiting a game's outcome, is released. Many gamblers value this adrenaline-induced state of excitement more than the money they may win. Others claim that gambling is simply a way to pass time-just another form of entertainment-and that they spend no more time and money on gambling than they would devote to some other form of entertainment if they did not gamble.

Various theories exist on why people gamble. There is a debate over whether the motivation to gamble is positive or negative. A common view is that gambling is negative. In particular, this view is common among religious groups. Others argue that the motivation is positive, that people enjoy gambling and it is usually a harmless diversion. Because people enjoy it they continue to gamble in spite of losing.

Any discussion of the motivation of gambling usually starts with the natural comparison to life. Life is a gamble. Everyday, people are faced with situations which involve risk and chance. Gambling activities are extensions of the risk and chance in life. The activity of gambling becomes play, it becomes a game. Gambling allows the person the choice of engaging in the activity, the amount of risk and, in many cases, the stakes. The stakes are a necessary element for many people. It turns the bet into not just an opinion but a commitment. There are six main reasons as to why people become addicted to gambling. They are :

1. To cope with traumatic life experiences. People gamble to avoid uncomfortable or painful emotions, including sadness (passing of loved ones), guilt, shame, anger, humiliation, failure and frustration. Gambling provides the illusion of control in uncontrollable situations and environments.
2. To chase losses. Chasing past losses is the result of experiencing early wins. After a winning streak, the gambler is deluded with the idea of the "Big Win." The gambler thinks that through this Big Win, his or her bills will be paid and problems solved. The Big Win is an illusion that keeps the gambler in action until he or she admits there is a problem and seeks help.
3. As a form of stimulation. Some people knowingly or unknowingly seek the risky, naughty, thrill-seeking lifestyle through gambling. They surrender to their impulses and gambling acts as a drug that is as powerful as any available.
4. As a coping mechanism. Problem gamblers seek the "zone out" or "in my own little world" effect. They feel the need to escape the stress, demands, and pressures of daily life.
5. To climb the economic ladder. Some gamblers have the illusion, indoctrinated through advertising, that gambling can provide a steady income and money will solve their problems.
6. To self-sabotage their success. Some problem gamblers feel uncomfortable with, guilty about, or undeserving of their success. Gambling is used as a tool for self-destruction. The reasons behind their need for self-destruction are rooted in the past which psychology can explain.

The Risk Factors of Gambling Addiction

Problem and pathological gambling may be an invisible or silent disease but it is not a costless disease. Social costs are the costs borne by society as a whole that result from the behaviour of the problem gambler. Social costs includes such items as fraud, theft, bad loans, bad checks, lost work time, unemployment and welfare benefits, insured or publicly supported medical costs, and criminal justice system costs. Those types of social costs are easier to quantify than other types of social costs that result from gambling such as increased rates of suicide, car accidents, and incidence of child abuse. Another view is that social costs should include lost productivity of spouses, impaired judgment and efficiency on the job, divorces, added administrative costs for unemployed, and the costs of depression and physical illness related to the stress and lower quality of family life. Gambling is not just an economic problem; it is also a moral problem. Some of the risk factors of gambling addition are discussed below.

Coping Strategies : Coping strategies are life management skills, which include the ability to deal with the demands of every day life at home and work. These life management skills are required to effectively deal with stress, interpersonal relationships, setbacks, disappointments, and obstacles that are a part of life. Problem gamblers often lack coping strategies and gamble to escape adversity and hardship.

Impulsivity : Impulsivity is the inability of the gambler to control their behaviour even when they know the consequences beforehand through experience. Impulsivity stems from personality. One trait of impulsivity is the inability to delay decision-making.

Stress : Most relapses are the result of reaction to stress. Stress can come from work, home, and relationships with acquaintances and friends.

Depression : The debate is still ongoing as to whether gambling addiction is the cause of depression or depression is the cause of gambling addiction.

Gender : Men and women tend to have different preferences in their gambling. Men are more likely to gamble in games such as blackjack and lotteries and women are more likely to engage in bingo and raffles. It used to be that men were more likely to become problem gamblers than women were. However, that is no longer the case. Women are now being targeted by the gambling industry just as men are. Men who become problem gamblers start gambling as teenagers, while women develop the problem later in life, usually 30 years old and up.

Socio-economic Status : As measured by percentage, people with low income lose more to gambling than those with high income. Those making less than ₹20,000 a year are likely to lose about 3% of their income to gambling, while those making over ₹30,000 will lose about 2% of their income to gambling. People with low income tend to view gambling as a quick ticket out of poverty. Studies have found community size to be positively related to gambling propensity, meaning that residents of big cities gamble more.

Family History : If you have a parent or anyone in your immediate family who is a problem gambler, you are at a much higher risk of becoming one yourself. This could be due to hereditary (genetics) or social learning (behaviours you learned from family members and peers growing up). Single, divorced, and separated people were found to be more likely to gamble than married people.

Irrational Thinking : Irrational thinking is also known as the "Gambler's Fallacy." Problem gamblers tend to look for patterns in random events. For example, if they lose 3 times in a row, then they think a win is due soon. They expect their luck to change if things are going against them. However, in reality any game of chance is fixed according to mathematical probabilities and losing 3, 4, 5, 6, or even 10 times in a row does not guarantee that a win is around the corner.

The consequences of gambling addiction are listed below :

- Large debts that take a long time to pay off
- Relationship problems with spouse, friends, and family members
- Loss of reputation in the community
- Loss of employment
- Poor performance in school; if problem persists, dropping out of school
- Problems with the law if fraud, embezzlement and other crimes are involved.
- Problems with mood swings and physical problems such as losing or gaining weight.
- Thoughts of suicide. Gambler addicts are more likely to commit suicide than drug addicts are.

Types of Gamblers

There are six types of gamblers identified by Dr. Robert L. Custer.

1. Action gamblers
2. Escape gamblers
3. Compulsive gamblers
4. Professional gamblers
5. Anti-social or personality gamblers
6. Casual social gamblers

Action gamblers : Action gamblers are mostly male who started their gambling careers as teenagers. Their addiction usually lasts anywhere from 5 to 30 years. Action gamblers prefer games that are perceived to have elements of skill involved(in reality, there is very little skill involved in any game of chance). Their games of choice include poker, online poker, blackjack, craps, sports betting, horse racing, or any game where they are presented with the opportunity to beat another player or the house. They usually develop gambling systems that are effective in the short-run but ineffective in the long-run, since the house or casino always has the mathematical advantage in the long-run. Action gamblers have personality traits that are domineering, manipulative, controlling, and egotistic. They tend to have above average IQ's (over 120 versus the average of 100). They are known to be friendly, sociable, generous, assertive and confident. But despite some of their good character traits, they have very low self-esteem. Action gamblers gamble to obtain the euphoric high that cocaine addicts crave. After prolonged exposure to gambling, action gamblers get that high just by thinking about gambling or planning when they'll gamble next. Like all problem gamblers, action gamblers eventually become detached from reality and become very good at lying; they eventually even lie to themselves.

Escape gamblers : Escape gamblers are mostly (over 65%) female who develop the problem later in life, usually after 30 years old. They prefer games that are perceived to have no elements of skill; these games include video poker, lottery, bingo, and slot machines. These games provide the relief from emotional (or sometimes physical) pain. While gambling, escape gamblers are numbed by the drug-effect of gambling and find themselves in a hypnotic-like state—their problems disappear while they're "in action." Escape gamblers are known to be reserved, but they can also be manipulative when the occasion calls for it (obtaining money to gamble). As an adult, they tend to be nurturing and are afraid of confrontation. Some escape gamblers were physically or verbally abused as children. The escape gambler, like the action gambler, has low-self esteem.

Compulsive gamblers : For compulsive gamblers, thoughts are preoccupied with gambling. Gambling becomes a means to escape from problems in life. Compulsive gamblers become co-dependent—relying on others for money and emotional needs. They failed in attempts to curtail or quit gambling, become restless and irritable when they can't control their impulse to gamble. They suffer from guilt, depression, helplessness, and anxiety become triggers for gambling. They chase loss until credit cards, bank cards are overextended and loans are used up. Compulsive gambler denies and lies to friends and family on the magnitude of the problem. He commits crime such as fraud, forgery, embezzlement, and theft to finance gambling. Gambling becomes more important than career, education, relationships, and important opportunities in life.

Professional gambler : A professional gambler is someone who makes their living off of games offered in a casino setting. A majority of the population believes that it is impossible to make a living from gambling, but they would be mistaken. Professional gamblers gamble for economic gain. Professional gamblers concentrate on those casino games for which a competitive long-term mathematical advantage can be gained where skill and research always guarantees winning. A majority of professional gamblers are poker players. Second most popular among professionals is gambling on horse racing or sports. It is also possible, but less common, to make a living at Blackjack or even Video Poker. The casino's advantage margin is so small in these games, the player has the best chance of winning. To be a professional gambler, all they do is practice - "Practice is the best master" (Latin Proverb).

Anti-social or personality gamblers : Gambling is considered a deviant behaviour (anti-social in nature) if someone continues to gamble despite knowing the negative consequences to themselves and those in their lives. Some anti-social behaviours include: lying to or conning others for fun or for personal benefit, being irresponsible, not holding down a job or paying back money, being impulsive and not considering the results of a certain behaviour, breaking laws, ignoring the safety of self or others, lacking remorse, not worrying about hurting other people and picking on other people or getting into fights.

Casual social gamblers : Casual social gambling" means gambling in which all of the following conditions are present : (1) Players compete on equal terms with each other; (2) No player receives, or becomes entitled to receive, anything of value or any profit, directly or indirectly, other than the player's personal gambling winnings; (3) No other person, corporation, unincorporated association, or entity receives or becomes entitled to receive, anything of value or any profit, directly or indirectly, from any source, including but not limited to permitting the use of premises, supplying refreshments, food, drinks, service, lodging or entertainment; (4) It is not conducted or played in or at a hotel, motel, bar, nightclub, cocktail lounge, restaurant, massage parlor, billiard parlor, or any business establishment of any kind, public parks, public buildings, public beaches, school grounds, churches or any other public area; and (5) None of the players is below the age of majority.

The Phases of Gambling Addiction

1. The winning phase : The winning phase is when the gambler experiences the early wins. They brag and exaggerate about their winnings and exude the image of the "big shot." They are very optimistic about winning. They think (falsely) that gambling will provide them with the good life without hard work, patience, and diligence.

2. The losing phase : The losing phase is when the losing streaks begin; the true odds of gambling are being reflected and the gambler doesn't stand a chance regardless what system he or she uses. At this point the gambler begins to miss work, asks for bailouts, neglects bills, borrows to gamble, lies to cover losses, and manipulates to get money for gambling. In the losing phase, the gambler thinks he's facing a temporary period of bad luck. However, it's not luck he's up against but the mathematics (probabilities) designed into every game of chance to ensure that the house or casino always wins.

3. The desperate phase : In the desperate phase the gambler faces isolation from friends and family. He feels remorse for the money he's lost and for his inability to control his behaviour. As a result panic ensues and he becomes desperate. He now has to decide whether to continue to gamble or quit. His reputation with those around him is affected, but rather than look inward he'll blame others, out of frustration and anger, for the predicament he's in. He is unwilling to accept responsibility and the consequences of his behaviour. At this point, some problem gamblers will resort to illegal activities to finance their gambling.

4. The critical phase : During the critical phase, the gambler realizes he is not making any progress in life and gambling isn't going to provide him with the good life he's imagined. He tries to be more responsible and is hopeful for the future. He is a more productive employee at work and becomes realistic about his money goals. Slowly he puts his pride aside and reaches out for help, either through counseling, Gamblers Anonymous, books, or a combination of the three, and begins to examine his spiritual needs.

5. The rebuilding phase : The building phase is about self-examination. Here the gambler accepts his strengths and weaknesses. He begins to pay his bills, develops new hobbies, and seeks new goals. He becomes more patient, less irritable and spends more time with his friends and family. Semblances of a normal life begin to return.

6. The growth phase : The growth phase is when the gambler develops a higher consciousness. He seeks self-understanding through studying psychology, philosophy, and spirituality. He learns the power of forgiveness and its role in putting the past behind in order to live in the present. Through self-examination he is able to forgive himself and others. He becomes more generous and understands his own needs and wants and the needs and wants of other people. He also deals with problems as they arise, accepts the hardships of life, and becomes realistic rather than overly optimistic or pessimistic about situations. He takes life one day at a time and deals with whatever life throws at him. He is comfortable in his own skin and doesn't feel the need to control, manipulate, or lie. He makes progress as a human being and begins to contribute to the community in which he or she lives.

Helping Problem Gamblers

It is hard to help someone unless they want to help themselves. It's hard enough to change who we are, so trying to change someone else is a daunting, if not an impossible task. It has always been a difficult task for many to search for ways on how to stop gambling. It is harder to even implement it on one's self. The main problem is that people who often suffer from gambling addiction deny that there is a problem. Their perception is that it's a small thing that will not go out of hand and they would just like to have fun. What they fail to realize is that the small thing can lead to bigger problems such as hurting other people aside from themselves because of major money and behavioural problems. The gambler would not choose to stop gambling unless nothing is left for them to bet with. Eventually, the first step on how to stop gambling is to admit that they have a gambling problem. The next motivation is for themselves, not for other people. The desire to stop should come from them. A good suggestion on how to stop gambling would be to stay away from places that obviously flaunt the escaped vice. He should have the will power to turn his back on the places he often goes to gamble to until the need, the desire to gamble will cease. A gambler has to remind himself that this is for his own good. It would also help to orally recite this mantra every day to convince himself that this is for the best.

A good schedule or daily regimen on how to stop gambling is most effective to stick to. Having a sure plan every day and every hour to keep a gamblers mind from incessant thoughts would help divert his attention to more important things. Many gambling addicts develop their problem out of boredom. He may combat his gambling problem by finding an enjoyable new hobby or activity. He may already enjoy a hobby that he didn't have time for because of gambling. One may make model airplanes, garden, play a musical instrument, get involved in religious rituals, or try creative writing and may just find something else he loves doing. If possible, he may choose an activity that involves others who can encourage him and make him feel like a part of a community.

The gambler will need to gain some support from someone he trusts. He must realise that he is responsible for the trouble he is in and take control of his finances. He may ask someone to look after his bank cards for a period of time and may destroy his credit cards. This way one will prevent himself from gambling more.

A support group with people who have overcome the problem or trying to rid of the problem can help contribute and inform of the different processes on how to stop gambling. Gambling support groups and members of civil society provide resources and programs to help gambling addicts overcome their problem. Programs are often free. They provide a nonjudgmental supportive atmosphere in which addicts can tell their stories and relate to others with the same addiction. This can help you stop gambling and rebuild your life. Self conviction and motivation by other people are typically a better regimen to stop all the problems and pains caused by gambling.

Avoiding stress is key to overcoming any addiction, and gambling is no exception. People under stress often resort to familiar destructive behaviours such as gambling or alcohol abuse, even though these things create even more stress. They need to consider meditation, yoga, deep breathing, and other relaxation exercises to avoid stress. They need to relax, enjoy the outdoors, and spend time with people they love to avoid patterns of stress.

One needs to limit available funds for possible use in gambling. Gambling requires money, and limiting access to money quickly diminishes one's gambling capacity. One may seek support from family and friends to help him limit spending and sticking to the budget.

The best way to help someone who is addicted to gambling is to make suggestions or offer advice and counseling. Help them indirectly to set goals and work out a strategy with them to reach those goals. For example, if they gamble to escape problems in life, buy them a book that teaches life management and coping skills. You can also offer your own life experiences that they can relate to. If they gamble to get the high, give them a gym pass or several bungy jumping passes. You can come up with many creative solutions without becoming an enabler or putting his needs and wants before your own. The perspective lenders must stop lending them money for bailouts or lying to cover for them. They may be engaged in community service.

References

Brown, R. (1987). Gambling addictions, arousal, and an affective/decision-making explanation of behavioural reversions or relapses. International Journal of the Addictions. 22 (11), 1053-1067.

Custer, R. (1984). Profile of the pathological gambler. Journal of Clinical Psychiatry, 45 (12, Sect 2), 35-38.

Ferris, J.,Wynne, H. & Single, E. (1999). Measuring Problem Gambling in Canada: Final Report – Phase 1. For the Inter-Provincial Task Force on Problem Gambling, Submitted on behalf of the ccsa, April 22,

Gabrielle A. and Reuven Brenner. (1991) "Gambling: Shaping an Opinion," in Eadington & Cornelius, eds., Gambling and Public Policy, University of Navada, p. 493.

Goodman Robert, (1994) "Legalized Gambling as a Strategy for Economic Development," United States Gambling Study, March, p. 158.

Griffiths, M. (1995). Towards a risk factor model of fruit machine addiction: A brief note. Journal of Gambling Studies, 11 (3), 343-346.

Grinols, E. L. and J. D. Omorov, (1995) "Development or Dreamfield Delusions? Assessing Casino Gambling's Costs and Benefits," (Champaign, IL: University of Illinois, p. 13.

Hunter Donald and Ernest Bleinberger, (1995) "Gaming in America," Economic Development Commentary vol. 19, no. 1 (Spring), p. 10.

Kindt John, (1994) "The Economic Costs of Legalized Gambling Activities," Drake Law Review vol. 43, no. 1, p. 72.

Lesieur, H. & Rosenthal, R. (1991). Pathological gambling: A review of the literature Journal of Gambling Studies, 7 (1), 5-39.

Patrick T. R, (1994) "There are No Sure Bets in Figuring Out Why We Wager," Chicago Tribune, 9 December, p. 36.

Rosenthal, R. (1987). The psychodynamics of pathological gambling: A review of the literature. In Galski, Thomas, et al. (Eds.). The Handbook of Pathological Gambling (pp. 41-70). Springfield, IL: Charles C Thomas.

Sharpe, L. & Tarrier, N. (1993). Towards a cognitive-behavioural theory of problem gambling. British Journal of Psychiatry, 162, 407-412.

Skinner, B.F. (1953). Science and Human Behaviour. New York: Macmillan.

Stein, S. (1989). A developmental approach to understanding compulsive gambling behaviour. In H. Shaffer (Ed.), Compulsive Gambling: Theory, Research, and Practice. Lexington Books, MA.

Upflod Darryl (2011) An Introduction to Conceptual Models of Problem Gambling, Centre for Addiction and Mental Health, Toronto.

Waiman P. M and Joseph H, (1994) "Age and Gambling Behaviour: A Declining and Shifting Pattern of Participation," in Eadington and Cornelius, eds., pp. 56-59.

Welles, Chris (1996) "America's Gambling Fever," U. S. News and World Report, 15 January, p. 55.

■■■

CHAPTER 15

Drug Addiction

Learning Objectives :

- ❒ **To Understand the concept of drug addiction.**
- ❒ **To comprehend the causes and the effects of drug addiction.**
- ❒ **To be acquainted with the types of commonly abused drugs.**
- ❒ **To discuss the drug related crime.**

Introduction

Drug addiction, also called substance dependence or chemical dependency, is a disease that is characterized by a destructive pattern of drug abuse that leads to significant problems involving tolerance to or withdrawal from the substance, as well as other problems that use of the substance can cause for the sufferer, either socially or in terms of their work or school performance. More than 2.6% of people suffer from drug addiction at some time in their life.

The term dual diagnosis refers to the presence of both a drug-abuse or dependence issue in addition to a serious mental-health problem in an individual. Substance abuse or dependence unfortunately occurs quite commonly in people who also have severe mental illness. This is important given that people with a serious mental illness are far more at risk of engaging in violence, being incarcerated, or contracting infection with the human immunodeficiency virus (HIV). Individuals with dual diagnosis are also at higher risk of being noncompliant with treatment.

An authoritative definition of drug addiction is propounded by the World Health Organization: "Drug addiction is a state of periodic and chronic intoxication detrimental to the individual and to society, produced by the repeated consumption of a drug (natural or synthetic). Its characteristics include: (1) An overpowering desire or need (compulsion) to continue taking the drug and to obtain it by any means; (2) A tendency to increase the dose; (3) A psychic (psychological) and sometimes a physical dependence on the effects of the drug." This definition of drug addiction includes many drugs such as hypnotic and sedative drugs (barbiturates, etc.) alcohol, amphetamine, mescaline (peyote).

Drug addiction is a complex brain disease. It is characterized by compulsive, at times uncontrollable, drug craving, seeking, and use that persist even in the face of extremely negative consequences. Drug seeking becomes compulsive, in large part as a result of the effects of prolonged drug use on brain functioning and, thus, on behaviour. For many people, drug addiction becomes chronic, with relapses possible even after long periods of abstinence. Scientific advances have offered remarkable insights into how the human brain works and how it molds behaviours that affect drug addiction, say the directors of the National Institute on Drug Abuse (NIDA) and the National Institute on Alcohol Abuse and Alcoholism (NIAAA), National Institutes of Health, in a newly published article. Building on these foundations, scientists can now investigate issues that were previously inaccessible, such as how environmental factors and genes affect how the brain responds to drugs of abuse to drive the process of addiction. The report, by NIDA Director Dr. Nora D. Volkow, and NIAAA Director Dr. Ting-Kai Li, is published in the December 2004 issue of Nature Reviews Neuroscience. "Drug addiction is a brain disease," says Dr. Volkow. "Although initial drug use might be voluntary, once addiction develops this control is markedly disrupted. Imaging studies have shown specific abnormalities in the brains of some, but not all, addicted individuals. While scientific advancements in the understanding of addiction have occurred at unprecedented speed in recent years, unanswered questions remain that highlight the need for further research to better define the neurobiological processes involved in addiction."

Recent studies have increased our knowledge of how drugs affect gene expression and brain circuitry, and how these factors affect human behaviour. They have shed new light on the relationship between drug abuse and mental illness, and the roles played by heredity, age, and other factors in increased vulnerability to addiction. Drug addiction is also called substance abuse. Substance abuse can simply be defined as a pattern of harmful use of any substance for mood-altering purposes. Medline's medical encyclopedia defines drug abuse as "the use of illicit drugs or the abuse of prescription or over-the-counter drugs for purposes other than those for which they are indicated or in a manner or in quantities other than directed."

But the broad range of substance abuse in today's society is not that simple. There are substances that can be abused for their mood-altering effects that are not drugs at all — inhalants and solvents — and there are drugs that can be abused that have no mood-altering or intoxication properties, such as anabolic steroids. Generally, when most people talk about substance abuse, they are referring to the use of illegal drugs. Most professionals in the field of drug abuse prevention argue that any use of illegal drugs is by definition abuse. Those drugs got to be illegal in the first place because they are potentially addictive or can cause severe negative health effects; therefore, any use of illegal substances is dangerous and abusive. Others argue that casual, recreational use of some drugs is not harmful and is merely use, not abuse. The most vocal of the proponents of recreational drug use are those who smoke marijuana. They argue that marijuana is not addictive and has many beneficial qualities, unlike the "harder" drugs.

But recent research has shown that even marijuana may have more harmful physical, mental, and psychomotor effects than first believed, and the National Institute on Drug Abuse reports that marijuana users can become psychologically dependent, and therefore addicted. Illegal drugs are not the only substances that can be abused. Alcohol, prescription and over-the-counter medications, inhalants and solvents, and even coffee and cigarettes, can all be used to harmful excess. Theoretically, almost any substance can be abused. When it comes to illegal substances, society has determined that their use is harmful, and has placed legal prohibitions on their use in order to protect the individual and to protect society from the costs involved with healthcare resources, lost productivity, the spread of diseases, crime and homelessness.

CAUSES OF DRUG ADDICTION

When people take drugs, either for medical purposes or recreation, there is a benefit or reward that they are trying to achieve. For example, pain medication is intended to bring relief to an injured or stressed area of their body. The unintended consequences of that is their need to take more and more of the drug to get the same result. To understand what causes drug addiction we need to first look at the effect of the drug on the individual. Of course, there are many variables to consider, but we will start by looking at what happens when a person takes drugs.

There are three theories on the causes of drug addiction. One theory offers a scientific explanation of when drug addiction occurs, stating, "Physical addiction appears to occur when repeated use of a drug alters reward pathways in your brain. The addicting drug causes physical changes to some nerve cells (neurons) in your brain. Neurons use chemicals called neurotransmitters to communicate."

Another theory sees the causes of drug addiction as being caused partially by the self-esteem of the individual. "People who have low self-concepts, who feel bad about themselves, have a higher rate of addiction. People with low self-concepts use psychoactive substances either to enhance or create pleasure in their lives, or to decrease the constant emotional pain they live with. The better a person feels about himself, the less likely he will be to use or abuse psychoactive substances. A society that has easy access to drugs that has a population that is "addiction-prone" due to genetics or emotional pain, and that has pro-use or unclear norms, is a society prone to addiction."

A third view is held by Dr. Greg Baer, an ophthalmologist, who asserts that the primary cause of addiction is a response to pain, emptiness and the lack of love. He does not ascribe to the theory that addiction is a disease. "After extensive interviews with thousands of addicts," says Dr. Baer, "less than one percent of them fail to respond enthusiastically to this definition of addiction and the clarification of the causes of drug addiction, whereas most of them have had serious problems with the usual definitions and explanations."

The beginning stages of drug addiction causes us to crave more and to use more. The unintended consequences of that is our need to take more and more of the drug to get the same result. Drug addiction causes the pathways inside the brain to be altered. Physical changes in the nerve cells are brought on by the drug. These cells (neurons) communicate with each other releasing neurotransmitters into the gaps or synapses between the nerve cells. This makes some drugs more addictive than others. Think of drug addiction as a progression. A person uses drugs and at some point the pathways inside the brain are altered. Physical changes in the nerve cells are brought on by the drug. These cells (neurons) communicate with each other releasing neurotransmitters into the gaps or synapses between the nerve cells. This makes some drugs are more addictive than others.

There are several other factors that contribute to drug addiction. The major factors are one's genetic makeup, personality and peer pressure.

Genetic/Inherited : We are all a product of our parents. If one's parents have addiction struggles, he/she is more susceptible to addiction. That's why drug addiction is more common in some families than in others. If one's parents smoke, chances are good that child will smoke. If one's parents used alcohol, the offspring will probably follow and use that drug in much the same way. If the father was an alcoholic, one has a predisposition to abusing that drug. Drug addiction causes one generation to pass it on to the next.

Personality : Aside from the inherited factors, some people have a personality that is more likely to become drug dependent. People are curious, so that alone can lead a person to try a drug. People experiment and see what happens. People look to relax and have pleasure. People want to feel good, and people are by nature impatient. Drugs give us an instant gratification that other things do not, so for that moment or hour of for whatever time-frame, they feel good. Someone diagnosed with depression, attention deficit disorder, or hyperactivity tends to be drug prone. Maybe there has been some stress, or anxiety in their life. Whatever the case, these are contributing factors. Even some common personality characteristics, such as aggression, may be a factor. Children who do not have confidence, healthy self-esteem may be prone to turning to drugs to fill the void. Drug addiction causes negative changes in personality that can lead to an even more destructive behaviour.

Peer Pressure/Social : People are basically wired to have relationships, and sometimes those relationships cause them to give in to something they otherwise would avoid in order to maintain the relationship. Peer pressure is huge and nowhere is this greater than during our teenaged years. Kids want to be cool. It begins as a social action, to take the drugs to be a part of the group, to be accepted. It's not just teenagers, as peer pressure takes so many different forms. There is social etiquette, for example, to take a drink during a party. "I'm a social drinker." Some people actually believe that drug addiction causes one to be accepted and part of the 'popular' group.

Easy Access : Drugs are available and accessible everywhere. Drug addiction causes people to sell drugs to the most vulnerable population, children. It's not just the stereotypical poor sections of the inner city that serve as the hotbed for drugs. Drugs are found in suburban shopping malls, rural schools, with beetle shop owners, well-to-do private school, on the job in factories, offices and remote job sites.

Race, Ethnicity : There is enough cause to support any claim that one race of people or any particular cultural group is more prone to drug addiction than another. Alcohol is offered as a mark of respect to the invited guests. Drug addiction is a human problem and crosses all boundaries.

Loneliness, Depression : People want to feel good physically and emotionally. Sometimes drugs are the substitution for a healthy life experience. The person in pain and they want to numb the pain. The drug numbs the pain and for a moment they don't feel as poorly. The person needs to escape the pain of the life experience, and for a short while, the drug takes them away and they feel "better."

Anxiety : Sometimes people need some help coping with life. Everyday life becomes a struggle and simple things become too much to handle. Drugs are used to deal with it. In the case of addiction, we are not talking about the use of medication, under the care and observation of a doctor. People who have been clinically diagnosed with anxiety can lead a very good life. We're talking here about people who just need to escape. Their drug of choice facilitates that escape.

COMMONLY ABUSED TYPES OF DRUGS

Virtually any substance whose ingestion can result in a euphoric ("high") feeling can be abused. While many are aware of the abuse of legal substances like alcohol or illegal drugs like marijuana (in most states) and cocaine, less well known is the fact that inhalants like household cleaners are some of the most commonly abused substances. Many different types of drugs can be abused: not only illegal drugs such as heroin, cannabis, cocaine or ecstasy, but also prescription drugs such as tranquilizers, analgesics (painkillers), and sleeping pills. Even medicines that can be bought from the market can be abused (such as cough mixtures or herbal remedies) and the abuse of alcohol is a major area of concern. On one hand, drugs such as cannabis and opiates, which have been in use over a period of time, may be termed traditional drugs. On the other, drugs such as heroin, mescaline, LSD and Angel Dust, which are relatively recent in origin, may be called modern drugs. Apart from this, the form of drugs prevalent in rural areas in India differs from that in urban areas. It follows that the use of psychotropic drugs in the country is not uniform (Khan and Krishna, 1982). All drugs can be divided into seven broad categories :

- Cannabinoids (e.g., hashish and marijuana)
- Stimulants (e.g., amphetamines and cocaine)
- Depressants (e.g., Xanax and Qualudes)
- Narcotics (aka opioids and morphine derivatives, e.g., heroin, opium, Vicodin)
- Hallucinogens (e.g., LSD and mescaline)
- Dissociative anesthetics (e.g., PCP)
- Other compounds (e.g., steroids and inhalants)

The following are many of the drugs and types of drugs that are commonly abused and/or result in dependence :

1. Alcohol: Although legal, alcohol is a toxic substance, particularly to a developing fetus when a mother consumes this drug during pregnancy.
2. Amphetamines: This group of drugs comes in many forms, from prescription medications like methylphenidate (Ritalin, Concerta) and dextroamphetamine and amphetamine (Adderall) to illegally manufactured drugs like methamphetamine ("meth"). Overdose of any of these substances can result in seizure and death.
3. Anabolic steroids: A group of substances abused by bodybuilders and other athletes, this group of drugs can lead to terrible psychological effects like aggression and paranoia, as well as devastating long-term physical effects like infertility and organ failure.
4. Caffeine: While it is consumed by many, coffee, tea and soda drinkers, when consumed in excess this substance can produce palpitations, insomnia, tremors and significant anxiety.

5. Cannabis: More commonly called marijuana, the scientific name for cannabis is tetrahydrocannabinol (THC). In addition to the negative effects the drug itself can produce (for example, infertility, paranoia, lack of motivation), the fact that it is commonly mixed ("cut") with other substances so drug dealers can make more money selling the diluted substance or expose the user to more addictive drugs exposes the marijuana user to the dangers associated with those added substances. Examples of ingredients that marijuana is commonly cut with include baby powder, oregano, embalming fluid, PCP, opiates, and cocaine.
6. Cocaine: A drug that tends to stimulate the nervous system, cocaine can be snorted in powder form, smoked when in the form of rocks (crack cocaine), or injected when made into a liquid.
7. Ecstasy: Also called MDMA to denote its chemical composition (methylenedioxymethamphetamine). This drug tends to create a sense of euphoria and an expansive love or desire to nurture others. In overdose, it can increase body temperature to the point of being fatal.
8. Hallucinogens: Examples include LSD and mescaline, as well as so-called naturally occurring hallucinogens like certain mushrooms, these drugs can be dangerous in their ability to alter the perceptions of the user. For example, a person who is intoxicated with a hallucinogen may perceive danger where there is none and to think that situations that are truly dangerous or not. Those misperceptions can result in dangerous behaviours (like jumping out of a window because the individual thinks they are riding on an elephant that can fly).
9. Inhalants: One of the most commonly abused group of substances due to its accessibility, inhalants are usually contained in household cleaners, like ammonia, bleach, and other substances that emit fumes. Brain damage, even to the point of death, can result from using an inhalant just once or over the course of time, depending on the individual.
10. Nicotine: The addictive substance found in cigarettes, nicotine is actually one of the most habit-forming substances that exists. In fact, nicotine addiction is often compared to the intense addictiveness associated with opiates like heroin.
11. Opiates: This group is also called narcotics and includes drugs like heroine, codeine, Vicodin, Percocet, and Percodan. This group of substances sharply decrease the functioning of the nervous system. The lethality of opiates is often the result of the abuser having to use increasingly higher amounts to achieve the same level of intoxication, ultimately to the point that the dose needed to get high is the same as the dose that is lethal for that individual by halting the person's breathing (respiratory arrest).
12. Phencyclidine: Commonly referred to as PCP, this drug can cause the user to feel extremely paranoid, become quite aggressive and to have an unusual amount of physical strength. This can make the individual quite dangerous to others.
13. Sedative, hypnotic, or antianxiety drugs: As these substances quell or depress the nervous system, they can cause death by respiratory arrest of the person who either uses these drugs in overdose or who mixes one or more of these drugs with another nervous system depressant drug (like alcohol or an opiate).

The National Institute on Drug Abuse published a list of commonly abused drugs and their street names, along with intoxication effects and adverse health consequences. All of the drugs pose a high potential for addiction. Following table shall make it clear (Kaur R and JK Gulati, 2007 : 390-92).

TABLE 15.1 Intoxication Effects of Commonly Abused Drugs

Drug Name/ Class	Commercial and / or Street Names	How Taken	Intoxication Effects	Adverse Health consequences
Marijuana Cannabinoid	Dope, pot, joints,grass, reefer, weed,etc.	Smoked, swallowed	Euphoria, slowed thinking and reaction time, confusion, impaired balance and coordination	Cough, frequent respiratoryinfections; impaired memory and learning; increased heart rate, anxiety; panic attacks
Cocaine Stimulant	Blow, bump,candy, charlie,coke, crack, rock,snow, toot	Snorted, smoked, injected	Increased heart rate, blood pressure, temperature, metabolism; feelings of exhilaration, energy, increased mental alertness	Rapid or irregular heart beat;reduced appetite, weight loss, heart failure, nervousness, insomnia, chest pain, respiratory failure,nausea, abdominal pain, strokes, seizures, headaches, malnutrition,panic attacks
Amphet-amines Stimulant	Biphetamine, Dexedrine; bennies, blackbeauties, crosses,hearts, speed, truckdrivers, uppers	Swallowed, smoked, injected, snorted	Increased heart rate, blood pressure, metabolism; feelings of exhilaration, energy, increased mental alertness, rapid breathing	Rapid or irregular heart beat;reduced appetite, weight loss, heartfailure, nervousness, insomnia,tremor, loss of coordination;irritability, anxiousness, restless-ness, delirium, panic, paranoia,impulsive behaviour, aggressiveness, psychosis
Methampheta-mine Stimulant	Desoxyn; chalk,crank, crystal, fire,glass, go fast, ice,meth, speed	Snorted, swallowed, smoked, injected	Increased heart rate, blood pressure, metabolism; feelings of exhilaration, energy, increased mental alertness; aggression, violence, psychotic behaviour	Rapid or irregular heart beat;reduced appetite, weight loss, heartfailure, nervousness, insomnia;memory loss, cardiac and neurological damage; impaired memory and learning

Drug Name/ Class	Commercial and / or Street Names	How Taken	Intoxication Effects	Adverse Health consequences
MDMA Stimulant	Adam, clarity, ecstasy, Eve,lover's speed,peace, STP, XXTC	Swallowed	Increased heart rate, blood pressure, metabolism; feelings of exhilaration, energy, increased mental alertness; mild hallucinogenic effects, increased tactile sensitivity, empathic feelings	Rapid or irregular heart beat; reduced appetite, weight loss, heartfailure, nervousness, insomnia;impaired memory and learning,hyperthermia, cardiac toxicity,renal failure, liver toxicity
Barbiturates Depressant	Amytal, Nembutal,Seconal, Phenobarbital; barbs, reds, redbirds, phennies,tooies, yellows,yellow jackets	Swallowed, injected	Reduced anxiety; feeling of well-being; lowered inhibitions; slowed pulse and breathing; lowered blood pressure; poor concentration, sedation, drowsiness	Fatigue; confusion; impaired coordination, memory, judgment;depression, unusual excitement,fever, irritability, slurred speech,dizziness, life-threatening withdrawal; respiratory depression and arrest,death
Benzodi-azepines Depressant	Ativan, Halcion,Librium, Valium,Xanax; candy,downers, sleepingpills, tranks	Swallowed, injected	Reduced anxiety; feeling of well-being; lowered inhibitions; slowed pulse and breathing; lowered blood pressure; poor concentration, sedation, drowsiness	Fatigue; confusion; impairedcoordination, memory, judgment;sedation, drowsiness; dizziness;respiratory depression and arrest, death
Heroin Narcotic (Opioid)	Diacetyl-morphine;brown sugar, dope,H, horse, junk,skag, skunk, smack,white horse	Injected, snorted, smoked	Pain relief, euphoria, drowsiness, unsteady gait	Nausea, constipation, confusion,sedation, respiratory depression and arrest, tolerance,unconsciousness, coma, death
LSD	Hallucinogen Lysergic aciddiethylamide; acid, blotter, boomers,cubes, microdot,yellow sunshine	Swallowed, absorbed through mouth tissues	Altered states of perception and feeling; nausea; increased body temperature, heart rate, blood pressure; loss of appetite, sleeplessness, numbness, weakness, tremors; persistent mental disorders	Persisting perception disorder(flashbacks)

Drug Name/ Class	Commercial and / or Street Names	How Taken	Intoxication Effects	Adverse Health consequences
PCP Dissociative anesthetics	Phencyclidine; angel dust, boat,hog, love boat,peace pill	Smoked, swallowed, injected	Increased heart rate and blood pressure, impaired motor function, possible decrease in blood pressure and heart rate; panic, aggression, violence	Memory loss; numbness;nausea/vomiting; loss of appetite,depression
Inhalants	Solvents (paintthinners, gasoline,glues), gases(butane, propane,aerosol propellants, nitrousoxide), nitrites(isoamyl, isobutyl,cyclohexyl); laughing gas,poppers, snappers,whippets	Inhaled through nose or mouth	Stimulation, loss of inhibition; headache; nausea or vomiting; slurred speech, loss of motor coordination; wheezing	Unconsciousness, cramps, weight loss, muscle weakness, depression, memory impairment, damage to cardiovascular and nervous systems, sudden death
Anabolic steroids	Anadrol, Oxandrin,Durabolin, Depo-Testosterone, Equipoise; roids,juice	Injected, swallowed, applied to skin	No intoxication effects	Hypertension, blood clotting and cholesterol changes, liver cysts and cancer, kidney cancer, hostility and aggression, acne; in adolescents,premature stoppage of growth; in males, prostate cancer, reduced sperm production, shrunkentesticles, breast enlargement; in females, menstrual irregularities,development of beard and other masculine characteristics

EFFECTS OF DRUG ADDICTION

In order to be diagnosed with drug abuse, an individual must exhibit a destructive pattern of drug abuse that leads to significant problems or stress. Recurrent drug use results in a lack of meeting important obligations at work, school, or home. There might be recurrent legal problems as a result of drug use. Continued drug use may be dangerous despite continued or repeated social or relationship problems as a result of the drug's effects. Recurrent drug use may seriously affect one's health, the family, the economy, the Law and our society.

Effects of drug addiction on health

If left unchecked, the drug is going to win. Drug abuse is a disease of the brain, and the drugs change brain chemistry, which results in a change in behaviour. Aside from the obvious behavioural consequences of addiction, the negative effects on a person's health are potentially devastating. While addicts use drugs to *"feel better,"* the unintended consequences include but are not limited to overdose, HIV/AIDS, stroke, cardiovascular disease and a host of related maladies. Suicide is also a common effect of drug addiction. *Depression* is also an effect of addiction.

Effects of drug addiction on the family

One of the saddest aspects of the insidious nature of drug addiction is that by the time an addict realizes he/she has a problem, that problem has already taken a heavy toll on the family. Parents in treatment centers tell counselors and therapists that they want to *"get their kids back,"* as drug addiction has taken over to the point where the courts have been forced to remove the children from the home. Husbands and wives, brothers and sisters, and sadly children are all impacted. Families can be sources of strength and support, or they can passively enable the addiction to advance. Families can share in the victory over drug addiction, or they can be the victims of it.

Effects of drug addiction on the economy

Beyond the personal health issues, beyond the devastating effect on families, beyond community crime statistics, drug addiction has a major impact on the American economy. The National Institute on Drug Abuse reported that some **$67 billion per year** is the impact that drug addiction has in USA. This total includes the cost of law enforcement, incarceration, treatments, traffic injuries, lost time in the workplace, etc. Drug addiction causes impaired reasoning, and therefore the crime rate is dramatically impacted by drug use. Addicts have a much higher likelihood of committing crimes than others.

Effects of drug addiction on our society

The National Library of Medicine estimates that some 20% of all people in the United States have used prescription medication for non-medical purposes. This is not talking about cocaine, heroin or methamphetamine use, but doctor-prescribed medication. Illegal drug use and prescription drug misuse, have given rise to a huge problem in the society.

Effects of drug addiction on the Law

The news media reports daily struggles with theft, drive-by shootings, drug busts, illegal trafficking and manufacturing of drugs, and arrests for crimes ranging from child neglect to murder. Look closer and chances are great that you will uncover a drug addiction component to any of these stories.

DRUG ADDICTION IN RELATION TO CRIME

This is a difficult subject to address as the relationship between the two is complex and thought-provoking. We know that many addicts resort to crime to pay for their habit but there also some people who are addicted to the criminal act itself. So we have people who wouldn't normally commit crime but have only turned to it out of an act of desperation and then there are those people who have already committed crime and then use this to fund their habit. One of the compelling reasons why more rational methods of dealing with drug addicts must be devised is the close relationship between drug addiction and crime. The compulsion for the drug makes every drug addict a law violator and a criminal. Mere possession of a narcotic drug which the addict must have to ward off withdrawal distress is a violation of the narcotic laws. Thus, every drug addict is subject to arrest by the police, and as we have seen, the arrests of addicts and of narcotic law violators have gone up by leaps and bounds. Addicts guilty of no other crime than illegal possession of narcotics are filling the jails, prisons and penitentiaries of the country. However, this is only a part of the distressing picture of the relationship between narcotic addiction and criminality. For most narcotic addicts, predatory crime (larceny, shoplifting, sneak thievery, burglary, embezzlement, robbery, etc.), is a necessary way of life.

The New York University and the Chicago studies on drug addiction support the notion that drug addiction necessarily leads to predatory crime as a way of life. For example, Chein and Rosenfeld (1957:53-54) make the following comments based on their studies of juvenile addicts: "Drug use leads to a criminal way of life. The illegality of purchase and possession of opiates and similar drugs makes a drug user a delinquent ipso facto. The high cost of heroin, the drug generally used by juvenile users, also forces specific delinquency against property for cash returns. The average addicted youngster spends almost forty dollars a week on drugs, often as much as seventy dollars. He is too young and unskilled to be able to support his habit by his earnings. The connection between drug use and delinquency for profit has been established beyond any doubt." A Chicago study comes to a similar conclusion: "...Almost without exception addicts resort to theft to obtain money for the purchase of the drugs. The compulsion of the addiction itself coupled with the astronomically high cost of heroin leads the addict inescapably to crime. For the addict there is very simply no alternative."

There has been considerable debate as to whether the criminality of the addict preceded or is merely a consequence of the drug addiction. The answer to the question of whether the addict was a delinquent or criminal prior to addiction largely depends upon the particular groups of addicts studied. The conclusion from the Chicago and New York studies is inescapable that "delinquency both preceded and followed addiction to heroin." "Persons who became users," stated the Chicago report, "were found to have engaged in delinquency in a group habitual form either prior to their use of drugs or simultaneously with their developing interest in drugs.

Organized criminals, indulging in illegal and violent activities over time with the motive of profit making have changed their strategies with the changing times. They are using modern technology, and highly specialized forms of organization. Most importantly, organized crime is not limited to the national boundaries but has acquired a transnational scope in terms of connections and linkages. Drug traffickers become their companions in breaching legal barriers. Today drug trafficking and organized crime have become the most profitable enterprise all over the world. According to an Interpol report the biggest wrongdoers are the Turkish syndicates and the Italian mafia, which control 80 per cent of the international drug trade.

The crime - drug nexus works at three levels. At the first level, criminals work as middlemen for drug trafficking. They purchase drug products from one country, use their own country as a transit point, and sell it in another country. Drugs being a rare commodity they work on huge profit margins. Indian criminals from northeastern India transfer drugs from the Golden Triangle of Myanmar, Laos and Thailand to the developed and affluent countries of Europe and America.

Secondly, criminals work as a link between drug and arms traffickers. As criminals require modern arms for committing heinous crimes they find drugs to be a useful exchange commodity. This is very convenient for arms traffickers as they get huge profits from selling illegal drugs. Most of the criminals from northern India work at this level, which provides them illegal arms from Afghanistan.

Lastly, criminals themselves indulge in drug smuggling and selling within and across borders. At this level drug trafficking provides a financial support system. Criminals on the western coast of India have specialized in this practice. One of the Dawood brothers is in charge of the drug trafficking business from South Asia to West Asia. This is the most dangerous trend, because control of all related illegal activities provides tactical advantages to the criminal group.

The impact of the crime-drug nexus is multidimensional. It leads to increase in drugs distribution and consumption in society, destroying productive human resources. In the transit areas of India like Nagaland and Manipur the drug addiction problem has assumed alarming proportions. Drug addiction also leads people to commit criminal activities. A study in the US has shown that the crime rate is directly proportional to drugs consumption, since drug addicts are available for criminal acts, as they need drugs at any cost. The crime - drugs nexus also increases weaponisation of society as these are smuggled along with drugs in transit areas. Illegal drugs smuggling has implications for the economy. In fact, drugs have become the backbone of the parallel economy along with gold and diamond. The most serious threat from this nexus is for national security. This nexus links criminals with transnational networks and out of the reach of enforcement agencies. Once part of the transnational networks, this nexus provides a base for other anti-national elements like terrorist groups. This became visible in incidents like the Bombay blasts where Dawood Ibrahim provided the logistics support to Pakistan's Inter-Services Intelligence (ISI) and its agents, and also participated in executing these anti-national plans.

Unfortunately, no national or international agency is seriously concerned about this menace. The political classes have themselves become a link in this networked business. Educated public opinion, active non-governmental agencies and international cooperative measures are some of the ways to check this growing challenge.

INDIAN SCENARIO

India too is caught in this vicious circle of drug abuse, and the numbers of drug addicts are increasing day by day. According to a UN report, One million heroin addicts are registered in India, and unofficially there are as many as five million. What started off as casual use among a minuscule population of high-income group youth in the metro has permeated to all sections of society. Inhalation of heroin alone has given way to intravenous drug use, that too in combination with other sedatives and painkillers. This has increased the intensity of the effect, hastened the process of addiction and complicated the process of recovery. Cannabis, heroin, and Indian-produced pharmaceutical drugs are the most frequently abused drugs in India. Cannabis products, often called charas, bhang, or ganja, are abused throughout the country because it has attained some amount of religious sanctity because of its association with some Hindu deities. The International Narcotics Control Board in its 2002 report released in Vienna pointed out that in India persons addicted to opiates are shifting their drug of choice from opium to heroin. The pharmaceutical products containing narcotic drugs are also increasingly being abused. The intravenous injections of analgesics like dextropropoxphene etc are also reported from many states, as it is easily available at 1/10th the cost of heroin. The codeine-based cough syrups continue to be diverted from the domestic market for abuse.

Drug abuse is a complex phenomenon, which has various social, cultural, biological, geographical, historical and economic aspects. The disintegration of the old joint family system, absence of parental love and care in modern families where both parents are working, decline of old religious and moral values etc lead to a rise in the number of drug addicts who take drugs to escape hard realities of life. Drug use, misuse or abuse is also primarily due to the nature of the drug abused, the personality of the individual and the addict's immediate environment. The processes of industrialization, urbanization and migration have led to loosening of the traditional methods of social control rendering an individual vulnerable to the stresses and strains of modern life. The fast changing social milieu, among other factors, is mainly contributing to the proliferation of drug abuse, both of traditional and of new psychoactive substances. The introduction of synthetic drugs and intravenous drug use leading to HIV/AIDS has added a new dimension to the problem, especially in the Northeast states of the country. Drug abuse has led to a detrimental impact on the society. It has led to increase in the crime rate. Addicts resort to crime to pay for their drugs. Drugs remove inhibition and impair judgment egging one on to commit offences. Incidence of eve- teasing, group clashes, assault and impulsive murders increase with drug abuse. Apart from affecting the financial stability, addiction increases conflicts and causes untold emotional pain for every member of the family. With most drug users being in the productive age group of 18-35 years, the loss in terms of human potential is incalculable. The damage to the physical, psychological, moral and intellectual growth of the youth is very high. Adolescent drug abuse is one of the major areas of concern in adolescent and young people's behaviour. It is estimated that, in India, by the time most boys reach the ninth grade, about 50 percent of them have tried at least one of the gateway drugs. However, there is a wide regional variation across states in term of the incidence of the substance abuse. For example, a larger proportion of teens in West Bengal and Andhra Pradesh use gateway drugs (about 60 percent in both the states) than Uttar Pradesh or Haryana (around 35 percent). Increase in incidences of HIV, hepatitis B and C and tuberculosis due to addiction adds the reservoir of infection in the community burdening the health care system further. Women in India face greater problems from drug abuse. The consequences include domestic violence and infection with HIV, as well as the financial burden. Eighty seven per cent of addicts being treated in a de-addiction center run by the Delhi police acknowledged being violent with family members. Most of the domestic violence is directed against women and occurs in the context of demands for money to buy drugs. At the national level, drug abuse is intrinsically linked with racketeering, conspiracy, corruption, illegal money transfers, terrorism and violence threatening the very stability of governments.

India has braced itself to face the menace of drug trafficking both at the national and international levels. Several measures involving innovative changes in enforcement, legal and judicial systems have been brought into effect. The introduction of death penalty for drug-related offences has been a major deterrent. The Narcotic Drugs and Psychotropic Substances Act, 1985, were enacted with stringent provisions to curb this menace. The Act envisages a minimum term of 10 years imprisonment extendable to 20 years and fine of ₹1 lakh extendable up to ₹2 lakhs for the offenders. The Act has been further amended by making provisions for the forfeiture of properties derived from illicit drugs trafficking. Comprehensive strategy involving specific programmes to bring about an overall reduction in use of drugs has been evolved by the various government agencies and NGOs and is further supplemented by measures like education, counseling, treatment and rehabilitation programmes. India has bilateral agreements on drug trafficking with 13 countries, including Pakistan and Burma. Prior to 1999, extradition between India and the United States occurred under the auspices of a 1931 treaty signed by the United States and the United Kingdom, which was made applicable to India in 1942. However, a new extradition treaty between India and the United States entered into force in July 1999. A Mutual Legal Assistance Treaty was signed by India and the United States in October 2001. India also is signatory to the following treaties and conventions :

- 1961 U.N. Convention on Narcotic Drugs
- 1971 U.N. Convention on Psychotropic Substances
- 1988 U.N. Convention Against Illicit Traffic in Narcotic Drugs and Psychotropic Substances
- 2000 Transnational Crime Convention

The spread and entrenchment of drug abuse needs to be prevented, as the cost to the people, environment and economy will be colossal. The unseemly spectacle of unkempt drug abusers dotting lanes and by lanes, cinema halls and other public places should be enough to goad the authorities to act fast to remove the scourge of this social evil. Moreover, the spread of such reprehensible habits among the relatively young segment of society ought to be arrested at all cost. There is a need for the government enforcement agencies, the non-governmental philanthropic agencies, and others to collaborate and supplement each other's efforts for a solution to the problem of drug addiction through education and legal actions. Practically every country has its own substance abuse problem to face.

References

Chein I and Rosenfeld E "Juvenile Narcotics Use," Law and Contemporary Problems, Duke University School of Law, Vol. 22, No. 1, 1957 (pp. 53-54)

Chicago University, Special Report on Heroin Addiction in Chicago, 1957 (p. 43).

Kaur R and JK Gulati (2007) Drug Abuse: Trends and Issues, Proceedings of International Marketing Conference on Marketing & Society, 8-10 April.

■■■

CHAPTER 16 Alcoholism

Learning Objectives :

- **To Understand the concept, history and the etymology of alcoholism.**
- **To comprehend the symptoms, causes and effects of alcoholism.**
- **To discuss the recent trends of alcoholism.**
- **To be acquainted with the stages of growth of the disease of alcoholism.**
- **To grasp the prevention strategies and the policy prescriptions.**

Introduction

The term 'alcoholism' is a widely used term. The problem of alcoholism, until a few decades ago, was considered a moral problem and a sign of social irresponsibility. After the introduction of the prohibition policy in some states, it was viewed as an illegal act. Now, it is considered by some scholars more as a complicated, chronic and immensely costly disease than a type of a deviant behaviour. The Geneva Partnership on Alcohol: Towards a Global Charter 2000 documented "The drinking of beverage alcohol is and has been a part of many cultures throughout the world; however, the ill effects of inappropriate alcohol consumption are widely recognized....All cultures and societies place restrictions on alcohol consumption, either by law or custom, varying from place to place." Drinking patterns are undergoing significant transformation, with traditional and contemporary values sometimes coming into conflict. Where drunkenness is viewed as inappropriate, it tends to be avoided. Where it is deemed heroic, masculinist or desirable, it tends to be embraced. Alcoholism has much in common with the problem of drug abuse.

The Concept of Alcoholism

Alcoholism, also known as alcohol dependence, (Wikipedia) is a disabling addictive disorder. It is characterized by compulsive and uncontrolled consumption of alcohol despite its negative effects on the drinker's health and social standing. Similar to other drug addictions, alcoholism is medically defined as a treatable disease. In professional and research contexts, the term "alcoholism" sometimes encompasses both alcohol abuse and alcohol dependence (Tracy 2005) and sometimes is considered equivalent to alcohol dependence. Alcohol abuse refers to repeated use of alcohol despite recurrent adverse consequences. Alcohol dependence means alcohol abuse combined with tolerance, withdrawal, and an uncontrollable drive to drink.

The term "alcoholism" has contesting definitions. The World Health Organisation uses the term alcohol dependence syndrome rather than alcoholism. The concept of "harmful use" (as opposed to "abuse") was introduced in 1992 to minimize underreporting of damage in the absence of dependence. Alcoholism is the cyclic presence of tolerance, withdrawal, and excessive alcohol use; the drinker's inability to control such compulsive drinking, despite awareness of its harm to his or her health, indicate that the person might be an alcoholic. An alcoholic is different from an 'occasional drinker'. Any person who takes alcohol is a 'drinker', while a 'compulsive drinker' who cannot live without taking alcohol is called an 'alcoholic'. An alcoholic is an excessive drinker whose dependence upon alcohol has reached such a degree that it results in a noticeable mental disturbance or an interference with his bodily and mental health, his interpersonal relations, and his smooth social and economic functioning; or one who shows the early signs of such developments. Clinebell (1956: 17) has defined an 'alcoholic' as one whose drinking interferes frequently or continuously with any of his important life adjustments and interpersonal relationships.

Alcoholism is a condition in which an individual loses control over his alcohol intake in that he is constantly unable to refrain from drinking once he begins (Johnson, 1973:519). According to Keller and Efron (1955:619-644), alcoholism is characterized by the repeated drinking of alcoholic beverages to an extent that exceeds customary use or compliance with the social customs of the community and that adversely affects the drinker's health or interferes with his social or economic functioning. Broadly speaking, alcoholism has been characterized by four factors : (1) excessive intake of alcoholic beverages, (2) individual's increasing worry over his drinking, (3) loss of the drinker's control over his drinking, and (4) the disturbance in functioning in his social world.

Blum (1973:508) has referred to drinking in two contexts : (i) in the context of prescribed social pattern where drinking is integrated in the culture of the society and it is perceived as part of everyday life, (for example, in Italy, the United States) and people do not find any psychological potential in it; and (ii) in the context of perceiving alcohol-use as disruptive to culture and society and people find addictive potential in it (as in India) and view drinking as a means of seeking pleasure and escape. Drinkers have been classified as 'non-addicts', 'addicts' and 'chronic alcoholics'. Non-addicts are categorized as 'experimenters' and 'regulars'. Don Cahalan (1969) has given a five-fold classification of alcohol drinkers on the basis of the frequency of drinking (and not the quantity of alcohol taken) : (1) Rare users, who drink once or twice a year. (2) Infrequent users, who drink once or twice in two-three months, that is, less than once a month. (3) Light drinkers, who drink once or twice a month. (4) Moderate drinkers, who drink three or four times in a month. (5) Heavy drinkers, who drink every day or several drinks during the day. The last category of drinkers is also described as 'hard-core' drinkers.

History and Etymology

Alcohol has a long history of use and misuse throughout recorded human history. Biblical, Egyptian and Babylonian sources record history of abuse and dependence on alcohol. In some ancient cultures alcohol was worshiped and in others its abuse was condemned. Excessive alcohol misuse and drunkenness were recognised as causing problems thousand of years ago. However, the defining of habitual drunkenness as it was then known as and its adverse consequences were not well established medically until the 1700s. In 1647 a Greek monk was the first to document that chronic alcohol misuse was associated with toxicity to the nervous system and body which resulted in a range of medical disorders such as seizures, paralysis and internal bleeding.

Historically the name dipsomania was coined by German physician Dr. C. W. Hufeland in 1819 before it was superseded by alcoholism. The term "alcoholism" was first used in 1849 by the Swedish physician Magnus Huss to describe the systematic adverse effects of alcohol (Tracy2005).

Alcoholics Anonymous (2001: 28) describes alcoholism as an illness that involves a physical allergy and a mental obsession (Ibid:23). However, "allergy" used in this context is not the same as used in modern medicine. Alcoholics suffer from a "(physical) craving beyond mental control." A 1960 study by E. Morton Jellinek is considered the foundation of the modern disease theory of alcoholism. Jellinek's definition restricted the use of the word "alcoholism" to those showing a particular natural history. The modern medical definition of alcoholism has been revised numerous times since then. The American Medical Association currently uses the word alcoholism to refer to a particular chronic primary disease.

A minority opinion within the field, notably advocated by Herbert Fingarette and Stanton Peele, argue against the existence of alcoholism as a disease. Critics of the disease model tend to use the term "heavy drinking" when discussing the negative effects of alcohol consumption. On the other hand, McGovern and White said "We have never called alcoholism a disease because, technically speaking, it is not a disease entity. For example, there is no such thing as heart disease. Instead there are many separate heart ailments, or combinations of them. It is something like that with alcoholism. Therefore we did not wish to get in wrong with the medical profession by pronouncing alcoholism a disease entity. Therefore we always called it an illness, or a malady — a far safer term for us to use" (2003:7).

Symptoms of Long Term Alcohol Misuse

Alcoholism is characterised by an increased tolerance of and physical dependence on alcohol, affecting an individual's ability to control alcohol consumption safely. These characteristics are believed to play a role in impeding an alcoholic's ability to stop drinking. Alcoholism can have adverse effects on mental health, causing psychiatric disorders to develop and an increased risk of suicide (Dunn:1999). Alcoholism is a disease that includes the following four symptoms :

1. Craving—A strong need, or urge, to drink.

2. Loss of control—Not being able to stop drinking once drinking has begun.

3. Physical dependence—Withdrawal symptoms, such as nausea, sweating, shakiness, and anxiety after stopping drinking.

4. Tolerance—The need to drink greater amounts of alcohol to get "high."

Other symptoms include Physical symptoms and Psychiatric symptoms.

1. Physical Symptoms : Long term alcohol abuse can cause a number of physical symptoms, including cirrhosis of the liver, pancreatitis, epilepsy, alcoholic dementia, heart disease, nutritional deficiencies, and sexual dysfunction, and can eventually be fatal. Other physical effects include an increased risk of developing cardiovascular disease, liver disease, and cancer. Damage to the central nervous system and peripheral nervous system can occur from sustained alcohol consumption. Women develop long-term complications of alcohol dependence more rapidly than do men. Additionally, women have a higher mortality rate from alcoholism than men. Examples of long term complications include brain, heart, and liver damage and an increased risk of breast cancer. Additionally, heavy drinking over time has been found to have a negative effect on reproductive functioning in women. This results in reproductive dysfunction such as decreased ovarian mass, problems or irregularity of the menstrual cycle, and early menopause.

2. Psychiatric Symptoms : Long term misuse of alcohol can cause a wide range of mental health problems. The most prevalent psychiatric symptoms are anxiety and depression disorders. Psychiatric symptoms usually initially worsen during alcohol withdrawal, but typically improve or disappear with continued abstinence. Psychosis, confusion, and organic brain syndrome may be caused by alcohol misuse, which can lead to a misdiagnosis such as schizophrenia. Panic disorder can develop or worsen as a direct result of long term alcohol misuse. Psychiatric disorders differ depending on gender. Women who have alcohol-use disorders often have a co-occurring psychiatric diagnosis such as major depression, anxiety, panic disorder, bulimia, post-traumatic stress disorder (PTSD), or borderline personality disorder. Men with alcohol-use disorders more often have a co-occurring diagnosis of narcissistic or antisocial personality disorder, bipolar disorder, schizophrenia, impulse disorders or attention deficit/hyperactivity disorder.

Social effects: Drug-related Crime

Alcohol is not a stimulant; it is a depressant or inhibitor on the central nervous system. Alcohol relaxes the customary controls on behaviour and the drinker becomes less restrained and feels more free. But taking alcohol even once in a while leaves the possibility of a habit-forming phenomenon open and the drinker may start taking it frequently and in large quantity which could have tragic and disastrous effects. It may affect him physically, destroy his ability to work and earn, ruin his family life, and demoralize him utterly. An innocent recreation, thus, may come to spoil the whole life of the drinker.

The various health problems associated with long-term alcohol consumption are generally perceived as detrimental to society, for example, money due to lost labor hours, medical costs, and secondary treatment costs. Alcohol use is a major contributing factor for head injuries, motor vehicle accidents, violence, and assaults. Beyond money, there are also significant social costs to both the alcoholic and their family and friends (McCully, 2004). For instance, alcohol consumption by a pregnant woman can lead to fatal alcohol syndrome, an incurable and damaging condition.

Estimates of the economic costs of alcohol abuse, collected by the World Health Organization (2004), vary from one to six percent of a country's GDP.

The biological mechanisms underpinning alcoholism are uncertain, however, risk factors include social environment, stress, mental health, genetic predisposition, age, ethnic group, and sex (Agarwal 2000;Chen et.al 2009). Long-term alcohol abuse produces physiological changes in the brain such as tolerance and physical dependence. Such brain chemistry changes maintain the alcoholic's compulsive inability to stop drinking and result in alcohol withdrawal syndrome upon discontinuation of alcohol consumption (Hoffman:1996). Alcohol damages almost every organ in the body, including the brain; because of the cumulative toxic effects of chronic alcohol abuse, the alcoholic risks suffering a range of medical and psychiatric disorders (Cann:2002). Alcoholism has profound social consequences for alcoholics and the people of their lives.

The social problems arising from alcoholism are serious, caused by the pathological changes in the brain and the intoxicating effects of alcohol (McCully, 2004). Alcohol abuse is associated with an increased risk of committing criminal offences, including child abuse, domestic violence, rape, burglary and assault. Alcoholism is associated with loss of employment, which can lead to financial problems. Drinking at inappropriate times and behaviour caused by reduced judgment, can lead to legal consequences, such as criminal charges for drunk driving or public disorder, or civil penalties for tortuous behaviour, and may lead to a criminal sentence. An alcoholic's behaviour and mental impairment, while drunk, can profoundly impact those surrounding them and lead to isolation from family and friends. This isolation can lead to marital conflict and divorce, or contribute to domestic violence. Alcoholism can also lead to child neglect, with subsequent lasting damage to the emotional development of the alcoholic's children.

Alcohol withdrawal

As with similar substances with a sedative-hypnotic mechanism, such as barbiturates and benzodiazepines, withdrawal from alcohol dependence can be fatal if it is not properly managed. With repeated heavy consumption of alcohol, these receptors are desensitized and reduced in number, resulting in tolerance and physical dependence. When alcohol consumption is stopped too abruptly, the person's nervous system suffers from uncontrolled synapse firing. This can result in symptoms that include anxiety, life threatening seizures, delirium tremens, hallucinations, shakes and possible heart failure.

Acute withdrawal symptoms tend to subside after one to three weeks. Less severe symptoms (e.g. insomnia and anxiety) may continue as part of a post withdrawal syndrome gradually improving with abstinence for a year or more. Withdrawal symptoms begin to subside as the body and central nervous system restore alcohol tolerance.

Causes of Alcoholism

Alcoholism can be explained only on the basis of a holistic approach rather than a single factor approach. A complex mixture of genetic and environmental factors influence the risk of the development of alcoholism (Enoch, 2006). Genes which influence the metabolism of alcohol also influence the risk of alcoholism, and may be indicated by a family history of alcoholism. One published article has found that alcohol use at an early age may influence the expression of genes which increase the risk of alcohol dependence. Individuals who have a genetic disposition to alcoholism are also more likely to begin drinking at an earlier age than average. The age of onset of drinking is associated with an increased risk of the development of alcoholism, and about 40 percent of alcoholics drink excessively by their late adolescence. Severe childhood trauma is associated with a general increase in the risk of drug dependency. Lack of peer and family support is associated with an increased risk of alcoholism developing (Ibid).

Genetic variation : Genetic differences exist between different racial groups which affect the risk of developing alcohol dependence. For example, there are differences between African, East Asian and Indo-racial groups in how they metabolize alcohol. These genetic factors are believed to in part, explain the differing rates of alcohol dependence among racial groups. Native Americans however, have a significantly higher rate of alcoholism than average. Other risk factors such as cultural environmental effects e.g. trauma have been proposed to explain the higher rates of alcoholism among Native Americans compared to alcoholism levels in Caucasians (Szlemko and Thurman, 2006; Gabbard,2001).

Pathophysiology : Alcohol's primary effect is the increase in stimulation of the central nervous system depression. With repeated heavy consumption of alcohol, these receptors are desensitized and reduced in number, resulting in tolerance and physical dependence. The amount of alcohol that can be biologically processed and its effects differ between sexes. Equal dosages of alcohol consumed by men and women generally result in women having higher blood alcohol concentrations (BACs). This can be attributed to many reasons, the main being that women have less body water than men (Karrol,2002). A given amount of alcohol, therefore becomes more highly concentrated in a woman's body. A given amount of alcohol causes greater intoxication for women due to different hormone release compared to men.

Parental attitude : One psychological view is that practically all alcoholics show the mark of deprivation of emotional needs during childhood. Clinebell (1956:45) reports four main types of parental attitudes which happen to be associated with alcoholism in adulthood and all of which tend to produce trauma and emotional deprivation in the child : (1) authoritarianism, (2) overt rejection, (3) moralism, and (4) success worship. That these factors are the key ones in the formation of an insecure personality resulting falling prey to alcohol is indicated by the fact that psychological studies of alcoholics repeatedly mention the following personality traits : a high level of anxiety in interpersonal relationships, emotional immaturity, ambivalence towards authority, low frustration tolerance, low self-esteem, feelings of isolation and guilt (Clinebell, 1956:49). These psychological traits are not the result of alcoholism but are the causes of alcoholism. They are often present in many alcoholics before they take to excessive drinking.

Personality factor : According to some scholars, there seems to be a definite connection between alcoholism and personality maladjustment. Initially, a person drinks to seek refuge for his problems of life or to find a temporary respite from his troubles. Gradually he starts drinking more and more frequently until he becomes utterly dependent on it. However, psychologists maintain that only those people take to frequent drinking who are emotionally immature and lack self-confidence. The personal problems of adjustment, anxiety, tension, guilt and frustration may lead a person to turn alcoholic. The current approach is that alcoholism is to be understood in terms of character and motivation. An alcoholic is a sick man. He is not to be looked upon with ridicule, condemnation and blame. He has fallen a victim to a set of complexes, attitudes and habits which bind him until the process of self-destruction is inevitable.

Sociological reasons : The sociological reasons for taking alcohol are essentially the same as for taking drugs. However, a distinction can be made in the causes of drinking alcohol and taking illicit drugs. Since alcohol is more socially acceptable than drugs, drinking reduces a person's fears, worries and anxieties. Besides, alcohol is more easily available than drugs. It is also cheaper than

many drugs like heroin, cocaine and LSD. The main sociological causes of taking alcohol are : (1) environmental pressures, (2) peer pressure, and (3) a dominant sub-culture.

While drinking is portrayed as a consequence of poverty, a detailed study of global consumption suggests that it is also associated with relative affluence. The patterns of drinking are changing rapidly with major changes in economic policies, liberalization, the market economy and growing consumerism in India. The break-up of the traditional joint family system, changes in values and attitudes, including those towards consuming alcohol, have also had an impact on its pattern of use. Cultural and religious controls, which prevented people from drinking alcohol, have been weakening and alcohol use is now seen in all sections of society. Alcohol use is also increasing among groups who were traditionally abstainers such as women, teenagers and the rural rich. The changing economy has brought with it changes in lifestyles and culture.

Problem of early detection : Attitudes and social stereotypes can create barriers to the detection and treatment of alcohol abuse. Lack of early detection increases the possibility of one becoming a habitual drunkard. Fear of stigmatization may lead alcoholics to deny that they are suffering from a medical condition, to hide their drinking, and to drink alone. This pattern, in turn, leads family, physicians, and others to be less likely to suspect that a person they know is an alcoholic. In contrast, reduced fear of stigma may lead men to admit that they are suffering from a medical condition, to publicly display their drinking, and to drink in groups. This pattern, in turn, leads family, physicians, and others to be more likely to suspect that a man they know is an alcoholic (Karrol, 2002).

Disease of Alcoholism : Stages of Growth

There are, and have been, many theories about alcoholism. The most prevailing theory, and now most commonly accepted, is called the Disease Model. Its basic tenets are that alcoholism is a disease with recognizable symptoms, causes, and methods of treatment. In addition, there are several stages of the disease which are often described as early, middle, late, treatment and relapse stages. It is useful to understand these stages in terms of how the disease presents itself.

1. The Early or Adaptive Stage of Alcoholism : The early or adaptive stage of alcoholism is marked by increasing tolerance to alcohol and physical adaptations in the body which are largely unseen. This increased tolerance is marked by the alcoholic's ability to consume greater quantities of alcohol while appearing to suffer few effects and continuing to function. This tolerance is not created simply because the alcoholic drinks too much but rather because the alcoholic is able to drink great quantities because of physical changes going on inside his or her body.

The early stage is difficult to detect. By appearances, an individual may be able to drink a great deal without becoming intoxicated, having hangovers, or suffering other apparent ill-effects from alcohol. An early stage alcoholic is often indistinguishable from a non-alcoholic who happens to be a fairly heavy drinker. In the workplace, there is likely to be little or no obvious impact on the alcoholic's performance or conduct at work. At this stage, the alcoholic is not likely to see any problem with his or her drinking and would scoff at any attempts to indicate that he or she might have a problem. The alcoholic is simply not aware of what is going on in his or her body.

2. The Middle or Dependence Stage of Alcoholism : There is no clear line between the early and middle stages of alcoholism, but there are several characteristics that mark a new stage of the disease. Many of the pleasures and benefits that the alcoholic obtained from drinking during the early stage are now being replaced by the destructive facets of alcohol abuse. The drinking that was done for the purpose of getting high is now being replaced by drinking to combat the pain and misery caused by prior drinking.

One basic characteristic of the middle stage is physical dependence. In the early stage, the alcoholic's tolerance to greater amounts of alcohol is increasing. Along with this, however, the body becomes used to these amounts of alcohol and now suffers from withdrawal when the alcohol is not present.

Another basic characteristic of the middle stage is craving. Alcoholics develop a very powerful urge to drink which they are eventually unable to control. As the alcoholic's tolerance increases along with the physical dependence, the alcoholic loses his or her ability to control drinking and craves alcohol.

The third characteristic of the middle stage is loss of control. The alcoholic simply loses his or her ability to limit his or her drinking to socially acceptable times, patterns, and places. This loss of control is due to a decrease in the alcoholic's tolerance and an increase in the withdrawal symptoms. The alcoholic cannot handle as much alcohol as they once could without getting drunk, yet needs increasing amounts to avoid withdrawal.

Another feature of middle stage alcoholics is blackouts. Contrary to what you might assume, the alcoholic does not actually pass out during these episodes. Instead, the alcoholic continues to function but is unable to remember what he or she has done or has been. Basically, the alcoholic simply can't remember these episodes because the brain has either stored these memories improperly or has not stored them at all. Blackouts may also occur in early stage alcoholics.

Impairment becomes evident in the workplace during the middle stage. This will become apparent at work in terms of any or all of the following: increased and unpredictable absences, poorly performed work assignments, behaviour problems with co-workers, inability to concentrate, accidents, increased use of sick leave, and possible deterioration in overall appearance and demeanor. This is the point where the alcoholic employee may be facing disciplinary action.

3. The Late or Deteriorative Stage of Alcoholism : The late, or deteriorative stage, is best identified as the point at which the damage to the body from the toxic effects of alcohol is evident, and the alcoholic is suffering from a host of ailments. An alcoholic in the final stages may be destitute, extremely ill, mentally confused, and drinking almost constantly. The alcoholic in this stage is suffering from many physical and psychological problems due to the damage to vital organs. His or her immunity to infections is lowered, and the employee's mental condition is very unstable. Some of the very serious medical conditions the alcoholic faces at this point include heart failure, fatty liver, hepatitis, cirrhosis of the liver, malnutrition, pancreatitis, respiratory infections, and brain damage, some of which is reversible.

In the early stage, the alcoholic does not consider himself or herself sick because his or her tolerance is increasing. In the middle stage, the alcoholic is unknowingly physically dependent on alcohol. He or she simply finds that continuing to use alcohol will prevent the problems of withdrawal. By the time an alcoholic is in the late stage, he or she is often irrational, deluded, and unable to understand what has happened. In addition to the effects of these changes, the alcoholic is faced with one of the most powerful facets of addiction: denial. An alcoholic will deny that he or she has a problem. This denial is a very strong force. If an alcoholic did not deny the existence of a problem, he or she would most likely seek help when faced with the overwhelming problems caused by drinking. While denial is not a diagnosable physical symptom or psychiatric disorder, it is an accurate description of the state of the alcoholic's behaviour and thinking and is very real.

4. Treating Stage of Alcoholism : An alcoholic will rarely stop drinking and stay sober without outside help. Also, he or she usually will not stop drinking without some kind of outside pressure. This pressure may come from family, friends, clergy, other health care professionals, law enforcement or judicial authorities, or an employer. For example, a spouse may threaten divorce, or the alcoholic may be arrested for driving under the influence. There was at one time a widespread belief that alcoholics would not get help until they had "hit bottom." This theory has generally been discredited as many early and middle stage alcoholics have quit drinking when faced with consequences such as the loss of a job, a divorce, or a convincing warning from a physician regarding the potentially fatal consequences of continued drinking.

There are obvious advantages to getting the alcoholic into treatment earlier rather than later. One advantage is that, the earlier treatment is begun, the probability of having less expensive treatment, such as outpatient care, is increased. There is also a greater likelihood of success in treatment with an individual who has not yet lost everything and still has a supportive environment to return to, including an intact family, good health, and a job. In addition, the employer has a stake in the early treatment of alcoholism, since the employee will have a greater chance of returning sooner to full functioning on the job if the disease is arrested at an earlier point. Early treatment is simply less disruptive and can help the alcoholic avoid further misconduct and poor performance. If an alcoholic doesn't get help until very late in the disease, there may have been irreparable harm done.

The alcoholic does not initially have to want to get help to go into treatment. Many people go into treatment because of some kind of threat such as loss of a job, divorce or death or possible incarceration. However, even the individual that is forced will eventually have to personally accept the need for treatment for it to be effective. Employers are a very potent force in getting the alcoholic into treatment. The threat of the loss of a job is often the push the alcoholic needs to enter treatment. There are various kinds of treatment and programs for alcoholism. Though some alcoholics do stop drinking on their own, this is rare. Most alcoholics require some type of treatment or help.

5. Relapse Stage : An important and frustrating facet of treating alcoholism is relapse or a return to drinking and is common. An alcoholic often relapses due to a variety of factors including:

- inadequate treatment or follow-up,
- cravings for alcohol that are difficult to control,
- failure by the alcoholic to follow treatment instructions,
- failure to change lifestyle,
- use of other mood altering drugs, and
- other untreated mental or physical illnesses.

Relapses are not always a return to constant drinking and may only be a one time occurrence. However, relapses must be dealt with and seen as a sign to the alcoholic that there are areas of his or her treatment and recovery that need work. Relapse prevention is an area in the treatment field that is receiving increased attention and research. A basic part of any effective treatment program will include relapse prevention activities.

The Recent Trends

Substance use disorders are a major public health problem facing many countries. "The most common substance of abuse/dependence in patients presenting for treatment is alcohol (Gabbard, 2001)." In the United Kingdom, the number of 'dependent drinkers' was calculated as over 2.8 million in 2001. About 12% of American adults have had an alcohol dependence problem at some time in their life. The World Health Organization estimates that about 140 million people throughout the world suffer from alcohol dependence. In the United States and western Europe 10 to 20 percent of men and 5 to 10 percent of women at some point in their lives will meet criteria for alcoholism. The World Health Organisation estimates that there are 140 million alcoholics worldwide in 2003.

Alcoholism has a higher prevalence among men, though in recent decades, the proportion of female alcoholics has increased. Current evidence indicates that in both men and women, alcoholism is 50–60 percent genetically determined, leaving 40–50 percent for environmental influences. Most alcoholics develop alcoholism during adolescence or young adulthood (Enoch,2006).

The most common cause of death in alcoholics is from cardiovascular complications. There is a high rate of suicide in chronic alcoholics, which increases the longer a person drinks. This believed to be due to alcohol causing physiological distortion of brain chemistry, as well as social isolation. Suicide is also very common in adolescent alcohol abusers, with 25 percent of suicides in adolescents being related to alcohol abuse. Approximately 18 percent of alcoholics commit suicide, and research has found that over 50 percent of all suicides are associated with alcohol or drug dependence. The figure is higher for adolescents, with alcohol or drug misuse playing a role in up to 70 percent of suicides.

Indian Alcohol Consumption Trend (2005)

Indian Liquor : Branded spirits market in India is 119 million cases out of which Whisky accounts for 55%, Rum 27%, Brandi 14%, Gin 3%, Vodka 1%. North and west India consumes more of whisky and South, more of Brandy and Rum. The UB Group (35%) has a market share of 48% after merger of Spirits Division of Shaw Wallace (13%), Jagajit (7%), Mohan Meakins (7%) and IDL (3.5%). The key MNC's operating in the India include Diageo, Seagram, Bacardi. About 80% of whisky is sold at below Rs.200 per bottle and 1% above Rs.600. AP (24), TN (19.3), Karnataka (11.2), Kerala (10.9) & Rajasthan (5.7) consume 60 % of the total IL, which is growing at about 11%.

Country Liquor : About 115 million cases of country liquor is sold in the country and the growth rate is 0.5%. The country liquor market is highly regulated, sold generally as commodity, often dominated by cartels, traders get disproportionate share of MRP. Country liquor companies predominantly are state centric e.g. Saraya & Radico in UP, GM Brewery & Vasant Dada in Maharashtra, IFB Agro in West Bengal, Som Distillers and Kedia in MP, Glenn in Haryana etc. A.P.,

Karnataka and Tamil Nadu have abolished RS based country liquor in favour of cheap ENA based IMFL. UP (19.5), Maharashtra (19.3), Karnataka (13.6), Punjab (10.2) & Haryana (9.4) consume 63% of the total country liquor.

Beer : The Indian beer market is 93 million cases and has grown at 5.54% p.a. over 2003-05. Several parts of India are showing much higher growth rates. Five States, Andhra Pradesh (18.3), Maharashtra (16.2), Tamil Nadu (9.0), Karnataka (9.0), & Rajasthan (5.6) account for 63% of total beer consumption and top 10 States for 85%. Indian beer industry due to merger and acquisition is dominated by two top players (i.e. UB and SAB Miller together have a market share of 75%) and provide attractive profit margins due to the consolidated nature of the industry. The Chinese beer market in contrast is marked by intense competition with 400 brewers. The top 10 brewers account for only 45% of the market. Lower taxes on beer, falling distribution margin and prices have contributed to the surge in the consumption of beer in India.

Alcoholic Treatments

Treatments are varied because there are multiple perspectives of alcoholism. Those who approach alcoholism as a medical condition or disease recommend differing treatments than, for instance, those who approach the condition as one of social choice. Most treatments focus on helping people discontinue their alcohol intake, followed up with life training and/or social support in order to help them resist a return to alcohol use. Since alcoholism involves multiple factors which encourage a person to continue drinking, they must all be addressed in order to successfully prevent a relapse. An example of this kind of treatment is detoxification followed by a combination of supportive therapy, attendance at self-help groups, and ongoing development of coping mechanisms. The treatment community for alcoholism typically supports an abstinence-based zero tolerance approach; however, there are some who promote a harm-reduction approach as well (Gabbard, 2001).

Detoxification : Alcohol detoxification or 'detox' for alcoholics is an abrupt stop of alcohol drinking coupled with the substitution of drugs, such as benzodiazepines, that have similar effects to prevent alcohol withdrawal. Individuals who are only at risk of mild to moderate withdrawal symptoms can be detoxified as outpatients. Individuals at risk of a severe withdrawal syndrome as well as those who have significant or acute comorbid conditions are generally treated as inpatients. Detoxification does not actually treat alcoholism, and it is necessary to follow-up detoxification with an appropriate treatment program for alcohol dependence or abuse in order to reduce the risk of relapse.

Group therapy and psychotherapy Alcoholics Anonymous : Various forms of group therapy or psychotherapy can be used to deal with underlying psychological issues that are related to alcohol addiction, as well as provide relapse prevention skills. The mutual-help group-counseling approach is one of the most common ways of helping alcoholics maintain sobriety. Alcoholics Anonymous was one of the first organizations formed to provide mutual, nonprofessional counseling, and it is still the largest. Others include LifeRing Secular Recovery, SMART Recovery, and Women For Sobriety.

Rationing and moderation : Rationing and moderation programs such as Moderation Management and DrinkWise do not mandate complete abstinence. While most alcoholics are unable to limit their drinking in this way, some return to moderate drinking. A 2002 U.S. study by the National Institute on Alcohol Abuse and Alcoholism (NIAAA) showed that 17.7 percent of individuals diagnosed as alcohol dependent more than one year prior returned to low-risk drinking. This group, however, showed fewer initial symptoms of dependency. A follow-up study, using the same subjects that were judged to be in remission in 2001–2002, examined the rates of return to problem drinking in 2004–2005. The study found abstinence from alcohol was the most stable form of remission for recovering alcoholics. A long-term (60 year) follow-up of two groups of alcoholic men concluded that "return to controlled drinking rarely persisted for much more than a decade without relapse or evolution into abstinence."

Medications : A variety of medications may be prescribed as part of treatment for alcoholism. Medications currently in use are Antabuse (disulfiram), Temposil (calcium carbimide), Naltrexone, Campral (acamprosate), and Topamax (topiramate). Medications may be problematic for the dual addicts who use drugs along with alcohol.

Role of Family : Involving an alcoholic's family in his treatment and rehabilitation enhances the chances of success by 75 to 80 per cent. The family members do not preach; nor do they blame or condemn the alcoholic. They minimize his problems, offer him sincere and unselfish help and guidance, and never abandon him.

Changing Values through Education : Some voluntary organizations undertake educational and information programmes to alert the alcoholics to the dangers of excessive drinking. Social workers help the drinkers in coping with life and changing the social values and attitudes about drinking.

Prevention and Policy Prescription

Targeting adolescents and young adults is regarded as an important step to reduce the harm of alcohol abuse. Increasing the age at which licit drugs of abuse such as alcohol can be purchased, the banning or restricting advertising of alcohol has been recommended as additional ways of reducing the harm of alcohol dependence and abuse. Credible, evidence based educational campaigns in the mass media about the consequences of alcohol abuse have been recommended. Guidelines for parents to prevent alcohol abuse amongst adolescents, and for helping young people with mental health problems have also been suggested. A strong social support system and early medical or psychiatric intervention may also help prevent the escalating consumption of alcohol so characteristic of alcoholism.

There is increasing recognition that consumption of alcohol is a major contributor to the burden of disease in India and the developing world, and is indeed a major public health concern. The complex relationships between policies, economics and politics of alcohol and public health and between governments, industry and individuals call for a thorough review of the current situation. While public health implications of policies concerning alcohol have long been accepted, the failure to implement many of these policies demands a more balanced and nuanced approach to the problem.

Benegal (2005:1051-56) makes a critical assessment of an alcohol policy for the 21st century in the context of Indian Society. A combination of (a) a population-based approach reducing overall consumption and (b) a high-risk approach targeting high-risk behaviours is essential to reduce the impact of hazardous alcohol use in the country (Ibid:1054).

This requires an urgent shift to a public health paradigm in the approach to alcohol use. Health systems, especially at primary care levels, must be geared to play a greater role in the early detection and prevention of alcohol-related harm, perhaps through brief, cost effective interventions that have been demonstrated to be useful. The social welfare system and the criminal justice system, often the first to come into contact with alcohol-related problems, can be sensitized in identifying and assisting individuals and families at risk from heavy drinking and acting as early referral systems. Extensive opportunities exist to lessen alcohol problems through community education and the prevention of drink driving, domestic violence, public disorder, unintentional injuries and criminal damage. Community programmes supporting healthier lifestyles, mass media campaigns that present the advantages of reduced consumption rather than the dangers of heavy alcohol and community development in general (job creation, skills development and upgrading infrastructure or recreational facilities in communities with high levels of alcohol abuse) should be utilized to encourage alternatives to drinking among the young and disadvantaged. Community action can also serve to shape attitudes, values and norms about drinking. Recently, several effective temperance campaigns have been led by heads of certain Hindu religious orders, although the impact has been limited to their immediate followers. Existing legislation relating to controls on availability and marketing, restrictions on advertising, and especially minimum drinking age, need to be enforced rigorously. Simulations have demonstrated that implementing a nation-wide legal drinking age of 21 years can achieve about 50–60% of the alcohol consumption reducing effects of prohibition. Each year that drinking is delayed significantly reduces the likelihood of developing alcoholism and the life-time risk of alcohol abuse. Dialogue and a degree of consensus between the health and revenue arms of government (i.e. federal and state agencies who are virtually opposing players) are crucial to reduce the mutually antagonistic directions of profit and welfare. A proportion of the considerable revenue from alcohol should be ploughed back into treatment and research. The priority for alcohol researchers in India is surely to focus on the public health dimensions of alcohol misuse and to focus on appropriate interventions. It is equally important for research findings to be published and publicized. Hopefully, the imminent centralized system of value added taxation will reduce the inequities in alcohol taxes between states, and thus diminish interstate smuggling, which nullifies the potential benefits of increasing alcohol prices to reduce consumption. Volume-based taxation (wherein a litre of beer is taxed equally as a litre of spirits) that promotes excess consumption of spirits requires rationalization and stricter control on the entry of illicit (often toxic) alcoholic beverages into circulation is essential (Ibid).

There is also a need to consider public health interests in issues relating to trade agreements. Investment in alcohol production and distribution by transnational companies in India and the removal of tariff barriers is likely to increase availability and consumption and limit the effect of prevention programmes. For all this to happen, a vital first step is for health planners and other stakeholders to debate and draft an explicit and rational alcohol policy appropriate for India as it marches into the uncertain future of the 21st century (Ibid).

The goal of a sustainable and effective alcohol policy can only be achieved by coordinated action by multiple stakeholders. The divergent frameworks used result in confusion and inaction. Multiple agencies including the Ministries of Law, Industry, Agriculture, Revenue, Health, Home, Customs and Enforcement, non-governmental organizations, and medical associations should be involved. Indian society and governments need to take a longer term view of issues and plan a coordinated, comprehensive and balanced approach. While complete prohibition has been shown to be a failure, the current permissiveness without the enforcement of regulations also represents a lack of responsibility from a public health perspective. The WHO's global strategy for alcohol and governmental and non-governmental efforts can be used synergistically to produce sustained action.

A Joint Working Group (JWG) on rationalization of Excise Policy and Taxation with representatives of all concerned central ministries and state governments was constituted by the Ministry of Food Processing Industries (Nodal Minister), Government of India, in December 2001.

Some general observations made by the Joint Working Group are :

1. Prohibition is not a viable policy.
2. Though the play of market forces in a free trade regime leads to optimization of resources, this may not be the best recipe for alcoholic products because of its very negative nature, moreover, after reaching saturation level in consumption in the developed countries, there is a very real danger of the surplus production in these countries being "marketed" and dumped into developing countries. The Union and State Governments should therefore, take necessary measures to protect citizens from this onslaught.
3. Scientific studies have proved the efficacy of the physical controls in controlling consumption, in spite of their known negative side effects (Illicit distillation, trafficking etc.) Further to minimize such side effects, control regime should be reasonable and transparent and give least discretion to the operators.
4. The tax element in the price structure will have to be high so long as the cost price of liquor continues to below. In the long run, with the increase in the cost price, due to increased demand of alcohol for industrial and fuel use, the element of taxes can be bought down.
5. A Model Excise Policy should ensure that the objectives of minimization of consumption and maximization of revenue are achieved without putting unreasonable restrictions on the individual's freedom of choice and unreasonably high taxation.
6. Raising rates beyond the threshold instead of generating higher revenue and discouraging consumption would lead to evasion of taxes smuggling and illicit distillation.

References

Agarwal-Kozlowski, K.; Agarwal, DP. (2000). Genetic predisposition for alcoholism. Ther Umsch 57 (4): 179–84.

Ahuja Ram (2005) Indian Social Problems, Jaipur, Rawat Publications.

Benegal V (2005) India: alcohol and public health, Addiction, Society for the Study of Addiction, 100, pp.1051–1056.

Blum Richard H. (1973) Drug Dealers—Taking Action, Jossey-Bass.

Caan, Woody; Belleroche, Jackie de, eds (2002). Drink, Drugs and Dependence: From Science to Clinical Practice (1st ed.). Routledge. pp. 19–20.

Chen, CY, Storr, CL and Anthony, JC. (2009). "Early-onset drug use and risk for drug dependence problems.". Addict Behav 34 (3): 319–22.

Clinebell HJ (1956) Understanding and Counseling the Alcoholic (New York : Abingdon Press, p.17).

Don Cahalan (1969) American Drinking Practices, New College & Univ Press.

Dunn, N; Cook (1999). "Psychiatric aspects of alcohol misuse.". Hospital Medicine 60 (3): 169–72.

Enoch, MA. (2006). "Genetic and environmental influences on the development of alcoholism: resilience vs. risk.". Ann N Y Acad Sci 1094: 193–201.

Gabbard, Glen O. (2001). Treatments of psychiatric disorders (3 ed.). Washington, DC: American Psychiatric Press.

Gitlow, Stuart (2006). Substance Use Disorders: A Practical Guide (2nd ed.). USA: Lippincott Williams and Wilkins. pp. 52 and 103–121.

Glavas MM, Weinberg J (2006). "Stress, Alcohol Consumption, and the Hypothalamic-Pituitary-Adrenal Axis". in Yehuda S, Mostofsky DI. Nutrients, Stress, and Medical Disorders. Totowa, NJ: Humana Press. pp. 165–183.

Holder, Harold D. (2001) Prevention of Alcohol Problems in the 21st Century: Challenges and Opportunities, The American Journal on Addictions, Volume 10, Issue 1, pp. 1–15.

Hoffman, PL.; Tabakoff, B. (1996). "Alcohol dependence: a commentary on mechanisms.". Alcohol Alcohol 31 (4): 333–40.

Indian Alcohol Policy Organization (2011) : http://www.indianalcoholpolicy.org/alcohol_atlas.

Jellinek, E. M., (1960) *The Disease Concept of Alcoholism*, Hillhouse, (New Haven).

Johnson, D. A (1973) The side-effects of fluphenazine decanoate. British Journal of Psychiatry, vol. 123, 519-522.

Karrol Brad R. (2002). "Women and alcohol use disorders: a review of important knowledge and its implications for social work practitioners". Journal of social work 2 (3): 337–356.

Keller M. and V. Efron, (1955) "The Prevalence of. Alcoholism," Quarterly Journal of Studies on Alcohol, 16, 619-644.

McCully, Chris (2004). Goodbye Mr. Wonderful. Alcohol, Addition and Early Recovery.. London: Jessica Kingsley Publishers.

Ogborne, AC. (2000). "Identifying and treating patients with alcohol-related problems.". CMAJ 162 (12): 1705–8.

Spillane, NS.; Smith, GT. (2007). "A theory of reservation-dwelling American Indian alcohol use risk.". Psychol Bull 133 (3): 395–418.

Stivers, Richard (2000). Hair of the dog: Irish drinking and its American stereotype. New York: Continuum.

Szlemko, WJ.; Wood, JW.; Thurman, PJ. (2006). "Native Americans and alcohol: past, present, and future.". J Gen Psychol 133 (4): 435–51.

Tracy, Sarah J. (2005). Alcoholism in America: from reconstruction to prohibition. Baltimore: Johns Hopkins University Press. pp. 31–52.

WHO (2004) "Global Status Report on Alcohol 2004" (PDF). World Health Organization.

■■■

CHAPTER 17

Cyber Crime

Learning Objectives :

- **To understand the concept and the scope of cyber crime.**
- **To comprehend the theories of classification of cyber crime.**
- **To discuss the types of cyber crime and the current trends.**
- **To grasp the policy implications.**

Introduction

Cyber crime is a global problem today that affects computer users and others in all developed and developing countries that rely heavily on computing and communications technologies. The need to respond effectively has, therefore, been a common theme throughout the networked world, although the level of concern and the impetus for reform have been closely correlated with the degree of usage and hence the risk of victimisation. Those countries with the highest usage levels tend to have taken greater steps with respect to regulation and law reform. 'Cyber crime' has, however, taken on a life of its own to some extent, and entered popular discourse today. Although cyber crime overlaps with a number of other areas of economic crime and white-collar crime, there are some aspects that are unique. Some forms of illegality simply could not have been committed prior to the introduction of computers and so, in this sense, cyber crimes (or at least elements of them) are dissimilar from conventional crimes. In addition, with cyber crime the level of apprehension of victimisation felt by many businesses is far greater than what appears to be the actual incidence of offences. The presence of computers in the commission of crime has enhanced the seriousness with which these crimes are viewed, at least in the minds of those at risk of victimization and those who have actually suffered from these crimes. It remains to be seen whether the courts deal with persons convicted of cyber crimes in ways similar to, or dissimilar from, the ways in which they treat those convicted of conventional crimes.

Cyber crime is a term used broadly to describe activity in which computers or computer networks are the tool, target, or theme of criminal activity. These categories are not exclusive and many activities can be characterized as falling in one or more categories. Many types of cyber crime are simply extensions of existing criminal activities, with the computer and internet severing them from the perpetrator's geographical location, providing anonymity and protection from law enforcement. While computers and the internet are valuable technological advances that benefit contemporary society in numerous ways, like all tools they can be used for evil as well as good. For those who put their self-centered desires above the good of others, and of society as a whole, they offer seemingly unlimited possibilities for criminal activity.

Defining and Measuring Cyber Crime

Before proceeding with our substantive discussion it is important to examine the definition, nature and scope of cyber crime. Clearly, digital technologies lie at the heart of cyber crime and these include computers, communication technologies and networked services. References to computers and digital technologies are used interchangeably.

There is, at present, a wide range of adjectives used to describe computer crime virtual, online, cyber, digital, high-tech, computer-related, Internet-related, telecommunications related, computer assisted, electronic, and 'e- crime'. In the same way that the term 'white-collar crime' sparked crucial discussion and controversy, these terms coined to delimit the scope of computer related misconduct are likely to be similarly problematic. 'Cyber crime' is used generically to describe a range of criminal offences, only some of which specifically relate to computers and the telecommunications infrastructure that supports their use. In this sense, it is similar to terms such as 'fraud', (such as 'obtaining financial advantage by deception'), all of which contain an element of dishonesty. Similarly, 'cyber crime', spelt as a single word in the titles of some recent pieces of legislation such as the Australian Cyber crime Act 2001 and the Council of Europe's Convention on Cyber crime, is a way of describing conduct that could entail a range of offences, many of which have nothing to do with computers in their legislative descriptions. Defining the term 'cyber crime' raises conceptual complexities. The term 'cyberspace' was first coined by William Gibson in his novel Neuromancer (1984) to describe a high-tech society in which people inhabit a virtual world divorced from terrestrial life. It has been used since then in a wide range of contexts to describe almost anything to do with computers, communications systems, the Internet, or, indeed, life in the twenty-first century. Chatterjee (2001, p. 81) reviews the many ways in which the term 'cyberspace' has been used, as well as the disparate other terms used to describe computer-related activities including those that infringe criminal laws.

Some have argued that virtual crime should be characterised as separate from and less serious than terrestrial crime, although Williams (2001, pp. 152–3) believes that 'the "real" and the "virtual" are not separate experiences and as such the nature of online communication enables a perpetrator to inflict recognisable levels of harm upon a victim via textual slurs and abuse'. The term 'cyber crime' encompasses any proscribed conduct perpetrated through the use of, or against, digital technologies, the pursuit or harassment of a victim by means of computers. This does not normally entail any new type of crime; the only new element is the means by which it is committed. Similarly, the theft of funds electronically is no different in terms of financial loss from the theft of currency from a bank, and the display of obscene images online (whether involving real human actors or images of people created electronically) involves the same affront to those who view the images as when they see them in a magazine. There may be differences in the extent and scale of the impact, but the effects of the acts themselves remain the same. Recently enacted laws targeting offences specifically related to computers, such as unauthorised access or modification of data, but the majority involves conventional crimes such as theft and other regulatory offences. The future will undoubtedly see new criminal laws enacted that have particular relevance to computers and new technologies.

Cyber crime is distinct from the traditional crime in that :

- The offender is not present at the crime scene
- The primary victim is most often an institution
- Offender characteristics and motive are heterogeneous
- Control involves global, technologically sophisticated, and political and indirect strategies.
- Cyber crime is harder to detect than traditional crime and as a result most cyber criminals are difficult to be caught, especially in transnational situations.
- What distinguishes cyber crime above and beyond other features is the intangible environment within which such offenses are committed which creates unlimited opportunities of offenders.

Scope of Cyber Crime

The involvement of computers in the commission of crime can extend from being clear and direct to being peripheral and of minor importance. The computer crime is 'an offence in which a computer network is directly and significantly instrumental in the commission of the crime. Computer interconnectivity is the essential characteristic.' Such a definition would exclude many instances involving non-networked computers and is overly restrictive. Clearly, there are certain crimes in which computing and communications technologies facilitate the commission of the offence but are not essential to its commission. Examples include the theft of funds by creating fictitious invoices on a company's computer, an offence which could just as easily be committed on paper. The opportunity to manipulate paper accounts might not, however, be immediately apparent to a potential offender, whereas the theft of funds electronically might seem more likely to be successful and not as easily detected. Hence there is a need to consider cases in which the use of a computer is of peripheral relevance.

A good example of the subtle differentiation between crimes in which computers are instrumental and crimes in which computers are incidental can be seen in instances of sexual abuse of children whom an offender has located online. For example, the offender met a number of young girls under the age of sixteen through an Internet chat room and, after winning their confidence, arranged to meet them in person. He then persuaded them to engage in various acts of indecency in return for money, cigarettes or alcohol which he provided. He also took obscene photographs of the girls, again in return for money and goods. Although such conduct could have been carried out without the use of the Internet – for example, following a chance meeting with a young person in any public place – it seems probable that chat rooms provided an easy and efficient way in which to meet potential victims and to cultivate their interest. The court in sentencing this offender impose a heavy penalty because of the physical acts perpetrated rather than because the contact had been initiated electronically. There are other cases in which computers are involved in the actus reus (physical

elements) of the crime as the object of the illegality, such as cases involving theft of computer hardware or components, but which have no other connection with digital technologies in terms of how the offence was perpetrated or the ultimate effect of the illegality. These cases may include theft of mobile telephones or hand-held personal organisers, or interference with automatic teller machines, vending machines or even modern motor vehicles which have computers on board. There are points of convergences and divergences between cyber crime and other forms of crime such as economic crime, white-collar crime, corporate crime, electronic commerce, copycat crime, intellectual property infringement, cross-border cyber crime and civil wrongs as discussed below.

1. Cyber Crime and Economic Crime

The interrelationship between cyber crime and economic crime raises difficult definitional questions. This is in part because most property crimes involving fraud or dishonesty that have been committed in recent years have involved the use of computers, simply because modern businesses rely so heavily on digital technologies for accounting purposes and for transfer of funds. Moreover, some (though not all) economic crime overlaps with 'white-collar' crime, distinguished by its relative sophistication and the background of its usual perpetrators. Cyber crime has connections with both white-collar crime and economic crime. Only a subset of cyber crimes has no economic component, that is, no financial benefit sought to be derived from the activity. Included are cases of cyber stalking or cyber vandalism in which computers are used to threaten or to harass. Many cases of hacking are also carried out for non-economic reasons. The dissemination of offensive content could be either financially motivated (for example, online businesses' distribution of child pornography for a fee) or not (for example, the publication of other offensive content such as racist material generally, with no fee demanded for access).

2. Cyber Crime and White-Collar Crime

The traditional definition of white-collar crime focused on crimes committed by persons of high status and social repute in the course of their occupation (Sutherland, 1940). Included in this definition were crimes committed by company officers, public servants, and professional people such as doctors and lawyers. The original emphasis was on economic crime, although over time, white-collar crime has come to include any acts of occupational deviance involving a breach of the law or ethical principles. As such, it has been suggested that white-collar crime now includes almost any form of illegality other than conventional street crimes (Freiberg 1992). Technological developments over the past decade have created further complexities surrounding the types of persons able to commit white-collar crime. The perpetrator of an online fraud, for example, might just as easily be a self-taught teenager using a personal computer at home as an educated professional in the workplace.

A simple categorisation distinguishes crimes committed by specified types of offenders (mainly professionals and individuals employed by corporations) from crimes perpetrated in specified ways (mainly economic crimes that involve sophistication, planning, or the use of technology in their commission). The essence of white-collar crime, however, remains rooted in abuse of power and breach of trust, usually involving the pursuit of financial gain as a motive. Clearly, not all white-collar crimes involve the use of digital technologies, although in recent times the vast majority have. Examples of those which do not include acts of violence committed in the workplace, such as sexual assault of patients by doctors, and some environmental. The category of fraud, or financial crimes of dishonesty, intersects with white-collar crime, economic crime, and cyber crime. Overlying the other concepts are the categories of property crime and corporate crime. Property crime is sometimes used synonymously with economic crime, although trespass, for example, or acts of vandalism would not be economic but nonetheless clearly property-related. Problems also arise in relation to the types of property protected by the criminal law. Notably, at least until recently, information usually has been regarded as being outside the scope of criminal prosecutions, dealt with instead by a range of intellectual property regimes such as copyright, patents, designs, and protection of confidential information.

3. Cyber Crime and Corporate Crime

Corporate crime can be defined in numerous ways. Examples of corporate crimes and other forms of corporate illegality include infringements of corporations laws, taxation and revenue offences, noncompliance with occupational health and safety laws and anti-discrimination legislation, breach of environmental protection laws, consumer protection offences relating to deceptive practices and the sale of dangerous or unhealthy products, infringement of trade practices and competition legislation, intellectual property piracy, bribery and corrupt practices in dealing with government agencies, and various economic offences concerning employees such as breaches of industrial awards and non-payment of wages and superannuation (Grabosky 1984). Clearly some of these activities would fall within the definition of cyber crime as conduct that involves the use of computers, although to date corporations have rarely been prosecuted for such activities, although their officers have.

4. Cyber Crime and Electronic Commerce

Electronic commerce encompasses a wide range of activities but essentially involves the use of computing and communications technologies to advertise, trade in and pay for goods and services. Various technologies can be used for electronic commerce including electronic mail, facsimile transfers, and a variety of web-based systems for the sharing and exchange of information. Acts of dishonesty, deception and misrepresentation relating to any of these technologies are included within the scope of electronic commerce.

5. Cyber Crime and Copycat crime

A copycat crime is a criminal act that is modelled or inspired by a previous crime that has been reported in the media or described in fiction. It has been shown that most of the persons who mimic crimes seen in the media (especially news and violent movies) have in most cases prior criminal records, prior severe mental health problems or histories of violence suggesting that the effect of the media is indirect (more affecting criminal behaviour) rather than direct (directly affecting the number of criminals) (Surette, 2002: 46-69). Technology breeds false familiarity, blurs fantasy and reality, and provides a virtual realm that mediates conscience. This has important implications for the study of criminal behaviour. Media and computer technology have changed the nature of social life in such profound ways that no behaviour is immune to its influence.

Early references to the copycat phenomenon appeared in the 1800s involving behaviours thought to be inspired by books. Sociologists in the 1970s examined the copycat phenomenon with respect to suicide suggesting that the suicide rate increases with the level of media coverage of suicide committed by a famous person. The criminological literature has been surprisingly silent on the subject of copycat crime in recent years. Copycat crime and cyber-crime are likely to become a significant part of the crime landscape in the 21st century. Copycat crime and cyber crime are subtypes that can cut across all of the major crime categories while maintaining distinct features. In some respects, copycat and cyber-crime represent more the process by which criminal behaviour occurs rather than a type of crime. Both copycat and cyber-crime can be violent, sex, economic, public order, or political crimes. Copycat and cyber crime are unique in that technology shapes their nature and presentation.

6. Cyber Crime and Intellectual Property Infringement

Infringement of intellectual property laws can entail both civil and criminal consequences, and in the majority of cases computers are the primary means by which infringements take place. Infringements include crimes committed through the use of computing and telecommunications technologies (such as unauthorised copying of digital information using computers and unlawful dissemination of data such as trade secrets, trade marks and logos using e-mail attachments, or the use of computers to fabricate documents); crimes directed at computing and communications technologies (such as direct theft of software or infringement of patents, designs, trade marks or circuit layouts relating to computer hardware); and crimes in which computing and communications technologies are incidental, but nonetheless involved in the commission of crime (such as the use of encryption to prevent the disclosure of communications between offenders, the use of steganography to conceal the transmission of infringing material, or the use of databases to record the movement of stolen data). As telecommunications systems are invariably controlled by computers, theft of services provides a clear example of crime directed at computers as the targets of illegality.

7. Cross-Border Cyber Crime

In addition to involving computers, the other feature of cyber crime is that it often takes place across jurisdictional boundaries. Cyberspace has been characterised as divorced from terrestrial life, existing in digital data streams that are transmitted via telephone lines and satellites. Often criminal conduct is committed remotely with offenders, victims, and others implicated in the offences located in different countries. Examples include various frauds committed by Russian criminals against banks and other businesses in the West, or the online stalking of a victim on the other side of the globe. In another recent example, offenders in Romania allegedly used e-mail messages to attempt to extort funds by threatening to sell hacked details of the computerised control systems of an Antarctic research facility (FBI 2003). The use of the Internet and other networked services clearly makes such activities possible and creates jurisdictional questions about the most appropriate place in which legal proceedings should be brought.

8. Cyber Crime and Civil Wrongs

In delimiting the scope of cyber crime, it is also important to distinguish between incidents that might result in criminal proceedings and those for which civil action for compensation may be taken.

Where both criminal and civil remedies are available, there may be uncertainty as to the most appropriate legal response, and in some cases, argument over whether the public authorities or the affected private interest should bear the cost of legal proceedings. A particularly complex area is the infringement of intellectual property rights, such as copyright and trade marks, where the same conduct can have both civil and criminal consequences.

Choosing the appropriate response raises difficult questions for victims, police, and prosecutors alike. Criminal proceedings have been brought in a number of cases involving copy- right infringement, particularly where software has been made available via the Internet. A recent example concerns the prosecution of a 21-year-old man in the United States for making 142 software programs available from his website. On 30 January 2001, he was sentenced to three years' probation, including 180 days of home confinement and a tether, as well as 40 hours of community service. He was ordered to make restitution to the software manufacturers, and prohibited from engaging in Internet activity without approval of the Probation Department. (United States Department of Justice 2001). Where computers are involved in the commission of acts infringing intellectual property laws or other civil matters and criminal proceedings have been brought, scholars include these within the scope of our discussion of cyber crime.

Theories of Classification of Cyber Crime

There are three categories of views with regard to classification of cyber crime. Within the first category are cases involving dissemination of offensive material electronically, online fraud and financial crime, electronic manipulation of share markets, and the dissemination of misleading advertising information, to name but a few. Also included are traditional crimes such as fraud or deception in which the involvement of computers constitutes a statutory aggravating element.

The second category include unauthorised access to computers and computer networks (so-called 'hacking' or 'cracking'), crimes involving vandalism and invasion of personal space, such as cyber stalking and denial of service attacks, and theft of telecommunications and Internet services.

The third category involves conduct that has been described as 'computer-supported crime' (Kowalski 2002, p. 6). This includes the use of encryption (the translation of data into secret code) or steganography (in which information is embedded within other, seemingly harmless data such as pictures) to conceal communications or information from law enforcement. It also includes the use of electronic databases to store and to organise information concerning proposed or completed criminal activities.

The issues raised for investigators therefore generally involve access to evidence rather than specifically proscribed conduct. These are just some examples, and the full range of potential conduct is limited only by the extent of one's criminal imagination.

Types of Cyber crime

Cyber crime ranges from the fundamental invasion of privacy, both personal and private, to crimes where the perpetrator hides in the anonymity of the internet, to attacks on the internet or computer network itself.

1. Child Pornography : The issue of child pornography has become more important due to its widespread distribution and availability on the internet. Child pornography is widely seen as a form

of child abuse and its production, distribution, and possession is outlawed in most countries. Prohibition may include images or videos of children engaged in sexual acts or of nude children. Many American states have made efforts to pass legislation outlawing its transmission via the internet.

Of the 184 members of The International Criminal Police Organization (Interpol), 95 do not have comprehensive laws covering child pornography. This can be seen as simply a lack of concern in those countries or also as a reflection of the varying norms of sexuality around the world, where in some countries child brides are still seen as ideal. The lack of laws against child pornography in such countries has led to the use of internet hosts there for such material, safe from prosecution.

2. Computer Viruses and other Malware : Malware is software designed to infiltrate or damage a computer system without the owner's informed consent. It is a portmanteau of the words "malicious" and "software." The expression is a general term used by computer professionals to mean a variety of forms of hostile, intrusive, or annoying software or program code. The more specific term "computer virus" is however used in common parlance, and often in the media, to describe all kinds of malware. Software is considered malware based on the perceived intent of the creator rather than any particular features. It includes computer viruses, worms, trojan horses, spyware, dishonest adware, and other malicious and unwanted software. In law, malware may also be referred to as a "computer contaminant."

Many early infectious programs, including the internet worm and a number of MS-DOS viruses, were written as experiments or pranks generally intended to be harmless or merely annoying rather than to cause serious damage. Since the rise of widespread broadband internet access, more malicious software has been designed for a profit motive. For instance, since 2003, the majority of widespread viruses and worms have been designed to take control of users' computers for black market exploitation. Infected "zombie computers" are used to send e-mail spam, to host contraband data such as child pornography, or to engage in distributed denial-of-service attacks as a form of extortion. [Daniel Tynan, "Zombie PCs: Silent, Growing Threat," PC World (July 9, 2004)]. Another strictly for-profit category of malware has emerged in spyware—programs designed to monitor users' web browsing, display unsolicited advertisements, or redirect affiliate marketing revenues to the spyware creator. Spyware programs do not spread like viruses; they are generally installed by exploiting security holes or are packaged with user-installed software.

3. Denial of Service Attacks : A denial-of-service attack (DoS attack) is an attempt to make a computer resource unavailable to its intended users. Although the means to, motives for, and targets of a DoS attack may vary, it generally comprises the concerted, malevolent efforts of a person or persons to prevent an Internet site or service from functioning efficiently or at all, temporarily or indefinitely. Perpetrators of DoS attacks typically—but not exclusively—target sites or services hosted on high-profile web servers. Denial of service attacks are one form of computer sabotage whereby people can effectively ruin their target's operations for what could be a lengthy period of time.

4. File Sharing : File sharing is the practice of making files available for other users to download over the internet and smaller networks. The duplication of copyrighted material, such as music and movies, is well understood as illegal. However, the status of sharing computer files containing such data has been less clear. The Recording Industry Association of America (RIAA) was able to force Napster to close its music file sharing operation based on the fact that the shared files were stored on a central Napster computer. Subsequently, Napster and others have set up subscription based legal file sharing for music files.

More controversial file sharing follows the peer-to-peer (P2P) model, where the files are stored on and served by personal computers of the users. Most people who engage in file sharing are also downloading files that other users share. Sometimes these two activities are linked together. P2P file sharing is distinct from file trading in that downloading files from a P2P network does not require uploading, although some networks either provide incentives for uploading such as credits or force the sharing of files being currently downloaded.

Under U.S. law, "the Betamax decision" (Sony Corp. of America v. Universal City Studios, Inc.), case holds that copying "technologies" are not "inherently" illegal, if substantial non-infringing use can be made of them. This decision, predating the widespread use of the internet applies to most data networks, including peer-to-peer networks, since distribution of correctly licensed files can be performed. These non-infringing uses include sending open source software, public domain files, and out of copyright works. Other jurisdictions tend to view the situation in somewhat similar ways.

In practice, many of the files shared on peer-to-peer networks are copies of copyrighted popular music and movies. Sharing of these copies among strangers is illegal in most jurisdictions. This has led many observers, including most media companies and some peer-to-peer critics, to conclude that the networks themselves pose grave threats to the established distribution model. The research that attempts to measure actual monetary loss has been somewhat equivocal. Whilst on paper the existence of these networks results in large losses, the actual income does not seem to have changed much since these networks started up. Whether the threat is real or not, both the Recording Industry Association of America (RIAA) and the Motion Picture Association of America (MPAA) have spent large amounts of money attempting to lobby lawmakers for the creation of new laws, and some copyright owners pay companies to help legally challenge users engaging in illegal sharing of their material.

Countries outside of the United States have served as havens for the hosting of file sharing software as the U.S. copyright laws are often difficult to enforce in these countries. [Thomas Mennecke, "Supreme Court Rules Against P2P Companies," Slyck News (2005)]. For example, the program Kazaa is owned by the Australian company Sharman Holdings, incorporated in Vanuatu, and developed by two Dutch software engineers. The online index of bit torrents The Pirate Bay is hosted in Sweden with backup servers in Russia.

5. Hacking : A "hacker" is someone who gains illegal access to computers. This action can be done maliciously or with the intent of pointing out possible security risks. Microsoft Corporation, the American multi-national computer technology corporation, and the Department of Defense are among those large organizations that have been the target of hackers. Hacking poses a risk to society as hackers can gain access to sensitive financial, personal, or security information that can then be used for extortion or exploited for an attack. Hacking can be understood as the cyber-version of industrial espionage.

6. Identity Theft : The term "identity theft" encompasses a broad range of identification-based crimes. It can be sub-divided into four categories : financial identity theft (using another's name and

Social Security number to obtain goods and services), criminal identity theft (posing as another when apprehended for a crime), identity cloning (using another's information to assume his or her identity in daily life), and business/commercial identity theft (using another's business name to obtain credit). There are a number of methods used by identity thieves including stealing mail or rummaging through trash, eavesdropping on public transactions to obtain personal data, stealing personal information in computer databases (Trojan horses, hacking), infiltration of organizations that store large amounts of personal information, impersonating a trusted organization in an electronic communication ("phishing"), and the use of spam. The unlawful acquisition of legally attributed personal identifiers is made possible by serious breaches of privacy. For consumers it is usually due to personal naiveté in who they provide their information to or carelessness in protecting their information from theft (such as vehicle break-ins and home invasions).

7. Internet Fraud : The term "internet fraud" generally refers to any type of fraud scheme that uses one or more online services—such as chat rooms, e-mail, message boards, or websites—to present fraudulent solicitations to prospective victims, to conduct fraudulent transactions, or to transmit the proceeds of fraud to financial institutions or to others connected with the scheme. The Federal Bureau of Investigation and police agencies worldwide have people assigned to combat this type of fraud; according to figures from the FBI, U.S. companies' losses due to internet fraud in 2003 surpassed US$500 million.

Internet fraud is committed in several ways. In some cases, fictitious merchants advertise goods for very low prices and never deliver. However, that type of fraud is minuscule compared to criminals using stolen credit card information to buy goods and services. Some common types of internet fraud include identity theft, purchase scams, counterfeit money orders, "phishing" for sensitive information, and "clickfraud" whereby false hits are generated for websites in order to generate advertising money.

Nigerian penal code includes they would be thieves e-mail people claiming to have a large sum of money they need to wire out of the country for some reason (large tax burden, political persecution, civil war, and so forth). They request for their victims to send them the fees for transferring this fortune to the victim's account in exchange for a portion of the wealth. The victim, of course, never receives the money transfer and loses the fee money they have already sent. This type of scam originated with schemes in the early 1980s in Nigeria, hence the alternate name of Nigerian Money Transfer Fraud.

8. Spam : "Spamming" is the abuse of electronic messaging systems to send unsolicited bulk messages, which are generally undesirable. While the most widely recognized form of spam is e-mail spam, the term is applied to similar abuses in other media: instant messaging spam, Usenet newsgroup spam, Web search engine spam, spam in blogs, mobile phone messaging spam, internet forum spam, and junk fax transmissions. Spam is also called junk mail. Some e-mail clients or servers have spam filters, which try to delete or move the spam messages.

Spamming is economically viable because advertisers have no operating costs beyond the management of their mailing lists, and it is difficult to hold senders accountable for their mass mailings. Because the barrier to entry is so low, spammers are numerous, and the volume of unsolicited mail has become very high. The costs, such as lost productivity and fraud, are borne by the public and by Internet service providers, which have been forced to add extra capacity to cope with the deluge. Spamming is widely reviled, and has been the subject of legislation in many jurisdictions.

E-mail and other forms of spamming have been used for purposes other than advertisements. Many early Usenet spams were religious or political. A number of evangelists have spammed Usenet and e-mail media with preaching messages. A growing number of criminals are also using spam to perpetrate various sorts of fraud (see "internet fraud" above) and in some cases have used it to lure people to locations where they have been kidnapped, held for ransom, and even murdered. E-mail worms or viruses may be spammed to set up an initial pool of infected machines, which resend the virus to other machines in a spam-like manner. The infected machines can often be used as remote-controlled zombie computers, for more spamming, to phish for bank account details, or other attacks.

Spam's direct effects include the consumption of computer and network resources and the cost in human time and attention of dismissing unwanted messages. In addition, spam has costs stemming from the kinds of spam messages sent, from the ways spammers send them, and from the "arms race" between spammers and those who try to stop or control spam. The European Union's Internal Market Commission estimated in 2001 that "junk e-mail" cost internet users •10 billion per year worldwide. In addition to direct costs are the indirect costs borne by the victims—both those related to the spamming itself, and to other crimes that usually accompany it, such as financial theft, identity theft, data and intellectual property theft, virus and other malware infection, child pornography, fraud, and deceptive marketing.

Cyber Crime Trends

A 2009 report by Gartner Research states that financial fraud was up by 47% in 2008 from 2007, with 687 data breaches reported. What does that translate to in dollars ? No one seems to know, although Chris Hoofnagle, Senior Fellow with the Berkeley Center for Law and Technology, says in an article that he wrote for the Fall 2007 issue of the *Harvard Journal of Law and Technology* that it's probably in the tens of billions : "Currently we don't know the scope of the problem.... We do know that it is a big market for identity theft prevention. In January 2009, Heartland Payment problem and that the losses are estimated in the tens of billions. Without reporting, we cannot tell whether the market is addressing the problem. Reporting will elucidate the scope of the problem and its trends, and as explained below, create a real Systems revealed that it was the victim of the largest data breach in history, involving more than 130 million accounts. No one really knows for sure because hackers had five months of uninterrupted access to Heartland's secure network before the breach was discovered."

Organized crime syndicates from Russia, Japan, Hong Kong, and the U.S. are consolidating their influence in the underground world of cyber crime because the risk-reward ratio is so good. Although law enforcement agencies are making sustained progress in cyber crime detection and enforcement—such as Operation DarkMarket, an FBI sting that resulted in the arrest of 56 individuals worldwide, more than $70 million in potential economic loss prevented, and recovery of 100,000 compromised credit cards—cyberspace is still a crime syndicate's dream environment for making a lot of money with little to no risk.

Indian Scenario (NCRB 2009 Report) :

Cyber Crimes are a new class of crimes rapidly increasing in India due to extensive use of Internet and I.T. enabled services. IT Act, 2000 was amended in 2009. Cyber-crime Offences & Relevant Sections under IT Act, 2000 are :

Tampering with Computer source documents	Sec.65
Hacking with Computer systems, Data alteration	Sec.66
Publishing obscene information	Sec.67
Un-authorized access to protected system	Sec.70
Breach of Confidentiality and Privacy	Sec.72
Publishing false digital signature certificates	Sec.73

Section 65 relates to source code. This is the most important asset of software companies. "Computer Source Code" means the listing of programmes, computer commands, design and layout. Its ingredients are knowledge or intention, concealment, destruction and alteration. Computer source code is required to be kept or maintained by law. The offense is cognizable and non-bailable. Punishment includes imprisonment up to three years, and / or fine up to ₹ 2 lakh.

Section 66 relates to hacking. Its ingredients are intention or knowledge to cause wrongful loss or damage to the public or any person, destruction, deletion, alteration, diminishing value or utility or injuriously affecting information residing in a computer resource. The offense is cognizable and non-bailable. Punishment includes imprisonment up to three years, and / or fine up to ₹2 lakh. Section 66 also covers data theft as well as data alteration.

Sec. 67 is about Pornography. Ingredients of pornography are publishing or transmitting or causing to be published in the and electronic form and obscene material. Sending pornographic or obscene e-mails are punishable under Section 67 of the IT Act. An offence under this section is punishable on first conviction with imprisonment for a term,which may extend to five years and with fine, which may extend to One lakh rupees. In the event of a second or subsequent conviction the recommended punishment is imprisonment for a term, which may extend to ten years and also with fine which may extend to Two lakh rupees. E-mails that are defamatory in nature are punishable under Section 500 of the Indian Penal Code (IPC), which recommends an imprisonment of upto two years or a fine or both. Threatening e-mails are punishable under the provisions of the IPC pertaining to criminal intimidation, insult and annoyance (Chapter XXII), extortion (Chapter XVII). E-mail spoofing is covered under provisions of the IPC relating to fraud, cheating by personation (Chapter XVII) and forgery (Chapter XVIII). The offense is cognizable and non-bailable.

The collection of data for 2009 has been done under pre-amended IT Act. Hence data under certain clauses introduced after amendment is not available in this Chapter. The Information Technology (IT) Act, 2000, specifies the acts which are punishable. Since the primary objective of this Act is to create an enabling environment for commercial use of I.T., certain omissions and commissions of criminals while using computers have not been included. Several offences having bearing on cyber-arena are also registered under the appropriate sections of the IPC with the legal recognition of Electronic Records and the amendments made in several sections of the IPC vide the IT Act, 2000. The statistics on Cyber Crimes are collected under the following heads :

(i) Offences registered under the Information Technology Act 2000.

(ii) Offences under the IPC (with use of Computers)

Cyber Crimes – Cases of Various Categories under IT Act, 2000, 420 cases were registered under IT Act during the year 2009 as compared to 288 cases during the previous year (2008) thereby reporting an increase of 45.8% in 2009 over 2008. 23.1% cases (97 out of 420 cases) were reported from Karnataka followed by Kerala (64), Maharashtra (53), Andhra Pradesh (30) and Punjab (28).

33.1% (139 cases) of the total 420 cases registered under IT Act 2000 were related to obscene publication / transmission in electronic form, normally known as cyber pornography. 141 persons were arrested for committing such offences during 2009. There were 233 cases of hacking with Computer Systems during the year wherein 107 persons were arrested. Out of the total (233) Hacking cases, the cases relating to Loss / Damage of computer resource/utility under Sec 66(1) of the IT Act were 49.4% (115 cases) whereas the cases related to Hacking under Section 66(2) of IT Act were 50.6% (118 cases). Maharashtra (25), Andhra Pradesh (21) and Kerala (15) registered maximum cases under Sec 66(1) of the IT Act out of total 115 such cases at the National level. The criteria and approaches currently vary across member states. Out of the total 118 cases relating to Hacking under Sec. 66(2), most of the cases (91 cases) were reported from Karnataka followed by Tamil Nadu (8) and Madhya Pradesh (6). 27.1% of the 288 persons arrested in cases relating to IT Act, 2000 were from Maharashtra (78) followed by Kerala (47). The age-wise profile of persons arrested in Cyber Crime cases under IT Act, 2000 showed that 64.6% of the offenders were in the age group 18 – 30 years (186 out of 288)and 28.8% of the offenders were in the age group 30 - 45 years (83 out of 288).

Maharashtra (6) and Kerala (4) reported offenders whose age was below 18 years. Crime head-wise and age group wise profile of the offenders arrested under IT Act, 2000 reveals that 49.0% (141 out of 288) of the offenders arrested were under 'Obscene publication /transmission in electronic form' of which 68.1% (96 out of 141) were in the age-group 18 –30 years. 57.9% (62 out of 107) of the total persons arrested for 'Hacking with Computer Systems' were in the age-group of 18 - 30 years.

Incidents of Cyber Crimes registered under IPC

TABLE 17.1

Computer Related Crimes under IPC and Special Laws

1.	Sending threatening messages by e-mail	Sec-503 IPC
2.	Sending defamatory messages by e-mail	Sec-499, 500 IPC
3.	Forgery of electronic records	Sec-463, 470, 471 IPC
4.	Bogus websites, cyber frauds	Sec-420 IPC
5.	E-mail spoofing	Sec-416, 417, 463 IPC
6.	Online sale of Drugs	NDPS Act
7.	Web - Jacking	Sec. 383 IPC
8.	Online sale of Arms	Arms Act

Some more offences dealt with under IPC

- Criminal breach of trust/Fraud- Sec. 405,406,408,409 IPC
- Destruction of electronic evidence-Sec.204,477 IPC
- False electronic evidence-Sec.193 IPC
- Offences by or against public servant-Sec.167,172,173,175 IPC

A total of 276 cases were registered under IPC Sections during the year 2009 as compared to 176 such cases during 2008 thereby reporting an increase of 56.8%. Maharashtra reported maximum number of such cases (108 out of 276 cases or 39.1%) followed by Chhattisgarh 16.7% (46 cases) and Punjab 10.1% (28 cases). Majority of the crimes out of total 276 cases registered under IPC fall under 2 categories viz. Forgery (158) and Criminal Breach of Trust or Fraud (90). Although such offences fall under the traditional IPC crimes, these cases had the cyber overtones wherein computer, Internet or its enabled services were present in the crime and hence they were categorised as Cyber Crimes under IPC. The Cyber Forgery (158 cases) accounted for 0.21% out of the 72,718 cases reported under Cheating. The Cyber Frauds (90) accounted for 0.55% of the total Criminal Breach of Trust cases under IPC (16,326).

The Crime head and State /UT-wise analysis of Cyber Crimes under IPC are very revealing. The Cyber Forgery cases were the highest in Maharashtra (67) followed by Chhattisgarh (32) and Gujarat (13). The cases of Cyber Fraud were highest in Maharashtra (30) followed by Punjab (19) and Gujarat & Tamil Nadu (11 each). A total of 263 persons were arrested in the country for Cyber Crimes under IPC during 2009. 61.2% offenders (161) of these were taken into custody for offences under 'Cyber Forgery', 30.0% (79) for 'Criminal Breach of Trust/Fraud' and 8.7% (23) for 'Counterfeiting'.

The States such as Maharashtra (89), Punjab (48), and Chhattisgarh (44) have reported higher arrests for Cyber Crimes registered under IPC. The age group-wise profile of the arrested persons under this category showed that 45.2% (119 out of 263) were in the age-group of 30 - 45 years and 41.8% (110 out of 263) of the offenders were in the age-group of 18-30 years. No offenders were below 18 years of age. Crime head-wise and agewise profile of the offenders arrested under Cyber Crimes (IPC) for the year 2009 reveals that offenders involved in Forgery cases were more in the age-group of 18 -30 (47.2%) (76 out of 161). 49.4% of the persons arrested under Criminal Breach of Trust / Cyber Fraud offences were in the age group 30-45 years (39 out of 79).

Incidence of Cyber Crimes in Cities 14 out of 35 mega cities did not report any case of Cyber Crime i.e, neither under the IT Act nor under IPC Sections during the year 2009. 20 mega cities have reported 178 cases under IT Act and 14 mega cities reported 168 cases under various section of IPC. There was an increase of 23.6% (from 144 cases in 2008 to 168 cases in 2009) in cases under IT Act as compared to previous year (2008), and an increase of 300.0% (from 42 cases in 2008 to 168 cases in 2009) of cases registered under various sections of IPC. Bengaluru (97), Ahmedabad (10), Bhopal, Coimbatore and Kochi (6 each) and Delhi City, Indore, Ludhiana and Pune (5 each) have reported high incidence of cases (145 out of 178 cases) registered under IT Act, accounting for more than half of the cases (81.5%) reported under the IT Act. Nasik has reported the highest incidence (68 out of 168 cases) of cases reported under IPC sections accounting for 40.5% followed by Mumbai (35 or 20.8%).

Tamil Nadu Landmark Cyber Case in record time

The Case of The State of Tamil Nadu vs Suhas Katti is notable for the fact that the conviction was achieved successfully within a relatively quick time of 7 months from the filing of the FIR. Considering that similar cases have been pending in other states for a much longer time, the efficient handling of the case which happened to be the first case of the Chennai Cyber Crime Cell going to trial deserves a special mention.

Mr S.Balu, the Assistant Commissioner of Police in charge of the Chennai's Cyber Crime Cell at the Commissioner's office acknowledged that the assistance provided by Naavi and the Cyber Evidence Archival Center was helpful in the speedy resolution of the case. The case was also notable since it silenced several critics about the inability of the Police in general to be capable of Cyber Crime investigation and more so production of satisfactory evidence to prove the case in a court of law. This case will therefore be considered as a land mark case in the history of Cyber Crime Management in India. The case related to posting of obscene, defamatory and annoying message about a divorcee woman in the yahoo message group. E-mails were also forwarded to the victim for information by the accused through a false e-mail account opened by him in the name of the victim. The posting of the message resulted in annoying phone calls to the lady in the belief that she was soliciting.

Based on a complaint made by the victim in February 2004, the Police traced the accused to Mumbai and arrested him within the next few days. The accused was a known family friend of the victim and was reportedly interested in marrying her. She however married another person. This marriage later ended in divorce and the accused started contacting her once again. On her reluctance to marry him, the accused took up the harassment through the Internet. On 24-3-2004 Charge Sheet was filed u/s 67 of IT Act 2000, 469 and 509 IPC before The Hon'ble Addl. CMM Egmore by citing 18 witnesses and 34 documents and material objects. The same was taken on file in C.C.NO.4680/2004. On the prosecution side 12 witnesses were examined and entire documents were marked.

The Defence argued that the offending mails would have been given either by ex-husband of the complainant or the complainant her self to implicate the accused as accused alleged to have turned down the request of the complainant to marry her. Further the Defence counsel argued that some of the documentary evidence was not sustainable under Section 65 B of the Indian Evidence Act. However, the court based on the expert witness of Naavi and other evidence produced including the witness of the Cyber Cafe owners came to the conclusion that the crime was conclusively proved.

The court has also held that because of the meticulous investigation carried on by the IO, the origination of the obscene message was traced out and the real culprit has been brought before the court of law. In this case Sri S. Kothandaraman, Special Public Prosecutor appointed by the Government conducted the case. Honourable Sri Arulraj, Additional Chief Metropolitan Magistrate, Egmore, delivered the judgement on 5-11-04 as follows :

"The accused is found guilty of offences under section 469, 509 IPC and 67 of IT Act 2000 and the accused is convicted and is sentenced for the offence to undergo RI for 2 years under 469 IPC and to pay fine of ₹ 500/-and for the offence u/s 509 IPC sentenced to undergo 1 year Simple imprisonment and to pay fine of ₹ 500/- and for the offence u/s 67 of IT Act 2000 to undergo RI for 2 years and to pay fine of ₹ 4000/- All sentences to run concurrently."

The accused paid fine amount and he was lodged at Central Prison, Chennai. This is considered the first case convicted under section 67 of Information Technology Act 2000 in India.

Policy Implications

Predictions are done about future threats. A recent report by the European Commission predicts: There is a 10% to 20% probability that telecom networks will be hit by a major breakdown in the next 10 years, with a potential global economic cost of around •193 billion ($250 billion). This could be caused by natural disasters, hardware failures, rupture of submarine cables (there were 50 incidents recorded in the Atlantic Ocean in 2007 alone), as well as from human actions such as terrorism or cyber attacks, which are becoming more and more sophisticated. The commission goes on to recommend an increased focus in key areas to counter future threats in cyberspace. These include :

Preparedness and prevention : Fostering cooperation of information and transfer of good policy practices between member states via a global forum, which will help businesses share experience and information with public authorities.

Detection and response : Supporting the development of a global information-sharing and alert system.

Mitigation and recovery : Stimulating stronger cooperation between member states via national and multinational contingency plans and regular exercises for large-scale network security incident response and disaster recovery.

International cooperation : Driving a world-wide debate to set priorities for the long-term resilience and stability of the Internet with a view to proposing principles and guidelines to be promoted internationally and establishing criteria for global critical infrastructure in the Information and Communication Technologies (ICT) sector.

Our examination of the concept and dimensions of cyber criminals has identified many fruitful avenues for policy reform. Although not all countries will agree on the most appropriate ways to respond to crimes of this nature, there are some issues that require urgent investigation, and on these all countries should, arguably, agree.

First, and of greatest importance, is the need for the process of legislative harmonisation to continue. There have been some important advances made already, such as the creation of the Council of Europe's Convention on Cyber crime and the United Nations Convention against Transnational Organised Crime. Individual countries need to continue their efforts to reform their own cyber crime legislation to ensure uniformity. This will help to facilitate the process of extradition as well as reducing inconsistencies in sentencing across countries. More importantly, the continuing process of legislative reform will help to ensure that unfortunate incidents, such as the failure of the prosecution in the 'Love Bug' virus case, will not be repeated. Sufficient numbers of cases have now come before the courts that most prosecutors are aware of the many problems that can be encountered in court. This knowledge, arguably, needs to be used to ensure that prosecutors are appropriately trained and equipped to respond to some of the tactics that defendants raise to avoid responsibility for their actions. Prosecution policies may also need to be revisited to enable appropriate decisions concerning the acceptance of new cases for prosecution. Exactly when, for example, should a 15-year-old be prosecuted for having disseminated malicious code, and when is online copy- right infringement serious enough to warrant the substantial costs associated with a multi-jurisdictional prosecution ? Once again, organisations such as the International Association of Prosecutors could assist in devising effective policy reforms in this area and in achieving uniformity across jurisdictions. Everyone involved in court proceedings, including prosecutors, defence counsel, solicitors, and jurors, needs to understand the sometimes complicated technological evidence that might be adduced in cases. In one Case in US, for example, counsel for the accused, on appeal, criticised the trial judge for not understanding the facts of the case and for displaying confused reasoning.

Some of the lawyers are not being familiar with the papers and are unable to answer questions about the case adequately. To overcome such problems, a specialist Bar might need to be established, and training courses offered for judges and lawyers. Similarly, the question of specialist units within prosecution agencies should be revisited, or a decision made as to whether it might be more cost-effective for the prosecution of some cyber crime cases to be outsourced to specially trained counsel in the private sector, rather than trying to maintain levels of expertise in this highly specialised area within public sector prosecution agencies. This problem, of course, principally arises for those jurisdictions in which prosecutors function solely within the public sector.

Resources may also need to be allocated to ensure that courtrooms are equipped with appropriate technologies for the presentation of electronic evidence in such a way as to make it readily understandable to jurors. Finally, in the most complex cases, jury trial may need to be made optional, in favour of trial by a judge assisted by a panel of trained assessors. At present, the number of cases that would benefit from reforms of this kind probably does not warrant considerable financial outlay, although as time progresses and more and more cyber crime cases reach the courts, the level of need will become clear.

Clearly, this idea is very much a preliminary overview of the prosecution and sentencing of cyber criminals, and one would be remiss without the suggestion of the possible issues and topics that could be worthy of attention by those who would follow in the path. These are discussed below.

Improved Knowledge Base : Improved standards of information with respect to the prosecution and trial of cyber criminals must lie at the heart of any research agenda. Better and more extensive research needs to be carried out not only locally, but also internationally. Definitions of cyber crime need to be carefully framed and problems of accurate quantification addressed. In addition, the rapidly changing nature of cyber crime must be considered. Government statistical and census agencies should play a key role in conducting surveys of the population, as they are well placed to undertake research objectively. Public sector offices of crime statistics could, for example, ask some questions about cyber crime and security issues in their regular surveys on household use of information technologies. Official statistical data collection should also seek to explore more useful categorisation of cyber crime when collecting data from police, courts and correctional agencies. At present, conventional crime statistics imply do not enable any differentiation between cyber crimes and other types of offence. Evidence also needs to come from those within industry. Persuading ISPs, financial institutions and merchants to be frank in disclosing cyber crime experiences is by no means simple, but some appropriate sharing of information must take place, even if conditions of confidentiality are imposed.

With improved levels of official information and more accurate surveys of cyber crime, we may then be in a position to direct resources appropriately to combat the problem. Courts and prosecutors should be better able to allocate resources and governments should be able to dedicate realistic levels of funding that would enable cases of this nature to be actively pursued through the courts.

Keeping Pace with Technological Developments : Of primary interest is the continuing game of 'leap-frog' being played by, on the one hand, those who perpetrate cyber crime, and, on the other, those who seek to control it. The pace of technological change is such that law and policy are subject to continuous challenge. With the increasing adoption of digital technologies and the increasing prevalence of computer literacy in the community, what once may have constituted 'special skills' under the United States Sentencing Guidelines are now mundane. Home computer users can counterfeit documents and establish their own fraudulent websites with ease.

Similarly, we have seen how post-release conditions imposed on offenders are now regarded as unrealistic and, indeed, counterproductive. In Western industrial societies at least, digital technology now pervades the lives of all but the most reclusive neo-luddites. Governments have set targets for electronic service delivery, which mean that those in the community will, except in extraordinary cases, be obliged to carry out their dealings with government electronically. And banks are now making it expensive and slow to transact business at a local branch with a teller. For courts to ban an offender from using computers, or place severe restrictions on their use, may make life close to unliveable. Certainly, the potential for rehabilitation could well be lost if workplace computers could not be used. Indeed, arguably the act of telephoning one's parole officer or calling a police station when on bail could infringe a ban on computer usage if one were to make use of a cellular telephone. Further research is, therefore, needed on how offenders deal with such conditional orders. In addition, it would be worthwhile to conduct interviews with a sample of those who have been convicted of cyber crimes, to evaluate the 'rational choice' hypothesis. It could, for example, be possible to find out how many offenders knew the details of the offences they were committing and the associated punishments that were likely, and why they decided to commit their crimes nonetheless. We could also find out whether particular sanctions are perceived as having greater deterrent potential than others. Does, for example, the prospect of being banned from using the internet produce a realistic deterrent to someone intent on dealing in child pornography online ?

Further Investigation of Cases : While some of the cases to which we have referred were relatively mundane, others raise complex questions about how they were able to be committed and why individual courts imposed the sentences they did. In another Case, for example, which involved an online 'sting' operation, the partial success of the fantasy defence, and contemporary challenges to the constitutionality of child pornography legislation in the United States, is one such case.

Explaining Attrition : One of the more interesting aspects of cyber crime prosecution is the attrition of cases before they get to the steps of the courthouse. The data show that 75 per cent of cases referred for prosecution to federal authorities in the United States as on 2009 were declined, primarily due to lack of evidence. An examination of attrition rates in a particular jurisdiction over time, or a comparison of jurisdictions with high and low attrition rates, might shed some interesting light on prosecution policies.

Police–Prosecutor Interactions : Prosecutors in some jurisdictions around the world are actively involved in the planning and execution of investigations, while elsewhere they are relatively passive recipients of briefs prepared and presented by the police. More intensive research on police – prosecutor relations in these jurisdictions could provide insight on how the demands of the courtroom are communicated back to the 'upstream' drivers of the investigation.

Defence Tactics : In addition, further research could be undertaken on the ways in which defence counsel approach their task of defending persons accused of cyber crime. Mann's (1988) study of white-collar defence counsel provides a useful model. How trial counsel manage their clients and their adversaries – the prosecutors who are pitted against them – would make an interesting contrast with existing work, including our own, which tends to represent the point of view of the state.

Global Cooperative Action : Finally, despite the difficulties in investigating cross-national cyber crime, the annals of law enforcement contain celebrated examples of successful law enforcement cooperation across national frontiers. Operation Cathedral, the investigation of a global child pornography ring, resulted in simultaneous raids and over 100 arrests in Australia, Austria, Belgium, Finland, France, Germany, Italy, Norway, Portugal, Sweden and the United States in September 1998. Operation Buccaneer, an international software piracy investigation, involved the simultaneous execution on 11 December 2001 of more than 65 search warrants against alleged software pirates in the United States, the United Kingdom, Australia, Finland, Sweden and Norway. Operations such as these present unique opportunities for comparative analyses of prosecution and sentencing processes. In the end, cases that come before the courts in the twenty-first century are likely to demonstrate many aspects of life as those of us in developed nations currently know it : activities that are heavily dependent on digital technologies, activities that involve participants in multiple countries, and activities that take place with an emphasis on speed. Prosecutors and courts need to be equipped with training and resources to respond effectively to such cases which, it may confidently be predicted, will become a regular feature in daily court listings. In order to maximise the efficiency of judicial processes, and to guard against the problem of history repeating itself, prosecutors and judges need to be made aware of the latest cases as soon as they come to light. Technology, which itself is deeply involved in the commission of modern crimes, will provide the best means of achieving such an objective. A century ago, the philosopher George Santayana said : 'Those who cannot remember the past are condemned to repeat it.' In today's digital environment, it may be said of prosecutors and judges that those who fail to anticipate the future are in for a rude shock when it arrives.

An Overview

Technology, media, and popular culture shape offender motivation, modus operandi, and play a role in neutralizing guilt and providing justification for offenders' actions. It is important to consider technology as a potential risk factor for criminal behaviour for some individuals. Technological influences on criminal behaviour exist along a continuum. Copycat crime and cyber-crime as two distinct subtypes of criminal behaviour that involve technological influence. Technology will play an increasingly salient role in influencing offender motivation and modus operandi in a segment of offenses and it is important that criminological theory and research examine technology as a risk factor for criminal behaviour. Cyber-crime has become a particularly relevant problem to police around the world and to society at large. The growing presence of the internet and computers in homes around the world means that more people are exposed to cyber-crime each year. The rise in popularity of broadband internet access also means a greater risk. In an era where all data is being digitized and stored on computers, protecting computers is integral to personal and national security.

References

Chatterjee, B. B. (2001), 'Last of the Rainmacs: Thinking About Pornography in Cyberspace', in Wall, D. S. (ed.), Crime and the Internet, Routledge, London, pp. 74–99.

Dreyfus, Suelette (1997). Underground: Tales of Hacking, Madness and Obsession on the Electronic Frontier. Mandarin.

Federal Bureau of Investigation (FBI) (2003), 'The case of the hacked South Pole': <http://www.fbi.gov/page2/july03/071803backsp.htm>.

Freiberg, A. (1992), 'Sentencing White-Collar Criminals', in Australian Institute of Judicial Administration (AIJA), Sentencing of Federal Offenders, Proceedings of a seminar for judges and magistrates held 1–2 November 1991, AIJA, Melbourne, pp. 1–19.

Gibson, W. (1984), Neuromancer, Harper Collins, London.

Grabosky, P. N. (1984), 'Corporate Crime in Australia: An Agenda for Research', Australian and New Zealand Journal of Criminology, vol. 17, pp. 95–107.

Graham, Paul (2004). Hackers and Painters. Big Ideas from Computer Age, O' Reilly Media, New York.

Kowalski, M. (2002), Cyber Crime: Issues, Data Sources and Feasibility of Collecting Police Reported Statistics, Canadian Centre for Justice Statistics, Ottawa.

Levy, Steven (1984). Hackers: Heroes of the Computer Revolution. Doubleday.

Mann, K. (1988), Defending White-Collar Crime, Yale University Press, New Haven.

NCRB (2009) Crime in India Compendium, Ministry of Home Affairs, GOI, New Delhi pp. 170-180.

Slatalla, Michelle and Joshua Quittner (1995). Masters of Deception: The Gang That Ruled Cyberspace. HarperCollins.

Smith Russell G. et, al. (2004) Cyber Criminals on Trial, Cambridge University Press.

Sterling, Bruce (1992). The Hacker Crackdown. Bantam.

Surette, R. (2002). Self-Reported Copycat Crime Among a Population of Serious and Violent Juvenile Offenders. Crime & Delinquency, 48(1), 46-69.

Turkle, Sherry (1984). The Second Self: Computers and the Human Spirit, new ed. Cambridge, MA: MIT Press.

United States Department of Justice (USDOJ) (2001), 'Man sentenced in Michigan for offering software programs for free downloading on "Hacker Hurricane" website', press release, 30 January 2001: <http://www.cybercrime.gov/baltutatsent.htm>.

Verton, Dan (2002). The Hacker Diaries : Confessions of Teenage Hackers. McGraw-Hill Osborne Media.

Williams, M. (2001), 'The Language of Cyber-crime', in Wall, D. S. (ed.), Crime and the Internet, Routledge, London, pp. 152–66.

■■■

CHAPTER 18

Prostitution

Learning Objectives :

- ❑ **To understand the meaning, history and the causes of prostitution.**
- ❑ **To comprehend the types of prostitution.**
- ❑ **To discuss the consequent dangers of prostitution.**
- ❑ **To be acquainted with the IPTA 1956.**
- ❑ **To grasp the role of law enforcement in combating prostitution.**
- ❑ **To grasp the prevention strategies.**

Introduction

Prostitution has been an institution which has made women a commodity throughout the social evolution of humanity around the world. The practice has been reported in virtually every culture and described throughout recorded history. As a form of deviance, prostitution has been of interest to sociologists as a reflection of various social processes and phenomena. Sociologists have studied prostitution as a form of sexual deviance and a reflection of the basic values, norms, and institutions within a society. Sociological studies of prostitution have been concerned with the function it serves in a society, the gender inequality and double standard implicit in the practice, and the social dynamics involved in becoming a prostitute. Prostitution represents a form of sexual deviance in that it is a sexual practice which is widely viewed as socially undesirable and degrading. Colloquial terms used to refer to prostitutes, such as hooker, hustler, and whore, carry negative connotations.

An important issue is defining what constitutes prostitution. Most definitions stress the relatively indiscriminate exchange of sexual activity for economic gain. Prostitution is one of the branches of the sex industry. A prostitute is a person, "who allows her body to be used for lewd purposes in return for payment." For the prostitute, the practice represents a means of supplementing an income. A person who trades sexual activity for a job promotion would not be labeled a prostitute by most observers, although this situation includes some of the same elements as prostitution. What separates prostitution from the previous example is the repeated, indiscriminate nature of the exchange. More than an isolated deviant act, prostitution entails a deviant profession.

A "prostitute" is derived from a composition of two Latin words: *pro* and *statuere* implying "to expose", "to place up front". A variety of terms are used for prostitution, some of which distinguish between different kinds, or imply a value judgment about them. The Greek word for prostitute is *porne.* The English word pornography, and its corollaries in other languages, are directly derivative of the Greek word *porne.* In ancient Rome, a registered prostitute was called a *meretrix* while the unregistered one fell under the broad category *prostibulae.* Common alternatives for prostitute include escort and whore, although not all professional escorts are prostitutes. The word "prostitute", or more frequently "whore", is also used as an insult towards a person (typically a woman or girl) who is perceived as being sexually promiscuous. Those seeking to remove the social stigma associated with prostitution often promote terminology such as sex worker, commercial sex worker (CSW), or sex trade worker. A hooker or streetwalker solicits customers in public places; a call girl makes appointments by phone.

In the report prepared by the Advisory Committee, League of Nations, Geneva (1943:8) the following observations were made which indicate tenacity with which prostitution persists in civilised communities: "Prostitution has outlived every social, economic, ethical system which the West has known since the time of the Greek City States. It has had its vicissitudes; but, flourishing or languishing public or clandestine, it has existed in large towns for the last years, thereby proving how deeply it is rooted in human social life."

Correctly or not, prostitute without specifying a gender is commonly assumed to be female; compound terms such as male prostitution or male escort are therefore used to identify males. Those offering services to female customers are commonly known as gigolos; those offering services to male customers are hustlers or rent boys.

Organizers of prostitution are typically known as pimps (if male) and madams (if female). More formally, they practice procuring, and are procurers, or procuresses.

The customers of prostitutes are known as johns or tricks in North America and punters in the British Isles. These slang terms are used among both prostitutes and law enforcement for persons who solicit prostitutes. In some places, men who drive around red-light districts for the purpose of soliciting prostitutes are also known as kerb crawlers.

Prostitution is commercialised sex and, as such, it involves two parties- seller and buyer. It can be described as indiscriminate sexual activity without the normal motives i.e procreation or pleasure, but is economically motivated so far as the seller is concerned. The seller generally is a woman and a man the buyer; the reverse roles are, however, not unknown. As per section 2(f) of The Immoral Traffic (Prevention) Act, 1956 "prostitution" means the sexual exploitation or abuse of persons for commercial purposes or for consideration in money or in any other kind, and the expression "prostitute" shall be construed accordingly. Also there is, though much rarer, male prostitution in which gratification is by one male to another. In the discussion which follows, the subject is in the most usual context, the commercialised heterosexual practice. Thus the most common form of prostitution involves women who sell sexual services to heterosexual men. The second most common group comprises male homosexual prostitutes who cater to gay men. The consistent pattern in society and throughout history is for the customers of prostitutes to be men. Male prostitutes who make themselves available to women (termed as gigolos) are uncommon, and lesbian prostitutes are considered extremely rare.

History of Prostitution

Prostitution is historically and culturally ubiquitous, with every culture adopting their own standards and attitudes. In Armenia the noblest families dedicated their daughters to the service of the goddess *Anaitis* in her temple at *Acilisena*. Prostitution was a part of daily life in ancient Greece. In ancient Greek society, prostitution was engaged in by both women and boys. In the more important cities, and particularly the many ports, it employed a significant proportion of the population and represented one of the top levels of economic activity. There were some commonalities between the Greek system and Roman system; but as the Roman Empire grew, prostitutes were often foreign slaves, captured, purchased, or raised for that purpose, sometimes by large-scale "prostitute farmers" who took abandoned children. In the ancient city of Heliopolis in Syria, there was a law that stated that every maiden should prostitute herself to strangers at the temple of Astarte. What has been said about western societies is true of many others. India had a system of licensing prostitutes and taxing them even before the Romans developed their system of regulating prostitution by the same tools.

In ancient India, there was a practice of having *Nagarvadhus*, "brides of the town". Famous examples include *Amrapali*, state courtesan and Buddhist disciple, described in *Vaishali Ki Nagarvadhu* by Acharya Chatursen and Vasantasena, a character in the classic Sanskrit story of *Mricchakatika*, written in the 2nd century BC by Sudraka. The Devadasis, who performed in temples, were described as "temple prostitutes". Kanhopatra is venerated as a saint in the Varkari sect of Hinduism, despite spending most of her life as a courtesan. Binodini Dasi started her career as a courtesan, and later became a Bengali theatre actress.

In the Mahabharata period, courtesans were not merely the fringe dwellers of the society. On occasions, they were invited by people of high social status. The word *prakasha-Sarvagmya* denotes prostitutes. Regarding their garments it is stated that they should put on red clothes, red garlands, and use ornaments of red gold. The red colour is regarded as colour as the symbol of *Yama*. Probably this rule was framed to distinguish easily the prostitutes or for warning the people about the dangerous effect of sexual union with a prostitute; the red colour indicates death. In Mahabharata courtesans of *Indra Puri* have been variously depicted. Mahabharata has recorded the name of forty two apsaras in all.

Epic Ramayana mentions courtesans as well as drinking bars. Where adultery and other immoral practices were indulged in. In Gorthi women of Vadini class used to provide entertainment for kings and nobles. There was the custom of engaging strumpets for entertaining distinguished guests. In the impending coronation of Rama, such women participated. Bharata ordered courtesans to join the welcome festival on the occasion of the return of the triumphant Rama. *Yajnavalkya* enumerates as one of the four tokens by which the police could trace a criminal that of living in house of ill fame.

Patliputra was at time of Chandra Gupta Maurya a flourishing center of prostitution and it was the first time the attention of the state was drawn to the colony of prostitutes for its effective control and to bring it under the obligation of a stabilized taxation system. Brothel keeping was looked upon as a source of Government revenue. Kautilya's Arthasastra contains rules for prostitutes and their activities and gives an account of how prostitutes should behave and how their lives be ordered. The women of the count were not common prostitutes. They were highly accomplished courtesans and were considered fit to give company to the cultured men of society.

The function of prostitutes was not only entertainment, but their services were also used for political purposes, especially in espionage work. *Kamasutra*, Vatsayayana, written probably in the early centuries of the Christian era, is the most important source of information about courtesans and prostitution in ancient India. The sixth chapter of Kamasutra deals exclusively with courtesans. In Vatsayayana's classification the lowest type of prostitute was the *Kumbhadasi* or common harlots. They used to entertain any one who could afford their low price. A *kulta* was a married women who practices prostitution. A *paricharika* or attendant was the daughter of courtesan who went through a form of brothel marriage. *Ganika* was the highest in rank of prostitutes and was well known for her good manners and deportment. Kamasutra Vatsayayana has devoted a number of pages about prostitution and prostitutes. Rules of conduct for the popular and successful practice of their trade have been prescribed.

Puranas have laid down that the women attached with body and soul to one husband was called *Pativarta*. Wherever one who shared her couch with two paramours was called *Kulata*, to three was a *Darshini*, to four was Pungaschalea and to five was called *veshya*. The women who gave herself to people more than five in number was known as *mahaveshya*.

The Mughal Empire (1526 -1857) in India also witnessed prostitution the word "tawaif" and *mujra* became common during this era. During the Mughal era in the subcontinent (1526 to 1857) prostitution had a strong nexus with performing arts. Mughals patronized prostitution which raised the status of dancers and singers to higher levels of prostitution. King Jahangir's harem had 6,000 mistresses which denoted authority, wealth and power. A tawaif was a courtesan who catered to the nobility of South Asia, particularly during the era of the Mughal Empire. These courtesans would dance, sing, recite poetry and entertain their suitors at *mehfils*. Like the geisha tradition in Japan, their main purpose was to professionally entertain their guests, and while sex was often incidental, it was not assured contractually. High-class or the most popular tawaifs could often pick and choose between the best of their suitors. They contributed to music, dance, theatre, film, and the Urdu literary tradition.

In Mughal period prostitution was a recognized institution. The *nawabs* and nobles were not feeling any shame in keeping concubines and prostitutes in their houses. They did not scruple to trample on the sanctity of the harem and turned their harem in private brothel. Akbar, the illustrious Mughal emperor maintained a seraglio in which there were 5000 women and it had a supreme staff of women officers who looked after its management. Emperor Akbar made some regulations so that the services of prostitutes might not be available very easily to the public. The prostitutes were confined to a place outside the capital city . This place was known as Saitanpura or the devils quarter.

A large number of women were attached to the Mughal courts for dancing and singing. The Mughal kings were great patrons of fine arts of music and dance, which flourished during their regime.

Women even accompanied the Mughal army and moved with them from place to place for the entertainment of the army personnel.

In Goa, a Portuguese colony in India, during the late 16th and 17th centuries, there was a community of Japanese slaves, who were usually young Japanese women and girls brought or captured as sexual slaves by Portuguese traders and their South Asian lascar crewmembers from Japan.

Even during the British era, prostitution flourished the famous kamathipura, a red light area in Bombay was built during this era for the refreshment of British troops and which was later taken over by Indian sex workers. During the British East India Company's rule in India in the late 18th and early 19th centuries; it was initially fairly common for British soldiers to frequently visit local Indian nautch dancers. Likewise, Indian lascar seamen taken to the United Kingdom frequently visited the local British prostitutes there.

The entry of Europeans into India profoundly altered the socio-economic and political milieu. The presence of sailors in port towns now rendered the climate morally vulnerable. After their long and dull some voyage they wanted to have some relaxation during their halts in ports and they got the pleasure by attending dance parties and vesting places of interest. They also used to visit houses of prostitutes which were like honeycombs in areas adjoining sea ports.

Conditions continued to deteriorate and prostitution thrived on a large commercial scale. Almost all unmarried Europeans and even married one kept native concubines. Patronage to prostitution and singing and dancing girls were extended to them by the *Zamindars, talukedars* and *nawabs.*

In absence of state control and regulation, the vice of prostitution became rampant and British rulers felt the necessity to control and regulate it. Law dealing with immoral traffic was enacted by British rulers. The enactment of the Contagious Disease Act 1868 opened up a new epoch in the history of prostitution in India. Certain important factors such as the increase in incidence of Venereal diseases in the presidency capitals and for enacting rules for controlling the activities of prostitutes in the towns neighboring to cantonments were behind this legislation.

Important feature of this Act was that it sanctioned the establishment of brothels and allowed prostitution in a regularized form. Every brothel keepers and prostitute were provided with identity cards with details of registration, which was subjected to be produced on demand. Every prostitute was required to be periodically medically checked, that she was not infected with venereal diseases. This medical checking was having the aim to protect the customers from contracting sexually transmitted diseases from infected prostitutes.

To protect the health of prostitutes was not main aim of the medical checkup provided by Contagious Disease Act 1868. The Contagious Disease Act was abolished in 1888 A.D. It was in 1923 that the Suppression of Immoral Traffic Act (SITA) was passed for Culcutta, Madras and Bombay presidencies. In Uttar Pradesh the Nayak Girls Protection Act and Minor Girls Protection Act were passed in 1923. Attempt to regulate and control prostitution was started in real legislative sense only after passing of these acts. In the 19th and early 20th centuries, thousands or even millions of women and girls from continental Europe and Japan were trafficked into British India, where they worked as prostitutes serving British soldiers and local Indian men.

In the 19th century, legalized prostitution became a public controversy as France and then the United Kingdom passed the Contagious Diseases Acts, legislation mandating pelvic examinations for suspected prostitutes. This legislation applied not only to the United Kingdom and France, but also to their overseas colonies. A similar situation did in fact exist in the Russian Empire; prostitutes operating out of government-sanctioned brothels were given yellow internal passports signifying their status and were subjected to weekly physical exams. Leo Tolstoy's novel Resurrection describes legal prostitution in 19th-century Russia.

Originally, prostitution was widely legal in the United States. Prostitution was made illegal in almost all American states between 1910 and 1915 largely due to the influence of the Woman's Christian Temperance Union which was influential in the banning of drug use and was a major force in the prohibition of alcohol. In 1917, President Woodrow Wilson wanted all prostitution ended near any military and naval base as America prepared to enter World War I.

In 1956, the United Kingdom introduced the Sexual Offences Act 1956, which would partly be repealed, and altered, by the Sexual Offences Act 2003. While this law did not criminalise the act of prostitution itself, it did prohibit such activities as running a brothel and soliciting.

Beginning in the late 1980s, many states in the US increased the penalties for prostitution in cases where the prostitute is knowingly HIV-positive. These laws, often known as felony prostitution laws, require anyone arrested for prostitution to be tested for HIV, and if the test comes back positive, the suspect is then informed that any future arrest for prostitution will be a felony instead of a misdemeanor. Penalties for felony prostitution vary in the states that have such laws, with maximum sentences of typically 10 to 15 years in prison. An episode of COPS which aired in the early 1990s detailed the impact of HIV/AIDS among prostitutes; this episode is deemed as part of HIV/AIDS awareness.

Commercial sex exploitation and consequent AIDS epidemic raised alarm world over including India. The need for change in SITA was continuously felt and some significant changes were introduced by an amendment. A new name the Immoral Traffic (prevention) Act 1956 (ITPA), colloquially called PITA, was given to the amended legislation in India. Under the new Act the definition of prostitution has been widened to mean 'the sexual exploitation or abuse of persons for commercial purposes'. The word 'person' has replaced the expressions 'women' and 'girls' making it possible to include males as well in the definition of prostitution, the male customers of the prostitutes also being the victims of commercial exploitation. The Act classifies the persons into three categories viz. children, minors, and adults; the former two being less than 18 and 16 years respectively. Under Section 5, procuring, inducing or taking for the sake of prostitution is made punishable with varying periods of imprisonment, greater punishment being prescribed for offences against minors and children.

Causes of Prostitution

The denunciation of prostitution is also as old and pervasive as the practice itself. The causes are to be appreciated in the context of the needs of the seller as well as of the buyer. At the outset, two popular beliefs may be dispelled. Firstly, it is generally believed that consumer of commercialised sex are impelled to go to prostitutes since no alternative is available to them for the gratification of their sexual urges. The second belief is that women become prostitutes due to economic reasons or motivation.

Alternative means of sexual gratification : While it is true that many persons patronise prostitutes because of lack of other means of sexual gratification, common instances being of bachelors, armymen away from their families and sailors. Those who visit prostitutes are sex-starved. To quote Kingsley Davis: "The common and ignorant assumption that prostitution exists to satisfy the gross sexuality of the young unmarried man, and that if he is taught to bridle gross sexual impulse or induced to marry early the prostitute must be idle, is altogether incorrect. The prostitute is something more than a channel to drain off superfluous sexual energy, and her attraction by no means ceases when men are married, for a large number of men who visit prostitutes, if not the majority, are married. And alike whether they are married or unmarried the motive is not one uncomplicated lust. The craving for variety, for perverse gratification, for mysterious and provocative surroundings, for intercourse free from entangling cares and civilised pretence, all play their part"(1937:746-55).

Gender difference in sexual urge : From the beginning of civilisation, ways and means have been devised to discipline the sexual urge of human beings and marriage is the most potent universally recognised institution devised to attain the goals. But even the institution of marriage cannot alter some of the basic truths. To appreciate the reasons for demand of extramarital sex by men, it is essential to keep in mind the difference between males and females in terms of their sexual urge. Man tends to be more promiscuous by nature than woman (Siddique,2005:459). This characteristic is also found among lower male mammals. Further, while a male may find it more difficult to go without sex for any substantial period of time, for a female it is generally easier to lead a discontinuous sex life. Partly, the difference in male and female sexual attitudes may also be due to social and cultural determinants; according to the traditional mores no laxity is permissible so far as women are concerned but male indiscretions are not only tolerated and condoned but are also taken as signs of virility. The fact, therefore, is that there are many men who need and manage, sex outside wedlock. As regards procurement of sex outside marriage, the easiest and sometimes the only way to get it is to buy it from those who are willing to sell it. To quote Kingsley Davis again : "When all other sources of gratification fail, due to defects of person or circumstances, prostitution can be relied upon to furnish relief. None of the exacting requirements of sex attraction and courtship are necessary. All that is needed is the cash, and this can be obtained in a thousand ways. Prostitution is the most malleable, the most uninvolved form of physical release"(Ibid).

Compelling situations : As regard the other common belief that women become prostitutes due to economic motivations, it is not true of at least those women who take up prostitution involuntarily; those compelled by their abductors or 'buyers, or even by family members including the husband. Further today, human trafficking is primarily for prostituting women and children. It is described as "the largest slave trade in history" and is the fastest growing criminal industry, set to outgrow drug trafficking.

Affluence : In the case of those women, who opt for a life of prostitution, economic factors may be the precipitatory cause; other factors like alienation from the family or the decay of moral values could be at the root of the problem (Bilimoria Rani, 1983). If poverty is the factor that is meant by the words 'economic reasons', then that does not, as such explain the emergence of an elite class of prostitutes in big cities the world over. In contemporary society there is a wide range of varieties of prostitutes. At the highest strata in the system are expensive and somewhat exclusive call girls or party girls, highly conscious of their 'superiority' and of being 'different' from their fellow professionals.

In London, for instance, prostitute of a more exclusive area like Mayfair may be very different from the one operating in Soho. Researching about American call girls, Greenwold (1958) found that none of the girls included in his study would dream of charging less than twenty dollars since doing so would make them feel like "common prostitutes." One of the girls interviewed asserted that she was "strictly a fifty-dollar girl". What distinguishes the members of this elite class is their attractiveness, financial standing, dress, manners, residence in an affluent neighborhood and even their political connections. It is evident that prostitution at the higher level of the social scale is basically is response to the mechanism of the consumer society. A few girls even of relatively well-off families indulge in prostitution, casual or regular, clandestine and exclusive manner in order to achieve higher standards living. In bigger cities of India even some girls studying in elite colleges and living in hostels are reported to be in the profession.

Social Evils : While there may be some other factors for drifting into prostitution in an individual case, poverty is the determining factor in most of the cases in India. Some of the causes peculiar to India which indirectly contribute towards the vice are social evils like the dowry system, child marriage and the religious and social constraints on widow remarriage. The committee on 'Status of Women' appointed by the Government of India in its report of 1975 observed: "...recruitment to this profession is easy and girls from the middle class are also joining this profession and these women practise prostitution often with the connivance of their parents or husbands in order to secure huge sums of money to keep up an appearance of affluence. Educated and outwardly respectable, these women are prompted to take to prostitution because of the undue emphasis on the values of affluence....Indeed it is the cult of consumption which encourages this pattern of prostitution."

The other evils include ill treatment by parents, bad company, family prostitutes, social customs, inability to arrange marriage, lack of sex education, media, prior incest and rape, early marriage and desertion, lack of recreational facilities, ignorance, and acceptance of prostitution under tradition.

Economic causes : Economic causes include poverty and economic distress. Prostitution is practiced in brothels by hapless victims of exploitation or poverty. Economics is the pervasive form of prostitution and this reality is indicated by the fact that money is mentioned as a motivating factor in virtually all of the literature relating to prostitution. One type of literature represents prostitutes as wretched creatures forced into prostitution by extreme economic deprivation, opposing this depiction, is a body of research which indicates that prostitutes choose prostitution as an occupational alternative that provides them relatively greater opportunity for earning money that affords them the highest attainable standard of living. Kingsley Davis observed about economic cause of prostitution and related it with female attractiveness : "Furthermore the division of labor by sex inevitably makes women dependent to some extent on their sexual attractiveness and puts men in control of economic means. Since the economic means are distributed unequally between classes but female attractiveness is not, some women of lower economic means can exploit their attractiveness for economic gain"(1971:345).

Havelock Ellis depicted that the higher wages are earned in prostitution in comparison to wages paid of ordinary women. This higher wage is actual motivating factor initiating the profession of prostitution. "No practicable rise in the rate of wages paid to women in ordinary industries can possibly compete with the wages which fairly attractive women of quite ordinary ability can earn by prostitution"(quoted in PK Singh, 2004:25). Esselstyne find out that causative factor of prostitution is that prostitution provides materially better living standard to women who are entering in this profession : "Women are attracted to prostitution in contemporary America because the income is high and because it affords an opportunity, to earn more, buy more, and live better than would be possible by any other plausible alternative."

Greenwald studied 26 call girls and reported that "Not one of the girls I interviewed attempted to explain her choice of her profession in terms of desperate economic needs." Pomeroy studied 175 prostitutes 83 to 93 percent of whom were motivated by economic factors, he noted that "the gross income from prostitution is usually larger that could be expected, from any other type of unskilled labor"(quoted in PK Singh, 2004:25). Jannifer James study points out that only 8.40 percent of prostitutes claimed to have started prostitution because of economic necessity, while 56.49 percent were motivated by a desire for money and material goods. Some researchers claim to find an abnormal, perhaps even neurotic materialism among prostitutes. O'Toll Jackman and Geis state that : "The rationalization by prostitutes violating social taboos against commercial sex behaviour take the form of exaggerating other values, particularly those of financial success, and for some the unselfish assumption of financial burden of people dependent upon them" (quoted in PK Singh, 2004:26).

Psychological causes : Psychological causes include desire for physical pleasure, greed, and dejection. Extreme deprivation seems to be the cause of prostitution in India rather than sexual and psychological aberrations. Inadequate socialization or over protection by parents, limited self-control and waywardness are significantly associated with psychological dimension of prostitution.

3. Types of Prostitution :

3.1 Religious Prostitutes in India : Religious prostitutes are known by different names e.g., Devadasis, Khudikar, Murloi, Basavi bhavin, Devali, Naikin and Jogins etc. They all start at an early age i.e. at puberty and prepuberty stages. Brief description of categories of religious prostitutes is given below :

Devadasis : Originally, devadasi were celibate dancing girls used in temple ceremonies and they entertained members of the ruling class. But sometime around the 6th Century, the practice of "dedicating" girls to Hindu gods became prevalent in a practice that developed into ritualized prostitution. Devadasi literally means God's (Dev) female servant (Dasi), where according to the ancient Indian practice, young pre-pubertal girls are 'married off', 'given away' in matrimony to God or Local religious deity of the temple. Her duties comprise a combination of Propriety, Ritual and Entertainment to assert positive fertility and prosperity. They on account of being married to God or goddess, such as Yellamma or Mathamma were called 'Nitya Sumangali'. The marriage usually occurs before the girl reaches puberty and requires the girl to become a prostitute for upper-caste community members. The system of devadasi started only after the fall of Buddhism and records about them start appearing around 1000 A.D. [Bharatiya Sanskruti Kosh, IV, 448]. It is viewed that the Devadasis are the Buddhist nuns who were degraded to the level of prostitutes after their temples were taken over by Brahmins during the times of their resurgence after the fall of Buddhism. According to the 1934 Devadasi Security Act, this practice is banned in India. The prostitution continued from ancient and medieval India and has taken a more gigantic outlook in modern India, the devadasi system still continues, according to a report of National Human Rights Commission of the Government of India, "after initiation as devadasis, women migrate either to nearby towns or other far-off cities to practise prostitution." The practice of dedicating devadasis was declared illegal by the Government of Karnataka in 1982 and the Government of Andhra Pradesh in 1988. However the practice is still prevalent in around 10 districts of north Karnataka and 14 districts in Andhra Pradesh. Districts bordering Maharashtra and Karnataka, known as the "devadasi belt," have trafficking structures operating at various levels. The women here are in prostitution either because their husbands deserted them, or they are trafficked through coercion and deception. Many are devadasi dedicated into prostitution for the goddess Yellamma.

This ban was reinforced again in 1980s but the law is broken every day. Poverty and 'Untouchability' contribute to the persistence of this terrible practice. Reference to dancing girls in temples is found in Kalidasa's "Meghadhoot". The popularity of devadasis seems to have reached its pinnacle around 10th and 11th century CE. The rise and fall in the status of devadasis can be seen to be running parallel to the rise and fall of Hindu temples. The devadasi system was mostly prevalent in southern India and it reached its height during the Chola Empire. Though government has taken adequate steps in order to combat with the problem of devadasi, even the devadasi prohibition act was not fully successful in solving the problem in India. Most important reason still being poverty, ignorance and hunger which are forcing them to this kind of exploitation. Now this was the old story or the beginning of prostitution apart from these there is also references of prostitution in Kama sutra written by Vatsyayana sometime between the second and fourth centuries A.C.E.

The system presently exists in Andhra Pradesh, Maharashtra and Karantaka and some parts of Goa. A girl whose mother is devadasi usually becomes a devadasi.

Jogins : The prevalence of Jogins, a similar practice like devadasi continues in large part of Andhra Pradesh. In this system the girls are 'Married' to God before puberty and they enter into prostitution when they reach puberty. In 1992 there were about 5,00 jogins in Nizamabat and 30,000 elsewhere in the state. Ganga a scheduled caste girl was initiated as a jogin around the year 1960 at the age of six. After eight years of initiation at the age of 14 a local landlord come and took her away at night and in a span of few years burdened her with a couple of children and paid a paltry sum of ₹ 200 to her father by way of compensation and maintenance. After that she was passed on from one landlord to another till she died at the age of 30. The children left by Ganga are begging and the landlord has taken another jogin. There are about 10,000 jogins in the area.

Basavi : It means 'femal bull' and it connotes the bulls freedom to wonder. Historically Basavis are forbidden to marry and expected to spend their lives performing religious duties. The system is known to exist in Andhra Pradesh and in scattered form in some areas of Karnataka. The Basavis do not immediately enter into life long career of prostitutes. They are still required to perform certain ritual duties. Since such work does not yield sufficient income they turn to prostitution.

3.2 Street Walkers : In street prostitution the prostitute solicits customers while waiting at street corners or "walking the street". The prostitute solicits customers while waiting at street corners, sometimes called "the track" by pimps and prostitutes alike. They usually dress in skimpy, provocative clothing, regardless of the weather. Street prostitutes are often called "streetwalkers" while their customers are referred to as "tricks" or "johns." Servicing the customers is described as "turning tricks." The sex is usually performed in the customer's car, in a nearby alley, or in a rented room. Motels and hotels which accommodate prostitutes commonly rent rooms by the half or full hour.

The street walkers may function independently or through pimps. Generally affiliated to brothels, hotels, cinema halls, etc. these women are vulnerable to attacks and pressures of both clients and the police, with little or no support to combat them. They have no security and are time and again exploited economically, emotionally and sexually either by the clients or others. They avoid concentrating in particular areas in order to escape the eyes of the law enforcing authorities. They have their unique methods of soliciting clients. Their income may range from the low of ₹ 5/- to the high of ₹ 500/.

3.3 The Cage Brothel Prostitute : This is an unique practice reported about Bombay, Delhi. Such prostitutes are attached to a brothel, which has a landlord who in turn rents the premises to a brothel keeper. The brothel keeper runs the brothel and appoints a manager who supervises and keeps a watch on all the women, pimps, procurers and henchmen. The caged prostitutes are basically minors, below 18 years of age. Their earnings may range between ₹ 50/- to ₹ 100 per day. All their earnings go to the brothel keeper till such time that the brothel keeper's investment made in procuring her has been recovered. The brothel keepers do not release these girls as long as they get regular and rich clients. Young girls and virgins are very much in demand and therefore their earnings during earlier years are high.

3.4 Brothel Prostitutes : Brothels are establishments specifically dedicated to prostitution, often confined to special red-light districts in big cities. Other names for brothels include bordello, whorehouse, cathouse, knocking shop, and general houses. Prostitution also occurs in some massage parlours, and in Asian countries in some barber shops where sexual services may be offered as a secondary function of the premises. Such prostitutes stay and practice in brothels or brothel like outfits under the rules and regulations of the brothel which are set by the people managing the same. They are basically dependent on brothel keepers, pimps, etc, and they provide sexual favors to their clients in exchange of money. A good amount of their earnings is shared among various categories of persons including brothel keepers, pimps, etc.

Unlike street walkers they are economically secure in a sense that they are not cheated by their clients. As they work through established setups they are easily identifiable. The concentration of such prostitutes is considerably high in the respective brothels and their earnings may vary from ₹ 10 to ₹ 50/- per client. The clientele is mainly from the lower or lower middle class groups, which include rickshaw pullers, laborers, etc. These prostitutes also belong to low socio-economic group families. The number of the clients they entertain daily is quite high and the localities where the brothels are situated are either densely populated or have high frequency of floating population. e.g. G.B. Road (Delhi), Kamathipura (Bombay), Khidirpur (Calcutta), etc.

3.5 Singing and Dancing Girl : This group of prostitutes do not start and end with sex. They normally start with singing and dancing to entertain the client and end with sexual favors. It is important to mention that these women are not entirely dependent on prostitution. They may or may not resort to prostitution regularly. They may have pimps and contact persons and have access to a part of the income along with the person who fetches them their clients. Unlike brothel prostitutes they have some degree of freedom in terms practicing prostitution. They are mainly drawn from lower or middle class groups. Majority of them operate through prostitution prone areas whereas there are a few who operate through brothels[38]. It was observed that a good many of present day prostitutes belonged to this category earlier but because of loss of patronage have taken to direct prostitution.

3.6 Lot Lizard : A "lot lizard" is a trashy, street-level, female prostitute who frequents truck-stop parking lots and rest areas at night, looking for customers. Most lot lizards openly "advertise" using Citizen Brand (CB) radios; others boldly walk from truck to truck randomly knocking on doors. CB, is a public, two-way personal radio service in portable communications using handheld transceivers with operation taking place within a narrow band of frequencies near 27 MHz. It is a commonly-encountered special case of street prostitution. Lot lizards mainly serve those in the trucking industry at truck stops and stopping centers. Prostitutes will often proposition truckers using a CB radio from a vehicle parked in the non-commercial section of a truck stop parking lot, communicating through codes based on commercial driving slang, and then join the driver in his truck.

3.7 Barmaids : This group earns anything between ₹ 2000 to ₹ 3500/- per month. All barmaids do not indulge in prostitution but quite a few of these girls do indulge in the practice after the bar is closed. Most women/girls in this group maintain that they are neither exploited the management nor are they overtly harassed by customers.

3.8 Theatre/Cinema Girls : A special category of prostitutes reported by the Bombay study are those who are 'Company' to men and accompany them of cinema halls. They do not necessarily indulge in sexual activities except when some men take them to hotels for the purpose after watching movies[40]. In such situations there women earn a little extra money.

3.9 Massage Parlours and Health Centre Attendants : They basically work secretively and under some garb. They too, like singing and dancing girls, do not start and end with sex. They start with massaging or exercising or checkups and finally land up with sexual intercourse. They either work in these setups on full time or part time basis or work through these setups. These women cater to the clients of middle and upper income groups and so have high incomes which they share with the person who fetch them a client. They may or may not be solely dependent on prostitution. Prostitutes in this category are also from middle and upper class families. This phenomenon is increasing day by day and the women are found to practice at some pick up point convenient both to the client and prostitutes.

3.10 Independently Operating Prostitutes : They normally belong to lower and lower middle class groups of families and serve to clients of the some class. They may or may not have pimps (majority of times they do not work through pimps). Normally found in the highly congested areas of big cities, like slums resettlement colonies, etc, they have a unique modus operandi. Majority of them including working women resort to prostitution regularly or intermittently as a means a supplementing their monthly incomes. Their earnings are quite low because of the type clientele they serve.

3.11 Call Girls : A call girl or escort is a sex worker who (unlike a street walker) is not visible to the general public; nor does she usually work in an institution like a brothel, although she may be employed by an escort agency. The client must make an appointment, usually by calling a telephone number. Call girls often advertise their services in small ads in magazines and via the Internet, although an intermediary advertiser, such as an escort agency, may be involved in promoting more up market escorts, while some may be handled by a pimp. Call girls may work either in-call, where the client comes to them, or out-call, where they go to the client. They function depending upon the situation of promises of greater payment.

It is a comparative recent phenomenon and it has been found that a good number of call girls belong upper middle class families. They may own apartments with telephones and other amenities, wherein they practice the profession. They may also go out to posh hotels and other places for the purposes of prostitution. They either operate independently or through a contact/manager. Some of them may have regular clients who may be willing to pay higher fees for exclusive services. They, too, share their income with their contacts. In many cases they have a choice to decide about their clients. Management to their business is either their own or their managers' affair. They operate from areas inhabited by middle and upper class families. Their gradually increasing numbers are difficult to check because of the constant changes in the socio-economic situation of the city and growing commercialization of profession.

3.12 Call Girl and Escort Agency : Escort agencies typically advertise over the Internet, in regional publications and even telephone listings like the Yellow Pages. Many maintain websites with photo galleries of the employees. An interested client contacts an agency by telephone and offers a description of what kind of escort they are looking for. The agency will then suggest an employee who might fit that client's need. However, this kind of advertising often entails bait and switch practices in which photographs of attractive models who have no connection with the agency are used to attract customers, who are then sent whatever escort is available to the agency at the moment. The agency collects the client's contact information and calls the escort. Usually, to protect the identity of the escort and ensure effective communication with the client, the agency arranges the appointment. Sometimes it may be up to the escort to contact the client directly to make arrangements for location and time of an appointment. If the agency does not supply transport to and from the client, the escort is also expected to call the agency upon arrival at the location and again upon leaving to assure his or her safe completion of the booking.

The purpose of discretion is to attempt to protect the escort agency (to some degree) from prosecution for breaking the law. If the employee is solely responsible for arranging any illegal aspects of their professional encounter the agency could try to maintain plausible deniability should an arrest be made. However in practice, the use of undercover police evidence or the use of links to reviews of the agencies escorts usually results in this failing.

Typically, an agency will charge their escorts either a flat fee for each client connection or a percentage of the prearranged rate. Most transactions occur in cash, and optional tipping of escorts by clients. Independent escorts, also known as providers, have differing fees depending on many factors. For example; different seasons bring about different costs (and differing levels of demand), as do regular and semi-regular customers. Some may charge by the hour, half hour or even in 15 minute blocks. Time extensions (if offered or requested) are usually priced at the same rate as the original booking. Some escorts pay another individual to act as their personal security, thus providing a level of protection to themselves from violent or abusive clients. An escort who works less often may be able to command a premium for his or her exclusivity. One who sees several clients each day may charge less, but earn more in the end. Independent escorts might see clients for extended meetings involving dinner or social activities, whereas escorts who work through agencies generally provide only sexual services.

3.13 Online Prostitution : Some prostitutes use the Internet to find customers (Siegal, 2005). A prostitute may use adult boards or create a website of their own with contact details, such as e-mail addresses. Adult contact sites, chats and on-line communities are also used. In many parts of the developed world, the Internet is one of the main ways in which buyers find and contact prostitutes. This, in turn, has brought increased scrutiny from law enforcement, public officials, and activist groups toward online prostitution. Reviews of the services of individual prostitutes often can be found at various escort review boards worldwide. These online forums are used to trade information between potential clients, and also by prostitutes to advertise the various services available. However, many other prostitutes are critical of such sites, as threats of a negative review are sometimes used by clients used to extort extra services or unsafe sex from sex workers. Sex workers, in turn, often use online forums of their own to exchange information on clients, particularly to warn others about dangerous clients. Participants in these forums also exchange information on law enforcement activity, working conditions, and other matters important to sex workers.

3.14 Sex Tourism : Sex tourism emerged in the late 20th century as a controversial aspect of Western tourism and globalization. Sex tourism is typically undertaken internationally by tourists from wealthier countries. Sex tourism is travelling for sexual intercourse with prostitutes or to engage in other sexual activity. The World Tourism Organization, a specialized agency of the United Nations defines sex tourism as "trips organized from within the tourism sector, or from outside this sector but using its structures and networks, with the primary purpose of effecting a commercial sexual relationship by the tourist with residents at the destination."

Often the term "sex tourism" is mistakenly interchanged with the term "child sex tourism". As opposed to regular sex tourism, which is often legal, a tourist who has sex with a child prostitute will usually be committing a crime in the host country, under the laws of his own country (notwithstanding him being outside of it) and against international law. Child sex tourism (CST) is defined as a travel to a foreign country for the purpose of engaging in commercially facilitated child sexual abuse. The term "child" is often used as defined by international law and refers to any person below the age of consent. Thailand, Cambodia, India, Brazil and Mexico have been identified as leading hotspots of child sexual exploitation.

3.15 The Chukri System : Also prevalent in Indian prostitution is the Chukri System, whereby a female is coerced into prostitution to pay off debts, as a form of bonded labor tound in Bangladesh and parts of West Bengal. The Chukri System is a type of debt bondage or forced labor system. In this system, the sex worker generally works without pay for one year or longer in order to repay a supposed debt to the brothel owner for food, clothes, make-up, and living expenses. The system creates a workforce of people virtually enslaved to their creditors, and constitutes one of the primary causes for women entering the sex trade. In this system, the prostitute generally works without pay for one year or longer to repay a supposed debt to the brothel owner for food, clothes, make-up and living expenses. In India, the Government's "central sponsored scheme" provides financial or in-kind grants to released bonded labourers and their family members, the report noted, adding over 2,850,000 people have benefited to date. Almost 5,000 prosecutions have been recorded so far under the Bonded Labor System (Abolition) Act of 1976.

3.16 Underage Sexwork : Many prostitutes are said to be underage, entering the sex trade as young as 12. Andhra Pradesh and West Bengal states together account for 26% of the total number of prostitutes in the country. Child prostitution is a major problem in India involving around 1.2 million children with at-least 100 million people were involved in human trafficking in India. Some women and girls are by tradition born into prostitution to support the family. The Bachara, for example, follow this tradition with eldest daughters often expected to be prostitutes. The Bachara are a tribe of people in the western part of Madhya Pradesh in central India. They are known for the tradition of child prostitution[citation needed], with families making their first daughters, as children, into prostitutes to support the family. The tradition is centuries old and is still practiced today.[citation needed] The documentary film Highway Courtesans was made in 2004 about this tribe. Their customers mostly are the truckers passing by the highway. Bachara women and their customers are subject to frequent extortion by the police for money. There have been very little efforts by the government for the welfare and upliftment of the tribe.[citation needed] They are looked upon with contempt by the other communities of the area.

3.17 Male Sex Worker : Male prostitution is increasingly visible in India. Gigolo service in India is growing. But there are cases of harassment of client women by gigolos. There are many "agencies" offering "handsome masseurs" in the classifieds of the newspapers. They offer both in and out services, although the facilities are usually very basic. Most western clients are visited at their hotels. Local middle-class Indians are also now using these services. Fees are discussed over the phone, typically ₹ 1000-3000. Safe sex and condom use is generally well understood. The workers typically do not speak the language the clients deserve. They are also found in Delhi's emerging gay night life scene, with several "one nighters" at various middle-class night clubs in the city.

In India, male homosexual acts are now legal but male prostitution is all but invisible and not much is currently known about the status of male sex workers. The stories they have to tell are similar: stories of missed opportunities, of poverty, hopelessness and exploitation. Due to the social stigma attached to homosexuality in India and the lack of legal protection, they tend to face higher risks than females. They are often faced with violence from the police, clients, and are often subjected to extortion from the police in order to carry on with their work. A large percentage of male sex workers are eunuchs or hijrahs. Most know of sexually transmitted diseases through experience, but there are few preventative measures, such as condoms, that are made available to them. Due to their legal status, no regimen of testing for HIV/AIDS or other diseases are made available. Men who pursue sex work in India are doubly criminalised. On the one hand, same sex practices are prohibited by article 377 of the Indian penal code. On the other hand, sex work is illegal. Male sex workers operate thus outside the law, which leaves them unprotected and defenseless. They are often victims of violence, unfair treatment and rape.

Prostitution- The Consequent Dangers

Prostitution has been denunciated in almost all societies for various reasons; the most fundamental reason being that it offends the elementary norms of decency and culture and involves human debasement of the lowest order. It has been condemned as an evil, albeit inevitable, by social reformers, religious thinkers and philosophers alike. Prostitution represents a social problem for several reasons. First, the prostitute is affected adversely by the pathological behaviour of systematic and indiscriminate promiscuity. Second, she is frequently a source for spreading venereal diseases. Third, she frequently is controlled by, and is a source of revenue for organized crime.

The prostitutes, who remain in this perverse craft for a long time deteriorates physically and psychologically. In addition to becoming infected on one or several occasions, she experiences sexual frigidity and attitudes of self-degradation. One study estimated that older prostitutes were sexually frigid in their contacts with their clients. And prostitutes, who had been sexually responsive at the outset, lost their responsiveness within several month of "professional activity". The Prostitute's past social relationships frequently become sundered and her ties with the parasitic panderer more intensely dependent. The brothel prostitute pursues a mode of life which segregates her from conventional girls. Sometimes her activities are watched by the panderer. Furthermore, the prostitute tends to drift in a day-to-day routine without any purpose or plan.

"Illegitimate sex is considered to be a great sin under all the religious systems and it should be too evident that the lowest level of illegitimate sexual conduct is reached in prostitution. What can be called more sordid, more void of modesty, more full of shame than prostitution brothels and every other evil of this kind ? Yet remove prostitution from human affairs, and you will pollute all things with lust; set them among honest matrons, and you will dishonour all things with disgrace turpitude", commented St. Augustine on prostitution. Thomas Acquinas expressed the same idea in the following words: (quoted in Siddique 2005:459), "Prostitution is like the filth in the sea, or a sewer in the palace. Take away the sewer, and you will fill the palace with pollution; and likewise with the filth (in the sea). Take away prostitutes from the world, ' you will fill it with sodomy."

The other adverse effects are the creation of nuisance on the streets due to solicitation by the prostitutes or their agents, spread of venereal diseases and the psychic or emotional disturbance to children and young persons belonging to the families of prostitutes or living in the brothel areas.

The problem of soliciting became so acute in London by the early fifties that it was one of the major reasons which led to the appointment of the Wolfendon Committee in 1954. Quite often the act of soliciting on the streets was responsible for trouble between prostitutes and their agents because of clash of business interests. A respectable non-prostitute woman could also be subjected to embarrassment when taken to be a prostitute by some prospective client in a locality of dubious nature.

Since prostitution involves systematic promiscuity, it is a pervasive source for the spread of venereal disease. Venereal disease afflicts over three million people yearly. Syphilis, gonorrhea, trichomoniasis and non-specific arthritis are some of the diseases which are spread through prostitution, a risk more pronounced in Africa and Asia where 80 to 97 per cent of venereal infections are contracted through prostitution. These diseases pale into insignificance when their consequences are compared to the lethal nature of 'AIDS' (Acquired Immune Deficiency Syndrome), now striking many countries.

In western countries people get more infections through 'good time girls' and casual acquaintances. It appears that while a person may be more careful in taking antibiotic medicines as a precautionary measure with regard to sex with prostitutes, he may not foresee the danger in case of non-prostitute females.

Health Problems : The prostitution leads to many health problems for the prostitutes like : Cervical cancer, Traumatic brain injury, HIV, STD and Psychological disorders. In a country like India where most of the people indulge themselves in unprotected sex with prostitutes. It is very difficult to eradicate the problem of aids. Historically, the AIDS epidemic in India was first identified amongst sex workers and their clients, before other sections of society became affected. The sex workers are themselves taking steps to combat with aids in some brothels in India for example, Sonagachi a brothel in Kolkata; where the sex workers are insisting their clients for use of condoms in order to avoid aids. But in all the other brothels in India social workers and NGO's are trying to acquaint the sex workers about the ill effects of AIDS and are insisting them for using condoms.

Prostitution is associated with the spread of sexually transmitted diseases (STDs) such as HIV: "One of the main reasons for the rapid spread of HIV in Asian countries is the massive transmission among sex workers and clients" (Rojanapithayakorn, 2006). As a result, prevention campaigns aimed at sex workers play a major role in restricting the spread of HIV. HIV is tied to prostitution in Africa as well. Typical responses to the problem are :

- banning prostitution completely.
- introducing a system of registration for prostitutes that mandates health checks and other public health measures.
- educating prostitutes and their clients to encourage the use of barrier contraception and greater interaction with health care.

Some think that the first two measures are counter-productive. Banning prostitution tends to drive it underground, making safe sex promotion, treatment and monitoring more difficult. Registering prostitutes makes the state complicit in prostitution and does not address the health risks of unregistered prostitutes. Both of the last two measures can be viewed as harm reduction policies.

In countries and areas where safer sex precautions are either unavailable or not practiced for cultural reasons, prostitution is an active disease vector for all STDs, including HIV/AIDS, but the encouragement of safer sex practices, combined with regular testing for sexually transmitted diseases, has been very successful when applied consistently. As an example, Thailand's condom program has been largely responsible for the country's progress against the HIV epidemic.

Consistently high HIV infection rates among sex workers (50% or more among Mumbai's female sex worker population since 1993), coupled with lack of information, failure to use protection, and the migrancy of their clients, may contribute to the spread of AIDS in the region and the country in India (Galwankar, 2002). Reaching women who are working in brothels has proven to be quite difficult due to the sheltered and secluded nature of the work, where pimps, Mashis, and brothel-keepers often control access to the women and prevent their access to education, resulting in a low to modest literacy rate for many sex workers.

Relation to Crime : Besides the harm caused to society described above, prostitution may also give rise to crimes like cheating, blackmail and breaches of peace in an indirect manner.

Children are sold into the global sex trade every year. Often they are kidnapped or orphaned, and sometimes they are sold by their own families. According to the International Labor Organization, the problem is especially alarming in Thailand, the Philippines, Sri Lanka, Vietnam, Cambodia, Nepal and India. One of the most serious problems associated with prostitution is the fact that the sex trade is surrounded by illegal, abusive and dangerous activities. One view insists that such situations occur because prostitution is kept illegal and the industry operates on the black market. Another, however, believes that legalizing and regulating prostitution does not improve the situation, but instead makes it worse.

Human Trafficking and Sexual Slavery : Sex trafficking and prostitution are inextricably intertwined. Sex trafficking is a form of modern day slavery that claims victims from around the world (most of them women and children). While many people may not fully understand the complexities of sex trafficking, its generally understood that it's a grave injustice that must be stopped. However, most people do not approach prostitution with similar sympathy and efforts to gain an understanding. Prostitution is highly stigmatized in our society, and often falsely portrayed in Hollywood and the media. Stereotypically prostitutes are seen as promiscuous, addicts, dishonest and criminals – not characteristics that generally garner much sympathy. But the reality is that prostitution and sex trafficking are linked on many levels.

PM Nair (2007:1) says "trafficking does not mean prostitution. They are not synonymous. In understanding trafficking, one should delink it from prostitution." As per the existing law, Immoral Traffic (Prevention) Act 1956 (ITPA) prostitution becomes an offence when there is commercial exploitation of a person. If a woman or child is sexually exploited and any person gains out of the same, it amounts to commercial sexual exploitation (CSE), which is a legally punishable offence wherein the culpability lies against all exploiters. Trafficking is the process of recruiting, contracting, procuring or hiring a person for CSE. Therefore, trafficking is a process and CSE is the result. The 'demand' in CSE generates, promotes and perpetuates trafficking. This is a vicious cycle. Trafficking could also be a means for other types of violations such as for developing pornographic material, for promoting sex tourism, for sexual exploitation under the facade of bar tending, massage parlours etc, or even for exploitative labor where sexual abuse may or may not coexist.

He further argues (Ibid) that ITPA envisages only trafficking for CSE. Commercial activity need not be in a brothel, but could also occur in places including a residential dwelling, a vehicle, etc. Therefore a police officer who is acting under ITPA has powers to take steps in all such situations where trafficking leads to or is likely to lead to CSE in any form, including those under the facade of massage parlours, bar tending, 'tourist circuit', 'escort services', 'friendship clubs', etc.

According to U.N. estimates, approximately 2.5 million people are being trafficked around the world at any given time, 80% of them women and children. Conservative estimates suggest that the sex industry generates some $32 billion annually. However, estimates of income generated from prostitution in one city, Las Vegas, are as high as $5 billion. Today, sex trafficking is a high-tech, globalized, electronic market, and predators are involved at all levels, using the same methods to control prostituted women that batterers use against their victims : minimization and denial of physical violence, economic exploitation, social isolation, verbal abuse, threats and intimidation, physical violence, sexual assault, and captivity. Despite the illogical attempt of some to distinguish prostitution from trafficking, trafficking is simply the global form of prostitution. Sex trafficking may occur within or across international borders, thus women may be either domestically or internationally trafficked or both. Young women are trafficked for sexual use from the countryside to the city, from one part of town to another, and across international borders to wherever there are men who will buy them.

Prostitution is widely socially tolerated, with the buyers socially invisible. Even today, many mistakenly assume that prostitution is sex, rather than sexual violence, and a vocational choice, rather than a human rights abuse. Although clinicians are beginning to recognize the overwhelming physical violence in prostitution, its internal ravages are still not well understood. There has been far more clinical attention paid to sexually transmitted diseases among those prostituted than to their depressions, lethal suicidality, mood disorders, anxiety disorders (including post-traumatic stress disorder) dissociative disorders, substance abuse, and traumatic brain injury. Regardless of its legal status or its physical location, prostitution is extremely dangerous for women. Homicide is a frequent cause of death.

Prostitution is an institution akin to slavery, one so intrinsically discriminatory and abusive that it cannot be fixed–only abolished. At the same time, its root causes must be eradicated as well : sex inequality, racism and colonialism, poverty, prostitution tourism, and economic development that destroys traditional ways of living. The conditions that make genuine consent possible are absent from prostitution : physical safety, equal power with johns and pimps, and real alternatives. It is a cruel lie to suggest that decriminalization or legalization will protect anyone in prostitution. Until it is understood that prostitution and trafficking can appear voluntary but are not in reality free choices made from a range of options, it will be difficult to garner adequate support to assist those who wish to escape but have no other choices. Enforcement of international agreements challenging trafficking and prostitution can aid in this effort as can laws challenging men's purchase of sex.

It is important to address men's demand for prostitution. Acceptance of prostitution is one of a cluster of harmful attitudes that encourage and justify violence against women. Violent behaviours against women have been associated with attitudes that promote men's beliefs that they are entitled to sexual access to women, that they are superior to women and that they are licensed as sexual aggressors. Those concerned with human rights must address the social invisibility of prostitution, the massive denial regarding its harms, its normalization as an inevitable social evil, and the failure to educate students in the mental health and public health professions. Trafficking and prostitution can only exist in an atmosphere of public, professional and academic indifference.

Establishments engaged in sexual slavery are the highest priority targets of law enforcement actions against prostitution. It has been suggested that human trafficking is the fastest growing form of contemporary slavery and is the third largest and fastest growing criminal industry in the world.

Poverty, social exclusion and war are at the heart of human trafficking. Some women are hoodwinked into believing promises of a better life, sometimes by people who are known and trusted to them. Traffickers may own legitimate travel agencies, modeling agencies and employment offices in order to gain women's trust. Others are simply kidnapped. Once overseas it is common for their passport to be confiscated by the trafficker and to be warned of the consequences should they attempt to escape, including beatings, rape, threats of violence against their family and death threats. It is common, particularly in Eastern Europe, that should they manage to return to their families they will only be trafficked once again.

Prostitution of Children : Regarding the prostitution of children the laws on prostitution as well as those on sex with a child apply. If prostitution in general is legal there is usually a minimum age requirement for legal prostitution that is higher than the general age of consent (see above for some examples). Although some countries do not single out patronage of child prostitution as a separate crime, the same act is punishable as sex with an underage person.

In India, the federal police say that around 1.2 million children are believed to be involved in prostitution. A CBI statement said that studies and surveys sponsored by the ministry of women and child development estimated that about 40% of all India's prostitutes are children. Thailand's Health System Research Institute reported that children in prostitution make up 40% of prostitutes in Thailand. Some adults travel to other countries to have access to sex with children, which is unavailable in their home country. Cambodia has become a notorious destination for sex with children. Thailand is also a destination for child sex tourism. Several western countries have recently enacted laws with extraterritorial reach, punishing citizens who engage in sex with minors in other countries. As the crime usually goes undiscovered, these laws are rarely enforced.

Violence against prostitutes : Female prostitutes are at risk of violent crime, as well as possibly at higher risk of occupational mortality than any other group of women ever studied. For example, the homicide rate for female prostitutes was estimated to be 204 per 100,000 (Potterat et al., 2004), which is considerably higher than that for the next riskiest occupations in the United States during a similar period (4 per 100,000 for female liquor store workers and 29 per 100,000 for male taxicab drivers) (Castillo et al., 1994). However, there are substantial differences in rates of victimization between street prostitutes and indoor prostitutes who work as escorts, call girls, or in brothels and massage parlors (Weitzer 2000, 2005). While women who work on the streets are the most likely to be victimized, attacks and even murders of prostitutes have also occurred in legal and licensed brothels (such as in the German brothel *Pascha*).

Combating Prostitution

Three kinds of societal responses to prostitution are discernible i.e. suppression, regulation and abolition. Suppression refers to the efforts to be made to wipe out the institution of prostitution. Regulation, on the other hand, accepts the inevitability of prostitution and efforts are directed to prevent and minimise its evil effects. It is sought to be achieved subject to certain safeguards like licensing, registration, enforcing health directives and prescribing limits in the city or town where the profession could be carried on. The term 'abolition' for the third response is, however, somewhat misleading. It does not mean that prostitution is to be abolished but stands for just the opposite. According to this approach the inevitability of prostitution and the futility of the efforts to regulate or end it must be recognized and all restrictions and control on prostitution must, therefore, be abolished.

It must, however, be noted that the approach adopted by a country does not necessarily fit neatly into one of these three responses. Strategies of suppression may not be ambitious or adequate enough for total suppression nor has regulation a universal meaning in all contexts.

Following the recommendations of the Wolfendon Committee, the Street Offences Act, 1959 was enacted in the U.K. The object of the Act is to prevent the nuisance aspect of prostitution and not to ban it. Section I of the Act provides : "It shall be an offence for a common prostitute to loiter or solicit in a street or public place for the purposes of prostitution." Besides the Act of 1959 there are some other local and special laws like the Vagrancy Act, 1824 and the Universities Act, 1825 (enacted for the University of Oxford but subsequently extended to Cambridge University) which also takes care of the problems arising out of prostitution. The objects of all these legislations are to eliminate the nuisance aspect of prostitution and to keep it away from some special areas.

The USA represents the system based on suppression. Prostitution is prohibited in most American jurisdictions. The only jurisdiction where prostitution may be permitted by law is the State of Nevada where counties have the option to permit it if they so desire. The Churchill County employed a referendum to decide the issue and about two-thirds of the people expressed themselves in favor of licensed brothels. In the early years of the decade begining 1970 public opinion polls indicated that about half of the U.S. population was in favor of decriminalising prostitution, more men than women favoring it.

In India, the main legislation dealing with the problem is the Suppression Immoral Traffic Act, 1956 (SITA). The Act does not prohibit prostitution as such but seeks to remove some of the conditions which promote prostitution. The provisions are directed mainly towards those who either organise prostitution by running brothels or induct women into the sex trade. Some of the relevant provisions in the Penal Code are : Sections 366, 366-A, 372 and 373. There are some provisions in the Children's Acts also as per definition of a neglected child in Section 2(iv) of the Central Act of 1960.

Some of the changes introduced in Immoral Traffic (Prevention) Act of 1956 (ITPA) by amending SITA are as given below :

1. Licensing authorities have been empowered to suspend licenses of hotels where children or minors are detected to be used for Prostitution.
2. Sections 10 and 12 of SITA providing for the release of convicted persons on probation of good conduct or after due admonition and on security from habitual offenders for good behaviour have been omitted from the Act.
3. Provision has been made for medical examination of persons removed from a brothel after a search has been effected under the relevant provisions of the Act.
4. Women or girls removed in pursuance of a search are to be interrogated only by women police officers and in the presence of woman social worker if no female police officer is available.
5. The Central Government has been empowered to set up special courts with jurisdiction to try offences having inter-State ramifications.

The Immoral Trafficking Prevention Act, 1956 ("ITPA")

Before going into the provisions of PITA, a comprehensive definition of trafficking needs to be brought in order. Trafficking is the recruitment, transportation, transfer, harbouring or receipt of persons, by means of the threat or use of force or other forms of coercion, of abduction, of fraud, of deception, of the abuse of power or of a position of vulnerability or of the giving or receiving of payments or benefits to achieve the consent of a person having control over another person, for the purpose of exploitation. *(Source: The UN Protocol to Prevent, Suppress and Punish Trafficking in Persons, 2000)*

The Immoral Trafficking Prevention Act, 1956 ("ITPA"), the main statute dealing with sex work in India, punishes acts by third parties facilitating prostitution like brothel keeping, living off earnings and procuring, even where sex work is not coerced.

Section 3. Punishment for keeping a brothel or allowing premises to be used as a brothel.

(1) Any person who keeps or manages, or acts or assists in the keeping or management of, a brothel, shall be punishable on first conviction with rigorous imprisonment for a term of not less than one year and not more than three years and also with fine which may extend to two thousand rupees and in the event of a second or subsequent to conviction with rigorous imprisonment for a term of not less than two years and not more than five years and also with fine which may extend to two thousand rupees.

(2) A any person who,

(a) Being the tenant, lessee, occupier or person in charge of any premises, uses, or knowingly allows any other person to use, such premises or any part thereof as a brothel, or

(b) Being the owner, lessor or landlord of any premises or the agent of such owner, lessor or landlord, lets the same or any part thereof with the knowledge that the same or any part thereof is intended to be used as a brothel, or is willfully a party to the use of such premises or any part thereof as a brothel, shall be punishable on first conviction with imprisonment for a term which may extend to two years and with fine which fine which may extend to two thousand rupees and in the event of a second or subsequent conviction, with rigorous imprisonment for a term which may extend to five years and also with fine.

(2-A) For the purposes of sub-section (2), it shall be presumed, until the contrary is proved, that any person referred to in clause (a) or clause (b) of that subsection, is knowingly allowing the premises or any part thereof to be used as a brothel or, as the case may be, has knowledge that the premises or any part thereof are being used as a brothel, if,

(a) A report is published in a newspaper having circulation in the area in which such person resides to the effect that the premises or any part thereof have been found to be used for prostitution as a result of a search made under this Act; or

(b) A copy of the list of all things found during the search referred to in clause (a) is given to such person.

Section 5. Procuring, inducing or taking person for the sake of prostitution.

(1) Any person who-

(a) Procures or attempts to procure a person whether with or without his/her consent, for the purpose of prostitution; or

(b) Induces a person to go from any place, with the intent that he/she may for the purpose of prostitution become the inmate of, or frequent, a brothel; or

(c) Takes or attempts to take a person or causes a person to be taken, from one place to another with a view to his/her carrying on, or being brought up to carry on prostitution; or

(d) Causes or induces a person to carry on prostitution; shall be punishable on conviction with rigorous imprisonment for a term of not less than three years and not more than seven years and also with fine which may extend to two thousand rupees, and if any offence under this sub-section is committed against the will of any person, the punishment of imprisonment for a term of seven years shall extend to imprisonment for a term of fourteen years:

Provided that if the person in respect of whom an offence committed under this sub-section,

(i) Is a child, the punishment provided under this sub-section shall extend to rigorous imprisonment for a term of not less than seven years but may extend to life; and

(ii) Is a minor; the punishment provided under this sub-section shall extend to rigorous imprisonment for a term of not less than seven years and not more than fourteen years.

So it can be seen that both the sections namely section 3 and section 5 punishes only the acts of the 3rd party and same does the other sections in the Act and so new legislation shall be passed as to punish the client who are visiting the prostitutes.

Clauses in the ITPA relating to living of the earnings are being challenged in court, together with criminalisation of brothels, prostitution around a notified public place, soliciting and the power given to a magistrate to evict sex-workers from their home and forbidding their re-entry. Other groups are lobbying parliament for amendments (Deccan Herald 7th March 2011).

The only situations for which a prostitute can be punished are :

1. Where prostitution is carried on in premises which are within a distance of 200 yards of any place of public religious worship, educational institution, hotel, hospital, nursing home or such other public place of any kind as may be notified in this behalf by the commissioner of police or district magistrate.
2. Where a prostitute makes positive attempts to seduce or solicit persons for purposes of prostitution.

One important object of the Act is to rescue girls in need of help and provision has been made for protection homes.

Enforcement of Laws

The position of prostitution and the law varies widely worldwide, reflecting differing opinions on victimhood and exploitation, inequality, gender roles, gender equality, ethics and morality, freedom of choice, historical social norms, and social costs and benefits.

Legal themes tend to address four types of issue : victimhood (including potential victimhood), ethics and morality, freedom of choice, and general benefit or harm to society (including harm arising indirectly from matters connected to prostitution).

Prostitution may be considered a form of exploitation (e.g. Sweden, Norway, Iceland, where it is illegal to buy sexual services, but not to sell them — the client commits a crime, but not the prostitute), a legitimate occupation (e.g., Netherlands, Germany, where prostitution is regulated as a profession) or a crime (e.g. many Muslim countries, where the prostitutes face severe penalties).

The legal status of prostitution varies from country to country, from being legal and considered a profession to being punishable by death. Some jurisdictions outlaw the act of prostitution (the exchange of sexual services for money); other countries do not prohibit prostitution itself, but ban the activities typically associated with it (soliciting in a public place, operating a brothel, pimping etc), making it difficult to engage in prostitution without breaking any law; while in a few countries prostitution is legal and regulated.

A dichotomy of prostitution in India is that it is legal but not regulated. The experience has almost been universal that the effective enforcement of legislation regarding prostitution is an impossible task. This is true irrespective of the fact whether the aim of law is suppression or regulation. The two main reasons for the failure of law are : the act involved is one of those violations which have been termed as 'crimes without victims' and the corruption of enforcement personnel. Realising the inherent difficulties in enforcement, United Kingdom gave up regulation in 1886; Norway abolished licensed houses in 1890; Denmark and Finland abandoned regulation in 1906 and 1907 respectively (Flexner, 1914:104).

A further difficulty is that the nature of the activity being what it is, it may not be possible to establish or even identify the vice in a given situation. The institution of singing and dancing girls' in India presents an interesting example. If genuinely so, these women are not involved in prostitution but are in a way concerned with purely cultural activities. 'Tawaif, a woman of this kind, at some time represented a very high level of culture and often had claims to literary attainments also. It is reported that in the olden days members of the elite class of Lucknow and other centres sent their sons to these women in order to be trained in fine etiquette and manner and for developing cultural appreciation pertaining to music and dance. Since the implementation of SITA many prostitutes camouflage the true nature of their profession by adopting the label of 'dancing' or 'singing' girls.

The idea of rescue homes has not been properly implemented because of the acute shortage of such homes. For a vast country like India there are only 81 rescue homes in all; most of the homes are very ill-managed and ill-equipped, specially those under government management.

There have been practical difficulties in the fulfillment of the formalities the implementation of certain provisions of SITA. Section 15(2) of the act empowers the special police officer, when he has reasonable grounds to believe that an offence in respect of a girl is being committed and a search with warrant will cause undue delay to search the premises without a warrant in the presence of two or more respectable persons of the locality, of whom one at least is to be a woman, and remove the girl and produce her forthwith before the magistrate. The insistence on the presence of two respectable persons of the locality is based on sound public policy in order to prevent the harassment of innocent persons by the police. More often than not, it is however not possible to manage two respectable persons willing to associate themselves with the proceeding in connection with activities of a dubious nature.

Recent Trends

According to the paper "Estimating the prevalence and career longevity of prostitute women" (Potterat et al., 1990), the number of full-time equivalent prostitutes in a typical area in the United States (Colorado Springs, CO, during 1970–1988) is estimated at 23 per 100,000 population (0.023%), of which fraction some 4% were under 18. The length of these prostitutes' working careers was estimated at a mean of 5 years. A follow-up paper entitled "Prostitution and the sex discrepancy in reported number of sexual partners" (Brewer et al., 2000) goes on to estimate a mean number of 868 male sexual partners per prostitute per year of active sex work, and offers the conclusion that men's self-reporting of prostitutes as sexual partners is seriously under-reported.

In 2003 it was estimated that in Amsterdam, one woman in 35 was working as a prostitute, compared to one in 300 in London. In the United States, a 2004 TNS poll reported 15% of all men have paid for sex at least once in their life. In Australia, a survey conducted in the early 2000s, showed that 15.6% of men aged 16–59 have paid for sex at least once in their life, and 1.9% had done so in the past year(Rissel, 2003).

A number of reports over the last few decades have suggested that prostitution levels have fallen in sexually liberal countries, most likely because of the increased availability of non-commercial, non-marital sex. Prostitutes have long plied their trades to the military in many cultures. For example, the British naval port of Portsmouth had a flourishing local sex industry in the 19th century, and until the early 1990s there were large red light districts near American military bases in the Philippines. The notorious Patpong entertainment district in Bangkok, Thailand, started as an R&R location for US troops serving in the Vietnam War in the early 1970s.

Prostitution is a thriving business in cities and towns across India, although, it is currently a contentious issue. In 2007, the Ministry of Women and Child Development reported the presence of over 3 million female sex workers in India, with 35.47 percent of them entering the trade before the age of 18 years. Human Rights Watch puts the figure of sex workers in India at around 20 million, with Mumbai alone being home to 200,000 sex workers, the largest sex industry centre in Asia. The number of prostitutes has doubled in the last decade.

Much new knowledge on sex work in India came from the first major survey, in April 2011. This was performed by the Center for Advocacy on Stigma and Marginalisation (CASAM), which is part of SANGRAM, a major NGO that deals with sex workers (Sahni & Shankar, 2011). The largest and best-known red-light districts are Sonagachi in Kolkata, Kamathipura in Mumbai, G. B. Road in New Delhi, Reshampura in Gwalior and Budhwar Peth in Pune host thousands of sex workers. In recent years red-light centers across various parts of India are common place for international sex tourism. Earlier, there were other centres such as Maduahdiah in Varanasi, Naqqasa Bazar in Saharanpur, Chaturbhuj Sthan in Muzaffarpur, Peddapuram and Gudivada in Andhra Pradesh. Kabadi Bazar of Meerut is a strange place where public life goes on normally with brothels quietly working of the roofs of the busy market.

Feminism, Prostitution and Legal Issues

The issue relates to the debate over whether to legalize or decriminalize the prostitution. Prostitution is a significant issue in feminist thought and activism. Many feminists are opposed to prostitution, which they see as a form of exploitation of women and male dominance over women, and as a practice which is the result of the existing patriarchal societal order. These feminists argue that prostitution has a very negative effect, both on the prostitutes themselves and on society as a whole, as it reinforces stereotypical views about women, who are seen as sex objects which can be used and abused by men. Other feminists hold that prostitution can be a valid choice for the women who choose to engage in it; in this view, prostitution must be differentiated from forced prostitution, and feminists should support sex worker activism against abuses by both the sex industry and the legal system.

In 1949, the UN General Assembly adopted a convention stating that trafficking persons for the purposes of prostitution is incompatible with human dignity, requiring all signing parties to punish pimps and brothel owners and operators and to abolish all special treatment or registration of prostitutes. As of January 2009, the convention was ratified by 95 member nations including France, Spain, Italy, Denmark, and not ratified by another 97 member nations including Germany, the Netherlands, the United Kingdom and the United States.

Roughly speaking, there are two possible attitudes such as (1) "Prostitution should be made to disappear" and (2) "Prostitution should be tolerated by society". The 'disappearance' argument is further subdivided into three categories : prohibitionism, abolitionism and neo-abolitionism. The 'tolerance' argument is further subdivided into two categories : regulation and decriminalization.

1. "Prostitution should be made to disappear" : Disappearance Argument

(a) Prohibitionism : Prostitution must be prohibited because both prostitutes and clients are criminalized and are seen as immoral, they are considered criminals. This attitude prevails nearly everywhere in the United States, with a few exceptions in some rural Nevada counties. Prohibition can save from otherwise social danger and health disorder.

(b) Abolitionism : Abolitionism suggests that prostitution itself is not prohibited, but most associated activities are illegal, in an attempt to make it more difficult to engage in prostitution, prostitution is heavily discouraged and seen as a social problem. Prostitution (the exchange of sexual services for money) is legal, but the surrounding activities such as public solicitation, operating a brothel and other forms of pimping are prohibited, the current situation in the United Kingdom, France and Canada among others.

(c) Neo-abolitionism : Prostitution is a form of violence against women, it is a violation of human rights, and the clients of the prostitutes exploit the prostitutes. Prostitutes are not prosecuted, but their clients and pimps are, which is the current situation in Sweden, Norway and Iceland (in Norway the law is even more strict, forbidding also having sex with a prostitute abroad).

2. "Prostitution should be tolerated by society": Tolerance Argument

(a) Regulation : Prostitution may be considered a legitimate business; prostitution and the employment of prostitutes are legal, but regulated; the current situation in the Netherlands, Germany and parts of Nevada. The degree of regulation varies very much, for example in Netherlands prostitutes are not required to undergo mandatory health checks while in Nevada the regulations are very strict.

(b) Decriminalization : Prostitution is labor like any other. "Sex industry premises should not be subject to any special regulation or laws", the current situation in New Zealand; the laws against operating a brothel, pimping and street prostitution are struck down, but prostitution is not regulated nearly at all. Proponents of this view often cite instances of government regulation under legalization that they consider intrusive, demeaning, or violent, but feel that criminalization adversely affects sex workers.

In some countries, there is controversy regarding the laws applicable to sex work. For instance, the legal stance of punishing pimping while keeping sex work legal but "underground" and risky is often denounced as hypocritical; opponents suggest either going the full abolition route and criminalize clients or making sex work a regulated business.

Some people opine that prostitution shall be made legal in India and accept them as a part of society because the problem of prostitution is inevitable. The benefit of legalizing prostitution in India will be that atleast we will have a track record of sex workers as for example when dance bar in Bombay were closed most of the bar dancers migrated to Gujarat and Karnataka and other neighboring state and started their business undercover. Legalising prostitution will see these women, who live life on the edge everywhere, gaining access to medical facilities, which can control the spread of AIDS. There is a very strong need to treat the sex industry as any other industry and empower it with legal safeguards. The practical implications of the profession being legal would bring nothing but benefits for sex workers and society as a whole. Keeping prostitution illegal also contributes to crime because many criminals view prostitutes and their customers as attractive targets for robbery, fraud, rape, or other criminal acts. The criminals realize that such people are unlikely to report the crimes to police, because the victims would have to admit they were involved in the illegal activity of prostitution when the attacks took place, now if it is legal then they will easily go and report this to police. Benefits of legalizing prostitution are :

- Legalization of prostitution and the sex industry will stop sex trafficking.
- Legalization of prostitution will control the sex industry.
- Legalization of prostitution will decrease clandestine, hidden, illegal and street prostitution.
- Legalization of prostitution will protect the women in prostitution as they will have rights.
- Women in systems of prostitution want the sex industry legalized as they are the one who suffers the most as they don't have any rights.
- Legalization of prostitution will promote women's health as they can have easy access to medical facilities which they don't have when it is illegal.
- Recognizing prostitution as an economic activity, thus enabling women in India to obtain working permits as "sex workers".

While some emphasize on legalization, the other school emphasizes on decriminalization. Though the words, "legalization" and "decriminalization" are used interchangeably, they have a slight difference in meaning. To legalize something is to "authorize" and "to make legal." When prostitution is legalized in a country, prostitutes are recognized as professionals. Therefore, they have to register or licensed as professionals, which allows a heavy intervention of government authorities over the industry. In countries like Germany, Australia, Nevada, where prostitution is legalized, prostitutes are required to submit fingerprints, photographs, and personal information to the police station.

To decriminalize, other hand, is to "eliminate criminal penalties for or remove legal restrictions against." Advocates for decriminalization of prostitution argue that the system allows the prostitutes to exercise their choice to sell their bodies with their voluntary consents to the customers. Advocates for decriminalization of prostitution apply "the same criteria to private consenting adult sex," to the principle of regulating sex industry. Therefore, they argue that prostitution should not be prohibited if it was performed in a privacy of one's home or hotel, and the money is freely exchanged with consent. Advocates also argue that decriminalization of prostitution will encourage prostitutes to seek help from the police men since prostitution will no longer be illegal. They also assert that decriminalization will reduce crimes against the prostitutes.

The Supreme Court in India (AIR1997SC3021, 1997(2)ALD(Cri)199, 1997(2)Crimes40(SC), JT1997(6)SC305, 1997(4)SCALE657, (1997)8SCC114, [1997]Supp2SCR105) came to conclusion that the country is better off to legalize prostitution. Despite of fierce criticism by the abolitionists, some people however proposed that decriminalization of prostitution might be a better fit for the India's case of sex industry. Granted, advocates for both decriminalization and legalization of prostitution are basing their arguments on the examples of other developed countries. New Zealand, for instance, is frequently used as an example of a country, which successfully decriminalized prostitution. Similarly, Germany is used as an example for country which legalized prostitution. However, the advocates for decriminalize or legalize prostitution in India fail to understand the problems which New Zealand or Germany faces because of decriminalized or legalized prostitution. Melissa Farley, a scholar who has done numerous researches on prostitution and sex industry summarizes the review of New Zealand government on prostitution industry since the decriminalization saying decriminalization of prostitution industry has only exacerbated the crime rates in New Zealand. Further in case of India, many prostitutes are illiterate and came for the low castes class. Whether prostitution becomes decriminalized, their understanding of their legal rights as well as their view of themselves will rarely be affected by it. Their illiteracy already prevents them from understanding their constitutional rights, and their self-image is deeply controlled by the attitude of Hindu culture towards people in the low castes classes. Hence, if decriminalization affects prostitution in India, it will only add certain problems that New Zealand is experiencing because of decriminalization to the current issues that India is already facing in prostitution industry.

Thus either India shall legalize prostitution which is the most suitable step that can be taken or shall make such deterrent laws as to curve up the problem of prostitution. Laws should not be such as to just remove the prostitutes but also to change the mentality of people who are interested in paid sex by punishing them in such a manner that people of same mentality will dare to indulge themselves in similar activities. The causes of prostitution in India are poverty, ill treatment by parents, bad company, family prostitutes, social customs, caste system, inability to arrange marriage, lack of sex education, media, prior incest and rape, early marriage and desertion, lack of recreational facilities, ignorance, and acceptance of prostitution. They are the byproducts of the pathological social system, which fundamentally prevent certain group of people in Indian society from receiving proper education and rising above the poverty line. Majority favor the suggestion of the Supreme Court authority rather than the views held by proponents for decriminalizing prostitution.

Prevention

Steps that should be taken in order to fight with prostitution can be :

- Formal education should be made available to those victims who are still within the school going age, while non-formal education should be made accessible to adults.
- The Central and State Governments in partnership with non-governmental organizations should provide gender sensitive market driven vocational training to all those rescued victims who are not interested in education.
- Rehabilitation and reintegration of rescued victims being a long-term Recruitment of adequate number of trained counsellors and social workers in institutions/homes run by the government independently or in collaboration with non-governmental organizations.

- Awareness generation and legal literacy on economic rights, particularly for women and adolescent girls should be taken up.
- Adequate publicity, through print and electronic media including child lines and women help lines about the problem of those who have been forced into prostitution.
- Culturally sanctioned practices like the system of devadasis, jogins, bhavins, etc. which provides a pretext for prostitution should be addressed suitably.

References

Abraham Sion, (1977) Prostitution and the Law, Edinburgh Books, United Kingdom.

Banarji, S.C. and Banarji R, (1989): 'The Castaway of Indian Society', Puntni Pustak, Calcutta.

Beckman, Karen Redrobe, (2003) Vanishing Women: Magic, Film, and Feminism, Duke University Press, pp. 31–3.

Beranand Jessie : (1957) Social Problems at Mid-century, Published by the Darden Press New York.

Bilimoria Rani, (1983) Female Criminality, New Delhi, Eastern Book Co.

Brewer DD, Potterat JJ, Garrett SB, et al.. (2000) "Prostitution and the sex discrepancy in reported number of sexual partners". Proc. Natl. Acad. Sci. U.S.A. 97 (22): 12385–8.

Bullough, Vern and Bullough, Bonnie (1978) Prostitution: An Illustrated Social History. New York: Crown Publishers.

Campbell, Russell, (2005) Marked Women: Prostitutes and Prostitution in the Cinema, University of Wisconsin Press.

Castillo DN, Jenkins EL (1994) "Industries and occupations at high risk for work-related homicide". J Occup Med 36 (2): 125–32.

Davis Kingsley, (1937) The Sociology of Prostitution, American Sociological Review, vol. 2, October, pp. 746-55.

Davis Kingsley: (1971) "Prostitution in Merton and Nisbet (ed)' contemporary Social Problems, Narcourt Brace Janmichlnc, New Yourk.

Encyclopedia Britannica, (2005) Britanntica Ready Reference Encyclopedia. vol 8, New Delhi.

Fisher, Michael H, (2007) "Excluding and Including "Natives of India": Early-Nineteenth-Century British-Indian Race Relations in Britain", Comparative Studies of South Asia, Africa and the Middle East 27 (2): 303–314.

Flexner, Abraham, (1914) Prostitution in Europe, New York. The Century Co.

Galwankar S, (2002) Sexual behaviours in migrant male workers from Mumbai: a need for sex education for the uneducated, Int Conf AIDS. Jul 7-12; 14: Abstract no. WePeD6342.

Gathia Joseph, (1998) 'Child Prostitution in India', Concept Publishing Company, New Delhi.

Greenwold, W, (1958) The Call Girl: A Social and Psychoanalytical Study, New York.

Gurumukh Ram Madan, (1966) "Indian Social Problems', Allied Publishers Private Ltd. Bombay, Delhi, Calcutta, Madras, vol.-I.

Gurumukh Ram Madan : (1967) "Indian Social Problems', Allied Publishers Private Ltd. Bombay, Delhi, Calcutta, Madras, vol.-II.

Hastings, James, (1952) "Encyclopedia of Religion and Ethics", vol.X, New York.

James Frazer, (1922) The Golden Bough, 3e, Chapter 31: Adonis in Cyprus.

Jenness, Valerie (1990). "From Sex as Sin to Sex as Work" Social Problems, 37(3), 403-420, 1990.

Joseph M. Patrssano, (2001) "World of Sociology" vol.II.

Kapour, P, (1975) "The Life and Work of Call Girls in India" Sahibabad Publishing House.

Keire, Mara L. (2010) For Business and Pleasure: Red-Light Districts and the Regulation of Vice in the United States, 1890-1933 (Johns Hopkins University Press); 248 pages.

Magill N. Frank, (1995) "International Encyclopedia of Sociology, vol. II. London Chiago.

Matthews Roger and Nill O. Maggie, (2003) "Prostitution"(ed) Published by Dart Mouth Publishing Company Ashgate Publishing Ltd., Gower House.

McCall, Andrew, (2004) Medieval Underworld, Sutton Publishing.

Motichand, (1973) "The World of Courtesans", Vikash Publishing House Pvt. Ltd. New Delhi.

Mukerji Santosh Kumar, (1996) "Prostitution in India", Inter India Publications, New Delhi.

Mukherjee K.K. and Das Deepa, (1996) "Prostitution in Metropolitan cities of India", A Study by: Central Social Welfare Boarol, Samaj Bhavan, New Delhi.

Nair PM, (2007) Trafficking Women and Children: Handbook for Law Enforcement Agencies in India, UNIFEM, New Delhi.

Nair PM, (2007) Human Trafficking, Konark Publications, New Delhi.

Raj M. Sunder, (1993) "Prostitution in Madras, A Study in Historical Perspective", Konark Publishers Pvt. Ltd. Delhi.

Rissel, Chris E. (2003) "Sex in Australia: Experiences of commercial sex in a representative sample of adults". Australian and New Zealand Journal of Public Health 27: 191.

Rojanapithayakorn W, (2006) "The 100% condom use programme in Asia" Health Matters 14 (28): 41–52.

Roper L, Luther on sex, marriage and motherhood.". The University of Warwick.

Rozayio Rita, (1988) "Trafficking in Women and Children in India", Uppal Publishing House, New Delhi.

Sahni R & VK Shankar, (2011) The First Pan-India Survey of Sex Workers: A summary of preliminary findings, Center for Advocacy on Stigma and Marginalisation (CASAM), Kolkata, April.

Siddique Ahmad, (2005) Criminology — Problems perspectives, Eastern Book Co., Fourth edition.

Siegal, Larry J. (2005). Criminology: The Core Second Edition. Thompson.

Singh, P.K. (2004) "Brothel Prostitution in India", University Book House, Pvt. Ltd. Jaipur.

Singh, S.N. and Basu N.K., (1994) "History of Prostitution in India", Cosmmy Publications, New Delhi.

Thakur, A.P. and Banerjee Anjana, (2005) "Encyclopedia of the Theoretical Sociology", vol.2, New Delhi.

Venkatchaliah, M.N. (1998) "Child Prostitution in India", Concept Publishing Company, New Delhi.

Weinberg S. Kirson, (1960) Social problems of our times, Prentice Hall Inc Engkwood Cliffs.

■■■

CHAPTER 19 Penology : Concept & Dimensions

Learning Objectives :

- **Acquainting the Learners with the concept of penology and punishment.**
- **To understand history, Reasons and types of punishment.**
- **To have a critical look at judicial process of sentencing offenders.**
- **To comprehend the Capital Punishment Controversy in India.**

Introduction

Concepts of Penology

Penology refers to the scientific study of penalties (from the Latin poena, "punishment"), although in its broadest sense it is also concerned with the consequences and merits of attempting to deal with various kinds of conduct by criminal prohibition ('criminalizing'). It includes not only the study of penal codes of law, but also the investigation of ways in which penal codes are applied by courts in practice, and the manner in which each type of penal measure is applied. For example, even when a penal code appears to oblige courts to pronounce a sentence (such as imprisonment for life in the case of murder), there are ways of avoiding this (such as convicting the offender of a less serious charge of homicide); most penal systems provide legal devices by which a sentence of imprisonment can be terminated before its nominal end.

Punishment is the practice of imposing something unpleasant or aversive on a person or animal, usually in response to disobedience, defiance, or behaviour deemed morally wrong by individual, governmental, or religious principles. Penology comprises penitentiary. Penitentiary may refer to Apostolic Penitentiary, a tribunal of mercy, responsible for issues relating to the forgiveness of sins in the Roman Catholic Church or Prison or a correctional facility. A prison is a place in which people are physically confined and, usually, deprived of a range of personal freedoms. Other terms are penitentiary, correctional facility, and jail, although in the United States "jail" and "prison" refer to different subtypes of correctional facility.

Penologists are interested in all such expedients, and in the criteria which are used by courts, administrators and other personnel to make distinctions between offenders, whether for such purposes or for other reasons. Other reasons may include the belief that certain types of offenders are more likely than others to respond to certain regimes, or that some prisoners are so dangerous that they must be given special sentences, detained longer than is normal for the offence, or given freedom only under especially strict conditions.

The study of "penology is the science of prison administration and rehabilitation of criminal." Contemporary penology concerns itself mainly with prison management and criminal rehabilitation. Rehabilitation means; To restore to useful life, as through therapy and education or To restore to good condition, operation, or capacity. The word seldom applies to theories and practices of punishment in less formal environments such as parenting. Parenting is the process of promoting and supporting the physical, emotional, social, and intellectual development of a child from infancy to adulthood.

An important task of penologists is to provide answers to the question of effectiveness of the measure taken. Effectiveness is usually assessed by reconvictions or rearrests, although this is not without problems. For example, it cannot take account of offences of which the offender is not suspected; the follow-up period must be substantial; in some jurisdictions rearrests or reconvictions for minor offences are not recorded centrally. The most serious problem, however, is the difficulty of being sure that offenders who remain free of rearrests during the follow-up period would not have remained free if otherwise dealt with: for example, if merely discharged without penalty. In consequence, follow-up studies must usually be content with comparing the reconviction rates after different measures. Even so, they have to take into account the fact that courts are selective, and do not allocate offenders randomly to different measures (a few random allocation studies have been achieved, but only for rather specific groups of offenders or offences (Farrington, 1983).

The criteria used to allot offenders to different measures may themselves be associated with higher or lower reconviction rates. For instance, the more previous convictions that individuals have, the more likely they are to be reconvicted, quite apart from any effect which a sentence may have on them. Again, offenders whose offences usually involve theft, burglary, drunkenness or exhibitionism are more likely to be reconvicted than those who commit serious sexual offences or personal violence. Statistical devices have to be used to allow for this, for example, by subdividing samples into high-, medium- and low-risk groups. When such precautions are taken, the differences between reconviction rates following such different measures as imprisonment, fines and probation have tended to disappear, and the choice of sentence therefore seem to make no difference to a person's likelihood of reconviction; or not enough difference to justify expensive measures. But this is now regarded as an exaggeration : some types of offenders do respond to some techniques (McLaren 1992).

In any case, other possible aims of penal measures have to be taken into account. Psychiatrists, for example, usually regard themselves as primarily concerned with the mental health of those committed to their charge by criminal courts; and social workers—including many probation officers—regard their clients' financial and family problems as more important than their legal transgressions.

Whether these views are accepted or not, some penal measures are valued as general deterrents, in the belief that even if they do not often affect the conduct of those who have experienced them, they discourage potential offenders who have not yet committed offences (Beyleveld 1980). The efficacy of general deterrents has been exaggerated, for example, by the supporters of capital punishment : statistical comparisons of jurisdictions which have abolished or retained the death penalty, or of decades in the same jurisdiction preceding and following abolition, suggest that the substitution of long periods of imprisonment for the death penalty does not affect rates of intentional homicide. In plain terms, potential murderers who think before they kill are as likely to be deterred by 'life' as by death. Whatever the penalty, however, its deterrent efficacy depends to a great extent on people's own estimates of the probability of being detected and punished. For some people this seems immaterial; but they tend to be those who commit impulsive or compulsive crimes.

Another aim of some penal measures is simply to protect other people against a repetition of the offence by the offender concerned, usually by some degree of incapacitation. Incapacitation may take the form of long detention, disqualification from certain activities (such as driving or engaging in certain occupations) or surgery (for example, castration for rapists). The more severe types of incapacitating measures are controversial, the chief objection being that the probability of offenders' repeating their offence seldom approaches certainty, and is often less than 50:50 (Floud and Young 1981).

This illustrates a more general tendency to acknowledge the relevance of jurisprudence for penology. Scepticism about the efficacy of corrective or deterrent measures, together with the excessive use of very long detention in the name of therapeutic treatment, has revived the classical emphasis on the need for penalties to reflect the culpability of the offender. The underlying Kantian morality of this was never quite abandoned by jurists in the former West Germany; but the revival of it in the USA and Scandinavia is an important phenomenon, although lacking the sophistication of German jurists (Von Hirsch 1976). Penology and justice studies emphasize description and analysis of antecedents of criminal behaviour and outcomes of consequences imposed by criminal justice on the criminal behaviour. Wilkenson (2004) notes that overall heterogeneity of a society may provide a meta-explanation for the variance in incarceration rates : There may be a multi-directional causality where close-knit societies are least likely to offend against one another.

English judges—and, quite independently, Durkheimian sociologists—have contributed yet another notion. Without necessarily accepting the retributive view (which has both difficulties and dangers) they hold that penalties have an important expressive or symbolic function, declaring publicly the moral disapproval with which most people regard harmful offences (Walker 1978). Some English judges have even stated that an important task of sentencers is to lead public opinion, although this seems to exaggerate the attention and respect which the public pay to sentences (Walker and Marsh 1984). More tenable is the proposition that sentences reflect people's disapproval : the question is whether sentencers are selected or trained so as to be sure of reflecting the views of the law-abiding public, particularly in societies with heterogeneous moralities.

Other subjects in which penologists have interested themselves are the rights of offenders, especially those recognized by conventions (such as those of the United Nations or European

Union); the protection of offenders against avoidable stigma; and the rights of victims to compensation, whether from the state or the offender, and to other forms of care. Contemporary penology concerns itself mainly with Prison reform, Prisoner abuse, and Recidivism, as well as theories of the purposes of punishment (such as Deterrence, Rehabilitation, Retribution, Utilitarianism and prison management).

Punishment

The theory of punishment is based on the notion that punishment is to be inflicted on offenders so as to reform them, or rehabilitate them so as to make their re-integration into society easier. Punishment for crime includes the following elements : (1) it must involve pain or other consequences normally considered unpleasant; (2) it must be administered for some offence against legal rules; (3) it must be administered to an actual or supposed offender for his offence; (4) it must be intentionally administered by persons other than the offender; and (5) it must be imposed and administered by an authority constituted by a legal system against which the offence is committed. Punishment is intended to ultimately produce some sort of justifiable suffering of the offender for retribution, deterrence, protection, and reformation and rehabilitation. According to Max Greenhut in his Penal Reform : A Comparative Study (P.3), three components must bc present "if punishment Is to act as reasonable means of checking crime." First "speedy and inescapable detection and prosecution must convince the offender that crime does not pay. Second, after punishment; the offender must have "a fair chance of a fresh start." And third, "the state which claims the right of punishment must uphold superior values which he (offender) can reasonably be expected to acknowledge."

The concept of punishment is that of inflicting some sort of pain on the offender for his violation of law. This is an instrument of public justice. To illustrate, if a thief is prosecuted and brought before a Court, his case heard, punishment awarded by the Court and finally executed by the State, then this become punishment in legal sense. It will not be a punishment in law, if father beats his son for committing a theft in his house or killing of Naxalites by the State for their anti-national activities without prosecuting them. Accordingly following points may need to be brought in order for a fuller understanding of the concept of punishment.

First, punishment is a privation (evil, pain, disvalue). Second, it is coercive. Third, it is inflicted in the name of State; it is "authorised". Fourth, punishment presupposes rules, their violation, and a more or less formal determination of that, expressed in a judgment. Fifth, it is inflicted upon an offender who has committed a harm, and this presupposes a set of values by reference to which both the harm and the pain are ethically significant. Sixth, the extent or type of punishment is in some defended way related to the commission of the harm, and aggravated or mitigated by reference to the personality of the offender, his motives and temptation.

Reasons of Punishment

Administration of Criminal Justice is a policy which studies the means by which the volume of harmful conduct in the society can be limited. The subjects of investigation in Criminal Justice includes : (a) appropriate measures of social organization whereby harmful activities may be prevented and (b) the treatment to be accorded to those who have caused harm. However, the last mentioned investigation involves decisions as to which appropriate sanctions can be imposed under Criminal Law, and these range from various degrees of imprisonment down to such measures as probation or mere admonition. The degree of punishment depends on (a) length of sentence, (b) personality (a stable personality means lesser influence), (c) relationship with outside world, (d) refusal to accept blindly the codes of prison, population, (e) seriousness in work and recreative activities, and (f) the character of associates. The various objects of punishment being :

Retributive

Retributive refers to deserved punishment. The retributionists called it to be the most immediate and animalistic reaction. Often it is found that being dissatisfied with the functioning of law and justice specially of police, people react spontaneously and mercilessly, beat or even kill the criminal apprehended by them. However, a tight crime and justice system on the other hand, can sublimate or suppress this animal feelings aroused.

Expiative

Expiative is an early and ancient justification of punishment. Earlier, the logic was that a criminal is a source of infection to the society and therefore he deserved to be expiated. The various methods of early expiation that were adopted included stoning to death, putting an iron rod on the back, dismemberment of limbs and other severe physical tortures. In expiation, often the criminal is taken as a scapegoat for outletting the outraged feelings created by the offence. However, the penologists are against this form of expiatory punishment. In "The Universal Declaration of Human Rights, Article 5, it is stated that no one shall be subjected to torture or cruelty or inhuman treatment of punishment. The Universal Covenant on Civil and Political Rights has mentioned in Article 6 that : (i) No one shall be deprived of his life arbitrarily, (ii) Death sentence is to be abolished or if allowed should be restricted to some serious crimes, (iii) There shall be no genocide, (iv) There shall be right to seek pardon, (v) There shall be no death sentence on persons below 18 years or on pregnant woman.

Expiatory sentence in present day world is considered as an act of brutality. Often this type of punishment is an expression of utter internal dissatisfaction and often disproportionate to the crime caused. So, there is the probability of having no rational in the proportion of punishment made to an offence.

Deterrence

Punishment is often said to have a deterrent effect on people other than the offenders and it is said that the deterrent aspect of punishment over the non-offenders are more than that of the offenders, so much so, that those who note the offences, are terrified to commit the offence themselves. However, the deterrent effect of punishment is an observable variable, dependent upon the following six factors : (I) The social structure and set of social values, (II) The particular population group, (III) The type of law that the punishment is intended to uphold, (IV) The type of severity of the penalty, (V) The degree of certainty of apprehension and punishment, (VI) The Individual's understanding of the law and its prescribed penalties as well as his perception of the total situation.

Protection

One of the main objects of punishment is protecting the society from the act of criminals. Most of the philosophy of punishment has been construed on the basis of this thesis. It is believed, that confined within walls, a criminal becomes harmless to the society. The criticism of this theory of punishment being that it has failed to distinguish individual impact on punishment, the psychic pressure of punishment and individual reaction of mass treatment.

Reformation

The modern attitude to punishment is that it is an individualized treatment process and responses to every individual events of crime. The question therefore is who the offender is and what is his offending phenomenon. Modern criminologists who support the reformatory object of punishment argue that crime is a deviation from a social norm. There are several factors responsible for such deviation, since, when a crime occurs it takes place due to culmination of variety of factors and that a good many crimes are the result of socio-economic milieu. These factors may be social and economic, may be due to erosion of moral values by parental neglect, stress of circumstances or doing a criminal activity in spur of heat of a moment.

Reformation and social rehabilitation of criminals are the two important aspects concerning the objectives of punishment in Criminal Justice administration nowadays. These two objectives can best be fulfilled if we effectively try the system of Probation and Parole in Criminal Justice Administration. Indian Government declared 1971 as "Proabation Year" in the backdrop of the alarming increase in crime, especially among teenagers and ideological breakers of law. Having regard to the significance of Probation as a potent instrument accepted in all civilized countries for achieving social rehabilitation of the offender and the prevention of recurrence of crime, all serious administrators, judges, jurists, students of criminology and the heads of institutions of social welfare, regard probation as an integral part of social defence and correctional therapy.

The spiritual principle behind probation in the proper functioning of Criminal Justice Administration in India is that " All persons are born equal and no man is born a criminal". God after all has made man in his own image. The great truth which inspires the Probation system is that if a criminal is not born but made, and made by socio-economic circumstances, then it is the duty of the society to treat the individual offender to wean him away from the criminal path. Criminologists have taken the view that it is social neglect and domestic stress, not to speak of prison contamination that makes the first offender a die-hard delinquent. With proper opportunities for rehabilitation and incentives to conform to the needs of social behaviour, that the society may hope to salvage the fresh men in the field of crime.

History of Punishment

The history of penology is the history of punishment. The ancient logic was that if a man killed, he in turn should be killed; the offender should always be injured in the same way or to the same extent that he had injured his victim. Blood vengeance became a tribal matter when members of different tribes were involved, hi many societies, early Babylonian, Greek, Roman, and Chinese civilizations, for example, parents had the right to kill their children for disobedience. Historically, punishment was harsh and corporal punishment was predominant. It was common, for example, to mutilate of offender's hands for forgery or perjury. In Arabia, even in recent times, the foot and/or hand of a thief could be dismembered. In various societies the tearing of the scalp, severing of the nose, whipping, benting, and other forms of torture were frequently practiced, both before and after the accused was declared guilty. Even in the United States, prisoners in Delaware could be whipped as recently as 1952.

In some cultures, wrongdoers were burned alive or beheaded the body was drawn and quartered, and facial mutilation and branding for the purpose of criminal identification was common. In some instances, offenders were put into boiling water or oil. These were all justified means to curb criminal behaviour and frighten potential lawbreakers. Such cruel forms of punishment it should be noted, also served as community entertainment. Often the tortures, beheadings, and hangings were held in public for the musement of the throngs.

In most cases, it seems that the harshest punishment had an adverse effect; crime increased instead of decreased. England's adopted the most harsh and brutal criminal code to that date. More than 350 crimes were designated as -capital crimes as means to deter crime. But crime flourished. After many years of trial and practice this punitive policy was revised. Today, England has the most liberal penal system in the world. Many penal reforms have been affected, including the abolition of capital punishment. England's crime rate today is substantially lower than it was in the past; it is also substantially lower than the crime rates in other modem countries at present.

Many societies practiced the punitive measure known as transportation -banishment or exile. Convicted criminals were sent to distant areas, often to geographically inhospitable regions, to live in a penal colony. If such prisoners survived, they often became outcasts or members of minority-groups. The best -known penal colony, to Americans, is Devil's Island, off French Guiana. The notorious penal island of Andaman, in India's Bay of Bengal, is a classic example. English felons were sent to the American colonies as early as 1597. Until the American Revolution, about 2000 were sent over annually. When opposition developed in American and Tanzania until the entire practice was halted in 1867.

Banishment is sometimes considered the equivalent to death. Socrates, for instance, was offered the choice of exile or death by poison. It was common among primitives to order undesirable members out of the group or local, knowing that the outcast would likely meet death from the hostile environment or from an enemy tribe. In modern states, another process called deportation has been of wide use. Because of the high degree of mobility of people between countries, many residents are aliens. If an alien becomes a criminal, the host country expels him and he is returned to his state of citizenship. Deportation of political criminals is most common, but persons are frequently deported for immoral acts, use of narcotics, and labor law violations.

The system of detention in jail and prison was not well established before the 18th century. As a result of urban development following the Industrial Revolution, most of the European nations experienced an increase in crime. Rising unemployment, poverty, and hunger turned the most severely affected person to a life of crime. Jails became over crowded. One measure to alleviate this problem was the system of transportation.

Penal philosophies usually undergo a development from retaliation to brutality, to restraint, to deterrence, to treatment and to rehabilitation. In the United States, Pennsylvania adopted England's severe penal code in 1676. Six years later the Pennsylvania Assembly passed, "The Great Law", relaxing the criminal code to a more workable and humane discipline. Despite reforms, most States practiced severe punishment. Criminals were treated inequitably for some 200 years, including present practices, in terms of discrimination in arrests, convictions and penalties depending on one's social status or political affiliation and power. Inhumane treatment of prisoners continued long after the Declaration of Independence.

Penal philosophies differ widely between East and West. This difference is primarily attributable to the fundamental differences between the two cultures in social values. To Orientals, crime is caused by neglecting the natural order. Man's action is regarded as part of the complex network of casual filaments within the Universe. Any social violation is interpreted as an ominous disturbance of the natural order, and thus is subject to punishment by heaven through designated agencies such as the family, clan or state.

In China, Korea, or Japan, if a very serious crime were committed the entire street in the area of the crime could be destroyed as a governmental punishment Such practices, persisting for centuries, produced subsocial system in which everyone observed the actions of others. If anybody were aware of the possibility of a crime, neighbors or relatives or friends would intervene to prevent the occurrence of the crime as a means of self-protection. Another scheme to deter crime involved the obligation of the family and its honour. When officials could not locate the offender, someone in his family was made to take the punishment. In some cases, the entire family would be enslaved. In extreme cases, the first or second of kin might be executed. The Chinese law, however, stated that no one would be subject to punishment if the accused did not confess his own act.

Throughout the long history of China, Korea and Japan, the methods of punishment employed were similar to those of the West. Penalties included death and torture, as well as whipping, beating, imprisonment, transportation, degradation, castration, amputation, tattooing, forced suicide, ostracism and ridicule, banishment, and disinheritance by family and clan groups. To a large extent, the scientific investigation of crime was initiated by the Chinese, and later supplemented by Japanese techniques. The Chinese developed fingerprinting more than 3000 years ago. Psychological methods were frequently used to obtain confessions from law-breakers.

In earlier times, an offender was punished as his captor saw fit. Frequently, retaliation was carried out on the spot; often, a victim was avenged by family members or relatives. Two dominating views were behind the spirit of retaliation. First, the offender had damaged or caused suffering to an individual or group. Second, the individual or his group must respond with an appropriate degree of revenge. In addition to an instinctual theory, the above is based on the view that man is rational and can choose his course of action. Since a free choice is made to break the law and thereby produce harm, revenge is justified, and perceived as even necessary, by the individual, group, or society.

For centuries, until the recent past, the ancient doctrine of Lex Taliones (Law of Retaliation) has served as the punitive function. However, with the origin of criminal laws is in that part of human instinct, which wants to retaliate. "An eye for an eye and tooth for a tooth", is the most ancient of human laws, he contends. This concept appeared in the Code of Hamurabi, around 2000 B.C., as well as throughout the Old Testament (There are, however, many passages in the Old Testament that supersede crude vengefulness, basing human relations including victim - offender -within a context of forgiveness and grace), "As people became more enlightened, the practice of individual retaliation for wrongs suffered was no longer sanctioned. Any action that might endanger the sovereign or good order in the realm became an offence against the kind, until practically all offences against property or persons were considered as public rather than private matters. Punishment was still dealt out on the basis of retribution, but it was inflicted by the state, not by the individual wronged, and within the framework of law, not under the vague sanction of ancient tradition and custom." However, cruel treatment of prisoners was normal. Mankind eventually revolted against these atrocities because such practices were ineffective in curtailing crime, and not because of humanitarian reasons. In the hope of balancing the extent of the crime with the extent of the punishment, imprisonment and fines were substituted for executions and torture.

Historical theories of punishment were based on the notion that fearful consequences would discourage potential offenders. Example of this principle can be found in the Draconian law of Ancient Greece and the Bloody Code which persisted in Renaissance England, when (at various times) capital punishment was prescribed for over 200 offenses. Similarly, certain hardened offenses under Sharia hadith tradition may incur fearful penalties.

Modern theories of the punishment and rehabilitation of offenders are broadly based on ancient scriptural texts of the Christian Bible and the Moslem Qur'an, which principles only reappeared in Europe some time after the publication of a seminal pamphlet "On Crimes and Punishments" published by Beccaria in 1764. It is based on the notion of proportionality. Previously, (for example, under England's Bloody Code) the penalty of theft had been the same regardless of the value stolen, giving rise to the English expression "It is as well to be hanged for a sheep or a lamb."

Subsequent development of the ideas of Beccaria made non-lethal punishment more socially acceptable. Consequently convicted prisoners had to be re-integrated into society when their punishment was complete.

In Europe, and many westernized legal systems, criminal punishments now tend to be in accordance with a popular theory once expressed by Martin Luther King around 1958 : "Hate begets hate; violence begets violence; toughness begets a greater toughness. We must meet the forces of hate with the power of love...Our aim must never be to defeat or humiliate the white man (e.g. our enemies), but to win his friendship and understanding."

A movement to abolish cruel treatment of prisoners began during the Age of Enlightenment and continued throughout the 19th and 20th centuries. However, there have been continual arguments for severe punishments, perhaps increasing somewhat in the early years of the 21st century. Contemporary justifications for such punishment often revolve around the "rights of the victims". Often underlying these perspectives are opinions that stress the vindictive eye-for-the-eye notions of the Old Testament and Qur'an, over the notion that the primary goal of incarceration should be the reform and reeducation of prisoners to facilitate their re-integration into society. Aslam Abdullah (2006, http://www.beliefnet.com) in Demystifying Muslim Justice observes that " most Americans' impression of Islamic justice is a rather barbaric image of retribution harshly and violently administered." Ask even educated Americans to explain the law in Muslim countries, and they'll inevitably talk about hands and heads summarily being severed. In fact, Islamic justice shares much with Christianity and Judaism. These similarities are not surprising, considering that our penal systems are influenced by common scriptural foundations. The Qur'an's most basic passage pertaining to punishment is a familiar one to Christians and Jews alike : "And We prescribed for them therein: The life for the life, and the eye for the eye, and the nose for the nose, and the ear for the ear, and the tooth for the tooth, and for wounds equality." That said, even in Talmudic times, this was never intended for Jews to be taken literally, and has always been interpreted by the Rabbis to indicate that appropriate compensation be paid in the form of money.

Within the framework of penology, the trend toward increasing the severity of punishments is reflected in publications such as Block's (1997, p. 12) advocacy of policy initiatives aimed at increasing the unpleasantness of prison life that would likely be "a cost-effective method of fighting crime" and Arpaio and Sherman's 1996 book claiming that the increase in the severity of treatment of prisoners will result in decrease in recidivism.

Penologists have consequently evolved occupational and psychological education programs for offenders detained in prison, and a range of community service and probation orders which entail guidance and aftercare of the offender within the community. Penalties world over can only be imposed by a criminal court and include fines, damages, unpaid compensatory work, and mandatory training as well as physical detention.

The opinion that increasing the severity of punishments will result in decrease of recidivism is not supported by some meta studies of this issue. Contrary to this popular opinion, the majority of research studies indicates that penal policies stressing rehabilitation over punishment result in lower recidivism rates. Most empirical studies consistently find that the employment status after the release from prison is the strongest predictor of recidivism. Thus, e.g., Pennsylvania's 2000 Legislative Report on Recidivism concludes that "most studies found that boot camps have not been very successful in achieving the goal of reducing crime" and that the fact that "employed offenders are almost three times as likely to succeed indicates that providing vocational training and employment opportunities for offenders should be a high priority."

Arpaio and Sherman proposed to increase the severity of imprisonment by the construction of tent prison camps in the Mojave Desert where summer temperatures reach 120 degrees Fahrenheit, by serving prisoners foul-tasting food, by humiliating prisoners by cross-dressing, and by reinstatement of the chain gangs. Mauer (1999, pp. 92–93) documents some other the measures used to implement the increasing the unpleasantness of prison life policies that include shooting around prisoners to keep them moving, forced consumption of milk of magnesia, placing naked inmates in strip cells, and handcuffing inmates for long periods of time.

Types of Punishment

Severe punishments (such as beatings, prolonged sleep deprivation, sensory deprivation, chaining) have been often inflicted on prisoners. There are many reasons given for justification of such punishment. In the 16th century, the Bishop of Trier, Binsfeld, in his Tractatus de Confessionibus Maleficorum (1596) claimed that since the sinfulness of the world increases, God also allows increasing the severity of punishments. Various forms of punishments have been evolved and applied in different societies through the ages. Tortures, sadistic forms of executing death sentences and all sorts of cruelties in prisons were some of the distinguishing features of the penal philosophy all over the world till relatively recent time. The theory of punishment is loaded with a wagon of concepts which are charted below.

(a) Incapacitation

Incapacitation in the context of sentencing philosophy refers to the effect of a sentence in terms of positively preventing (rather than merely deterring) future offending. Imprisonment incapacitates the prisoner by physically removing them from the society against which they are deemed to have offended. Long term imprisonment with the intention to incapacitate is often used by criminal justice systems against habitual criminals who recidivate. Examples of incapacitation include back-to-back life sentences, three-strikes sentencing, and other habitual offender laws. Cutting off a hand of a thief is also an example; this acts to prevent further thefts in a drastic manner, in addition to it having a perceived deterrent effect on others.

(b) Denunciation

Denunciation in the context of sentencing philosophy refers to the disapproval of an act by society that is expressed by the imposition of a sentence. This can be considered as one of the purposes of sentencing, as well as being a possible justification for the imposition of a sentence.

Denunciation arguments can be used to justify more serious sentences than are required by the principles of retribution and deterrence. Denunciation arguments can also be used to justify the existence of laws which are never in practice enforced; they stand as statements of a society's values rather than working parts of a criminal justice system.

(c) Incarceration

Incarceration is the detention of a person in jail —typically as punishment for a crime. People are most commonly incarcerated upon suspicion or conviction of committing a crime, and different jurisdictions have differing laws governing the function of incarceration within an larger system of justice. Incarceration serves two essential purposes with regard to criminals :

1. to punish criminals for committing crimes
2. to isolate criminals to prevent them from committing more crimes

Incarceration rates, when measured by the United Nations, are considered distinct and separate from the imprisonment of political prisoners and others not charged with a specific crime. Historically, the frequency of imprisonment, its duration, and severity have varied considerably. There has also been much debate about the motives for incarceration, its effectiveness and fairness, as well as debate regarding the related questions about the nature and etiology of criminal behaviour.

(d) Corporal Punishment

Before the rise of humanitarianism in the penal philosophy, mutilation, branding, flogging were well-recognized modes of punishment. In India, corporal punishment of whipping which was first added and regulated the Whipping Act of 1864 (repealed and replaced by the Whipping Act of 1909) was abolished in 1955. In England, flogging was abolished quite some time ago. Yet, if this type of punishment is itself a dead letter, the issue whether or not to reintroduce. It is very much a living one, and as recently 1960 the Advisory Council on the Treatment of Offenders was asked to consider the desirability of such a course (Fitzgerald,1962: 228) In the U.S.A., Delaware: the only State still using the whipping post while the State of Maryland has reserved it for wife-beaters only.

To prove the efficacy of whipping, it is argued frequently by some people that flogging by parents or school teachers in childhood have a constructive effect. Such an argument ignores two aspects of whipping as a correctional measure. Firstly, the whipping administered by a parent or even a teacher is to be viewed in the context of the emotional relationship between the child and the pain-inflicting source. That certainly is lacking in the case of a person doing it on behalf of the police, jail or some other authority. Secondly, the whipping done by a parent or school teacher inflicts pain generally within a very short time after the delinquent act is committed by the child and such promptness in the punishment is bound to heighten the effect of the chastisement.

(e) Fines

Fines as an additional or alternative form of punishment have been increasingly favored by the law as well as judicial authorities. They are singularly more appropriate in offences relating to traffic, employment of persons unauthorized by law and violation of laws regarding manufacture-and distribution of goods. They are very frequently imposed in relation to property crimes like embezzlement, fraud, theft, violations of lottery and gambling laws and minor offences like loitering and disorderly conduct. In the Indian Penal Code the provision for fines, as punishment, was justified by its framers thus :

"Fine is the most common punishment in every part of the world and it is a punishment the advantages of which are so great and obvious that we propose to authorise the courts to inflict it in every case.... Imprisonment, transportation, banishment, solitude, compelled labor are not equally disagreeable to all men. With fine the case is different. In imposing a fine it is always necessary to have regard to the pecuniary circumstances of the offender, as to the character and magnitude of the offence. The mulet which is ruinous to the laborer is easily borne by a tradesman and is absolutely unfelt by a rich zemindar."

(f) Forfeiture and Confiscation of Property

Forfeiture of property which is the subject-matter of the offence, is a possible mode of punishment under the Sections 126, 127 and 169 of the Indian Penal Code in certain circumstances. But forfeiture of the whole of the property of criminal is not possible according to the present law. Such a punishment was possible under the original Penal Code but the provisions were repealed in 1921. Recently the Law Commission invited opinions on the basis of a questionnaire relating to the question of introduction of some forms of punishment in the Code. One of the items in the questionnaire was regarding the confiscation of the entire property of the criminal. The opinions received by the Commission were largely against the introduction of confiscation as a punishment in the Code. The Commission too was of the view that "this harsh punishment, which will fall not only on the criminal but on his dependant family, is not to be commended"(42nd Report.p.48). It may be observed here that the Commission does not appear to have applied its mind to the problem in depth. Simply saying that the family and dependants would be affected is probably not enough.

Such a punishment is certainly called for in cases of smugglers as per Smugglers and Foreign Exchange Manipulators (Forfeiture of Property) Act, 1976 and black marketeers where prima facie the source of income or property acquired by the offender may be illegal As regards hardships to the family, the same is caused in varying degrees by all forms of punishment. When a killer is sentenced to death or life imprisonment, hardship is inevitable to his family if he happens to be the sole or the most important breadwinner for the family. In a country, which claims to be a welfare State, the family, if hit hard, must be provided financial relief by government agencies and the family's possible suffering need not be used as an argument for withholding a punishment from an offender who deserves it otherwise. The punishment was retained in the Indian Penal Code Bill of 1972, Clause 19(vi).

(g) Banishment

This punishment is one of the most ancient ones. The object is to eliminate criminals from society by sending them to far-off places and sometimes even rebels, revolutionaries and reformers were also taken care of by this process. In India, transportation meant the dispatch of more dangerous criminals to 'Kala Pani' i.e. to Andaman and Nicobar Islands. The draftsmen of the Indian Penal Code, while proposing this punishment, took note of the extraordinary fear among Indians of the sea, particularly those living far away from it and felt that the punishment caused more terror in the Indian minds than what was actually warranted by the actual punishment. It was, therefore, thought to have it in the Penal Code because of its peculiar deterrent value. As observed by Hari Singh Gaur, it had an additional dimension for Hindus because going beyond the seas involved the forfeiture of one's caste (1972:380).

(h) Imprisonment

Imprisonment in its pure and simple form is a kind of punitive reaction, its object being primarily to deprive the offender of his liberty which is the most serious damage which can be caused to a human being, next only to deprivation of life by the death sentence. In recent times, the correctional idea has also been brought in to introduce some changes in prison philosophy but even now, imprisonment remains the most common mode reflecting the punitive approach and its fundamental character remains the same.

Imprisonment as a form of punishment is of relatively recent origin and got prominence as a result of the decreasing use of capital punishment and transportation in most countries during the last century or so. The traditional concept of prisons was different from the one in which it has been understood since the middle of the last century.

Fitzgerald says (1962: 233)"Traditionally it was held that imprisonment should be used only for the custody of offenders until such time as they could conveniently be dealt with, and this view prevailed in Europe from the time of the Roman Emperor Justinian for the next thousand years or so. In medieval England prisons were places where suspects were detained until the royal Judges came round on circuit with a commission of 'gaol delivery', to empty the jails and inquire into the alleged crimes."

One important reason why all forms of punishments have been quite ineffective is the popular belief that only a small percentage of all criminal offenders are punished. The law is no deterrent because people know that immediate arrest, conviction, and a definite sentence will not absolutely follow the commission of a crime. Because of the doubt, the great uncertainty of apprehension and conviction following a crime, the criminal allows this to outweigh his fear of punishment.

An objective of punishment is the prevention of further wrong doing by making the original criminal act injurious to the offender as well as to the offended, by making the way of the transgressor unpleasant. If punishment is to be effective in curtailing crime, it must first be enforced as soon as possible after the commission the crime. The association between the crime and the administration of punishment grows weaker with the passage of time; punishment loses whatever deterrent value might have had, when it is long delayed. Second, the punitive practices of the past should be a constant reminder that severe and harsh punishments do not, and not, suppress crime. Habitual criminals should not be treated in the same manner, as non-habitual or situational criminals. The former should be given extended sentences. Third, reformation and not punishment alone should be the goal of sentencing. Four, the limitations of punishment must be recognized. The thought punishment, even if it is a certain consequence of the crime, may hinder some potential criminals, but it will not deter the hardened, dedicated determined professional. Finally, humane punishment helps to reassure the law-abiding citizen and gives him a sense of security.

Judicial Discretion in Sentencing Offenders :

Guilt once established, the sentencing dilemma begins. A wide discretion is given to the judges in sentencing the offenders. The sentences awardable to the offenders are mainly of three types : (a) death, (b) imprisonment (including imprisonment for life,) and (c) fine (which also includes forfeiture of property). The determination of appropriate sentence for the convicted person is as important as the adjudication of the guilt of the accused in the modern sentencing system. The significance of the modern sentencing system lies in the individualisation of punishment and consequently to the rehabilitation of the offenders. That is the reason that the Indian Penal Code and the other penal laws normally indicate the maximum punishment awardable for an offence and then leave it to the discretion of the Court to pass a suitable sentence within such maximum limit.

Bentham mentions the following circumstances in mitigation of punishment : (1) absence of bad intention; (2) provocation; (3) self-preservation; (4) preservation of some near friends; (5) transgression of the limit of self-defence; (6) submission to the menaces; (7) submission to authority; (8) drunkenness; (9) childhood. (Jag Mohan Singh v. State of U.P AIR 1973 S.C 947: 1973 Cr. L.J 370,1973 SCC (Cri) 162). In fact, in many cases considerable delay in the disposal of the case is treated as a factor in awarding lesser punishment.

The policy of the law in giving a very wide discretion in the matter of punishment to the judge has its origin in the impossibility of laying down standards. Take for example, the offence of criminal breach of trust punishable u/ s 409 I.P.C. The maximum punishment prescribed for the offence is imprisonment for life. The minimum could be as low as one day's imprisonment and fine. It is obvious that if any standards were to be laid down with regard to several kinds of breaches of trust by the person referred in that section that would be an impossible task. All that could be reasonably done by the legislature is to tell the judges that between the maximum and minimum prescribed for an offence, they should on balancing the aggravating and mitigating circumstances as disclosed in the case, judicially decide what would be the appropriate sentence (Jag Mohan Singh V. State UP quoted supra.).

"Though according to new section 354(3) of the Code of Criminal Procedure, 1, when the conviction if for an offence punishable with death, or in the alternative, with imprisonment for life or imprisonment for a term of years, the judgment shall state the reasons for the sentence awarded, and, in case of sentence of death, the special reasons for such sentence. This section provides for recording of reasons and special reasons but it does not provide any guidelines for recording such reasons." In Bachan Singh v. State of Punjab (A.I.R 1980 SC 898 at 935-36) the Supreme Court rejected the contention that aggravating and mitigating circumstances for imposing death or life imprisonment should be given as a guideline to the court. The Supreme Court observed that the expression "special reasons" in the content of S. 354 (3), means "exceptional reasons founded on the exceptionally grave circumstances of the particular case relating to the crime as well as criminal. Thus the extreme penalty should be imposed only in extreme cases. It should not be imposed save in the rarest of rare cases when the alternative option is unquestionably foreclosed." What are the rarest of rare cases, and extreme cases are not clearly defined and discretion is left to the sentencing Judge which can be corrected by the appellate courts. In Balwant Singh v. State of Punjab, it was observed by the Supreme Court in this context that it is unnecessary and it is also not possible to make a catalogue of the special reasons which may justify the passing of a death sentence in a case.

Now section 361 of the Code of Criminal Procedure read with S.354 (4) requires the Court to give special reasons in its judgment if sentence is imposed. This section lays down an injunction as distinguished from discretion under section 360 of Cr. P.C, 1973 and sections 3 and 4 of the Probation of Offenders Act, 1958. Though this section (5.361) restricts the discretion of the Court in respect of all the offenders irrespective of age and lays emphasis on special reasons but does not provide for the source of information for recording special reasons like section 6(2) of the Probation of Offenders Act, which requires calling of a report from the Probation Officer for recording special reasons, even otherwise the object of this section is more laudable than the Probation of Offenders Act of 1958 as is clear from the observation of the Joint Select Committee on the code of criminal procedure Bill, 1970. XXV (1972) which proposed the insertion of this section. The Committee in its report stated :

"It has come to the notice of the Committee that the salutory provisions of the Probation of Offenders Act, 1958, the Children Act, 1960 and other similar laws intended for the treatment of youthful offenders are being applied by the Courts only rarely although it was expected that the provisions of these Acts would be applied liberally by the Courts. To ensure that they are so applied the new provision has been inserted requiring the Court to give special reasons in the judgment for not applying the provisions of these special laws whenever they may be applied."

The cumulative legislative effects of sections 235 (2), 248 (2), 354 (3), 354 (4), 360 and 361 of the Code of Criminal Procedure, 1973 ; Sections 3,4 and 6 of the Probation of Offenders Act, by and large is that the Court is to ensure greater awareness in examining each case with a view to determining the most appropriate sentence or for passing other post conviction orders including probation and the Court is required to record reasons and special reasons in the cases stated above when a sentence is passed.

Sentencing Power of the Courts

The sentencing Court, after adjudging the guilt of the accused, has to consider what is proper length of imprisonment to be imposed for a particular offence. In a majority of offences the I.P.C. and other penal laws provide punishment of imprisonment of varying terms. The law normally prescribes maximum punishment to be awarded in respect of an offence and except in a very few cases, does not prescribe the minimum term of imprisonment. The penal policy is to give wide discretion to the Court in awarding the appropriate sentence after considering the various factors as stated above. However, the courts are given sentencing powers according to their status.

While considering the quantum of sentence where an accused person is convicted of several offences at one trial, the total quantum of punishment shall be determined according to the provisions of Sec. 31 of the Code of Criminal Procedure, 1973. It reads :-

"31. Sentence in case of conviction of several offences at one trial–

(1) When a person is convicted at one trial of two or more offences, the Court may subject to the provisions of Section 71 of the Indian Penal Code (45 of 1860), sentence him for such offences, to the several punishments prescribed there for which such court is competent to inflict; such punishments when consisting of imprisonment to commence the one after the expiration of the other in such order as the Court may direct, unless the Court directs that such punishments shall run concurrently.

(2) In the case of consecutive sentences, it shall not be necessary for the Court by reason only of the aggregate punishments for the several offences being in excess of the punishment which it is competent to inflict on conviction of a single offence, to send the offender for trial before a higher Court:- provided that–

(a) in no case shall such person be sentenced to imprisonment for a longer period than fourteen years;

(b) the aggregate punishment shall not exceed twice the amount of punishment which the Court is competent to inflict for a single offence.

(3) For the purpose of appeal by a convicted person, the aggregate of the consecutive sentences passed against him under this section shall be deemed to be a single sentence. In other cases sentences which may be passed by criminal courts are given in section 28 and 29 of the Code of Criminal Procedure, 1973.

Habitual Offenders and Enhanced Sentence

Enhanced punishments are prescribed against persons who are persistent and habitual offenders. Habitual offenders are those who repeatedly commit crimes and earlier punishments given to them do not make any deterrent or reformative effect on them. According to Dr. D.J West, habitual offenders may be either mentally normal human beings or mental deviants. There are two kinds of mental deviants i.e., those who are active aggressive deviants possessing psychopathic characteristics and the other type who are called passive inadequate deviants (Institute of Criminology, University of Cambridge, 1962). In case of mentally abnormal offenders, enhanced punishments may not be appropriate measure. In these cases, the solution lies in the therapeutic diagnosis and treatment.

Enhanced sentences are governed by section 75 of the Indian Penal Code. The Section lays down that persons once found guilty of any offence under Chapter XII or XVII of the Code and sentenced to a term of three years or more imprisonment, may be awarded life imprisonment or upto a ten years term if again found guilty of any offence under anyone of the two chapters. These two chapters deal with offences relating to stamps and coins and property like theft, extortion, robbery and dacoity.

The object of this section is to enable the Courts to pass a longer sentence on a persistent habitual offender than would be merited by the offence. The object of the enhanced punishment is the protection of the public rather than the punishment of the offenders. The law commission of India has recognised the utility of section 75 of the Penal Code for protecting the public against persistent offenders.(42nd Law Commission's Report, pp.78-80). The commission observed that there was no reason to confine the application of the provision to offences relating to coin and stamps and offences against property and, therefore, recommended the extension of section 75 to all the offences under the Code punishable with imprisonment upto three years or more. They also recommended that section 75 ought not to be applied where there was no imprisonment on the last conviction but some other causes like fine or probation was adopted. It was further recommended that the provision for enhanced punishment be applied only if the subsequent offence was committed by him.

Capital Punishment Controversy in India

At one time or another, societies have used the following Capital punishments: Beating (using mace to crush the head), Beheading, Burning Crucifixion, Drowning, Destruction by wild beasts, Flaying, Hanging, Impaling upon a sharp stick , Precipitation from a height , Stoning, Strangling, Hanging, Electrocution, Gas chamber, Death by Guillotine. Capital punishment is one of those subjects of human concern which give rise to an endless debate without producing any conclusions which can be scientifically tested to make them convincing to both the parties to the debate. To abolish or not to abolish is the problem which has been faced in many countries. Capital punishment is provided for the following offences :

- Waging war against the Government or attempting to wage such war or abetting of such war (Sec. 121 IPC).
- Abetment of mutiny committed by the officers of defence forces with mutiny being committed in consequence of that abetment (Sec. 132 IPC).
- Giving or fabricating false evidence with intent to procure conviction of a person, of a capital offence, if an innocent person is thereby convicted and executed in consequence of such false evidence (Para 2 of Section 194 IPC).
- Murder (Section 302 IPC).
- Abetment of suicide of a person under eighteen years of age, an insane person, delirious person and idiot person or one in a state of intoxication (Section 305 IPC).
- An attempt to murder by a person under sentence of imprisonment for life if as a result of such attempt hurt is caused (Para 2 of Section 307 IPC).

In addition, there are some other categories of cases of constructive liability to death penalty :

Where an act which constitutes an offence punishable with death is done by several persons in furtherance of common intention of all, each of such person is to be sentenced to death. (Section 34 IPC). If five or more persons conjointly commit dacoity and anyone of them commits murder in so committing the dacoity every one of those persons is punishable with death (Section 396 IPC).

In certain circumstances, abetment of offences punishable with death, is also punishable with death (Secs. 109 to 119 IPC).

The Armed Forces laws also have a provision for the sentence of capital punishment for certain offences under those laws after trial by a Court-martial. In case of above noted provisions of the I.P.C. two options are available to the Courts. Either to sentence the accused to death or to impose on him a sentence of imprisonment for life.

Over the years, emphasis has shifted from brutal punishments to humane treatment of law-breakers. The debate regarding the retention of the death sentence has been going on for quite some time now. A Bill was introduced in the Lok Sabha in 1956 which was rejected by the House. Efforts were also made in the Rajya Sabha to move the resolution for the abolition of capital punishment in 1958 and in 1962 but they were withdrawn each time after some debate in the House.

The Law Commission in its Thirty-fifth Report (1967:354) concluded that the risks involved in the abolition of capital punishment could not be undertaken in the present state of the country. They summed up their impressions as follows :

"The issue of abolition or retention has to be decided on a balancing of the various arguments for and against retention. No single argument of abolition or retention can decide the issue. In arriving at any conclusion on the subject, the need for protecting society in general and individual human beings must be borne in mind. It is difficult to rule out the validity of or the strength behind many of the arguments for abolition. Nor does the Commission treat lightly the argument based on irrevocability of the sentence of death, the need for a modern approach, the severity of the punishment and the strong feelings shown by certain sections of public opinion in stressing deeper questions of human values.

Having regard, however, to the conditions in India, to the variety of the social upbringing of its inhabitants, to the disparity in the level of morality and education in the country, to the vastness of its area, to the diversity of its population and to the paramount need for maintaining law and order in the country at the present juncture, India cannot risk the abolition of capital punishment. Arguments which would be valid in respect of one area of the world may not hold good in respect of another area in this context. Similarly even if abolition in some parts of India may not make a- material difference, it may be fraught with serious consequences in other parts. On a consideration of all the issues involved, the Commission is of the opinion that capital punishment should be retained in the present state of the country."

In Jagmohan Singh v. State of U.P, the constitutional validity of capital punishment was challenged before the Supreme Court. It was argued that the 'right to live' was basic to the freedoms guaranteed under Article 19 of the Constitution. The Supreme Court rejected the contention and held that capital punishment cannot be regarded as unreasonable per se or not in the public interest and hence could not be said to be violative of Article 19 of the Constitution.

The question of constitutionality of capital punishment per se and its application as an alternative punishment under Section 302 came up once again before the Supreme Court in Eachan Singh v. State of Punjab (1980, 2 SCC 684: 1980 SCC (Cri) 580). Three main contentions were advanced in a number of petitions filed before the Court, some of which had been made in Jagmohan Singh earlier and were rejected by the Court :

1. Capital punishment is violative of Article 19 of the Constitution because the freedoms guaranteed therein cannot be enjoyed without the basic right to life and that it defiles the dignity of the individual vouchsafed in the preamble of the Constitution. Regarding the inherent nature of the death sentence it was contended that it served no social purpose.
2. Capital punishment as an alternative punishment in Section 302 of the Penal Code violates Article 21 of the Constitution inasmuch as the legislative policy laid down therein and in Section 354(3) of the Criminal Procedure Code of 1973 give too wide a discretion to the judges without proper and adequate legislative guidelines.
3. India being a party to the Stockholm Declaration of 1977 was committed to abolish capital punishment.

The Court through the majority of four judges, (Bhagwati, J. dissenting) rejected all the above contentions and did not find capital punishment to be unconstitutional or unreasonable per se. Regarding the alleged violation of Article 19, the Court held that 'in pith and substance, the penal laws do not deal with the subject-matter of rights enshrined in the Article and that Section 302 of the Penal Code does not have to stand the test of Article 19'. The Court after surveying all the relevant case-law right from A.K. Gopalan v. State of Madras (AIR 1950 SC 27) to Maneka Gandhi v. UOI, (1978, 1 SCC 248) concluded that a law or an order made there under will be hit by Article 19 if the direct and inevitable consequences of such a law or order is to abridge or take away any freedom under Article 19 e.g. a law or order relating to preventive detention. It is not so if the effect and operation of the statute, by itself, upon a person's fundamental right is remote or dependent upon factors which may or may not come into play, e.g. the penal laws dealing with murder, rape or theft.

As to the second argument; the Court mentioned that the discretion permitted under Section 354(3) of the Code of Criminal Procedure was to be exercised judicially having regard to various factors in a given situation and rigid standardization in this regard was neither possible nor even desirable. It was not attempted to be achieved by the legislature and cannot be expected from courts either. Article 21 was therefore not violated through Section 302 of the Penal Code and Section 354 of the Criminal Procedure Code.

Finally, after examining the relevant provisions of the International Covenant on Civil and Political Rights, which provided the basis to the Stockholm Declaration, 1977, that the Covenant did not stand for the abolition of death sentence but required that the penalty ought not to be awarded arbitrarily and must be confined to extremely heinous crimes only, the Court was satisfied that the Indian position was identical to the Covenant by virtue of Articles 20 and 21 of the Constitution and Section 354(3) of the Criminal Procedure Code.

The Supreme Court did not also agree with the contention that capital punishment did not serve any purpose and after surveying substantial literature on the controversial subject concluded that the punishment might have some significant deterrent effect; a position taken by many eminent personalities and the Law Commissions of India and the United Kingdom.

The Court noted that in India very few scientific studies in regard to crime and punishment in general and capital punishment in particular have been made. The data used by some abolitionists from within the country and also from abroad was stale and incomplete. It also noted that the various attempts made in India from time to time to abolish or restrict the death sentence to specified kinds of murders had not met with any success in Parliament. The Court was also impressed by the fact that not only did many countries still retain capital punishment but in many others the punishment has either been reintroduced or efforts have been made to revive it.

Yet one more, rather very desperate, futile attempt was made in Shashi Nayar v. Union of India (1992, 1 SCC 96: 1992 SCC (Cri) 24) to get capital punishment declared unconstitutional. Raj Gopal Nayar, the petitioner's husband, was sentenced to death for killing his father and stepbrother. After having failed to get the death sentence struck off through all the judicial and mercy channels, the constitutionality of the death sentence was challenged before the Apex Court employing almost the same arguments as in Jag Mohan and Bachan Singh earlier. Besides invoking Article 21 of the Constitution and asserting that capital punishment does not serve any social purpose, it was argued that the Law Commission's 35th Report of 1967, which the majority opinion cited in support of the capital punishment in Bachan Singh, ought not to continue to guide the Court because much water had flowed under the bridge since then and that there was no empirical study before the Court that the situation prevailing in 1967 was still prevailing. The Court rejected the contentions and held :

"The death penalty has a deterrent effect -and it does serve a social purpose. No material has been placed before us to show that the view taken in Bachan Singh requires reconsideration. Further, a judicial notice can be taken of the fact that the law and order situation in the country has not only not improved since 1967 but has deteriorated over the years and is fast worsening today. The present is, therefore, the most inopportune time to reconsider the law on the subject." Similar observations were also made earlier in Jumman Khan v. State of U.P., (1991) 1 SCC 752: 1991 SCC (Cri) 283 and the Court had declined to reconsider the constitutionality of capital punishment.

As already explained that the capital punishment is the exception and not the rule in Indian Law. Courts awarding capital sentence has to explain that the case falls within the definition of "rarest of rare case". The Court explained the concept of 'rarest of rare case' in Mohd. Chaman v. State (2000, 2 SCC 29) The Court out rightly rejected the idea of laying down standards and norms before an act of murder takes place. The Court said :

"....Such standardization is wellnigh impossible. Firstly, degree of culpability cannot be measured in each case; secondly, criminal cases cannot be categorised, there being infinite, unpredictable and unforeseable variations; thirdly, on such categorisation, the sentencing process will cease to be judicial, and fourthly, such standardization or sentencing discretion is a policy matter belonging to the legislature beyond the court's function."

The Supreme Court further explained the awarding of death sentence in Ramdeo Chauhan v. State of Assam : "It is true that in a civilised society a tooth for a tooth and a nail for a nail or death for death is not the rule out it is equally true that when a man becomes a beast and a menace to society he can be deprived of his life according to the procedure established by law. The awarding of lesser sentence only on the ground of the appellant being a youth at the time of occurrence can not be considered as a mitigating circumstance in view of one finding that the murders committed by him were most cruel, heinous and dastardly." Hence the request for referring the matter to a larger Bench was rejected.

References

Beyleveld, D. (1980) A Bibliography on General Deterrence Research, Westmead.

Deb, R. (1975) Reforms in Criminal Law, Some Suggestions, Journal of Indian Law Institute ,vol. 17 Oct. Dec, 603.

Farrington, D.P.F. (1983) 'Randomised experiments on crime and justice', in M.Tonry and N.Morris (eds) Crime and Justice, vol. 4, Chicago.

Fitzgerald P.J (1962) Criminal Law and Punishment, Clarendon Law Series.

Gaur Hari Singh (1972) Penal Law of India, vol. I.

Floud, J. and Young, W. (1981) Dangerousness and Criminal Justice, London.

McLaren, K. (1992) Reducing Reoffending: What Works Now, Wellington, NZ.

NPA (2008) Precis on Criminology, National Police Academy, Hyderabad.

Sirohi JPS (1992) Criminology and Criminal Administration, Allahabad, Allahabad Law Agency.

Von Hirsch, A. (1976) Doing Justice, the Choice of Punishments: Report of the Committee for the Study of Incarceration, New York.

Walker, N. (1978) 'The ultimate justification', in C.F.H. Tapper (ed.) Crime, Proof and Punishment: Essays in Memory of Sir Rupert Cross, London.

Walker, N. and Marsh, C. (1984) 'Do sentences affect public disapproval?', British Journal of Criminology.

■■■

CHAPTER 20

Prisons Conditions in India

Learning Objectives :

- **Acquainting the Learners with the definition and the type of prisons in India.**
- **To understand the prison conditions and the history of prison reform in India.**
- **To comprehend the situation of prison offenses in India.**

Introduction

Concept of Prison

The word prison is of French origin *'prisoun'* and is understood as a place in which people are physically confined and, usually, deprived of a range of personal freedoms. Other terms are penitentiary, correctional facility, and jail (or gaol), a jailhouse, pokey, poky, slammer, clink or a Correctional institution.. Prisons are conventionally institutions which form part of the criminal justice system of a country, such that imprisonment or incarceration is a legal penalty that may be imposed by the state for the commission of a crime. Accordingly, prison is a correctional institution used to detain persons who are in the lawful custody of the government (either accused persons awaiting trial or convicted persons serving a sentence). It is also called a penal institution, where persons are confined for punishment and to protect the public.

A criminal suspect who has been charged with or is likely to be charged with criminal offense may be held on remand in prison if he is denied or unable to meet conditions of bail, or is unable or unwilling to post bail. A criminal defendant may also be held in prison while awaiting trial or a trial verdict. Prisons may be used for internment of those not charged with a crime. Prisons may also be used as a tool of political repression to detain political prisoners, prisoners of conscience, and "enemies of the state", particularly by authoritarian regimes. In times of war or conflict, prisoners of war may also be detained in prisons. A prison system is the organizational arrangement of the provision and operation of prisons, and depending on their nature, may invoke a corrections system. Amongst the facilities that prisons may have are :

1. A main entrance, which may be known as the 'gatelodge' or 'sally port' (stemming from old castle nomenclature).
2. A religious facility, which will often house offices and facilities for counseling of individuals or groups.
3. An 'education facility', often including a library, providing adult education or continuing education opportunities.
4. A gym or an exercise yard, a fenced, usually open-air-area which prisoners may use for recreational and exercise purposes.
5. A healthcare facility or hospital.
6. A segregation unit (also called a 'block' or 'isolation cell'), used to separate unruly, dangerous, or vulnerable prisoners from the general population, also sometimes used as punishment (solitary confinement).
7. A section of vulnerable prisoners (VPs), or protective custody (PC) units, used to accommodate prisoners classified as vulnerable, such as sex offenders, former police officers, informants and those that have gotten into debt or trouble with other prisoners.
8. A section of safe cells, used to keep prisoners under constant visual observation, for example when considered at risk of suicide.
9. A visiting area, where prisoners may be allowed restricted contact with relatives, friends, lawyers, or other people.
10. A death row in some prisons, a section for criminals awaiting execution
11. A staff accommodation area, where staff and corrections officers live in the prison, typical of historical prisons.
12. A service/facilities area housing support facilities like kitchens.
13. Industrial or agricultural plants operated with convict labor.
14. A recreational area containing a TV and pool table.

The available accounts indicate that imprisonment is comparatively a modern method of dealing with the offenders. Although there had been a practice of imprisoning the wrongdoers for a particular period of time, those places were not the concrete structures meant for keeping the offenders. Such places could be part of a palace, a church or any other place declared as prison by the rulers at that time.

Prison was initially regarded as temporary holding den rather than a primary institution of punishment and rehabilitation. Imprisonment was used to compel the offender to pay the fine imposed on him, or to spend the period between arrest and trial or between the conviction for a capital offence and its execution. Torture, mutilation of limbs and outlawry were the common modes of punishment in the early days. Cellars and other places used as detention houses were kept in extremely inhuman conditions.

The period, which saw the rise in cruelties in the criminal law and its administration, also witnessed the birth and growth of the 'enlightenment' era in England and Europe, which ultimately paved the way for more humane approach towards criminals. Prominent among these philosophers of enlightenment were John Locke 1632-1704), David Hume (1711-1756), Jeremy Bentham (1748-1832), Voltaire (1694-1728) and Charles de Montesquieu (1689-1775). Due to the emergence of humanitarian movement, transition from corporal punishment to prison as punishment took place in the 18th century. John Howard may be rightly called the pioneer in the prison reform movement. He published his classic work, State of Prisons, in the year 1777, which greatly influenced prison reforms in Europe and in the United States of America. He advocated proper food, clothing and clean atmosphere for prisoners. He recognized the importance of religious and moral instructions and the utility of learning trades and crafts in prison. The Pentoville prison, constructed in 1842, was one of the earliest of the modern prisons constructed in England.

In the United States of America two different types of prison systems evolved by the end of eighteenth century. In the Pennsylvania system, the inmates were kept in complete seclusion all through the sentence. They were not given any work. Exposure to religious sermons was emphasized as a means of reforming the offenders. The Auburn system provided for association of inmates during the day time only in connection with hard work for maximum production. At night, the inmates were kept separate from one another. The prisoners were not allowed to speak to one another and strict discipline was maintained inside the jail.

Number and types of Prisons in Indian States/UTs as on 2008

NCRB New Delhi in their Fourteenth Report on Prison Statistics 2008 have given a detailed description about Number of Prisons in Indian States/UTs till the end of 2008. Prison and its administration is a State subject as it is covered by item 4 under List II in Schedule VII of the Constitution of India. Prison establishments in different States/UTs comprise of several tiers of Jails. The most common and standard Jail Institutions which are in existence in the States/UTs are better known as Central Jails, District Jails and Sub Jails. The other types of jail establishments are Women Jails, Borstal Schools, Open Jails and Special Jails. Total number of jails in the States/UTs is that Maharashtra has the highest number (213) of jails among the States/UTs followed by Andhra Pradesh (143), Tamil Nadu (133), Madhya Pradesh (120), Rajasthan (108).

Central Jail

The criteria for a jail to be categorised as a Central Jail differs from State to State, however, the common features observed in all the States/UTs are that the prisoners sentenced to imprisonment for a longer period are confined in the Central Jails, which have larger capacity in comparison to other jails. These jails also have rehabilitation facilities. Meghalaya, Odisha, Uttaranchal, Andaman & Nicobar Islands, Chandigarh, Dadra & Nagar Haveli, Daman & Diu and Lakshadweep do not have any Central Jail in their territories. Madhya Pradesh, Maharashtra, Rajasthan, Tamil Nadu and Delhi have the highest number of 9 Central Jails each followed by Andhra Pradesh and Punjab (8 each). The available information regarding capacity for Prison Inmates in Central Jails in respect of States/UTs indicate that Tamil Nadu (14,093), Maharashtra (14,771), West Bengal (11,457) and Bihar (12,339) have comparatively larger capacity followed by Punjab (8,337), Madhya Pradesh (10,703), Rajasthan (8,118), Andhra Pradesh (7,399), Uttar Pradesh (7,090), Karnataka (6,104) and Delhi

(4,800). The capacity to accommodate female prisoners in Central jails is comparatively higher in Madhya Pradesh (567), West Bengal (405), Maharashtra (426), Punjab (453), Kerala (249), Karnataka (416), Andhra Pradesh (262) and Rajasthan (190).

District Jail

District Jails serve as the main Prisons in some of the States/UTs. States which have considerable number of District Jails are Uttar Pradesh (51), Rajasthan (25), Bihar (32), Maharashtra (24), Madhya Pradesh (22), Jharkhand (19), Assam (18), Odisha (13), Harayana and West Bengal (12 each). A&N Islands, Delhi and Pondicherry have one District jail each among UTs. The District Jails in Uttar Pradesh (33,878), Bihar (16,208), Haryana (7,729), West Bengal (5,803), Maharashtra (5,272), Odisha (4,567), Rajasthan (4,929), Madhya Pradesh (5,506) and Jharkhand (6,218) have the capacity of lodging a large number of inmates with an average capacity of 664, 507, 703, 327, 484, 443, 250, 220 and 197 inmates per jail respectively. Comparatively higher accommodation for female inmates was also provided in District Jails in Uttar Pradesh (1,652), West Bengal (528), Bihar (507), Haryana (495), Madhya Pradesh (419), Maharashtra (292), Jharkhand (280) and Assam (254).

Sub Jail

10 States have reported comparatively higher number of sub-jails revealing a well organized prison setup even at lower formation. These States are Maharashtra (172), Andhra Pradesh (121), Tamil Nadu (99), Madhya Pradesh (90), Karnataka (74), Odisha and Rajasthan (59 each), West Bengal (30), Kerala (26) and Bihar (16). Madhya Pradesh had the highest capacity of inmates (7,221) in various Sub-Jails followed by Odisha (6,215), Andhra Pradesh (4,532), Rajasthan (3,611), Bihar (3,255), Tamil Nadu (2,995), Karnataka (2,703), West Bengal (1,901), Maharashtra (1,739) and Gujarat (1,491).

Women Jail

Women jails exist only in 13 States/UTs and function exclusively for women prisoners. Tamil Nadu has 3 women jails followed by Andhra Pradesh, Rajasthan and West Bengal (2 each), Bihar, Karnataka, Kerala, Maharashtra, Odisha, Punjab, Tripura, Uttar Pradesh and Delhi (1 each). The total capacity of women inmates was highest in Tamil Nadu (1,013) followed by West Bengal and Delhi (400 each), Andhra Pradesh (380), Rajasthan (350), Punjab (320), Maharashtra (262), Uttar Pradesh (90), Bihar (83), Karnataka (69), Odisha (55), Kerala (48) and Tripura (30).

Borstal School

The primary objective of Borstal Schools is to ensure care, welfare and rehabilitation of young offenders in a different environment suitable for children and keep them away from contaminating atmosphere of the prison. Borstal schools cannot be treated as either a miniature jail or a substitute for it. The delinquents detained in Borstal Schools are provided various vocational trainings. They are also given education with the help of trained teachers. Ten States namely, Andhra Pradesh, Haryana, Himachal Pradesh, Karnataka, Kerala, Madhya Pradesh, Maharashtra, Punjab, Rajasthan and Tamil Nadu each have one Borstal School in their respective jurisdiction. Tamil Nadu had the highest capacity for keeping 633 inmates followed by Punjab (489) and Haryana (355). Haryana (195) and Himachal Pradesh (15) have reported capacity for lodging female inmates in their Borstal Schools. None of the UTs has the existence of Borstal School.

Open Jail

Prisoners with good behaviour satisfying certain norms prescribed in the prison rules are admitted in open prisons. Minimum security is kept in such prisons and prisoners are trained in agriculture. Only 13 States have the Open Jails in their jurisdiction. Amongst these States, Rajasthan reported the highest number of 13 open jails. Maharashtra had 4 followed by Andhra Pradesh, Gujarat, Kerala and Tamil Nadu (2 jails each). The remaining 7 States – Assam, Himachal Pradesh, Karnataka, Odisha, Punjab, Uttarakhand and West Bengal have one Open Jail each. Such jail was not established in any of the UTs. The highest capacity of inmates in open jails was reported from Maharashtra (922), Kerala (530), Rajasthan (529), Andhra Pradesh (430), Uttarakhand (300), Tamil Nadu (110), Assam, Gujarat and Odisha (100 each), Himachal Pradesh and Karnataka (80 each), West Bengal (70) and Punjab (50).

Special Jails

Special Jail means any prison provided for the confinement of a particular class or particular classes of prisoners which are broadly as follows :

- Prisoners who have committed serious violations of prison discipline.
- Prisoners showing tendencies towards violence and aggression.
- Difficult discipline cases of habitual offenders.
- Difficult discipline cases from a group of professional organised criminals.

Tamil Nadu has the highest number of Special Jails (7) out of the 9 States having Special Jails followed by Kerala (5), West Bengal (4), Gujarat, Karnataka, Odisha and Uttar Pradesh (2 each), Assam and Maharashtra (1 each). As far as the available capacity in these jails is concerned, the highest capacity for keeping the prisoners was available in Odisha (1,073) followed by West Bengal (841), Uttar Pradesh (714), Tamil Nadu (590), Assam (372), Kerala (298),Karnataka (250), Maharashtra (246) and Gujarat (188) in their Special Jails. Provision for keeping female prisoners in these special jails was available in Tamil Nadu (418), West Bengal (95), Odisha (30), Kerala (27), Gujarat (18), Assam and Karnataka (12 each) and Maharashtra (3).

History of Prison Reforms in India

In order to fully appreciate the magnitude of the problem and the parameters relevant to reforms in the context of human rights, it would be desirable to look at the evolution of prison administration over the years.

Prisons in India at the time of takeover of the country by the East India Company were in a terrible condition. Obviously, the East India Company was not interested in investing any money in non-profit welfare programs like improvement of jails. It was only in 1836 that a committee was appointed by the Government to report on the existing conditions in jails in the country and to suggest steps for future administration of prisons. As there was a lot of scepticism regarding the possibility of reformation of criminals in England at that time, the committee recommended the exclusion of all reforming influences such as moral and religious teaching or education. The committee, however, suggested the construction of central prisons where the convicts might be engaged in some dull, monotonous, and wearisome work.

The first modern prison in the country, a central prison, was constructed in 1846 at Agra. This was followed by construction of central prisons at Bareilly and Allahabad in 1848, at Lahore in 1852, at Chennai in 1857, at Bombay, Alopore, Benaras and Fetehgarh in 1864 and at Lucknow in 1867.

The Second Jail Committee of 1864 recommended a certain minimum space for each prisoner inside the prison, better clothing and food, and regular medical check-up of prisoners. Both reports of 1834 and 1864 indicate that the British regime was interested in prisons only from the point of view of administration and discipline.

A conference of experts held in 1877 resulted in a Draft Bill being prepared governing the principles and practices of prison management. The recommendations made by the Third Jail Committee proved to be unproductive but the reports of the committees appointed in 1889 and 1892 led to the passing of the Prison Act of 1894. The Act laid down rules about the management and administration of prisons in India. However, the penal policy as reflected through the Act continued to be far from reformative. But the Act ultimately did not materialise. In 1888, the Fourth Jail Commission was appointed and from its object and scope, it is clear that even after a lapse of over a century solutions for the same problems are still being sought. This includes translating principles into effect in various jails, the cost of maintaining prisons, ensuring sanitary conditions, prison discipline, etc. However, the Prison Act of 1894, which is in vogue even today was the result of the 1888 Commission.

This Act sought to streamline prisons administration and put it on a uniform footing throughout the country. It provided for separation of prisoners based on age, their civil or criminal status and on the basis of whether they were unconvinced or convicted criminals. The Medical Officers was required to visit the prison daily and examine prisoners confined in the cells for more than 24 hours. The Act also restricted employment of criminal prisoners sentenced to rigorous imprisonment to no more than nine hours on any day. The Medical Officer was made responsible for ensuring that the prisoner's health was not injured by the work in which they were employed. No officer's subordinate to the Superintendent was empowered to award punishments. Female and civil prisoners were specially excluded from the punishment of handcuffing or fetters or whipping. Incidentally, whipping was abolished by the Abolition of Whipping Act, 1950.

The Act was "largely based on deterrent principles concerned more with prison management than with the treatment of prisoners and gave more consideration to prison offences and punishment than to their effect."

Modern prison reform in the country can be said to emanate from the Indian Jails Committee of 1919-20. For the first time this report identified reformation and rehabilitation as the true objective of prison administration. The Committee recommendations that the care of criminals should be entrusted to adequately trained staff selected after careful scrutiny. It also rejected the idea of excessive employment of convict overseers and recommended the induction of technical staff in jail services.

The Committee made the important recommendations that separate jails should be earmarked for various categories of prisoners, prescribing a minimum area of 75 square yards per inmate within the jail walls. It took strong objection to the presence of children in jails meant for adults. It recommended the creation of special courts for hearing of cases of juvenile delinquents and their housing in remand homes. The Committee also made valuable suggestions regarding education for convicts and aftercare programs for helping ex-convicts. It urged the holding of a conference of Inspectors General of Prison staff every alternate year. Many of the recommendations were not implemented on the ground that the subject of prisons was within the purview of the provincial governments. It is ironical that even today that one of the major stumbling blocks in ensuring uniformity in prison conditions all over India is the fact that prison administration is a state subject.

Immediately after Independence the Government of India sought to achieve further reforms in prisons. Dr. W. C. Reckless, an expert of the U.N.O. was invited in 1951 to make his recommendations. His recommendations resulted in the revival of the conference of Inspector Generals of Prisons after a lapse of 17 years. Later on, a committee was appointed to prepare an All India Jail Manual in 1957 on the basis of the recommendations of Dr. Reckless and the All India Conference of Inspectors General of Prisons. An All India Jail Management Committee submitted its report in 1960. This resulted in the settings up of the Central Bureau of Correctional Services, which was later redesignated as the National Institute of Social Defence. A working group in 1973 suggested that Government should make effective use of alternatives to imprisonment as a policy measure and also highlighted the desirability of proper training of prison personnel and improvements in their service conditions. Further, it made important recommendations with regard to the classification and treatment of offenders and laid down principles of follow-up and after-care procedures. It said that developments of prisons and correctional administration should no longer be divorced from the national development process and prison administration should be treated as an integral part of the social defence component of national planning. The committee recommended abolition of solitary confinement as a mode of punishment. It also recommended classification of prisoners according to their needs in terms of personality. The other recommendations included greater use of probation system, creation of a Central Bureau of Correctional Services, and establishment of After-care Organizations in every state.

Persistent criticism about the manner in which the prison system was functioning and the fact that it did not measure up to the test of law and international standards of human dignity and preservation of fundamental human rights of prison inmates, resulted in the setting up of the All India Committee on Jail Reform (1980 chaired by Mr. Justice A.N. Mulla (The Mulla Committee examined all aspects of prison administration and made wide-ranging recommendations, which if implemented would go a long way to make prison administration efficient, humane end professional. The recommendations of the Mulla Committee touched upon legislative, operational, security aspects besides matters like classification of prisoners, living conditions in prison, medical and psychiatric services, treatment programs, vocational training for prison inmate, problems related to undertrials and other unconvicted prisoners, problems of women prisoners etc. The report laid emphasis on the management of prisons to be entrusted to a cadre of professional.

Justice A. N. Mulla recommended a draft national policy on jail reforms. The following are some of the important recommendations of the committee :

- Creation of Department of Prisons and Correctional Services in each State.
- No unnecessary detention of under-trial prisoners.
- New alternatives to imprisonment such as community service, forfeiture of property, and compensation of victims.
- Improvement of living conditions in prisons.
- Prison service to be developed as a professional career service - Indian Prisons and Correctional Service. Proper training facilities for prison personnel to be developed at the national, regional and State levels.
- To open prisons to positive and constructive public discernment.
- Probation, aftercare, rehabilitation, and follow-up of offenders, to form an integral part of the functions of the Department of Prisons and Correctional Services.

The committee also recommended the incorporation of the following rights for the prisoners in the proposed scheme of "National Prison Legislation".

1. Right to Human Dignity
2. Right to Minimum Needs
3. Right to Communication
4. Right to Access to Law
5. Right against Arbitrary Prison Punishments
6. Right of Meaningful and Gainful Employment
7. Right to be released on due date.

All the above rights are more or less already contained in the Constitution, jail manuals and judicial pronouncements but the real challenge is regarding their implementation.

The National Police Commission (1977-80) looked into issues like arrest, detention in custody, interrogation of women, and delay in investigation (which contributes to the undue detention in custody of non-convicted persons). Besides highlighting the need to adhere to the provisions of law, it made wide ranging suggestions to amend laws and procedures to cut down on delays at the investigation and trial stages, and avoid custodial violence and lock-up illegalities to inspect police lock-ups and report on them.

The report of the National Commission for women on "Custodial Justice for Women" (1993) merits attention The Krishna Iyer Committee from the basis of this Report. The following are some of the more important aspects, many of which do not cast any financial burden for their implementation.

1. Women prisoners – like men – should be informed of their rights under the law.
2. Women constables should conduct searches.
3. Medical check-ups of women prisoners or undertrials, should be done by women doctors as soon as they come to prison.
4. Women prisoners should be allowed to contact their families and communicate with their lawyers, women social workers, and voluntary organisations.
5. Women prisoners should be allowed to keep their children with them.

6. Voluntary organisations of women should be encouraged to be associated with women prisoners.
7. Separate jails should be provided for women.
8. Special prosecution officers should be available to present the case of women prisoners.

The National Human Rights Commission (NHRC) in its first Annual Report (1993-94) has expressed its deep concern about the "appalling conditions of over crowding, lack of sanitation, poor medical facilities, inadequate diet and the like, in most of the jails of the country. These serious deficiencies are compounded by unconscionable delays in the disposal of cases for various reasons and mismanagement in the administration of jails, all of which need to be remedied."

As an immediate undertaking, the Commission is in touch with competent judicial and executive authorities in Delhi with a view to :

- Expediting the trial of cases including those of some 3000 foreign nationals in various jails in the country;
- Convening meeting of the Sentence Revising Board for the release, whenever possible, of those serving life sentence and who have already completed the maximum of 13 ½ years of their term.
- Segregating juveniles' prisoners sentenced for minor offences from those serving longer terms for heinous crimes.

The courts in India have also laid down specific rules and guidelines in regard to matters like the right to physical protection; protection against physical assault (Sunil Batra's case); restrictions on handcuffing and bar fetters (in Prem Shukla's case); on solitary confinement (in Sunil Batra's as well as Kishore Singh's case); the right to a speedy trial (in Hussainara Khattum's case); freedom of expression (in P. Pandurang Sansgui's case); and press interviews (in Prabhu Dutt's case), etc.

The Supreme Court issued directions regarding the procedure to be followed when a person is arrested. In Joginder Kumar vs. State of UP and others (1994), the Court refereed to the National Police Commission's finding that 60% of all arrests were either unnecessary or unjustified and laid down four requirements to be strictly followed :

- The right of the arrested person to request that a friend, relative or other persons be informed of his arrest and the place where he is detained.
- The duty of the police officer to inform the arrested person of this right.
- An entry to be made in the police station diary as to who was informed of the arrest.
- The duty of the Magistrate before whom the arrested person is produced, to satisfy himself that these requirements have been complied with.
- A police officer making an arrest should record in the case diary the reasons for making the arrest, implying there by that every arrest by the police has to be justified.

The Mulla Committee, the National Police Commission, Justice Krishna Iyer's Committee, and the National Human Rights Commission have made numerous valuable recommendations to bring about not only improvements and reform in the jail administration but in the entire criminal justice system itself. Unless there is comprehensive reforms of the criminal justice system in its entirely, there is unlikely to be decisive change. Various commissions and committees have examined problems relating to different elements of the criminal justice system (CJS). But what is required is a detailed look at the CJS as a whole. Perhaps this matter could be remitted to a Criminal Justice Commission, drawing upon talents from all the branches of the CJS. Such an effort is long overdue and would be an essential and urgent step towards reform of vital spheres of public administration affecting human rights and human dignity. There are problems concerning such issues right from the stage of recording the FIR, during investigation (which often involves search, seizure, arrest, detention and interrogation), prosecution, trial, sentencing, jail life, parole, review, remission and rehabilitation, not to mention recidivism and relapse. Unless the government agencies dealing with specific aspects of these processes and matters work in co-ordination and their efforts are complementary to each other, there cannot be harmonious and purposeful results. In the current processes severe damage is caused to basic humanitarian considerations, the rule of law and public confidence in the credibility of the entire system has been shaken. The results can be and in fact are very disturbing the Society is losing faith in the system of justice. Sensitivities in regard to human sufferings and the inescapable disregard of law have been dulled. Because of these, the foundations of a free and democratic society are in serious jeopardy.

Prison Conditions in India

Kamayani Bali of Human Rights Law Network, New Delhi has made a review of prison conditions in India suggesting the problems faced and solutions thereto. The report of the All India Committee on Jail Reforms (1980-83) chaired by Mr. Justice A N Mulla, had observed that "prison administration in India has been of and on, a subject of criticism in the Press, the Parliament and the Judiciary". "Over-crowded prisons, prolonged detention of undertrial prisoners, unsatisfactory living conditions, lack of treatment programs and allegations of an indifferent and even in human approach of prison staff have repeatedly attracted the attention of critics over the year." Unfortunately, nothing much seems to have changed even during the intervening decade and more and there has been no worthwhile reforms affecting basic issues of great relevance to prison administration in India. Broadly speaking following are the major problem areas, which afflict the prison system and need priority attention.

1. Overcrowding

This is the most visible problem and yet no long term or short-term remedies have been found. Prisons in the country have prisoners far in excess of their capacity. In Delhi, Tihar Jail holds 8700 prisoners against a stipulated capacity of 2200. The reasons for overcrowding in jail are many. Inordinate delays in trials result in many undertrials having to be detained in jail for unduly long periods – in many cases extending to years. This, together with the routine new additions, literally clogs the system. In many cases, prisoners who are facing charges of grave, professional, violent crimes are outnumbered by others like suspected drug offenders, ticketless travellers, Railway alarm-chain pullers, and a variety of others who have technically violated law. Many of them are in jail only because they could not pay the fines imposed on them by courts. In some cases, prisoners prefer to continue in jail because they just cannot afford even a single meal a day outside ! Then there are prisoners who prefer to spend a couple of months in jail then to pay "maintenance" to their wives as ordered by courts. Under these circumstances, the problem of overcrowding can be solved or at least reduced by a variety of measures.

There has to be a conscious policy not to overcrowded prisons by finding alternative methods of dealing withstanding non-criminal offenders like ticketless travellers and alarm chain-pullers, apart from unsatisfactory prison management, could lead to release of violent offenders and professional criminals. This has happened in some countries like the USA.

The National Police Commission pointed out that 60% of all arrests were either unnecessary or unjustified. This has resulted in overcrowding and accounts for undue increase of the expenditure of jails. Therefore, restraint by the police in resorting to unwarranted arrests by following the guidelines laid down by the Supreme Court would go a long way to reduce overcrowding in jails.

2. Delay in Trial

Trials delay in the courts has assumed very serious proportions. Even though this problem has been highlighted by the Mulla Committee, National Police Commission and thorough police interest litigation (in the Hussainara Khattun's case), there has been no relief at all. On the contrary, the situation seems to be getting worse, what with cases mounting in courts. In 1978, the pitiable plight of undertrials of Bihar Jails, primarily due to enormous congestion in courts brought to light by K.F. Rustomji resulted in the matter being noted by the Supreme Court, thanks to a Public Interest Litigation initiated at the instance of Ms. Kapila Hingorani. But even today, the situation remains far from satisfactory. On the contrary, we continue to read in newspaper of children of under trial women's growing to adulthood in prison. During a visit to the Tihar Jail in Delhi, the undertrial prisoners spoke with one voice and with deep anguish, frustration, and helplessness about the enormous injury caused to them due to inordinate delays in courts. The National Human Rights Commission has also taken note of this problem of delay affecting undertrials, including about 3300 foreign nationals in various jails in the country. No one aspect of prison administration has affected the human rights of prisoners as delays in trial, for which the police, judiciary and the legal profession are all to blame. It is only by the joint efforts of all these three links that there can be any semblance of improvement that undertrial account for over 50% of the total jail population. In many places like Andhra, Assam, Goa, Karnataka, Madhya Pradesh, Maharashtra, Manipur, Mizoram, Nagaland and Odisha is above 60%. The Mulla Committee also made pointed reference to this matter.

Delay commences at the investigation stage itself. In many cases, charge sheets are filed by the police very late leading to a long chain reaction. The next bottleneck occurs in the course of service of summons to witnesses, often due to collaboration between the process server, police, and the witness(es). This stage of the judicial process has lent itself to corrupt practices, which need to be set right.

The legal requirement of having to give copies of relevant documents to the accused needs to be streamlined. At present this contributes delay. This is unpardonable, particularly because modern copying machines are now available and can easily replace the old, time-consuming practice of making hand written copies, which are often illegible. However, the maximum delay takes place when evidence is to be recorded. In some cases, it has taken more than eight years to examine witnesses and record evidence. Official witnesses, lawyers, and public witnesses can attribute this delay to procedural complexities and absenteeism in one form or another. Defence and prosecution lawyers make their own contribution to the prolongation of trials. The situation has further deteriorated after the separation of prosecution from the police leading to dilution in accountability and loss of effectiveness.

There are also lapses in producing undertrials in courts on the dates of hearing. On many an occasion, they are either not produced or produced late, with the result that the hearing gets adjourned causing avoidable delay. This is due to the district police authorities not making available police escorts on the ground of non-availability of manpower. In many districts, the sanctioned manpower is inadequate. On their part, the courts are also not without blame. Even though law requires that trials should be conducted from day to day till completed, in practice this rarely happens. Cases are adjourned for a couple of months at a time, which further aggravates delay.

A large number of recommendations have been made to reduce delays. The Reports of the Law Commission, National Police Commission and the annual conference of chief Justices must have also discussed this matter a number of times. Numerous seminars have been held, the latest is one seminar on the Criminal Justice System organized by the Law Commission and the Bar Council of India. But the fact is that no one has yet cut the Gordian knot !

3. Privacy and Communication

The arrangements for facilitating communication between prisoners and their relatives, friends and legal advisors leave much to be desired. Many of these aspects have been dealt with in the Mulla Committee Report and deserve immediate implementation.

A connected issue, though more relevant in the context of providing legal aid is the near total absence of any contribution from the legal community.

4. Hygiene

Overcrowding has aggravated the problem of hygiene. In many jails conditions are appalling. At the tehsil level jails not even rudimentary conveniences have been provided. In many cases, such jails are no better than an unsatisfactory lock up. Latrine and bathroom facilities are inadequate to cater to the daily minimum needs of prisoners.

5. System of Jail Visitors

Even though this system exists on paper, it has not proved effective in practice. Visitors, including judicial and non official members, pay cursory attention to peripheral matters and do not take pains to go into the details of major problems, like the long and overdue detention of undertrials, inadequacies of medicare, problems faced by female prisoners, etc.

The system of jail visitors is to be an effective and useful instrument, the trials should examine all aspects of prison life, and with particular emphasis on these having a bearing on human rights and human living conditions. There should be a system of accountability for visitors, who are privileged to look at and review the actual working of jails, behind high walls !

6. Sentence Revising Board

The Model Prison Manual for granting remission was circulated to all States in 1960. Inspite of this, there is no uniformity. The Mulla Committee had noted that the remission system generally operates in an arbitrary manner with little regard to individual differences and the merits of each case. The Committee also observed that the grant of remission constitutes an area highly prone to corrupt practice if the discretion in this regard is not exercised judiciously.

The Delhi Sentence Remission Board has for example been virtually non-functional. In some other States, the Boards have not been fully constituted. Because of these omissions, prisoners due to be considered for premature release have been denied this privilege. The matter has attracted the attention of the National Human Rights Commission, which of the National Human Rights Commission, which plans to take it up with the concerned authorities.

7. Open Prisons

Numbers of Open prisons have not been adequate in the country as reported by the Mulla Committee. In open prison there is no armed guard, bars, confining walls and locks. The prisoners are allowed to move around freely and to earn their livelihood by means of labor. In India, a beginning in the open prison system was made in 1952 under the inspiration of Dr. Sampurnanand when a camp was opened at Chakia in Varanasi district. The inmates were allowed to work on their own or in local factories and wages were paid to them. The best thing about the open prisons is that the system counters 'prisonization' of the convict.

8. Administrative Matters

Organisational structure and personal matters are being inadequately and inappropriately dealt in the absence of the requisite expertise of officers in prison administration to assist the Government in taking appropriate policy decision. For all practical purposes annual conference of IGs Prisons is not being held regularly. This has denied jail administrators the opportunity to meet and discuss problems of common interest and evolve strategies to introduce uniform practices, to the extent possible, all over India. However, occasional conferences have been held to discuss specific issues and developments of current interest. Mixing together of various types of prisoners, for example, convicted offenders and undertrial prisoners, adults and juveniles and mentally stable and mentally challenged are matters of concern. A "Model Prison Manual" was prepared as far back as 1970 and circulated to all States. The Mulla Committee noted that this had not been adopted and the position has not changed since.

9. Role of NGOs

Support sought from civil society by the prison administration has not been adequate. There is considerable scope for NGOs and voluntary agencies to work in the field of treatment, after care and rehabilitation of offenders. If the services of well organized NGOs could be obtained on a regular basis, it would be desirable to assign a role for them as visualised by the Mulla Committee in regard to the following: -

- Services of experts in education, medicine, psychiatry, law, social work, the media etc. could be utilised in the formulation of correctional policy by associating them with advisory committees at the national and state levels.
- Appointments of eminent citizens interested in correctional work as visitors to prisons. They could also be appointed as non-official members on sentence reviewing boards for district and central prisons.

- In conducting adult education and free legal aid programs;
- Running health camps;
- Organising recreational and cultural activities;
- Individual coaching to inmates pursuing higher studies; and
- The NGOs should play a greater role in creating the right kind of awareness about the prison administration and the core problems of correctional sub-systems with a view to eliciting public co-operation.

There is increasing public awareness in regard to human rights and the need to adhere to humanitarian laws. Issues like custodial violence and deaths, unduly long detention of persons under special laws like Tada, COFEPOSA etc have drawn widespread public attention as well as judicial notice. Recently NHRC cases of persons under detention under TADA have been reviewed by the various state Governments and their number brought down drastically. In specific cases courts of laws have given relief. The Supreme Court has upheld the validity of TADA but has given strict guidelines for the enforcement agencies to follow. Continued efforts to create public awareness particularly by social workers and NGOs would help immensely in ensuring that the law enforcement officials strictly follow the procedural aspects laid down by law and that they are accountable for that in any regard.

There have been attempts by individual officers to improve the living conditions in prisons as well as to introduce a variety of reforms in jail administration. In a number of jails in India the system of "Jail Panchayat" is being followed whereby the inmates of prisons themselves take active part and interest in improving their lot and in taking up issues of common interest with the jail administration. In many prisons measures have been initiated to improve educational facilities for prison inmates. In Madhya Pradesh attempts are being made to achieve 100% literacy among prisons in some select prisons. In most of the prisons steps have been taken to impart vocational training with a view to help in rehabilitation of prisoners after their release from jail. There is also growing awareness among prison officials to maintain decent standards of hygiene and healthy environment within jail premises. Sometimes ago, a lot of initiative was taken by the prison in Imphal apart from improving health care, teaching of yoga, meditation, etc. Ample opportunities are being made available to the prisoners to organise and take part in the cultural activities throughout the year. But there is no organized or systematic efforts to bring about reforms in a sustained and institutionalised manner because of a variety of reasons like lack of formal policy and commitment of the Government in this regard, inadequate budget, non availability of required professional skills among the persons in charge of jail administration and the failure to fully take advantage of available resources and facilities with NGOs. Even though the Mulla Committee made strong recommendations about the need to have more open prisons, the progress in this regard is very little. In many jails the prisoners do not get the benefit of free legal aid and assistance and the legal profession has not played its part fully in this regard.

The prisons have programs for moral, spiritual, educational, and vocational training. The programs for vocational training include weaving, power looms and handlooms, tailoring, carpet making, carpentry, printing, tent making, black smithy, etc. The vocational programs may vary from one prison to another and also from one State to the other.

10. Separation of Mentally Challenged

There is another category of prisoners viz. non-criminal mentally ill persons who are held in various jails in India. In the normal course such persons should have been sent to Mental Hospitals rather than to prisons. An expert committee was set up by the Supreme Court of India, whose Report has brought out that increasing number of poor people are seeking police assistance in throwing the kith and kin behind the bars. The committee has found close nexus between police, magistracy, jail administration, and relatives in this regard. The Report pointed out that mental hospitals completely lacked medical facilities and living conditions inside the jails were beyond one's comprehension. Between July and September as many as 209 non-criminal mentally ill inmates were screened in UP jails and they were discharged. In Gujarat the jail administration has issued instructions not to accept such persons in jail. Instead, they are being directed to mental hospitals.

11. Lack of Recognition

Jail administration in India, even though an important limb of the criminal justice system has suffered neglect and lack of recognition. A lot has been talked about the police, a little less about the courts and almost nothing about prisons and prisoners. The problems of prison administration need to be highlighted to focus public attention on this very vital sphere of social concern.

12. Problem of being State Subject

Prison administration is a state subject and this often cited as the main reason for the Centre not being able to implement the recommendations of the Mulla Committee. Similar reasons used to be given for not enacting an All India Children's Act till the then Prime Minister put his full weight behind the Centre enacting legislation for the purpose. This only shows that if there is political will, there shall be no difficulty in the Centre taking an active and direct interest in prison administration. If it of making prison administration a central subject should be seriously considered. This will also help in ensuring uniformity all over the country besides making it possible and feasible to have a single all India cadre of jail administration.

The judiciary in India adopted a status quo jurisprudence and shown a lack of appreciation and concern by its 'hand-off' approach to the operations of prisons. A major break through in prison rights jurisprudence came in 1974 in D.B.M. Patnaik (AIR 1974 SC 2092) and the court asserted that the mere detention does not deprive the convicts of all the fundamental rights they otherwise possess.

The court encouraged the undertrials and convicts to "appeal confidently to the court brutalities of jail authorities". In 1977 in Hiralal (AIR 1977 SC 2237) and Mohd. Giasuddin (AIR 1977 SC 1926), the Supreme Court stressed for the first time the need for rehabilitation of prisoners. It was Maneka Gandhi (AIR 1978 SC 597) case, which generated a strong current and converting the right to life and personal liberty in Art. 21 into a great shield against deprivation of human rights. The Supreme Court reiterated the principle that "imprisonment does not spell farewell to fundamental rights" in Charles Shobhraj. (AIR 1978 SC 1514).

Sunil Batra (AIR 1978 SC 1675) is another important milestone in the field of prison justice and prisoner's rights. Delivering the far-reaching judgment, the court held that "the fact that a person is legally in prison does not prevent the use of Habeas Corpus to protect his other inherent rights". Sunil Batra displayed judicial concern for the miserable conditions of the prisoners to such an extent that for the first time in the history of the Court, the Chief Justice of India, M.H. Beg along with Justice Krishna Iyer and Kailasam visited the Tihar Jail on the 23rd January, 1978 to ascertain the existing conditions. The Court also permitted the Citizens for Democracy (CFD), a human rights group to formally intervene in the case.

In Prem Shankar (AIR 1980 SC 1535), the Supreme Court observed that "no prisoners shall be handcuffed or fettered routinely for the convenience of the custodian's escort." In Francis Mullin (AIR 1991 SC 747), the Supreme Court held that "a necessary component of the right to life, the prisoner or detainee will be entitled is to have the interview with members of his family and friends." Kunnikal Narayanan (AIR 1973 Ker 97) challenged the prison authorities who prevented him from receiving "Mao's literature" in the Kerala High Court held that the prisoners have to be paid reasonable remuneration and minimum wages for labor performed in prison. The Supreme Court in Hoskot (AIR 1978 SC 1548), Moti Ram (AIR 1978 SC 1594) ,Hussainara (1980) 1 SCC pp.88, 91, 93, 98, 108 and 115-16 cases respectively, Sheela Barse (AIR 1983 SC 378), and many other cases not only articulated new rights, but also developed new techniques for dealing with complaints of prisoners and their demands for humane treatment, legal assistance and justice. The attitudinal change of judiciary in recent times is in consonance with the changing premises behind imprisonment emphasizing on rehabilitative aspects and treating prisons as correctional institutions.

13. Privatization of prisons

This issue is being widely debated in some Western countries. Proponents of privatization argue that prisons can be run more efficiently and economically by private agencies. However, those opposed to such a move, hold the view that handing over the control of prisons to private parties may result in disastrous consequences. Complete privatization may indeed be harmful. The private parties may become hand in glove with the prisoners and indulge in criminal activities or the prisoners may be ruthlessly exploited to maximize profits. However, entrusting some aspects of management of prisons to private parties may lead to better administration of prisons.

14. Classification of Offenders

Individualization of treatment of offenders means that the personality of each offender is assessed and prison programs are designed to meet the individual requirements as far as possible. For that, it is of utmost importance that offenders are classified. Classification of offenders involves consideration of two aspects - the type of prison the offender is to be sent to and the type of treatment to be given to the prisoner within the prison to which he is sent. As regards the classification of prisons, one way to do it is in terms of their security. For classifying convicts in a given prison, it is necessary to have the details of their backgrounds for diagnosis. This requires the co-operation of persons of diverse fields like social workers, doctors, psychiatrists, psychologists, and criminologists. In the U.S.A. and some other advanced countries, the work of diagnosis and treatment of the offender is done by committees of these experts. However, in India the classification work is done by court convicting the offender. The courts do not apply their mind while classifying the offenders. A clerk of the court fills up the classification slip and the presiding officer of the court clears it as a matter of routine.

As per the rules in the jail manuals of some States, prisoners are classified on the basis of age, sex, mental health, nature of offence and whether a habitual or casual prisoner. Casual prisoners are sub-classified into (a) star, and (b) general; and habitual prisoners into (a) non-professional, and (b) professional. A casual prisoner is "one who is first offender and who lapses into crime not because he has a criminal mentality but on account of his surroundings, physical disability or mental deficiency." Casual prisoners whose previous character was good, antecedents not criminal and whose crime does not indicate grave cruelty and moral turpitude or depravity of mind, are classified into the 'star' sub-category. Non-professional habitual prisoners are those offenders who lapse into crime owing to their surroundings, or some physical or mental defect, and who are not first offenders. In the other sub-category are included all other habitual prisoners.

The All India Jail Reforms Committee (1980-83) recommended that prisoners might be classified in terms of the following factors :

(a) sex; (b) age; (c) the kind of law, civil or criminal, under which confined; (d) stage of investigation and trial; (e) nature of sentence; simple, rigorous, short, medium or long term of imprisonment, life imprisonment or death sentence; (f) casual or habitual offender; (g) disease or infirmity; and (h) socio-economic status i.e., A, B, or C class. Based on the segregation factors, the Committee suggested a variety of institutions to meet the needs of different categories.

Administration of prisons under our constitution is the responsibility of State governments and the Central Government performs only advisory functions. Therefore, it is up to the State governments to accept and implement any of the recommendations made by any commission. Dr. Kiran Bedi recently proposed comprehensive prison reforms. These are still under consideration.

15. Other Problems of prisons in India

The following are some of the other problems of prisons in India.

- Bad condition of jails - lack of proper ventilation, drainage etc.
- Poor quality of food supplied to the prisoners..
- Lack of individualized treatment of prisoners.
- Lack of proper medical attention to prisoners.
- High handed behaviour of prison staff.

16. Prison Offences

- Use of insulting or threatening language.
- Criminal assault or use of Criminal Force.
- Immoral or indecent or disorderly behaviour.
- Wilful refusal to work.
- Wilful idleness or negligence of work by a Rigorous Imprisonment (R.I) Prisoner.
- Feigning illness.

- Wilfully bringing a false accusation against an officer or prisoner.
- Omitting or refusal to report conspiracy, attempt to escape, as soon as it comes to his knowledge.
- Conspiracy to escape/ assisting in escape.

Thus, the problems afflicting prisons are many, prolonged neglect and the imperatives of prison reforms do not create further delay. The number of prisoners in India is not much, if they are cared for and administered in a systematic and humane manner, the jail population may even come down further. This is a social goal worth achieving.

References

Beyleveld, D. (1980) A Bibliography on General Deterrence Research, Westmead.

Deb, R. (1975) Reforms in Criminal Law, Some Suggestions, Journal of Indian Law Institute ,vol. 17 Oct. Dec, 603.

Farrington, D.P.F. (1983) 'Randomised experiments on crime and justice', in M.Tonry and N.Morris (eds) Crime and Justice, vol. 4, Chicago.

Fitzgerald P.J (1962) Criminal Law and Punishment, Clarendon Law Series.

Gaur Hari Singh (1972) Penal Law of India, vol. I.

Floud, J. and Young, W. (1981) Dangerousness and Criminal Justice, London.

McLaren, K. (1992) Reducing Reoffending: What Works Now, Wellington, NZ.

NPA (2008) Precis on Criminology, National Police Academy, Hyderabad.

Sirohi JPS (1992) Criminology and Criminal Administration, Allahabad, Allahabad Law Agency.

Von Hirsch, A. (1976) Doing Justice, the Choice of Punishments: Report of the Committee for the Study of Incarceration, New York.

Walker, N. (1978) 'The ultimate justification', in C.F.H. Tapper (ed.) Crime, Proof and Punishment: Essays in Memory of Sir Rupert Cross, London.

Walker, N. and Marsh, C. (1984) 'Do sentences affect public disapproval?', British Journal of Criminology.

■■■

CHAPTER 21 Theories of Penology

Learning Objectives :

- ❐ **Acquainting the Learners with the concept and history of deterrence.**
- ❐ **To understand the types of deterrence.**
- ❐ **To comprehend the role of police in deterrence.**
- ❐ **To make a critical appreciation of deterrence and deterrence theory as a military strategy.**

DETERRENCE THEORY

Concept of Deterrence

The concept of deterrence is the omission of a criminal act because of the fear of sanctions or punishment. While not the entire premise, deterrence is certainly an important foundation of the criminal justice system. Law enforcement exists both to apprehend wrongdoers and to convince would-be wrongdoers that there is a risk of apprehension and punishment if they commit a crime. Laws that provide a host of different sanctions for the commission of criminal offenses (fines, probation, imprisonment) serve notice that criminal statutes contain a credible threat that—it is hoped—will inhibit those who have been punished from committing additional crimes (specific deterrence) and those who have not yet offended from committing crimes at all (general deterrence). Particular criminal justice policies such as police "crackdowns," enhanced enforcement in and surveillance of high crime areas or "hot spots," mandatory minimum sentences, sentencing enhancements for firearm possession, "three-strike" laws, and others are all enacted with the expectation that they will successfully engineer the crime rate down through deterring offenders and would-be offenders. It is reasonable to argue that a belief or expectation that sanction threats can deter crime is at the very heart of the criminal justice system.

As it relates to the criminal justice system, deterrence goes to great lengths in an attempt to help keep the citizens of our communities—our families—safe, and on many levels it succeeds. Free will enables human beings to purposely and deliberately choose to follow a calculated course of action. If crime is to be deterred, punishment (pain) must exceed the pleasures gained from the fruits of crime. Based on the theories of Beccaria and Bentham, deterrence theory assumes potential and actual punishment can deter crime. Cesare Beccaria, an eighteenth-century philosopher, was among the first to conduct information gathering on the correlation between the imposed punishment of crimes and complaint behaviour of society (Keel, 2005). Beccaria believed that criminal decisions were based on a few simple factors, being that humans have free will (they have the power to act upon their own accord); humans are rational creatures and able to weigh prospective outcomes of their actions, seeing which may benefit or detract from the quality of their lives; human decisions are based on the simplest views of man (primarily, pleasure is preferable over pain); finally that an organized system of laws and punishments which catered to these human traits is necessary to help keep society compliant (Winfree & Abadinsky, 2003). As these traits heavily involve human choice and rationality, Beccaria's theories fall into the Rational Choice school of criminal theory and are closely related to rational choice itself, being that humans are notorious for considering their actions and weighing the pros and cons as it relates to themselves before acting. Beccaria's studies into deterrence theory waned, and for over a century criminology theorists focused on the criminal mind instead of preventing the mind from acting.

The idea of deterrence has two components : (1) specific punishment dispensed to the wrong-doer to prevent them from recommitting the crime and (2) the presumed affect that fear of punishment will prevent others from committing similar crimes. The deterrence theory finds no justification for action in a past offence, which has more than a certain evidential importance, and it depends upon consequences of punishment other than the immediate satisfaction given to victims of offences and others (as in retribution theory). It need not ignore these satisfactions given. It rightly explains them of relatively less value. What is taken as important factor is that punishment prevents further offences. By deterrence is meant the use of punishment to prevent others from committing crimes. To achieve this purpose, the accused is punished so that he will be an example of warning to those who are thinking to violate the law. The general assumption is that this will curb the criminal activities of others in addition to the real offenders.

Punishment, is therefore, justified to control individual crime and to have a deterring effect on other criminals. According to Bentham, general prevention ought to be the chief end of punishment as it is its real justification. If we consider an offence which has been committed as an isolated fact, the like of which would never recur, punishment would be useless. It would be only adding one evil to another. But when we consider that an unpunished crime leaves the path of crime open, not only to the same delinquent but also to all those who may have the same motives and opportunities for entering upon it, we perceive that punishment inflicted on the individual becomes a source of security to all. That punishment which considered in itself appeared base and repugnant to all generous sentiments is relevant to the first rank of benefits when it is regarded not as an act of wrath or vengeance against a guilty or unfortunate individual who has given way to mischievous inclinations, but as an indispensable sacrifice to the common safety.

Paternoster, Raymond (2010:765-824) presents a brilliant review of modern concepts and theories of deterrence. Most of the time, people are not compelled to commit criminal acts but do so when the expected benefits outweigh the expected gains. A key term in deterrence theory, therefore, is the notion of *utility*, which simply refers to the total satisfaction that is derived from a course of action or consumption. A very simple equation to express the basic premise of deterrence theory is that an individual will decide to offend if :

U(Crime) = p_1 (Benefits of Crime) + p_2 (Costs of Crime) + p_3 (Benefits of Non-Crime) + p_4 (Costs of Non-Crime)

This equation states that the utility of crime is equal to the sum of the benefits of crime (easy money, the "high" of drug use, prestige, or reputation), the costs of crime (getting arrested, getting convicted, going to jail, loss of family respect, not seeing one's children while in jail or prison), the benefits of non-crime (a steady and safe income from a job, physical security from not being shot or falling ill from ingesting bad drugs), and the costs of non-crime (low wages, feeling like a failure from a low-prestige job). The *p*'s reflect the probability of each event occurring. The utility of crime, then, is determined not only by the outcomes, but by the probability of those outcomes.

One of the costs of crime is the possible legal punishment, including being arrested for committing the act, getting convicted of the crime, and being subjected to some kind of punishment as a result of conviction (jail, probation, prison, or, in some jurisdictions for some crimes, death). Deterrence theorists presume that punishment by the legal system will increase the cost of—and therefore inhibit—crime. There are three properties of legal punishment that are related to its cost, the (1) certainty, (2) severity, and (3) celerity (or swiftness) of punishment. Other things being equal, a legal punishment is more costly when it is more certain (more likely than not to be a consequence of crime), severe (greater in magnitude), and swift (the punishment arrives sooner rather than later after the offense). This leads to the three key hypotheses that can be deduced from deterrence theory, which have served to guide empirical research over the past fifty years :

H1: The greater the certainty of legal punishment, the lower the crime rate

H2: The greater the severity of legal punishment, the lower the crime rate.

H3: The greater the celerity of legal punishment, the lower the crime rate.

The important deterrence variables are, then, certainty, severity, and celerity. There are two levels of these punishment properties, an objective level and a subjective (or perceptual) level. In general, the criminal justice system is premised in part on the expectation that having rigorously enforced laws and appropriate penalties that are applied with reasonable swiftness will lower crime by deterring punished persons from offending again and deterring would-be offenders from committing new crimes. How certainly, severely, and swiftly a jurisdiction *actually responds* to crime constitutes the objective properties of punishment. The greater the celerity of legal punishment, the lower the crime rate.

History of Deterrence

Deterrence theory has played a role in societies throughout history. One of the earliest examples of deterrence theory was during the Feudalist era between the ninth and fourteenth centuries. The Feudalist era was part of the Middle Ages and characterized by lords (men of power and asset), vassals (individuals loaned land by the lords) and fiefs (lands). In exchange for the lands, the vassal would pledge military allegiance to the lord loaning the lands. This period of English history inherited a relatively unchanged judicial system that came from the Anglo Saxon era dating back to 600 AD, often referred to as the Dark Ages (Brittain Express, 2006). The Anglo Saxon period was brutal and the strong dictated to the weak. Laws during the Feudalist era were designed, for the most part, to keep the peasants in their place. Laws were enacted to deter the peasants from attempting to rise above their station in life or challenge the status quo. Punishment for crimes during this period was not enforced unless the person receiving the punishment was considered a subordinate to the person upon whom the crime was committed.

The earliest known example of deterrence as it relates to crime and punishment is the Code of Hammurabi. This ancient text was the first of its kind, a document set forth by a ruling party which categorized crimes and their corresponding punishments. This code, carved upon a large stone slab and placed in public view, was intended to educate members of society on what is expected of their conduct and what would happen should they violate those expectations (King, 1998). The Code of Hammurabi also made it clear that ignorance of the law is no excuse for its breaking, another step along the path of crime deterrence.

When Henry VII (1457 - 1509) took the throne at the beginning of the Tudor period, he began to move belief toward the idea of royal absolutism and the divine right of kings to be omnipotent, answering to no one. This totalitarian perspective was firmly imprinted on the system of justice. Incarceration was not a common form of punishment. Prisons were simply holding areas where prisoners were kept until trial and sentencing. Trials were one sided in favor of the prosecution. Defendants were not allowed legal counsel and justice was swift and brutal.

During the reign of Henry VIII (1491 –1547), it was believed that extremely cruel and unusual punishment, in a public setting, would deter wrongdoers from committing additional civil discord (if they lived through it)(Bartollas, 2002). The affects of the judicial system were directly connected to an individual's station in life. Someone wealthy or influential could often commit offenses with impunity, while someone of lesser birth would be severely punished for the same offense. Public flogging, hangings and beheadings were common features in the public arena. Some individuals were hanged until almost dead, then taken down and drawn and quartered. Henry VIII had two of his many wives, Anne Boleyn and Catherine Howard, beheaded on the grounds of adultery. These examples of punishment dealt to those who displeased the king acted as strong deterrents to those who served in the Royal Court.

During the reign of Queen Elizabeth I (1533 –1603), depending on the crime, the offender might be sent to the galley service (the beginnings of community service concept?), where the prisoners were held as slaves and chained four men to an oar on great ships of the realm. These were criminals convicted of committing less than capital crimes, but they would end up serving an equivalent to life imprisonment. The following quote from Queen Elizabeth I, offers a reference to both deterrence theory and the idea of community service, though the "service" in this case was more about serving as an example to others as well as punishing the offender and one might be hard pressed to see this punishment as tempered with clemency and mercy.

"Except when convicted of wilful Murder, Rape and Burglary," a criminal might be reprieved from execution and sent to the galleys, "where in, as in all things, our desire is that justice may be tempered with clemency and mercy...and the offenders to be in such sort corrected and punished that even in their punishment they may yield some profitable service to the Commonwealth."

The brutality and severeness of the judicial system was designed to not only deter individuals from committing crimes, but to deter them from challenging or speaking against those in power. It was rarely fair, unevenly dealt and ripe with corruption.

A unique opportunity for making as evaluation of the deterrent approach to crime was presented by Danish experience during the second world war. In 1944, the German occupying forces deported the Danish police, and for some time the country had only a local guard force invested with police authority. There followed an immense rise in the number of robberies, thefts, frauds etc. but no comparable increase in murder or sexual crimes. While this experience does show that crime is reduced very considerably by the prospect of detection and presumably punishment, it suggests that deterrent methods are of less value reducing the incidence of those crime in which strong passions or deep psychological problems are involved.

One may say that this is a situation of war times which is one of behaviour in an abnormal situation. But even then, it may be said that punishment has a deterrent effect with respect to many offences. It may deter many potential offenders. Punishment is justified only under certain conditions, one of which is that it must deter to a certain extent. We then have the principle that a man's punishment is morally justified if the following conditions are satisfied. It does indeed deter and in, so doing (1) causes less distress than would occur if it were not imposed, also (2) there is no other punishment that would deter as effectively at a cost of less distress. Punishments which satisfy these conditions have been called economical deterrents or punishments that defer economically. It is also to be noticed that punishment must be sufficiently severe to deter effectively. It has sometimes been true that offenders regarded the penalties for their offences as like licences: an unavoidable feature of conduct of a certain kind and not so undesirable as to be a different.

General and Specific Deterrence

While it has always been in place, interest surfaced once more and governments began to invest more time and effort into acting on these theories. Initially, two forms of deterrence consumed theorists: general, and individual (or specific). Later on, theorists have added two more major types of deterrence to the fray, being absolute deterrence and restrictive deterrence (Winfree & Abadinsky, 2003). All four types of deterrence target various types of potential offenders. One thing to keep in mind is that deterrence is a very difficult concept to measure as it produces no solid, tangible primary results. All results seen from deterrent methods are dependant upon something else (secondary).

The concept of general deterrence is the most proactive of all as it seeks to target potential crimes before they happen. This branch of theory is a starting point in the deterrence continuum and often targets the crime in general, not the offender. One may ask how a rule deters a crime instead of a person. I believe that general deterrence does so by issuing blanket knowledge that if one commits a crime then there will be punishment. As first time offenders do not have a criminal history, they could be anyone. General deterrence is the going about of issuing a law and making it known that it is not permitted, and that there will be consequences should one commit such an act. For example, a "No Trespassing" sign which quotes a state law, such as the State of Washington's, is aimed at the general public, reminding or educating them that if they "knowingly enter or remain unlawfully in or upon premises of another" it is viewed as Criminal Trespass in the second degree (Chapter 9A.52 RCW: Burglary and Trespass, 2007).

Many examples of general deterrence we see everyday do not even have to quote the law itself: Electronic article surveillance is a popular deterrence method in many grocery and department stores, ominous tags attached to clothing and causing pillars to sound each time tags pass between them (this also appeals to the human aversion to public shaming); video surveillance cameras which watch our every move in retail establishments, convenience stores, banks, and government buildings; uniformed personnel. All of these security measures help deter the commission of crimes, and are aimed at the general public. It is also important to note that general deterrence is the most common, widespread form of this theory.

Individual deterrence, the second branch of deterrence theory and often a following step after general deterrence, is a bit more intimate when it comes to dealing with crime. Instead of being aimed at the general public and deterring crime in a general manner, it is used to target new offenders, often in their first stages of following a potentially criminal path. Individual deterrence is usually decided by the probation or parole system, often by the power of a judge or similar presiding power. Whereas the general branch informs the public of the law and punishment associated with their misdeed, individual deterrence helps to promote better decision making by giving the courts a measure of discretion when dealing with first time offenders. For example, a group of youths who string toilet tissue around an ill-favored teacher's home using the ever popular tactic of "teepeeing" have technically committed the crime of defacing private property. Other crimes may be committed depending on youth curfews and when the event took place, if any real damage was done to the property, etc. While defacing private property may be punishable by a fine and, say, 15 days in jail, a judge is not likely to tie up valuable resources in the justice system on this. Taking into account that all of the youths present are fairly successful students, have never been in any trouble with the law before, and are remorseful after the fact, the judge may issue community service on a graffiti cleaning crew in a nearby public park, six months probation, and a formal letter of apology to the victim. This discretionary sentence seeks to show the youths what kind of impact property defacement has on those victimized, and helps to show them that crimes—no matter how fun—are punishable, and steer them from a life of further crime.

An example on the opposite side of the individual deterrence spectrum may be a repeat DUI offender : Being caught three times driving under the influence a person has shown that they likely face alcoholism issues and are unable or unwilling to seek help on their own, or they may convey through their actions that they simply do not care about the safety of other drivers and pedestrians on the road. Taking their prior offences for the same crime and their apparent lack of seriousness for the situation, a judge may take it upon him or herself to show the offender how serious they have erred. A suggested sentence for this may be attending a DUI victim's panel, mandatory alcohol treatment, and serving thirty days in county jail. This example shows individual deterrence at its strongest form, meaning harsher sanctions against those who continually commit a type of crime.

Absolute and Restrictive Deterrence

Since the time of Beccaria, social scientists have added two other forms of deterrence to this theory : Absolute and restrictive deterrence. Absolute deterrence is the knowledge of what will happen when one commits a crime and their unwillingness to do so in the future (Winfree &

Abadinsky, 2003). For example, a man who has previously robbed a convenience store has been released from prison after serving six years of his life for this act. Upon being released he secures an entry level job with a small moving company making low wages. As a mover he has access to property and valuables, however knowing that crime has gotten him in trouble before, and despite the fact that stealing from clients could go virtually unnoticed he decides to resist the thefts. He has decided that crime, even though his original crime was much more serious and garnered more time of his life than simple theft would, is not worth the price that he may pay should he be caught. This is absolute deterrence.

Absolute deterrence does not need to involve an offender who has been punished for any crime to exist (Winfree & Abadinsky, 2003). Take for example a man who usually solicits prostitutes in a certain area of town when he is unable to find an upscale escort service. He knows of a few friends at work who have regularly frequented this area for sex over the past few years along with him. At one point a friend of his is arrested, his behaviours emerge to his friends and family, and he enters a period of life involving the consequences of this action. Our subject decides that the risk is too great to continue looking for sex in this or any area and cannot imagine what would happen to his life, his family, and his children should he be arrested for this crime. Even though he was not the target of this punishment he decides that both prostitution and so called "escort" services are not worth the risk of being involved in. This is also absolute deterrence.

Restrictive deterrence, the fourth and final concept, does not involve the satiation of crime by a previous offender, only a modification of one's habits in order to avoid being caught. This may involve modifying the types of crimes one commits in order to remain free and reap the benefits of their actions (Siegel & Senna 2006). Take for example an offender who seeks monetary gain without the traditional actions. She prefers to make small amounts of money without rousing suspicion so she passes off several bad checks in another person's name. After being caught for this fairly quickly due to the surge in measures relating to identity theft, ordering to pay it all back, and serving sixty days in county jail she embarks on other criminal endeavors with the same goal but less risk. She instead turns to an elaborate scheme where she purports to be a caretaker for the elderly all while helping herself to their bank accounts by having legitimate access. Her ends have not changed as she is still stealing from others' bank accounts; however her means has in order to reduce the risk of further punishment.

The Role of the Police in Deterrence

Role of police is very significant in controlling and preventing crime and more so as the agency to show off deterrence. In 1994, President Clinton signed the Violent Crime Control and Law Enforcement Act, which initiated many policies aimed at increasing the deterrent bite of the law, including new firearms restrictions, tougher sentencing laws, and a plan to put 100,000 new police officers on the street. There is evidence that, as a result of federal funding, there was a noticeable increase in the number and per capita rate of police officers in the United States beginning in the early 1990s. Although it is difficult to say for sure, one estimate indicated that there was an increase of approximately 50,000 to 60,000 police officers in the United States (a per capita expansion of about 14%). The intended effect of putting more police officers on the streets was to increase the objective probability or certainty of arrest, with a subsequent deterrent effect on the crime rate. In addition to a sheer increase in numbers, however, police experiments with more targeted strategies of enforcement such as using technology to identify geographic areas of high crime or "hot spots," and they engaged in "crackdowns" that targeted particular kinds of crime, such as drug dealing and weapons offenses, with enhanced enforcement. It is possible that the rather dramatic decrease in crime observed in the United States since the 1990s was due to the combination of more police and better policing.

With respect to the role of the police, many scholars have conclude that at least some of the decline in crime rates can be attributed to an increase in the number of police and the increased certainty of punishment associated with that increase. The effect of aggressive police patrolling practices (having patrol units on the street stopping cars and issuing moving citations) can lower robbery with increased arrest ratios. What the police do can affect the objective certainty of punishment.

A Critique of Deterrence

In spite of its central importance, and the very high expectation we have that legal punishment and criminal justice policies can inhibit crime, we do not have very solid and credible empirical evidence that deterrence through the imposition of criminal sanctions works very well. It is easy enough to see that criminal deterrence frequently and effectively influences our actions. We all "throw out the anchor" on the highway when we spot a patrol car. We generally do not park in handicapped-only parking spots. We do not light up a joint at the movies. Burglars like to break into unoccupied homes. Drug deals are typically made in the shadows and not out in the open where police can see. But while we know that deterrence works in the absence of doing nothing about crime, we are decidedly less certain about relative or marginal deterrent effects. For example, does increasing the number of police officers reduce crime ? Does adding three years to a prison sentence for the use of a gun deter firearm-related felonies ? Does an increased use of incarceration deter crime better compared with previous levels of incarceration ? Criminal deterrence is very weak today in the sense : (1) it is very difficult to isolate and measure a deterrent effect precisely because a great many things must happen before deterrence can occur; (2) it just may be very difficult to generate strong deterrent effects through the legal system because the system is unable to exploit human rationality effectively.

The general ill-repute in which deterrence theory was held in the field of criminology before the mid-twentieth century was nicely captured by Hans von Hentig (1938). There are large groups of persons who are simply immune to the deterrence-like threats of the law including those motivated by love or maternal instincts, the young and women who tend to be impetuous, those motivated by ideology, the "have-nots," and the feeble-minded. People are moved by the consequences of their actions, the pursuit of pleasure and avoidance of pain, "appear[s] to be unreal and simple-minded, when we think of the enormous multiformity and complication of life." According to Hentig (1938:559), deterrence is doomed to fail because the pleasure of committing a criminal act is a "near object" while the cost provided by the legal system is a "long-distance danger," very much in the remote future and therefore unable to offset the more immediate advantages of crime.

It is also argued that psychological learning principles are inconsistent with the idea that punishment could inhibit behaviour under normal circumstances. Punishment is essentially an ineffective way to control or to eliminate the behaviour of the punished person. Eliminating crime by means of more effective law enforcement and heavier penalties, is at first repugnant to the democratic citizen and smacks of the police state. Criminologists today think that punishment is archaic and barbaric, that crime is really due to some underlying individual pathology, and that the only way to reduce it is through the rehabilitation of individual offenders. They articulate a comprehensive research agenda. Two very critical points in that agenda were that it is very likely that any empirical deterrent effect for punishment : (1) depends upon the certainty of punishment, and (2) requires an understanding of would-be offenders' perceptions of punishment rather than the objective properties of punishment. Similarly a law can have no deterrent influence upon a potential criminal if he is unaware of its existence.

There are, three features about legal punishment that prevent it from being an even more effective deterrent : (1) legal punishments are generally very uncertain, (2) they are only imposed long after the crime has been committed and so have low celerity, and (3) the pleasures of crime are immediate and so carry greater weight than the delayed costs of crime in the would-be offender's calculus.

Thus, deterrence is a deeply rooted concept in our society, and goes to great lengths to help keep society in working order. Many studies have factored in previous crime statistics before a deterrence idea has been implemented, such as red light cameras. Others have gone to the source of the criminal behaviour, and relied upon offenders to say what deters them from future criminal activity. Unfortunately lack of measurable and solid results means the concept is difficult to gauge, as human word and explanations of what may be coincidence due not prove a concept to be true. Deterrence theory has its roots in prehistory when it operated simply on the basis of strength and skill over weakness and lack of skill. As humans evolved, deterrence theory evolved in that people found more and more brutal ways of attempting to deter others from taking away power, property or position. Even with all this "creativity" in play, crime still occurred and still occurs today. Deterrence seems to be much like treating symptoms rather than curing disease. What is true is that deterrence is a concept that humans are familiar with on their very basest of human emotions, that good is better than bad, a concept that will likely always work in one form or another, and will continue to work towards society's benefit.

Deterrence Theory as a Military Strategy

Deterrence theory as a military strategy developed during the Cold War. It is especially relevant with regard to the use of nuclear weapons, and figures prominently in current United States foreign policy regarding the development of nuclear technology in North Korea and Iran.

The term is also used more casually to suspect a party in any field of potential conflict of being prepared to inflict unacceptable damage on an aggressor, and making sure the potential aggressor is aware of the risk so that they may refrain from aggression. Establishing a credible and known deterrence is at the opposite end of the spectrum from a policy of appeasement.

Deterrence, thus, is a strategy by which governments threaten an immense retaliation if attacked, such that aggressors are deterred if they do not wish to suffer great damage as a result of an aggressive action. Weapons of mass destruction (WMDs), conventional weapons strength, economic sanctions, or any combination of these can be used as deterrents. Mutually Assured Destruction (MAD) is a form of this strategy, which came to prominence during the Cold War when it was used by the US to characterize relations between the United States and Soviet Union. Both nations were prepared to fight a full scale nuclear and conventional war, but were not willing to risk the carnage of a full scale nuclear war [Yost, D. 2007. "Analysing International Nuclear Order." International Affairs 83:3, 549-574].

Deterrence by denial is a strategy whereby a government builds up or maintains defense and intelligence systems with the purported aim of neutralizing or mitigating attacks. Aggressors are deterred if they choose not to act, perceiving the cost of their action to be too high in relation to its likelihood of success.

Criticism of Deterrence Theory as a Military Strategy

Deterrence theory is criticized for its assumptions about opponent rationales : first, it is argued that suicidal or psychotic opponents may not be deterred by either forms of deterrence.

Second, diplomatic misunderstandings and/or opposing political ideologies may lead to escalating mutual perceptions of threat, and a subsequent arms race which elevates the risk of actual war.

Thirdly, a military build-up increases a country's risks of budget deficits, restrictions on civil liberties, the creation of a military-industrial complex, and other such potentially-undesirable measures.

Fourthly, a number of new or nuanced criticisms of "traditional" deterrence theory emerged around group psychology. Studies of the specific group psychology of several leaders and leader groups, including the Israeli and Arab leaders in 1973 and the Kennedy Administration during the Bay of Pigs Invasion and Cuban Missile Crisis, indicated that in many cases executive groups use poor decision-making techniques and improperly assess available information. These errors can and often do preclude truly rational end-behaviour in deterrence situations (Psychology and Deterrence, Ed. Robert Jervis, Johns Hopkins University Press; 1989).

Finally there is an intellectual dilemma as well. The intellectual constructs within deterrence theory are responsible for at least one unshakeable dilemma, examined here. Two positive attributes of deterrence credibility are frequently cited : the virtues of certainty and the virtues of ambiguity. The principle behind certainty is that deterrence works best if the potential aggressor understands clearly and fully the unavoidable consequences of a retaliatory response. There is no room for ambiguity. The principle behind ambiguity is that deterrence works best when the aggressor is confounded by uncertainties in enemy response, which frustrate attack planning and contingency planning. Risks are incalculable in the presence of total ambiguity. Both principles have obvious values but are mutually-exclusive and contradictory. They cannot exist simultaneously and the defending nation must decide which principle is to prevail. In any given situation, how is a choice made between principles ? To summarize the dilemma, there is no reliable way to determine whether ambiguity is best or certainty is best. This dilemma undercuts the very roots of deterrence theory.

References

Bureau of Justice Statistics (1981, 1981-1996, 1998). Crime and Justice in the United States and in England and in Wales."

David Kopel, Paul Gallant & Joanne Eisen (2001). Britain: From Bad to Worse.Americas first freedoms p.26

Disaster Center (2010). US and DC Crime Rates, Source: FBI UCR index.

Hentig, Hans von (1938) *The Limits of Deterrence*, 29 J. AM. INST. CRIM. L. & CRIMINOLOGY.

Jacob Scullum (2000). Guns Down Under reason, Australia, p. 10.

Keel, Robert (2005). Rational Choice and Deterrence Theory Lecture Notes. University of Missouri, St. Louis.

King, L.W. (1998). Ancient History Sourcebook: Code of Hammurabi, c. 1780 BCE, Fordham University web site:http://www.fordham.edu/halsall/ancient/hamcode.html.

Liptak, Adam (June 29, 2008). Gun Laws and Crime:A Complex Relationship:http://www.nytimes.com/2008/06/29/weekinreview/29liptak.html?_r=2.

National Rifle Association Institute for Legislative Action (2010). Gun Laws, Culture, Justice & Crime In Foreign Countries Fairfield, VA

Paternoster, Raymond (2010), How Much Do We Really Know About Criminal Deterrence ? The Journal of Criminal Law & Criminology, vol.100 (3) pp. 765-824.

Schalleger (2002) Evolution of Deterrence Crime Theory, Http://www.associatedcontent.com/article/31600

Siegel, L, & Senna, J (2006). Introduction to Criminal Justice. Mason: Thomson Wadsworth.

State of Washington (2007). Chapter 9A.52 RCW: Burglary and trespass, web site: http://apps.leg.wa.gov/RCW/default.aspx?cite=9A.52.

Summerfield, Morgan (2006), Evolution of Deterrence Crime Theory Associated ContentWebsite:http://www.associatedcontent.com/article/32600/evolution_of_deterrence_crime_theory.html.

Wikipedia (2006). Code of Hammurabi, from web site: http://en.wikipedia.org/wiki/Code_of_hammurabi.

Winfree, L.T., & Abadinsky, H (2003). Understanding crime: Theory and practice.Belmont: Thomson Wadsworth.

■■■

CHAPTER 22 Theory of Retributive and Restorative Justice

Learning Objectives :

- ❒ **Acquainting the Learners with the history and sub-types of Retributive justice.**
- ❒ **Acquainting the Learners with the history and practices of Restorative justice.**
- ❒ **To make a critical appreciation of Retributive justice.**

Introduction

Concepts of Retributive Justice

Retributive justice is a theory of justice that considers that punishment, if proportionate, is a morally acceptable response to crime, with an eye to the satisfaction and psychological benefits it can bestow to the aggrieved party, its intimates and society. In ethics and law, "Let the punishment fit the crime" is the principle that the severity of penalty for a misdeed or wrongdoing should be reasonable and proportionate to the severity of the infraction. The concept is common to most cultures throughout the world. Its presence in the ancient Jewish culture is shown by its inclusion in the law of Moses, specifically in Deuteronomy 19:17-21, and Exodus 21:23-21:27, which includes the punishments of "life for life, eye for eye, tooth for tooth, hand for hand, foot for foot." That phrasing in turn resembles the older Code of Hammurabi. Many other documents reflect this value in the world's cultures. However, the judgment of whether a punishment is appropriately severe can vary greatly between cultures and individuals.

Proportionality requires that the level of punishment be scaled relative to the severity of the offending behaviour. However, this does not mean that the punishment has to be equivalent to the crime. A retributive system must punish severe crime more harshly than minor crime, but retributivists differ about how harsh or soft the system should be overall.

Traditionally, philosophers of punishment have contrasted retributivism with utilitarianism. For utilitarians, punishment is forward-looking, justified by a purported ability to achieve future social benefits, such as crime reduction. For retributionists, punishment is backward-looking, and strictly for punishing crimes according to their severity(Cavadino and Dignan,1997: 39). Depending on the retributivist, the crime's level of severity might be determined by the amount of harm, unfair advantage or moral imbalance the crime caused.

History

In the early period of all systems of law the redress of wrongs takes precedence over the enforcement of contract rights, and a rough sense of justice demands the infliction of proportionate loss and pain on the aggressor as he has inflicted on his victim. Hence the prominence of the "lex talionis" in ancient law. The Bible is no exception: in its oldest form it included the "lex talionis," the law of "measure for measure" (this is only the literal translation of middah ke-neged middah). In the 19th century, philosopher Immanuel Kant wrote in The Metaphysical Elements of Justice of retribution as a legal principle : "Judicial punishment can never be used merely as a means to promote some other good for the criminal himself or for civil society, but instead it must in all cases be imposed on him only on the ground that he has committed a crime"(Martin, 2005:174). Immanuel Kant regards punishment as a matter of justice. He states that if the guilty are not punished, justice is not done.

Subtypes

There are two distinct types of retributive justice. The classical definition embraces the idea that the amount of punishment must be proportionate to the amount of harm caused by the offence. A more recent version advocated by the philosopher Michael Davis dismisses this idea and replaces it with the idea that the amount of punishment must be proportionate to the amount of unfair advantage gained by the wrongdoer. Davis introduced this version of retributive justice in the early 1980s, at a time when retributive justice was making a resurgence within the philosophy of law community, perhaps due to the practical failings of reform theory in the previous decades.

Criticism

According to some theories of ethics, punishment (or proportionate punishment) is self-contradictory in the sense that "Two wrongs don't make a right". Some hold that the motive behind the Christian sanction for interpersonal relations ("turn the other cheek" before seeking retribution for a wrong), and the motive behind the sanctions for social magistrates (which include the application of retributive justice, e.g., "just stonings"), conflict. On the other hand, the motives for the social sanctions can be attributed to other justifications beyond simple retaliation.

Many more jurisdictions following the retributive philosophy, especially in the United States, follow a set tariff, where judges impose a penalty for a crime within the range set by the tariff. As a result, some argue that judges do not have enough discretion to allow for mitigating factors, leading to unjust decisions under certain circumstances. In the case of fines, the financial position of an offender is not taken into account, leading to situations where an unemployed man and a millionaire could be forced to pay the same fine, creating an unjust situation; either the fine would be too punitive for the unemployed offender, or not large enough to punish the millionaire (Martin, 2005, pp. 174–175). Alternatives to retributive measures include psychiatric imprisonment, restorative justice and transformative justice. A general overview of criminal justice puts each of these ideals in context.

Theory of Restorative Justice

Restorative justice often called "reparative justice", is an approach to justice that focuses on the needs of victims and offenders, instead of the need to satisfy the abstract principles of law or the need of the community to exact punishment. Victims are given an active role in a dispute and offenders are encouraged to take responsibility for their actions, "to repair the harm they've done-by apologizing, returning stolen money, or (for example) doing community service." It is based on a theory of justice that focuses on crime and wrongdoing as acted against the individual or community rather than the state (Marty, 2001). Restorative Justice can involve a fostering of dialog between the offender and the victim show the highest rates of victim satisfaction, true accountability by the offender, and reduced recidivism (repeat offences). Restorative justice is defined as :

"... a broad term which encompasses a growing social movement to institutionalize peaceful approaches to harm, problem-solving and violations of legal and human rights. These range from international peacemaking tribunals such as the South Africa Truth and Reconciliation Commission to innovations within the criminal and juvenile justice systems, schools, social services and communities. Rather than privileging the law, professionals and the state, restorative resolutions engage those who are harmed, wrongdoers and their affected communities in search of solutions that promote repair, reconciliation and the rebuilding of relationships. Restorative justice seeks to build partnerships to reestablish mutual responsibility for constructive responses to wrongdoing within our communities. Restorative approaches seek a balanced approach to the needs of the victim, wrongdoer and community through processes that preserve the safety and dignity of all."

Restorative justice is very different from either the adversarial legal process or that of civil litigation. "Court-annexed ADR (alternative dispute resolution) and restorative justice could not be philosophically further apart" because lawyers seek to reduce issues between offenders and victims to only legally relevant ones and to protect their offending client, whereas restorative justice seeks "expanding the issues beyond those that are legally relevant, especially into underlying relationships"(Braithwaite, 2002:249).

Citing Greif, Liebmann has written that " a way of looking at restorative justice is to think of it as a balance between a number of different tensions : a balance between the therapeutic and the retributive models of justice, a balance between the rights of offenders and the needs of victims and a balance between the need to rehabilitate offenders and the duty to protect the public" (Liebmann, 2007:33).

"Restorative justice posits a paradigm shift that is best understood by asking the oft-quoted "three questions." The more common three questions for a system of justice to ask are "1. What laws have been broken?, 2. Who did it ?, 3. What do they deserve ?" Restorative justice asks, "1. Who has been hurt ?, 2. What are their needs ?, 3. Whose obligations are these ?" (Zehr, 2002).

History

Restorative approaches to crime date back thousands of years. In North America, the first traces of restorative justice have been attributed to the First Nations communities. In Israel, the Pentateuch specified restitution for property crimes. In Babylon, the Code of Hammurabi (c. 1700 BC) prescribed restitution as a sanction for property offenses. In Rome, the Twelve Tables (449 BC) ordered convicted thieves to pay double the value of stolen goods. In Ireland, under the Brehon Laws (first recorded in the Old Irish period) compensation was the means of restitution for most crimes. In Germany, tribal laws promulgated by King Clovis I (496 AD) called for restitutional sanctions for both violent and nonviolent offenses. In England, the Laws of Ethelbert of Kent (c. 600 AD) included detailed restitution schedules. In New Zealand/Aotearoa, prior to European contact, Maori had a well-developed system of custom and process that ensured the protection of individuals, the stability of social life and the integrity of the group i.e. a system of restorative justice. Retributive justice began to replace this system following the Norman invasion of Britain in 1066 A.D. William the Conqueror's son, Henry I, issued laws detailing offenses against the "king's peace." By the end of the 11th century, crime was no longer perceived as injurious to persons, but rather was seen as an offense against the state (Wikipedia).

Restorative justice takes many different forms, but all systems have some aspects in common. In criminal cases, victims have an opportunity to express the full impact of the crime upon their lives, to receive answers to any lingering questions about the incident, and to participate in holding the offender accountable for his or her actions. Offenders can tell their story of why the crime occurred and how it has affected their lives. They are given an opportunity to make things right with the victim—to the degree possible—through some form of compensation (Leo, 2006).

In social justice cases, impoverished people such as foster children are given the opportunity to describe what they hope for their futures and make concrete plans for transitioning out of state custody in a group process with their supporters. In social justice cases, restorative justice is used for problem solving(Braithwaite, 2002). In criminal cases, types of compensation include, but are not limited to: money, community service in general, community service specific to the deed, self-education to prevent recidivism, and/or expression of remorse.

Restorative justice sometimes happens in the context of a courtroom, and sometimes within a community or nonprofit organization. In the courtroom, the process might look like this : For petty or first-time offenses, a case may be referred to restorative justice as a pretrial diversion, with charges being dismissed after fulfillment of the restitution agreement. In more serious cases, restorative justice may be part of a sentence that includes prison time or other punishments. In the community, concerned individuals meet with all affected parties to determine what the experience and impact of the crime were for all. Those called out for offenses listen to others' experiences first, preferably until they are able to reflect and feel what those experiences were for the others. Then they speak to their experience : how it was for them to do what they did. A plan is made for prevention of future occurrences, and for the offender to heal the damage to the injured parties. All agree. Community members hold offender accountable for adherence to the plan.

Most academics and government definitions of restorative justice restrict that definition to those programs that involve an encounter between the offender and the victim. Some grassroots organizations, like the Mennonite Central Committee Canada, define restorative justice programs less on who the clientele of the program is, and more on the programs values. This means that programs that only serve victims (or offenders for that matter), but have a restorative framework, are considered a restorative justice program. Restorative justice pioneer Howard Zehr was honored as the recipient of the 2006 Community of Christ International Peace Award.

Many Libertarians support restorative justice because it is a victim-centric rather than state-centric approach to law enforcement. Victim-offender mediation, or VOM (also called victim-offender dialogue, victim-offender conferencing, victim-offender reconciliation, or restorative justice dialogue), is usually a meeting, in the presence of a trained mediator, between the victim of a crime and the person who committed that crime. This system generally involves a few participants, and often is the only option available to incarcerated offenders. VOM originated in Canada where it formed part of an alternative court sanction in a 1974 Kitchener, Ontario case involving two accused vandals who met face-to-face with their many victims.

Recidivism

In a criminal context, reduction of recidivism is a secondary goal of restorative justice (Hennessey, 2005), achieved incidentally through the primary goal of restoration of offenders (Jaimie, 2005). Proponents of restorative justice argue that it can prevent reoffending(Hennessey, 2005), and deter other potential criminals (John, 1999). Critics have argued, on the other hand, that these practices have no significant influence in lowering crime rates (Jaimie, 2005 and John, 1999). The response to such arguments tends to focus on whether restorative justice is superior to traditional methods in this regard, rather than on whether it leads to an absolute decrease in crime rates (Jaimie, 2005). The majority of the arguments on both sides, however, appear to be predominantly theoretical, as the use of restorative practices in their current form is relatively recent and has not yet become widespread (Hennessey, 2005).

The results of those studies which have compared recidivism rates to those achieved by traditional court procedures have been varied (Hennessey, 2005). Some studies have demonstrated a relative reduction in recidivism subsequent to restorative procedures, but the decrease has not always been dramatic (Katherine, 2004 and James, 2006). An example of such research is a study of 1,298 juveniles conducted throughout the late 1980s and 1990s, which found a 32% reduction in recidivism relative to a comparison group. A 2001 report by Department of Justice Canada found that "restorative justice programs, on average, yielded reductions in recidivism compared to non-restorative approaches to criminal behaviour." These findings were supported by a 2007 Cambridge University report. On the other hand, some studies have indicated a relative increase in reoffending after restorative procedures, while still others have found no significant difference between the results of either retributive or restorative justice processes (Hennessey, 2005). These results appear insufficient to draw any universal conclusions (John, 1999).

Practices

The restorative practices concept has its roots in restorative justice. The International Institute for Restorative Practices has been developing a comprehensive framework for practice and theory that expands the restorative paradigm beyond its origins in criminal justice. Restorative practices is an emerging field of practice and study devoted to building social capital and achieving social discipline through participatory learning and decision-making.

The field of restorative practices offers a common thread to tie together theory, research and practice in fields such as education, counseling, criminal justice, social work and organizational management. Individuals and organizations in many fields are developing models and methodology and performing empirical research that shares the same fundamental hypothesis.

The fundamental unifying hypothesis of restorative practices is that human beings are happier, more cooperative and productive, and more likely to make positive changes in their behaviour when people do things with them, rather than to them or for them. This hypothesis maintains that the punitive and authoritarian to mode and the permissive and paternalistic for mode are not as effective as the restorative, participatory, engaging with mode.

These various fields employ different terms, all of which fall under the rubric of restorative practices : In the criminal justice field the phrase used is "restorative justice" in social work the term employed is "empowerment"; in education, talk is of "positive discipline" or "the responsive classroom"; and in organizational leadership "horizontal management" is referenced. Use of restorative practices is spreading worldwide, in education, criminal justice, family and youth and-serving and workplace applications.

Restorative justice has been applied to property offences, as well as civil and criminal offences. However, it has been deemed as unsuitable for drug offences, sexual assault and domestic violence. South Australia and New Zealand are among two of the areas which have dealt with juvenile sexual offence. Indigenous regions of Canada have implemented different approaches, such as circle sentencing, to tentatively deal with domestic violence. Advocates believe that it may be applicable to these indigenous communities because of the different level of regard and effectiveness they have for a punitive court system compared to non-indigenous persons. However, it is acknowledged that restorative justice has no real, hard and fast rules and regulations, and so caution should be exercised in applying it to these communities.

Theory

Thus "Restorative Justice" has developed to have several different meanings and to be practiced in a number of ways. At its inception some theoreticians wondered why the basic mechanism worked. David Moore in Australia on exposure to the work of Silvan Tomkins and Donald Nathanson formulated that the theoretical psychological foundations of why these "scripts" worked so well was due to mechanisms laid out in "Affect Psychology" and had much to do with Tomkins' view of "shame" and "interest." While by no means universally accepted many in Restorative Justice use these ideas to articulate particularly the idea that there is no reintegrative "shaming" there is only reintegrative "shame." That is a general recognition of the shared community burden of shame.

References :

Braithwaite, J. Restorative Justice & Responsive Regulation 2002, Oxford University Press, p. 249.

Denborough, D. (Ed.) (1996). Beyong the prison: Gathering dreams of freedom. Adelaide, South Australia: Dulwich Centre Publications.

Denton, D. (1998). Horizontal management. Lanham, MD: Rowman and Littlefield.

Hennessey Hayes, (2005) 'Assessing Reoffending in Restorative Justice Conferences' Australian and New Zealand Journal of Criminology, 38(1) 77.

Hopkins, B., (2004) Just Schools: A Whole School Approach to Restorative Justice, Jessica Kingsley Publishers, 208 pages.

Jaimie Beven et al., (2005) 'Restoration or Renovation: Evaluating Restorative Justice Outcomes' Psychiatry, Psychology and Law, 12(1) p. 194.

James Bonta et al., (2006) 'Restorative Justice and Recidivism: Promises Made, Promises Kept ?' in Dennis Sullivan and Larry Tifft (eds), Handbook of Restorative Justice: A Global Perspective, p.108.

John Braithwaite, (1999) 'Restorative Justice: assessing optimistic and pessimistic accounts' in Michael Tonry (ed), Crime and Justice: A Review of Research (Volume 25).

Katherine Basire, (2004) 'Taking Restorative Justice Seriously' 13 Canterbury Law Review 31.

Leo Zaibert, (2006) Punishment And Retribution Ashgate Publishing.

Liebmann, M. Restorative Justice: How it Works, London: Jessica Kingsley Publishers, p. 33.

Marty Price, J.D. "Personalizing Crime," Dispute Resolution Magazine. Fall 2001.

McCold, P. (2003). A survey of assessment research on mediation and conferencing. In L. Walgrave (Ed.), Repositioning Restorative Justice (pp. 67–120). Devon, UK: Willan Publishing.

McCold, P., & Wachtel, T. (2003). In pursuit of paradigm: A theory of restorative justice. Paper presented at the XIII World Congress of Criminology, Rio de Janeiro, Brazil. Retrieved October 25, 2007.

Nelsen, J. (1996). Positive discipline (2nd Ed.). New York: Ballantine Books.

O'Connell, T. (1998). From Wagga Wagga to Minnesota. Paper presented at the First International Conference on Conferencing, Minneapolis, MN, USA.

O'Connell, T., Wachtel, B., Wachtel, T. (1999) Conferencing Handbook: The New Real Justice Training Manual. Pipersville, PA: The Piper's Press.

Simon, B. (1994). The empowerment tradition in American social work. New York: Columbia University Press.

Zehr, Howard, (2002) The Little Book of Restorative Justice Intercourse, PA: Good Books.

■■■

CHAPTER 23

Reformatory and Rehabilitative Theory of Punishment

Learning Objectives :

- **Acquainting the Learners with the Reformatory Theory of Punishment.**
- **Acquainting the Learners with the theory of rehabilitation of prisoners.**
- **To expand the concepts of Probation, Parole and Furlough.**
- **To make a critical appreciation of the Concept of Pardon in India.**

Introduction

Modern society considers various objectives in order to control crime and it considers imprisonment a means to attain the twin aims, i.e., reform and treatment of the criminals so that they will commit no crime after their release. Society also seeks protection from criminals and for this purpose prison isolates criminals from the community for a certain time. All these objectives-reformation, retribution and deterrence, within the prison result in cross conflict. This conflict between reformative, deterrent and retributive measures has been a controversial issue from the point of correctional administration as well as treatment of the offender. Some criminologists have even gone to the extent of suggesting that introducing new reformative techniques in the correctional field should ultimately eliminate prisons. Special mention may be made of the Probation, fine, collective labor, etc. With the advent of social science to the arena of punishment, however, a new clearly defined school of thought has arisen whose insistence on the reform of the convict as the central theme of criminal theme of criminal sanctions excludes or subordinates all other ends of punishment. Certainly the ideal of complete reform has not been reached, although there is wide agreement on the value of this goal of punishment.

The history of reformation has been a tale of the search of mankind for the absolute correctional technique and of failure. Again and again the idea of reformation through punishment has appeared in one form or the other, as an expression of an idea higher than mere retribution or deterrence. "*Can punishment be in form of reform*": an age-old controversy ? '*Hate the crime and not the criminal*' very rightly said by Mahatma Gandhi. Whatever be the ultimate aim of punishment in the first instance it is the imposition on an evil. Punishment as Benthem said is itself an evil; it is a negative concept. Any power of punishment to reform is widely and strongly denied on the dual grounds of principle and of experience. Experiments have shown that punishments ordinarily do not reform. On the contrary, it often degrades, coleuses and brutalizes. Men commonly come out of prison worse, than they went in. Punishment tends to search for an answer but more often ends up raising more questions.

It should, however, be borne in mind that reformative idea must be kept within sensible limits. In its extreme application, the reformative idea will not secure its goal and might lead, as farce reformation should go hand in hand with retribution and deterrence. The majesty of law must be maintained through punishments, which has to be exemplary also. A certain amount of terror is also desirable, and at times, evens necessary. So the idea of deterrence cannot be ignored. Deterrence cannot be wholeheartedly and completely substituted by reformation pure and sole. The perfect blend of deterrence and reformation can be seen in Indian Jurisprudence. Some of the legislative provisions, which deal with reformative concept, are as follows : Reformative Schools Act, 1897, Juvenile Justice Act, 2000 (amended 2006), Probation Offender Act, 1958 and Parole rules.

Reformatory School Act, 1897 : One of the earliest outcome of the Reformatory Theory of Punishment is Reformatory School Act, 1897. The kind of approach taken by this act is curative. As this act basically deals with children and the focus of this act is to prevent first youthful offender, whose antecedent are not shown to be bad to sent to ordinary jails, which may have the effect of making them hardened criminal, the law as curative measure provide other form of punishment.

First youthful offenders are sent to Reformatory School run by government. There they have every basic facility of food, clothing, education and opportunity for exploring themselves. These all measures are under Government control and scrutiny of head of Reformatory School. Offenders are restricted within a particular boundary. This act of imposing restriction acts punishment on them.

However, it does not mean that all the offence of youth is always guided by Reformatory School Act. Certain limitations of this act are : (1) The offence of murder which is punishable to death or imprisonment of life, do not include the offence to be dealt with Reformatory Schools Act (2). A sentence of mere fine is not sufficient to order detention is Reformatory School. Further, the Juvenile Justice Act incorporates and supplements the object and aim of the Reformatory Schools Act in true spirit and sense.

The problem of social reformation and rehabilitation of prisoners is, with the passing of years, assuming larger dimensions in almost every country of the world. This is so because the moral consciousness of human being is more actively manifest in the anxiousness of the society to secure the redemption and rehabilitation of those who by committing less serious offence, have strayed away for the time being from the normal walks of social life. The main aim of social reformation and rehabilitation of prisoners can be achieved by any of the following ways : (i) Parole and (ii) After-care of released prisoners.

Parole is a form of conditional release granted to the prisoners after they have served a portion of their sentences. The conditional release involves a service which includes the control, assistance, and guidance the offenders need as they serve the remainder of their sentences within the free community.

Parole is related to but somewhat different from probation. Like probation, parole also ideally includes treatment in the form of supervision, guidance, and assistance. But while in the case of probation the suspension of sentence is made at the court level without the offender having served any part of the sentence in penal confinement while on the other hand, parole is granted by an executive board or the institution itself only after a prisoner has served a portion of his sentence. Parole is granted by administrative or an executive and is always preceded by serving part of sentence. Parolees are considered as undergoing both punishment and treatment. By parole it meant conditional release of a prisoner from prison but not from legal custody of the state. It should also be noted that, although in the transformatory age parole is granted in most of the instance but it couldn't be claimed by the prisoner as a right.

After-care of released prisoners is one of the most effective means to look after the financial and family burden of the prisoners in terms of providing vocational training for creating employment opportunities of the prisoner and their families. This constitutes one of the most important integral part of correctional programmes of prisoners, since immediately after release from prison an offender has to confront with a lot of social and personal problems, such as loss of family contacts, lack of suitable employment opportunities, social stigma of prison sentence. It is for the solution of these problems that a discharged offender needs the society's help, care, sympathy and solace. The significance of after-care help being without the provision for these services, a prisoner in all probability, will find no other alternative and to revert back to the crime.

Despite the usefulness of after-care services, there has been very little progress in this major area of correctional services in India. Some sporadic efforts to render a part of the service to the released offenders have, however, been made in some of the states. Yet, the actual after-care work that has been done so far in this country falls far from the short of the requirements. Lack of funds, ignorance of the psychological and economic basis of crimes, and general apathy are the major factors standing in the way.

The choice between punishment and reformation is a difficult one and it requires a judicious application of mind by the person dealing with such offender. This decision must be guided by the ultimate aim of providing justice to the victim, society and the offender. Justice is supreme and justice ought to be beneficial for the society so that the society is placed in a better-off situation. Law courts exist for the society and ought to rise up to the occasion to do the needful in the matter, and as such ought to act in a manner so as to sub serve the basic requirement of the society. The greatest virtue of the law is its flexibility and its adaptability; it must change from time to time so that it answers the cry of the people, the need of the hour and the order of the day.

In T.K. Gopal v. State of Karnataka (AIR 2000 SC 1669) the Supreme Court observed : "In the matter of punishment for offence committed by a person, there are many approaches to the problem. On the commission of the crime, three types of reactions may generate. The traditional reaction of universal nature is known as punitive approach. The other approach is the therapeutic approach; the third is the preventive approach. Under the punitive approach, the rationalization of punishment is based retributive and utilitarian theories. Deterrent theory which also part of the punitive approach proceeds on the basis that the punishment should act as a deterrent not only to the offender but also to others in the community. The therapeutic approach aims at curing the criminal tendencies, which were the product of a diseased psychology. The therapeutic approach has since been treated as an effective method of punishment which not only satisfies the requirements of law that a criminal should be punished and the punishment prescribed must be meted out to him, but also reforms the criminal through various processes, the most fundamental of which is that inspite of having committed a crime, may be a heinous crime, he should be treated as a human being entitled to all the basic human rights, human dignity and human sympathy."

Similarly, in State of M.P v G.Singh (AIR 2003 SC 3191) the Supreme Court observed : "The law regulates the social interests, arbitrates conflicting claims and demands security of persons and property of the people is an essential function of the state. It could be achieved through instrumentality of criminal law. Undoubtedly, there is a cross cultural conflict where living law must find answer to the new challenges and the courts are required to mould the sentencing system to meet the challenges. The contagion of lawlessness would undermine social order and lay it in ruins. Protection of society and stamping out criminal proclivity must be the object of the law, which must be achieved by imposing appropriate sentence. Therefore, law as a cornerstone of the edifice of "order" should meet the challenges confronting the society. In operating the sentencing system, law should adopt the corrective machinery or the deterrence based on factual matrix. The sentencing process should be stern where it should be, and tempered with mercy where it warrants to be. The facts and given circumstances in each case, the nature of the crime, the manner in which it was planned and committed, the motive for the commission of the crime, the conduct of the accused, the nature of weapons used and all other attending circumstances are relevant facts which would enter into the arena of consideration. The undue sympathy to impose inadequate sentence would do more harm to the justice system. It is, therefore, the duty of every court to award proper sentence having regard to the nature of the offence and the manner in which it was executed or committed, etc. The imposition of sentence without considering its effects on the social order may be in reality a futile exercise. The social impact of the crime, e.g. where it relates to offences against women, dacoity, kidnapping and other offences involving moral turpitude or moral delinquency which have great impact on social order and public interest, cannot be lost sight of and per se require exemplary treatment."

The following factors are relevant for determining whether a deterrent or reformative approach should be adopted :

(1) the facts and circumstances of each case,
(2) the nature of the crime,
(3) the manner in which it was planned or committed,
(4) the motive for the commission of the crime,
(5) the conduct of the accused,
(6) the nature of the weapons used,
(7) prior criminal record of the offender,
(8) age of the accused,
(9) home life of the accused,
(10) emotional and mental condition of the accused,

(11) the prospects for rehabilitation,

(12) the possibility of return to normal life in the community,

(13) the possibility of treatment and training of the offender,

(14) the social impact of the crime, and

(15) all other attending circumstances.

These factors are, however, illustrative and not exhaustive in nature. Thus, there is no fixed formula or guidelines which can be used to determine whether the offender should be punished by sentencing him to imprisonment or he should be given a chance to reform himself and to start a fresh life. Each case has to be decided on its own facts and circumstances. It is submitted that unless the court is undoubtedly of the opinion that the offender is a threat and menace to the society, he should be given a chance to reform himself and to contribute positively towards the betterment and development of the society at large. Further, whenever the court is divided on the issue of punishment or reformation, the benefit of such a situation should be extended to the offender and he should be given a chance to correct and reform himself. The aim of this section under the chapter on penology deals with three fundamental concepts related to the theory of reformation : Probation, Parole and Furlough.

Probation

The earlier penological approach held imprisonment, that is, custodial measures to be the only way to curb crime. But the modern penological approach has ushered in new forms of sentencing whereby the needs of the community are balanced with the best interests of the accused : compensation, releases on admonition, probation, imposition of fines, community service are few such techniques used. Through this paper, the advantages of probation are highlighted along with how it could be made more effective in India. Modern Criminal Jurisprudence recognizes that criminals are not born but made, since when a crime occurs it takes place due to culmination of variety of factors and that a good many crimes are the result of socio-economic milieu. These factors may be social and economic, may be due to erosion of moral values by parental neglect, stress of circumstances or doing a criminal activity in spur of heat of a moment. So, whenever a criminal activity takes place it is not always that the person so involved had prior intention to commit the crime; thus sentencing such a man to prison term may often embitter him and when he comes out of the prison bars he becomes an enemy of society. Thus, it is absolutely necessary to rekindle him and bring him mainstream of life. This type of rekindling can be best tried through parole and probation, community service and creative and employment opportunities. Thus probation service is becoming increasingly popular for many reasons, the most important being the economic cost of maintaining offenders in prison (Chakrabarti, ,1999:6-7). A compact definition of Probation can be stated as "Probation is a method of treating (correcting) suitably selected offender by releasing him into the community upon certain conditions prescribed by the court on conviction, before sentencing (offender) generally upon supervision of Probation Officer."

Probation as a method of correctional service evolved after the criminologists and legal jurists thought of means of reformation to criminals by giving them a chance to prove "their worth" and not confine them to the prison bars. The word "probation" is derived from the latin word "probatus" meaning "tested" or "proved". The first Probation officer in this world is said to be JOHN AUGUSTUS, a Boston Cobbler who is regarded as the "Father of Probation" who in 1841 volunteered to assist offenders if the court would release them to his care. As John Augustus was a cobbler operating in and around the courts of Boston, he daily used to watch the court proceedings of the criminals and used to wonder on the alternative methods available for the rectification of criminals other than imprisonment, a method enabling the criminals to rectify their mistakes and lead a normal societal life far away from the dark side of criminal world. John Augustus also had the self-belief that young culprits convicted of criminal offence if sentenced to prison would make them more strong, hard and fast criminals by mixing with other old criminals. Thus the entire life of these culprits would be ruined if they are made to stay in jails without giving them a chance to reform themselves or rectify their mistakes. This thinking led John Augustus to formulate the concept of Probation who later in 1841 appealed to the court to voluntarily assist the offenders and lead them to correctional methods if the court would release them to his care. He first started his work by granting bail to a man charged with being a "common drunkard". The man was released when he sweared in front of the court that he would never taste intoxicating liquors once he is not prisoned. In this manner Augustus started assisting offenders who came to him seeking relief on account of probation of good conduct, granted by a decree of the Court.

Correctional services like Probation and Parole is also important in the administration of Criminal Justice and in the method of mainstreaming criminals as it is an integral part of the total structure of the punishment system in a contemporary legal world. It also serves as an alternative to imprisonment especially of short-term and has now taken within its purview all the offences except those punishable with death or imprisonment for life. Legal statutes like Probation of Offenders Act, 1958 and the Code of Criminal Procedure, 1973 have made probation a more viable method of dealing with offenders than imprisonment, because the judge is required to record special reasons in the judgment of a criminal case other than capital punishment and imprisonment for life stating why probation is not granted to all the eligible offenders irrespective of their ages.

Probation in the Criminal Justice system postulates a proposition that to a large extent crime is the result of individual pathology- a deviance that can be dealt with through the use of Psycho-social treatment methods. The criminologists who propagate the correctional philosophy of punishment argue that "every saint has a past and every sinner a future". In spite of these arguments of criminologists, however the basic object of Probation as enlisted in the Probation of Offenders Act,1958 is the protection of society by preventing the crime through rehabilitation of the offender in the society without curbing his freedom, not subjecting him to rigorous prison life and depriving him of his social and economic obligations. The Supreme Court of India has also emphasized the reformatory and rehabilitative aspects of sentencing. Even there are several currents running through the waters of administration of criminal justice in these days and to quote a line from Justice R.C.Patnaik's judgment in Saradhkar Sahee v. State of Odisha "Let not the wind of change pass us without inspiring us."

Correctional services like Probation and Parole is a milestone in the progress of modern liberal trend of reform in the field of penology. It is the result of the recognition of the principle that the purpose of criminology is more to reform the criminal than to punish him. It prevents the turning of offenders into hardened criminal by providing them educative and reformative treatment in the community. It is needed is to treat prisoners therapeutically, so that they may be healed of their criminal activities and become good citizens. To make an offender a non-offender is the purpose of penology. In this background, Probation has a great practical value of socializing the convict and making him a responsible citizen.

The Growth of the Development of Probation :

Ancient Period

Dr. P.K. Sen, a renowned criminologist in his book Tagore Law Lecture on 'Penology Old and New' points out that the idea of releasing an offender after due admonition is not borrowed in India from USA or England but the ancient Hindu Law givers laid down that punishment must be regulated by consideration of the motive and nature of the offence, the time and place, the strength, age, conduct, learning and economic position of the offender and above all, by the fact whether the offence was repeated. All these ideas were envisaged by the Smriti writers as early as 300 B.C.

The ancient Smriti writers were aware of the complexities of human nature and duly paid attention to individuality of the offender. The foresights of Smriti writers were remarkable since in their writings there was direct reference of the release of offenders on account of good conduct and integrity of character, which seems to support the modern concept of Probation.

Manu, the ancient law giver in Manusmriti pondered the idea of releasing an offender after gentle admonition. About 2000 years ago, Brahaspati in his Dandabheda Vyavastha referred to admonition as punishment. According to Brahaspati, a gentle admonition should be administered to a man for light offence.

Yajnavalkya, laid down that having ascertained the guilt, the place and time, as also the capacity, the age and means of the offender, the punishment should be determined and if possible the man should be released if after putting into observation maintains good character and conduct. Kautilya, in his Arthashastra advised the king to award punishment which should neither be mild nor severe.

In ancient Indian mythology, Vishnu said that King should pardon no one for having offended twice. In Brahmabaibarta Purana, Lord Mahadev pronounced that if people commit offence it is the duty of pious man to forgive him. Even Maurya rulers were in favor of mild punishment. One of the edicts of the emperor Ashoka contains provision for remission of punishment. He advised his officers to examine and reduce punishment awarded to prisoners after due consideration of circumstances which substantially coincides with those mentioned by Smriti writers. Thus, the philosophy of Probation is not entirely new in Indian Criminal Law and views of our ancient law-givers had support for the modern Probation System.

Medieval Period

The administration of Criminal Justice in India during the time of Muslim reign in India upto the advent of British rulers was based on Islamic Criminal Law which did not recognise principles of correctional method or admonition. But we do find traces of principles of probation during Maratha rule and Peshwa period where cases of an offender who committed a crime was not at once awarded punishment but was given a chance to improve himself. The cases include :

(i) Case of Viswanath Bhatt Patankar– In 1775-76 one Vishwanath Bhatt Patankar was arrested for committing thefts. Since, he was unable to furnish security he was sent to prison. Janardhan Bhatt Bhide, a cousin of Vishwanath Bhatt stood surety for him promising that he would not again commit theft or any other offence.

(ii) Case of Janki Lagadin– In 1785-86, one Janki Lagadin was imprisoned at Fort Visapur for charges of adultery. Her father Shivaji Gaikwad prayed for her release. The prayer was granted on his standing as surety for her future good conduct.

The above cited two cases can be said to be the earliest cases of Probation in India.

Colonial Period

Although the probation system rooted in ancient Indian Criminology, found its legal recognition for the first time in 1898 in section 562 of the then Criminal Procedure Code(Cr. P.C.-Act XX of 1898). This section was actually taken from the English Probation of First Offenders Act,1887. Even S.562 was amended by the amendment of Cr.P.C. in 1923 which drastically changed the law of Probation in India. The old S.562 did not contain any specific provision empowering High Court as a court of Revision in Probation matters which was granted in new section. Also, under the new section 562 of 1923, a court may grant probation in case of offenders not under the age of 21 years for offences punishable with not more than 7 years and in respect of woman offender below 21 years for offences punishable with death or imprisonment for life.

During 1931, the Government of India circulated a proposed draft of Probation of Offender Bills to the then local governments, provinces and princely states for their views. In pursuance of the above suggestion some provinces enacted their own probation laws. The enactments include : (1) The C.P. & Berar Probation of Offenders Act, 1937, (2) The Bombay Probation of Offenders Act, 1938, (3) The U.P. First Offenders Act, 1938.

Modern Period

After India attained Independence, a joint committee was set up under the chairmanship of Mr. Hukum Singh on the bill to provide for release of offenders on Probation. This bill was presented in front of Lok Sabha on the 25th of February,1958. The descent views of the Bill were expressed by several members of the Lok Sabha like Rajendra Pratap Singh, Jagdish Awasthi, Yadav Narayan Jadhav. This bill was finally passed by the Legislature to give shape to an Act called Probation of Offenders Act,1958 on May 16,1958. The object of the Probation of Offenders Act,1958 is the protection of society by preventing the crime through rehabilitation of the offender in the society as its useful member without curbing his freedom, subjecting him to rigorous prison life and depriving him of his social and economic obligations. The Act also seems to accomplish this object by replacing punitive approach of punishment with reformatory ones. The Act is also significant in the modern liberal trend of reform in the field of penology. The Act also provides provisions for reformation, social rehabilitation of prisoners, helping prisoners families and in terms of removal of disqualification in employment opportunities.

Procedure for Probation Service

S. 4(2) and S. 6(2) of the Probation of Offenders Act provide that the judge would consider the report of the probation officer before deciding on whether to grant probation. S. 14 of the said Act lays down the duties of the Probation Officers.

The pre-sentence report of the Probation Officer is the fundamental document for the guidance of the Court whether to grant the benefit of probation to the accused or not. The object of the pre-sentence report is to appraise the court about the character of the offender, exhibit his surroundings and antecedents and throw light on the background which prompted him to commit the offence and give information about the offenders conduct in general and chances of his rehabilitation on being released on probation.

The judge may also pass a supervision order under section 4(3) of the Act, whereby the offender is placed under the supervision of a probation officer and certain conditions are imposed upon him. This is mostly in the form of regular visits to the supervising officer. Some of the conditions which must be followed have been laid down in S. 4(4). On the application of the probation officer such conditions may be varied- S. 8(2) and alco the offender may be discharged- S. 8(3). If the offender fails to follow the conditions laid down by the Court, the original sentence against him may be revived S. 9. The mechanics of probation are as under:--

(a) It begins with an investigation of the offender after his guilt has been established.

(b) The investigation is made by a probation officer and is usually ordered by the court whenever an offender is legally eligible for probation.

(c) Statutes commonly exclude from consideration persons convicted of serious offences such as dacoity and murder or those previously convicted of offences.

(d) The social investigation by the probation officer helps the court in determining whether the offender is suitable for probation and in outlining the corrective treatment.

(e) The investigation is primarily concerned with the offender as an individual, his personality and background education and employment, and his family and social environment.

(f) The report of the social investigation is submitted by the probation officer as a confidential report to the court.

(g) When probation is ordered by the court the offender is placed under the supervision of the probation officer and under conditions specified in the probation order.

(h) These conditions in the case of adults require that the probationer conducts himself properly, that he maintains his local residence, reports regularly to his probation officer, supports his family, pays restitution, avoids criminal association and disreputable places and abstains from drinking, he may be called upon to obtain medical or psychiatric care. The status of a probationer may carry other restrictions as to civil rights, freedom to act in certain matters without approval and access to social or employing organisations.

(i) The duration of probation is limited to some maximum term, early discharge by the court in recognition of good conduct is a common practice.

Role of Probation Officer :

(a) The probation officer's fundamental task is to help the probationer become a more responsible and better-adjusted person. The essence of probation is good casework and not mere supervision. The aim of the probation officer is to bring about the correction and rehabilitation of the individual probationer.

(b) Individual counselling and available social, occupational, educational and mental health resources are employed to help the probationer resolve specific problems, conflicts and attitudes that have led to his criminal or delinquent.

(c) The probation officer has to watch that the probationer observes the condition of probation and has to be watchful regarding further anti-social activities of the probationer.

(d) If the probationer violates the terms of his probation and commits a further offence during the period of probation, he may be brought back before the court for review or revocation of the original order of probation.

(e) When the probationer completes his period of probation successfully, he is discharged from other obligations to the court.

(f) Studies made in several countries show that 70% to 80% of all probationers successfully fulfil the terms of probation and are discharged. The success of probation is to be judged not from success at the expiry of the period of probation but by a follow-up after some years of the termination of probation.

However, probation in India is mostly dependent on the policies of the State rather than a uniform Central Policy. In Karnataka a State level Probation Advisory Committee has been constituted with High Court Judge as Chairman with official and non-officials as members. A District level Probation Advisory Committee has been constituted in each district consisting of the District and Sessions Judge as Chairman with official and non-officials as members. After Care Programmes have been set up to improve the lives of those released on probation. The After Care Programme, in Kerala, is intended to rehabilitate released prisoners and probationers coming under the supervision of District Probation Officers. By utilizing this amount they can engage in small scale income generating activities. The amount of assistance is ₹ 10,000/- per head. If the amount is insufficient for meeting the expenses this can be attached with some bank loan. Department of Juvenile Welfare and Correctional Services was set up in Andhra Pradesh in 1990. It gives the following probation services taking care of probationers released by the courts and ex-convicts, released juveniles, after-care work, counseling and guidance to reform themselves and not to revert to crime and for their rehabilitation through Govt. Welfare Agencies.

Criticisms against the Concept of Probation and Their Counter

There are some critics who look at probation as a form of leniency towards the offenders. To quote Dr. Walter Reckless, probation like parole, seems to the average laymen a sap thrown to the criminal and a slap at society. Probation is still generally perceived as a lenient approach rather than a selective device for the treatment of offenders who are no threat to public safety. Probation system lays greater emphasis on the offender and in the zeal of reformation the interests of the victim of the delinquents are completely lost sight of. This obviously is against the basic norms of justice.

Keeping in view the increasing crime rate and its frightening dimensions, it is assumed that undue emphasis on individual offender at the cost of societal insecurity can hardly be appreciated as a sound penal policy. Some criticize probation because it involves undue interference of non-legal agencies in the judicial work which hampers the cause of justice.

Further, when non-custodial correctional measures are used arbitrarily, without being resorted to on objective grounds, there is danger of men of means taking undue advantage and abusing the system as against those who would really deserve but have no advocacy or support, and of the whole approach becoming counter-productive and coming into public disrepute.

The answers to these criticisms would lie in the fact that the aim of the criminal justice system is to correct the offender and for some offences this would be best done outside the prison. Further, laying down strict guidelines to determine when probation should be awarded would defeat the very purpose of the concept. The broad parameters laid down age of the offender; surrounding circumstances, nature of the offence, etc. provide a broad framework for the judge to apply his discretion. It would also defeat the purpose if probation has to be granted when certain conditions are satisfied, if for example, the facts on record show clear premeditation to do a wrongful act.

Responding to the other criticism, it is essential that non-legal agencies, namely probation officers, interference is only meant for smooth functioning, and also it is not mandatory for the judge to consider using the probation officer always. He may not ask for a pre-sentence report, may not put the offender under supervision.

Problems in the Practical Implementation of Probation in India

S. 6 of the Probation of Offenders Act, which makes it easier for a person below 21 years of age to benefit from probation. This is regardless of their antecedents, personality and mental attitude. It might lead to recidivism because many of them may not respond favorably to this reformative mode of treatment. Also, in many cases it is difficult to ascertain whether the delinquent is a first offender or a recidivist.

The Probation of Offenders Act, in sections 4(2) and 6(2), lays down that report of the probation officer is considered before awarding probation. But, the Courts generally have shown scant regard for the pre-sentence report of the probation officer because of lack of faith in integrity and trustworthiness of the Probation Officers. In their view calling for the pre-sentence report would mean unnecessary delay, wastage of time, undue exploitation of the accused by the probation officer and likelihood of biased report being submitted by him, which would jeopardize the interest of the accused and would be contrary to the object envisaged by the correctional penal policy.

On personal interview with some judicial officers and probation officers, conducted by Abdul Hamid, it has come to light that neither judicial officers feel it desirable to get report from the probation officers, nor the probation officers feel it obligatory on their part to submit their reports in the courts unwarranted.

Section 4 of the Probation of Offenders Act does not make supervision of a person released on probation mandatory when the court orders release of a person on probation on his entering into a bond with or without sureties. This is not in accordance with the probation philosophy, which considers supervision essential in the interests of the offender, against corrective justice.

The lower judiciary in India has not at all taken into consideration the objects and reasons of this act, while applying its discretion in regard to grant of probation. In an umpteen number of cases the accused had to move the High Court and even the Supreme Court to get the relief of probation. If an accused gets relief of probation only in the High Court or the Supreme Court after passing through the turmoil of a long and cumbersome judicial process, he would, psychologically, be diverted towards hardenedness and the whole purpose of the Act would be forfeited. Variation or discharge of the probationer is based solely on the report of the probation officer; this leaves the probationer at the mercy of the Probation Officer.

The after probation services are not very effective. Thus, even considering that a sentence of probation has been passed and the offender is placed under supervision it is nothing more that a regular visit to the officer. There is no scientific process of rehabilitation and the Probation Officers aren't adequately trained. They are recruited between 20 and 26 years of age. They are grouped into districts and supervised by a state/provincial chief. There is no in-service training and occasional refresher courses, and thus they are not adequately trained.

Further, often there is a lack of interest for social service among the probation personnel. Lack of properly qualified personnel, want of adequate supervision and excessive burden of casework are attributed as the three major causes of inefficiency of the probation-staff.

Effectiveness of Probation :

The following issues are important in the effectiveness of the probation service :

1. Who should be treated under probation supervision ?
2. What type of supervision is most conducive to successful rehabilitation ?
3. How does a probationer adjust himself in the community ?
4. What are the factors leading to failure of probation ?
5. Is probation effective ?

1. Who should be treated under probation supervision ?

The selection of right type of offender for probation has not been an easy task. The young first offender is the obvious choice. However, there are others for whom probation will be equally effective. A careful pre-sentence investigation by the probation officer is vital in determining which offender should or should not be granted probation. Such report should not only mention facts, but should sense the feeling behind these facts. The data on present offence, criminal history, attitude towards crime, family ties, earning capacity, play a major role, in the decision making process.

2. Role of supervision

The two important characteristics of supervision are its quantity and its quality. The quantitative aspect includes the size of the caseload. A workload of 50 units, calculated on the basis of one unit for each probationer and five units for each pre-sentence report, is an ideal load. During well meaning supervision many things can go wrong such as admonishments, needless warnings and punitive watching. The client may show hostility and anxiety. Some offenders will succeed under supervision regardless of the type of service, while others will violate no matter how much treatment they receive. The probationers need intensive supervision, which can be gradually relaxed. Those who stay clear of the trouble during this initial period, generally improve their chance of success.

For most offenders lengthy and intensive probation supervision is not only unnecessary but wasteful. For fruitful supervision during probation, the family, school and the employer can be heavily utilized. Generally the offenders have poor family ties and if these can be improved it shows positive results. Sometimes the whole family require the treatment. In case of juveniles who had been attending schools earlier, they should be encouraged to continue with their studies. The help of their teacher can also be sought.

The key point in re-socialization through probation is regular employment.

3. Adjustment in society

Under proper supervision, the probationers develop self control, responsibility and interest in work. The families of offenders are generally helpful. Even though some of them do not think themselves criminal, they are willing to seek the help of the probation officer.

4. Training of Probation Officer

Emotional maturity, common sense, capacity to learn by experience and interest in welfare of human beings are the basic requirements. He should be trained in criminology, corrections and counselling. He should know the conditions which cause crime and what type of offenders constitute his case load, the resources in the community and reactions of the community to the offenders, For effective supervision, he has to combine in himself the roles of guide and a counsellor.

5. Effectiveness of Probation

Probation is very effective in juveniles, drunkards and women offenders. It can be effectively used in the case of adults, particularly casual offenders. There are several advantages to probation. The main advantages are its cost and the opportunity for the offender to remain a productive member of the society. The satisfaction of being able to look after his family, receive love and care of the family members are other great factors which help the criminal in re-socialization.

The success rate varies between 60 to 75 %. About 50% are self-correcting who improve regardless of the quality of service. The contribution of the service can be gauged only from the higher success rates.

Suggestions to Make Probation Service Fulfill Its Purpose

A few suggestions may be implemented at the legislative and the administrative level, which would make probation effective in India. Changes that could be brought about in the law are enumerated below. These changes are mostly applicable to the Probation of Offenders Act as it is more widely applicable than S.360 of the Code.

Due importance must be given to the reports of the probation officers by making necessary amendments in section 4(2) and section 6(2) of the Act. Probation must be based on thorough investigation into the case history of the offender and the circumstances associated with his crime. United Nations Standard Minimum Rules for Non-Custodial Measures also provides that the judicial authority must avail of such a report.

Recidivists have often proved a failure in the process of probation. It has, therefore, been generally accepted that probation should only be confined to the cases of juveniles, first offenders and women offenders. Though S.360 of CrPC lays down that only first offenders will be granted this benefit, if they are not below 21 years of age, no such condition has been laid down in the Act. Necessary amendment may be done to incorporate the same.

It must be made mandatory for offenders to be placed under supervision of a probation officer, by amendment under S.4(3) of the Act, as that would best serve the philosophy of probation. If the officer feels that the offender would not commit a crime, he could then submit to the court an application for the offenders discharge. Also, it has been left to the discretion of the Probation Officer to decide and inform the Court about necessity to vary an order of probation or to discharge the probationer, so there must be a complaint mechanism provided is a probationer wants to complain against a decision concerning the implementation made by the implementing authority, or the failure to take such a decision.

The proviso to S(4) of the Act lays down that probation would be granted only after the offender or his surety, have fixed place of abode or regular occupation. A large segment of offenders consists of the poor, the illiterate and the unskilled. It would not be possible for them to fulfill the conditions in all cases, hence the proviso should be amended to not make it mandatory, and leave it at the jurisdiction of the Court. Amendment could be made to The Code of Criminal Procedure to include the provisions for pre-sentence report and supervision. To make the judiciary more responsive, an amendment could be brought about in The Probation of Offenders Act which would make it mandatory for the judiciary to lay down the grounds as to why the benefit of probation must not be given, on the lines of S.361 of the Code.

The provisions under the Probation of Offenders Act and the Code of Criminal Procedure could be amended to be similar to the Juvenile Justice (Care and Protection of Children) Act, where more detailed procedures are laid down, like for the setting up of observation homes, report of the probation officer. India, being a developing country can't spend heavily on correctional measures, as its emphasis would be more on economic improvement. Due to lack of economic resources most developing countries violate the UN Standard Minimum Rules. It wouldn't be possible for India to adopt all of the measures prescribed by the UN, but India could adopt a few of the measures. Changes could be brought about in the way administration deals with probation. Some of them are enumerated below.

The first among them must be to have trained probation personnel. This isn't there today because the task of the probation officers is not given much importance in India. It is considered to be a mere formality, but if utilized well they would be most effective. The quality of probation service must be improved by making the service conditions of the probation staff more lucrative. This will attract well-qualified and competent persons to the profession. The probation personnel ought to be specially trained so that they can discharge their duty as probation officer competently.

A nation-wide uniform scheme of training for probation personnel with emphasis on social-work and rehabilitative techniques would serve a useful purpose to improve the efficacy of probation service in India. Guidelines for the training of Probation officers as have been laid down in the United Nations Standard Minimum Rules for Non-Custodial Measures, may be followed to the extent possible.

Thus, it can be said that the concept of Probation would be effective only where the judiciary and the administration work together there must be a common understanding between the Magistrate (or) Judge and the Probation Officer. Probation would be effective only when there is a sincere attempt made to implement it. It would be of great benefit for a country like India, where the jails are often overcrowded, with frequent human rights violations which would harden the human inside a person. Probation is an affirmation of the human inside every being and it must be given due importance.

■

PAROLE

Alexander Maconochie, a Scottish geographer and captain in the British Royal Navy, introduced the modern idea of parole when, in 1840, he was appointed superintendent of the English penal colonies in Norfolk Island, Australia. He developed a plan to prepare the prisoners for eventual return to society that involved three grades. The first two consisted of promotions earned through good behaviour, labor, and study. The third grade in the system involved conditional liberty outside of prison while obeying rules. A violation would return them to prison and starting all over again through the ranks of the three grade process. Maconochie "was a deeply religious man, of generous and compassionate temperament, and convinced of the dignity of man." His two basic principles of penology were that :

- as cruelty debases both the victim and society, punishment should not be vindictive but should aim at the reform of the convict to observe social constraints, and
- a convict's imprisonment should consist of task, not time sentences, with release depending on the performance of a measurable amount of labor.

Black's Law Dictionary says "The release of a prisoner from imprisonment before the full sentence has been served, Parole is granted for good behaviour on the condition that, parolee regularly reports to a supervising officer for a specified period". "Conditional release of prisoner : the early release of a prisoner, conditioned on good behaviour and regular reporting to the authorities for a set period of time" (Mitra's Dictionary) or Parole can be defined as a form of conditional pardon by which the convicts is released before the expiration of his term. *Article 14 of the constitution* Strikes at the Arbitrariness in state action and ensures fairness and equality of treatment.

Parole may have different meanings depending on the field and judiciary system. All of the meanings originated from the French word parole meaning "voice", or "spoken word". Following its use in late-resurrected Anglo-French chivalric practice, the term became associated with the release of prisoners based on prisoners giving their word of honor to abide by certain restrictions. One proposed reform is that parole bonds be used to incentivize defendants not to re-offend (Reynolds, 2000). In criminal justice systems, parole is the supervised release of a prisoner before the completion of their sentence in prison. This differs from amnesty or commutation of sentence in that parolees are still considered to be serving their sentences, and may be returned to prison if they violate the conditions of their parole. A specific type of parole is medical parole or compassionate release which is the release of prisoners on medical or humanitarian grounds. Conditions of parole often include things such as obeying the law, refraining from drug and alcohol use, avoiding contact with the parolee's victims, obtaining employment, and maintaining required contacts with a parole officer.

Some justice systems, such as the United States federal system, place defendants on supervised release after serving their entire prison sentence; this is not the same as parole. Parole should not be confused with probation, as parole is serving the remainder of a sentence outside of prison, where probation is given instead of a prison sentence and as such, tends to place more rigid obligations upon the individual serving the term. In US there are also difference between parole and mandatory supervision and between parole and "good time". Some states in the US have what is known as "mandatory supervision," whereby an inmate is released prior to the completion of their sentence due to legal technicalities which oblige the offender justice system to free them. In some states such as Texas, inmates are compensated with "good time," which is counted towards time served. For example, if an inmate served five years of a ten year prison term, and also had five years of "good time," they will have completed their sentence "on paper," obliging the state to release them. Where parole is granted or denied at the discretion of a parole board, mandatory supervision does not involve a decision making process : one either qualifies for it or does not. Mandatory supervision tends to involve stipulations that are more lenient than those of parole, and in some cases place no obligations at all on the individual being released.

A variant of parole in US is known as "time off for good behaviour", or, colloquially, "good time". Unlike the traditional form of parole – which may be granted or denied at the discretion of a parole board – time off for good behaviour is automatic absent a certain number (or gravity) of infractions committed by a convict while incarcerated (in most jurisdictions the released inmate is placed under the supervision of a parole officer for a certain amount of time after being so released). In some cases "good time" can reduce the maximum sentence by as much as one-third. It is usually not made available to inmates serving life sentences, as there is no release date that can be moved up. Similarly, in the United States, current policy prohibits US soldiers who are prisoners of war from accepting parole. The Code of Conduct for the US Armed Forces states : "I will accept neither parole nor special favors from the enemy." This position is reiterated by the Department of Defense. "The United States does not authorise any Military Service member to sign or enter into any such parole agreement."

In US immigration law, the term parole has three different meanings.

A person who does not meet the technical requirements for a visa may be allowed to enter the US for humanitarian purposes. Persons who are allowed to enter the US in this manner are known as parolees.

Another use related to immigration is advance parole, in which a person who already legally resides in the US needs to leave temporarily and return without a visa. This typically occurs when a person's application for a green card (permanent residency) is in process and the person must leave the US for emergency or business reasons. In the wake of September 11, 2001, there has been greater scrutiny of applications for advance parole.

A person who goes out of the country on "advanced parole" has to go through the following process : Canada by road : US Immigration officers will require you to submit one parole document to them. They will stamp your passport and another parole document, and issue a new I-94 form.

Parole in India

Parole, thus, in India is a legal sanction that lets a prisoner leave the prison for a short duration, on the condition that s/he behaves appropriately after release and reports back to the prison on termination of the parole period. It must be noted that a parole is different from a "furlough." While parole is granted to a prisoner detained for any offence irrespective of the duration of imprisonment, a furlough is only granted to prisoners facing long sentences, five years or more.

Under Indian jail laws, a prisoner may be released on parole in certain emergency situations, which includes the following health-related concerns :

- Complete and incurable blindness
- Advanced pulmonary tuberculosis, which incapacitates the prisoner from committing further crimes of the nature for which s/he was sentences;

- Is dangerously ill, and is likely to have a better recovery outside prison;
- Has become mentally unstable, and requires treatment in a asylum;

Additionally, a prisoner in India may be granted parole in the following exceptional cases :

- To perform funeral rites;
- To visit a sick or dying member of the family;
- To attend important functions, such as marriage of son, daughter, brother or sister;
- To construct a house or repair a badly damaged house.

Under Section 3 of the Temporary Release Act 1988, the State Government has the authority to release prisoners temporarily on some basic grounds as stated in the jail manuals. Thus, it is well known as a conditional release.

Parole is mere suspension of the sentence for the time being keeping the quantum of sentence intact. Release on parole is designed to afford some relief to the prisoners in certain specified exigencies. Parole is granted by the consent of the Divisional Commissioner for a maximum duration of one month. Generally parole is granted to those prisoners who have a good character according to the concerned authorities. Essence of parole is release from prison before completion of sentence on the condition that the prisoners abide by certain rules during balance of sentence. Convicted prisoners sentenced to life imprisonment can be granted parole after they undergo a minimum period of 3 years subject to the condition that their conduct.

There may be provision of escort as well. Escort parole is a parole granted to a prisoner to attend to emergencies like death, funeral etc., of near relatives under police protection for 3 days by the Superintendent of Jail, Government also may grant it.

Rejection of parole on valid grounds is not unconstitutional. Therefore, the right of prisoners when taken away in compliance with Section 3 (3) shall not be unconstitutional. Parole and furlough are not fundamental rights, such rights should be considered emanates from states own executive or statutory rules.

Person on parole is said to be in custody and is entitled to federal habeas corpus relief. Custody is a very elastic term and may mean actual physical detention in jail and the same is synonymous with restraint on liberty. Thus, persons released on probation or parole has been held to be in custody for the purpose of habeas corpus proceedings (Nirmaljit Kaur vs. State of Madhya Pradesh 2004 (7) SCC 558)

Following are the differences between parole and probation.

1. Parole is a conditional release of the convict from the prison while probation refers to suspended sentence.
2. Probation may be granted to a person while facing trial but parole is for convicts.
3. Probation is jural based on the order of court and parole is non-jural based on executive order.
4. Probation is a pre-care reformatory service and parole is a after care correctional service.

Conditions on which a parole can be granted

There are certain grounds on which a parole can be granted to the prisoners. They are as follows :–

(a) A member of the prisoner's family has died or is seriously ill or the prisoner himself is seriously ill; or

(b) The marriage of the prisoner himself, his son, daughter, grandson, granddaughter, brother, sister, sister's son or daughter is to be celebrated; or

(c) The temporary release of the prisoner is necessary for ploughing, sowing or harvesting or carrying on any other agricultural operation on his land or his father's undivided land actually in possession of the prisoner; or

(d) It is desirable to do so for any other sufficient cause.

(e) Parole can be granted only after a portion of sentence is already served (Seresh Chandra v. State of Gujrat, AIR 1976 SC 2462).

(f) If conditions of parole are not abided by the parolee he may be returned to serve his sentence in prison, such conditions may be such as those of committing a new offence.

(g) Parole may also be granted on the basis of aspects related to health of convict himself.

Duration of Parole

The period for which a prisoner may be released shall be determined by the State Government so as not to exceed–

(a) Where the prisoner is to be released on the grounds of illness, three weeks;

(b) Where the prisoner is to be released on the grounds of family function, four weeks;

(c) Where the prisoner is to be released on the grounds of agriculture, six weeks.

Provided that the temporary release under clause (c) can be availed more than once during the year, which shall not, however, cumulatively exceed six weeks. The period of release under this section shall not count towards the total period of sentence of a prisoner i.e. the period of parole has to be excluded in reckoning the period of detention (Avtar Singh vs. State of Haryana, 2002-(003)-SCC -0018 –SC). Where on his request the detenue is released on parole for specified period, it is generally appropriate that the court leaves it to administrative or jail authorities to prescribe the terms and conditions of parole. As held by the apex court in case of Suresh Chandra vs. State of Gujrat (AIR 1976 SC 2462), a person is released from a correction institution after portion of sentence is served. Parole is generally given to people who are young in age.

Exceptions to Parole

In India, the parole system is being misused to a great extent. A recent report by the Comptroller and Auditor General (CAG) stated that the parole granted to prisoners in India is misused by many of them as a convenient escape route. A random review by the CAG in two districts of Punjab, found that out of the 82 prisoners released under the parole system between May 2001 and 2008, 51 did not report back and 31 overstayed their parole period by as much as 1 year. The CAG has blamed inaction of the state police for this bleak scenario. The following are the exceptions to parole:- (a) Prisoners who are mentally ill, (b) One's with homicidal tendencies, (c) Parole cannot be granted in the cases of Conservation of Foreign Exchange and Prevention of Smuggling Act (COFEPOSA), and (d) In case of preventive detention, parole cannot be granted.

Parole under Different Laws

1. NDPS ACT 1985 : There exists no bar to affect the bar created by the provisions affecting the power of high court to suspend a grant parole under NDPS Act 1985 (Amarjit Singh vs State

1993 JCC 196).Though the offence under the NDPS Act 1985 are very serious in nature and affects entire society, thus they have to be dealt with the heavy hand and the question that whether the court are still powerless or without authority in granting small indulgences to even a convict under provisions of NDPS.

2. ARMY ACT 1950 : Under section 164,165, 179, 182, 185, by virtue of section 179 convict can inter-alia apply for his release on parole.

Distinctions between Bail and Parole

Bail and parole both have different connotations in law.

Bail is granted to a person who has been arrested in a non bailable offence or has been convicted of an offence after trial. Effect of granting bail is to release the accused from interment custody though the court would still retain constructive control over him through sureties as such literal meaning of word bail is surety.

Parole has a connotation other than bail, even though substantial legal effect of bail and parole may be the release of a person from detention of custody. Parole in fact is a temporary release from custody which does not suspend the sentence or period of detention which provides conditional release from custody and changes the mode of undergoing sentence (Sunil Falchand Shah vs. Union of India 2000 AIR SC 1023).

Decisions of Courts on Concept of Parole

A. SUPREME COURT

1. Provision for grant of parole to pregnant women : R.D Upadhyay vs. State of Andhra Pradesh AIR 2006 (3) 1946.

Herein it was held that women under trial prisoners and women convict when having a child birth in prison then such a prisoner has a suitable option of arrangements for temporary release or parole to be made to enable an expectant prisoner to have delivery outside prison and only in exceptional cases such a facility can be denied.

2. Computation of period of parole : State of Haryana vs. Karambir Singh 2001(5) SCC 427 Period of parole granted to the convict cannot be counted for the purpose of sentence.

3. Parole does not amount to suspension, remission or commutation of sentence : Dadu vs. State of Maharashtra 2000 AIR SC 3203 Parole does not amount to suspension remission or commutation of sentence.

4. Pendency of trial in parole : State of Haryana vs. Nauratta Singh 2000 AIR SC 1179 Expression parole or furlough not to be stretched to the period during which person is enlarged on bail during pendency of trial appeal or revision.

B. DELHI HIGH COURT

1. Grant of parole for repair of house : Rajesh vs. State 2006(1) JCC 284

While granting the accused a four week parole sought for repairing the house and arranging marriage of the sister and on account of his tendency of jumping on previous occasions his parole consisted of a condition requiring him to report to the concerned police station on every Monday of the week.

2. Grant of parole any number of times : Lalit Kumar vs. State of NCT of Delhi 2004 (3) JCC 1324

Herein petitioner had already undergone 12.5 years of sentence was released on parole on four occasions. Petitioners never misused the liberty of parole. In this case parole was granted for four weeks. On execution of personal bonds of ₹ 25000 with one surety.

FURLOUGH

Furlough is a conditional release of a prisoner who is sentenced to long term imprisonment. Under Section 4 of the Temporary Release Act is a brief release from prison. It is conditional and is given in the case of long term imprisonment. Release on furlough is not granted when it is hazardous to public interest. Furlough is also a temporary form of release. The period of sentence spent on furlough by the prisoners need not be undergone by him as was done in the case of parole. Furlough is granted as a good conduct remission. It is granted by the Inspector General of Prisons after the prisoner undergoes a period of 2 years imprisonment subject to the condition that his conduct is good in the jail and Police report does not indicate any negative reaction from his locality. Furlough is granted for a maximum period of 14 days only.

Conditions of temporary release under furlough under section 4 are more regress and a prisoner shall not be entitled to such temporary release unless he fulfils the conditions laid down under section 4. Certain classes of prisoners cannot get benefit of furlough. Convicts who are on bail cannot be granted furlough (State of Haryana vs. Mohinder Singh).

Temporary release of prisoners on Furlough

Basic essentials for release of prisoners on furlough are the following :

The State Government or any other officer authorised by it in this behalf may, in consultation with such other Officer as may be appointed by the State Government, by notification, and subject to such conditions and in such manner as may be prescribed, release temporarily, on Furlough, any prisoner who has been sentenced to a term of imprisonment of not less than four years, and who :

(a) has immediately before the date of his temporary release, undergone continuous imprisonment for a period of three years, inclusive of the presentence detention, if Any;

(b) has not during such period committed any jail offence (except an offence punished by a warning) and has earned at least three annual good conduct Remissions.

Exceptions to Furlough

(a) Furlough cannot be granted when it is hazardous to public interest.

(b) Furlough cannot be granted in case of short term imprisonment.

Rights of Life Convicts

Life convicts literally speaking must remain in jail for rest of their life. It is in this context that there release from jail of and on for short period has to be considered an opportunity have to be afforded to them not only to solve their personal and family problems but also to maintain their links with society. They too must breathe fresh air for atleast some time provided they maintain good conduct consistently during incarceration and show a tendency to reform them and become good citizens. Thus redemption and rehabilitation of such prisoners for good of societies must receive due weight while they are undergoing sentence of imprisonment. Looking from this angle should be released on parole or furlough subject of course to their undergoing 14 years of actual imprisonment. Thus consequently it cannot be said by any stretch of reasoning that concept of furlough must be given a go by once and for all. Thus, it cannot be said any stretch of reasoning concept of furlough must be given ago by once in for all so far as life convicts hit by 433-a are concerned.

Key Distinction between Parole and Furlough

Both Parole and Furlough are conditional release.

1. Parole can be granted in case of short term imprisonment whereas in Furlough it is granted in case of long term imprisonment.
2. Duration of parole extends to one month whereas in the case of Furlough; it extends to 14 days maximum.
3. Parole is granted by Divisional Commissioner and furlough is granted by Deputy Inspector General Prison.
4. For parole specific reason is required whereas furlough is meant for breaking the monotony of imprisonment.
5. The term of imprisonment is not included in the computation of the term of parole, whereas it is vice-versa in furlough.
6. Parole can be granted a number of times whereas there is limitation in the case of furlough.
7. Since furlough is not granted for any particular reason it can be denied in the interest of the society(State of Maharashtra vs. Suresh Panduram, 2006, AIR SC 2471).

In the landmark Judgement of **Charanjit Lal vs. State of Delhi (**1985CriLJ1541, 1985(1)Crimes1115, 28(1985)DLT92, 1985(9)DRJ135)**,** it was held:-

1. The four main objectives which a state intends to achieve by punishing an offender are – Deterrence, Prevention, Retribution and Reformation.
2. Life convicts' release from jail off and on for short periods has to be considered and opportunities have to be afforded to them not only to solve their personal and family problems but also to maintain their links with society.
3. They must breathe fresh air for at least sometime provided, of course, they maintain good conduct consistently during incarceration and they show a tendency to reform them and become good citizens.
4. Redemption and rehabilitation of such prisoners for the good of the society must receive due weight while they are undergoing sentence of imprisonment.

Balbir Singh Versus **State of NCT of Delhi IN THE HIGH COURT OF DELHI (**W.P.(Crl.) 1327/2010, SHIV NARAYAN DHINGRA, J : January 04, 2011)

This petition under Section 482 Cr.P.C has been preferred by the petitioner against an order dated 12th August, 2010 passed by competent authority declining an application of the petitioner for grant of furlough on the basis of adverse police report.

The police report showed that the petitioner was a resident of Trilok Vihar, where his mother was residing. The mother of the petitioner had already entered into an agreement to sell the residence and she could leave the place any time. There was every possibility that if the petitioner was released, he may escape the clutches of law and may not report back to undergo imprisonment. It was also reported that the Mohalla where his mother used to live was under terror of the petitioner and many people were not wiling to talk due to fear of the petitioner. The people living in that Mohalla ultimately told the ASI that it would be danger to the peace of the area if the petitioner was allowed to live there.

The petitioner was granted parole only about two months before making application for furlough. The petitioner's request for parole and furlough at such short interval is not understood. Parole and furlough are granted to maintain contact with the family but they are not to be granted so frequently that a criminal, who is undergoing sentence for serious offence like murder becomes danger to the society. I, therefore, consider that the competent authority rightly dismissed the application of the petitioner. The petition is hereby dismissed.

An Overview

Four main objects which punishment of an offender by the state is intended to achieve are deterrence, prevention, retribution and reformation. There has been substantial diversion from the previously existing popular concept of retribution. Of late the focus has shifted upon the concept of reformation. The earlier criminal law concept of an eye for eye and a tooth for tooth has been replaced by a more humane concept which emphasizes upon the re-allocation of an accused into the society.

It is easier to put away the offender in prison and to forget him than to work with him in the community. Erroneously society feels safer if the offender is isolated, removed from the scene and put out of circulation. It is vaguely hoped that when he comes out of prison, he will have learned to behave better. This, however, is often belied. It is now realized that the society enjoys only a false sense of security, because every offender is going to return to society some day more angry, socially handicapped disabled, outdated in skills, less employable and as such less capable of rehabilitation. Although the prisons are repulsive and revolting many prisoners come out only to return quickly to them. Having been removed from society for a long time, they are beset with tremendous difficulties upon re-entry into larger society. The community also turns a cold shoulder refusing them reintegration. All these considerations re-enforce the concept of community based treatment. Hence grew the concept of probation.

The concept of probation may be thought of as the application of modern scientific casework to specially selected offenders who are placed by the court under the personal supervision of Probation Officer, sometimes under conditional suspension of punishment, and given treatment aimed at their complete and permanent social reformation. As a sentence probation represents a judicial disposition, which establishes the defendants legal status under which his freedom in the community is continued subject to supervision of a probation officer and subject to the condition imposed by the court. As an organization, probation is a service, agency, designed to assist the court and to execute certain services in the administration of criminal justice. As a process, probation involves the pre-sentence investigation for the court and the supervision of the persons in the community.

To quote James Vorenberg, "there are fashions in everything and today's fashion in criminal justice is mandatory minimum sentence." The concept of parole and furlough are in fact a step towards the accomplishment of this very purpose. The provisions of parole and furlough provide for a humanistic approach towards those lodged in jails. Main purpose of such provisions is to afford to them an opportunity to solve their personal and family problems and to enable them to maintain their links with society. While exercising the discretion in granting parole and furlough the courts should not be too lenient, as great number of crimes are committed by offenders who have been put back on the street after convictions, thus opportunity of parole and furlough may be afforded on their maintaining a consistent good conduct during incarceration and on their showing a tendency to reform themselves to become good citizens.

In instances where the convicts (released on parole and furlough) do not maintain good conduct or in instances where they fail to follow the condition attached to their protocols or where they pose a threat to the peace and well being of the society then in large public interest the law reserves with it the right to cancel such temporary release.

THE CONCEPT OF PARDON IN INDIA

A pardon is the forgiveness of a crime and the penalty associated with it. It is granted by a head of state, such as a monarch or president, or by a competent church authority. For a clear understanding of the concept of pardon, it is worthwhile to list out some other related concepts such as Amnesty, Commutation, Remission, Reprieve and Clemency. These terms, however differ subtly from country-to-country.

(a) Amnesty : 'Forgetting' the crime, e.g. if a car thief witnesses a murder, he will often be granted amnesty for his crime in order to allow him to testify against the murderer, or after a civil war a mass amnesty may be granted to absolve all participants of guilt and 'move on'. Weapon amnesties are often granted so that people can hand in weapons to the police without any legal questions being asked as to where they obtained them/why they had them/etc.

(b) Commutation : Commutation is an associated term, meaning the lessening of the penalty of the crime without forgiving the crime itself. This is substituting the penalty for a crime with the penalty for another, whilst still remaining guilty of the original crime (e.g., someone who is guilty of murder may have their sentence commuted to life imprisonment rather than death)

(c) Remission : Complete or partial cancellation of the penalty of a crime, whilst still being considered guilty of said crime (i.e., reduced penalty).

(d) Reprieve : A reprieve is the temporary postponement of punishment. This is temporary postponement of a punishment, usually so that the accused can mount an appeal (especially if he or she has been sentenced to death).

(e) Clemency : Clemency is a general term encompassing all of the above, or just referring to amnesty and pardons. Clemency is often requested by foreign governments that do not practice capital punishment when one of their citizens has been sentenced to death by a foreign nation that does practice it.

Today, pardons are granted in many countries when individuals have demonstrated that they have fulfilled their debt to society, or are otherwise deserving (in the opinion of the pardoning official) of a pardon. Pardons are sometimes offered to persons who claim they have been wrongfully convicted. Some believe accepting such a pardon implicitly constitutes an admission of guilt, so in some cases the offer is refused (cases of wrongful conviction are nowadays more often dealt with by appeal than by pardon).

In India, Article 72 of the Constitution of India empowers the President of India to grant pardon, reprieve, respite or remission of punishment, or to suspend, remit or commute the sentence of any person convicted of any offence, particularly in cases involving capital punishment. The president cannot act as per his own whims and fancies and in this process he is to be guided by the Home Minister and the council of ministers. The power to pardon rests on the advice tendered by the executive to the president, who subject to the provisions of Article 74(1) must act in accordance with such advice. More so it has been sufficiently established in Maru Ram vs. Union of India (AIR 1980 SC 2147) and Kehar Singh vs. Union of India (AIR 1989 SC 653) that the exercise of the executive prerogative is subject to a judicial review.

In Maru Ram vs. Union of India, the scope of Article 72 was examined. The court opined that faith was to be instilled in the vision of the founding fathers and for such a task it is necessary that rules must be drafted, laying down the guidelines for the exercise of such vast prerogative. The court stated that the proportional rise in power should be accompanied with a proportional rise in responsibility and that any such power must be protected from the capacity of being abused. The Honorable judges provided that the pardoning power can be subject to a review where an executive decision has been made on altogether irrational, arbitrary, unreasonable or mala fide grounds such as discrimination on the basis of religion, caste, colour or political loyalty. In Kehar Singh vs. Union of India, Pathak, CJ observed "The power of pardon is part of a constitutional scheme and we have no doubt that it should be so treated also in the Indian Republic. It has been reposed by the people through the Constitution in the head of the state and enjoys high status. It is a constitutional responsibility of great significance, to be exercised when occasion arises in accordance with the discretion contemplated by the context." The court also justifies the existence of a "Pardon", by acknowledging the fallibility of human judgment being undeniable even in a supremely legally trained mind and therefore any such errors can be remedied by entrusting power to a higher authority, which shall " scrutinize the validity of the threatened denial of life or the continued denial of personal liberty."

A similar & parallel power vests in the Governors of each State under Article 161. However, a convention has developed that the Governor's powers is exercised for only minor offenses, while requests for pardons and reprieves for major offenses and offenses committed in the Union Territories are deferred to the President. It also depends upon other provisions of law i.e., section 54 & 55 of Indian penal code, Sections 432,433 and 433A of criminal procedure code of Indian criminal justice system and also the sentencing policies of state.

The Law Commission [xxix]also stressed on the need for the existence of an executive Pardoning Power citing the following as reasons :

1. Facts not placed before the court.
2. Facts placed before the court but not in a proper manner.
3. Facts disclosed after the passing of the sentence.
4. Events which have developed after the passing of the sentence.

The law commission stated that there is a plethora of other reasons which do not lend themselves to codification and therefore the law commission insisted on retaining the Scope of the pardon prerogative.

In State of Rajasthan vs. Union of India (1978 1 SCR 1 at pp 80-81), it was held that whenever a question arises whether an authority under the constitution has acted within the limits of its power or exceeded it, it can certainly be decided by the court. Indeed, it would be its constitutional obligation to do so. The constitution entrusts the court with the responsibility to interpret the constitution and determine the validity or constitutionality of any executive action.

However, it is also to be understood that there is no inherent right in the condemned person to insist on an oral hearing. "The manner of consideration of the petition lies within the discretion of the president and it is for him to decide how best he can acquaint himself with all the information that is necessary for its proper and effective disposal." Mr. Ram Jethmalani, the counsel on behalf of Kehar singh, had requested an oral presentation before the president which was subsequently denied by the Secretary to the president stating that it was not in accordance with "the well established practise in respect of consideration of mercy petition."

Recently the case for granting pardon to Mohammed Afzal, convicted of conspiracy in the December 2001 attack on the Indian Parliament has come to the fore. He was sentenced to death by the Supreme Court of India in 2004. The sentence was to be carried out on 20 October 2006. The sentence is now stayed. Political groups from Kashmir are insisting on a pardon for the sake of maintenance of the peace process in the valley. Various human rights group are also alleging denial of a fair and free trial to Mohd. Afzal. In particular, Arundhati Roy, in an scathing article has castigated the trial and deemed it as one denying natural justice. Earlier the mercy petition on behalf of Dhanonjoy Chatterjee, convicted of rape and murder and sentenced to death, was denied by the president in spite of widely staged and much publicized efforts of Human rights and Anti-capital punishment groups.

An Overview

Executive pardon since the ages has undergone massive change, and so has unbridled executive power. It only bodes well for a democratic nation that the ambit of a free and fair judiciary claims victory over a politically motivated executive. However, propagating absolute denial of clemency to the executive will not lend itself to better circumstances as an already overburdened judiciary tries to deal with the immense backlog of disputes. An executive seeking to reach a just end for the accused without any other considerations and a strict and ever watchful judiciary, looking over the exercise of such vast amplitude of power can guarantee a balanced and legitimate use of this. Furthermore, there should be equity before law and equal opportunity to all. For this the poor and the illiterate should be provided assistance in drafting, and pursuing their mercy petitions. Lastly and most importantly, it should be made a matter of policy that those prisoners who seem to have atoned and reformed should be pardoned and suitably rehabilitated. This shall encourage the criminal offenders to turn their lives around and start afresh. The president should use his wisdom and discretion well in order to take a just decision based on objective criteria to grant pardons.

THE THEORY OF REHABILITATION

Ever increasing crime is one of the major problems of modern society. A lot of people get involved in the vicious circle of crime and their lives change dramatically. They get caught by the law and are then sent to prison. The prisoners have a lot of problems in their lives that lead them to crime. The causes can be addiction, psychological problems; stressful relationships etc. Rehabilitation programs are carried out in jails for prisoners to make the prisoner a part of mainstream society and get employment after their release. These programs help them to go back to normal life and earn livelihood ethically. The scenario does have a light at the end of the tunnel. There seems to be a good reason why some former inmates do not return to jail: it appears that their time in incarceration was spent productively, changing some vital aspect of their personality.

The Authorities are taking efforts to accommodate the rising number of prisoners in spite of new prisons built every year. Because of that many people's criminal record are being away and they are allowed to mix with the common people. Research indicates that after indulging criminal activities, approximately 70% of the released prisoners return to jail within a few years. Most of them fall in four broad areas : Violent offending, Sex offending, Cognitive skills and Addictions.

The goal of the prisoner's rehabilitation programs is to reform the prisoners from committing crimes and assisting them in starving normal lives without any criminal activities. The best rehabilitation programs for prisoner will not only helpful for the criminal but also the society at great. Many people say that prisoner's can not be reformed and whatever punishment they get for their crimes, irrespective of that they always return to their old habits. It has been seen that after undergoing rehabilitation programs, many prisoners are reforming and returning to normal life.

Prisoner rehabilitation programs are also offered by some social organizations. They help these people to refrain and reform from criminal activities. The program identifies the skills and potential that the person has and gives training to improve their skill so that they earn something for their decent living. The faith based prisoner rehabilitation programs try to find out the spiritual values of the person. The program offers religious education and methods of healing and slowly makes them regret their criminal activities by converting them to lead towards normal life.

Because of this, most of these programs are aimed to import basic education to them. After achieving this, they have given some vocational education and training. Research and studies shows that encouraging prisoners with education provide better results. After under going literacy and training, many former offenders have returned to the normal life.

Another common problem with many prisoners is the drug addiction. Large numbers of people indulge in the criminal activities only for the sake of fulfilling their drug addiction needs. Many prisoners rehabilitation programs are working on this issue. They create awareness about abuse of drug and it's bad effects for individual with the society in which he lives. Also this program offers counseling and treatment for health recovery.

When a prisoner is admitted in a jail, the authorities assess the individual and find out the type of rehabilitation that needs to be given. This is an essential and inseparable part of the rehabilitation program carried for prisoners. Education and vocational training are the two key areas focused on in the rehabilitation program. This will help them in getting a job after completing their punishment. A lot of techniques are used to rehabilitate the prisoners according to their needs. As every prisoner's problems are different and they have to be handled subjectively.

Criminal rehabilitation can help to solve the problem of overcrowding in most prisons. The criminal population continues growing, as the death penalty has been abolished, and the state would need to spend more on facilities to house criminals.

In criminal rehabilitation, prisoners are given opportunity to increase their content knowledge base. This is essential as studies show that many inmates do not have basic grade school education. This would severely impede their success of acquiring jobs, thus many had to turn to a life of crime. Basic criminal rehabilitation programs ensure that there is a standard level of literacy amongst the inmates who sign up for the course.

Rehabilitation also ensures that inmates are socially well adjusted. Psychological assessments are being meted to test for mental or physical disabilities that led to their incarceration in the first place. Should the inmates be ready and willing to accept counseling and assessment, many of them are able to return to society as relatively well-balanced individuals. For drug addicts, this is a pertinent issue, as many of them are struggling with addiction problems. Counseling would help to balance inner dynamics that led to the addiction, and possibly the criminal behaviour that financed the addiction.

Criminal rehabilitation has many positive benefits and can impact the lives of many inmates as well as their families. It can help with wider social issues as well, such as reducing discrimination and stigma.

Here are some of the techniques used in prisoner rehabilitation :

- The technique varies from person to person. It depends on the crime committed as well as the age, gender, nature and background of the prisoner.
- The root cause of the crime has to be found out. Once that is known, the right program can be implemented for the prisoner based on his needs.
- If the offender has some psychological problems then he or she is given counseling. This helps in making the prisoner a relaxed, stress free individual.
- A prisoner who is an addict has to undergo a drug or alcohol rehabilitation program. The prison can come back to normalcy after finishing the program with help of medication, therapies and counseling.
- Vocational training techniques different from each individual depending on his or her interests and talents. Pottery, carpentry, weaving, designing are the areas in which training is provided.
- Education teaches them skills that they will require in the job market as well as in the outside world. The program is designed in such a manner that the prisoner comes out of the jail as a skilled, proficient and an independent individual.

The help is given to the prisoners after the rehabilitation in form of contacts for jobs, counseling, discussion boards every month. This helps the prisoner is getting the confidence and support that he or she has someone to rely on. This feeling makes them strong and they start a new life.

References

Braithwaite, J. Restorative Justice & Responsive Regulation 2002, Oxford University Press, p. 249.

Bureau of Justice Statistics (1981, 1981-1996, 1998). Crime and Justice in the United States and in England and in Wales."

Cavadino, M & Dignan, J. (1997). The Penal System: An Introduction (2nd ed.), p. 39. London: Sage.

Chakrabarti, Nirmal Kanti,1999, Probation Services in the Administration of Criminal Justice, 1st edn., Deep & Deep Publication Pvt. Ltd, New Delhi

David Kopel, Paul Gallant & Joanne Eisen (2001). Britain: From Bad to Worse.Americas first freedoms p.26

Denborough, D. (Ed.) (1996). Beyong the prison: Gathering dreams of freedom. Adelaide, South Australia: Dulwich Centre Publications.

Denton, D. (1998). Horizontal management. Lanham, MD: Rowman and Littlefield.

Disaster Center (2010). US and DC Crime Rates, Source: FBI UCR index.

Hennessey Hayes, (2005) 'Assessing Reoffending in Restorative Justice Conferences' Australian and New Zealand Journal of Criminology, 38(1) 77.

Hentig, Hans von (1938) *The Limits of Deterrence*, 29 J. AM. INST. CRIM. L. & CRIMINOLOGY.

Hopkins, B., (2004) Just Schools: A Whole School Approach to Restorative Justice, Jessica Kingsley Publishers, 208 pages.

Jacob Scullum (2000). Guns Down Under reason, Australia, p. 10.

Jaimie Beven et al., (2005) 'Restoration or Renovation: Evaluating Restorative Justice Outcomes' Psychiatry, Psychology and Law, 12(1) p. 194.

James Bonta et al., (2006) 'Restorative Justice and Recidivism: Promises Made, Promises Kept ?' in Dennis Sullivan and Larry Tifft (eds), Handbook of Restorative Justice: A Global Perspective, p.108.

John Braithwaite, (1999) 'Restorative Justice: assessing optimistic and pessimistic accounts' in Michael Tonry (ed), Crime and Justice: A Review of Research (Volume 25).

Katherine Basire, (2004) 'Taking Restorative Justice Seriously' 13 Canterbury Law Review 31.

Keel, Robert (2005). Rational Choice and Deterrence Theory Lecture Notes. University of Missouri, St. Louis.

King, L.W. (1998). Ancient History Sourcebook: Code of Hammurabi, c. 1780 BCE, Fordham University web site:http://www.fordham.edu/halsall/ancient/hamcode.html.

Leo Zaibert, (2006) Punishment And Retribution Ashgate Publishing.

Liebmann, M. Restorative Justice: How it Works, London: Jessica Kingsley Publishers, p. 33.

Liptak, Adam (June 29, 2008). Gun Laws and Crime:A Complex Relationship:http://www.nytimes.com/2008/06/29/weekinreview/29liptak.html?_r=2.

Mandelbaum, David G.,1998, Society in India, vol-I, Popular Prakashan Ltd, Bombay, p.82.

Martin, Jacqueline (2005). The English Legal System (4th ed.), p. 174. London: Hodder Arnold.

Marty Price, J.D. "Personalizing Crime," Dispute Resolution Magazine. Fall 2001.

McCold, P. (2003). A survey of assessment research on mediation and conferencing. In L. Walgrave (Ed.), Repositioning Restorative Justice (pp. 67–120). Devon, UK: Willan Publishing.

McCold, P., & Wachtel, T. (2003). In pursuit of paradigm: A theory of restorative justice. Paper presented at the XIII World Congress of Criminology, Rio de Janeiro, Brazil. Retrieved October 25, 2007.

National Rifle Association Institute for Legislative Action (2010). Gun Laws, Culture, Justice & Crime In Foreign Countries Fairfield, VA

Nelsen, J. (1996). Positive discipline (2nd Ed.). New York: Ballantine Books.

O'Connell, T. (1998). From Wagga Wagga to Minnesota. Paper presented at the First International Conference on Conferencing, Minneapolis, MN, USA.

O'Connell, T., Wachtel, B., Wachtel, T. (1999) Conferencing Handbook: The New Real Justice Training Manual. Pipersville, PA: The Piper's Press.

Paternoster, Raymond (2010), How Much Do We Really Know About Criminal Deterrence ? The Journal of Criminal Law & Criminology, vol.100 (3) pp. 765-824.

Schalleger (2002) Evolution of Deterrence Crime Theory, Http://www.associatedcontent.com/article/31600

Siegel, L, & Senna, J (2006). Introduction to Criminal Justice. Mason: Thomson Wadsworth.

Simon, B. (1994). The empowerment tradition in American social work. New York: Columbia University Press.

State of Washington (2007). Chapter 9A.52 RCW: Burglary and trespass, web site: http://apps.leg.wa.gov/RCW/default.aspx?cite=9A.52.

Summerfield, Morgan (2006), Evolution of Deterrence Crime Theory Associated ContentWebsite:http://www.associatedcontent.com/article/32600/evolution_of_deterrence_crime_theory.html.

Wikipedia (2006). Code of Hammurabi, from web site: http://en.wikipedia.org/wiki/Code_of_hammurabi.

Winfree, L.T., & Abadinsky, H (2003). Understanding crime: Theory and practice.Belmont: Thomson Wadsworth.

Zehr, Howard, (2002) The Little Book of Restorative Justice Intercourse, PA: Good Books.

■■■

CHAPTER 24 Social Defence

Learning Objectives :

- ❐ **To understand the concept and the content of social defence.**
- ❐ **To know the characteristics of the history of social defence.**
- ❐ **To grasp the interrelationship between crime trends and social defence.**
- ❐ **To understand police as a social service agency.**
- ❐ **To know the approaches to promotion of social defence.**

Introduction

Social scientists should give no policy advice, but should rather point out, publicly, the shortcomings of policies- *Morton A. Kaplan [and Philip Green] et al., Strategic Thinking and its Moral Implications, University of Chicago Center for Policy Study, 1973, p.67.*

Social defence in common parlance means the protection of society against crime through a systematically organized and coherent action by both the State, civil society people's group. Though this term has long been in use in the criminological and penological literature, the modes and modalities of achieving its inherent objective have been shifting with the advancement in social sciences and behavioural disciplines. Even today, because of the complexity of issues involved, it has not been possible to evolve a wholly satisfactory theoretical framework for policy formulation and programme development in this field.

Whereas the initial interpretation of the notion of social defence implied the protection of society through the repression of crime, its modern conception visualizes a system that emphasizes control of crime through social defence. Therefore social defence not only endeavours to perfect the system that protects society against criminal acts but also goes much beyond the ambit of criminal law in extending measures to forestall criminogenic situations and to treat offenders through appropriate remedial, educational and rehabilitative services.

Apart from its meaning in criminology as a strategy for crime prevention and control, the term has now a wider meaning and social defence refers not to an elaborate school of thought but to a vigorous movement sustained by a variety of organizations sharing certain values and methods in dealing with phenomena of crime, in wisely dealing with a dangerous offender, or the efficient protection of society accomplished without persistently violating the offenders' rights. Hence social defence has acquired a paradigmatic character over a period of time. It connotes several meanings and denotes several names. It has the narrow and the broad definitions.

In a narrow sense social defence is a nonviolent alternative to military defence. It is based on widespread protest, persuasion, noncooperation and intervention in order to oppose military aggression or political repression. It uses methods such as boycotts, refusals to obey, strikes, demonstrations and setting up alternative government. Social defence is nonviolent defence of the vital features of society — including human rights, local autonomy, and participation - against all oppressive forces. There are actually several different names that all mean about the same thing. The main ones are social defence, nonviolent defence, civilian-based defence and civilian defence.

In a broad sense social defence is nonviolent community resistance to aggression. This includes defence against exploitation, injustice, subjugation and discrimination like defence against male violence against women. Social defence often refers to nonviolent defence based on grassroots initiatives. Those favouring the broad definition can readily provide examples from struggles by feminists, environmentalists, peace activists, minorities and umpteen others.

Conceptual Dichotomies

In the backdrop of the aforesaid discussion of the concept it is understood that social defence is one term with several meanings and one meaning with several terms. A discussion of the following dichotomies shall further clarify the content and character of social defence namely Social Defence and Civilian-based Defence, Social Defence and Military Defence, Social Defence and Social Offence and Protest as a Form of Social Defence.

(a) Social Defence and Civilian-based Defence

There are actually several different names that all mean about the same thing. The main ones are social defence, nonviolent defence, civilian-based defence and civilian defence. It is usually unfruitful to get into discussions about names, except with people promoting social defence who need to agree about what they are going to call it. The different names do have different connotations. The expression "civilian-based defence" usually refers to nonviolent defence operating under direction of a government, whereas the expression "social defence" often refers to nonviolent defence based on grassroots initiatives.

(b) Social Defence and Military Defence

1. Social defence provides a way for everyone to take responsibility for defence, unlike military methods.
2. Social defence contains no military capability and thus nuclear attack and aerial bombardment become pointless and harder to justify.
3. There are cases of military coup and toppling of Government. With social defence, this problem does not arise, since social defence is based on popular participation and so removes the dependence on a professional defence force. The nonviolent methods used against a foreign aggressor can also be used against local military forces that try to take power.

4. Social defence is much more compatible with a society based on equality and wide political participation. Military forces are based on hierarchy and obedience. They train people to kill on command. This is contrary to the equality, questioning, mutual respect and dialogue that help promote a democratic society. The influence of military systems often inhibits or thwarts greater participation in the rest of society.
5. A militarised society is one based on command and obedience. A social defence society would be one based on self-reliance and independent action.

(c) Social Defence and Social Offence

The most fruitful way to develop social defence is by promoting social offence. There needs to be a vast expansion of action by people supporting nonviolent opponents of repression The philosophy behind this approach resides in the familiar saying that, "the best defence is a good offence." Most people are genuinely altruistic and are willing to act against repression elsewhere, as testified by the success of Amnesty International. The advantage of this approach is the possibility of mobilising people and giving them experience in relevant nonviolent action, as well as developing international solidarity.

The challenge for social defence is to demonstrate that it is the most effective way to resist. Otherwise some people will head for the hills and possibly end up being massacred as well as helping to justify violent attacks on the nonviolent opposition.

A social defence system will offer plenty of opportunities for people who want to make courageous and potentially dangerous stands. In fact, some of the rebels in present society, who are often in trouble with the authorities, could well become the heroes in a nonviolent resistance. Perhaps a social defence system should provide real glamour for certain types of resistance, while trying to remove the romantic image of violence. Without practical skills and a deep commitment for national security, violent resistance is a losing proposition. Social offense becomes the final option in dealing with violence and innocent killings by guerrillas or insurgents. Social defence is a tool that can be used for the wrong cause. Activists need to carefully study the situation before intervening. Social defence by its nature does not give rise to any threat of violence towards an enemy, and so avoids the problem that 'defence' may also potentially be used for offence. There is also an immediate need to realistically book corrupt politicians, businessmen and government functionaries in order to restore the faith and confidence of masses in the credibility of the government in affected areas. Such steps must be taken in a time bound manner with a clear and implementable approach.

(d) Protest as a Form of Social Defence

The idea of 'protest' typically evokes images of a dissident minority taking a public stand. Protest is usually associated with groups that are outside the mainstream, that lack inside connections with the wielders of power. Protest is often 'against' something or other, an attempt to stop a policy or practice which would otherwise go ahead unquestioned. To many people, protesters have a bad image: the rabble in the streets. Although the vast majority of protest activity in liberal democracies is nonviolent in reality and intent, an aura of actual or potential violence commonly accompanies media presentations and popular perceptions of protest. These images are part of an overall view which balances the 'right to protest' against a need for 'law and order.'

The conventional presentation of protest by the media, by government and by academics focuses on political activity by a particular segment of the population. Protest activity is assumed to be problematical, whereas other political processes are taken as less so.

The dominant political actors in liberal democracies are the state, corporations and professions. The actions of these groups are not usually called protest, though. The term protest is applied to actions of groups that are painted as outside the mainstream. When trade unions go on strike that is recognised as a form of political protest—and often stigmatised—but when corporations redirect investments out of a particular area (a 'capital strike') that is taken to be a normal exercise of corporate prerogatives. When peace activists purposefully break a regulation to block a train carrying nuclear materials, that is civil disobedience. When government departments fail to provide information by mandatory deadlines, when corporations continually flout environmental regulations, when prison warders beat prisoners or when ex-husbands refuse to pay child support, that is seen as cause for concern but is not categorised as civil disobedience. Whereas violence by 'protesters' is invariably condemned and often called terrorism, violence by the police or military is usually seen as legitimate. Direct police and military repression of social movements has occurred when these movements have become threats to dominant interest groups. For example, in the US from about 1880 to 1920, the police were regularly used to smash worker opposition to capitalist exploitation (Bowden, 1978). In the 1970s and 1980s, massive police operations in France and West Germany were used to disrupt protests against nuclear power, a technology heavily promoted by the governments of those countries. A conclusion frequently drawn from such experiences is that protest is allowed so long as it doesn't have much impact. Once protest on a massive scale develops in way that threatens powerful interests in or allied to the state, police or military force is brought to bear against it. That protest is part of a wider political struggle is apparent in this double standard. During and after World War I, the powerful socialist movements in Europe were crushed by the governments there, a process enabled by the mobilisation of force against all those who opposed the war effort. After World War II, the cold war provided the context in the US for a witchhunt of dissidents in trade unions, the media, education and elsewhere.

Defining protest in a narrow fashion is one way to ensure that it poses no threat to established institutions and social relations. Those who argue for a broader view of which sorts of protest are legitimate are entering the struggle in a way that supports those groups for whom protest is a method of overcoming powerlessness in orthodox channels. Those who assume the primacy and legitimacy of the state invariably take a narrow view of civil disobedience; those who question the state take a broader view.

The usual point of view of the dominant groups is that people should leave social problems to the authority, elites and experts. If action isn't happening quickly enough, then they should work through 'normal channels.' The trouble with these channels is that they are biased in favor of privileged groups.

Protest as a form of social defence however presupposes non-violence and respect for the rule law. 'Nonviolence' refers to actions which do not by themselves cause physical harm to humans, whereas 'violence' refers to those which do. Protest, mostly nonviolent, played a major role in the ending of slavery, extension of the franchise, curtailing ruthless aspects of the exploitation of labor and extending rights to women and minorities. Gene Sharp, the leading researcher on nonviolent action, has identified 198 different types of nonviolent action and given examples of each one. Sharp divides the methods of social defence into three categories: symbolic actions, noncooperation, and intervention and alternative institutions.

It is a common misconception that nonviolence is passive. The expression "passive resistance" has been used to describe a type of nonviolent resistance. Its core is nonviolent action, and this includes strikes, fraternisation and setting up alternative institutions. There are also offensive measures to be taken, such as communications to undermine international and domestic support for the aggression. Social defence does not mean just sitting there and accepting whatever the aggressor inflicts.

History of Social Defence

The concept of Social Defence, as adopted by the United Nations in 1948, relates to the entire gamut of activities and programmes in the sphere of the prevention of crime and the treatment of offenders. While the ultimate object of social defence might still be the same, the modes and modalities of achieving this goal have considerably changed in the light of the development of behavioural science and a deeper understanding of the phenomenon of crime.

The International Society of Social Defence was founded in the aftermath of the Second World War. Its aim was to provide scientific answers –at the international level– to problems linked to criminality and its prevention by means of legal and social analysis and a humane criminal policy. The purpose of the International Society of Social Defence is to study, in co-operation with the existing specialized associations, the best means of combating crime, in particular, by utilising the findings of the human sciences. The Society is composed of individual members from many countries. Members from the same country may form national groups, and this has been done in several countries.

An obvious approach to understand accurately the commendable movement for social defence is to examine critically the theme discussed at one of the international congresses organised by the International Society of Social Defence. No longer can criminal law abruptly arrive at the legal status of offenders by following only legal criteria but it should in addition give intense regard to bio-psycho-social criteria.

Maturing and maturity can be exhaustively discussed from different points of view. Apart from biological maturing and maturity there is also psychological and social maturity. In practice, it is altogether impossible to separate those formally, because the intricate development of an individual results in an integrated and dynamic bio-psycho-social complex.

The psychiatrist, the psychologist, the physician and the social worker should methodically explore the potent factors intimately associated with maturing and vividly indicate the significant manner in which particular types of crime obviously correspond to particular stages of development. Their pertinent findings can ultimately form the scientific basis for judicial, legal and administrative personnel to relate confidently the legal status and legal procedure to the age of offenders. 'Comparative studies primarily based on experiments made in individual countries are materially important as legislation is increasingly differentiating the legal system according to age. Such illuminating studies are bound to lead to the uninterrupted evolution of criminal justice and unique methods of treatment and rehabilitation. Based on the degree of maturing and maturity, which duly provides the 'real' age of offenders, the desired form of punishment, the effective system of treatment and the proper conditions of release would be differentiated.

Further social defence in the form of nonviolent resistance to aggression can be traced to a number of writers, including Henry David Thoreau, Leo Tolstoy, Elihu Burritt (a Christian pacifist), William James and Bertrand Russell. The campaigns led by Gandhi in South Africa and India were important in developing the idea of a nonviolent alternative to war. Gandhi himself began advocating defence by nonviolent resistance in the 1930s. A number of writers were inspired by Gandhi and developed his ideas. In the 1930s, advocates of a nonviolent substitute for war included Richard Gregg, Bart de Ligt, Kenneth Boulding, Jessie Wallace Hughan and Krishnalal Shridharani.

Perhaps the first fully-fledged description of a national social defence system was by Stephen King-Hall, a British writer and former naval officer, in his book *Defence in the Nuclear Age* published in l958, King-Hall thought that British parliamentary democracy could be better defended from communism if the military were abolished and replaced by organised nonviolent resistance. King-Hall's treatment moved social defence onto the agenda as a pragmatic rather than just a moral alternative.

Shortly after this, the idea of social defence was developed by various researchers including Theodor Ebert in West Germany, Johan Galtung in Norway, Adam Roberts in Britain and Gene Sharp in the United States. Gene Sharp, the leading researcher on nonviolent action, has identified 198 different types of nonviolent action and given examples of each one. Sharp divides the methods of social defence into three categories : symbolic actions, noncooperation, and intervention and alternative institutions. Since then, these and other researchers have worked on the idea, inspired both by historical writing about nonviolent struggles and by contemporary use of nonviolent action in a variety of campaigns.

The idea of social defence has really only been around since the 1950s and 1960s, and it is not yet widely known among the general public. There is a relatively small degree of formal endorsement of social defence. Many green parties, such as the West German green party, endorse social defence, as do a number of activist groups. On the other hand, a large number of groups — such as environmental groups and social justice groups — make very sophisticated and conscious use of the methods of nonviolent action. The development of these skills and experience in nonviolent action lays a good foundation for the furthering of development of social defence. A discussion of characteristics of social defence can now be brought to order.

Characteristics of Social Defence

1. The concept of **Social Defence** relates to the entire gamut of activities and programmes in the sphere of the prevention of crime and the treatment of offenders.

2. The social defence approach is based on the premise that the criminal justice system by itself cannot undo such aberration of the wider socio-economic system as are associated with

crime. Even the reduction in crime as a more realistic goal can be achieved only by extending crime prevention and control measures beyond the criminal justice system and by building these into a broader social action to curb conditions which produce crime.

3. More effective is the social defence, more the society tending to be normal. However, war against crime has to be waged, if not to win, atleast to be ensured that it is not being lost. Thus, the social defence approach aims not only at perfecting the system that deals with offenders but also at forestalling conditions that generate criminality.

4. The social defence approach takes best care of poverty, inequality, injustice, interests of the vulnerable groups and the treatment of offenders.

5. Social defence is a nonviolent alternative to military defence. It is based on widespread protest, persuasion, noncooperation and intervention in order to oppose discrimination and political repression.

6. Social defence is nonviolent community resistance to aggression. This includes defence against any form of exploitation, subjugation and discrimination like defence against male violence against women. Social defence is nonviolent defence of the vital features of society — including human rights, local autonomy, and participation — against all oppressive forces. Other examples may include struggles by feminists, environmentalists, peace activists and others.

7. It is a common misconception that nonviolence is passive. The expression "passive resistance" has been used to describe a type of nonviolent resistance, but usually it is better to avoid it since it gives the wrong impression. Social defence is not passive. Its core is nonviolent action, and this includes strikes, fraternisation and setting up alternative institutions. There are also offensive measures to be taken, such as communications to undermine international and domestic support for the aggression. Social defence does not mean just sitting there and accepting whatever the aggressor inflicts. Social defence is based on the principle that no regime — whether democracy or military dictatorship — can survive without the passive support or nonresistance of a large fraction of the population. In other words, all societies are built on consent, cooperation and obedience.

8. Social defence is already part of the defence policy of some governments. Sweden has a policy of "total defence", and this includes military defence, civil defence, economic self-reliance, psychological defence — and social defence. In several other countries — most notably Switzerland — the general population is considered to be essential to military defence. Although the focus is on military resistance, nonviolent resistance would inevitably be part of this. In quite a number of countries, including the Netherlands, Denmark and Austria, there have been official government investigations into the possibilities for social defence. We expect that government interest will expand in coming years.

The Department of Social Defence of Government of Tamil Nadu is providing services for the development of children found in difficult circumstances and the girls and women requiring care, treatment and rehabilitation. The Department is also providing institutional and non-institutional services for the rehabilitation of the children entering the institutions, besides involving the Non-Governmental Organisations in all the programmes to ensure protection of the rights of the children and their all round development. The institutions under the Department provide care, treatment (correctional), training (educational and vocational) and rehabilitation to the children (neglected and children in conflict with law) coming under the provisions of the Juvenile Justice (Care and Protection of Children) Act, 2000 and girls and women institutionalised under the Immoral Traffic (Prevention) Act, 1956. The Department also provides aftercare services to the boys, girls and women discharged from the institutions. The Commissionerate of Social Defence Co-ordinates drug abuse control and prevention programme of the Ministry of Social Justice and Empowerment, Government of India in Tamil Nadu. An information cell is functioning in the Commissionerate of Social Defence headed by the Commissioner. Information about the services available for neglected children, children in conflict with law and women in moral danger are obtained between 3 to 4 p.m. in the office of the Commissioner of Social Defence, 300, Purasawalkam High Road, Chennai-10. Information about institutional programmes, leave, discharge on license, absolute discharge etc. in respect of the children are made available from the information cell. Public Relation Officer/Personal Assistant to Commissioner can also be contacted for such information services.

Complaints Redressal Procedure is available in the Commissionerate of Social Defence, The Deputy Director (Administration) of the Office of Director of Social Defence is the Vigilance Officer to receive the complaints either in person or by the post in name cover. Reply about the action taken will be communicated to the complainant within 30 days from the date of receipt of complaint. The complaints will be received on all working days between 3 to 5 p.m. in the office of the Commissioner of Social Defence of Government of Tamil Nadu.

9. Social defence is not an instant alternative. To be fully effective, advance planning, preparation and practice are essential. Instead of disarmament, which is the standard appeal of the peace movement, the aim would be transarmament: a gradual switch from military defence and deterrence to social defence and deterrence. Most important is people's understanding of their own capabilities and of the dynamics of nonviolent action. For example, workers must understand the power of strikes, telecommunications workers must understand the power of information and everyone must understand the power of symbolic action. In general a social defence system will be stronger when a society is more self-reliant and, in particular, not dependent on or subject to groups of elites.

10. One cannot expect the government to introduce social defence on its own. Therefore, the best approach is to develop the capacity of individuals, groups and communities to use nonviolent action to defend themselves and the things they believe in. Campaigns for social defence can be linked to campaigns by workers, women, peace groups, minorities and others. The actual introduction of social defence is not likely to occur as a process of normal policy. It is impossible to graft social defence onto a society that it is not suited for it. The best approach is to promote campaigns to change society to become more self-reliant, participatory and equal. When society has changed sufficiently in this direction, then introduction of social defence will be a natural process. People will only defend their society when it is worth defending. At present, energy systems such as electricity production are quite centralised and hence is takenover. A population would be more self-reliant if there were a move to greater energy efficiency and decentralisation. Communities with energy-efficient buildings with solar heaters and local wind generators are more resistant to threat than communities dependent on supplies of fuel and electricity. The same applies to transport. Communities designed to maximise reliance on walking and cycling are less vulnerable than ones built around either mass transit or the automobile. Similarly for food. The greater the production of food in local gardens, the less the vulnerability to disruptions in the food supply.

Crime Trends and Social Defence

It is true that crime is essentially an offshoot of the disorganizational process of society itself and, as such, it is difficult to visualize a system that could completely neutralize an 'aggression' which originates from within. Then, in the absence of any fool-proof theory of crime causation, it is beyond human capacity to create a society entirely free from crime. Moreover, the definition of crime itself may vary from country to country and within the same country from time to time, in keeping with the changes in social structure, cultural values and normative expectations and no single system can hold good for different situations. Despite this position, there is a candid awareness that the objective behind social defence cannot be realized merely by focussing on individuals who are recognized as offenders; it inevitably requires also reaching out to those who are vulnerable to crime. Thus, the social defence approach aims not only at perfecting the system that deals with offenders but also at forestalling conditions that generate criminality. In this process, it closely interacts with various sectors of socio-economic development in creating an environment conducive to the prevention of crime and in mobilizing resources appropriate for the treatment and rehabilitation of offenders in the totality of their life situations.

Further, in order to maintain an orderly functioning of the society in transition, there is an exaggerated need for an effective enforcement of the existing laws, and even the enactment of new laws, to adequately cope with the emerging forms and trends of crime. There are enough indications that, in the years to come, with the development of information technology and telecommunications and the acceleration of economic activities within and across national borders, organized crime is bound to acquire a much more volitional and disruptive form. The emergence of terrorism, environmental crime, cyber crime, etc., are most ominous for a developing country like India. Such a reality only makes it imperative that the crime prevention and control strategies must extend beyond the criminal justice system and integrate into sectors in which individuals are born and they live and grow. From this viewpoint, social defence as a comprehensive approach towards ameliorating conditions responsible for social maladjustment, deviance and crime gains significance.

Crime and Development

The nexus between crime and development, especially in the wake of globalization, liberalization of trade and commerce and free market economy, has been clearly recognized at various international fora. The world community is gravely concerned about the baneful effect of crime on the peace, progress and prosperity of nations. Many countries have seen as to how crime thwarts the development process, undermines human dignity and disrupts the well-being and welfare of people. It is invariably found that the development process, if not property monitored, tends to become criminogenic.

From the standpoint of social defence, the development process must be geared towards ensuring social justice, protecting human rights and providing for an equitable sharing of socio-cultural and economic opportunities by one and all. Any development process which destroys the self-generating and self-sustaining capacity of the people, alienates certain sections from the mainstream of social and economic life, widens the gap between the rich and the poor, intensifies ethnic and caste conflicts, and erodes public confidence in the rule of law, is socially destructive and, thus, breeds crime. History shows how an imbalanced development can stir social strife, racial discriminations, religious bigotry and fragmentation of society, and how authoritarian governance deprives people of a life compatible with human dignity. While the government tends to rely mainly on a trickle-down approach, civil society has to work from bottom upward in helping the poor to shape their destiny and to secure their place in society with dignity.

Poverty Factor

The importance of poverty factor in crime could hardly be overlooked. While poverty per se cannot be taken as a direct cause of crime, it does make individuals in stark deprivations more prone than the others to social maladjustment and to their coming in conflict with law.

The situation is further compounded when, in the wake of industrialization and consequential urbanization, the poor migrants in search of livelihood, are found to cluster around slums and squatter dwellings and to live in a state of social marginality and economic neglect. It is, therefore, encouraging that the Government of India has provided a major thrust to poverty alleviation in the national development plans. As a result of various measures, the poverty ratio is reported to have considerably declined in the recent years. The present trend augurs well for the creation of just society committed to the rule of law. It is well accepted that the strategies for poverty alleviation have to evolve within the framework of social justice for which the rule of law is a pre-requisite. With the globalization of national economy the obligation of multinationals, business houses and financial institutions towards protecting the rights and interests of the poor has to be clearly spelt out. Of course, in a free society, the media has to serve as a powerful agent in promoting a social climate conducive to a solidarity with the poor in their fight against poverty.

Vulnerable Groups

Among various groups subjected to social inequalities, cultural discrimination and economic handicaps, the condition of women below poverty line is much more precarious than that of their male counterparts. Therefore, in the planning of programmes for social defence, a vigorous drive has to be launched towards women's empowerment and gender justice and the criminal justice system has to act relentlessly against unscrupulous elements degrading their status.

Doubtlessly, the economic and social marginalization of the poor deprives a vast population of children in the country of their right to grow normally in body and mind. As children constitute the supreme national asset for the making of tomorrow, the failure of society to bring them up as socially healthy individuals not only multiplies poverty but also leaves them extremely fragile to withstand the onslaught of anti-social elements. In several cities, a large number of poor children are found to be living or working on the street in search of livelihood through odd jobs in extremely sub-human and hazardous conditions. A rising trend of the abuse of children for unconscionable gain and their instrumental use in crime, and of their transportation beyond national borders for nefarious purposes under the garb of adoption, marriage or employment, is also a matter of grave concern.

India has a first-hand experience of setting in motion a host of democratic processes to resolve the problems of social inequality and class divide, including the formulation of special laws to provide a protective umbrella to the weaker sections of society as are oppressed for centuries.

The overall strategy is to secure distributive justice and allocation of resources to support programmes for the social, economic and educational advancement of the weaker sections in general and Scheduled Castes and Scheduled Tribes, Backward Classes and Minorities in particular. Indeed, the criminal justice system which is founded on the principles of fairness and equity has to prove its credibility of being equally fair and equitable to the poor as to the rich, in actual operation.

Public Participation

As crime is a social phenomenon, no system for its prevention and control could ever be conceived without an active participation of the public. In fact, public participation is an inseparable ingredient of the process that defines a behaviour as crime and strives to tackle it. Whereas the critical attitude of the public that abhors crime and cries for the offender to be so punished as to become a deterrent for the others, is clearly discernible, the positive role of the public in preventing conditions which precipitate crime and in facilitating the offender to mend his behaviour and to reintegrate into society has yet to be fully recognized. Of course, the public opinion towards crime manifests in an ambivalent manner: while on one side, demanding for a stringent action against those who offend, on the other hand, pleading for the powers of those who administer punishment to be restrained and curbed. It is, therefore, logical that civil society is encouraged to take a balanced view and to guide the public in subscribing to a system that protects society against crime without impinging on the human rights of all those involved, whether as offenders or victims. The edifice of social defence can stand only on an enlightened participation of the public as the harbinger, the means and the end of the process.

The need for public participation in the treatment and rehabilitation of offenders in equally well established. The future of various non-custodial measures is tied with not only the offender's responsiveness to community correction but also the extent to which the public is prepared to extend its helping hand in the correctional process and to accept the offender within its fold. Besides adopting a positive attitude towards the rationale and efficacy of community-based treatment, the public can substantially contribute to the reformative process by providing social and material inputs. In the sphere of institutional treatment, the public can play a significant role in two broad ways : first, by supplementing correctional programmes in custodial institutions, and, secondly, by serving as a bridge for the offender's transition from custody to free society.

Civil Society

In the prevention of crime, Civil Society that spring from within society have a definite advantage over official agencies in making a dent on the problems that culminate into crime. A variety of situation- problem- and individual-oriented approaches to crime prevention can be devised and implemented through the self-help endeavours and collective initiatives of the people themselves. A situation-oriented strategy would be based on an optimum use of all such voluntary agencies as are engaged in protecting the vulnerable, like children, women and other economically weaker or socially disadvantaged sections in society against any form of abuse or exploitation which is likely to induce asocial reaction. A problem-oriented strategy would bring within its ambit all such activities and programmes as are initiated through voluntary action to tackle social problems which have a nexus with crime. An individual-oriented strategy would obviously require working with those who, under situational compulsions, have fallen to a socially deviant behaviour and still have a chance to redeem. In all these areas, voluntary organizations have to function in conjunction with families, communities and other social institutions which have a bearing on the process of resocialization. Apart from a direct intervention in situations and problems responsible for crime, voluntary organizations can be greatly instrumental in moulding public opinion and in mobilizing social support for an effective implementation of social legislation enacted to eradicate social evils such as dowry, child marriage, 'Sati', beggary, prostitution, etc. Most of these laws contain specific provisions for public participation in different forms.

It, therefore, needs to be emphasized that whereas any transfer or dilution of the responsibility that legitimately comes within the purview of the State would be rather hazardous, civil society has every right to know as to what transpires behind walls. However, it is distressing to note that, while a large number of voluntary organizations in social welfare are being financially supported by the government, very few are really coming forth to work for the mainstreaming of social deviants.

Criminal Justice and Correctional Strategy

The voluminous criminal code is framed to protect substantially the general interests of society. Those who do not flagrantly violate the code are frankly accepted by society and those who violate it criminally are liable for punishment. When an offender is vigorously subjected to a criminal trial, his regrettable action is officially made known to a wide circle of law-abiding citizens who strongly disapprove of his criminal conduct. The social defence movement is not keenly enthusiastic about the conventional form of criminal trial, which it would like to reduce drastically to the exclusive announcement of guilt. Those who support social defence approach say that the criminal courts may give reasons for selecting the sentences they impose, but there should be no objection to a correctional system providing scientific programme of treatment.

Correctional Strategy

In the backdrop of a rapidly changing crime situation, any action to interpret principles underlying social defence into actual practices has to start with the rationalization of sentencing policy. The penal policy should not only strive at balancing the interest of society with that of the individual involved in crime but also at addressing the plight of the victim. The range of dispositional alternatives has to be so widened as to enable the court to place a person found guilty in a setting which is most conducive to his mainstreaming. In the social defence approach not only the offender but also the offence has to be taken into consideration by the Judge, who should have appropriate guidelines for sentencing, based not on the importance of the deterrent effects and uniformity of treatment but on the necessity of uniformity in the approach. More importantly, the right of an accused to a speedy trial which, in the case of the poor, seems to have been severely trampled has to be restored by reinforcing the judicial system.

In India, as in many other countries, where most of the persons coming within the purview of the criminal justice system are involved in crime under various kinds of situational compulsions, correctional approach to crime control has to be pursued as an integral part of the social development process. An analysis of crime statistics would show that a large segment of offenders consists of the poor, the illiterate and the unskilled. Such offenders are seen to be victimized twice : once, when they are denied of their basic human needs in open society and forced to live in a sub-culture of social marginality, and, again, when they are grinded in the mill of criminal justice for having infringed the law. An increased investment on the provision of correctional services in relation to these persons would be most productive not only in reducing crime but also in improving the quality of life among the strata they come from and are ultimately to return to. In this regard, a priority attention needs to be given to raising the standards of diagnostic, educational and developmental programmes with all the necessary technical inputs, in close conjunction with community based welfare agencies. There is no dearth of success stories in this regard but most of these have so far been confined to a few bold experiments or to some individual initiatives. It is high time that correctional measures are directed towards developing the capacity, caliber and competence of persons in conflict with law in such a manner as to assure that, after their passage through the criminal justice system, they would be able to stand on their own as dignified and law-abiding citizens. Of course, the approach towards those who indulge in crime as a way of life and refuse to understand the language of correction will have to be differentiated.

Prison Reform

Doubtlessly, prisons constitute the oldest and most widely used mode of dealing with offenders. In the recent years, it has also been a subject of unprecedented criticism, scrutiny and debate from the viewpoint of its social defence role. It is well recognized that as long as certain types of offenders are to be segregated from society in the interest of public safety, and are expected to return as better human beings than what they were when incarcerated, the institution of prison will have to play an important in the dispensation of justice. However, there are several problems that prison administration is presently confronted with in discharging its public safety and reformative functions. In India, as in many parts of the world, imprisonment continues to be applied indiscriminately and excessively, either as a convenient way of dealing with all sorts of crime or because of the non-availability or limited range of effective alternatives. In the face of competing priorities, it is hard to mobilize adequate resources to bring in the desired systemic reforms in terms of the necessary infrastructure, scientific classification of inmates, diversification of prisons for various categories of offenders, provision of correctional services and duly qualified and professionally trained personnel to handle custodial and correctional tasks. The problem of overcrowding and a swelling proportion of undertrials among prison inmates has thrown the system haywire. As the judicial process further speeds up and the pendency of cases in courts decreases, prison population will correspondingly multiply, because a much larger number of accused persons are on bail and awaiting trial. There is also a real need to establish appropriate linkages between institutional programmes and community-based welfare resources to ensure that the processes of recovery, reeducation and rehabilitation initiated in prisons are systematically followed up till the discharged prisoners are able to reintegrate themselves into society. What is most urgently called for is the formulation of a national policy and bringing in a basic uniformity in laws governing prisons, so that imprisonment as defined in the Indian Penal Code has the same meaning, when actually executed, in every part of the country.

Apart from the imperativeness of bridging the gap that exists between societal expectations and operational realities, the prisons administration has to run on the premise that its rehabilitative function can be accomplished only in an atmosphere that fosters human rights of persons in custody and generates among them a will to improve their quality of life. In this respect, the Supreme Court of India, discarding its erstwhile 'hands off' doctrine in favor of a judicial intervention when the rights of prisoners are found in jeopardy, has already enunciated three basic principles : (i) a person in custody does not become a 'non person'; (ii) a prisoner is entitled to all human rights within the limitations of imprisonment, and, (iii) there is no justification for aggravating the suffering which is already inherent in the process of incarceration.

Accordingly, the Apex Court has issued a number of directives for prison authorities to afford to prisoners all such facilities for self-improvement and correctional therapy as are consistent with their conditions of imprisonment. Besides a detailed interpretation of the relevant Constitutional provisions, the principles embodied in various United Nations instruments, to which India is a party, have also been invoked in guiding this process. The framework laid down by the Supreme Court of India to protect the rights and interests of prisoners has far-reaching implications for prison reform in a futuristic perspective. It not only entails a thorough overhauling of the prison administration but also an enlightened participation on the part of civil society.

The current emphasis on the humanization of prisons as an essential condition for invigorating their reformative and rehabilitative role brings into focus the issue of torture. Despite a high level of denial among the concerned authorities, torture does exist in prisons broadly in three different ways. First, there is a form of torture which may be intentional in nature and resorted to against all canons of rules and regulations to 'discipline' a prisoner or to 'set him right'. It may happen sporadically but does take place when the custodian under pressure loses his balance and the prisoner is totally at his mercy. No civilized society would ever condone such an intentional torture. Secondly, there is torture incidental to the sub-human conditions prevailing in prisons, especially in developing countries. Such incidental torture has to be prevented by adhering to certain minimum standards of institutional care in terms of living conditions, basic needs and the necessary amenities and privileges to treat prisoners as human beings. And, thirdly, there is an element of torture inherent in incarceration itself, when the individual involved is deprived of his freedom and isolated from his family and the community he belongs to. This form of torture can also be considerably reduced, if not eliminated altogether, by using prisons more selectively only for offenders who endanger public safety, by enlarging the range of alternatives to imprisonment, and, even when a offender is justifiably imprisoned, by providing him with ample avenues to maintain ties with outside world, and a possibility for an early release as an incentive for good behaviour and responsiveness to correctional treatment. Abolition of torture other than what is consequential to lawful sanctions is fast emerging as a vital issue of prison reform to be addressed squarely.

Positive Supervision of Crime and Individualization of Punishment

Punishment confidently presupposes guilt whereas precautionary measures of security are based on the inevitable dangerousness of crime. The latter are helpfully preventive in nature and have social defence as their well-defined objective. Unlike punishment, which is substantially based on responsibility, security measures can be indeterminate and cover those who cannot be held responsible, or have even been acquitted or placed under restricted liberty with supervision. An individual held not responsible need not be punished, though as a protective measure he may be carefully segregated. Punishment has its essential basis in responsibility though the inadequate criterion used in its diligent application is individualization.

Social defence persistently demands, constant review *of* the factual basis on which prolonged detention, ordered in the interests of public safety of offenders who are considered responsible, rests. Generally, unwarranted presumptions arc made regarding the extraordinary risk to the public. Accurate data based on careful assessment of the individual are not always available. Further, the re-education and complete rehabilitation of the individual is not always seen as a clear objective to be speedily realized in processing the offender. Close collaboration between judicial personnel and members of interested professions will boldly lead to methods of positive supervision of crime.

Even where the criminal law abundantly provides long-term detention for persistent offenders, persons actually detained need not be the most dangerous among offenders. Legitimately giving extra punishment on the basis of previous offences for which he has already suffered, may lead to the adverse criticism of punishing twice for the same offence. However, it is not manifestly inconsistent with social defence to provide increased sentences especially when the offender is mentally abnormal. For social defence, it is hopelessly inadequate if offenders, whose mental abnormality materially reduced their degree of responsibility, receive attenuated sentences, or if the same concession is extended to those suffering from less serious types of abnormality. Legal provisions should unhesitatingly make indeterminate duration of institutionalisation possible, effectively strengthen guarantee against releases doomed premature, sufficiently emphasize the rights of the defendant, properly institute mandatory legal aid by competent lawyers, studiously outline the procedure of confining a person for a period of observation, and amply protect the rights of the institutionalized person. Psychiatrists, psychologists, physicians, criminologists and social workers should competently work in enduring collaboration with judicial and legal personnel to decide impartially on the type of institutional placement, transfer to other specialized facilities and the date of release.

Social defence has an irresistible interest in the psychopathic personality who may dismally experience immense difficulties in his environment and also habitually commit crimes. While general rules of procedure may be effective, it is also absolutely necessary to assess an individual case repeatedly in terms of the intellectual qualities of the offender and the difficult nature of the offence in which he is involved. These offenders undeniably constitute a social danger, and therefore, need early and continued treatment to enable them to return to society as productive members.

Punishment is awarded to protect society adequately against the imminent dangers constantly posed by criminals. If the offender is mentally ill or retarded, imprisonment is irrelevant from the therapeutic and social points of view, and admittedly harmful to the offender since he is made to suffer mutely in order to protect society. Instead of ordinary prisons, social defence requires institutions, which provide scientific treatment programmes needed to process such offenders. In the penetrating observation, prolonged detention and vigilant discharge of abnormal offenders, the lawyer providing indispensable assistance should be unconditionally free to take appropriate measures on behalf of the client even without consulting him. After care in such cases should not rigidly restrict itself to social and economic aspects but should specifically provide other related professional assistance.

Individualization of punishment is an arduous duty for the courts to perform conscientiously as they cannot clearly evaluate all aggravating and mitigating circumstances, sometimes of a contradictory nature. From the standpoint of social defence even to judge the dangerousness of the offender, the court needs more relevant criminological data regarding him. Individual prevention based on social defence is viewed formidably impractical as judges cannot minutely examine all accompanying circumstances during a short trial, and in addition, are not sufficiently trained in criminology. All that the court can now do is to make extensive use of experts as a part of healthy social defence.

Probation is a form of sentence in its own right as it is not the suspension of official processing of the offender but one of professionally implementing a scientific programme of treatment in the community. The considered treatment under probation is beneficially extended to some offenders who are dispassionately considered likely to benefit from it. Breach of conditions leads to subsequent imprisonment. Creatively using the probation officer's report, the judge consciously takes decisions on grant of probation and the conditions to be imposed on the offender. The probationer receives continuous guidance and assistance not only from the professional probation service but also from voluntary workers especially interested in helping offenders placed on probation. Social defence is vividly realized in probation.

Non-Custodial Measures

Whereas the rationale behind the segregation from society of certain types of offenders in the public interest and their treatment in closed institutions is firmly established, the correctional potential of non-custodial measures has yet to be fully utilized in the administration of justice. Though probation as a form of non-institutional treatment of offenders under conditions of good behaviour, with or without supervision, has been in practice since the British period, the country has yet to provide a sound basis for its application on an extensive scale. Probation is still generally perceived as a lenient approach rather than a selective device for the treatment of offenders who are no threat to public safety. In fact, for want of scientifically evolved criteria to be safely relied upon in the placement of offenders in a non-institutional setting, the range of community corrections remains limited and imprisonment continues to be followed as the most convenient course, even for offenders whose institutionalization for short periods has no therapeutic value. It is true that when non-custodial correctional measures are used arbitrarily, without being resorted to on objective grounds, there is real danger of men of means taking undue advantage and abusing the system as against those who would really deserve but have no advocacy or support, and of the whole approach becoming counter-productive and coming into public disrepute. It is, therefore, necessary that a ground is prepared for community correction to prove its credentials to function, if not more, as effectively as custodial correction in reforming and rehabilitating offenders.

For this purpose, the United Nations Standard Minimum Rules for Non-Custodial Measures which seek to strike a proper balance between the interests of the individuals involved in crime and those of society at various stages of the criminal justice process, offer a blue-print for action and strategy. In keeping with the principles of the observance of human rights, the requirements of social justice, and the rehabilitation needs of offenders, the rules spell out a wide range of disposition modalities at the pre-trial, sentencing and post-sentencing stages. Apart from empowering the police and prosecution agencies to discharge offenders under specified conditions, the suggested sentencing alternatives include : verbal sanctions, such as admonition, reprimand and warning; conditional discharge; status penalties, economic sanctions and monetary penalties; confiscation or an expropriation order; restitution to the victim or a compensation order; suspended or deferred sentence; probation and judicial supervision; community service order; referral to an attendance centre; house arrest; and any other mode of non-institutional treatment or combination of various measures. With a view to avoiding institutionalization and to secure early reintegration of offenders into society, the post-sentencing dispositions to be tried out are : furlough and half-way houses; work or education release; parole; remission and pardon. Obviously, the implementation of such non-custodial measures would call for the development of a scientific basis for the selection, placement and supervision of cases, treatment processes, staffing resources, community participation, etc. However, all these approaches are worth experimenting with so as to assess their suitability to indigenous socio-cultural and economic conditions.

Social Defence and Police

As a social service agency, police generally act as most easily available agency with 24 hours duty and on weekends for the impoverished, the sick, the old and the lower socio-economic classes. Police can provide counseling to the juvenile delinquents, school boys, unemployed youths, drug addicts, and such others. Police can also educate members of the society to protect themselves and their property and on such matters as will increase society's understanding of the functions of the police, their limitations and areas where public can help police. Police service can provide assistance to those who are in state of disorganized mental state or are in need for psycho-emotional stress requiring immediate care.

As an agency for social defence it has to play its part in crime prevention, civil defence measures, internal security schemes and village defence measures in border areas. Police play an important part in providing rescue and relief operations during floods, droughts, famines and other natural calamities or disasters. The police must accept their service oriented role to the community and must strive to improve the quality of social life by initiating a continuous programme to infuse in the public security awareness, acquaint them of safety measures and precautions and demonstrate various emergency facilities available to the community. Police can play its service oriented role by stepping up their involvement in helping people in distress such as drug addicts, alcoholics, misguided youths and other suitable referrals.

The role played by the police in a society that has converted to social defence is a serious issue. Very few activities of the police require arms. But some situations require it. There is a need for further development of nonviolent methods to control individuals who become dangerously violent. It should be remembered that if a social defence is based on community self-reliance, there would be greater community responsibility for "policing", too.

Scholars argue that the police probably cause more crime than they prevent. Criminologists know that the crime rate has little connection with the level of policing or imprisonment. Most prisons breed crime, and most police forces breed corruption. If social defence is a viable alternative to the military, then surely it can be extended to deal with crime. After all, what is war except organised crime controlled by governments ? In the words of sociologist Charles Tilly, "If protection rackets represent organized crime at its smoothest, then war making and state making — quintessential protection rackets with the advantage of legitimacy — qualify as our largest examples of organized crime."

While the prevention of crime would necessitate forging of constructive linkages between the formal system and various sectors of socio-economic development, the treatment of offenders would inevitably entail a progressive refinement of the criminal justice processes. With newly emerging forms and trends of criminality, some of which are much more volitional and disruptive in nature, a holistic strategy has to be worked out jointly by various criminal justice including police, social welfare and development agencies. Such a concept of social defence warrants not only a thorough reorganization of the traditionally operative crime control mechanisms but also innovation of and experimentation with new approaches in coping more effectively with the changing crime scenario.

Of interest to social defence is the neglect even by criminology of the intricate problem of impunity. Insecurity continually develops in a society where offenders enjoy freedom or safety from necessary consequences of crime. Services in the general area of social defence are thus rendered inaccessible to such offenders. When a community becomes slowly conscious of impunity, it suffers from anxiety, which is aggravated by more crimes. Stem silence at the level of society prodigiously breeds impunity. Measures of social defence undeviating aim at making all offenders persistently realize the certainty of detection. The police can be most effective in analysing the problem of impunity, and public agencies should involve themselves directly in the investigation of all offences.

Promotion of Social Defence

The social defence movement itself is willingly undergoing perceptible change. Quantitative and qualitative analysis of crime in countries with differently structured social and economic organizations definitely indicate new forms in the phenomena of crime. This is all the more true if studies are directed towards cost of crime, offenders on the road, drug offenders, etc. These need constant examination and substantial revision through special classifications even at the state level before our renewed efforts to deal competently with crime can take us nearer the desired goal of social defence.

There has been relatively little on how to go about promoting social defence in present-day society. For this it is more fruitful to look at the efforts of those who have actually promoted social defence. Two distinct approaches stand out. Although they are not mutually exclusive, the distinction between these two approaches is useful here to move the discussion further. Their characteristics are given in the table below.

Table 24.1 Two Approaches to the Promotion of Social Defence.

Parameter	Elite Reform	Grassroots Action
Implementation	Governments	Mass action
Key target audience	Government and military officials	Social movements
Domain of defence	National	Local, national, transnational
Social context	Social defence as a functional alternative to military defence	Social defence as part of wider social change
Key promoters	Academic researchers	Activists
Argument, justification	Rational superiority of social defence state sponsored violence	Commitment to non-violence, participation, social justice

The first approach - exemplified by the work of Gene Sharp - is based on convincing key decision-makers in government of the superiority of civilian-based defence. The second approach is based on grass-roots action, usually linked to social movements. Many who adopt this latter approach are activists in groups not formally dealing with social defence, such as environmental groups, although there are also some groups specifically promoting social defence.

So far, no method of promoting social defence has had any notable successes. Only a few governments have shown interest in social defence, and none has taken major steps toward replacing its military forces by non-violent popular resistance. (There have been inquiries in, for instance, Sweden, Denmark and the Netherlands, but with little continuing consequence.) Similarly, no substate community has converted itself to non-violent resistance in a way that poses a comprehensive alternative to military defence. Given this lack of obvious successes, the discussion of prospects for social defence relies heavily on theoretical arguments, analogies and interpretation of historical struggles, most of which were not consciously linked to non-violence.

Here is examined the two approaches to promoting social defence by looking at the problem of formulating a convincing scenario.

The main problem with the reformist approach, according to its critics, is that it does not deal with social structures, in particular with the vested interests in present political systems. It can be argued that most arrangements in society are based not on the logic of human needs (such as security) but on the interests of social groups in power, wealth and status. According to this view, the present political systems are in place because they serve the interests of national elites, corporate elites, etc. Some government leaders may have the best of intentions to change the system, but they are unable to overcome some latent commitments in their functions that keep them in power. For example, in capitalist societies, the law enforcement is the ultimate defender of the ownership of private property, and is used in crisis situations to break strikes and smash militant workers' movements. A defence system based on popular non-violent action would be without the usual repressive sanction for property and bureaucratic privilege (among other systems of inequality). Any scenario, such as the social defence reform scenario, that ignores this issue is unrealistic. The advocates of the reform approach have not explained how they expect to avoid this fate.

The grass-roots action approach to social defence suffers a different problem in terms of scenarios. There are some examples of dramatic popular non-violent action which seem to hold the potential for a power equivalent to the state machinery. But, according to critics, the results of such action are often pitifully weak or disastrously misguided.

It is wise not to be overly committed to generalizations in this area, because no research has been done on the relative effectiveness of different methods of promoting social defence, nor are the potential criteria for evaluating different methods even spelled out, much less agreed upon. Because there has been so little experience in promoting social defence, and so little overt progress towards it, it is premature to rule out any method that seems compatible with social defence itself.

In conclusion, it may be reiterated that as no formal system has a complete answer to the problem of crime, integrated efforts are needed to tackle the problem at its very source. The social defence approach is based on the premise that the criminal justice system by itself cannot undo such aberration of the wider socio-economic system as are associated with crime. However, war against crime has to be waged, if not to win, atleast to be ensure that it is not being lost. Even the reduction in crime as a more realistic goal can be achieved only by extending crime prevention and control measures beyond the criminal justice system and by building these into a broader social action to curb conditions which produce crime. Such a perspective requires focussing on two major areas : (i) prevention of crime by protecting various vulnerable groups within the framework of social justice; and, (ii) treatment of offenders in a just, fair and equitable manner, with due regard to their human rights, and on the basis of a differential handling of individuals who violate law under various kinds of situational compulsion vis-a-vis those who perpetrate crime in an organized manner.

References

Bowden, Tom, (1978). Beyond the Limits of the Law: A Comparative Study of the Police in Crisis Politics. Harmondsworth: Penguin.

Charles Tilly, (1985) "War making and state making as organized crime", in Peter B. Evans, Dietrich Rueschemeyer and Theda Skocpol (eds.), Bringing the State Back In (Cambridge: Cambridge University Press, pages 169-191.

Chomsky, Noam and Edward S. Herman, (1979), *The Political Economy of Human Rights.* Boston: South End Press.

Gamson, William A, (1975), *The Strategy of Social Protest.* Homewood, Illinois: Dorsey.

Larson, Margali Sarfatti, (1977), *The Rise of Professionalism: A Sociological Analysis.* Berkeley: University of California Press.

Martin, Brian, (1991), "Social defence: arguments and actions", in Shelley Anderson and Janet Larmore (eds.), *Nonviolent Struggle and Social Defence* London: War Resisters' International, pp. 115-119.

Martin Brian, (1994), Protest in a Liberal Democracy, Philosophy and Social Action, vol. 20, Nos. 1-2, January-June, pp. 13-24.

Mauss, Armand L, (1975). *Social Problems as Social Movements.* Philadelphia: Lippincott.

NPA, (2008), Precis in Criminology National Police Academy, Hyderabad

Peters, Charles and Taylor Branch, (1972), *Blowing the Whistle.* New York: Praeger.

Plate, Thomas, and Andrea Darvi, (1983), *Secret Police.* London: Sphere.

Sharp Gene, (1973), *The Politics of Nonviolent Action* Boston: Porter Sargent.

Singh Hira (Dr.), (2007), Social Defence: Vision 2020, National Institute of Social Defence, New Delhi.

Zashin, Elliot M, (1972), Civil Disobedience and Democracy. New York: Free Press.

Zinn, Howard, (1968), Disobedience and Democracy: Nine Fallacies on Law and Order. New York: Vintage.

■■■

CHAPTER 25

Victimology

Learning Objectives :

- **To Understand the meaning, history and the importance of studying victimology.**
- **To comprehend the types of victims and those needing special assistance.**
- **To discuss the concept of victim under broad system of criminal justice.**
- **To be acquainted with the Indian scenario.**

Introduction

Victimology is a branch of criminology that studies victims, survivors and their relationships with the people who victimized them. It scientifically examines the degrees of suffering caused by offenders. Victimology researchers investigate the physical, emotional and financial impact of crime.

Concept of crime is intimately linked with the existence of the victim. When a crime takes place, a large number of factors combine and both the offender and the victim get involved in more ways than one. It is, therefore, hardly justifiable to pay attention only to the offender and bypass the existence of the victim. The occurrence of an offence is viewed as a human phenomenon in which social concern should be not only with the bringing of the offender to book and the maintenance of law and order in society, but also with the victim who has immediately borne the brunt as a result of the offence. If we go in to the legal proceedings, the importance of the role of victim seems to have been highlighted and acknowledged even by criminal law. The current viewpoint holds that reaction to crime on the part of the state should not be solely determined by the interests and needs of the victim. In fact provocation on the part of the victim has been treated for many centuries as a mitigating factor in the law, particularly in the case of homicide.

The emphasis on focusing on the victim and his problem are not without justification. If the victim in the wake of an offence is left to himself, in many cases it may lead to situations, which would hardly be in the long-range interest of peace and tranquility in society. The likelihood cannot be overlooked that the fallout of the offences spawns conditions conducive to anti-social or criminal behaviour on the part of the victim(s). It is hardly necessary to add that in the event of violations of law the criminal justice system is not only constrained against to intervene but also is obliged to shoulder the responsibility of such law-violators, during the period when they are in custody or undergoing imprisonment. This underscores the need for timely and effective preventive measures including succor to victims of crime.

History

In early days, a victim was a person or animal sacrificed to honor a God or leader. The concept of victim dates back to ancient cultures and civilizations, such as the ancient Hebrews. Its original meaning was rooted in the idea of sacrifice or scapegoat — the execution or casting out of a person or animal to satisfy a deity or hierarchy. As time went on, the concept of a victim was expanded to include other meanings. Initially, victimologists defined victims as "hapless dupes who instigated their own victimization." Later, scholars opposed this notion and redefined the concept to remove blame from victims.

Victimology or the study of victims was developed in the 1940s and 1950s. Notable criminologists Hans von Hentig and Benjamin Mendelsohn began studying the interactions between victims and perpetrators, and started drawing links between their actions. They are considered the "fathers of the study of victimology." Early victimology argued that there was a correlation between a victim's carelessness, conduct, behaviour, personality or misunderstanding and the crime committed. In essence, it said some victims were partially at fault for the crimes. The study of victimology used the information to recognize those cases and develop cost-saving and risk-reducing law enforcement practices.

The term "crime victim" generally refers to any person, group, or entity who has suffered injury or loss due to illegal activity. The harm can be physical, psychological, or economic. The legal definition of "victim" typically includes the following :

* A person who has suffered direct, or threatened, physical, emotional or pecuniary harm as a result of the commission of a crime; or in the case of a victim being an institutional entity, any of the same harms by an individual or authorized representative of another entity. Group harms are normally covered under civil and constitutional law, with "hate crime" being an emerging criminal law development, although criminal law tends to treat all cases as individualized.

Besides "primary crime victims", there are also "secondary crime victims" who experience the harm second hand, such as intimate partners or significant others of rape victims or children of a battered woman. It may also make sense to talk about "tertiary crime victims" who experience the harm vicariously, such as through media accounts or from watching television.

Many victims feel that defining themselves as a "victim" has negative connotations, and choose instead to define themselves as a "survivor." This is a very personal choice that can only be made by the person victimized. The term "survivor" has multiple meanings; e.g. survivor of a crime, "survivor benefits." It remains to be seen whether this terminology for victims of crime will endure.

"Victim defenses" have recently emerged in cases of parricide (killing one's parents) and homicide of batterers by abused spouses. Advocates for battered women were among the first to recognize the issue, and promote the "battered woman syndrome" to defend women who killed or seriously injured a spouse or partner after enduring years of physical, emotional and/or sexual abuse. Attorneys have also drawn upon theories of Post-traumatic Stress Disorder to defend their client's behaviour. From time to time, media attention to these defenses becomes intense, and certain "high profile" cases tend to influence public opinion and spread confusion over who is the "victim" and who is the "victimizer." One of the goals of victimology as a science is to help end this state of societal confusion.

Today's victimology combines criminology with several other disciplines, including psychology, social work, sociology, economics, law and political science. The most commonly studied areas of victimization are murder, rape, spousal abuse, elder abuse, child abuse and kidnapping. As the field expands identity theft, terror attacks, hate crimes and disability discrimination are also being researched.

The Study of Victimology

Before we can understand victimology, we need to appreciate that it is a fairly new subfield or area of specialization within criminology. Criminology is a rather broad field of study that encompasses the study of law making, law breaking, and societal reactions to law breaking. Victimology, much like criminal justice, falls into the third of these areas. Victimology doesn't have any subfields within itself; in fact, there are few theories, and little or no schools of thought. Going back to criminology, there are four subfields : penology (and the sociology of law); delinquency (sometimes referred to as psychological criminology); comparative (and historical) criminology; and victimology.

Andrew Karmen broadly defined victimology as "The scientific study of victimization, including the relationships between victims and offenders, the interactions between victims and the criminal justice system — that is, the police and courts, and corrections officials — and the connections between victims and other societal groups and institutions, such as the media, businesses, and social movements." From this definition, we can see that victimology encompasses the study of :

- victimization
- victim-offender relationships
- victim-criminal justice system relationships
- victims and the media
- victims and the costs of crime
- victims and social movements

The method to study victimology, are discussed below :

1. Define the problem : Find the asymmetry, analyze responsibility, explore the kinds of harm. The first goal of victimology is to understand crime. Studying details about victims also can provide information about criminals.This includes what causes it, who is responsible for it and what harm it causes. While the criminal justice system looks into these factors, it does so to determine appropriate punishment. Victimology does it from the side of the victim, trying to understand how the crime impacts the life of the victim and those around him.

2. Study Crime-victim relationship : One of the key factors looked at by victimologists is the relationship status of the criminal and the victim. It questions whether the two were friends, family members, romantically involved, acquaintances or perfect strangers. From there, the focus turns to how the particular victim was chosen and if he was targeted for a reason. Victimology began in Europe after World War II as a way to understand the relationship between criminals and victims.

3. Measure true dimension of the problem : Analyze statistics, see what kind of people are involved, accurately gauge extent of harm. Another important area of interest for victimologists is how often certain crimes occur and where they take place. They seek to explain why the crime rate varies in different age groups, gender groups, socio-economic classes, race groups and ethnicities. By grouping people and studying their lifestyles, victimologists hope to learn enough to prevent crime on an annual and individual lifetime basis.

4. Investigate how CJ system handles the problem : Look at what CJ system ignores, ask what victim wants, analyze effects, chronicle emergence of victim's movement. Victimologists also study the ways in which legal professionals treat victims as part of a trial. Many advances in victims' rights have come from research showing the injured party's needs being overlooked. Victimology experts study the criminal justice system to see what is missing. In many cases, this means analyzing what the victim wants in regards to punishment and compensation, how the victim is affected by the crime and whether this has an effect on the community in general.

5. Examine societal response to problem : Look at issues of constitutional rights, analyze proposed legislation, analyze media reaction, see if anyone is cashing in on the problem.

The third International Symposium on victimology held at Munstar, West Germany, in September 1979, has classified victims under five heads : (a) completely innocent victims like children and people who are the victims of crime when they are completely unconscious; (b) victims with minor guilt and victims of ignorance like women who prefer miscarriage and pay for it with their lives; (c) voluntary victims like the ones who commit suicide or are killed by Euthenasia; (d) victims who are more guilty than offenders like persons who provoke others to commit crime; and (e) the most guilty type of victims who commit offences against others and get killed by the others in self-defence. Sometimes, it is a matter of mere chance as to who becomes the criminal and who the victim. Indeed, the two roles may be exchanged either successively or simultaneously, i.e. one of the parties may be first the criminal and afterwards the victim, as in the case of a swindler who is killed by his angry victim, or of two swindlers who cheat each other at the same game. The situation may throw the victim into utter despair and lead him to crime.

The role of the victims in specific categories of offences has brought to light several important points. Some of them are as under :-

(i) The distinction between criminal and victim, which in former days appeared clear out as that of Black and White, actually often becomes vague and blurred in individual cases. The longer and the more deeply the actions of the persons involved are scrutinized, the more doubtful will it occasionally be who is to blame for the tragic outcome. This is especially so in certain cases of homicide, where attention may be drawn to the intricate nature of the relationships between the 'penal couple' (criminal and victim).

(ii) There are certain individuals who either because of their personal characteristics or as members of certain occupations can be described as 'latent victims', i.e. they are more prone than others to become the object of crimes. Obvious examples of such victim proneness are comparatively defenseless persons such as young, old, females, drunks, feeble-minded, illiterate, members of minority groups and so on. Among members of the victim prone occupations are taxi-drivers, prostitutes, and all persons who have to handle large sums of money. Moreover, there are some psychological possibilities as in the case of masochists who deliberately provoke physical aggression to satisfy their own unconscious wishes.

(iii) The victim is not necessarily an individual, it may also be a collective entity, a firm or a whole race or nation. The provisions of the substantive criminal law will usually indicate whom they are intended to protect. For example, in case of genocide, victims are not only those individual members of the national, racial, religious groups whose destruction had been the object of the crime, but also the groups as such.

(iv) The details of the victim-offender relationships have also been studied particularly for those types of crime where personal factors are most likely to play an important part e.g. homicide, sex offences and offences of an economic character.

(v) In the case of sex crimes against children, the distinction between participating and accidental victims has become generally accepted. Participant victims, as compared with others, were found usually to be girls who had known the offender for some time, had more than one sexual experience with him, had received some form of remuneration from him and kept the relationship secret from their parents. All of them showed considerably emotional disturbance and various other personality traits distinguished them from the accidental victims, such as much stronger needs for security and love. A number of them were severe neurotics, felt rejected by their parents and had considerable anxiety about sex. In many cases, their parents too had sexual conflicts and marital difficulties.

(vi) In the field of economic crime we have to make distinction according to the type of property which the offenders violate. If the victim is rich, or a big firm, or the 'Government', or an insurance company, scruples of any kind are entirely unknown, whereas the property of the small man is usually respected. The many ways in which property owners, especially when insured, facilitate the work of thieves and burglars are a well known source of concern to the police. The indifferent attitude of large stores to shoplifting may also serve a good example. Art forgeries have become the flourishing business which are at present largely through the ignorance, snobbery and speculative habits of certain well-to-do patrons.

Victimology Theory

Over the centuries, the word victim came to have additional meanings. During the founding of victimology in the 1940s, victimologists such as Mendelson and Von Hentig tended to use textbook or dictionary definitions of victims as hapless dupes who instigated their own victimizations. Over the years, ideas about victim precipitation have come to be perceived as a negative thing; "victim blaming" it is called. Research into ways in which victims "contribute" to their own victimization is considered by victims and victim advocates as both unacceptable and destructive. Yet a few enduring models and near-theories exist.

1. Luckenbill's (1977) Situated Transaction Model : This one is commonly found in sociology of deviance textbooks. The idea is that at the interpersonal level, crime and victimization is a contest of character. The stages go like this : (1) insult - "Your Momma"; (2) clarification - "Whaddya say about my Mother"; (3) retaliation - "I said your Momma and you too"; (4) counter retaliation - "Well, you're worse than my Momma"; (5) presence of weapon - or search for a weapon or clenching of fists; (6) onlookers - presence of audience helps escalate the situation.

2. Benjamin & Master's Threefold Model : This one is found in a variety of criminological studies, from prison riots to strain theories. The idea is that conditions that support crime can be classified into three general categories : (1) precipitating factors - time, space, being in the wrong place at the wrong time; (2) attracting factors - choices, options, lifestyles (the sociological expression "lifestyle" refers to daily routine activities as well as special events one engages in on a predictable basis); (3) predisposing factors - all the socio-demographic characteristics of victims, being male, being young, being poor, being a minority, living in squalor, being single, being unemployed.

3. Cohen & Felson's (1979) Routine Activities Theory : This one is quite popular among victimologists today who are anxious to test the theory. Briefly, it says that crime occurs whenever three conditions come together : (1) suitable targets - and we'll always have suitable targets as long as we have poverty; (2) motivated offenders - and we'll always have motivated offenders since victimology, unlike deterministic criminology, assumes anyone will try to get away with something if they can; and (3) absence of guardians - the problem is that there's few defensible spaces (natural surveillance areas) and in the absence of private security, the government can't do the job alone.

4. Greg Cooper and Mike King gave the theory victimology continuum principle. (http://www.iois.net/victimology/victimology.htm). They have suggested four types of victimology continuum principle.

Victimology Continuum Principle #1 :

"The lower the victim's risk level to the associated crime, the higher the probability that there was pre-association between the victim and the offender."

Victimology Continuum Principle #2 :

"The lower the victim's risk level to the associated crime, the higher the consideration given that the victim was targeted by the offender."

Victimology Continuum Principle #3 :

"The higher the victim's risk level is to the associated crime, the higher the probability that the offender is a stranger to the victim."

Victimology Continuum Principle #4 :

"The higher the victim's risk level is to the associated crime, the higher the probability that the victim is one of opportunity."

This is shown in the following diagram :

Figure 25.1

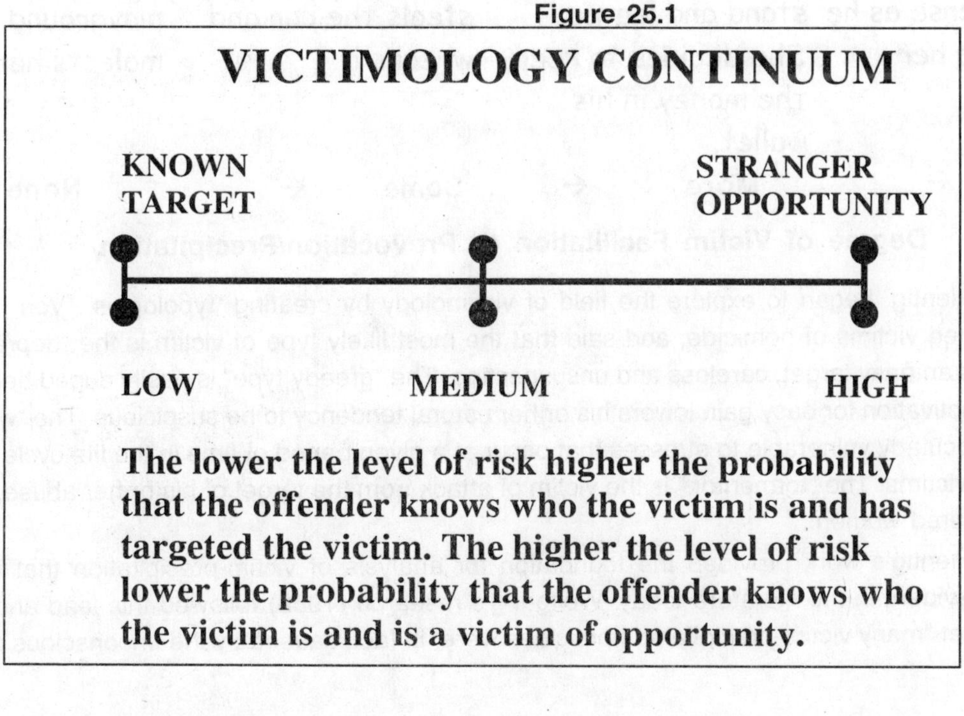

The phenomena that criminals and victims often have the same socio-demographic characteristics (e.g., are in relatively the same age group) is known as the propinquity hypothesis; and that criminals and victims often live in physical proximity to one another is called the proximity hypothesis.

5. Mendelsohn's Theory of Victim Precipitation : Mendelsohn proposed a theory that most victims have certain characteristics that make them more prone to be victimized. He identified six types of victims, the first of which, "the innocent," is in the wrong place at the wrong time. The remaining five types describe victims who somehow contribute to their own harm. Mendelsohn called that "victim precipitation." In 1948, von Hentig developed similar theories when he studied homicide victims. He put them into categories such as the "depressive type" who is a careless, easy target, and the "greedy type" who is easily fooled because his lust for gain is more powerful than his suspicion.

Victim precipitation theory has been most contentious when it is applied to rape. Here below is illustrated Four Scenarios Illustrating the Degree of Victim/Offender Responsibility According to Victim Precipitation Theory.

Figure 25.2

Degree of Criminal Intent of the Perpetrator

None →	Some →	More →	Much
Victim Provocation	Equal Responsibility	Victim Facilitation	Victim Innocent
A woman who has suffered years of abuse stabs and kills her husband in self-defense as he is beating her again.	Victim using the services of a prostitute leaves his wallet on the bed stand and leaves. She decides to keep the money in his wallet.	Victim leaves keys in his car while he runs into a store. A teenager impulsively steals the car and wrecks it.	A sex offender kidnaps a screaming young girl from a playground and molests her.
Much ←	**More** ←	**Some** ←	**None**

Degree of Victim Facilitation or Provocation/Precipitation

Von Hentig, began to explore the field of victimology by creating "typologies". Von Hentig (1948) studied victims of homicide, and said that the most likely type of victim is the "depressive type" who is an easy target, careless and unsuspecting. The "greedy type" is easily duped because his or her motivation for easy gain lowers his or her natural tendency to be suspicious. The "wanton type" is particularly vulnerable to stresses that occur at a given period of time in the life cycle, such as juvenile victims. The "tormentor," is the victim of attack from the target of his or her abuse, such as with battered women.

Von Hentig's work provided the foundation for analysis of victim-precipitation that is still somewhat evident in the literature today. Wolfgang's research (1958) followed this lead and later theorized that "many victim-precipitated homicides were, in fact, caused by the unconscious desire of the victims to commit suicide." Schafer's theoretical work (1968) also represented how victimology invested a substantial amount of its energy to the study of how victims contribute - knowingly or unknowingly — to their own victimization, and potential ways they may share responsibility with offenders for specific crimes. In fact, Schafer's book, The Victim and His Criminal, from this approach, is supposed to be a corrective to Von Hentig's book, The Criminal and His Victim.

Several other victimologists in the 1950s and 60s continued to theorize that victims were somehow responsible, whether consciously or unconsciously, of causing their own victimization.

In the 1970s, the field shifted its focus to studying victimization as a means of preventing increases in future crime and helping victims get what they need to move on. Victimology sought to give law enforcement and counselors tools to better handle cases and ultimately speed recovery and resolution.

This notion of "victim precipitation" was vigorously attacked by scholars in the 1980s, and was replaced by the notion of victims as anyone caught up in an asymmetric relationship or situation. "Asymmetry" means anything unbalanced, exploitative, parasitical, oppressive, destructive, alienating, or having inherent suffering. In this view, victimology is all about power differentials. Today, the concept of victim includes any person who experiences injury, loss, or hardship due to any cause. Also today, the word victim is used rather indiscriminately; e.g., cancer victims, holocaust victims, accident victims, victims of injustice, hurricane victims, crime victims, and others. The thing that all these usages have in common is an image of someone who has suffered injury and harm by forces beyond his or her control.

Typology of Victims

Just as certain persons are thought to have a high probability of indulging in criminal behaviour, so also some others may have a greater likelihood of being victimised. Von Hentig made the first ever study of the role of victims in crime and found some general characteristics among them which may be summarised as follows :

1. The poor and ignorant immigrants and those who are requisitive or greedy are the victims of offences involving frauds.
2. Quite often, the victims of larceny (theft) are intoxicated or sleeping persons.
3. The depressed or apathetic person is a victim because he is "deprived of warning posts" and is indifferent to harm or injury "in prospect".
4. Wanton or sensual persons may become victims due to situations precipitated by themselves.
5. A lonesome and heartbroken person may become especially vulnerable because of the loss of critical faculties in him.

Among "general classes of victims", Von Hentig includes the young, females, the old, the mentally defective and the intoxicated, immigrants, members of minority groups and the 'dull normal'.

Mendelsohn studied victims on the basis of their contribution to crime and classified them into the following categories :

1. Completely innocent victims, e.g., children, persons in sleep.
2. Victims with minor guilt and victims of ignorance such as pregnant women who go to quacks for procuring abortion.

3. Voluntary victims, such as the ones who commit suicide or are killed by euthanasia.
4. Victims who are more guilty than the offenders, such as persons who provoke others to commit crimes.
5. The criminal type of victims who commit offences against others and are killed or hurt by others in self-defence.

Victims Needing Special Attention

The work done by the pioneers mentioned above, and some others, has gradually led to some interest and concern for the actual and potential victims of crime in many countries around the world, India being also one of such countries. Certain categories of vulnerable persons and victims need special and greater attention.

1. Elderly Victims

In Western countries elderly persons tend to be in lower income groups and have, therefore, to live in dubious and criminal neighborhoods. Again, many of them live alone and repeated victimisation is possible in their case. These factors may not be much relevant in traditional countries like India where elderly people by and large, live with their children and they do not generally have to face any peculiar problems of physical insecurity. In the USA, the problem of crimes against elderly persons is sufficiently serious to have drawn the attention of the White House Conference on Ageing held in 1973. Besides recommending quite a few specific measures, the conference suggested top priority to the protection of the elderly and recommended that a portion of federal funds for the prevention of crime allocated to the States or local communities should be earmarked for this purpose.

2. Child Victims

The problem requires attention regarding offences involving violence in general and sexual abuse in particular. They need special attention because inept handling by the law enforcement agencies may prove to be even more damaging than the crime committed against the child; a factor which may discourage the initiation of proceedings by the parents of the child victims. Some special measures are, therefore, called for which may do away with the appearance and cross-examination of the child in the law court. But the problem is that such a cause may violate the ordinary principles of criminal procedure and evidence and, therefore, deny the accused his rights such as the right to cross-examine the prosecution witnesses. An innovation in this area has been introduced in Israel regarding the reception of the evidence of child victims of sex offences by the Evidence Revision (Protection of Children) Law of 1955. The purpose of the law is to protect the child from the undesirable effects of testimony in the police station or the court. Under this law, a child is examined by a young 'interrogator' who decides as to under what conditions the child should, if at all, appear before the court. Where it is not desirable for the child to appear before the court, the evidence of the child may reach the court via the youth 'interrogator' who under ideal conditions should be a combination of a lawyer and psychologist; such a combination being rare, the youth 'interrogator' is usually a graduate and experienced in social work. The 'interrogation' or rather the examination may take place either in the child's home or in the interrogator's office and the parents of the child mayor may not be allowed to be present. The system does away with some of the fundamental principles of evidence like the rule against hearsay evidence and the courts always insist on corroboration of the child's evidence tendered before the youth 'interrogator' which now is not necessary in case of ordinary witnesses.

3. Victims of Sex Offences

Offences against women, in particular serious offences such as that of rape, have been increasing everywhere including in traditional societies like India and greater attention is now being given to the problem of the victims vis-a-vis the criminal justice system. Perhaps this class of victims, irrespective of the age factor, deserves the maximum consideration in view of the emotional, psychological and human problems involved. The police and court proceedings may be as traumatic as the offence itself which led to the proceedings. The feeling is almost universal among the victims that instead of being treated as victims, they are treated by the police and law agencies as if they themselves are the culprits. The rules of criminal law and evidence, for all practical purposes, are tilted against the victim as evident by the requirements given below :

(i) It is for the prosecution to prove the lack of consent on the part of the victim; an element not easily provable especially when the accused is not a stranger to the victim. The courts often insist that proof be given of the resistance offered by the victim.

(ii) The credit of a rape victim may be impeached by showing that she was of a generally immoral character. The evidence of immoral character of the accused can, however, not be given unless good character is asserted by him in his defence. Highly humiliating and scandalous questions are often put to the victim despite the legal bar against such questions. Lately some legislative measures have been introduced in order to alleviate the sufferings of the rape victims. The changes have been made in the substantive penal law, procedural law and law of evidence.

Section 376 of the Penal Code has been amended by introducing punishments of longer duration. The minimum punishment now prescribed for the offence of rape is (Section 228-A, Penal Code) 10 years; the minimum is 10 years and the maximum is life imprisonment if the offender happens to be a police officer, public servant, jail superintendent or a hospital employee in whose custody the victim was put when the offence was committed. Two new provisions have also been added in the Penal Code. Disclosing the identity of the victim of the offence of rape has been made punishable.? Proceedings in a rape trial are to be held in camera. A new provision has been introduced in the Evidence Act laying down the presumption that there is no consent of a victim of a rape if the offence was committed by the husband during judicial separation from the victim and in cases where the offence was committed while the victim was in the custody of the police or in a hospital or in a rehabilitation home (Section 114-A, Evidence Act).

In the USA, one significant development has been the creation of "rape crises centres". These centres have all female participants including some rape victims, whose common concern is the problem of sex crime and they aim at providing counseling and other therapeutic measures to the victims.

They may operate through a telephone "hot line" or from an office maintained for the purpose. Two different kinds of centres have emerged as regards their attitude towards criminal justice administration agencies; some have cooperative while others have an antagonistic relationship with these agencies. In general, the views of the antagonistic variety have been summed up as follows (William MacDonald (Ed.), Criminal Justice and the Victim (1976), pp. 24-25) :

1. The sex victim does not benefit from and only stands to suffer further degradation by reporting her victimisation to the police and cooperating with the prosecution of the case.
2. Nothing is lost by not reporting the victimisation to the police because the criminal justice system is grossly ineffective in apprehending, successfully prosecuting, or even incarcerating dangerous sex criminals.
3. Psychiatry, the behavioural and social sciences and the "helping" professions have nothing to offer the female victim of sex crimes. They do not know how to protect her by rehabilitating sex offenders and they do not know how to treat her as a victim.
4. The best therapist for the female sex victim is another victim.

4. Female Victims

In Western societies, the issues regarding criminality against women generally pertain to and are confined to sexual offences but in the Indian setting, women are exposed to gang rapes by policemen and dominant caste groups, sati, wife-beating, prostitution and even occasional witch-hunting. Legislation exists and is reinforced from time to time to meet the challenges posed by some of the above-mentioned maladies but effective enforcement of the laws is not forthcoming; the defective legislation being only a marginal cause in the overall socio-political milieu of the country. Crimes against women, particularly dowry deaths, are increasing at a rapid pace. The situation has worsened to such an extent that a central legislation against eve-teasing is on the anvil.

Most of the crimes against women in India are directly traceable to their inferior and subservient status as sanctified by social values and ideas based on religion, real or imagined. While there might have occurred a change among small sections of the elite and truly educated persons, no genuine change in perception is, however, discernible among the masses as regards their traditional attitudes towards women and their role and status in the family and society. Politicisation in certain situations only further aggravates the situation. Sati, for instance, may not be a common phenomenon in our contemporary society but the fairly recent sordid episode of Deorala (On 4th September, 1987, a young Rajput woman was burnt to death on the pyre of her husband in a small town in Rajasthan. It is doubtful whether it was a voluntary sati or the woman was forced to do it) has brought into focus some very disturbing features regarding women's lot, especially the plight of Hindu widows. It also manifested the complete grip of the outmoded and, sometimes, imagined religious sanctions on substantial chunks of the population, including even the so-called educated people and the marked apathy and indifference of some of the politicians and police force regarding certain vital issues involved in such criminal situations. The virtues of sati are not extolled by villagers and illiterate persons alone but even the noted art critic Ananda Coomaraswami has expressed laudatory views about the practice of sati (The Dance of Shiva (1945), quoted by V.N. Datta in Sati : Widow Buming in India, 1988, p.231). He considers the sacrifice of sati spiritually significant and equates her with a patriot who "must be unmoved by an appeal to self-interest or a merely utilitarian demonstration of futility".

Dowry deaths caused by the husbands and in-laws of the hapless women, who are unable to fulfill their husband's or in-laws' unreasonable expectations based on greed and avarice, are quite often given the colour of suicide. Despite some well-meaning legislation to help the prosecution,(Sections 113-A and 113-8 of the Indian Evidence Act) it is necessary to secure conviction of the offenders in a majority of the cases.

5. Minority Groups and Weaker Sections

Members of the ethnical, religious or linguistic minorities in pluralistic societies may be especially vulnerable to crimes. In conflicts resulting from socio-economic imbalances and political factors, they are generally the worst sufferers. The phenomenon is almost universal and a typical example is provided by India where communal strifes owing to a variety of causes, pose a perennial problem. A number of caste and communal riots occur each year in the country leading to the murder, rape and destruction and looting of property on a large scale in which the main sufferers obviously are those belonging to minority and weaker sections and this is irrespective of the fact as to who was responsible for the flare-up in the first instance. Hardly any administrative or legal action is possible or even tried in situations involving a large number of persons and, paradoxical as it may sound, the greater the tragedy, the lesser is the possibility of the protection of the victims and of punishment to the perpetrators of the ghastly crimes.

Consumers of various goods and services are quite often the victim of unscrupulous practices of manufacturers, traders, contractors and members of other professional and occupational groups. Though there are civil and criminal laws in plenty, as are also the various 'warranties' and 'guarantees' given to protect the consumer, they do not serve much useful purpose in actual practice.

The Consumer Protection Act, 1986 is the most significant and recent legislation on the subject. It is very comprehensive and all kinds of goods and services are brought under it. Under the Act, advisory and adjudicating bodies known as Consumer Protection Councils are to be set up at the Central and State levels. In order to serve the purpose they have been assigned, these councils have to protect the various rights of consumers : the right to be informed about the quality, quantity, potency, purity, standard and price of goods, right to be assured access to a variety of goods at competitive prices; right to be led at appropriate forums; right to seek redressal against unfair trade or unscrupulous exploitation of consumers; and right to consumer education.

Importance of Studying Victimology

Victimology is a crucial part of crime analysis. Through it the investigator tries to evaluate why this particular person was targeted for a violent crime. Very often, just answering this question will lead the investigator to the motive, which will lead to the offender. Victimology is an essential step in arriving at a possible motive. If investigators fail to obtain complete victim histories, they may be overlooking information that could quickly direct their investigations to motives and to suspects.

The analysis of socio-economic and personal background of the victim may serve as a guide to construct a comprehensive victimology. It is noted that not every crime requires an exhaustive review of the victim; however the general rule applies : The more violent, complex, bizarre and/or perplexing the crime, the greater direction a victimology will provide.

As we analyze the socio economic and personal background of the victim, we may attempt to answer the related questions of what, when, where, why and how. This will reveal the interdependent and dynamic relationship between each referenced item while leading the investigation toward the final product... the most probable type of suspect.

The next most provocative question to ask is "why" was this victim the subject of the crime? The answer leads the investigation toward the criminal motivation. Often, the offender will inadvertently reveal himself through behaviour committed during a crime. This diagnosis assists the investigator during the initial stages of formulating a probable motive and reasonable suspects. Additional pertinent questions include :

Why was the victim selected by the offender ?
Why was the crime committed in this location ?
Why was the crime committed in this manner ?
Why did the offender approach the victim in a particular manner ?
Why did the offender assault the victim in this particular manner ?
Why did the offender inflict these particular wounds on the victim ?
Why did the offender use/choose a particular weapon ?
Why was the victim left in a particular position ?

While processing the foregoing questions on victimology, the investigator becomes acutely aware of the dynamics occurring between each of them. Each element and their interdependent association formulate a progressive system which can effectively unravel mysteries. This approach can begin to incrementally lead the investigation toward an increased probability of successful resolution.

Victimology consists of the application and integration of asking the appropriate diagnostic questions, acquiring accurate data and then assimilating it through the filters and principles designed to lead the investigation. By using this approach, the investigator is better equipped to focus his energy and efforts to the reasonable probabilities rather than the chance of possibilities.

The foregoing insights illustrate how effectively an investigation can be guided by conducting a thorough victimology. By surgically applying the appropriate diagnostic questions, the investigator is better prepared to accurately "prescribe" scarce investigative resources more efficiently and effectively. The application of these principles and their benefits extend beyond the investigative approach. Victimology is the key to crime analysis. It is the foundation to understanding the crime, the offender, the criminal motive and the victim. The integration of victimology has multidimensional impact and multi-disciplinary considerations.

With a thorough victimology, the investigator will more effectually construct a guided investigation, as well as an interview or interrogation focus, and a prosecutor will be better equipped to formulate a strategy and court presentation reflecting the *corpus delecti*.

Victim Trauma

Psychological trauma impairs the ability and/or willingness of crime victims to cooperate with the criminal justice system. Victims must be treated better by the criminal justice system because it cannot accomplish its mission without the cooperation of victims.

Victims whose crime-related fear makes them reluctant to report crimes to police or who are too terrified to testify, effectively make it impossible for the criminal justice system to accomplish its mission. Thus, it is important to understand :

- Victims' crime-related mental health problems.
- What aspects of the criminal justice system process are stressful to victims.
- What can be done to help victims with their crime-related mental health problems.
- What can be done to help victims cope with criminal justice system-related stress.

Effective partnerships among the criminal justice system, victim assistance personnel, and trained mental health professionals can help victims with crime-related psychological trauma and with criminal justice system-related stress. By helping victims through such partnerships, the criminal justice system also helps itself become more effective in curbing and reducing crime.

How can the Criminal Justice System Address the Needs of Traumatized Crime Victims ?

- Treat victims as human beings, not as evidence.
- Always provide victims with information about case status and prepare them for what will happen at trial.
- Pay close attention to any psychological trauma the victim may be experiencing.
- Arrange for someone to be present at the trial whom the victim can count on for emotional support.
- Inquire about any specific fears or concerns the victims may have about trial and testimony.
- Inform and consult with victims about potential plea-bargain procedures.
- Give victims opportunity for input into proceedings when possible, including the opportunity to make a victim impact statement.
- Refer victims who need help with stress management to mental health professionals specifically trained to provide it.
- Tell victims you are sorry that the crime happened and ask how you can help.

Victim Expectations

Mendelsohn is credited with being the first to study the relationship between the victim and the doer (offender) and taken together, he termed the two as the "penal couple". In his victimology he focussed his attention on the possible role of the victim on the circumstances surrendering him in the commission of the crime. He identified four factors in the context of the crime of rape and other offences against morality which could be responsible for lessening of the resistance on the victim's part :

1. The familial, authoritative or hierarchical relations existing between the accused and the victim.
2. The temperament of the victim, which may obscure the reasoning faculty.
3. The liberative social surroundings of the victim.
4. The superiority of the social milieu of the accused in relation to that of the victim.

Most crime victims think that the criminal justice system should be responsible for providing them with counseling for crime-related psychological trauma.

Victim Compensation

Compensation refers to payment by the State to the victim of a crime or to his dependents .to make up for the loss incurred. It may be mentioned in passing that the term 'compensation' is not used universally. In some of the countries, it is known as restitution, damage or reparation. The term 'restitution' is closely related to compensation. Restitution is agreed to by the offender willingly or on being compelled by the court. Reparation is obtained from the offender for the victim of crime. On the other hand, the term "damage" is civil in character and is awarded mostly by civil courts. Thus, compensation differs from reparation in that it does not involve offenders, from restitution and damages as it involves neither the Criminal Court nor the Civil Court. The idea that the persons who suffer have a claim upon society is not altogether new. There lies the welfare of the state in the well-being of the citizens, and the two tend to rise and fall together. Moreover, should the victims be left to fend for themselves, it may be reiterated, many of them may digress to anti-social path. Thus, while the idea of providing compensation to the victims of crime may not be questioned, it would be conceptually unrealistic and administratively impracticable to provide compensation en-block to the victims of all offences. From this stems the necessity of evolving norms, which would be socially and economically reliable and valid which, in turn, implies that different types of victims of crime, their background and circumstances may be studied systematically. In recent years the administrators and the policy makers of the criminal justice system all over the world have been reflecting over the injustice to the victim of crime and the need to develop schemes of payment of compensation for his loss.

State victim compensation programs are designed to provide financial assistance to victims and, in some cases, to family members and other eligible persons. Usually, a victim must have suffered actual physical harm or other tangible loss. Keep track of your losses such as destroyed or stolen property and cost of any emotional counseling. This includes medical bills and lost wages as a result of the abusers actions.

In most states, a victim must fully cooperate with law enforcement and prosecution efforts in order to qualify for compensation. Restitution, on the other hand, is ordered by the court or in some states, by the paroling authority making the offender pay for the financial loss of the victim. However, as a rule, neither victim compensation nor restitution include punitive damages for injury or loss suffered by the victim. A lawsuit in civil court is usually required to recover punitive damages.

Some studies have been conducted and issues have been discussed, focusing on compensation to the victims of crime. The First International Symposium on Victimology (Jerusalem 1973) has further strengthened the victim compensation programme. Section 4 of the proceedings of the Symposium deals with victim compensation. Papers presented to the symposium centre around mainly on the effectiveness of programmes to compensate victims of crime, help for victims of robbery, compensation to victims, a new approach to the victim of crime and so on.

In India, criminal law is offender-oriented. The criminal justice is concerned with the offender. The Constitution lays down: "No person shall be deprived of his life or personal liberty except according to procedure established by law" (Art. 21). Accordingly, it is the responsibility of the state to protect its people.

Section 357 of the Criminal Procedure Code deals with compensation to the victims of crime. The Research Centre of the Maharashtra Criminal Investigation Department undertook a study of

Pune city in the year 1960-61, and found that out of 510 cases of violence reported, only 182 prosecutions took place, and only in 13 cases was the aggrieved party awarded compensation. Indeed this shows the incapacity of Section 357 of the Criminal Procedure Code. Singh and Jatar (1978) in the study of compensation to the victims of dacoity gangs have found that no compensation but ex gratia assistance limited to only Rs.500.00 was given to the victims. This help to the victims was available only in about 15 percent of the cases.

Compensation involves three major questions : (i) who should be paid compensation; (ii) how much; and (iii) what should be the mechanism. As regards the first question, Glasebrook (1962) recommends compensation only for the victims of crime of violence. According to Dean and Feency (1968) compensation may be given to these three categories of victims also, (a) where prosecution has not taken place, (b) where the offender has been acquitted and (c) in the cases where the offender was insane or a minor below the age of 10 years. The compensation may be paid to the victim himself or to his surviving dependents.

The second question involves the determination of how much compensation is to be paid to the victims. The compensation for victims of crime may cover loss or injury suffered by victims or their surviving dependents. It should include payment for property loss, physical injury, incapacitation or disability, loss of income, mistaken imprisonment and for legal fees and similar expenses. The judgment in a civil suit if any should also be taken into consideration.

Such provisions may also tempt people to make false claims, or to demand inflated amounts. The pattern of opportunity structure may even change the behaviour of the individual. To check these, a mechanism may be evolved. This may be in the form of a tribunal, compensation board or a special court to deal with claims for compensation. While these provisions may add to the burden of the tax payer, those are morally and socially more than justifiable. Suitable legislation may be enacted to provide for compensation to the victims of crime, defining eligibility for compensation and prescribing both the mode of assessment of damage and the quantum of compensation. This would, in turn, involve the establishment of suitable machinery charged with the function of determining compensation.

In Indian context, there are a few areas in which the existence of the victim and his problems have been recognized. For example, in the case of an air-accident all the bonafide passengers are compensated. So is the case with the passengers on the Indian Railways and state transport buses like the U.P., State Road Transport Corporation and the Delhi Transport Corporation. The Motor Vehicle Act also enjoins that the vehicle would carry a compulsory 'Third-party insurance' and should they run over and hurt or kill a pedestrian, the insurance company is required to pay the compensation. While such provision go a long way towards providing relief to those who have suffered, all these hardly suffice institutional arrangements for providing compensation to the victims of crime.

In India, civil laws do provide for such eventualities. The victim or his next-of-kin may file civil suits and claim compensation from those who have been responsible for the damage. However, perpetrators of crime are not always identified; in other cases they may not be solvent; and, above all, the civil suits are usually a long-drawn process. Taken together, such provisions in the civil law are far from being satisfactory of adequate. In a multi-religious country, group disturbances or riots are almost a recurring feature. Following an eruption, destruction of property and loss of limb or life is not unusual. In recent times, a few states in the country have also seen much eventualities, the victims have been given ex-gratia grants. It would not, however, take long to see that those are ad-hoc measures and, therefore, their application is not predictable and certain.

The State which lays down rules of behaviour should not only punish the crime-doers but also take cognizance of the victim. Nevertheless, a concerted attention on the victim and his problems is yet to be paid in the country. This is not to say that the criminal justice system in the country has been altogether unresponsive to these issues. Section 357 of the Criminal Procedure Code, 1973 lays down :

1. "When a court imposes a sentence of fine or a sentence (including a sentence of death) of which fine forms a part, the court may, when passing judgment, order the whole or any part of the fine recovered to the applied - (a) in defraying the expenses properly incurred in the prosecution; (b) in the payment to any person of compensation for any loss or injury caused by the offence, when compensation is ,in the opinion of the court, recoverable by such person in a civil court; and when any person is convicted of any offence for having caused the death of another person or of having abetted the commission of such an offence; (d) when any person is convicted of any offence which includes theft, criminal misappropriation, criminal breach of trust, or cheating, or of having dishonestly received or retained, or of having voluntarily assisted in disposing of stolen property knowing or having reason to believe the same to be stolen, in compensating and bona fide purchaser of such property for the loss of the same if such property is restored to the possession of the person entitled thereto.
2. If the fine is imposed in a case which is subject to appeal, no such payment shall be made before the period allowed for presenting the appeal has elapsed, or if an appeal be presented, before the decision of the appeal.
3. When a court imposes a sentence, of which fine does not form a part, the Court may, when passing judgment, order the accused person to pay, by way of compensation, such amount as may be specified in the order to the person who has suffered any loss or injury by reason of the act for which the accused person has been so sentenced.
4. An order under this section may also be made by an Appellate Court or by the High Court or Court of Session when exercising its powers of revision.
5. At the time of awarding compensation in any subsequent civil suit relating to the same matter, the court shall take into account any sum paid or recovered as compensation under this section.

Similar provisions have also been made under Section 5 of the Probation of Offenders Act, 1958. However, it would be noticed that these legal provisions barely touch the fringe of the problem. The victim is eligible to receive full or partial compensation in only those cases where a fine has been imposed on the offender; and in those cases where the offender has not been identified, apprehended, convicted or is insolvent, the victim would remain unprovided for. Sometimes ago, in respect of victims of crime who belong to Scheduled Castes or Tribes, a proposal was mooted under Section 357 of the Cr.P.C. that a fine must be imposed upon the guilty and the victim must be provided compensation, by one of the States. It could not be, however, acceded to by the Government of India because it was selective.

Victim's Basic Rights

Strange and ironical as it may sound, the administration of criminal justice is not much concerned with the victims of crime except that in a few cases some marginal action is possible to render limited or token help to the sufferer. The entire focus of the criminal justice system is on the offender : to punish him or to seek his reformation and rehabilitation with all the resources and goodwill available through the courts and other governmental and non-governmental agencies. Efforts are made to understand his personality and the individual and social factors which might have contributed to his criminal behaviour; the victims of the crimes are, on the other hand, 'the forgotten people in the system'. President Gerald R. Ford sent the following message to the American Congress in 1975 :

"For too long, the law has centred its attention more on the rights of the criminal than on the victims of the crime. It is high time we reversed this trend and put the highest priority on the victims and potential victims."

Since about 1981 (the start of Victims Rights Week), there has been numerous legislation, conferences, and task forces. Some so-called victim's rights (such as denial of bail, anti-suppression of evidence, and victim-initiated appeals) clearly are anti-defendant and pro-prosecutor to the extent that they undermine cherished principles that an accused person is considered innocent until proven guilty, and that the burden of proof falls on the state.

The United National General Assembly in 1985 adopted a declaration. This declaration is known as "Basic Principles of Justice for Victims of Crime and Abuse of Power". The declaration treated as Magna Carta of Rights of victims globally. This declaration deals with certain important aspects of problems of victims of crime including victims of abuse of power. Some of the important suggestions discussed in the declaration are as under :

(a) Victim should be treated with compensation and respect for their dignity. They are entitled to access to the mechanisms of justice and to promote redress, as provided for by National Legislation, for the harm they have suffered.

(b) Judicial and Administrative mechanism should be established and strengthened where necessary to enable victims to obtain redress through formal and informal procedures that are expeditious, fair, inexpensive and accessible. The victims should be informed about their rights in seeking redress through such mechanisms.

(c) Providing proper assistance to victims through the legal process.

(d) Taking measures to minimise inconvenience to victims, protect their privacy where necessary and ensure their safety as well as that of their families and witnesses on their behalf, from intimidation and retaliation.

A Victim's Basic Rights Are :

1. The right to attend and/or participate in criminal justice proceedings.

The victim can attend the trial, sentencing, and/or the parole. Many states also allow the victim to make an oral or written statement to be considered by the court or parole board at such proceedings. Some of you may remember the televised statements from families of victims of serial killer Jeffrey Dahmer. Victims are increasingly being given the right to attend and sometimes address the court at other critical proceedings in the criminal justice process, such as :

(a) Hearings on bail or pre-trial release of the offender;

(b) Entry of plea agreements;

(c) Post-trial relief or release hearings;

(d) Probation hearings;

(e) Commutation or pardon hearings.

2. The right to notification of the stages/proceedings in the criminal process.

This may not only include notification to the victim or victim's family of scheduled criminal proceedings and their outcomes, but also advance notice of proceedings where the victim has the right to attend and/or make a statement, as well as when hearings have been canceled and rescheduled. The right to be notified is a crucial one, because without it, victims cannot adequately pursue their other rights.

3. The right to notification of other legal remedies.

Victims may also have the right to be informed of the option to sue the offender for money damages in the civil justice system, to collect witness fees for their testimony, as well as other rights.

4. The right to protection from intimidation and harassment.

The right to protection from intimidation and harassment by the offender or the offender's family or associates may be extended to the victim's family members, as well as the victim. In the event you receive threats, bribes, or other attempts to persuade or intimidate you into testifying untruthfully, to forget, or to make yourself unavailable as a witness, report it immediately to the Sheriff's Office.

If you are the victim of domestic violence (violence within the family) or repeat violence (two incidents), you can file an injunction for protection with appropriate Court. Other kinds of protection include :

(i) Police escorts to and from court;

(ii) Secure waiting areas separate from those of the accused and his/her family during court proceedings;

(iii) Witness stands that are shielded from the direct view of the offender; especially if the victim is a child, in which case many courts now allow video taped testimony to be used to protect the child from the trauma of the courtroom and further exposure to the accused.

(iv) Closing the courtroom to those who are not parties to the case; and

(v) Residence relocation.

5. The Right to Notification of Employer and Creditors.

At your request, the State Attorney's Office or Sheriff's Office will inform your employer that your cooperation in their investigation and prosecution of the case may necessitate your absence from work. At your request, they can also contact your creditors to seek their consideration if you are unable, temporarily, to continue payments as a result of the crime. Credit card companies are usually willing to suspend interest and payments if notified of the situation by an authority.

6. The right to confidentiality of records.

Unlike many other criminal cases, police and court records are not public record if they involve a juvenile or if the case deals with sexual assault or rape. These records are usually only available to the attorneys and parties to the case.

7. The right to speedy trial provisions.

Usually, this Constitutional right is used as a tactic by defense attorneys to rush the prosecution to court before they have all their ducks in a proverbial row. But it's a two-edged sword. As a victim, YOU have the right to a speedy trial also.

8. The right to prompt return of the victim's personal property seized as evidence from the offender.

This can include photos, clothing, recordings, letters etc.

Victim and Criminal Justice

One important and basic factor in the administration of criminal justice is the victim's decision as to whether he should invoke the judicial process. There are a number of motives and factors responsible for the wide gap between the actual volume of the crime and the reports made to the police about it.

Crime reporting to the police would obviously depend to a great extent on the victim's perception regarding efficiency of the police and the likely response from them. In India, more than the inefficiency of the police, real or perceived, it is the unlikelihood of police interest and response and corruption on their part as perceived by a majority of the citizens which makes them cynical and somewhat convinced about the futility of approaching the police. The police is rather regarded as being in league with criminals of certain varieties. Some of the self-evident propositions in this regard are :

1. If the seriousness of the crime is greater, the probability of reporting is greater.
2. If the victim has more positive attitudes towards the police, is more trusting of the police, then the probability of reporting is greater.
3. If the victim believes that the police and other law enforcement institutions are effective, then the probability of reporting is greater.
4. If the victim is more involved in the community, more integrated in to the community, then the probability of reporting is greater.
5. In the event of property loss, if the lost items were insured, then the probability of reporting is greater.
6. If the offender is a stranger, then the probability of reporting is greater.

The police may not be the only inhibiting factor constituting obstacles to the invocation of the criminal law process by the victims. The victim continues to suffer at all the stages subsequent to the reporting to the police, during investigation and processing of the case by the police and even during the trial by the court. The suffering of a victim's witness in terms of harassment and humiliation, various kinds of inconveniences and the financial losses may be quite formidable. The convenience and interests of judges and others connected with the court (prosecution and defence lawyers, court clerks, etc.), as can be expected, are given priority over those of the victim. The delays and adjournments are inevitable, resulting in unnecessary running around and avoidable repeated and futile trips by the victim to the court.

An interesting area of recent researches is the determination of the extent to which the administrative and judicial action is influenced by the conduct and personal factors and characteristics of the victim. Some American studies have examined the police discretion in charging the alleged offenders in this context. Not surprisingly, it has been found that the police are more likely to bring charges in an assault case if the victim and offender are strangers and less likely to do so if they are related or known to each other. The greater the prestige of the victim, the lesser the chance of the case being dropped. Older victims being regarded as defenseless, it has been found that cases in which the homicide victim was sixty or older were less likely to be dismissed by the prosecutor.

Judicial attitude, as reflected in sentencing, also appears to be amenable to the personal factors and characteristics of victims. More severe judicial action is discernible in cases of helpless victims, e.g., elderly persons, children and females. Lesser severity is reflected in the sentencing in victim precipitated crimes. Interestingly enough, the American studies also reveal that the attractiveness of the victim and defendant also affected sentences passed by the court, greater sentence passed where the victim happened to be attractive and lesser sentence in case of attractive defendants.

The Indian Scenario

Section 357 of the Code of Criminal Procedure, 1973 is the main provision dealing with compensation to crime victims (Sections 237, 250 and 358 of the CRPC). Section 545 of the old Criminal Procedure Code dealt with the same subject-matter though it was somewhat narrower in scope. Section 357(1) lays down inter alia :

"357(1). Whenever under any law in force for the time being a criminal court imposes a fine...or a sentence of which fine forms a part, the court may, when passing judgment, order the whole or any part of the fine recovered to be applied :

(a) in defraying expenses properly incurred in the prosecution;

(b) in the payment to any person of compensation for any loss or injury caused by the offence when compensation is, in the opinion of the court, recoverable by such person in a civil court;

(c) when any person is convicted of any offence for having caused the death of another person of having abetted the commission of such an offence in paying compensation to persons who are, under the Fatal Accidents Act, 1855, entitled to recover damages from the person sentenced for the loss resulting to them from such death."

Sub-section (3) of Section 357 was added, as recommended by the Law Commission in its Forty-first Report, in the new Criminal Procedure Code of 1973 and it provides :

"357(3). When a court imposes a sentence of which fine does not form a part, the court may, when passing judgment, order the accused person to pay, by way of compensation such amount as may be specified in the order, to the person who has suffered any loss or injury by reasons of the act for which the accused person has been sentenced."

The court has very limited discretion under Section 357(1); it can give compensation only out of the fine if imposed on the offender. The court has, however, much more discretion under sub-section (3) of Section 357; though only if fine does not form a part of the sentence. Theoretically, the power of the court is unlimited, though practical considerations would prevail. A magistrate can order for higher compensation than the amount of fine he can impose.

In Sarwan Singh vs. State of Punjab (1978, 4 SCC 111), it was said by the Court that in awarding compensation the court should not just consider what compensation ought to be awarded to the heirs of the deceased and then impose a fine which is higher than the compensation. The Court laid down that the amount of fine should be determined on the basis of various factors including the nature of crime, number of injuries, and the paying capacity of the offender. In this case the two appellants were fined ₹ 3500 each along with rigorous imprisonment of 5 years in view of sufficient funds being available with them.

The courts have generally been averse to the imposition of fine along with a severe or even substantial punishment by way of imprisonment. In Mohammed Shah vs. Emperor, (AIR 1934 Lah 519), the offender was convicted under Sections 32, 149, and 148 of the Penal Code. He was awarded one year's imprisonment and a fine of ₹ 500 out of which ₹ 400 was awarded to the heirs of the victim. The Lahore High Court held imprisonment to be substantial and, therefore, fine to be unwarranted. The court further held that compensation in any case would not have been payable to the heirs in view of the blameworthiness on the part of the deceased. He was himself the aggressor and had encroached upon the land of the offender.

The Supreme Court has expressed its disapproval of combining the punishment of fine with a death sentence and even with life imprisonment (1977 SCC (Cri) 397). In the instant case, the High Court had reduced the punishment of death sentence to life imprisonment awarded by the trial court and imposed a fine of ₹ 20,000 on the offender payable to the heirs of the deceased. The Supreme Court reduced the fine and compensation to ₹ 3000 and made the following observations :

"Though for the offence of murder, courts have the power to combine a sentence of death with a sentence of fine, that power is sparingly exercised because the sentence of death is an extreme penalty to impose and adding to that grave penalty a sentence of fine is hardly calculated to serve any social purpose..., even a sentence of life imprisonment is seldom combined with a heavy sentence of fine. Before imposing the sentence of fine, particularly a heavy fine, along with the sentence of death or life imprisonment, one must pause to consider whether the sentence of fine is at all called for and, if so, what is the proper or adequate fine to impose in the circumstances of the case."

The judicial attitude is, however, reflected somewhat differently in Guruswami v. State of T.N. (1979 3 SCC 799), where it was held that in a case of murder it is only fair that proper compensation should be provided for the dependants of the deceased. It was a case where the appellant had murdered his father and brother as a result of some family feud and was sentenced to death. On appeal, the Supreme Court reduced the punishment to life imprisonment and imposed a fine of ₹ 10,000 on the offender to be paid to the heirs of the deceased. A perusal of a subsequent case, Baldev Singh v. State of Punjab (1995 6SCC 593: 1995SCC Cri 1132) also indicates that quite often the Supreme Court prefers to substitute a severe punishment given to the offender in award of compensation to the victim in deaths resulting due to family feud. In the instant case, the appellants were - guilty of culpable homicide of a family and awarded life imprisonment by the High Court. On appeal, the Supreme Court observed that in appropriate cases, award of compensation to the victim served better than the deterrent punishment to the offender. The two appellants were altered to pay ₹ 35,000 each to the widow of the deceased and her children and the sentence of life imprisonment against the affluents was reduced to the period of imprisonment already undergone by them.

In quite a few cases of death caused by rash or negligent act, punishable under Section 304-A of the Penal Code, the courts have reduced the period of imprisonment to the one already undergone but imposed a severe fine in order to provide substantial relief to the dependants of the deceased (Sukhdev Singh v. State of Punjab 1982 2 SCC 439) With lie same object, 'an innovative use of the probation technique is being made "compensate the dependants of the victim of death caused by rash or negligent act as would be evident by two cases decided by the Punjab and Haryana High Court (Sunder Lal v. State of Punjab and Gurmit Singh v. State of Punjab; Chandigarh Crime Cases, p. 126 and p. 335.) The killer is released on probation if, besides other favourable factors, the offender is willing to pay compensation to the heirs of the victim. Though probation was denied in Dr Jacob George v. State of Kerala (1994 SCC (Cri) 774) where a homeopath attempting to procure an abortion by operating upon a woman caused her death, the Supreme Court reduced the imprisonment to the two months already undergone. The fine imposed upon the petitioner was increased from ₹ 5000 to one lakh required to nurse the child of the deceased reasonably well.

With the growth of "victimology" the Courts have taken the cognizance of the cases where they pay the compensation to the victims of crime especially where one of the parties are state. In Chairman Railway Board v. Chandrima Das (2000, 2 SCC 465), the court awarded a compensation of ₹ 10 lakh to the victim who was victim of rape at Rail Yatri Niwas. They made it clear that running of railway is a commercial activity and the court made a distinction between sovereign and commercial activity in this case. A new category of the victims have emerged during the recent years. They are the victims whose death is caused in custody. During the year 2000, 40 deaths in police custody were reported. All these persons were remanded to police custody by the Courts. This figure has shown an increase of 90.5% over the year 1999. Besides this, 38 deaths in police custody were reported who were not remanded to them by the courts during 2000. While as the number of such cases was 43 during the year 1993.(Crime in India, NCRB, p. 356.) and they are also awarded compensation. In another case, (Phoolwati v. NCT of Delhi, 2000 Cri LJ 1613) where the death of the deceased was caused in police custody the court awarded a compensation of ₹ 3 lakhs to the deceased's relatives.

The Supreme Court in State of Punjab v. Ajaib Singh (2 SCC 486: 1995) gave a new dimension to the idea of compensation of India. The Court held that the respondent police officer will deposit an amount of ₹ 5 lakhs even though he has been acquitted by the High Court.

It is evident that only marginal action is possible under Section 357 of the Code of Criminal Procedure to compensate the victims of crime. The various constraints and limitations may be summed up as follows :

1. Much depends upon the paying capacity of the offender and, in most cases, this acts as a bar against the victim getting any compensation.
2. There is a general reluctance on the part of the criminal courts regarding the use of the criminal law process for compensation.
3. The courts are reluctant to impose fine along with substantial imprisonment in serious offences and the scope of fine in any case is very limited in terms of quantum in minor offences.
4. Maximum fines have been laid down for various offences which were fixed a long time ago and their monetary value must now be a very small fraction of what they might have been at the time when these fines were introduced in the Penal Code.
5. Conviction is necessary for the payment of compensation. As is well known, conviction may not be possible in many cases irrespective of the merits of the case.

Besides the Code of Criminal Procedure, there are some other enactments which contain provisions for compensation to crime victims. Section 5 of the Probation of Offenders Act authorises a court releasing an offender on probation to direct him to pay such compensation as the court thinks reasonable for loss or injury caused to any person by the commission of the offence and such costs of the proceedings as the court thinks reasonable. The probation system may be quite conducive to restitutive justice and in the USA restitution is quite often the condition for release of an offender on probation. Compensation by the prisoners out of their earnings during imprisonment may also be a good strategy; an appropriate sanction in view of the fact that prisoners are given wages for the work done in prison.

Some laws provide for the imposition of punitive tax on the community as a whole in an area where riots occur or members of an unlawful assembly commit some offences. The Bombay Police

Act, 1951 is such an enactment. This remedy is perhaps the only effective remedy indicated in large-scale violence involving a large number of offenders and victims. This remedy is particularly useful in the context of communal riots in the country since hardly any person is convicted, not even prosecuted, regarding participation in communal disturbances due to various political and legal constraints and expediency. Similar legislation is needed for other parts of the country in view of the recurring nature of communal riots.

Compensation by the State to the victims of crime is the only proper remedy but it is very unlikely for such a scheme to materialize in India and other developing countries. The fact, however, remains that ineffective means of preventing and controlling crime are the main factors leading to victimisation. There is a great need for the constant evaluation and improvement of law and enforcement procedures in order to reduce crime victimisation to the minimum level. Government and non-governmental agencies have to perform effective roles in providing to victims both emergency and prolonged medical, psychiatric, psychological and social services which are altogether lacking at present in the country. In any future set-up in India, the following issues raised in an international symposium held in Israel must also be considered :

1. Should there be a maximum and/or minimum level for compensation ?
2. What is the nature of the losses that should be recompensed, e.g., direct damage, loss of earnings, pain and suffering ?
3. Should consideration be given to the victim's conduct at the time of the offence and/or to his general character in determining the question of compensation.
4. Should payment be by right and denied only for stated reasons by the court ?
5. Should present-day schemes be extended to include crimes against property ?
6. Should the State be entitled to claim reimbursement from the criminal and/or should the State be empowered to compel criminals to give part of their earnings to the State ?
7. Should States set up compulsory insurance schemes for certain professions whose exercise relies upon an element of fidelity and trust in order to cover damage caused by one of their members to stock exchanges ?
8. Should compensation schemes contain opportunities for appeal ?
9. Should bystanders attempting to aid victims be entitled to compensation for damage or losses suffered ?
10. Should the victim be entitled to immediate partial compensation in order to tide him over initial expenses, the determination of the final sum to be made subsequently by the compensation board ?
11. Should an accused person who is found not guilty be entitled to compensation for court costs incurred and/or for other losses ?
12. Should a judge in a criminal trial be entitled to order compensation by the State concurrently with his verdict ?
13. Should the office of ombudsman be set up to provide direct focus on the needs of the victim, with special concern for mitigating immediate trauma, prevention of further stress at the hands of society, as well as offering treatment for victim ?

References

Cohen, Lawrence E., and Marcus Felson, (1979) "Social Change and Crime Rate Trends: A Routine Activity Approach." American Sociological Review 44.4: 588.

Great Britain, Home Office, (1996) The victim's charter: a statement of the service standards for victims of crime, London: HMSO.

Greg Cooper and Mike King, (2002) Analyzing Criminal Behaviour, IQ Publishing,Institute of Investigative Science, Egypt.

Hentig, von, Hans, (1948) The Criminal and His Victim. New Haven: Yale U. Press.

Karmen, A., (1992) Crime Victims. Pacific Grove: Brooks/Cole.

Luckenbill DF (1977) Situated Transaction Model: Juvenile Homicide. Amazon Books

Maguire, M and J. Pointing (1988), Victims of crime: a new deal? Milton Keynes: Open University Press.

Mendelsohn, B, (1974), 'The origins of the doctrine of victimology' In I. Drapkin and E. Vicano (eds) Victimology, Lexington, IN: D.C. Heath and Co.

Newman, O, (1972) Defensible Space. NY: Macmillan.

Rajan, V.M, (1981) Victimology in India New Delhi, Allied Publisher, pp. 10-11.

Schwartz, M. & V. Pitts, (1995), "Exploring a Feminist Routine Activities Approach to Explaining Sexual Assault" Justice Quarterly 12: 1.

Shapland, J. et al, (1985) Victims and the criminal justice system, Aldershot: Gower.

Siddique A, (2007) Criminology Eastern Book Company, Lucknow, Fifth Edition.

Sparks, R. Genn, H. and Dodd, D, (1977) Surveying victims, Chichester: Wiley.

Walklate, S., (1989), Victimology: the victim and the criminal justice process, London: Unwin Hyman.

Walklate, S., (1998) Understanding criminology: current theoretical debates, (Chapter Seven 'Criminal victimisation, politics, welfare') Buckingham: Open University Press.

Walterc. Reckless, (1971), The Crime Problem New York: Appleton-Century-Crofts, Inc., 1950 (Indian Reprint, 1971), p. 138.

William MacDonald (Ed.), (1976), Criminal Justice and the Victim, Sage, Beverly Hills, Calif, pp. 24-25.

Young, A., (1996), Imagining crime, London: Sage.

Zedner, L. 1997, (2002), 'Victims' In M. Maguire, R. Morgan and R. Reiner (eds.) The Oxford Handbook of Criminology, 3rd Ed Oxford: Clarendon Press.

■■■

CHAPTER 26 Victims of Terrorism

Learning Objectives :

- **To Understand the conceptual, theoretical and the methodological dimensions of Victims of terrorism.**
- **To grasp the policy implications of treatment of of Victims of terrorism.**

Introduction

New Millenium has given rise to a new terrorism which has now not only fattened the scope of victimology but also has increased the concerns for the victims of gravest terror crimes. 9/11 terror attacks at WTC in 2001, Bali bombings of October 2002, Madrid terrorism of March 2004 , London bombings of 2005 & Mumbai attacks of 2006 & 2008 produced large scale devastations & victimization with the sole aim of terrorists to produce maximum public attention. Possibility of the use of chemical weapons by terrorists in future as against the present use of other weapons of mass destruction may give rise to a new form of terrorism called 'Catastrophic Terrorism'. Millions of people throughout the world suffer harm as a result of crime, the abuse of power, terrorism and other stark misfortunes. Literature on terrorists and terrorism is abundant, but very little is written on the victim. Whether the studies are published by psychologists or political scientists, by psychiatrists or sociologists, data on the victim are conspicuous by their absence.

There is hardly any mention of how victims of terrorism are chosen, their characteristics, the role they play, if any, in preparing or triggering the terrorist acts, their behaviour during the ordeal, and their post-victimization condition. While policy response in context of victims of terrorism, as reviewed in this paper, are yet in their infancy in the west. It is yet to be born in much terrorism affected countries of the world. Interestingly, the policy response to terrorism in parts of the world has – until recently – been primarily focused on adjusting police tactics and criminal procedural laws relating to organised crime to the new demands placed on law enforcement (Albrecht and Kilchling, 2005:5). Many countries go with traditional mindset acknowledging that victimization, terrorist or otherwise, is a normal and natural occurrence, a fact of life, portraying it instead as a pathological and abnormal phenomenon.Their rights and needs as victims of this harm have not been adequately recognized. The dearth of the empirical data renders the systematic study of the victims of terrorism particularly difficult. This paper presents some reflections on inter linkage between victimology and terrorism under four major heads : Conceptual, Theoretical, methodological and Praxiological aspects.

Terrorism : One Meaning Several Terms & One Term Several Meanings

Victimization and terrorism are the old realities but new rhetoric in the terminological discourse of criminology with its recent comrade-in-arm victimology. Concepts concerning victims in the context of terrorism are relatively abstract or include victims of terrorism under the broader heading of victims of crime in general. In addition, policies and legislation relating to victims of crime or victims of terrorism vary widely on the domestic level. First need is to sort out the definitional issues before entering into a discourse on theory and/or praxis. There are many concepts but fewer consensuses, many theories and fewer arrivals, many policies yet fewer results.

Both the concepts terrorism and victim are far from controversies and comments. They represent one term each with diverse meanings or one meaning endowed with diverse terms. The term "terrorism" comes from Latin terrere, "to frighten." The terror *cimbricus* was a panic and state of emergency in Rome in response to the approach of warriors of the *Cimbri* tribe in 105BC. The Jacobins cited this precedent when imposing a Reign of Terror during the French Revolution. After the Jacobins lost power, the word "terrorist" became a term of abuse. The French National Convention declared in September 1793 that "terror is the order of the day." The period 1793-94 is referred to as the regime de la terreur (Reign of Terror). Maximilien Robespierre, a leader in the French revolution proclaimed in 1794 that, "Terror is nothing other than justice, prompt, severe, inflexible." Common principles among legal definitions of terrorism provide an emerging consensus as to meaning and also foster cooperation between law enforcement personnel in different countries. Among these definitions there are several that do not recognize the possibility of legitimate use of violence by civilians against an invader in an occupied country. Other definitions would label as terrorist groups only the resistance movements that oppose an invader with violent acts that indiscriminately kill or harm civilians and non-combatants, thus making a distinction between lawful and unlawful use of violence. Ultimately, the distinction is a political judgment.

A precise definition of the word "terrorism" is often difficult as the term is at times politically and emotionally charged. Studies have found over 100 definitions of "terrorism" (Schmid et. al.1988) The concept of terrorism may itself be controversial as it is often used by state authorities to delegitimize political or other opponents, and potentially legitimize the state's own use of armed force against opponents (described as "terror" by opponents). A less politically and emotionally charged, and more easily definable, term is violent non-state actor. Terrorism has been practiced by a broad array of political organizations for furthering their objectives. It has been practiced by both right-wing and left-wing political parties, nationalistic groups, religious groups, revolutionaries, and ruling governments. Those labeled "terrorists" by their opponents rarely identify themselves as such, and typically use other terms or terms specific to their situation, such as separatist, freedom fighter, liberator, revolutionary, vigilante, militant, paramilitary, guerrilla, rebel, patriot, or any similar-meaning word in other languages and cultures. Jihadi, mujaheddin, and fedayeen are similar Arabic words which have entered the English lexicon. It is common for both parties to a conflict to describe each other as terrorists. The pejorative connotations of the word can be summed up in the aphorism, "One man's terrorist is another man's freedom fighter". In his book "Inside Terrorism" Bruce Hoffman wrote in Chapter One: Defining Terrorism that

"On one point, at least, everyone agrees: terrorism is a pejorative term. It is a word with intrinsically negative connotations that is generally applied to one's enemies and opponents, or to those with whom one disagrees and would otherwise prefer to ignore. 'What is called terrorism,' Brian Jenkins has written, "thus seems to depend on one's point of view. Use of the term implies a moral judgment; and if one party can successfully attach the label terrorist to its opponent, then it has indirectly persuaded others to adopt its moral viewpoint." Hence the decision to call someone or label some organization 'terrorist' becomes almost unavoidably subjective, depending largely on whether one sympathizes with or opposes the person/group/cause concerned. If one identifies with the victim of the violence, for example, then the act is terrorism. If, however, one identifies with the perpetrator, the violent act is regarded in a more sympathetic, if not positive (or, at the worst, an ambivalent) light; and it is not terrorism" (1998:32).

The focal point of terrorist activity tends to move around the world. In the meantime, the international community adopted a series of sectoral conventions that define and criminalize various types of terrorist activities. In addition, since 1994, the United Nations General Assembly has condemned terrorist acts using the following political description of terrorism: "Criminal acts intended or calculated to provoke a state of terror in the general public, a group of persons or particular persons for political purposes are in any circumstance unjustifiable, whatever the considerations of a political, philosophical, ideological, racial, ethnic, religious or any other nature that may be invoked to justify them."

Points of acrimonious debate in the United Nations in the search for a common legal definition of terrorism relate to (Friedrichs, 2006) the relationship (or lack thereof) between terrorism and anti-colonial and national liberation struggles and to the question whether or not the activities of States' armed forces in armed conflicts and in exercise of their official duties should be excluded from any definition of terrorism. For more than a decade the debate has been going on in the United Nations in New York (Letschert et al, 2010:40).

The United States Department of Defense defines terrorism as "the calculated use of unlawful violence or threat of unlawful violence to inculcate fear; intended to coerce or to intimidate governments or societies in the pursuit of goals that are generally political, religious, or ideological." Within this definition, there are three key elements—violence, fear, and intimidation—and each element produce terror in its victims. The FBI uses this : "Terrorism is the unlawful use of force and violence against persons or property to intimidate or coerce a government, the civilian population, or any segment thereof, in furtherance of political or social objectives." The U.S. Department of State defines "terrorism" to be "premeditated politically-motivated violence perpetrated against non-combatant targets by sub-national groups or clandestine agents, usually intended to influence an audience."

The USA PATRIOT Act defines terrorism activities as "activities that (A) involve acts dangerous to human life that are a violation of the criminal laws of the U.S. or of any state, that (B) appear to be intended (i) to intimidate or coerce a civilian population, (ii) to influence the policy of a government by intimidation or coercion, or (iii) to affect the conduct of a government by mass destruction, assassination, or kidnapping, and (C) occur primarily within the territorial jurisdiction of the U.S."

The Unlawful Activities (Prevention) Amendment Act, 2008 in India defines terrorist act as "Whoever does any act with intent to threaten or likely to threaten the unity, integrity, security or sovereignty of India or with intent to strike terror or likely to strike terror in the people or any section of the people in India or in any foreign country" :-

(a) By using bombs, dynamite or other explosive substances or inflammable substances or firearms or other lethal weapons or poisonous or noxious gases or other chemicals or by any other substances (whether biological radioactive, nuclear or otherwise) of a hazardous nature or by any other means of whatever nature to cause or likely to cause–

(i) Death of, or injuries to, any person or persons; or

(ii) Loss of, or damage to, or destruction of, property; or

(iii) Disruption of any supplies or services essential to the life of the community in India or in any foreign country; or

(iv) Damage or destruction of any property in India or in a foreign country used or intended to be used for the defence of India or in connection with any other purposes of the Government of India, any State Government or any of their agencies; or

(b) Overawes by means of criminal force or the show of criminal force or attempts to do so or causes death of any public functionary or attempts to cause death of any public functionary; or

(c) Detains, kidnaps or abducts any person and threatens to kill or injure such person or does any other act in order to compel the Government of India, any State Government or the Government of a foreign country or any other person to do or abstain from doing any act, commits a terrorist act. Explanation : For the purpose of this section, public functionary means the constitutional authorities and any other functionary notified in the Official Gazette by the Central Government as a public functionary.

Similarly the one meaning of terrorism is known through several names or terms. Rebels, insurgents, separatists, extremists, militants, guerrillas, insurrectionists, freedom fighters, fundamentalists are all called terrorists in some sense or the other.

The above-mentioned legal definitions and trajectory of terminologies only partly overlap with some of the more widely used sociological definitions of terrorism. Conceptualizing terrorism through different lenses can be useful in order to understand terrorism in all its forms and manifestations. Schmid developed, (2004:382), an academic consensus undisputed definition containing 16 elements. The definition reads as follows : Terrorism is an (1) anxiety-inspiring method of repeated (2) violent action, employed by (semi-) (3) clandestine individual, group, or state actors, for (4) idiosyncratic, criminal, or political reasons, whereby – in contrast to assassination – the direct targets of violence are not the main targets. The (5) immediate human victims of violence are generally chosen (6) randomly (targets of opportunity) or (7) selectively (representative or symbolic targets) from a target population, and serve as message generators. (8) Threat- and violence-based (9) communication processes between terrorist (organization), (imperilled) victims, and main targets are used to (10) manipulate the main target (audience(s)), turning it into a (11) target of terror, a (12) target of demand, or a (13) target of attention, depending on whether (14) intimidation, (15) coercion, or (16) propaganda is primarily sought.

Thus, terrorism has many different faces and features. The idea here is not to aggravate the controversies, but to bring the meaning and features to order. Despite the contesting definitions and the diversity of terrorist acts, it is more important to locate the differences of effect of the various forms of terrorism on victims, thus adhering to a victims' perspective. Important question to address is whether there is a real need to adopt specific standards for victims of terrorism, thereby implying that their needs might differ from victims of ordinary crime. Do these needs differ in degree, i.e. when the consequences of terrorism are more or less severe, does meeting the need in question become more or less important ? Are there indications that meeting a need of victims of terrorism requires additional efforts in implementation ?

Victim of Terrorism : Concept and Dimensions

Just as it is difficult to agree on a definition of terrorism, it also appears problematic to define the term 'victims' in the context of terrorist attacks. Similarly the meaning victim may be understood through several terms like injured party, wounded, bereaved, undue sufferers, the adversaries of the terrorist organization (usually one or several governments) and survivors and so on.

The term 'victim' in itself gives rise to controversies because it is considered to have multiple connotations as well. The term "crime victim" generally refers to any person, group, or entity who has suffered injury or loss due to illegal activity. The harm can be physical, psychological, or economic. The legal definition of "victim" typically includes the following : "A person who has suffered direct, or threatened, physical, emotional or pecuniary harm as a result of the commission of a crime; or in the case of a victim being an institutional entity, any of the same harms by an individual or authorized representative of another entity."

The theory of victim precipitation gives different connotation to victim that in the majority of cases of victimization, victims in some way, knowingly or unknowingly, passively or actively, influenced their victimization. Although the theory is often criticized as anti-victim, it leaves enough scope to distinguish between victims of street crime and victims of terrorist crime. The victim facilitation in case of victims of terrorist crime is near zero. What makes the term 'victim' in relation to terrorism difficult to define relates to the variety of audiences that are affected by the act. It remains to be seen whether they could be qualified as victimised entities and are all entitled to protection offered by international victims' rights instruments.

Most of these instruments define who falls under the scope of protection. The 1985 UN Declaration of Basic Principles of Justice for Victims of Crime and Abuse of Power contains the following definition in Articles 1 and 2 :

1. "Victims" means persons who, individually or collectively, have suffered harm, including physical or mental injury, emotional suffering, economic loss or substantial impairment of their fundamental rights, through acts or omissions that are in violation of criminal laws operative within Member States, including those laws proscribing criminal abuse of power.
2. The term "victim" also includes, where appropriate, the immediate family or dependants of the direct victim and persons who have suffered harm in intervening to assist victims in distress or to prevent victimization.

The UN Basic Principles and Guidelines on the Right to a Remedy and Reparation for Victims of Gross Violations of International Human Rights Law and Serious Violations of International Humanitarian Law use the same definition as the 1985 Declaration. Legal persons are not entitled to protection under these two documents, contrary to the scope of protection offered by the Rules of Procedure and Evidence for the purpose of the Statute of the International Criminal Court. Rule 85 notes that the notion of victims may also include 'organizations or institutions that have sustained direct harm to any of their property, which is dedicated to religion, education, art or science or charitable purposes, and to their historic monuments, hospitals and other places and objects for humanitarian purposes.'

The European Union (EU) Framework Decision on the standing of victims in criminal proceedings contains the following definition in Article 1 (a): 1 (a) 'victim' shall mean a natural person who has suffered harm, including physical or mental injury, emotional suffering or economic loss, directly caused by acts or omissions that are in violation of the criminal law of a Member State. The Framework Decision understandably, thus limits the scope of protection to natural persons who suffered harm directly caused by acts or omissions that are in violation of the criminal law of a Member State. Article 8, however, broadens the scope to some extent in the sense that Member States 'shall ensure a suitable level of protection for victims and, where appropriate, their families or persons in a similar position, particularly as regards their safety and protection of their privacy, where the competent authorities consider that there is a serious risk of reprisals or firm evidence of serious intent to intrude upon their privacy.' The reason for this limited scope can be derived from the heading of this Framework Decision; it concerns the standing of victims specifically in criminal proceedings. Lastly, legal persons do not fall under the scope of protection.

The Council of Europe (CoE) Guidelines on the Protection of Victims of Terrorist Acts mention in principle 1 that States should ensure that any person who has suffered direct physical or psychological harm as a result of a terrorist act as well as, in appropriate circumstances, their close family can benefit from the services and measures prescribed by these Guidelines. These persons are considered victims for the purposes of these Guidelines.

This definition does not include economic loss, as included in the UN Declaration and the EU Framework Decision. Viewing these definitions, it can be noted that not every category should qualify as victim entitled to victim's protection. Where does that leave those who were in another way confronted with the attack ? One could think of first responders like the fireman who was involved in the rescue operation in the aftermath of the attack or the man or woman who lost livelihood/business after devastation or bombing ? Or what about the people who were confronted with the terrorist attack through the media (mainly television images) and show symptoms of Post Traumatic Stress Disorder (PTSD) ?

The specific characteristics of terrorism, namely the intention to cause death or serious bodily harm to civilians and non-combatants or cause damage to property, with the purpose, inter alia, to intimidate the larger population, might justify the formulation of a broader definition. Thus a comprehensive definition amidst controversies needs to be brought in order.The following typology of victims made by Schmid (Ibid) is relevant in present context. It distinguishes between primary (also referred to as direct) victims, and secondary (also referred to as indirect or vicarious) victims of terrorism.

Primary victims

1. Those that are killed by terrorist kidnappers, hostage-takers, gunmen or bombers.
2. Those that are injured, mutilated or mentally tortured by terrorists but are ultimately released or liberated.
3. Those who are wounded or die in a counter-terrorist rescue operation at the hands of terrorists or armed first responders.
4. Those who become mentally or physically handicapped or die (in suicide) in a causal sequel to one or several terrorist events in which they were involved or of which they were direct witnesses.

Secondary victims

1. Those close to persons in the four categories of primary and direct victims: family, dependants, friends and colleagues.
2. Those whose names appear on 'death lists'.
3. Those that have otherwise a well-founded reason to fear that they might be a victim in the future.
4. Those first responders to acts of terrorism who become traumatized and experience "burn-out".
5. Those that experience income loss or property damage due to acts of terrorism.
6. Those whose normal lifestyle is changed by terrorist threats and counter-terrorism measures.

Tertiary victims

Based on this classification proposed by Schmid, and taking into account existing definitions in international instruments, Letschert et al (2010) suggest sub-dividing the large group of secondary victims into two groups, thereby introducing a separate category of tertiary victims (also referred to as vicarious victims) and propose the following definition of tertiary victims. All others not listed under primary and secondary victims could be considered tertiary victims e.g firemen, rescue team, media persons, PTSD patients who confronted with the terrorist attack through the media (mainly television images) and the counsel.

Theoretical Discourse

Only in the last two decades, the discipline of victimology has firmly become established on the academic scene. The decades of the 1990s and 2010s could easily be described as a period of consolidation, data gathering and theorization, with new legislation, victim compensation, redress and mediation, help, assistance and support to enable victims to recover from the negative effects of victimization. The wealth of data collected mainly through victimization surveys has led to various theoretical formulations. Models have been developed to offer plausible explanations for the variations in victimization risks, and for the clustering of victimization in certain areas and certain groups. They have also helped to unravel the intriguing phenomenon of repeat victimization. Theories available wheather it is *lifestyle model* developed by Hindelang, Gottfredson, and Garofalo (1978) or the *Routine Activity*

Approach developed by Cohen and Felson (1979) or the *opportunity model* (Cohen *et al.*, 1981) relate to victims of crime in general. Enough of theories are yet to be built on victims of terrorism.

Theoretical discourse begins with differentiating between victims of terrorism and other victims of crime Letschert et al (2010) especially in the context of their needs and rights. For the most part, the needs of direct victims of terrorism are similar to those of other victims of crime, differing not in kind but rather in degree or in possibilities for implementation. On average, the impact of terrorism in a financial, psychological and physical sense may be larger, but definitely not always. Often terror victims will require immediate medical and financial assistance, but this will be the case for some victims of crime as well. Similarly, victims of terrorism, like victims of crime, will need to be treated respectfully and provided with information about and participation in their case. Both will need reassurance of their safety, and will need to come to terms with feelings of anxiety and anger they are likely to have after the event.

The main difference appears to lie in the context in which terrorist victimization occurs, and its audience. Victims of terrorism, by definition, are attacked as representative of a larger group. Acknowledgment of their victimization entails recognizing this fact. This gains even more relevance in situations where the terrorist attack is framed in the context of war. Victims may then feel they are civil casualties of war rather than 'just' victims of crime.

The fact that terrorists use violence against direct targets to threaten, frighten and otherwise influence a wider group of indirect or vicarious victims, implies that the audience of the crime transcends the direct victims. Indeed, the effects on vicarious victims in absolute terms may outweigh those of the direct victims. It was shown that symptoms of post-traumatic stress disorder are found in members of the public not present at the site of terrorist attacks and unrelated to those who were. The increased levels of fear in the general public may result in various behavioural reactions, from lower levels of tourist activity, to decreased use of public transport systems and the occurrence of 'worrying well'. Moreover, research into terror management theory shows the effects of terrorism on public opinion and political preferences. The anger at the attack may lead to a process of vicarious retribution, which can involve backlash attacks on people who in one way or another resemble the terrorists and a general antipathy against those who are seen to belong to the same group.

Literature at present in the context of theorizing the need and rights of victims of terrorism may fall into three broad approaches : Theory of Compensation, Theory of Restoration and Theory of Victim Rights.

Theory of Compensation : Reparation entails the following concepts : restitution, compensation, rehabilitation, satisfaction and guarantees of non-repetition, often used in the framework of massive and systematic cases of grave human rights violations. The word compensation in literal sense means a thing that compensates or is given to compensate (for); a counterbalancing feature or factor; amends, recompense; spec. money given to compensate loss or injury, or for requisitioned property. When we talk about Compensation to the victims it means something given in recompense i.e. equivalent rendered. It is to be noted that the whole purpose of compensation is to make good the loss sustain by the victim or legal representative of the victim. Victims should receive fair, appropriate and timely compensation, which is easily accessible. This need applies as much to victims of ordinary crime as to victims of terrorism (needs in kind). The need for reparation in the broader meaning could be more apparent for victims of large-scale terrorist acts than for victims of ordinary crime. It is open to question whether normal procedures for compensation will be sufficiently swift in reaction to large-scale terrorist victimisation, resulting in a large group of both primary and secondary victims. Victims of terrorism being excluded from insurers' policies coverage is a major issue today. Another issue relating to differentiation in degree relates to the possibility that terrorist attacks will result in cross-border victims. As terrorists attacks have not infrequently targeted tourist places, many victims come from abroad. Apart from the other difficulties this may pose for victims and their families, it also adds to the cost of victimisation, making meeting the financial need even more important. For instance, one can have doubts about the adequacy of State compensation schemes in case of *large-scale* terrorist acts, with regard to procedural matters and with regard to the possibility to receive compensation for property damage.

Economic hardships and budgetary restraints have greatly limited the scope of compensation for mass victims of terrorism as well as the number of victims who receive help. The literature suggests that the sums victims get from the schemes are, for the most part, token amounts, and that the programs in reality fulfil no more than a symbolic function. Only a very tiny proportion of victims end up receiving any compensation, and for those who do, it is more often than not too little and too late.

Questions relating to compensation of victims of terrorism appear more complicated compared to compensation in cases of individual 'ordinary' crime. For instance, in many cases, the perpetrator either died (due to a suicide attack) or is difficult to capture; catastrophic terrorism produces large numbers of victims which makes it more difficult to come up with an effective and appropriate compensation scheme. Whether the needs of victims of terrorism in the field of victim support and assistance differ from victims of 'ordinary' crime also requires further study. It appears that victims of terrorism experience a higher level of distress and different needs, partially due to the magnitude and scope of the events and quite possibly also due to the media attention. It has also been noted that the cross-border or transnational character of many acts of terrorism and the diverse identity of their victims, poses new demands on procedures and organization of victim relief and support schemes.

Theory of Restoration : The following are the underlying assumptions of restorative justice programmes can be identified: the response to crime should repair as much as possible the harm suffered by the victim; offenders should be brought to understand that their behaviour is not acceptable and that it had consequences for the victim and the community; offenders can and should accept responsibility for their action; victims should have an opportunity to express their needs and to participate in determining the best way for the offender to make reparation, and the community has a responsibility to contribute to this process. The perception of restorative justice is to understand crime first of all as harm done to people and communities. It implies an inherent concern for victims' needs and their role in the criminal justice system and encourages offenders to understand the harm and the consequences of their behaviour. A further aim is that the offender accepts his responsibility and tries to repair the harm done to the victim. However, the application of restorative justice programmes in terrorist cases is still only rarely applied and developed. The main challenge is to consider terrorism as a possible case for a restorative justice approach. Restorative justice principles and values are indicators for the reaction to all types of crime that could also be applied for terrorism. In this respect, it could be envisaged to adopt and elaborate appropriate restorative justice practices and to implement them in cases of terrorism in relation to the criminal justice system.

Theory of Victim Rights : Terrorism as a form of manifest extremism has become a human rights issue. Human rights are relevant to terrorism as concerns both its victims and its perpetrators. Universal Declaration of Human Rights, established "recognition of the inherent dignity and inalienable rights of all members of the human family." The innocent victims of terrorism suffer an attack on their most basic right to live in peace and security. The rapid development of more stringent counter-terrorism efforts has pitched the issue of human rights and terrorism into high relief. This is true not only in the United States, but in a number of countries who have signed on as partners in a global coalition to crack down on terrorist activity.The suspected perpetrators of attacks also have rights, as members of the human family, in the course of their apprehension and prosecution. They have the right not to be subject to torture or other degrading treatment, the right to be presumed innocent until they are deemed guilty of the crime and the right to public trial. Fighting terrorism means shining a light into the darkest corners of human existence, and bringing a real promise of human rights to the entire world's people, so that desperate men and women are not driven to follow leaders whose only real message is vengeance.

The theories reviewed above are still being experimented and have raised more questions than they have answered, questions that need to be addressed in realizing the rights and meeting the needs of victims of terrorism.

Methodological Issues

Victimology in the context of Terrorism is relatively a new area of research. One of the primary tasks of theoretical victimology is to collect empirical data on crime victims. The main instrument used at present to collect this information is victimization surveys, conducted at a local, regional, national and international level. Worthy of notice among these surveys are the ones carried out on a regular basis, at regular intervals, in England, the U.S, Germany, Canada and Australia. Each of these surveys yields a wealth of information on crime victims. Although a wide variety of new programmes have been tried and dubbed as "promising" most of these have not been subjected to any form of empirical evaluation. Before these programme can be accepted as worthy of duplication, they must be carefully scrutinized over a sufficient time period (Dussich, 2005).

While activism to affirm victims' rights and to improve their lot has been in full swing on many fronts, animated by political and ideological considerations, research has been lagging behind, and in many instances totally lacking. Several areas, pivotal to the theory and practice of victim assistance, have hardly been investigated and are in dire need of solid empirical research. One has to wonder why it is that when the field of victim services is flourishing, research on the effects of victimization and on the impact of victim assistance is hard to come by. And yet it seems obvious that individualized care, individualized assistance, and personalized treatment or counselling require a profound knowledge of the differential impact of victimization and the differential needs of crime victims (Fattah, 1999: 193). Clearly, this is an area that offers golden opportunities for original empirical qualitative research, but it is not by any stretch of the imagination the only one. Being a young discipline, many areas of victimology remain virgin territory and have yet to be explored by inquisitive and adventurous researchers. The coming years will witness a growing realization that action not backed by research is a mere ideological exercise, and that practice not grounded in theory is dangerous and potentially harmful. The large sums spent on national victimization surveys could be better spent either to fund much needed qualitative research or to add to the subsidies of poorly-funded victim assistance programs and victim services. Further unexplored dimensions include terrorist financing cross-boarder and transnational terrorism, now branded as neo terrorism or catastrophic terrorism in the new millennium. Specific qualitative research studies need to be conducted on terrorism from victim's perspectives.

Praxiological Considerations

Policy precedes praxis. Major international instruments include EU Council Framework Decision of 15 March 2001 on the standing of victims in criminal proceedings, the EU Council Framework Decision of 13 June 2002 on combating terrorism, provisions of thc International Criminal Court (ICC), the 1985 UN Declaration of Basic Principles of Justice for Victims of Crime and Abuse of Power and Recommendations for Assistance to Victims of Acts of Terrorism at the Conference on Standards for Victims of Terrorism, organized by the European Forum for Restorative Justice in cooperation with the International Victimology Institute at Tilburg University, the Centre for the Study of Terrorism and Political Violence at the University of St Andrews, the Leuven Institute of Criminology at the Catholic University of Leuven and Victim Support the Netherlands on 10 and 11 March 2008.

As new programmes and new laws evolve some prove effective and others not. In the search for programmes and laws that fulfil the fundamental aims of the United National Declaration "... to be treated with compassion and respect for their dignity... to be provided with access to the mechanisms of justice and to prompt redress... to be informed of their rights... to be informed of their role and the scope, timing and progress of the proceedings and of the disposition of their cases... to be provided with proper assistance throughout the legal process... to have their privacy protected and insure their safety... to be considered for receipt of restitution... to be informed about receiving compensation..." These criteria determine the value of programmes and laws so that they can be evaluated and ultimately recommended as worthy of duplication. Some of these are listed below.

1. Law Enforcement

- In San Diego, California there is a partnership between the police and the YWCA which resulted in a Community Domestic Violence Resource Network. This has resulted in a major resource for all the police agencies in the community for accurate information about the availability of shelters at any given time.
- In Provo, Utah victims participate in crime solving, called "victim-assisted" investigations.
- In Orange County, California a group of five victim advocates working together in a non-profit programme work with police and prosecutors to ensure comprehensive services for victims of gang violence.

2. Prosecution

- In Kenosha, Wisconsin, a programme established by the district attorney established special prosecution units for domestic violence and sensitive crimes.
- In Pinellas County, Florida, the state's attorney's office established a special prosecutor to be responsible for all elder exploitation and neglect cases. This includes police training, community outreach and education for other prosecutors.
- In Philadelphia, Pennsylvania, the Victim Services Unit located in the District Attorney's Office, employs Vietnamese and Cambodian victim-witness coordinators to assist victims from Southeast Asia throughout their case process, including translating information and helping victims with emergency medical and financial assistance.

3. Judiciary

- In Tucson, Arizona, the Municipal Court established a partnership with the police, victim advocates, prosecutors and health care professionals to form a Community Domestic Violence Awareness Centre.
- In New York State, the Permanent Judicial Commission on Justice for Children was established to provide assistance to children in the courts so as to provide a special space for child care so that those children whose parents are in court have a safe haven during their stay.
- In Santa Clara County, California, the courts have established local family violence councils to provide a comprehensive response to domestic violence.

4. Corrections

- In Texas, the Department of Criminal Justice started a victim-offender mediation/dialogue programme for victims of severe violence and their incarcerated offenders.
- In California one of the best known victim-offender programmes is the Impact of Crime on Victims (IOC) initiated by the California Youth Authority. This programme is aimed at providing hours of education to inform offenders about how crimes affect victims and society.
- The US Federal Bureau of Prisons piloted victim awareness classes on drug and domestic violence crimes for offenders in halfway houses in Baltimore, Maryland, and Tampa, Florida.

5. State Actors

A number of countries have become leaders in the field of victimology and victim services. The most active and productive countries which have had a major international influence have been : The United States of America, Great Britain, The Netherlands, Canada and recently Japan. Each of these countries have produced model programmes in the area of services, laws, training, education and research :

(i) The USA's Office of Victims of Crime is one of the few governmental agencies in the world which services people in all countries by producing documents, funding for research and hosts conferences in the areas of victimology and victim assistance. Perhaps one of the most dramatic aspects of the USA's role in victim services is the large number of different types of victim service programmes throughout the country (roughly 20,00 separate programmes for victims of sexual assault, child abuse, elder abuse, domestic violence, etc.) In response to the UN Declaration for Victims the USA established a special partnership with the United Nations, the USA produced the two follow-up documents to the 1985 Declaration referred to above : the Handbook on Justice for Victims, and the Guide for Policy Makers, both published in 1999 by the UN Office for Drug Control and Crime Prevention.

(ii) Great Britain was the origin of a number of significant and innovative practices in victim services, such as the concept of victim reparations all called compensation by Margery Fry in 1957; started one of the first victim support services in the world in Bristol; nationwide victim support services in all major cities controlled by a national office in London; home of the journal International Review of Victimology and has over the last ten years provided leadership within the European community in the area of victim services.

(iii) The Netherlands has sponsored a number of very significant activities which continue to reflect their government's support for the world victim rights activities. Set up by the Ministry of Justice in 1987, which provided the early funding for the International Crime Victim Surveys working group which lead to a series of publications about victimization in many countries across the globe. In 1997 the Dutch hosted the 9th World Society of Victimology International Symposium and the website that was created for that event continues today as the most active and useful resource for international victimological materials. In addition to the establishment of victim services throughout the nation, The Netherlands also produces very high quality research in various aspects of victimology. The most recent contribution is the creation of a new research institute dedicated to victimology called the International Victimology Institute Tillburg (INTERVICT).

(iv) Canada has also established a wide variety of sophisticated victim service programmes throughout their country; hosted the 10th WSV International Symposium on Victimology; is home to a number of leading Victimologists, especially Irvin Waller the prime mover of the UN Declaration and hosts the International Center for the Prevention of Crime in Montreal, established in 1997, which contributes to the promotion of victim-friendly programmes and activities available to interested persons around the world.

(v) Japan entered the field of victimology early when Professor S. Nakata introduced Mendelsohn's concept of victimology in 1958 followed in 1966 with the publication of K. Miyazawa's dissertation on international victimology. One of Miyazawa's students, H. Morosawa established a degree programme for students interested in victimology at his university in Mito, Japan and in the last ten years this university has become internationally known for the innovations in victim services, victimological research, its hosting of WSV courses, and recently became the home of the new Tokiwa International Victimology Institute. Today this university now boasts a new Graduate School of Victimology supported by ten victimologists. The most dramatic recent event in Japan was the enactment of the new Fundamental Law for Victims of Crime.

(vi) Victim Assistance to Child Victims of Terrorist Violence in India: National Foundation for Communal Harmony, New Delhi has started a Flagship scheme Project Assist with the provision of providing assistance to the children rendered orphan or destitute in various communal, castes, ethnic or terrorist violence. The financial assistance is provided to such children for their care, education and training. Number of children getting assistance is given below :

- As on 31.03.2010, 10431children were covered under Project Assist in nineteen states and a union territory.
- A sum of ₹ 4.84 crore was released as assistance to these children during 2009-10.
- ₹ 35.06 crore has been paid as assistance up to 31.03.2010.
- 486 fresh cases have been approved and 4302 cases renewed for grant of assistance under Project Assist in the current financial year.

The Foundation collaborates with the following non-government organisations :

(i) Aashwas, a unit functioning under the Assam Police, Guwahati: Aashwas is an Assam Police Project which is helping the Foundation by identifying child victims of militancy in Assam who are being provided financial assistance.

(ii) Council for Rehabilitation of Widows, Orphans, Handicapped and Old Persons in Jammu & Kashmir, Srinagar: The Council is working for victims of militancy. The council has been helping the Foundation in identifying child victims of terrorism in the state of Jammu & Kashmir.

(iii) Society to Save Victims of Violence and Terrorism, Kokrajhar, Assam: The Society is collaborating with the Foundation in implementation of the Project 'Assist' in Assam. A system of random verification of the mode of utilization is also being carried out.

Besides recently Government of India (25.01.2010) devised a Scheme called the 'Central Scheme for Assistance to Civilian Victims of Terrorist, Communal and Naxal violence. The Scheme sets the eligibility as under :

"........ (iii) Families of the victims would be eligible to get assistance under the scheme even if they have received any other assistance, by way of payment of ex-gratia or any other type of relief from the Government or any other source except when a similar scheme is already being implemented by the Central Government.

(iv) Next of kin of employees of Central Government, CPSEs, Autonomous Institutions and other Government Organizations including State Governments / State PSEs and similar organizations of State Governments would be eligible to receive financial assistance of ₹ 3 lakhs in case of death/permanent incapacitation (50% and above) on account ofincidents of Terrorist/ communal/ naxalite violence.

(v) The cases received / recommended for assistance under SRE Scheme of MHA will not be covered under this scheme. The concerned DM / State Government will ensure that an applicant may receive the assistance either under this Scheme or under SRE Scheme of MHA. A certificate to this effect will also be given by DM in respect of each case.

(vi) Foreign Nationals and NRIs shall also be eligible / covered under the scheme w.e.f. 1.4.08 i.e. the date from which this scheme has been made effective.

6: Assistance

(i) An amount of ₹ 3 lakh would be given for each death or permanent incapacitation to the affected family under the scheme...."

While policy responses in context of victims of terrorism, as reviewed in this paper, are in their infancy in the west, it is yet to be born in much terrorism affected countries of the world. Interestingly, the policy response to terrorism in parts of the world has – until recently – been primarily focused on adjusting police tactics and criminal procedural laws relating to organised crime to the new demands placed on law enforcement (Albrecht and Kilchling, 2005:5). Many countries go with traditional mindset acknowledging that victimization, terrorist or otherwise, is a normal and natural occurrence, a fact of life, portraying it instead as a pathological and abnormal phenomenon. Indian policy instrument, of course, has been basically child centric. Despite enormous strides, a great deal remains to be done. And victims' services really serve official needs, not victims' needs. Overall, however, the two most important services provided to crime victims by victims assistance programs are information and moral support.

An Overview

The fight against terrorism is receiving increased awareness due to recent worldwide large-scale terrorist acts, and only since then has some attention been directed specifically to victims of terrorism. Existing legal instruments of international bodies like the European Union, the Council of Europe and the United Nations concerning victims of terrorism are relatively abstract or include victims of terrorism under the broader heading of victims of crime in general.

Concepts : Both the concepts terrorism and victim are far from controversies and comments. They represent one term each with diverse meanings or one meaning endowed with diverse terms. Despite the contesting definitions and the diversity of terrorist acts, it is more important to locate the differences of effect of the various forms of terrorism on victims, thus adhering to a victims' perspective. Important question to address is whether there is a real need to adopt specific standards for victims of terrorism, thereby implying that their characteristics, aspirations, needs and rights differ from victims of ordinary crime. Accordingly victims of terrorist crimes fall into three distinct categories : Primary, Secondary, and tertiary or vicarious victims.

The Theory : The theory of victim precipitation gives different connotation to victim that in the majority of cases of victimization, victims in some way, knowingly or unknowingly, passively or actively, influenced their victimization. Although the theory is often criticized as anti-victim, it leaves enough scope to distinguish between victims of street crime and victims of terrorist crime. Literature at present in the context of theorizing the need and rights of victims of terrorism may fall into three broad approaches : Theory of Compensation, Theory of Restoration and Theory of Victim Rights. The theories reviewed above are still being experimented and have raised more questions than they have answered, questions that need to be addressed in realizing the rights and meeting the needs of victims of terrorism.

Method : Methodologically, victimology in the context of terrorism is relatively a new area of research. The large sums spent on national victimization surveys could be better spent either to fund much needed qualitative research or to add to the subsidies of poorly-funded victim assistance programs and victim services. Further unexplored dimensions include terrorist financing cross-boarder and transnational terrorism, now branded as neo terrorism or catastrophic terrorism in the new millennium. Specific qualitative research studies need to be conducted on terrorism from victim's perspectives.

Policy and Praxis : While policy response in context of victims of terrorism, as reviewed in this paper, are in their infancy in the west, it is yet to be born in much terrorism affected countries of the world. Interestingly, the policy response to terrorism in parts of the world has – until recently – been primarily focused on adjusting police tactics and criminal procedural laws relating to organised crime to the new demands placed on law enforcement (Albrecht and Kilchling, 2005:5). Many countries go with traditional mindset acknowledging that victimization terrorist or otherwise, is a normal and natural occurrence, a fact of life, portraying it instead as a pathological and abnormal phenomenon. Indian policy instrument, of course, has been basically child centric. Despite enormous strides, a great deal remains to be done. And victims' services really serve official needs, not victims' needs. Overall, however, the two most important services provided to crime victims by victims assistance programs are information and moral support.

Healing, recovery, redress and prevention of future victimization are the primary objectives of most crime victims. And if the primary purpose of social intervention is to restore peace, redress harm, heal injury, and prevent repetition of the offense, then it is easy to foresee application of the restorative justice paradigm, with its constructive elements: mediation, reconciliation, restitution and compensation, as the way of the future (Fattah, 2000). It is thus to be expected that the policies reasoned out and practices advocated by victimologists from now on especially under the possibility of 'catastrophic terrorism', will be very different from those of yesteryears. If a safe prediction is to be made about victimology of the future, it will uphold its original intended character as a truly humanistic discipline and a truly scientific practice.

References

Albrecht, H.J. and Kilchling, M., (2005) Victims of Terrorism Policies and Legislation in Europe, An Overview on Victim related Assistance and Support', Max Planck Institute for Foreign and International Criminal Law.

Cohen, L. E., And Felson, M. (1979), "Social Change and Crime Rate Trends: A Routine Activities Approach", American Sociological Review 44: 588-608.

Cohen, L. E. Klugel, J. R., And Land, K. C., (1981), "Social Inequality and Predatory Criminal Victimization: An Exposition and Test of a Formal Theory", American Sociological Review 46 : 505-524.

Das, D.K. & Kratcoski, P.C., (2003), Terrorist Victimisation: Definition, Focus and Impact, in: Das & Kratcoski (eds.), Meeting the Challenges of Global Terrorism, Prevention, Control and Recovery, Lanham, Lexington Books, Boulder, New York, Oxford.

Dussich, John P. J, (2005), Victimology – Past, Present and Future, UNAFEI, 131st International Senior Seminar Visiting Experts' Papers, Japan.

Fattah, E. A., (1978), "Some Theoretical Developments in Victimology", Victimology: An International Journal 4 (2): 198-213.

Fattah, E.A., (1991), Understanding Criminal Victimization, Scarborough, Ont.:Prentice Hall, Canada.

Fattah, E. A. (1992), Towards a Critical Victimology, London: MacMillan. New York: St. Martins Press.

Fattah, E. A., (2000), Victimoology: Past, Pesent and Future, Sociologie, 33 (1).

Friedrichs, J., (2006), Defining the International Public Enemy: The political struggle behind the legal debate on international terrorism, Leiden Journal of International Law, vol. 19, pp. 69–91.

Harries, R., (1999), The cost of criminal justice, Home Office research findings, No. 103, London: Home Office.

Hindelang, M., Gottfredson, M. And Garofalo, J, (1978), Victims of Personal Crime, Cambridge, MA.: Ballinger.

Hoffman, Bruce, (1998), "Inside Terrorism" Columbia University Press. pp. 32.

Jones, A., (1999) Out of sight, out of mind? The experiences of homeless women, London: Crisis.

Klaus, K, (1994) Crime statistics, Congressional Digest, 73(6-7): 167-9.

Letschert, Rianne; Staiger, Ines; Pemberton, Antony (Eds.) (2010), Assisting Victims of Terrorism: Towards a European Standard of Justice, Springer Science, New York, XXXII+ 348 pp.

Maguire, M. And Pointing, J, (1988), Victims of Crime— A New Deal ?, Milton Keynes : Open University Press.

Maguire, M. , (1991), 'The needs and rights of victims of crime' In M. Tonry (ed) Crime and justice: a review of research, vol. 14:363-433, Chicago.

Mendelsohn, B, (1974) 'The origins of the doctrine of victimology' In I. Drapkin and E. Vicano (eds) Victimology, Lexington, IN: D.C. Heath and Co.

Mohanty, S & R. K. Mohanty, (2010), Police in India, Role Responsibility & Accountability, Cuttack, Legal Associate.

Phipps, A, (1988) Ideologies, political parties and victims of crime In M. Maguire and J. Pointing (eds.) Victims of crime: a new deal ? Milton Keynes: Open University Press.

Schmid, Alex, and Jongman, Albert, (1988) Political Terrorism: A new guide to actors, authors, concepts, data bases, theories and literature. Amsterdam ; New York : North-Holland ; New Brunswick: Transaction Books.

Schmid, Alex, (2004), Terrorism, The Definitional Problem, Case Western Reserve Journal of International Law, vol. 36, Nos. 2 and 3, pp.375–419, p. 382.

Shapland, J. et al (1985) Victims and the criminal justice system, Aldershot: Gower Sparks.

Genn, H.R. and Dodd, D., (1977), Surveying victims, Chichester: Wiley.

Stanko, E.A., Crisp, D., Hale, C., Lucraft, H., (1998), 'Counting the costs: estimating the impact of 'domestic' violence in the London Borough of Hackney', Crime Concern, London.

Von Hentig, H., (1948) The criminal and his victim, New York: Yale University Press.

Werthem, F., (1949) The show of violence, New York: Gollancz.

Wikipedia upto 2010 Wikipedia.com

Young, A., (1996) Imagining crime, London: Sage.

■■■